W9-AOD-788

ACTIVE NETWORK AND FEEDBACK AMPLIFIER THEORY

ACTIVE NETWORK AND FEEDBACK AMPLIFIER THEORY

Wai-Kai Chen

Department of Electrical Engineering
Ohio University

Hemisphere Publishing Corporation

Washington New York London

McGraw-Hill Book Company

New York St. Louis San Francisco Auckland Bogotá
Hamburg Johannesburg London Madrid Mexico
Montreal New Delhi Panama Paris São Paulo
Singapore Sydney Tokyo Toronto

ACTIVE NETWORK AND FEEDBACK AMPLIFIER THEORY

1 2 3 4 5 6 7 8 9 0 E B E B 7 8 3 2 1 0 9

Library of Congress Cataloging in Publication Data

Chen, Wai-Kai, date.
Active network and feedback amplifier theory.

Includes bibliographies and indexes.
1. Feedback amplifiers. 2. Electric networks, Active. 3. Electric network analysis. I. Title.
TK7871.2.C435 621.3815'3 79-16997
ISBN 0-07-010779-3

This book was set in Press Roman by Hemisphere Publishing Corporation. The editors were Lynne Lackenbach and Judith B. Gandy; the production supervisor was Rebekah McKinney; and the typesetter was Wayne Hutchins.
Edwards Brothers Incorporated was printer and binder.

To my mother Shui-Tan Hsia Chen and
to the memory of my father Yu-Chao Chen

CONTENTS

PREFACE

Since Bode published his classical text "Network Analysis and Feedback Amplifier Design" in 1945, very few books have been written that treat the subject in any reasonable depth. The purpose of this book is to bridge this gap by providing an in-depth, up-to-date, unified, and comprehensive treatment of the fundamentals of the theory of active networks and its applications to feedback amplifier design. The guiding light throughout has been to extract the essence of the theory and to discuss the topics that are of fundamental importance and that will transcend the advent of new devices and design tools. Intended primarily as a text in network theory in electrical engineering for first-year graduate students, the book is also suitable as a reference for researchers and practicing engineers in industry. In selecting the level of presentation, considerable attention has been given to the fact that many readers may be encountering some of these topics for the first time. Thus, basic introductory material has been included. The background required is the usual undergraduate basic courses in circuits and electronics as well as the ability to handle matrices.

The book can be conveniently divided into two parts. The first part, comprising the first three chapters, deals with general network analysis. The second part, composed of the remaining four chapters, is concerned with feedback amplifier theory. Chapter 1 introduces many fundamental concepts used in the study of linear active networks. We start by dealing with general n-port networks and define passivity in terms of the universally encountered physical quantities *time* and *energy*. We then translate the time-domain passivity criteria into the equivalent frequency-domain passivity conditions. Chapter 2 presents a useful description of the external behavior of a multiterminal network in terms of the indefinite-admittance matrix and demonstrates how it can be employed effectively for the computation of network functions. The significance of this approach is that the indefinite-admittance matrix can usually be written down directly from the network by inspection and that the transfer functions can be expressed compactly as the ratios of the first- and/or second-order cofactors of the elements of the indefinite-admittance matrix. In Chap. 3 we consider the specialization of the general passivity condition for n-port networks in terms of the more immediately useful two-port parameters. We introduce various types of power gain, sensitivity, and the notion absolute stability as opposed to potential instability.

Chapters 4 and 5 are devoted to a study of single-loop feedback amplifiers. We begin the discussion by considering the conventional treatment of feedback amplifiers based on the ideal feedback model and analyzing several simple feedback networks. We then present in detail Bode's feedback theory, which is based on the concepts of return difference and null return difference. Bode's theory is formulated elegantly and compactly in terms of the first- and second-order cofactors of the elements of the indefinite-admittance matrix, and it is applicable to both simple and complicated feedback networks, where the analysis by conventional method for the latter breaks down. We show that feedback may be employed to make the gain of an amplifier less sensitive to variations in the parameters of the active components, to control its transmission and driving-point properties, to reduce the effects of noise and nonlinear distortion, and to affect the stability or instability of the network. The fact that return difference can be measured experimentally for many practical amplifiers indicates that we can include all the parasitic effects in the stability study and that stability problems can be reduced to Nyquist plots.

The application of negative feedback in an amplifier improves its overall performance. However, we are faced with the stability problem in that, for sufficient amount of feedback, at some frequency the amplifier tends to oscillate and becomes unstable. Chapter 6 discusses various stability criteria and investigates several approaches to the stabilization of feedback amplifiers. The Nyquist stability criterion, the Bode plot, the root-locus technique, and root sensitivity are presented. The relationship between gain and phase shift and Bode's design theory is elaborated. Finally, in Chap. 7 we study multiple-loop feedback amplifiers that contain a multiplicity of inputs, outputs, and feedback loops. The concepts of return difference and null return difference for a single controlled source are now generalized to the notions of return difference matrix and null return difference matrix for a multiplicity of controlled sources. Likewise, the scalar sensitivity function is generalized to the sensitivity matrix, and formulas for computing multiparameter sensitivity functions are derived.

The book is an outgrowth of notes developed over the past several years while teaching courses on active network theory at the graduate level at Ohio University. There is little difficulty in fitting the book into a one-semester or two-quarter course in active network theory. For example, the first four chapters plus some sections of Chaps. 5 and 6 would be ideal for a one-semester course, whereas the entire book can be covered adequately in a two-quarter course.

A special feature of the book is that it bridges the gap between theory and practice, with abundant examples showing how theory solves problems. These examples are actual practical problems, not idealized illustrations of the theory. A rich variety of problems has been presented at the end of each chapter, some of which are routine applications of results derived in the text. Others, however, require considerable extension of the text material. In all there are 239 problems.

Much of the material in the book was developed from my research during the past few years. It is a pleasure to acknowledge publicly the research support of Ohio University through the Professional Leave Program, the Ohio University Baker Fund

Awards Committee, and the National Science Foundation. I am indebted to many graduate students who have made valuable contributions to this book. Special thanks are due to my doctoral students H. Elsherif and M. Jameel, who gave the complete book a careful and critical reading. I was deeply saddened by the unexpected death of Professor G. E. Smith, who kindly reviewed the manuscript and made valuable suggestions. However, I am especially indebted to my wife Shiao-Ling and children Jerome and Melissa for their patience and understanding throughout the course of this work.

Wai-Kai Chen

CHAPTER

ONE

CHARACTERIZATIONS OF NETWORKS

Over the past two decades, we have witnessed a rapid development of solid-state technology with its apparently unending proliferation of new devices. Presently available solid-state devices such as the transistor, the tunnel diode, the Zener diode, and the varactor diode have already replaced the old vacuum tube in most practical network applications. Moreover, the emerging field of integrated circuit technology threatens to push these relatively recent inventions into obsolescence. In order to understand fully the network properties and limitations of solid-state devices and to be able to cope with the applications of the new devices yet to come, it has become increasingly necessary to emphasize the fundamentals of active network theory that will transcend the advent of new devices and design tools.

The purpose of this chapter is to introduce many fundamental concepts used in the study of linear active networks. We first introduce the concepts of portwise linearity and time invariance. Then we define passivity in terms of the universally encountered physical quantities *time* and *energy*, and show that causality is a consequence of linearity and passivity. This is followed by a brief review of the general characterizations of n-port networks in the frequency domain. The translation of the time-domain passivity criteria into the equivalent frequency-domain passivity conditions is taken up next. Finally, we introduce the discrete-frequency concepts of passivity. The significance of passivity in the study of active networks is that passivity is the formal negation of activity.

1.1 LINEARITY AND NONLINEARITY

A network is a structure comprised of a finite number of interconnected elements with a set of accessible terminal pairs called *ports* at which voltages and currents may be measured and the transfer of electromagnetic energy into or out of the structure can be made. Fundamental to the concept of a port is the assumption that the instantaneous current entering one terminal of the port is always equal to the instantaneous current leaving the other terminal of the port. A network with n such accessible ports is called an *n-port network* or simply an *n-port*, as depicted symbolically in Fig. 1.1. In this section we review briefly the concepts of linearity and nonlinearity and introduce the notion of portwise linearity and nonlinearity.

Refer to the general representation of an n-port network N of Fig. 1.1. The port voltages $v_k(t)$ and currents $i_k(t)$ (where $k = 1, 2, \ldots, n$) can be conveniently represented by the *port-voltage* and *port-current vectors* as

$$\mathbf{v}(t) = [v_1(t), v_2(t), \ldots, v_n(t)]' \tag{1.1a}$$

$$\mathbf{i}(t) = [i_1(t), i_2(t), \ldots, i_n(t)]' \tag{1.1b}$$

respectively, where the prime denotes the matrix transpose. There are $2n$ port signals, n port-voltage signals $v_k(t)$, and n port-current signals $i_k(t)$, and each port is associated with two signals $v_k(t)$ and $i_k(t)$. The port vectors $\mathbf{v}(t)$ and $\mathbf{i}(t)$ that can be supported by the n-port network N are said to constitute an *admissible signal pair* for the n-port network. Any n independent functions of these $2n$ port signals, taking one from each of the n ports, may be regarded as the *input* or *excitation* and the remaining n signals as the *output* or *response* of the n-port network. In Fig. 1.1 we may take, for example, $i_1(t), i_2(t), \ldots, i_k(t), v_{k+1}(t), \ldots, v_n(t)$ to be the input or excitation signals. Then $v_1(t), v_2(t), \ldots, v_k(t), i_{k+1}(t), \ldots, i_n(t)$ are the

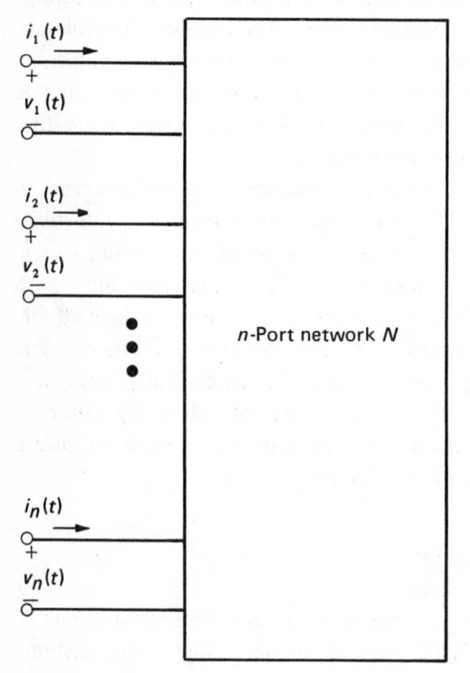

Figure 1.1 The general symbolic representation of an n-port network.

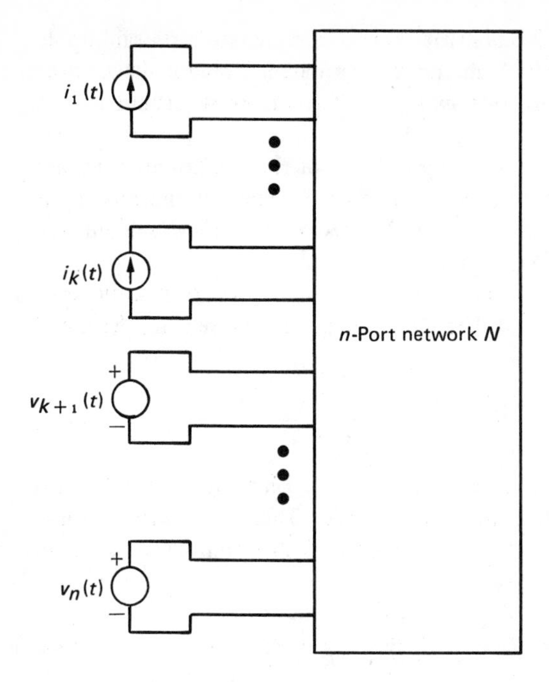

Figure 1.2 A specific input excitation of an n-port network.

output or response signals. This input-output or excitation-response situation is shown in Fig. 1.2. To facilitate our discussion, let $\mathbf{u}(t)$ be the *excitation vector* associated with the excitation signals, and $\mathbf{y}(t)$ the *response vector* associated with the response signals. For the excitation-response situation of Fig. 1.2, the excitation and response vectors are given by

$$\mathbf{u}(t) = [i_1(t), i_2(t), \ldots, i_k(t), v_{k+1}(t), \ldots, v_n(t)]' \tag{1.2a}$$

$$\mathbf{y}(t) = [v_1(t), v_2(t), \ldots, v_k(t), i_{k+1}(t), \ldots, i_n(t)]' \tag{1.2b}$$

respectively. When we speak of zero excitation of an n-port, we mean that every excitation signal is zero; that is, $\mathbf{u}(t) = \mathbf{0}$. On the other hand, a nonzero excitation is meant a set of n excitation signals, not all of them being zero; that is, $\mathbf{u}(t) \neq \mathbf{0}$.

Generally speaking, a network is said to be *linear* if the superposition principle holds. This implies that the response resulting from all independent sources acting simultaneously is equal to the sum of the responses resulting from each independent source acting one at a time. In this sense, any network comprised of linear network elements (linear resistors, linear inductors, linear capacitors, linear transformers, or linear controlled sources) and independent sources is a linear network. Thus, to verify the linearity of a network by this definition, we must have the complete knowledge of the internal structure of the network. For an n-port, the accessible part of the network may be only at its n ports. For this reason, the above definition of linearity may not be adequate for an n-port. For our purposes, we introduce, in addition to the above definition, the notation of portwise linearity.

Definition 1.1: Linearity and nonlinearity An n-port network is said to be *portwise linear* or simply *linear* if the superposition principle holds at its n ports. An n-port is *portwise nonlinear* or simply *nonlinear* if it is not portwise linear.

In other words, if $\mathbf{y}_a(t)$ and $\mathbf{y}_b(t)$ are the responses of the excitations $\mathbf{u}_a(t)$ and $\mathbf{u}_b(t)$ of an n-port, respectively, then the n-port is portwise linear if and only if for any choice of real scalars α and β, the vector $\mathbf{y}(t) = \alpha\mathbf{y}_a(t) + \beta\mathbf{y}_b(t)$ represents the response of the excitation $\mathbf{u}(t) = \alpha\mathbf{u}_a(t) + \beta\mathbf{u}_b(t)$.

The network of Fig. 1.3 is linear in the usual sense. Let us form a one-port from this network as shown in Fig. 1.4. The port voltage and current are described by the equation

$$v(t) = \left(R_1 + \frac{R_2R_3}{R_2 + R_3}\right) i(t) + \frac{ER_2}{R_2 + R_3} \tag{1.3}$$

Suppose that we take $i(t)$ to be the excitation and let $i_a(t) = i_b(t) = 1$ A be two excitations. Assume, for simplicity, that $\alpha = \beta = 1$. Then the corresponding responses $v_a(t)$, $v_b(t)$, and $v_{a+b}(t)$ of the excitations $i_a(t)$, $i_b(t)$, and $i_a(t) + i_b(t)$ are given by

$$v_a(t) = v_b(t) = R_1 + \frac{R_2R_3}{R_2 + R_3} + \frac{ER_2}{R_2 + R_3} \tag{1.4a}$$

$$v_{a+b}(t) = 2R_1 + \frac{2R_2R_3}{R_2 + R_3} + \frac{ER_2}{R_2 + R_3} \tag{1.4b}$$

Since $v_{a+b}(t) \neq v_a(t) + v_b(t)$, the one-port is nonlinear in the portwise sense. Instead of forming a one-port, suppose that we form a two-port from the network of Fig. 1.3. The resulting two-port network is shown in Fig. 1.5; its port voltages and currents are characterized by

$$\begin{bmatrix} v_1(t) \\ v_2(t) \end{bmatrix} = \begin{bmatrix} R_1 + R_2 & R_2 \\ R_2 & R_2 + R_3 \end{bmatrix} \begin{bmatrix} i_1(t) \\ i_2(t) \end{bmatrix} \tag{1.5}$$

It is straightforward to demonstrate that the two-port is now portwise linear. Thus, a portwise nonlinear network need not contain any nonlinear network elements and can often be rendered portwise linear by extracting internal sources at newly formed ports.

As another example, consider the one-port of Fig. 1.6, in which the capacitor is

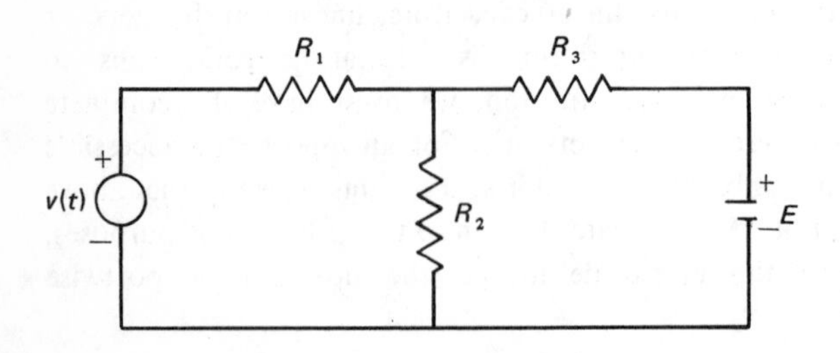

Figure 1.3 A linear network in the usual sense.

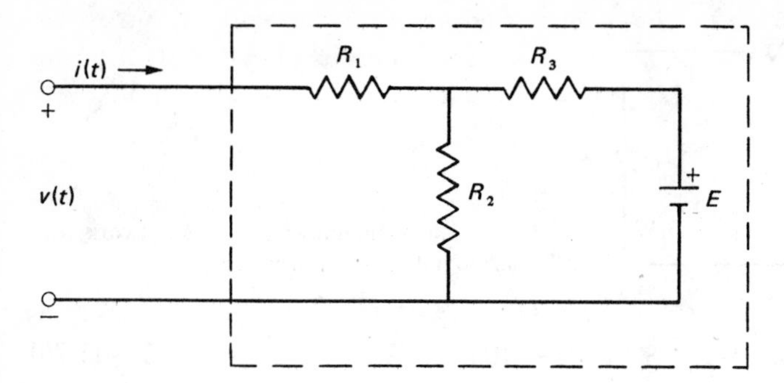

Figure 1.4 A nonlinear one-port network with $E \neq 0$.

initially charged to a voltage $v_C(0+) = V_0$. The terminal relation of the one-port is given by

$$v(t) = Ri(t) + \frac{1}{C}\int_0^t i(x)\, dx + V_0 \tag{1.6}$$

By following Eqs. (1.4), it is easy to confirm that the one-port is portwise nonlinear. Indeed, the presence of any independent sources or any initial conditions on the energy-storing elements in an n-port would render the n-port portwise nonlinear. On the other hand, an n-port network comprised of linear network elements with zero initial conditions and devoid of any independent sources is always portwise linear. For example, in the one-port networks of Figs. 1.4 and 1.6, if the independent source E and the initial voltage V_0 are set to zero, the resulting one-ports become portwise linear.

From the examples discussed above, it is clear that a portwise nonlinear n-port need not contain any nonlinear elements, but the presence of nonlinear elements does not necessarily imply that the n-port is portwise nonlinear. Figure 1.7 is a one-port comprised of two nonlinear resistors connected in series. The nonlinear resistors are characterized by the equations

$$v_\alpha(t) = i_\alpha(t) - i_\alpha^2(t) \tag{1.7a}$$

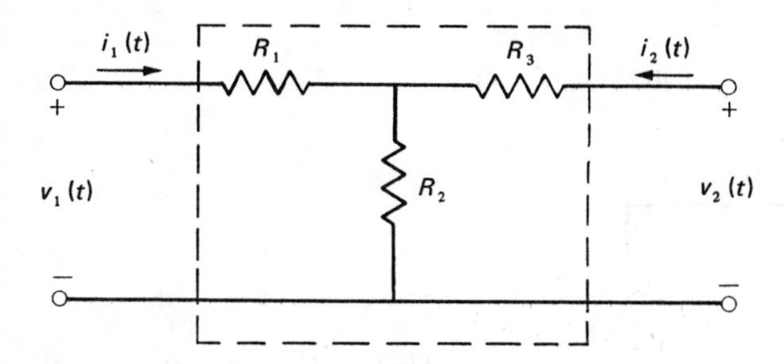

Figure 1.5 A linear two-port network.

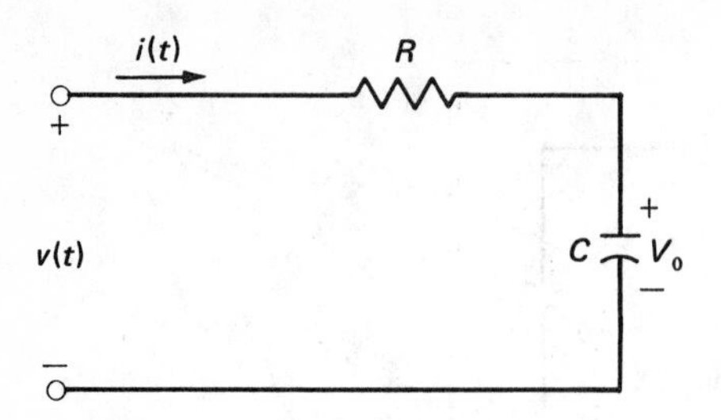

Figure 1.6 A nonlinear one-port network with nonzero initial capacitor voltage.

$$v_\beta(t) = i_\beta^2(t) \tag{1.7b}$$

The port voltage and current are related by the equation

$$v(t) = i(t) \tag{1.8}$$

showing that this one-port is equivalent to a 1-Ω resistor and thus is portwise linear. Suppose that a two-port is formed from this one-port by connecting two wires across one of the resistors as shown in Fig. 1.8. The resulting two-port becomes portwise nonlinear.

We emphasize the difference between the linearity of a network and the portwise linearity of an n-port. Throughout the remainder of this book, we are concerned mainly with portwise linearity. For simplicity, the word portwise will usually be dropped, as also indicated in Definition 1.1, and will be used only for emphasis.

1.2 TIME INVARIANCE AND TIME VARIANCE

A network is said to be *time-invariant* if it contains no time-varying network elements. Otherwise, it is called a *time-varying network.* Like those discussed in the preceding section, if the port behavior of a network is the major concern, the above definition may not be adequate for an n-port. For this reason, we define portwise time-invariance.

Definition 1.2: Portwise time-invariance and time variance An n-port network is said to be *portwise time-invariant* or simply *time-invariant* if, for every real finite constant τ,

$$\mathbf{u}_a(t) = \mathbf{u}_b(t - \tau) \tag{1.9a}$$

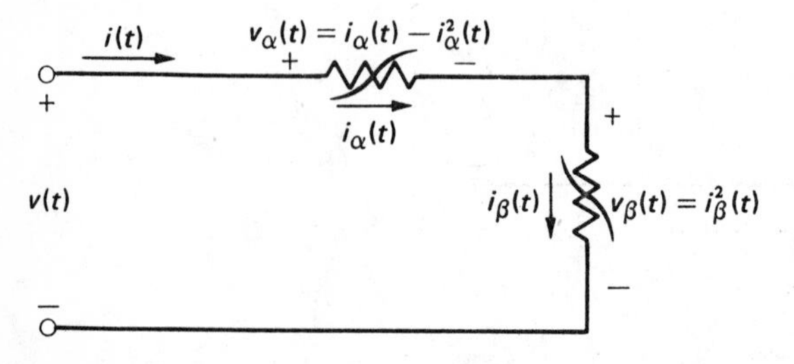

Figure 1.7 A linear one-port network comprised of two nonlinear resistors.

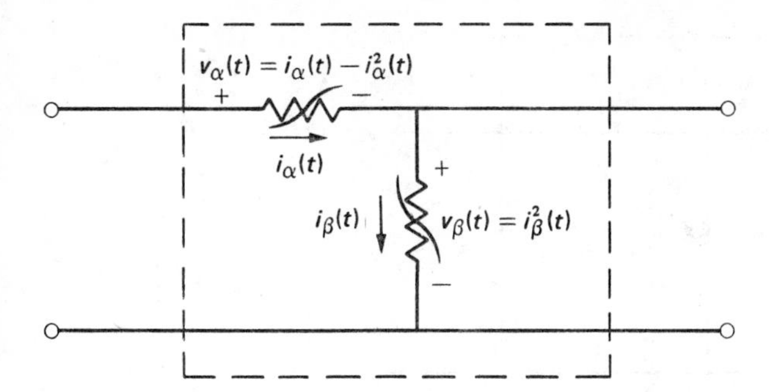

Figure 1.8 A nonlinear two-port network.

then

$$\mathbf{y}_a(t) = \mathbf{y}_b(t - \tau) \tag{1.9b}$$

where $\mathbf{y}_a(t)$ and $\mathbf{y}_b(t)$ are the responses of arbitrary excitations $\mathbf{u}_a(t)$ and $\mathbf{u}_b(t)$, respectively. An n-port network is *portwise time-varying* or simply *time-varying* if it is not portwise time-invariant.

In other words, an n-port network is time-invariant if its port behavior is invariant to a shift in the time origin. Putting it differently, a time-invariant n-port should produce the same response no matter when a given excitation is applied. Thus, any n-port network comprised only of time-invariant network elements and devoid of any initial conditions is always portwise time-invariant. The converse, however, is not necessarily true. It is quite easy to conceive of an n-port with time-varying elements that exhibits a port behavior that is time-invariant. Figure 1.9 shows a one-port comprised of a series connection of two time-varying resistors, whose input impedance is $2R$ ohms. Thus, the one-port is portwise time-invariant. As before, if a two-port is formed from this one-port by connecting two wires across one of the time-varying resistors, as shown in Fig. 1.10, the resulting two-port becomes portwise time-varying. Also, in general, n-ports with initial stored energies that affect port behavior must be considered to be portwise time-varying.

It is important to distinguish between a time-invariant network and a portwise time-invariant n-port. The network of Fig. 1.9 is time-varying, but it is a portwise time-invariant one-port.

In the remainder of this book, we shall mostly consider time-invariant

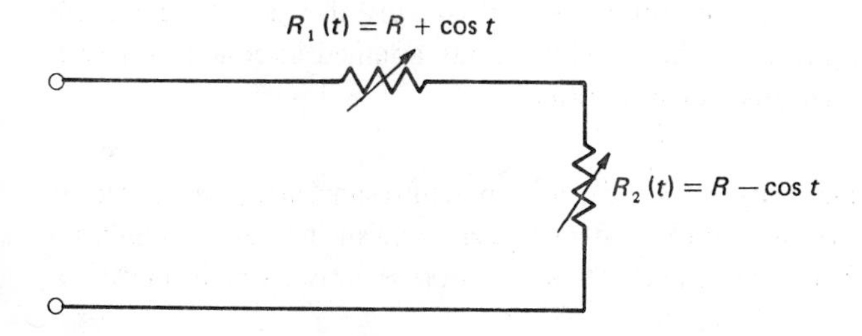

Figure 1.9 A time-invariant one-port network comprised of two time-varying resistors.

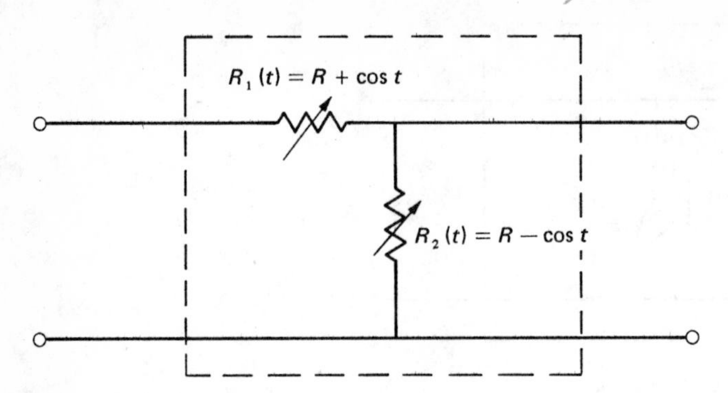

Figure 1.10 A time-varying two-port network.

networks. However, time-varying networks are important in modern engineering applications, having been used successfully in the design of parametric amplifiers and switching networks.

1.3 PASSIVITY AND ACTIVITY

The concept of passivity is not quite so well understood. For many, it simply means the absence of internal sources. A poorly designed transistor circuit may be passive in the sense that it is incapable of delivering energy at any time. A better definition of passivity, which is now widely accepted, is obtained by considering the energy delivery capability of a network.

Definition 1.3: Passivity and activity An n-port network is said to be *passive* if, for *all* admissible signal pairs $\mathbf{v}(t)$ and $\mathbf{i}(t)$, the total energy stored in and delivered to the n-port is nonnegative for all time t; that is,

$$\varepsilon(t) = \varepsilon(t_0) + \int_{t_0}^{t} \mathbf{v}'(x)\mathbf{i}(x)\, dx \geqslant 0 \tag{1.10}$$

for *all* initial time t_0 and *all* time $t \geqslant t_0$, where $\varepsilon(t_0)$ is the finite energy stored in the n-port at the initial time t_0. An n-port network is *active* if it is not passive. An n-port network is *strictly passive* if the equality is not attained in condition (1.10) for all nonzero admissible signal pairs $\mathbf{v}(t)$ and $\mathbf{i}(t)$.

In other words, a passive n-port is incapable of delivering energy at any time. To demonstrate activity, we need only find one excitation such that the condition (1.10) is violated for at least one time t. Thus, an n-port is active if and only if for *some* excitation

$$\varepsilon(t)=\varepsilon(t_0)+\int_{t_0}^{t}\mathbf{v}'(x)\mathbf{i}(x)\,dx<0 \tag{1.11}$$

for *some* initial time t_0 and *some* time $t \geqslant t_0$.

The linear time-invariant resistor, inductor, and capacitor with nonnegative element values are simple examples of passive one-ports. The two-port network of Fig. 1.5 is, as expected, also passive. To demonstrate this, we compute its terminal voltages in terms of its terminal currents, which are given by the equations

$$v_1(t)=(R_1+R_2)i_1(t)+R_2i_2(t) \tag{1.12a}$$

$$v_2(t)=R_2i_1(t)+(R_2+R_3)i_2(t) \tag{1.12b}$$

Substituting these in Eq. (1.10) yields

$$\varepsilon(t)=\varepsilon(t_0)+\int_{t_0}^{t}[v_1(x)i_1(x)+v_2(x)i_2(x)]\,dx$$

$$=\int_{t_0}^{t}[R_2(i_1+i_2)^2+R_1i_1^2+R_3i_2^2]\,dx \geqslant 0 \tag{1.13}$$

since the integrand is nonnegative, where $\varepsilon(t_0)=0$. Thus, the two-port network of Fig. 1.5 is passive.

The absence of internal sources does not always imply passivity. For example, a nonlinear one-port resistor characterized by the equation

$$v(t)=i(t)+i^2(t) \tag{1.14}$$

is active, because we can demonstrate that (1.11) can be fulfilled for some excitation, some initial time t_0, and some time t. To see this, let us apply a current source $i(t)=-e^t$ to the nonlinear resistor. Then from (1.14), (1.11) can be expressed as

$$\varepsilon(t)=\varepsilon(t_0)+\int_{t_0}^{t}(i+i^2)i\,dx=\frac{1}{2}e^{2t}-\frac{e^{3t}}{3}-\frac{1}{2}e^{2t_0}+\frac{e^{3t_0}}{3} \tag{1.15}$$

where $\varepsilon(t_0)=0$. For $t>\ln 1.5$ and $t_0 \leqslant \ln 1.5$, $\varepsilon(t)$ is negative, thus showing the nonlinear resistor to be active. In fact, any resistor, linear or otherwise, is passive if and only if its *vi*-characteristic lies solely in the first and third quadrants of the *vi*-plane. An unbiased tunnel diode, whose terminal *VI*-characteristic is as shown in Fig. 1.11*b*, is therefore a passive device. However, a properly biased tunnel diode of Fig. 1.12*a*, with its *vi*-characteristic as shown in Fig. 1.12*b*, is active in that it is capable of delivering energy for small signal levels; its justification is left as an exercise (see Prob. 1.1).

The definition of passivity as defined above cannot be stated solely in terms of the port variables $\mathbf{v}(t)$ and $\mathbf{i}(t)$; the initial stored energy $\varepsilon(t_0)$ must be taken into

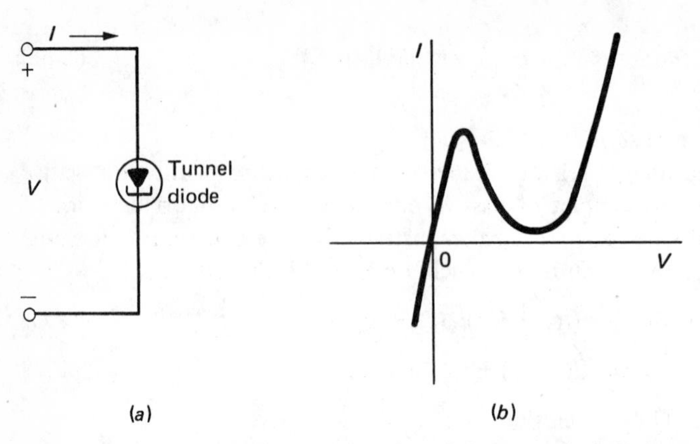

Figure 1.11 (*a*) An unbiased tunnel diode and (*b*) its terminal *VI*-characteristic.

account. Consider, for example, a linear time-invariant capacitor of capacitance C. To test passivity, we compute

$$\begin{aligned}\mathcal{E}(t) &= \mathcal{E}(t_0) + \int_{t_0}^{t} v(x)i(x)\,dx = \mathcal{E}(t_0) + C\int_{v(t_0)}^{v(t)} v\,dv \\ &= \mathcal{E}(t_0) + \tfrac{1}{2}Cv^2(t) - \tfrac{1}{2}Cv^2(t_0) = \tfrac{1}{2}Cv^2(t) \qquad (1.16)\end{aligned}$$

where $\mathcal{E}(t_0) = \frac{1}{2}Cv^2(t_0)$, showing that a linear time-invariant capacitor is passive if its capacitance is nonnegative and active if its capacitance is negative. Of course, negative capacitance is not physically realizable. However, it is possible to design an electronic circuit to attain a negative capacitance within a small operating range and a narrow frequency band. Now if the term corresponding to the initial stored

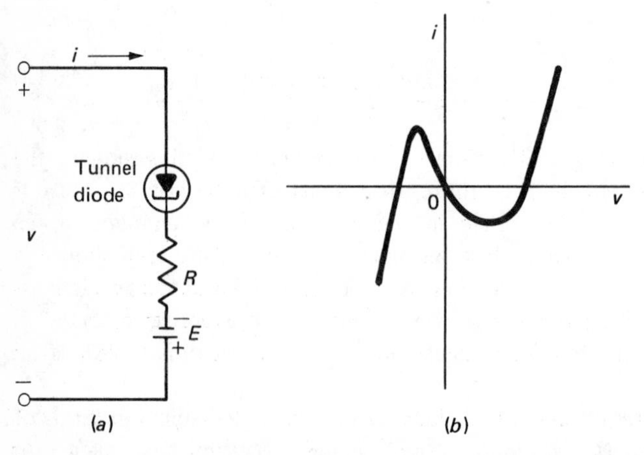

Figure 1.12 (*a*) A biased tunnel diode and (*b*) its terminal *vi*-characteristic.

energy $\mathcal{E}(t_0)$ is not included in Definition 1.3, Eq. (1.16) becomes $\mathcal{E}(t) = \frac{1}{2}Cv^2(t) - \frac{1}{2}Cv^2(t_0)$ and can then be made negative for some excitation, say, $v(t) = e^{-t}$ for positive C. This is totally unacceptable, and no one expects a linear time-invariant capacitor with positive capacitance to be active.

The above discussion of the capacitor brings out an important variation on the passivity of an n-port network. As is well known, the capacitor stores energy during part of the time and then returns it to the sources later. Thus, the total energy delivered to the capacitor from the very early time when the capacitor is completely quiescent starting at $t_0 = -\infty$ to the time $t = \infty$ must be identically zero. This gives rise to the concept of losslessness. An n-port network is lossless if all of the energy delivered into its ports is ultimately returned from the ports. More formally, we have the following definition.

Definition 1.4: Losslessness An n-port network is said to be *lossless* if, for *all finite* admissible port signals $\mathbf{v}(t)$ and $\mathbf{i}(t)$ that are square-integrable from t_0 to ∞, that is,

$$\int_{t_0}^{\infty} \mathbf{v}'(x)\mathbf{v}(x)\, dx < \infty \tag{1.17a}$$

$$\int_{t_0}^{\infty} \mathbf{i}'(x)\mathbf{i}(x)\, dx < \infty \tag{1.17b}$$

then

$$\mathcal{E} = \mathcal{E}(t_0) + \int_{t_0}^{\infty} \mathbf{v}'(x)\mathbf{i}(x)\, dx = 0 \tag{1.17c}$$

for *all* initial time t_0, where $\mathcal{E}(t_0)$ denotes the finite energy stored in the n-port at the initial time t_0.

The requirement that the port signals be square-integrable is necessary. Consider, for example, a linear time-invariant inductor of inductance $L = 1$ H. Clearly, $v(t) = e^t$ and $i(t) = e^t$ are admissible port signals for the inductor, but they are not square-integrable. The total energy stored in the inductor at any time t is $\mathcal{E}(t) = \frac{1}{2}e^{2t}$. As t approaches infinity, $\mathcal{E}(t)$ becomes unbounded and Eq. (1.17c) is not satisfied. Thus, in testing for losslessness, Eq. (1.17c) should be checked for all finite admissible port signals $\mathbf{v}(t)$ and $\mathbf{i}(t)$ that are square-integrable from any initial time t_0 to $t = \infty$. For the integrals in (1.17) to exist, it is necessary that $v(\infty) = 0$ and $i(\infty) = 0$. Also, for $t_0 = -\infty$, we must have $v(-\infty) = 0$ and $i(-\infty) = 0$. In the case of the inductor considered above, the total energy stored at any time t is $\mathcal{E}(t) = \frac{1}{2}Li^2(t)$, which approaches zero as t approaches infinity, showing that a linear time-invariant inductor with nonnegative inductance is lossless.

When an n-port is completely quiescent with no stored energy at some very early time starting $t_0 = -\infty$, an alternative and more compact definition of passivity becomes

$$\mathcal{E}(t) = \int_{-\infty}^{t} \mathbf{v}'(x)\mathbf{i}(x)\,dx \geqslant 0 \tag{1.18}$$

for *all* excitations and for *all* time t. Likewise, an initially relaxed n-port network is active if for *some* excitation

$$\mathcal{E}(t) = \int_{-\infty}^{t} \mathbf{v}'(x)\mathbf{i}(x)\,dx < 0 \tag{1.19}$$

for *some* time t, and is lossless if for *all finite* admissible port signals $\mathbf{v}(t)$ and $\mathbf{i}(t)$ that are square-integrable, as defined in (1.17a) and (1.17b) with t_0 being replaced by $-\infty$,

$$\mathcal{E} = \int_{-\infty}^{\infty} \mathbf{v}'(x)\mathbf{i}(x)\,dx = 0 \tag{1.20}$$

Example 1.1 Figure 1.13 represents a gyrator of gyration resistance r. The gyrator is a two-port device whose terminal voltages and currents are characterized by the equations

$$v_1(t) = -ri_2(t) \tag{1.21a}$$

$$v_2(t) = ri_1(t) \tag{1.21b}$$

The total energy delivered to the gyrator from $-\infty$ to t is given by

$$\mathcal{E}(t) = \int_{-\infty}^{t} (v_1 i_1 + v_2 i_2)\,dx = \int_{-\infty}^{t} (ri_1 i_2 - ri_1 i_2)\,dx = 0 \tag{1.22}$$

showing that the gyrator is a lossless two-port network.

Earlier, we demonstrated that the absence of internal sources in an n-port does not necessarily mean that the n-port is passive. However, the presence of internal sources, dependent or independent, does not guarantee that the n-port is always active. This is illustrated by the following example.

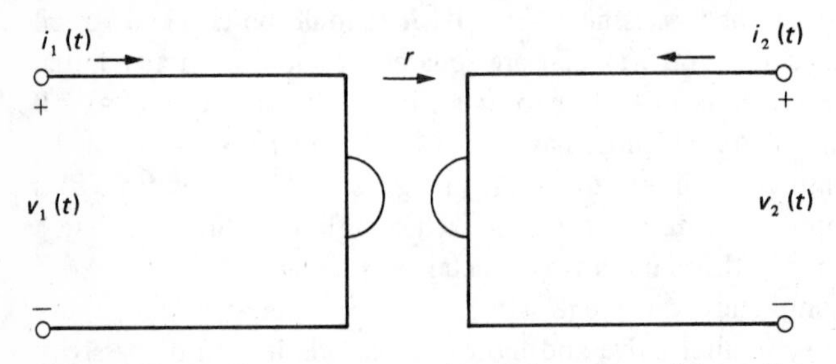

Figure 1.13 The diagrammatic symbol of the gyrator with gyration resistance r.

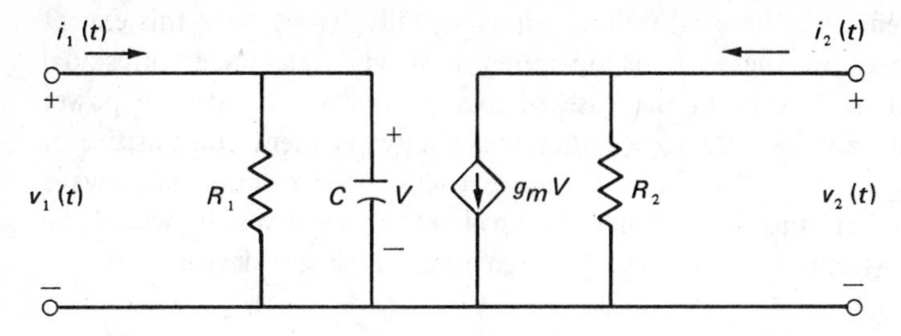

Figure 1.14 An equivalent network of a transistor.

Example 1.2 Consider the equivalent network of a transistor as shown in Fig. 1.14. The port currents and voltages are related by the equations

$$i_1(t) = C\frac{dv_1(t)}{dt} + \frac{v_1(t)}{R_1} \tag{1.23a}$$

$$i_2(t) = g_m v_1(t) + \frac{v_2(t)}{R_2} \tag{1.23b}$$

Substituting these in $\varepsilon(t)$ of (1.18) or (1.19) gives

$$\varepsilon(t) = \int_{-\infty}^{t} \left(Cv_1 \frac{dv_1}{dx} + \frac{v_1^2}{R_1} + \frac{v_2^2}{R_2} + g_m v_1 v_2 \right) dx \tag{1.24}$$

Choose port voltages $v_1(t)$ and $v_2(t)$ as the excitation and set

$$v_2(t) = -\frac{g_m R_2 v_1(t)}{2} \tag{1.25}$$

yielding the total energy delivered to the transistor at time t as

$$\varepsilon(t) = \frac{1}{2} Cv_1^2(t) - \frac{1}{4} R_2 \left(g_m^2 - \frac{4}{R_1 R_2} \right) \int_{-\infty}^{t} v_1^2(x)\, dx \tag{1.26}$$

The first term on the right-hand side of Eq. (1.26) denotes the energy stored in the capacitor C at the time t. Since the capacitor is lossless, the first term reduces to zero as t approaches infinity. Equation (1.26) shows that if

$$g_m^2 > \frac{4}{R_1 R_2} \tag{1.27}$$

the input energy $\varepsilon(t)$ to the transistor is always negative for sufficiently large t because the integrand in the second term is always positive for nonzero $v_1(t)$, indicating that the device is active. If, on the other hand, $g_m^2 \leqslant 4/R_1 R_2$, then the device becomes passive and behaves like an energy sink or absorber.

Before we turn our attention to another concept, we remark that it has been suggested (and used by many) that we use average power in defining the concept of passivity instead of energy. There is nothing fundamentally wrong with this except that we have to regard the network as operating at steady state under sinusoidal excitations and responses. Also, in the case of lossless n-ports, the average power into the n-port would always be zero whether the reactive elements are positive or negative. The average power flow into a pure capacitor, for example, is always zero regardless whether the capacitance is positive or negative. It would be inconceivable to call a capacitor with negative capacitance a passive device.

1.4 CAUSALITY AND NONCAUSALITY

The term *causality* connotes the existence of a *cause-effect* relationship. Intuitively, it states that a causal n-port network cannot yield any response until after the excitation is applied. In other words, a causal n-port is not anticipative; it cannot predict the future behavior of the excitation. Of course, all the physical networks must be nonanticipative, but the networks that we deal with and analyze are models made up by the elements that are idealizations of actual physical devices such as resistors, capacitors, inductors, transformers, and generators. We cannot always assume that they are nonanticipative. Furthermore, physical networks may be noncausal.

In this section, we formally introduce the concept of causality in terms of the excitation-response properties of an n-port and then show that any linear passive n-port with the exception of a few mostly trivial cases is causal. Since causality is defined in terms of excitation and response, we must first specify the excitation and response variables of the n-port under consideration. To keep our discussion general, let $\mathbf{u}(t)$ be the vector associated with the excitation signals, and $\mathbf{y}(t)$ the response signals. For the excitation-response situation of Fig. 1.2, the excitation vector $\mathbf{u}(t)$ and the response vector $\mathbf{y}(t)$ are given by

$$\mathbf{u}(t) = [i_1(t), i_2(t), \ldots, i_k(t), v_{k+1}(t), \ldots, v_n(t)]' \tag{1.28a}$$

$$\mathbf{y}(t) = [v_1(t), v_2(t), \ldots, v_k(t), i_{k+1}(t), \ldots, i_n(t)]' \tag{1.28b}$$

With these preliminaries, the intuitive notion of causality can now be stated formally as follows.

Definition 1.5: Causality and noncausality An n-port network is said to be *causal* if

$$\mathbf{u}_a(t) = \mathbf{u}_b(t) \qquad -\infty < t < t_1 \tag{1.29a}$$

then

$$\mathbf{y}_a(t) = \mathbf{y}_b(t) \qquad -\infty < t < t_1 \tag{1.29b}$$

for any t_1, where $\mathbf{y}_a(t)$ and $\mathbf{y}_b(t)$ are the responses of arbitrary excitations $\mathbf{u}_a(t)$ and $\mathbf{u}_b(t)$, respectively, of the n-port. An n-port network is *noncausal* if it is not causal.

The definition states in effect that two identical excitations over any time interval must result in identical responses over the same time interval for an n-port to be causal. To test for noncausality, it is sufficient to find *some* excitations $\mathbf{u}_a(t)$ and $\mathbf{u}_b(t)$ such that for *some* time t_1, $\mathbf{u}_a(t)=\mathbf{u}_b(t)$ for $t<t_1$ gives rise to $\mathbf{y}_a(t)\neq\mathbf{y}_b(t)$ for *some* $t<t_1$. Accordingly, an n-port is noncausal because it is either anticipative or the response is not a unique function of the excitation.

Examples of the causal networks are the one-port linear resistor, capacitor, and inductor under either voltage or current excitation. The two-port networks of Figs. 1.13 and 1.14 are also examples of causal networks under all excitations. For an example of a noncausal one-port, consider the nonlinear resistor of Eq. (1.14) under voltage excitation-current response. The vi-relation of this nonlinear resistor, as repeated here, is

$$v(t)=i(t)+i^2(t) \tag{1.30}$$

Choose two input voltages $v_a(t)$ and $v_b(t)$ such that, for $t<t_1$,

$$v_a(t)=v_b(t)=e^{-t} \tag{1.31}$$

For this input, there correspond two different responses

$$i_a(t)=-\tfrac{1}{2}+\sqrt{\tfrac{1}{4}+e^{-t}} \tag{1.32a}$$

$$i_b(t)=-\tfrac{1}{2}-\sqrt{\tfrac{1}{4}+e^{-t}} \tag{1.32b}$$

for $t<t_1$. Hence, this nonlinear resistor is noncausal under voltage excitation-current response. Now suppose that the excitation is a current source, and the response is the port voltage. Then for any two input currents $i_a(t)$ and $i_b(t)$ such that

$$i_a(t)=i_b(t) \qquad t<t_1 \tag{1.33}$$

the corresponding voltages $v_a(t)$ and $v_b(t)$ are equal,

$$v_a(t)=i_a(t)+i_a^2(t)=i_b(t)+i_b^2(t)=v_b(t) \qquad t<t_1 \tag{1.34}$$

showing that the resistor is causal under current excitation-voltage response.

Likewise, the unbiased tunnel diode of Fig. 1.11 is a causal one-port under voltage excitation-current response, and a noncausal one-port under current excitation-voltage response (see Prob. 1.3).

Example 1.3 Consider a two-port ideal transformer of turns ratio $k:1$, whose port voltages and currents are related by the equation (Fig. 1.15)

$$\begin{bmatrix} v_2(t) \\ i_2(t) \end{bmatrix} = \begin{bmatrix} \dfrac{1}{k} & 0 \\ 0 & -k \end{bmatrix} \begin{bmatrix} v_1(t) \\ i_1(t) \end{bmatrix} \tag{1.35}$$

If $\mathbf{u}(t)=[v_1(t),\ i_2(t)]'$ is taken as the excitation, then the response $\mathbf{y}(t)=[i_1(t),\ v_2(t)]'$ is uniquely determined through Eq. (1.35). Consequently, for any two identical excitations we obtain identical responses, meaning that the transformer is causal under the chosen excitation and response. However, if $\mathbf{u}(t)=[v_1(t),\ v_2(t)]'$ is

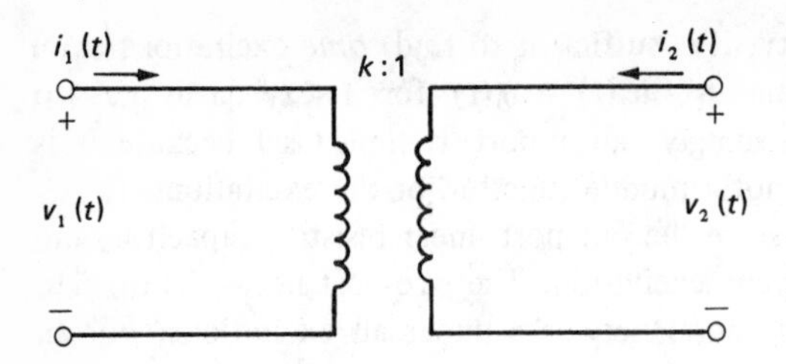

Figure 1.15 The two-port ideal transformer with turns ratio $k:1$.

taken as the excitation, then we can choose two inputs $\mathbf{u}_a(t)$ and $\mathbf{u}_b(t)$ such that, for all t,

$$\mathbf{u}_a(t) = \begin{bmatrix} kv(t) \\ v(t) \end{bmatrix} = \mathbf{u}_b(t) \tag{1.36}$$

$v(t)$ being arbitrary. The corresponding responses are

$$\mathbf{y}_a(t) = \begin{bmatrix} i(t) \\ -ki(t) \end{bmatrix} \quad \text{and} \quad \mathbf{y}_b(t) = \begin{bmatrix} \hat{i}(t) \\ -k\hat{i}(t) \end{bmatrix} \tag{1.37}$$

where $i(t)$ and $\hat{i}(t)$ are arbitrary. Choosing $i(t) \neq \hat{i}(t)$ for some t shows that two identical excitations do not result in identical responses. Thus, the ideal transformer is noncausal under voltage excitation-current response. Likewise, we can show that the transformer is noncausal under current excitation-voltage response.

From the above discussion, we emphasize that in characterizing an n-port as being causal or noncausal, we must specify the excitation and response. An n-port may be causal under one set of excitation and response and noncausal under a different set. In general, most linear networks are causal, and noncausal linear networks are very rare. We shall show shortly that with the exception of a few mostly trivial cases, any linear passive n-port network is causal.

It has also been suggested in the literature that causality or nonanticipativeness of an n-port network be defined in such a way that zero excitation over a time interval implies zero response over the same time interval; that is, $\mathbf{u}(t) = \mathbf{0}$ for $t \leqslant t_1$ implies $\mathbf{y}(t) = \mathbf{0}$ for $t \leqslant t_1$. This criterion for causality is not satisfactory in that some obviously causal and nonanticipative n-ports may be rendered noncausal and anticipative by this approach. Consider, for example, the one-port network of Fig. 1.16, comprised of a series connection of a resistor and a battery. Shorting the two

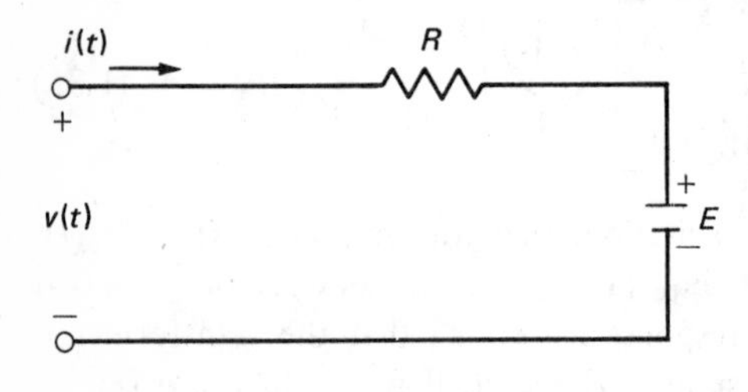

Figure 1.16 A causal and nonanticipative one-port network.

terminals gives rise to a nonzero current $i = -E/R$, indicating that a zero excitation $v(t) = 0$ does not imply a zero response $i(t) \neq 0$. It would be incorrect to say that this simple one-port is noncausal and anticipative.

Before we show that causality is a consequence of linearity and passivity, we review briefly the concept of linearity. As indicated in Sec. 1.1, an n-port is said to be *linear* if the superposition principle holds, that is, if $\mathbf{y}_a(t)$ and $\mathbf{y}_b(t)$ are the responses of the excitations $\mathbf{u}_a(t)$ and $\mathbf{u}_b(t)$, respectively, of an n-port, then for any choice of real scalars α and β, the vector $\alpha\mathbf{y}_a(t) + \beta\mathbf{y}_b(t)$ represents the response of the excitation $\alpha\mathbf{u}_a(t) + \beta\mathbf{u}_b(t)$. When we speak of an n-port that can support n *linearly independent* excitations, we mean that if $\{\mathbf{u}_k(t), \mathbf{y}_k(t);\ k = 1, 2, \ldots\}$ denotes the set of all excitation-response pairs that can be supported by the n-port, there exist n linearly independent excitation vectors $\mathbf{u}_k(t)$'s in the set.

Theorem 1.1 A linear passive n-port network that can support n linearly independent excitations for all time is also causal.

PROOF Let $\mathbf{y}(t)$ and $\mathbf{y}_0(t)$ be the responses of the excitations $\mathbf{u}(t)$ and $\mathbf{u}_0(t)$, respectively. Assume that for arbitrary finite $t_1 \geqslant t_0$,

$$\mathbf{u}(t) = \mathbf{0} \qquad \text{for } t < t_1 \tag{1.38}$$

Write, for any scalar α,

$$\hat{\mathbf{u}}(t) = \mathbf{u}_0(t) + \alpha\mathbf{u}(t) \tag{1.39}$$

Then $\hat{\mathbf{u}}(t)$ coincides with $\mathbf{u}_0(t)$ for $t < t_1$; that is,

$$\hat{\mathbf{u}}(t) = \mathbf{u}_0(t) \qquad \text{for } t < t_1 \tag{1.40}$$

Denote by $\hat{\mathbf{y}}(t)$ the response corresponding to the excitation $\hat{\mathbf{u}}(t)$. Then, by linearity,

$$\hat{\mathbf{y}}(t) = \mathbf{y}_0(t) + \alpha\mathbf{y}(t) \tag{1.41}$$

Appealing to passivity yields

$$\varepsilon(t) = \varepsilon(t_0) + \int_{t_0}^{t} \hat{\mathbf{u}}'(x)\hat{\mathbf{y}}(x)\,dx \geqslant 0 \tag{1.42}$$

for all t (see Prob. 1.5). Thus, for all $t < t_1$ and for any real scalar α,

$$\varepsilon(t_0) + \int_{t_0}^{t} \mathbf{u}_0'(x)\mathbf{y}_0(x)\,dx + \alpha\int_{t_0}^{t} \mathbf{u}_0'(x)\mathbf{y}(x)\,dx \geqslant 0 \tag{1.43}$$

Since $\varepsilon(t_0)$ is finite and α is any real scalar, for the inequality (1.43) to hold it is necessary that

$$\int_{t_0}^{t} \mathbf{u}_0'(x)\mathbf{y}(x)\,dx = 0 \qquad \text{for } t < t_1 \tag{1.44}$$

for, if not, we can choose an appropriate value of α to make the left-hand side of (1.43) negative. Equation (1.44) is possible for any $t<t_1$ only if

$$\mathbf{u}_0'(t)\mathbf{y}(t)=0 \qquad \text{for } t<t_1 \tag{1.45}$$

Let $\mathbf{u}_k(t)$, $k=1, 2, \ldots, n$, be n linearly independent excitations that can be supported by the n-port under consideration. Since in deriving Eq. (1.45) $\mathbf{u}_0(t)$ is an arbitrary excitation, Eq. (1.45) is also applicable to $\mathbf{u}_k(t)$. Applying it to $\mathbf{u}_k(t)$ successively gives

$$\mathbf{u}_k'(t)\mathbf{y}(t)=0 \qquad k=1, 2, \ldots, n \text{ and } t<t_1 \tag{1.46}$$

or, more compactly,

$$[\mathbf{u}_1(t), \mathbf{u}_2(t), \ldots, \mathbf{u}_n(t)]'\mathbf{y}(t)=0 \qquad \text{for } t<t_1 \tag{1.47}$$

Since the vectors $\mathbf{u}_k(t)$ are linearly independent, the coefficient matrix is nonsingular. Taking the inverse of this matrix yields

$$\mathbf{y}(t)=\mathbf{0} \qquad \text{for } t<t_1 \tag{1.48}$$

To complete our proof, let $\mathbf{y}_a(t)$ and $\mathbf{y}_b(t)$ be the responses of the n-port resulting from the excitations $\mathbf{u}_a(t)$ and $\mathbf{u}_b(t)$, respectively, and set

$$\mathbf{u}_a(t)=\mathbf{u}_b(t) \qquad \text{for } t<t_1 \tag{1.49}$$

Then we have

$$\mathbf{u}(t)=\mathbf{u}_a(t)-\mathbf{u}_b(t) \qquad \text{for } t<t_1 \tag{1.50}$$

Appealing once more to linearity, we get

$$\mathbf{y}(t)=\mathbf{y}_a(t)-\mathbf{y}_b(t) \qquad \text{for } t<t_1 \tag{1.51}$$

which according to (1.48) must be identically zero for $t<t_1$, giving

$$\mathbf{y}_a(t)=\mathbf{y}_b(t) \qquad \text{for } t<t_1 \tag{1.52}$$

This completes the proof of the theorem.

The ideal one-port short circuit is noncausal under voltage excitation-current response because it admits an infinitude of responses for the sole excitation $v(t)=0$. It fails the sufficient condition of the above theorem in that the one-port does not support any independent excitation. However, it is causal under current excitation-voltage response because at each instant of time all the current excitations are related by a multiplicative constant, meaning that the ideal one-port short circuit supports one independent excitation, thus fulfilling the sufficient requirement of the theorem. As for the ideal transformer considered in Example 1.3, if $\mathbf{u}(t)=[v_1(t), i_2(t)]'$ is taken as the excitation, the transformer clearly can support two independent excitations. One could be chosen, for example, as $[v_1(t), 0]'$ and the other as $[0, i_2(t)]'$ for arbitrary but positive $v_1(t)$ and $i_2(t)$. Thus, according to the theorem, the ideal transformer is causal under the specified excitation and response. Now suppose that $\mathbf{u}(t) = [v_1(t), v_2(t)]'$ is taken as the excitation. Then the transformer can support only excitations of the type

$$\mathbf{u}(t) = \begin{bmatrix} kv(t) \\ v(t) \end{bmatrix} = v(t) \begin{bmatrix} k \\ 1 \end{bmatrix} \tag{1.53}$$

$v(t)$ being arbitrary. This shows that the ideal transformer can support only one independent excitation and hence the sufficient condition of the theorem is not satisfied, and, as demonstrated in Example 1.3, the ideal transformer is noncausal.

From the definition of causality, it is obvious that a causal n-port network must have its response as a single-valued function of its excitation. The converse is, however, not necessarily true. Even a linear passive n-port for which the response is a single-valued function of the excitation may be noncausal (see Prob. 1.9). According to Theorem 1.1, the response can be a multiple-valued function of the excitation only if the linear passive n-port does not support n linearly independent excitations for all time. The following theorem demonstrates that the instantaneous power entering a linear passive n-port is uniquely determined by its excitation.

Theorem 1.2 The instantaneous power entering a linear passive n-port network is uniquely determined by its excitation.

PROOF Let $\mathbf{y}_a(t)$ and $\mathbf{y}_b(t)$ be the responses of two arbitrary excitations $\mathbf{u}_a(t)$ and $\mathbf{u}_b(t)$ of a linear passive n-port, respectively. It is sufficient to show that if $\mathbf{u}_a(t) = \mathbf{u}_b(t)$, then $\mathbf{u}_a'(t)\mathbf{y}_a(t) = \mathbf{u}_b'(t)\mathbf{y}_b(t)$.

Let $\mathbf{y}(t)$ and $\mathbf{y}_0(t)$ be the responses of the arbitrary excitations $\mathbf{u}(t)$ and $\mathbf{u}_0(t)$, respectively. Assume that $\mathbf{u}_a(t) = \mathbf{u}_b(t) = \mathbf{u}_0(t)$ and

$$\mathbf{u}(t) = \mathbf{u}_a(t) - \mathbf{u}_b(t) = \mathbf{0} \tag{1.54}$$

By linearity we have

$$\mathbf{y}(t) = \mathbf{y}_a(t) - \mathbf{y}_b(t) \tag{1.55}$$

Following the steps outlined in Eqs. (1.38)-(1.45) of Theorem 1.1 yields

$$\mathbf{u}_0'(t)\mathbf{y}(t) = 0 \tag{1.56}$$

Substituting (1.55) in (1.56) in conjunction with $\mathbf{u}_a(t) = \mathbf{u}_b(t) = \mathbf{u}_0(t)$ gives

$$\mathbf{u}_a'(t)\mathbf{y}_a(t) = \mathbf{u}_b'(t)\mathbf{y}_b(t) \tag{1.57}$$

This completes the proof of the theorem.

Theorem 1.2 indicates that even if the response of a linear passive n-port network is a multiple-valued function of its excitation, the instantaneous power entering the n-port is uniquely determined by its excitation, and Theorem 1.1 gives a sufficient condition under which the response can be uniquely determined by its excitation.

As an example, let us again consider the ideal transformer of Fig. 1.15 and take $\mathbf{u}(t) = [v_1(t),\ v_2(t)]'$ to be the excitation. Then the transformer can support only excitations of the type (1.53), which yields an arbitrary current response $\mathbf{y}(t)$ of the type

$$\mathbf{y}(t) = \begin{bmatrix} i(t) \\ -ki(t) \end{bmatrix} = i(t) \begin{bmatrix} 1 \\ -k \end{bmatrix} \tag{1.58}$$

where $i(t)$ is an arbitrary real function. The instantaneous power entering the transformer is always zero,

$$P(t) = \mathbf{u}'(t)\mathbf{y}(t) = v(t)i(t)(k-k) = 0 \tag{1.59}$$

showing that the excitation uniquely determines the instantaneous power, namely the zero power, entering the transformer.

1.5 MATRIX CHARACTERIZATIONS OF n-PORT NETWORKS

In the foregoing, we have presented a set of postulates characterizing the physical nature of a network. However, these descriptions, although very general, are not very useful because they are difficult to apply. In the present section, we discuss the frequency-domain characterizations of the class of linear time-invariant n-port networks, which are then used to derive the equivalent frequency-domain conditions of passivity. Since in the remainder of this book we deal exclusively with this class of networks, the adjectives *linear* and *time-invariant* will be omitted in the discussion unless they are used for emphasis. For our purposes, we shall deal with the Laplace-transformed variables and assume that all independent sources and initial conditions have been set to zero in the interior of the n-port network.

As indicated at the beginning of this chapter, any n independent functions of the $2n$ port variables (Fig. 1.1), taking one each of the two variables associated with each of the n ports, can be regarded as the excitation, and the remaining n variables as the response. In the present section, the restriction can be relaxed. Any n independent functions of the $2n$ port variables can be regarded as the excitation and the remaining n variables as the response. For example, both $i_1(t)$ and $v_1(t)$ can be taken as the excitation signals instead of one of them, as previously required. The reason is that in the preceding sections we require that $\mathbf{u}'(t)\mathbf{y}(t)$ represent instantaneous power entering the n-port. This is not possible if both variables such as $v_1(t)$ and $i_1(t)$ are chosen as the excitation signals.

For linear time-invariant n-port networks, their port behavior is completely characterized by giving the relationships among excitation and response signals. Let $\mathbf{y}(t)$ be the response of the excitation $\mathbf{u}(t)$. Then the square matrix $\mathbf{H}(s)$ of order n relating the Laplace transform $\tilde{\mathbf{y}}(s)$ of the response vector $\mathbf{y}(t)$ to the Laplace transform $\tilde{\mathbf{u}}(s)$ of the excitation vector $\mathbf{u}(t)$ is called the *general hybrid matrix* of the n-port, that is,†

$$\tilde{\mathbf{y}}(s) = \mathbf{H}(s)\tilde{\mathbf{u}}(s) \tag{1.60}$$

The elements of $\mathbf{H}(s)$ are referred to as the *general hybrid parameters.* We write

$$\tilde{\mathbf{u}}(s) = [\tilde{u}_1(s), \tilde{u}_2(s), \ldots, \tilde{u}_n(s)]' \tag{1.61a}$$

$$\tilde{\mathbf{y}}(s) = [\tilde{y}_1(s), \tilde{y}_2(s), \ldots, \tilde{y}_n(s)]' \tag{1.61b}$$

$$\mathbf{H}(s) = [h_{ij}(s)] \tag{1.61c}$$

From (1.60) it is quite clear that

$$h_{ij}(s) = \left.\frac{\tilde{y}_i(s)}{\tilde{u}_j(s)}\right|_{\tilde{u}_x(s)=0,\, x \neq j} \tag{1.62}$$

This indicates that $h_{ii}(s)$ represents the input immittance looking into the ith port when the excitation signals at all other ports are set to zero, and that $h_{ij}(s)$, $i \neq j$, denotes the transfer function from port j to port i when all the excitation signals except $\tilde{u}_j(s)$ are set to zero. By setting an excitation signal to zero, we mean the open-circuiting of a current source and the short-circuiting of a voltage source.

In the special situation, when all the excitation signals are currents, Eq. (1.60) becomes

$$\mathbf{V}(s) = \mathbf{Z}(s)\mathbf{I}(s) \tag{1.63}$$

where $\mathbf{V}(s)$ and $\mathbf{I}(s)$ denote the transforms of the port-voltage and the port-current vectors $\mathbf{v}(t)$ and $\mathbf{i}(t)$, as shown in Eq. (1.1). The matrix $\mathbf{Z}(s)$ is called the *open-circuit impedance matrix* or simply the *impedance matrix*, whose elements are referred to as the *open-circuit impedance parameters* or simply the *impedance parameters.* The reason is that the elements are obtained by open-circuiting the appropriate ports. Likewise, if all the excitation signals are voltages, Eq. (1.60) becomes

$$\mathbf{I}(s) = \mathbf{Y}(s)\mathbf{V}(s) \tag{1.64}$$

The matrix $\mathbf{Y}(s)$ is called the *short-circuit admittance matrix* or simply the *admittance matrix*, whose elements are referred to as the *short-circuit admittance parameters* or *admittance parameters.*

We illustrate the above results by the following example.

Example 1.4 Consider the high-frequency equivalent network of a bipolar transistor as shown in Fig. 1.17. Let $\tilde{\mathbf{u}}(s) = [I_1(s),\ V_2(s)]'$ and $\tilde{\mathbf{y}}(s) = [V_1(s),$

†Capital letters are customarily used to denote the Laplace transforms of the corresponding lowercase time-domain functions, but $\mathbf{Y}(s)$ is the standard symbol for the admittance matrix and U the U-function. Here we use the tilde to distinguish $\mathbf{y}(t)$ and $\mathbf{u}(t)$ from their transforms $\tilde{\mathbf{y}}(s)$ and $\tilde{\mathbf{u}}(s)$.

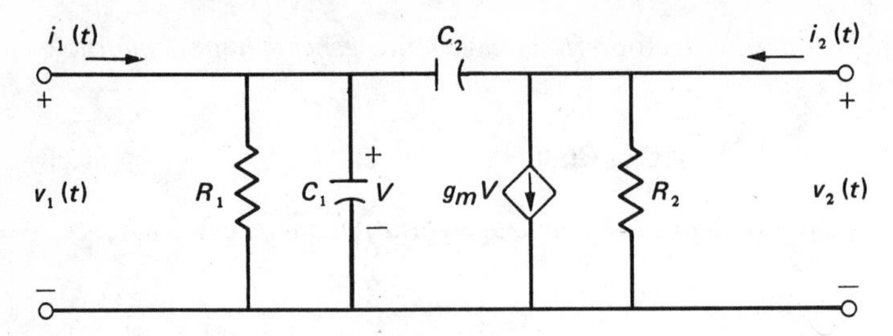

Figure 1.17 A high-frequency equivalent network of a bipolar transistor.

$I_2(s)]'$, where $V_k(s)$ and $I_k(s)$ are the transforms of $v_k(t)$ and $i_k(t)$, respectively. Then, from (1.62), we have

$$h_{11}(s) = \left.\frac{V_1(s)}{I_1(s)}\right|_{V_2(s)=0} = \frac{1}{G_1 + s(C_1 + C_2)} \tag{1.65a}$$

$$h_{21}(s) = \left.\frac{I_2(s)}{I_1(s)}\right|_{V_2(s)=0} = \frac{g_m - sC_2}{G_1 + s(C_1 + C_2)} \tag{1.65b}$$

where $G_1 = 1/R_1$. They are obtained from the network of Fig. 1.17 by short-circuiting the output port (port 2). Likewise, we compute $h_{12}(s)$ and $h_{22}(s)$. The resulting hybrid matrix is given by (see Prob. 1.8)

$$\mathbf{H}(s) = \frac{1}{G_1 + s(C_1 + C_2)} \begin{bmatrix} 1 & sC_2 \\ g_m - sC_2 & q(s) \end{bmatrix} \tag{1.66a}$$

with $G_2 = 1/R_2$, and

$$q(s) = (G_1 + sC_1)(G_2 + sC_2) + sC_2(G_2 + g_m) \tag{1.66b}$$

To compute the impedance matrix, which is conventionally written as

$$\mathbf{Z}(s) = \begin{bmatrix} z_{11}(s) & z_{12}(s) \\ z_{21}(s) & z_{22}(s) \end{bmatrix} \tag{1.67}$$

we let $\tilde{\mathbf{u}}(s) = [I_1(s), I_2(s)]'$ and $\tilde{\mathbf{y}}(s) = [V_1(s), V_2(s)]'$. From Eq. (1.62) we get

$$z_{11}(s) = \left.\frac{V_1(s)}{I_1(s)}\right|_{I_2(s)=0} = \frac{G_2 + sC_2}{q(s)} \tag{1.68a}$$

$$z_{21}(s) = \left.\frac{V_2(s)}{I_1(s)}\right|_{I_2(s)=0} = \frac{sC_2 - g_m}{q(s)} \tag{1.68b}$$

which are obtained from the network of Fig. 1.17 by open-circuiting the output port. In a similar manner, we can compute $z_{12}(s)$ and $z_{22}(s)$. The impedance matrix is given by

$$\mathbf{Z}(s) = \frac{1}{q(s)} \begin{bmatrix} G_2 + sC_2 & sC_2 \\ sC_2 - g_m & G_1 + s(C_1 + C_2) \end{bmatrix} \tag{1.69}$$

To determine the admittance matrix $\mathbf{Y}(s)$, conventionally written as

$$Y(s) = \begin{bmatrix} y_{11}(s) & y_{12}(s) \\ y_{21}(s) & y_{22}(s) \end{bmatrix} \tag{1.70}$$

we let $\tilde{\mathbf{u}}(s) = [V_1(s), V_2(s)]'$ and $\tilde{\mathbf{y}}(s) = [I_1(s), I_2(s)]'$. From Eq. (1.62) we have

$$y_{11}(s) = \left.\frac{I_1(s)}{V_1(s)}\right|_{V_2(s)=0} = G_1 + s(C_1 + C_2) \tag{1.71a}$$

$$y_{21}(s) = \left.\frac{I_2(s)}{V_1(s)}\right|_{V_2(s)=0} = g_m - sC_2 \tag{1.71b}$$

They are obtained by short-circuiting the output port of the two-port network of Fig. 1.17. In a similar fashion, we can compute $y_{12}(s)$ and $y_{22}(s)$. The resulting admittance matrix is given by

$$\mathbf{Y}(s) = \begin{bmatrix} G_1 + s(C_1 + C_2) & -sC_2 \\ g_m - sC_2 & G_2 + sC_2 \end{bmatrix} \tag{1.72}$$

Finally, suppose that we take $\tilde{\mathbf{u}}(s) = [V_2(s), -I_2(s)]'$ as the excitation and $\tilde{\mathbf{y}}(s) = [V_1(s), I_1(s)]'$ as the response. The hybrid matrix $\mathbf{H}(s)$, conventionally written as

$$\mathbf{H}(s) = \mathbf{T}(s) = \begin{bmatrix} A(s) & B(s) \\ C(s) & D(s) \end{bmatrix} \tag{1.73}$$

is called the *transmission* or *chain matrix*, whose elements are referred to as the *transmission* or *chain parameters.* They are also known as the *ABCD parameters.* Again, from Eq. (1.62) we have

$$A(s) = \left.\frac{V_1(s)}{V_2(s)}\right|_{I_2=0} = \frac{G_2 + sC_2}{sC_2 - g_m} \tag{1.74a}$$

$$C(s) = \left.\frac{I_1(s)}{V_2(s)}\right|_{I_2(s)=0} = \frac{q(s)}{sC_2 - g_m} \tag{1.74b}$$

which are obtained from the two-port by open-circuiting the output port. Likewise, we can compute $B(s)$ and $D(s)$ by short-circuiting the output port. The resulting transmission matrix is given by

$$\mathbf{T}(s) = \frac{1}{sC_2 - g_m} \begin{bmatrix} G_2 + sC_2 & 1 \\ q(s) & G_1 + s(C_1 + C_2) \end{bmatrix} \tag{1.75}$$

The above discussion indicates that there are many matrix representations for an n-port network. Depending on the applications, an n-port network can be represented by an impedance, admittance, hybrid, or transmission matrix. They are all, of course, special cases of the general hybrid matrix defined in Eq. (1.60). It is sometimes necessary to convert one set of parameters to another in order to achieve the desired result. The hybrid parameters, for example, may be specified for a transistor by the manufacturer, and we wish to use the admittance parameters in our design. Some conversions are straightforward, and others are not. Comparing Eqs. (1.63) and (1.64), for example, we see that if $\mathbf{Z}(s)$ is nonsingular, the admittance matrix $\mathbf{Y}(s)$ is simply the inverse of the impedance matrix and vice versa. In the following, we present a procedure that permits a simple conversion from one set of parameters to another.

Express Eq. (1.60) in a slightly different form as

$$[-\mathbf{1}_n, \mathbf{H}(s)] \begin{bmatrix} \tilde{\mathbf{y}}(s) \\ \tilde{\mathbf{u}}(s) \end{bmatrix} = \mathbf{0} \tag{1.76}$$

where $\mathbf{1}_n$ denotes the identity matrix of order n. Observe the pattern of the coefficient matrix. The vector $[\tilde{\mathbf{y}}(s), \tilde{\mathbf{u}}(s)]'$ contains all the $2n$ port variables of the n-port under consideration. Suppose that we premultiply the coefficient matrix of Eq. (1.76) by the negative of the inverse of the submatrix consisting of the columns corresponding to the variables chosen as the new response signals. Then the submatrix formed by the other columns in the resulting matrix product is the general hybrid matrix with respect to the new excitation signals. We illustrate this procedure by the following example.

Example 1.5 Consider the same equivalent network of a bipolar transistor of Fig. 1.17. Suppose that the admittance matrix (1.72) is known, and we wish to determine the hybrid matrix (1.66*a*). To this end, we write (1.72) as

$$\begin{bmatrix} -1 & 0 & G_1 + s(C_1 + C_2) & -sC_2 \\ 0 & -1 & g_m - sC_2 & G_2 + sC_2 \end{bmatrix} \begin{bmatrix} I_1(s) \\ I_2(s) \\ V_1(s) \\ V_2(s) \end{bmatrix} = \mathbf{0} \tag{1.77}$$

Since the new response signals are $I_2(s)$ and $V_1(s)$, we premultiply the coefficient matrix of (1.77) by the negative of the inverse of the submatrix consisting of second and third columns, which is

$$-\begin{bmatrix} 0 & G_1 + s(C_1 + C_2) \\ -1 & g_m - sC_2 \end{bmatrix}^{-1} = \frac{1}{G_1 + s(C_1 + C_2)} \begin{bmatrix} sC_2 - g_m & G_1 + s(C_1 + C_2) \\ -1 & 0 \end{bmatrix} \tag{1.78}$$

We obtain the new coefficient matrix as

$$\frac{1}{G_1 + s(C_1 + C_2)} \begin{bmatrix} g_m - sC_2 & -G_1 - s(C_1 + C_2) & 0 & q(s) \\ 1 & 0 & -G_1 - s(C_1 + C_2) & sC_2 \end{bmatrix} \tag{1.79}$$

The submatrix formed by the first and fourth columns of (1.79) is the desired hybrid matrix relating the response vector $[I_2(s),\ V_1(s)]'$ to the excitation vector $[I_1(s),\ V_2(s)]'$. By interchanging the positions of $I_2(s)$ and $V_1(s)$ in the response vector, which corresponds to the interchange of row 1 and row 2 in the hybrid matrix, we obtain (1.66*a*).

To compute the transmission matrix $\mathbf{T}(s)$ of (1.75), it is convenient to rewrite (1.77) as

$$\begin{bmatrix} G_1 + s(C_1 + C_2) & -1 & -sC_2 & 0 \\ g_m - sC_2 & 0 & G_2 + sC_2 & 1 \end{bmatrix} \begin{bmatrix} V_1(s) \\ I_1(s) \\ V_2(s) \\ -I_2(s) \end{bmatrix} = \mathbf{0} \tag{1.80}$$

Premultiplying the coefficient matrix of (1.80) by the negative of the inverse of the submatrix consisting of the first two columns, which is

$$\frac{1}{sC_2 - g_m} \begin{bmatrix} 0 & 1 \\ sC_2 - g_m & G_1 + s(C_1 + C_2) \end{bmatrix} \tag{1.81}$$

we obtain

$$\frac{1}{sC_2 - g_m} \begin{bmatrix} g_m - sC_2 & 0 & G_2 + sC_2 & 1 \\ 0 & g_m - sC_2 & q(s) & G_1 + s(C_1 + C_2) \end{bmatrix} \tag{1.82}$$

The submatrix formed by the last two columns is the desired transmission matrix.

We remark that in deriving the various representations for an n-port, we implicitly assume that the inverses of the various submatrices exist. It is clear that if any one of them is singular, the conversion is not possible, and the corresponding set of parameters does not exist. For example, the ideal transformer of Fig. 1.15 does not possess the impedance, admittance, or transmission representation except the hybrid representation of Eq. (1.35) or its inverse.

In addition to the port characterizations of a network, it is sometimes advantageous to consider a network with n accessible terminals. Terminal characterization of a network and its applications will be considered in the following chapter.

1.6 EQUIVALENT FREQUENCY-DOMAIN CONDITIONS OF PASSIVITY

The concept of passivity was introduced in Sec. 1.3 as a useful qualitative measure of the time-domain behavior of networks. As pointed out earlier, the definition, although very general, is difficult to apply. The examples we presented were either simple one-ports or two-ports. In this section, the time-domain passivity criterion is translated into frequency-domain passivity conditions for linear time-invariant n-port networks.

Let $\mathbf{y}(t)$ and $\mathbf{u}(t)$ be the response and the excitation of a linear time-invariant n-port network N. As in Secs. 1.3 and 1.4, we restrict that the elements of $\mathbf{u}(t)$ be taken one each of the two variables associated with each of the n ports, so that the product $\mathbf{u}'(t)\mathbf{y}(t)$ represents the instantaneous power entering the n-port. Specifically, we shall show that the linear time-invariant n-port network is passive if and only if its general hybrid matrix is a positive-real matrix. For this reason, we first introduce the concept of positive realness.

1.6.1 Positive-Real Matrix and Passivity

Let

$$\mathbf{A} = [a_{ij}] \tag{1.83}$$

be a matrix of order n. We write

$$\bar{\mathbf{A}} = [\bar{a}_{ij}] \tag{1.84a}$$

$$\mathbf{A}^* = \bar{\mathbf{A}}' \tag{1.84b}$$

$$\mathbf{A}_h = \tfrac{1}{2}(\mathbf{A} + \mathbf{A}^*) \tag{1.84c}$$

$$\mathbf{A}_s = \tfrac{1}{2}(\mathbf{A} + \mathbf{A}') \tag{1.84d}$$

where $\bar{a}_{ij}$ denotes the complex conjugate of a_{ij}. The matrix $\mathbf{A}_h$ is called the *hermitian part* of $\mathbf{A}$, and $\mathbf{A}_s$ the *symmetric part*, because $\mathbf{A}_h$ is hermitian and $\mathbf{A}_s$ is symmetric. The *r*ight *h*alf of the complex s-plane will be abbreviated as RHS. Likewise, the *l*eft *h*alf of the complex s-plane will be abbreviated as LHS. The open RHS is the region defined by $\text{Re}\, s > 0$ (Re denotes the real part of), and the closed RHS is the region defined by $\text{Re}\, s \geqslant 0$. When we say that $\mathbf{A}$ is analytic in a region, we mean that every element of $\mathbf{A}$ is analytic in the region. On the other hand, when we say that $\mathbf{A}$ has a pole at s_0, we mean that at least one element of $\mathbf{A}$ has a pole at s_0.

Definition 1.6: Positive-real matrix An $n \times n$ matrix function $\mathbf{A}(s)$ of the complex variable s is said to be a *positive-real matrix* if it satisfies the following three conditions:

1. $\mathbf{A}(s)$ is analytic in the open RHS.
2. $\bar{\mathbf{A}}(s) = \mathbf{A}(\bar{s})$ for all s in the open RHS.
3. Its hermitian part $\mathbf{A}_h(s)$ is nonnegative definite for all s in the open RHS.

Recall that a hermitian matrix $\mathbf{A}_h$ is said to be nonnegative definite if its associated hermitian form $\mathbf{X}^*\mathbf{A}_h\mathbf{X}$ is nonnegative; that is,

$$\mathbf{X}^*\mathbf{A}_h\mathbf{X} \geqslant 0 \tag{1.85}$$

for all complex n-vectors $\mathbf{X}$. The second condition of the definition states that each element of $\mathbf{A}(s)$ is real when s is real. If $\mathbf{A}(s)$ is a matrix of rational functions, condition 1 is redundant because it is implied by conditions 2 and 3. However, if general functions are considered, condition 1 must be included.

Definition 1.7: Positive-real function A positive-real matrix of order 1 is called a *positive-real function.*

We illustrate the above definitions by the following example.

Example 1.6 Consider the admittance matrix (1.72):

$$\mathbf{Y}(s) = \begin{bmatrix} G_1 + s(C_1 + C_2) & -sC_2 \\ g_m - sC_2 & G_2 + sC_2 \end{bmatrix} \tag{1.86}$$

Write $s = \sigma + j\omega$. Then the hermitian part of $\mathbf{Y}(s)$ is given by

$$\mathbf{Y}_h(s) = \tfrac{1}{2}\,[\mathbf{Y}(s) + \mathbf{Y}^*(s)] = \begin{bmatrix} G_1 + \sigma(C_1 + C_2) & \tfrac{1}{2}g_m - \sigma C_2 \\ \tfrac{1}{2}g_m - \sigma C_2 & G_2 + \sigma C_2 \end{bmatrix} \tag{1.87}$$

We now investigate the three conditions of Definition 1.6. Since $\mathbf{Y}(s)$ is devoid of poles in the open RHS and since $\bar{\mathbf{Y}}(s) = \mathbf{Y}(\bar{s})$, the first two conditions are satisfied. To show that $\mathbf{Y}_h(s)$ is nonnegative definite for $\sigma > 0$, we recall that a hermitian matrix is nonnegative definite if and only if all of its principal minors are nonnegative (see App. I). This is equivalent to requiring that

$$G_1 + \sigma(C_1 + C_2) \geqslant 0 \tag{1.88a}$$

$$G_2 + \sigma C_2 \geqslant 0 \tag{1.88b}$$

$$\det \mathbf{Y}_h(s) \geqslant 0 \tag{1.88c}$$

where $\sigma > 0$. The third inequality is equivalent to

$$4(G_1 + \sigma C_1)(G_2 + \sigma C_2) + 4\sigma C_2 G_2 + 4\sigma g_m C_2 - g_m^2 \geqslant 0 \qquad \sigma > 0 \tag{1.89}$$

By assuming that G_1, G_2, C_1, C_2, and g_m are all nonnegative and

$$4G_1 G_2 \geqslant g_m^2 \tag{1.90}$$

(1.88) and (1.89) are all satisfied, and the hermitian matrix $\mathbf{Y}_h(s)$ is nonnegative definite for all s in the open RHS. Consequently, the admittance matrix is positive real. On the other hand, if $4G_1G_2 < g_m^2$, for a sufficiently small value of σ the inequality (1.89) can be violated, showing that $\mathbf{Y}(s)$ is not positive real.

To use the definition to test for positive realness of a matrix is usually complicated. The first two conditions are relatively easy to check. The third constraint is somewhat difficult to verify, because the nonnegative definiteness of the hermitian part must be investigated for all s in the open RHS. In the case where a matrix is rational, each element being the ratio of two polynomials, an equivalent set of conditions can be stated that reduces the test for the points in the vast open RHS to the points on the boundary, the real-frequency axis ($j\omega$-axis). Its justification can be found in Chen (1976).

Theorem 1.3 An $n \times n$ rational matrix $\mathbf{A}(s)$ is positive real if and only if the following four conditions are satisfied:

1. $\bar{\mathbf{A}}(s) = \mathbf{A}(\bar{s})$.
2. $\mathbf{A}(s)$ has no poles in the open RHS.
3. Poles of $\mathbf{A}(s)$ on the real-frequency axis, if they exist, are simple, and the associated residue matrix $\mathbf{K}$ evaluated at each of these poles is hermitian and nonnegative definite.
4. Its hermitian part $\mathbf{A}_h(j\omega)$ is nonnegative definite whenever it is defined.

We remark that the associated residue matrix $\mathbf{K}$ evaluated at a pole is the matrix of residues of the elements of $\mathbf{A}(s)$ evaluated at the pole. We illustrate this by the following examples.

Example 1.7 Consider the hybrid matrix (1.66*a*) of Fig. 1.17, as repeated below:

$$\mathbf{H}(s) = \frac{1}{G_1 + s(C_1 + C_2)} \begin{bmatrix} 1 & sC_2 \\ g_m - sC_2 & q(s) \end{bmatrix} \tag{1.91a}$$

where
$$q(s) = (G_1 + sC_1)(G_2 + sC_2) + sC_2(G_2 + g_m) \tag{1.91b}$$

The matrix is analytic in the closed RHS because the only singularity is at the pole $s = -\sigma_0 = -G_1/(C_1 + C_2)$. Thus, conditions 1, 2, and 3 are all satisfied. To test condition 4, we first compute the real-frequency hermitian part $\mathbf{H}_h(j\omega)$, yielding

$$\mathbf{H}_h(j\omega) = \frac{1}{(C_1 + C_2)(\omega^2 + \sigma_0^2)} \times \begin{bmatrix} \sigma_0 & \frac{1}{2}g_m\sigma_0 + j\omega(C_2\sigma_0 + \frac{1}{2}g_m) \\ \frac{1}{2}g_m\sigma_0 - j\omega(C_2\sigma_0 + \frac{1}{2}g_m) & p(\omega) \end{bmatrix} \tag{1.92a}$$

where
$$p(\omega) = \omega^2(C_1G_2 + C_2G_2 + G_1C_2 - C_1C_2\sigma_0 + C_2g_m) + G_1G_2\sigma_0 \tag{1.92b}$$

For $\mathbf{H}_h(j\omega)$ to be nonnegative definite, it is necessary and sufficient that all of its principal minors are nonnegative (App. I). This is equivalent to requiring that for real frequencies ω,

$$\sigma_0 \geqslant 0 \tag{1.93a}$$

$$p(\omega) \geqslant 0 \tag{1.93b}$$

$$\det \mathbf{H}_h(j\omega) \geqslant 0 \tag{1.93c}$$

By assuming that G_1, G_2, C_1, C_2, and g_m are all nonnegative, Eqs. (1.93*a*) and (1.93*b*) are clearly satisfied, because $G_1C_2 - C_1C_2\sigma_0 \geqslant 0$. The third inequality is equivalent to

$$(\sigma_0^2 + \omega^2)(4G_1G_2 - g_m^2) \geqslant 0 \tag{1.94}$$

As in Example 1.6, if $4G_1G_2 \geqslant g_m^2$, Eq. (1.94) is satisfied and $\mathbf{H}_h(j\omega)$ is nonnegative definite. Consequently, the hybrid matrix $\mathbf{H}(s)$ is positive real. If $4G_1G_2 < g_m^2$, Eq. (1.94) can be violated for any value of ω on the real-frequency axis, indicating that $\mathbf{H}_h(j\omega)$ is not nonnegative definite and $\mathbf{H}(s)$ is not positive real. To appreciate the significance of Theorem 1.3, it is suggested that the reader use the definition to verify the above results.

Example 1.8 We use Theorem 1.3 to verify the results obtained in Example 1.6. The first two conditions are clearly satisfied by the matrix $\mathbf{Y}(s)$ of Eq. (1.86). To verify condition 3, we compute the associated residue matrix. The admittance matrix $\mathbf{Y}(s)$ has a pole at the infinity, which is customarily considered to be on the real-frequency axis. The associated residue matrix evaluated at this pole is given by

$$\mathbf{K} = \begin{bmatrix} C_1 + C_2 & -C_2 \\ -C_2 & C_2 \end{bmatrix} \tag{1.95}$$

which is hermitian and nonnegative definite. Thus, condition 3 is satisfied. For condition 4, we compute the real-frequency hermitian part of $\mathbf{Y}(s)$, giving

$$\mathbf{Y}_h(j\omega) = \begin{bmatrix} G_1 & \frac{1}{2}g_m \\ \frac{1}{2}g_m & G_2 \end{bmatrix} \tag{1.96}$$

which is nonnegative definite if and only if $4G_1G_2 \geqslant g_m^2$, confirming Eq. (1.90). Thus, the admittance matrix $\mathbf{Y}(s)$ is positive real if and only if $4G_1G_2 \geqslant g_m^2$.

With these preliminaries, we are now in a position to state the equivalent frequency-domain conditions of passivity. In order to exhibit the result succinctly, we relegate the proof to the next section.

Theorem 1.4 A linear, time-invariant n-port network possessing a general hybrid matrix,† which is analytic in the open RHS, is passive if and only if the general hybrid matrix is positive real.

The theorem does not require any specific hybrid matrix; any characterization will do. The only requirement is that the product $\mathbf{u}'(t)\mathbf{y}(t)$ of the associated excitation $\mathbf{u}(t)$ and response $\mathbf{y}(t)$ denote the instantaneous power entering the n-port network under consideration. Thus, the impedance matrix, the admittance matrix,

†The general hybrid matrix is restricted to the one that is defined by an excitation vector whose elements are taken one each of the two variables associated with each of the n ports.

and the hybrid matrix discussed in Example 1.4 are all permitted, but the transmission matrix is not. Although all the permissible characterizations would eventually result in the same conclusion, the amount of computation involved may be quite different. The availability of so many different characterizations permits one to choose a particular approach or a combination of approaches, so that the problem at hand can be solved in the simplest and most satisfying manner.

Consider, again, the small-signal high-frequency equivalent network of a bipolar transistor of Fig. 1.17. Three permissible matrix characterizations were derived in Example 1.4 and are given in Eqs. (1.66*a*), (1.69), and (1.72). To determine the conditions under which the transistor is passive, we need only derive conditions under which the general hybrid matrix is positive real. In Examples 1.7 and 1.8, we have shown that the hybrid matrix (1.66*a*) or the admittance matrix (1.72) is positive real if and only if $4G_1G_2 \geqslant g_m^2$. Thus, according to Theorem 1.4, the transistor is passive if and only if $4G_1G_2 \geqslant g_m^2$. Since activity is the formal negation of passivity, the transistor is active if and only if $4G_1G_2 < g_m^2$. The same conclusion can be reached from the impedance matrix (1.69), but the amount of work involved is considerable (see Prob. 1.10). Thus, to test for passivity or activity of a device, the choice of an appropriate matrix is as important as the test itself. The reader is urged to compare the amount of labor involved in the above three cases.

1.6.2 Outline of a Proof of Theorem 1.4

Let $\mathbf{y}(t)$ be the response of an excitation $\mathbf{u}(t)$ of an n-port network N. We first demonstrate that the n-port is passive if and only if

$$\varepsilon = \varepsilon(t_0) + \int_{t_0}^{\infty} \mathbf{u}'(x)\mathbf{y}(x)\,dx \geqslant 0 \tag{1.97}$$

for all initial time t_0 and for all excitations $\mathbf{u}(t)$.

The condition is clearly necessary, because Eq. (1.10) must be true for all time t and, in particular, for $t = \infty$, and because $\mathbf{u}'(t)\mathbf{y}(t) = \mathbf{v}'(t)\mathbf{i}(t)$. To show that Eq. (1.97) also implies Eq. (1.10), we assume that $\varepsilon(t)$ of (1.10) is negative for some $t = t_1$. We can now simulate this condition in (1.97) by setting

$$\mathbf{u}(t) = \mathbf{v}(t) \qquad t < t_1 \tag{1.98a}$$

$$= \mathbf{0} \qquad t \geqslant t_1 \tag{1.98b}$$

Then ε of (1.97) will also be negative, contradicting our hypothesis. Thus, $\varepsilon(t)$ of (1.10) must be nonnegative for all $t \geqslant t_0$.

Without loss of generality, we assume that the n-port under consideration is initially relaxed. The passivity condition becomes

$$\varepsilon = \int_{-\infty}^{\infty} \mathbf{u}'(x)\mathbf{y}(x)\,dx \geqslant 0 \tag{1.99}$$

for all excitations $\mathbf{u}(t)$.

Let $\tilde{\mathbf{y}}(s)$ and $\tilde{\mathbf{u}}(s)$ be the Laplace transforms of $\mathbf{y}(t)$ and $\mathbf{u}(t)$, respectively. Then $\tilde{\mathbf{y}}(s)$ and $\tilde{\mathbf{u}}(s)$ are related by

$$\tilde{\mathbf{y}}(s) = \mathbf{H}(s)\tilde{\mathbf{u}}(s) \tag{1.100}$$

The time function $\mathbf{y}(t)$ may be expressed as the convolution product

$$\mathbf{y}(t) = \int_0^t \mathbf{h}(t-\lambda)\mathbf{u}(\lambda)\, d\lambda \tag{1.101a}$$

where $\mathbf{h}(t)$ represents the response resulting from the unit impulse excitation; that is,

$$\mathbf{h}(t) = \mathcal{L}^{-1}\, \mathbf{H}(s)\mathbf{g} \tag{1.101b}$$

where $\mathbf{g}$ is a vector comprised only of 1's. Consider the function

$$e(w) = \int_{-\infty}^{\infty} \mathbf{u}'(x-w)\mathbf{y}(x)\, dx \tag{1.102}$$

and observe that

$$e(0) = \mathcal{E} \tag{1.103}$$

the total energy stored in the n-port. On the real-frequency axis, Eq. (1.100) becomes

$$\tilde{\mathbf{y}}(j\omega) = \mathbf{H}(j\omega)\tilde{\mathbf{u}}(j\omega) \tag{1.104}$$

Consider $\tilde{\mathbf{y}}(j\omega)$, $\tilde{\mathbf{u}}(j\omega)$, and $\mathbf{H}(j\omega)$ as the Fourier transforms of $\mathbf{y}(t)$, $\mathbf{u}(t)$, and $\mathbf{h}(t)$, respectively. Then through the convolution integral, the time-domain response can be written as

$$\mathbf{y}(t) = \int_{-\infty}^{\infty} \mathbf{h}(t-\lambda)\mathbf{u}(\lambda)\, d\lambda \tag{1.105}$$

Substituting (1.105) in (1.102) yields

$$e(w) = \int_{-\infty}^{\infty}\int_{-\infty}^{\infty} \mathbf{u}'(x-w)\mathbf{h}(x-\lambda)\mathbf{u}(\lambda)\, d\lambda\, dx \tag{1.106}$$

which is recognized as the convolution of the functions of the forms $\mathbf{u}'(-t)$, $\mathbf{h}(t)$, and $\mathbf{u}(t)$, whose Fourier transforms are $\tilde{\mathbf{u}}'(-j\omega)$, $\mathbf{H}(j\omega)$, and $\tilde{\mathbf{u}}(j\omega)$. Thus, the Fourier transform on both sides of Eq. (1.106) is given by

$$\mathcal{F}\, e(w) = \tilde{\mathbf{u}}^*(j\omega)\mathbf{H}(j\omega)\tilde{\mathbf{u}}(j\omega) \tag{1.107}$$

where $\tilde{\mathbf{u}}'(-j\omega) = \tilde{\mathbf{u}}'(\overline{j\omega}) = \bar{\tilde{\mathbf{u}}}'(j\omega) = \tilde{\mathbf{u}}^*(j\omega)$, so that

$$e(w) = \frac{1}{2\pi}\int_{-\infty}^{\infty} \tilde{\mathbf{u}}^*(j\omega)\mathbf{H}(j\omega)\tilde{\mathbf{u}}(j\omega)e^{jw\omega}\, d\omega \tag{1.108}$$

Finally, from Eq. (1.103) we see that the total energy ε stored in the n-port can now be expressed in terms of the frequency-domain excitation $\tilde{\mathbf{u}}(j\omega)$ and the characterization $\mathbf{H}(j\omega)$ by setting $w = 0$ in (1.108), yielding

$$\varepsilon = \frac{1}{2\pi}\int_{-\infty}^{\infty} \tilde{\mathbf{u}}^*(j\omega)\mathbf{H}(j\omega)\tilde{\mathbf{u}}(j\omega)\, d\omega \tag{1.109}$$

Necessity We show that if the n-port is passive, then the general hybrid matrix $\mathbf{H}(s)$ must be positive real. To this end, we apply a particular excitation of the form

$$\mathbf{u}(t) = [u_1(t), u_2(t), \ldots, u_n(t)]' \tag{1.110}$$

where
$$u_k(t) = 0 \qquad t < 0 \tag{1.111a}$$
$$= c_k \sigma_0^{1/2} e^{-\sigma_0 t} \cos(\omega_0 t - \phi_k - \phi_0) \qquad t \geqslant 0 \tag{1.111b}$$

with $k = 1, 2, \ldots, n$; where $s_0 = \sigma_0 + j\omega_0$ is an arbitrary point in the open RHS, and c_k, ϕ_k, and ϕ_0 are arbitrary real constants. Note that each excitation signal has the same associated σ_0, ω_0, and ϕ_0, but different c_k and ϕ_k. The Fourier transform $\tilde{u}_k(j\omega)$ of the signal $u_k(t)$ is given by

$$\tilde{u}_k(j\omega) = c_k \sigma_0^{1/2} \frac{(\sigma_0 + j\omega)\cos(\phi_k + \phi_0) + \omega_0 \sin(\phi_k + \phi_0)}{(\sigma_0 + j\omega)^2 + \omega_0^2} \tag{1.112}$$

Consider the product $\bar{\tilde{u}}_i(j\omega)\tilde{u}_k(j\omega)$, which can be simplified to

$$\bar{\tilde{u}}_i(j\omega)\tilde{u}_k(j\omega) = \frac{c_i c_k \sigma_0 A_1}{[(\sigma_0 + j\omega)^2 + \omega_0^2][(\sigma_0 - j\omega)^2 + \omega_0^2]} \tag{1.113a}$$

where

$$A_1 = (\omega^2 + \sigma_0^2)\cos(\phi_i + \phi_0)\cos(\phi_k + \phi_0) + \omega_0^2 \sin(\phi_i + \phi_0)\sin(\phi_k + \phi_0)$$
$$+ \omega_0 \sigma_0 \sin(\phi_i + \phi_k + 2\phi_0) + j\omega_0 \omega \sin(\phi_i - \phi_k) \tag{1.113b}$$

For our purposes, we express Eq. (1.113a) as

$$\bar{\tilde{u}}_i(j\omega)\tilde{u}_k(j\omega) = \tfrac{1}{4} c_i c_k (A_2 + A_3 + A_4 + \bar{A}_4) \tag{1.114a}$$

where
$$A_2 = \frac{\sigma_0 e^{j(\phi_i - \phi_k)}}{\sigma_0^2 + (\omega - \omega_0)^2} \tag{1.114b}$$

$$A_3 = \frac{\sigma_0 e^{-j(\phi_i - \phi_k)}}{\sigma_0^2 + (\omega + \omega_0)^2} \tag{1.114c}$$

$$A_4 = \frac{\sigma_0 e^{j(\phi_i + \phi_k + 2\phi_0)}}{[\sigma_0 - j(\omega - \omega_0)][\sigma_0 + j(\omega + \omega_0)]} \tag{1.114d}$$

The validity of (1.114a) can be verified by combining the terms on the right-hand side of (1.114a) and comparing the results with (1.113a). Expressing $\mathbf{H}(j\omega)$ explicitly as in (1.61c) and substituting (1.110) in (1.109) yield

$$\mathcal{E} = \frac{1}{2\pi}\int_{-\infty}^{\infty} \sum_{i=1}^{n} \sum_{k=1}^{n} \bar{\tilde{u}}_i(j\omega)h_{ik}(j\omega)\tilde{u}_k(j\omega)\, d\omega$$

$$= \frac{1}{8\pi} \sum_{i=1}^{n} \sum_{k=1}^{n} c_i c_k \left[\int_{-\infty}^{\infty} h_{ik}(j\omega)A_2\, d\omega + \int_{-\infty}^{\infty} h_{ik}(j\omega)A_3\, d\omega \right.$$

$$\left. + \int_{-\infty}^{\infty} h_{ik}(j\omega)(A_4 + \bar{A}_4)\, d\omega\right] \tag{1.115}$$

In the theory of functions of a complex variable there is a well-known integral called *Poisson's integral* [see, for example, Guillemin (1969)], which states that

$$F(s) = F(\sigma + j\omega) = \frac{1}{\pi}\int_{-\infty}^{\infty} \frac{\sigma F(jx)}{\sigma^2 + (\omega - x)^2}\, dx \tag{1.116}$$

is valid for any s in the closed RHS. Applying this integral to the first two integrals on the right-hand side of (1.115) gives

$$\mathcal{E} = \tfrac{1}{4}(\mathcal{E}_1 + \mathcal{E}_2) \tag{1.117a}$$

where

$$\mathcal{E}_1 = \tfrac{1}{2} \sum_{i=1}^{n} \sum_{k=1}^{n} c_i c_k \left[h_{ik}(\sigma_0 + j\omega_0)e^{j(\phi_i - \phi_k)} + h_{ik}(\sigma_0 - j\omega_0)e^{-j(\phi_i - \phi_k)}\right]$$

$$= \text{Re} \sum_{i=1}^{n} \sum_{k=1}^{n} (c_i e^{j\phi_i})h_{ik}(\sigma_0 + j\omega_0)(c_k e^{-j\phi_k}) = \text{Re}\, \mathbf{X}^*\mathbf{H}(s_0)\mathbf{X} \tag{1.117b}$$

in which, as before, $s_0 = \sigma_0 + j\omega_0$, and

$$\mathbf{X} = [c_1 e^{-j\phi_1}, c_2 e^{-j\phi_2}, \ldots, c_n e^{-j\phi_n}]' \tag{1.117c}$$

$$\mathcal{E}_2 = \frac{1}{2\pi} \sum_{i=1}^{n} \sum_{k=1}^{n} c_i c_k \int_{-\infty}^{\infty} h_{ik}(j\omega)(A_4 + \bar{A}_4)\, d\omega \tag{1.117d}$$

Since the n-port network is passive, $\mathcal{E}$ is nonnegative, showing that

$$\mathcal{E}_1 + \mathcal{E}_2 \geqslant 0 \tag{1.118}$$

In a similar manner, if we apply an excitation of the type (1.111) with $\phi_0 + \frac{1}{2}\pi$ replacing ϕ_0 in (1.111b), everything else being the same, we arrive at an expression similar to (1.117a) for the total energy stored in the n-port, which again must be nonnegative. The result is given by

$$\mathcal{E}_1 - \mathcal{E}_2 \geqslant 0 \tag{1.119}$$

The reason for this is that, by replacing ϕ_0 by $\phi_0 + \frac{1}{2}\pi$, Eqs. (1.114*b*) and (1.114*c*) remain the same and (1.114*d*) becomes $-A_4$. Thus, (1.117*b*) remains the same as before but (1.117*d*) is changed to $-\varepsilon_2$. Adding (1.118) and (1.119) shows that, for an arbitrary point $s_0 = \sigma_0 + j\omega_0$ in the open RHS,

$$2\,\varepsilon_1 = 2 \text{ Re } \mathbf{X}^*\mathbf{H}(s_0)\mathbf{X} = \mathbf{X}^*\mathbf{H}(s_0)\mathbf{X} + \mathbf{X}^*\mathbf{H}^*(s_0)\mathbf{X} = 2\mathbf{X}^*\mathbf{H}_h(s_0)\mathbf{X} \geqslant 0 \tag{1.120}$$

for all complex $\mathbf{X}$, because c_k and ϕ_k of (1.117*c*) are arbitrary. This, together with the fact that $\mathbf{H}(\bar{s}) = \bar{\mathbf{H}}(s)$ and that $\mathbf{H}(s)$ is analytic in the open RHS, shows that $\mathbf{H}(s)$ is positive real.

Sufficiency To show that positive realness is also sufficient, we write (1.109) as

$$\begin{aligned}\varepsilon &= \frac{1}{2\pi}\int_{-\infty}^{\infty} [\text{Re } \tilde{\mathbf{u}}^*(j\omega)\mathbf{H}(j\omega)\tilde{\mathbf{u}}(j\omega) + j \text{ Im } \tilde{\mathbf{u}}^*(j\omega)\mathbf{H}(j\omega)\tilde{\mathbf{u}}(j\omega)]\, d\omega \\ &= \frac{1}{2\pi}\int_{-\infty}^{\infty} \text{Re } \tilde{\mathbf{u}}^*(j\omega)\mathbf{H}(j\omega)\tilde{\mathbf{u}}(j\omega)\, d\omega \end{aligned} \tag{1.121}$$

where Im denotes the *imaginary part of.* The second integral of the first line is zero because $\text{Im } \tilde{\mathbf{u}}^*(j\omega)\mathbf{H}(j\omega)\tilde{\mathbf{u}}(j\omega)$ is an odd function of ω. Since $\mathbf{H}(s)$ is positive real, its hermitian part $\mathbf{H}_h(s)$ must be nonnegative definite for all s in the open RHS, or

$$\mathbf{X}^*\mathbf{H}_h(s)\mathbf{X} = \text{Re } \mathbf{X}^*\mathbf{H}(s)\mathbf{X} \geqslant 0 \qquad \sigma = \text{Re } s > 0 \tag{1.122}$$

for all complex $\mathbf{X}$. In the limit as σ approaches to zero, (1.122) becomes

$$\text{Re } \mathbf{X}^*\mathbf{H}(j\omega)\mathbf{X} \geqslant 0 \tag{1.123}$$

for all complex $\mathbf{X}$ and for all ω whenever $\mathbf{H}(j\omega)$ is defined. This shows that the integrand of (1.121) is nonnegative, and ε must be nonnegative. Thus, the n-port network is passive. This completes the proof of the theorem.

1.7 DISCRETE-FREQUENCY CONCEPTS OF PASSIVITY AND ACTIVITY

In the preceding section we showed that for a linear, time-invariant n-port network, the time-domain definition of passivity can be translated into the equivalent frequency-domain condition that states that the general hybrid matrix of the n-port must be positive real. In the present section we extend these results by introducing the discrete-frequency concepts of passivity and activity. These concepts are very useful for studying the behavior and limitations of active networks.

In showing that positive realness is a necessary condition for an n-port to be passive, we apply a particular signal of the type (1.111). Thus, the logical set of signals with which to connect directly the time and frequency domains are those of the form

$$u_k(t) = c_k e^{\sigma_0 t} \cos(\omega_0 t - \phi_k) = \text{Re}\,(\hat{c}_k e^{s_0 t}) \tag{1.124}$$

where

$$s_0 = \sigma_0 + j\omega_0 \qquad \sigma_0 \geqslant 0 \tag{1.125a}$$

$$\hat{c}_k = c_k e^{-j\phi_k} \tag{1.125b}$$

and c_k and ϕ_k are arbitrary real constants. To avoid the use of Re, which is not distributive over the product, we express (1.124) as

$$u_k(t) = \tfrac{1}{2}(\hat{c}_k e^{s_0 t} + \bar{\hat{c}}_k e^{\bar{s}_0 t}) \tag{1.126}$$

with each of these time-domain complex exponential signals corresponding to a simple pole of the signal transform at the point s_0 or $\bar{s}_0$ in the complex frequency s-plane. This class of signals is the most significant in the study of linear time-invariant networks.

Let $\mathbf{y}(t)$ and $\mathbf{u}(t)$ be the response and the excitation of an n-port. Assume that each component of $\mathbf{u}(t)$ is of the form (1.124) or (1.126). Then we can write

$$\mathbf{u}(t) = \tfrac{1}{2}(\mathbf{u}_0 e^{s_0 t} + \bar{\mathbf{u}}_0 e^{\bar{s}_0 t}) \tag{1.127}$$

where $\mathbf{u}_0$ is an arbitrary complex but fixed n-vector. Provided that we set up the appropriate initial conditions, the response of the n-port is given by (see Prob. 1.44 for an illustration)

$$\mathbf{y}(t) = \tfrac{1}{2}[\mathbf{H}(s_0)\mathbf{u}_0 e^{s_0 t} + \bar{\mathbf{H}}(s_0)\bar{\mathbf{u}}_0 e^{\bar{s}_0 t}] \tag{1.128}$$

where $\mathbf{H}(s)$ is the general hybrid matrix of the n-port as defined in (1.60). By substituting (1.127) and (1.128) in (1.10), the total energy stored in the n-port at any time t can be determined, and for $\sigma \neq 0$, with $\measuredangle$ denoting *the angle of*, we have

$$\begin{aligned}
\mathcal{E}(t) &= \mathcal{E}(t_0) + \int_{t_0}^{t} \mathbf{u}'(x)\mathbf{y}(x)\,dx \\
&= \mathcal{E}(t_0) + \frac{1}{2}\,\text{Re} \int_{t_0}^{t} [\mathbf{u}_0^*\mathbf{H}(s_0)\mathbf{u}_0 e^{2\sigma_0 x} + \mathbf{u}_0'\mathbf{H}(s_0)\mathbf{u}_0 e^{2s_0 x}]\,dx \\
&= \mathcal{E}(t_0) + \frac{1}{4} e^{2\sigma_0 t} \left\{ \frac{1}{\sigma_0}\,\text{Re}\,\mathbf{u}_0^*\mathbf{H}(s_0)\mathbf{u}_0 + \frac{1}{|s_0|}|\mathbf{u}_0'\mathbf{H}(s_0)\mathbf{u}_0| \right. \\
&\quad \left. \times \cos[2\omega_0 t + \measuredangle\underline{\mathbf{u}_0'\mathbf{H}(s_0)\mathbf{u}_0} - \measuredangle\underline{s_0}] \right\} + C_0 \geqslant 0
\end{aligned} \tag{1.129}$$

where the constant C_0 is introduced so that for $t = t_0$, $\mathcal{E}(t) = \mathcal{E}(t_0)$. Since the terms containing $\exp(2\sigma_0 t)$ grow exponentially without bound as t is increased, ultimately dominating the other terms, for (1.129) to hold for all $t \geqslant t_0$, it is necessary that

$$\text{Re}\,\mathbf{u}_0^*\mathbf{H}(s_0)\mathbf{u}_0 + \frac{\sigma_0}{|s_0|}|\mathbf{u}_0'\mathbf{H}(s_0)\mathbf{u}_0| \cos[2\omega_0 t + \measuredangle\underline{\mathbf{u}_0'\mathbf{H}(s_0)\mathbf{u}_0} - \measuredangle\underline{s_0}] \geqslant 0 \tag{1.130}$$

for sufficiently large t. Even so, if $\omega_0 \neq 0$, the cosine function can assume the value -1 for some t, necessitating that

$$\text{Re}\ \mathbf{u}_0^* \mathbf{H}(s_0)\mathbf{u}_0 - \frac{\sigma_0}{|s_0|} |\mathbf{u}_0' \mathbf{H}(s_0)\mathbf{u}_0| \geqslant 0 \tag{1.131}$$

For $\omega_0 = 0$, Eq. (1.130) becomes Re $[\mathbf{u}_0^* \mathbf{H}(\sigma_0)\mathbf{u}_0 + \mathbf{u}_0' \mathbf{H}(\sigma_0)\mathbf{u}_0] \geqslant 0$. Writing $\mathbf{u}_0 = \mathbf{u}_1 + j\mathbf{u}_2$ with $\mathbf{u}_1$ and $\mathbf{u}_2$ real, the inequality becomes $\mathbf{u}_1' \mathbf{H}(\sigma_0)\mathbf{u}_1 \geqslant 0$. But, since the imaginary part $\mathbf{u}_2$ of $\mathbf{u}_0$, being the real part of some other complex n-vector, must also satisfy this constraint, that is, $\mathbf{u}_2' \mathbf{H}(\sigma_0)\mathbf{u}_2 \geqslant 0$, we conclude that Re $\mathbf{u}_0^* \mathbf{H}(\sigma_0)\mathbf{u}_0 \geqslant 0$, which is implied by (1.131) (see Prob. 1.14). In the situation where $\sigma = 0$, a similar inequality can be derived. The only difference is that in (1.129) we replace $e^{2\sigma_0 t}/\sigma_0$ by t. In fact, we can show that (1.131) is valid for all $s_0 \geqslant 0$, $\omega_0 \neq 0$. The details are left as an exercise (see Prob. 1.15). Hence, we obtain the necessary condition (1.131) for $\omega_0 \neq 0$ and Re $\mathbf{u}_0^* \mathbf{H}(s_0)\mathbf{u}_0 \geqslant 0$ for $\omega_0 = 0$.

We remark that for $s_0 = \sigma_0$ there is no need to take $\mathbf{u}_0$ complex, because any excitation of the type (1.127) with $\mathbf{u}_0$ complex corresponds to an excitation $\hat{\mathbf{u}}_0 \exp(\sigma_0 t)$ with $\hat{\mathbf{u}}_0$ real, where $\hat{\mathbf{u}}_0 = \text{Re}\ \mathbf{u}_0$. Also, when we state that $\mathbf{u}(t)$ of (1.127) is the excitation and $\mathbf{y}(t)$ of (1.128) is the response, we implicitly assume that we have already set up appropriate initial conditions inside the n-port network such that when the excitation $\mathbf{u}(t)$ is applied at the time $t = t_0$, *no transient occurs* at t_0 and thereafter, and the voltages and currents appearing at the ports behave exactly as specified. The amount of energy required to set up the appropriate initial conditions is represented by $\varepsilon(t_0)$ in (1.129). These initial conditions need never be evaluated in this study; it is sufficient to know that they exist.

The condition (1.131) will now be employed in characterizing the single-frequency behavior of a linear, time-invariant n-port network.

Definition 1.8: Passivity at a complex frequency A linear time-invariant n-port network possessing a general hybrid matrix matrix† $\mathbf{H}(s)$ is said to be *passive at a complex frequency* s_0 in the closed RHS if, for all finite nonzero complex n-vectors $\mathbf{u}_0$,

$$\mathbf{u}_0^* \mathbf{H}_h(s_0)\mathbf{u}_0 - \frac{\sigma_0}{|s_0|} |\mathbf{u}_0' \mathbf{H}_s(s_0)\mathbf{u}_0| \geqslant 0 \tag{1.132}$$

where $\mathbf{H}_h(s)$ and $\mathbf{H}_s(s)$ denote the hermitian part and the symmetric part of $\mathbf{H}(s)$, respectively, and $s_0 = \sigma_0 + j\omega_0$. The second term on the left-hand side of (1.132) is defined to be zero for $s_0 = 0$. For $s_0 = \sigma_0$, Eq. (1.132) is to be taken for all finite nonzero real n-vectors $\mathbf{u}_0$.

The statement of the definition of passivity at a single point s_0 is not really meaningful for real networks, because passivity must always occur at the complex conjugate pair of the complex frequencies s_0 and $\bar{s}_0$. To avoid unnecessary complications and cumbersome notation, we shall be satisfied with the above

†See the footnote on page 29.

definition with the understanding that it must always occur at the complex conjugate pair.

Definition 1.9: Activity at a complex frequency A linear, time-invariant n-port network possessing a general hybrid matrix is said to be *active at a complex frequency* s_0 in the closed RHS if it is not passive at s_0.

The condition (1.132) is identical to (1.131) on recognition of the relations

$$\text{Re } \mathbf{u}_0^* \mathbf{H}(s_0)\mathbf{u}_0 = \mathbf{u}_0^* \mathbf{H}_h(s_0)\mathbf{u}_0 \tag{1.133a}$$

$$\mathbf{u}_0' \mathbf{H}(s_0)\mathbf{u}_0 = \mathbf{u}_0' \mathbf{H}_s(s_0)\mathbf{u}_0 \tag{1.133b}$$

To test activity for an n-port at s_0, it is sufficient to demonstrate that

$$\mathbf{u}_0^* \mathbf{H}_h(s_0)\mathbf{u}_0 - \frac{\sigma_0}{|s_0|} |\mathbf{u}_0' \mathbf{H}_s(s_0)\mathbf{u}_0| < 0 \tag{1.134}$$

for *some* finite complex n-vector $\mathbf{u}_0$.

We emphasize the difference of the concepts that an n-port is passive and that an n-port is passive at a closed RHS point s_0. An n-port is passive if and only if its general hybrid matrix $\mathbf{H}(s)$ is positive real, implying that the hermitian matrix $\mathbf{H}_h(s)$ is nonnegative definite for all s in the open RHS; that is,

$$\mathbf{u}_0^* \mathbf{H}_h(s)\mathbf{u}_0 \geqslant 0 \qquad \sigma = \text{Re } s > 0 \tag{1.135}$$

for all complex n-vectors $\mathbf{u}_0$. An n-port is passive at a single complex frequency s_0 in the closed RHS if

$$\mathbf{u}_0^* \mathbf{H}_h(s_0)\mathbf{u}_0 \geqslant \frac{\sigma_0}{|s_0|} |\mathbf{u}_0' \mathbf{H}_s(s_0)\mathbf{u}_0| \tag{1.136}$$

for all complex n-vectors $\mathbf{u}_0$. As demonstrated in Eq. (1.131), Eq. (1.136) is a consequence of positive realness of $\mathbf{H}(s)$. However, the converse is not necessarily true. An n-port may be passive at s_0 but it may not be a passive n-port in the sense of Definition 1.3. Definition 1.3 is applicable to any network and for any excitation, but Definition 1.8 is valid only for linear time-invariant n-ports and for the excitations of the type (1.127).

The definitions of passivity and activity at a single complex frequency s_0 in the closed RHS suggest that for any given n-port, the closed right half of the complex frequency s-plane can be partitioned into regions of passivity and activity. The *active region* is the set of points in the closed RHS at which the n-port is active, and the *passive region* is the set of points in closed RHS at which the n-port is passive. These regions are clearly mutually exclusive. We know a priori, then, that the device characterized by the n-port may be used as an amplifier only for signals with frequencies in the active region. In fact, if an n-port is passive at the complex frequency s_0, then the n-port may be replaced by a passive n-port that exhibits identical behavior with respect to all real signals of the form (1.127) comprised of an arbitrary combination of two complex exponential signals at the frequencies s_0

and $\bar{s}_0$. The realization generally involves a $2n$-port transformer and a nonreciprocal Darlington type-E section. We shall not discuss this general aspect of the subject any further, because it would take us far afield into network synthesis. As a result, the n-port may neither amplify nor oscillate at this complex frequency.

1.7.1 Single Complex Frequency Passivity and Activity of One-Port Networks

For one-port networks, the condition (1.132) for passivity at a single complex frequency can be greatly simplified. It is more usefully stated as

$$\operatorname{Re} H(s_0) - \frac{\sigma_0}{|s_0|} |H(s_0)| \geqslant 0 \tag{1.137}$$

for a one-port characterized by its immittance $H(s)$ to be passive at a closed RHS point s_0. On the σ-axis, Eq. (1.137) becomes

$$H(\sigma_0) \geqslant 0 \qquad \text{for } \sigma_0 \geqslant 0 \tag{1.138}$$

For the remainder of the closed RHS, we can use the equivalent condition

$$\left| \frac{\operatorname{Im} H(s_0)}{\operatorname{Re} H(s_0)} \right| \leqslant \left| \frac{\omega_0}{\sigma_0} \right| \tag{1.139}$$

to determine the region of passivity.

We illustrate the above results by the following example.

Example 1.9 Figure 1.18 is the equivalent network of a tunnel diode biased in the linear negative-resistance region. The input impedance of the device is given by

$$z(s) = R_1 + sL + \frac{R}{RCs - 1} \tag{1.140}$$

whose real and imaginary parts are computed as

$$\operatorname{Re} z(s) = R_1 + \sigma L + \frac{R(RC\sigma - 1)}{(1 - RC\sigma)^2 + R^2 C^2 \omega^2} \tag{1.141a}$$

$$\operatorname{Im} z(s) = \omega L - \frac{R^2 \omega C}{(1 - RC\sigma)^2 + R^2 C^2 \omega^2} \tag{1.141b}$$

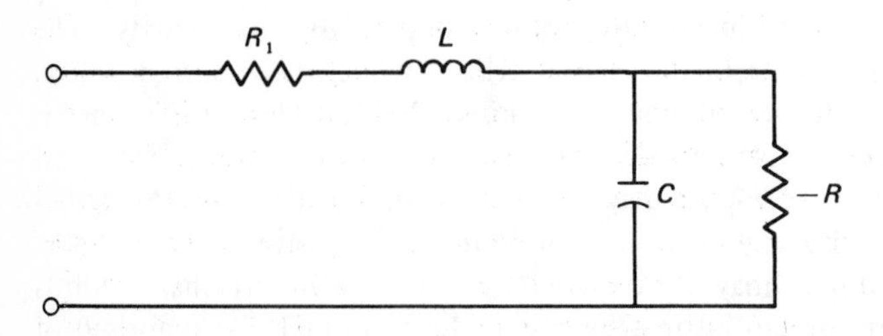

Figure 1.18 The equivalent network of a tunnel diode biased in the linear negative-resistance region.

To test passivity on the σ-axis we use Eq. (1.138), which yields

$$\frac{RLC\sigma^2 + (R_1RC - L)\sigma + R - R_1}{RC\sigma - 1} \geqslant 0 \qquad \text{for } \sigma \geqslant 0 \tag{1.142}$$

To satisfy (1.142), the numerator and the denominator on the left-hand side of (1.142) must have the same sign. For the tunnel diode, R is usually much larger than R_1. Assume that

$$(R_1RC - L)^2 < 4RLC(R - R_1) \tag{1.143}$$

meaning that the numerator polynomial has no real root and therefore is always positive. Thus, on the σ-axis the device is passive for all

$$\sigma \geqslant \sigma_m = \frac{1}{RC} \tag{1.144}$$

On the real-frequency axis, the condition (1.137) becomes

$$\text{Re } z(j\omega) = R_1 - \frac{R}{1 + R^2C^2\omega^2} \geqslant 0 \tag{1.145a}$$

requiring that

$$|\omega| \geqslant \omega_m = \frac{1}{RC}\sqrt{\frac{R}{R_1} - 1} \tag{1.145b}$$

For the remainder of the closed RHS, we apply (1.139), yielding

$$|\omega|\left|R_1 + \sigma L + \frac{R(RC\sigma - 1)}{(1 - RC\sigma)^2 + R^2C^2\omega^2}\right| \geqslant \left|\sigma\omega L - \frac{R^2\omega\sigma C}{(1 - RC\sigma)^2 + R^2C^2\omega^2}\right| \tag{1.146}$$

The boundaries of the passive and active regions occur at the points when (1.139) is satisfied with the equality sign. This gives

$$\omega \text{ Re } z(s) = \pm\,\sigma \text{ Im } z(s) \tag{1.147}$$

Choosing first the positive sign results in the equation

$$R_1 = \frac{R - 2R^2C\sigma}{(1 - RC\sigma)^2 + R^2C^2\omega^2} \tag{1.148}$$

As ω approaches zero, the curve intersects the positive σ-axis at

$$\sigma_1 = \frac{1}{RC} - \frac{1}{R_1C} + \frac{1}{R_1C}\sqrt{1 - \frac{R_1}{R}} \tag{1.149}$$

The partition of the closed RHS into the regions of passivity and activity is sketched in Fig. 1.19. The shaded area and the solid bar from $\sigma = 0$ to any $\sigma < 1/RC$ denote the active region, and the remainder of the closed RHS is the passive region. The boundary between active and passive regions is determined by (1.148).

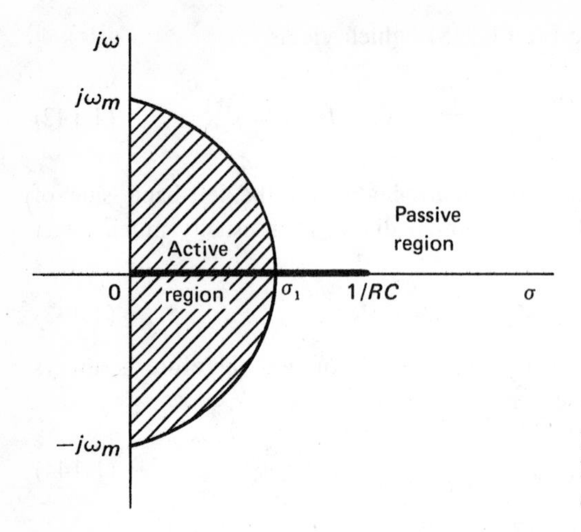

Figure 1.19 The partition of the closed RHS into the regions of passivity and activity by a properly biased tunnel diode.

In the case that we choose the negative sign in (1.147), the boundary of the active and passive regions is described by the equation

$$R_1 + 2\sigma L = \frac{R}{(1 - RC\sigma)^2 + R^2 C^2 \omega^2} \tag{1.150}$$

As ω approaches zero, this curve intersects the positive σ-axis at a point determined by the cubic equation

$$2R^2 C^2 L\sigma^3 + RC(R_1 RC - 4L)\sigma^2 - 2(R_1 RC - L)\sigma + R_1 - R = 0 \tag{1.151}$$

Suppose that we choose a point $s_0 = j\omega_0$, $\omega_0 > \omega_m$, on the real-frequency axis; according to (1.146) the device is passive at $j\omega_0$. This device may be replaced by an equivalent passive one-port comprised of the series connection of a resistor of resistance

$$r = R_1 - \frac{R}{1 + R^2 C^2 \omega_0^2} \geqslant 0 \tag{1.152a}$$

and an inductor of inductance

$$l = L - \frac{R^2 C}{1 + R^2 C^2 \omega_0^2} \geqslant 0 \tag{1.152b}$$

which exhibits identical behavior with respect to all real signals of the type (1.127). The general replacement formulas for a one-port will be discussed in the following example.

Example 1.10 Consider a one-port impedance $z(s)$ that is passive at a point s_0 in the closed RHS. The replacement of this one-port by a passive equivalent one-port that exhibits identical behavior with respect to all real signals of the type (1.127) may be effected in terms of one of the four simple passive one-ports, depending on the location of s_0 and the sign of Im $z(s_0)$.

For $s_0 = \sigma_0 > 0$, the impedance $z(\sigma_0)$ can be represented by a single resistor of resistance

$$r_1 = z(\sigma_0) \geqslant 0 \tag{1.153}$$

For $s_0 = j\omega_0$, $\omega_0 > 0$ and $\text{Im } z(j\omega_0) \geqslant 0$, or $\omega_0 < 0$ and $\text{Im } z(j\omega_0) \leqslant 0$, the impedance $z(j\omega_0)$ is the series combination of a resistor of resistance

$$r_2 = \text{Re } z(j\omega_0) \geqslant 0 \tag{1.154a}$$

and an inductor of inductance

$$l_2 = \frac{\text{Im } z(j\omega_0)}{\omega_0} \geqslant 0 \tag{1.154b}$$

For $s_0 = j\omega_0$, $\omega_0 > 0$ and $\text{Im } z(j\omega_0) < 0$, or $\omega_0 < 0$ and $\text{Im } z(j\omega_0) > 0$, $z(j\omega_0)$ is the series combination of a resistor of resistance

$$r_3 = \text{Re } z(j\omega_0) \geqslant 0 \tag{1.155a}$$

and a capacitor of capacitance

$$c_3 = -\frac{1}{\omega_0 \text{ Im } z(j\omega_0)} \geqslant 0 \tag{1.155b}$$

Finally, for $s_0 = \sigma_0 + j\omega_0$, $\sigma_0 > 0$ and $\omega_0 \neq 0$, the impedance $z(s_0)$ can be represented equivalently by the series combination of an inductor of inductance l_4 and a capacitor of capacitance c_4; that is,

$$z(s_0) = l_4 s_0 + \frac{1}{s_0 c_4} \tag{1.156a}$$

Substituting $z(s_0) = \text{Re } z(s_0) + j \text{ Im } z(s_0)$ in Eq. (1.156*a*) and solving for l_4 and c_4 yield

$$l_4 = \frac{1}{2\sigma_0} \text{Re } z(s_0) + \frac{1}{2\omega_0} \text{Im } z(s_0) \tag{1.156b}$$

and

$$c_4 = \frac{1}{(|s_0|^2/2\sigma_0) \text{ Re } z(s_0) - (|s_0|^2/2\omega_0) \text{ Im } z(s_0)} \tag{1.156c}$$

which according to (1.137) are nonnegative because $z(s)$ is passive at the point s_0. The replacement one-port networks are presented in Fig. 1.20.

Consider, for example, the equivalent network of a tunnel diode of Fig. 1.18. Some typical values of its parameters are given below:

$$R_1 = 1\ \Omega \qquad R = 50\ \Omega$$

$$L = 0.4 \text{ nH} \qquad C = 10 \text{ pF}$$

Using these values and the results obtained in Example 1.9, we conclude that the device is passive on the nonnegative σ-axis for all $\sigma \geqslant \sigma_m = 2 \cdot 10^9$ Np/s, and on the real-frequency axis for all $|\omega| \geqslant \omega_m = 14 \cdot 10^9$ rad/s. Suppose that we choose $s_0 = (2 + j10) \cdot 10^9$. It is straightforward to demonstrate that the device is passive

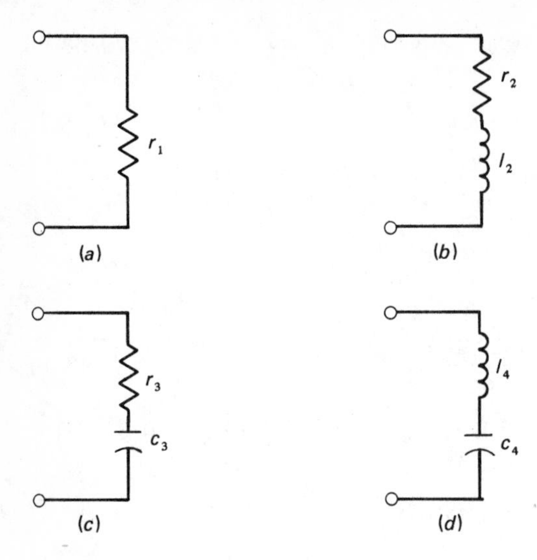

Figure 1.20 The replacement of a passive one-port impedance $z(s_0)$ by a passive equivalent one-port network with $s_0 = \sigma_0 + j\omega_0$. (*a*) $s_0 = \sigma_0 > 0$. (*b*) $s_0 = j\omega_0$, $\omega_0 > 0$ and $\operatorname{Im} z(j\omega_0) \geqslant 0$, or $\omega_0 < 0$ and $\operatorname{Im} z(j\omega_0) \leqslant 0$. (*c*) $s_0 = j\omega_0$, $\omega_0 > 0$ and $\operatorname{Im} z(j\omega_0) < 0$, or $\omega_0 < 0$ and $\operatorname{Im} z(j\omega_0) > 0$. (*d*) $s_0 = \sigma_0 + j\omega_0$, $\sigma_0 > 0$ and $\omega_0 \neq 0$.

at this point s_0. Thus, it can be replaced by an equivalent network comprised of the series connection of an inductor of inductance l_4 and a capacitor of capacitance c_4. From (1.156) we obtain $l_4 = 0.15$ nH and $c_4 = 12.82$ pF, where $z(s_0) = 1.8 - j6$. This equivalent one-port exhibits identical behavior with respect to all real signals comprised of two complex exponential signals at the complex conjugate pair of frequencies $s_0 = (2 + j10) \cdot 10^9$ and $\bar{s}_0 = (2 - j10) \cdot 10^9$, as shown in (1.127).

1.7.2 Active Networks

In the foregoing, we have demonstrated that by applying the definitions of passivity and activity at a single complex frequency, a given n-port can divide the closed RHS into regions of passivity and activity. The physical significance is that, in the passive region, the given n-port network can be replaced by a passive n-port that exhibits identical behavior with respect to all real signals of the type (1.127). In the active region, the n-port network can be made to achieve power gain or to oscillate by means of an appropriate passive imbedding network. In practical applications, we are concerned mainly with the behavior of the n-port on the real-frequency axis and, less frequently, on the positive σ-axis rather than the entire closed RHS. On these axes, the passivity condition (1.132) can be greatly simplified.

First, on the real-frequency axis, (1.132) becomes

$$\mathbf{u}_0^* \mathbf{H}_h(j\omega_0)\mathbf{u}_0 \geqslant 0 \tag{1.157}$$

for all finite nonzero complex n-vectors $\mathbf{u}_0$, showing that the hermitian part $\mathbf{H}_h(s)$ of the general hybrid matrix $\mathbf{H}(s)$ of the given n-port must be nonnegative definite at the point $s = s_0 = j\omega_0$. On the positive σ-axis, $\mathbf{H}(\sigma_0)$ is real and (1.132) becomes

$$\mathbf{u}_0^* \mathbf{H}_h(\sigma_0)\mathbf{u}_0 \geqslant |\mathbf{u}_0' \mathbf{H}_s(\sigma_0)\mathbf{u}_0| = |\mathbf{u}_0' \mathbf{H}_h(\sigma_0)\mathbf{u}_0| \tag{1.158}$$

because $\mathbf{H}_s(\sigma_0) = \mathbf{H}_h(\sigma_0)$. In computing the quadratic form $\mathbf{u}_0^*\mathbf{H}_h(\sigma_0)\mathbf{u}_0$ for real $\mathbf{H}_h(\sigma_0)$, there is no need to consider all complex $\mathbf{u}_0$; real $\mathbf{u}_0$ would be sufficient. To see this, we show that if

$$\mathbf{u}'\mathbf{H}_h(\sigma_0)\mathbf{u} \geqslant 0 \tag{1.159}$$

for all real $\mathbf{u}$, then

$$\mathbf{u}^*\mathbf{H}_h(\sigma_0)\mathbf{u} \geqslant 0 \tag{1.160}$$

for all complex $\mathbf{u}$, and vice versa. Clearly, (1.160) implies (1.159). To demonstrate that (1.159) also implies (1.160), we consider any complex n-vector $\mathbf{u}$ and write

$$\mathbf{u} = \mathbf{u}_1 + j\mathbf{u}_2 \tag{1.161}$$

where $\mathbf{u}_1$ and $\mathbf{u}_2$ are real n-vectors. Substituting (1.161) in $\mathbf{u}^*\mathbf{H}_h(\sigma_0)\mathbf{u}$ gives

$$\begin{aligned}\mathbf{u}^*\mathbf{H}_h(\sigma_0)\mathbf{u} &= \mathbf{u}_1'\mathbf{H}_h(\sigma_0)\mathbf{u}_1 + j\mathbf{u}_1'\mathbf{H}_h(\sigma_0)\mathbf{u}_2 - j\mathbf{u}_2'\mathbf{H}_h(\sigma_0)\mathbf{u}_1 + \mathbf{u}_2'\mathbf{H}_h(\sigma_0)\mathbf{u}_2\\ &= \mathbf{u}_1'\mathbf{H}_h(\sigma_0)\mathbf{u}_1 + \mathbf{u}_2'\mathbf{H}_h(\sigma_0)\mathbf{u}_2 \geqslant 0\end{aligned} \tag{1.162}$$

because $\mathbf{u}_1'\mathbf{H}_h(\sigma_0)\mathbf{u}_2 = \mathbf{u}_2'\mathbf{H}_h(\sigma_0)\mathbf{u}_1$. The second line follows directly from (1.159). The equivalence of (1.159) and (1.160) does not mean that their values are equal. In general, they are not:

$$\mathbf{u}^*\mathbf{H}_h(\sigma_0)\mathbf{u} \neq \mathbf{u}'\mathbf{H}_h(\sigma_0)\mathbf{u} \tag{1.163}$$

Consider, for example, the case where $\mathbf{H}(\sigma_0) = \mathbf{1}_2$, the identity matrix of order 2. Choosing $\mathbf{u} = [j, 1]'$ shows that the left-hand side is 2 whereas the right-hand side is 0.

Theorem 1.5 On the real-frequency axis or the positive σ-axis, a linear time-invariant n-port network is passive at a point $s_0 = j\omega_0$ or $s_0 = \sigma_0$ if and only if the hermitian part of its associated general hybrid matrix is nonnegative definite at s_0.

PROOF From Eq. (1.157) we see that the theorem is clearly true for all the points on the real-frequency axis. Thus, we consider the situation where $s_0 = \sigma_0 > 0$. If the n-port is passive at σ_0, meaning that (1.158) holds, $\mathbf{H}_h(\sigma_0)$ must be nonnegative definite. Conversely, if $\mathbf{H}_h(\sigma_0)$ is nonnegative definite, then for all real $\mathbf{u}_0$, $\mathbf{u}_0^*\mathbf{H}_h(\sigma_0)\mathbf{u}_0 = \mathbf{u}_0'\mathbf{H}_h(\sigma_0)\mathbf{u}_0$ and (1.158) follows. Note that on the σ-axis $\mathbf{H}(\sigma)$ is real, and there is no need in taking $\mathbf{u}_0$ complex, as indicated in Definition 1.8. This completes the proof of the theorem.

A problem of great practical interest is to determine the maximum frequency of sinusoidal oscillation. For this we define the following.

Definition 1.10: Maximum frequency of oscillation The largest ω on the real-frequency axis at which an n-port network is active is called the *maximum frequency of oscillation* of the n-port network.

In the piecewise-linear analysis of a regenerative oscillator, such as the blocking oscillator, a somewhat similar problem is posed by the question of designing a pulse circuit of minimum rise time. More precisely, given an active device, what is the smallest time constant for an exponentially increasing signal that the device can produce? For this we introduce the concept of fastest regenerative mode.

Definition 1.11: Fastest regenerative mode The largest σ on the nonnegative σ-axis at which an n-port network is active is called the *fastest regenerative mode* of the n-port network.

We illustrate the above results by the following examples.

Example 1.11 Consider the small-signal high-frequency equivalent network of a bipolar transistor of Fig. 1.21, whose admittance matrix was computed in Example 1.4 and is given by

$$\mathbf{Y}(s) = \begin{bmatrix} G_1 + s(C_1 + C_2) & -sC_2 \\ g_m - sC_2 & G_2 + sC_2 \end{bmatrix} \tag{1.164}$$

The hermitian part of $\mathbf{Y}(s)$ is obtained as

$$\mathbf{Y}_h(s) = \begin{bmatrix} G_1 + \sigma(C_1 + C_2) & \frac{1}{2}g_m - \sigma C_2 \\ \frac{1}{2}g_m - \sigma C_2 & G_2 + \sigma C_2 \end{bmatrix} \tag{1.165}$$

On the real-frequency axis, $\mathbf{Y}_h(j\omega)$ is nonnegative definite if and only if

$$G_1 G_2 \geqslant \tfrac{1}{4} g_m^2 \tag{1.166}$$

Thus, the device is passive for all real frequencies provided that (1.166) is satisfied. On the other hand, if

$$G_1 G_2 < \tfrac{1}{4} g_m^2 \tag{1.167}$$

the device becomes active for all real frequencies, and the maximum frequency of oscillation is therefore infinity. The situation is evidently physically impossible. The reason for this seemingly inconsistent result is that we assume the equivalent network of the transistor to be an adequate representation of the transistor for all

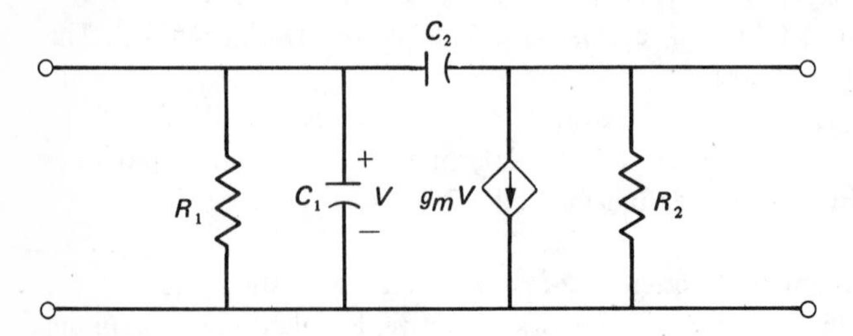

Figure 1.21 A small-signal high-frequency equivalent network of a bipolar transistor.

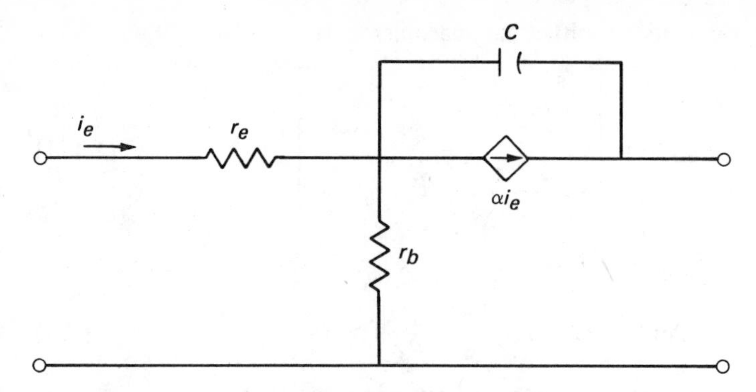

Figure 1.22 The small-signal T-equivalent network of a transistor.

frequencies. That this is not the case follows directly from the observation that at higher frequencies, parasitic effects must be taken into consideration, yielding a finite maximum frequency of oscillation.

On the positive σ-axis, $\mathbf{Y}_h(\sigma)$ is nonnegative definite if and only if

$$G_2 + \sigma C_2 \geqslant 0 \tag{1.168a}$$

$$G_1 + \sigma(C_1 + C_2) \geqslant 0 \tag{1.168b}$$

$$C_1 C_2 \sigma^2 + (C_1 G_2 + C_2 G_2 + C_2 G_1 + C_2 g_m)\sigma + G_1 G_2 - \tfrac{1}{4} g_m^2 \geqslant 0 \tag{1.168c}$$

Thus, if Eq. (1.166) is true, the device is passive for all nonnegative σ. Assuming that $G_1 G_2 < \frac{1}{4} g_m^2$ and setting the left-hand side of (1.168c) to zero yield the minimum value σ_m for which (1.168c) is satisfied:

$$\sigma_m = \frac{1}{2C_1 C_2} [\sqrt{b^2 + C_1 C_2 (g_m^2 - 4G_1 G_2)} - b] \tag{1.169a}$$

where

$$b = C_1 G_2 + C_2(G_1 + G_2 + g_m) \tag{1.169b}$$

Thus, the fastest regenerative mode of the device is given by (1.169a).† The device is passive for all $\sigma \geqslant \sigma_m$ and active for all $\sigma < \sigma_m$. In the piecewise-linear analysis of a regenerative oscillator, the value σ_m imposes a limitation on the period of relaxation oscillation.

Example 1.12 Figure 1.22 is the small signal T-equivalent network of a transistor. The two-port network is most conveniently characterized by its impedance matrix with

$$\mathbf{Z}(s) = \begin{bmatrix} r_e + r_b & r_b \\ r_b + \dfrac{\alpha}{sC} & r_b + \dfrac{1}{sC} \end{bmatrix} \tag{1.170}$$

†Strictly speaking, the fastest regenerative mode is $\sigma_m - \epsilon$, $\epsilon \to 0$.

On the real-frequency axis, its hermitian part becomes

$$\mathbf{Z}_h(j\omega) = \begin{bmatrix} r_e + r_b & r_b + \dfrac{j\alpha}{2\omega C} \\ r_b - \dfrac{j\alpha}{2\omega C} & r_b \end{bmatrix} \tag{1.171}$$

which is nonnegative definite if and only if

$$\det \mathbf{Z}_h(j\omega) = r_e r_b - \frac{\alpha^2}{4\omega^2 C^2} \geqslant 0 \tag{1.172}$$

Thus, the matrix remains nonnegative definite for all real frequencies

$$\omega \geqslant \omega_m = \frac{\alpha}{2C\sqrt{r_e r_b}} \tag{1.173}$$

showing that ω_m is the maximum frequency of oscillation† for the transistor.

On the nonnegative σ-axis, the hermitian part is obtained as

$$\mathbf{Z}_h(\sigma) = \begin{bmatrix} r_b + r_e & r_b + \dfrac{\alpha}{2\sigma C} \\ r_b + \dfrac{\alpha}{2\sigma C} & r_b + \dfrac{1}{\sigma C} \end{bmatrix} \tag{1.174}$$

which is nonnegative definite if and only if

$$\det \mathbf{Z}_h(\sigma) = 4C^2 r_e r_b \sigma^2 + 4C(r_b + r_e - \alpha r_b)\sigma - \alpha^2 \geqslant 0 \tag{1.175}$$

Choosing the equality and solving for σ yield the smallest value of σ at which $\mathbf{Z}_h(\sigma)$ is nonnegative, giving

$$\sigma_m = \frac{-(1-\alpha)r_b - r_e + \sqrt{[(1-\alpha)r_b + r_e]^2 + r_e r_b \alpha^2}}{2Cr_b r_e} \tag{1.176}$$

This is also the fastest regenerative mode for the transistor.

In the special situation where $\alpha = 0$, we have $\omega_m = 0$ and $\sigma_m = 0$. It can easily be shown that $\mathbf{Z}_h(s)$ is nonnegative definite for all s in the closed RHS, meaning that the device is passive at every point in the closed RHS. Thus, the impedance matrix $\mathbf{Z}(s)$ is a positive real matrix and the device is simply passive, as discussed in Sec. 1.6.1.

Example 1.13 Consider the two-port network of Fig. 1.21. We wish to partition the closed RHS into regions of passivity and activity by this two-port. To simplify the computation, let $C_2 = 0$. The admittance matrix becomes

$$\mathbf{Y}(s) = \begin{bmatrix} G_1 + sC_1 & 0 \\ g_m & G_2 \end{bmatrix} \tag{1.177}$$

†Strictly speaking, the maximum frequency of oscillation is $\omega_m - \epsilon, \epsilon \to 0$.

As indicated in Example 1.8, the two-port is passive, its admittance matrix being positive real, if $4G_1G_2 \geqslant g_m^2$. To exhibit activity in the closed RHS, we assume that $4G_1G_2 < g_m^2$. This does not mean that every point in the closed RHS is active; there is still a region of passivity.

We first investigate the behavior of the network for the positive σ-axis and the real-frequency axis points. According to Theorem 1.5, we need only test for nonnegative definiteness of the hermitian part of $\mathbf{Y}(s)$, which is given by

$$\mathbf{Y}_h(s) = \begin{bmatrix} G_1 + \sigma C_1 & \frac{1}{2}g_m \\ \frac{1}{2}g_m & G_2 \end{bmatrix} \tag{1.178}$$

Since $\mathbf{Y}_h(s)$ is independent of ω, on the real-frequency axis $\mathbf{Y}_h(j\omega)$ is not nonnegative definite, showing that the device is active for all real frequencies. On the positive σ-axis, $\mathbf{Y}_h(\sigma)$ is nonnegative definite if and only if

$$\sigma \geqslant \sigma_m = \frac{g_m^2 - 4G_1G_2}{4G_2C_1} > 0 \tag{1.179}$$

Thus, the device is active for all $\sigma < \sigma_m$ and passive for $\sigma \geqslant \sigma_m$.

For the remainder of the closed RHS, we must apply Eq. (1.136). Consider the points s_0 in the region defined by the vertical strip $0 < \sigma < \sigma_m$, as indicated in Fig. 1.23. At any of these points s_0, $\mathbf{Y}_h(s_0)$ is not nonnegative definite, implying that there exists a vector $\mathbf{u}_0$ such that $\mathbf{u}_0^*\mathbf{Y}_h(s_0)\mathbf{u}_0 < 0$. Clearly, for this $\mathbf{u}_0$ Eq. (1.136) cannot be satisfied, and the device is therefore active in this region. For $\sigma = \sigma_m$, $\mathbf{Y}_h(\sigma_m + j\omega)$ is singular and there exists a nontrivial solution $\mathbf{u} = [u_1, u_2]'$, $u_1 \neq 0$, such that $\mathbf{Y}_h(\sigma_m + j\omega)\mathbf{u} = \mathbf{0}$. Now we can write Eq. (1.136) as

$$|\sigma_m + j\omega|\mathbf{u}^*\mathbf{Y}_h(\sigma_m + j\omega)\mathbf{u} \geqslant \sigma_m|\mathbf{u}'\mathbf{Y}_h(\sigma_m + j\omega)\mathbf{u} + j\omega C_1 u_1^2| \tag{1.180}$$

which reduces to $|\sigma_m \omega C_1 u_1^2| \leqslant 0$. This inequality cannot be satisfied unless $\omega = 0$. Thus, the device is active along the vertical line $\sigma_m + j\omega$, $\omega \neq 0$. The results are sketched in Fig. 1.23.

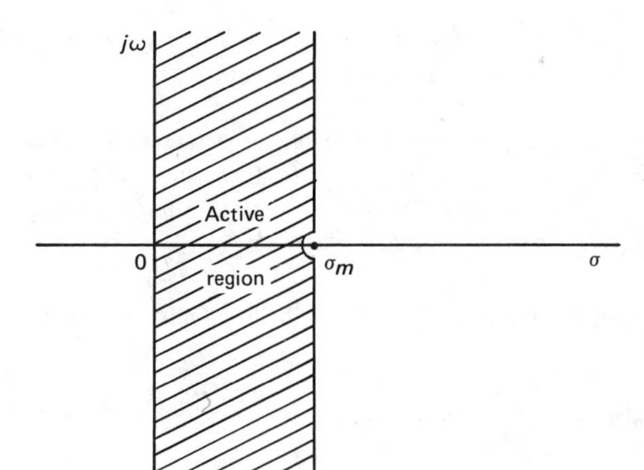

Figure 1.23 The partition of the closed RHS into regions of passivity and activity by the two-port network of Fig. 1.21.

1.7.3 Physical Significance of Single Complex Frequency Passivity and Activity

In introducing the concepts of passivity and activity at a single complex frequency s_0 in the closed RHS, we started from the time-energy definition (Definition 1.3) of passivity. We showed that if we apply an excitation of the type (1.127) to a given n-port network, which is known to be passive, then (1.131) or (1.132) follows. In the present section, we demonstrate that if an n-port is passive at a single complex frequency s_0, then for all excitations of the type (1.127) the energy into the n-port is always nonnegative. Furthermore, if an n-port is active at s_0, there exists an excitation source of the type (1.127) such that, given sufficient time, an arbitrarily large amount of energy can be extracted from the n-port.

Theorem 1.6 If a linear, time-invariant n-port network is passive at a single complex frequency s_0 in the closed RHS, then for all excitations of the type (1.127) the total energy delivered to the n-port network is nonnegative.

PROOF Let $\mathbf{H}(s)$ be the general hybrid matrix of the n-port, which is passive at $s_0 = \sigma_0 + j\omega_0$, $\sigma_0 \geqslant 0$. Assume first that $\sigma_0 \neq 0$ and $\omega_0 \neq 0$. Suppose that there exists an excitation of the type (1.127) such that the total energy delivered to the n-port is negative for some t, say t_1. Then we can obtain an expression for the total energy $\varepsilon(t_1)$ similar to that given in (1.129) with t_1 replacing t. If $\varepsilon(t)$ is negative at t_1, it is negative at $t_1 + k\pi/\omega_0$ for any positive integer k, however large. Since $\varepsilon(t_0)$ and C_0 in (1.129) are constants and, for $\sigma > 0$, the terms containing $\exp[2\sigma_0(t_1 + k\pi/\omega_0)]$ in $\varepsilon(t_1 + k\pi/\omega_0)$ can be made arbitrarily large. This implies that the sum of the terms inside the curl brackets of (1.129) is negative at $t_1 + k\pi/\omega_0$. Hence, the left-hand side of (1.131) must be negative. This contradicts the hypothesis that the n-port is passive at s_0. In a similar manner, we are led to the same conclusion for $\sigma = 0$ or $\omega_0 = 0$, the details being left as an exercise (see Prob. 1.18). The only difference is that for $\sigma_0 = 0$, $e^{2\sigma_0 t}/\sigma_0$ is replaced by t in (1.129), and the same argument follows. This completes the proof of the theorem.

The following theorem establishes a relation between activity at a single complex frequency and energy-delivering capability of an n-port network.

Theorem 1.7 If a linear, time-invariant n-port network is active at a single complex frequency $s_0 = \sigma_0 + j\omega_0$ with $\sigma_0 \geqslant 0$ and $\omega_0 \neq 0$, then there exists an excitation of the type (1.127) such that, given sufficient time, an arbitrary, large amount of energy can be extracted from the n-port network.

PROOF Since the n-port is active at s_0, there exists a fixed complex vector $\mathbf{u}_0$ such that for $\sigma_0 \neq 0$,

$$\frac{1}{\sigma_0}\mathbf{u}_0^*\mathbf{H}_h(s_0)\mathbf{u}_0 - \frac{1}{|s_0|}|\mathbf{u}_0'\mathbf{H}_s(s_0)\mathbf{u}_0| < 0 \qquad (1.181)$$

where $\mathbf{H}(s)$ is the general hybrid matrix of the n-port. Following Eqs. (1.127)-(1.129), we obtain, for $\sigma > 0$ and for large t,

$$4\varepsilon(t) \approx e^{2\sigma_0 t}\left[\frac{1}{\sigma_0}\text{Re }\mathbf{u}_0^*\mathbf{H}(s_0)\mathbf{u}_0 + \frac{1}{|s_0|}|\mathbf{u}_0'\mathbf{H}(s_0)\mathbf{u}_0| \cos(2\omega_0 t + \theta)\right] \tag{1.182}$$

where θ is the phase angle contributed by both the quadratic form $\mathbf{u}_0'\mathbf{H}(s_0)\mathbf{u}_0$ and s_0. Thus, if $t_k = (k+\frac{1}{2})\pi/\omega_0 - \theta/2\omega_0$, for sufficiently large integers k,

$$\begin{aligned}4\varepsilon(t_k) &\approx e^{2\sigma_0 t_k}\left[\frac{1}{\sigma_0}\text{Re }\mathbf{u}_0^*\mathbf{H}(s_0)\mathbf{u}_0 - \frac{1}{|s_0|}|\mathbf{u}_0'\mathbf{H}(s_0)\mathbf{u}_0|\right] \\ &= e^{2\sigma_0 t_k}\left[\frac{1}{\sigma_0}\mathbf{u}_0^*\mathbf{H}_h(s_0)\mathbf{u}_0 - \frac{1}{|s_0|}|\mathbf{u}_0'\mathbf{H}_s(s_0)\mathbf{u}_0|\right]\end{aligned} \tag{1.183}$$

Combining Eqs. (1.181) and (1.183) yields

$$4\varepsilon(t_k) \approx -\left|\frac{1}{\sigma_0}\mathbf{u}_0^*\mathbf{H}_h(s_0)\mathbf{u}_0 - \frac{1}{|s_0|}|\mathbf{u}_0'\mathbf{H}_s(s_0)\mathbf{u}_0|\right| e^{2\sigma_0 t_k} \tag{1.184}$$

or

$$\varepsilon(t_k) < -M \tag{1.185}$$

where M is any prescribed positive number, however large. For the case $\sigma_0 = 0$, the above reasoning still applies provided that we replace $e^{2\sigma_0 t}/\sigma_0$ by t. Thus, for sufficient time, an arbitrary, large amount of energy can be extracted from the given n-port network. This completes the proof of the theorem.

Thus, in the passive region, an active n-port network behaves like a passive n-port network and can be replaced by a passive n-port network. If an n-port is passive at every point in the closed RHS, the n-port is simply passive in the sense of Definition 1.3. In the active region, the n-port can be made to achieve power gain or to oscillate by means of an appropriate passive imbedding network. We illustrate these by the following examples.

Example 1.14 Figure 1.24 is the small-signal high-frequency equivalent network of a field-effect transistor, whose admittance matrix is given by

$$\mathbf{Y}(s) = \begin{bmatrix}(C_1 + C_2)s & -sC_2 \\ g_m - sC_2 & G + sC_2\end{bmatrix} \tag{1.186}$$

The hermitian part is found to be

$$\mathbf{Y}_h(s) = \begin{bmatrix}\sigma(C_1 + C_2) & \frac{1}{2}g_m - \sigma C_2 \\ \frac{1}{2}g_m - \sigma C_2 & G + \sigma C_2\end{bmatrix} \tag{1.187}$$

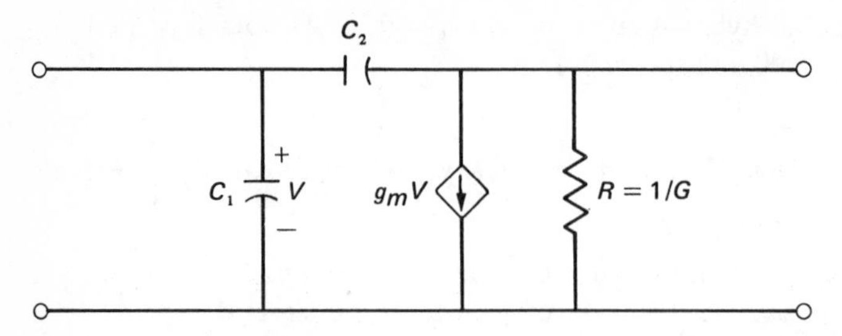

Figure 1.24 A small-signal high-frequency equivalent network of a field-effect transistor.

On the nonnegative σ-axis, $\mathbf{Y}_h(\sigma)$ is nonnegative definite if and only if det $\mathbf{Y}_h(\sigma) \geqslant 0$, which is equivalent to

$$C_1C_2\sigma^2 + (C_1G + C_2G + g_mC_2)\sigma - \tfrac{1}{4}g_m^2 \geqslant 0 \tag{1.188}$$

Setting det $\mathbf{Y}_h(\sigma) = 0$, we obtain the positive root of the above polynomial in σ, giving the fastest regenerative mode

$$\sigma_m = \frac{-(C_1 + C_2)G - g_mC_2 + \sqrt{(C_1G + C_2G + g_mC_2)^2 + g_m^2C_1C_2}}{2C_1C_2} \tag{1.189}$$

To demonstrate that this mode can be realized by an appropriate passive imbedding, we consider a gyrator with gyration resistance $2/g_m$, as shown in Fig. 1.25. The admittance matrix of the gyrator is given by

$$\mathbf{Y}_g(s) = \begin{bmatrix} 0 & \tfrac{1}{2}g_m \\ -\tfrac{1}{2}g_m & 0 \end{bmatrix} \tag{1.190}$$

If the transistor and the gyrator are connected in parallel as shown in Fig. 1.26, the admittance matrix of the overall two-port network is given by

$$\mathbf{Y}(s) + \mathbf{Y}_g(s) = \begin{bmatrix} (C_1 + C_2)s & \tfrac{1}{2}g_m - sC_2 \\ \tfrac{1}{2}g_m - sC_2 & G + sC_2 \end{bmatrix} \tag{1.191}$$

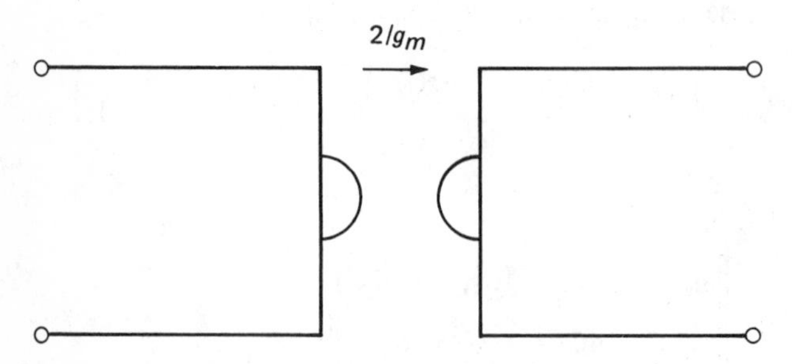

Figure 1.25 A gyrator with gyration resistance $2/g_m$.

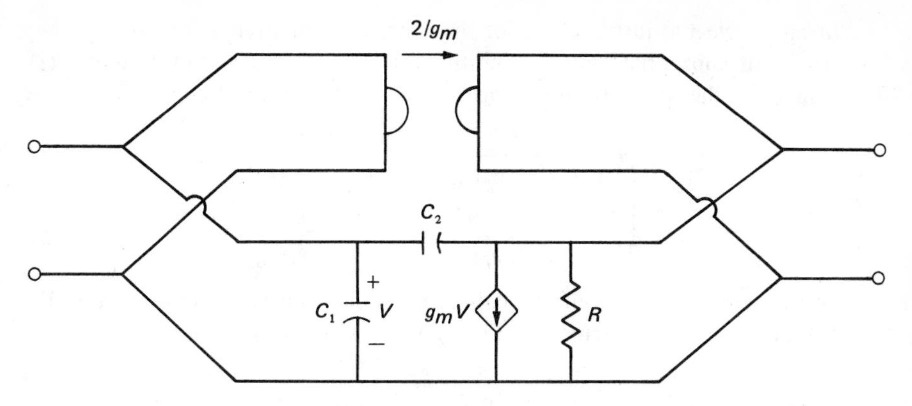

Figure 1.26 The parallel connection of a gyrator and a transistor.

The open-circuit natural frequencies of the resulting two-port network are the zeros of det $[\mathbf{Y}(s) + \mathbf{Y}_g(s)]$. Since Eqs. (1.187) and (1.191) are identical except that σ is replaced by s, the regenerative mode of the network is therefore σ_m, which is the largest σ for this transistor under arbitrary passive imbedding.

Example 1.15 In this example, we demonstrate that if a one-port network is active at a single complex frequency s_0 in the closed RHS, then it may be employed for energy gain with respect to all excitations of the type (1.127).

Let $z_p(s) = 1/y_p(s)$ be a positive real impedance characterizing a passive one-port network N_p. Assume that the one-port N_p is excited by a source combination as shown in Fig. 1.27*a* or 1.27*b* with $u(t) = v(t)$ or $i(t)$ and

$$u(t) = \tfrac{1}{2}(u_0 e^{s_0 t} + \bar{u}_0 e^{\bar{s}_0 t}) \tag{1.192}$$

where u_0 is an arbitrary complex but fixed constant, and $z_b(s) = 1/y_b(s)$ is another passive impedance. To be specific, we shall consider only the situation depicted in

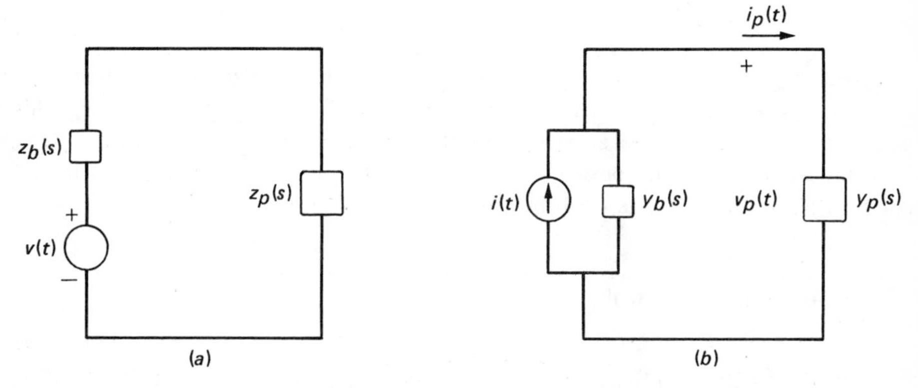

Figure 1.27 (*a*) A passive one-port network excited by a voltage source. (*b*) A passive one-port network excited by a current source.

Fig. 1.27*b* and leave the justification for the other configuration as an exercise (see Prob. 1.20). With appropriate initial conditions, the voltage $v_p(t)$ and current $i_p(t)$ at the terminals of the one-port admittance y_p in Fig. 1.27*b* are obtained as

$$v_p(t) = \frac{1}{2}\left(\frac{I_0 e^{s_0 t}}{y_b(s_0) + y_p(s_0)} + \frac{\bar{I}_0 e^{\bar{s}_0 t}}{y_b(\bar{s}_0) + y_p(\bar{s}_0)}\right) \tag{1.193a}$$

$$i_p(t) = \frac{1}{2}\left(\frac{I_0 y_p(s_0) e^{s_0 t}}{y_b(s_0) + y_p(s_0)} + \frac{\bar{I}_0 y_p(\bar{s}_0) e^{\bar{s}_0 t}}{y_b(\bar{s}_0) + y_p(\bar{s}_0)}\right) \tag{1.193b}$$

where $I_0 = u_0$. Then the total energy delivered to the initially relaxed one-port N_p up to any time t can be determined by (1.18) and is given by

$$\mathcal{E}(t) = \frac{|I_0|^2}{4|y_b(s_0) + y_p(s_0)|^2} e^{2\sigma_0 t}\left[\frac{1}{\sigma_0} \operatorname{Re} y_p(s_0) + \frac{1}{|s_0|} |y_p(s_0)| \cos(2\omega_0 t + \theta)\right] \tag{1.194a}$$

where
$$\frac{I_0^2 y_p(s_0)}{s_0 [y_b(s_0) + y_p(s_0)]^2} = \frac{|I_0|^2 |y_p(s_0)|}{|s_0| \, |y_b(s_0) + y_p(s_0)|^2} e^{j\theta} \tag{1.194b}$$

To maximize $\mathcal{E}(t)$ for a fixed $y_p(s_0)$, we minimize $|y_b(s_0) + y_p(s_0)|$. For this we choose

$$\operatorname{Im} y_b(s_0) = -\operatorname{Im} y_p(s_0) \tag{1.195a}$$

$$B_0 \equiv \operatorname{Re} y_b(s_0) = \frac{\sigma_0 |\operatorname{Im} y_p(s_0)|}{|\omega_0|} \tag{1.195b}$$

Equation (1.195*b*) represents the minimum value that the real part of $y_b(s)$ can have at the point s_0, because at this point (1.139) applies. Thus, the maximum energy that can be delivered to the one-port N_p up to any time t is given by

$$\mathcal{E}_m(t) = \frac{|I_0|^2}{4\sigma_0 [\operatorname{Re} y_p(s_0) + B_0]^2} e^{2\sigma_0 t}\left[\operatorname{Re} y_p(s_0) + \frac{\sigma_0}{|s_0|} |y_p(s_0)| \cos(2\omega_0 t + \theta)\right] \tag{1.196}$$

Now suppose that we have a one-port network characterized by its admittance function $y_a(s)$, which is active at the point s_0. We show that by using this admittance, we can achieve an energy input to the one-port N_p greater than $\mathcal{E}_m(t)$. For this we consider the network of Fig. 1.28. The total energy delivered to the one-port N_p up to any time t can be computed and is given by

$$\mathcal{E}_a(t) = \frac{|I_0|^2}{4\sigma_0 |y_a(s_0) + y_p(s_0)|^2} e^{2\sigma_0 t}\left[\operatorname{Re} y_p(s_0) + \frac{\sigma_0}{|s_0|} |y_p(s_0)| \cos(2\omega_0 t + \phi)\right] \tag{1.197a}$$

where
$$\frac{I_0^2 y_p(s_0)}{s_0 [y_a(s_0) + y_p(s_0)]^2} = \frac{|I_0|^2 |y_p(s_0)|}{|s_0| \, |y_a(s_0) + y_p(s_0)|^2} e^{j\phi} \tag{1.197b}$$

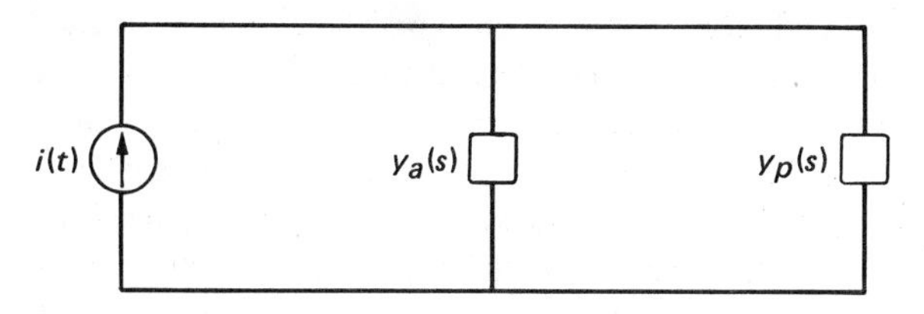

Figure 1.28 An active one-port network employed for energy gain.

Assume that we can choose a passive $y_p(s)$ so that $y_a(s_0) + y_p(s_0)$ is real. Under this situation, the phase angles θ and ϕ are equal, and the expression for net energy gain becomes

$$\frac{\mathcal{E}_a(t)}{\mathcal{E}_m(t)} = \frac{\{\operatorname{Re} y_p(s_0) + [\sigma_0 |\operatorname{Im} y_p(s_0)|/|\omega_0|]\}^2}{|\operatorname{Re} y_p(s_0) + \operatorname{Re} y_a(s_0)|^2} \tag{1.198}$$

Since $y_a(s)$ is active at s_0, from (1.137) we have

$$\operatorname{Re} y_a(s_0) - \sigma_0 \left| \frac{y_a(s_0)}{s_0} \right| < 0 \tag{1.199}$$

giving

$$|\omega_0| \operatorname{Re} y_a(s_0) < \sigma_0 |\operatorname{Im} y_a(s_0)| \tag{1.200}$$

Since by choice $\operatorname{Im} y_p(s_0) = -\operatorname{Im} y_a(s_0)$, the expression (1.198) is always indicative of energy gain greater than unity. Thus, we conclude that if a one-port network is active at a point in the closed RHS, then it can be employed for energy gain with respect to signals of the type (1.127).

1.8 SUMMARY

We began the chapter by introducing the concepts of portwise linearity and time invariance. A portwise linear n-port network need not be a linear network in the usual sense. Conversely, a linear network may not be a portwise linear n-port. Similar statements can be made for the time invariance. We then defined passivity and activity in terms of the universally encountered quantities time and energy, and we showed that causality is a consequence of linearity and passivity. From the definition of causality, a causal n-port network must have its response as a single-valued function of its excitation. The converse, however, is not necessarily true. In the situation where the response of a linear passive n-port network is a multiple-valued function of its excitation, we showed that the instantaneous power entering the n-port is uniquely determined by its excitation. This was followed by a brief review of the general characterizations of linear, time-invariant n-port networks in the frequency domain. The characterizations are stated in terms of the general hybrid matrix. The translation of the time-domain passivity criteria into the equivalent frequency-domain passivity condition was taken up next. We showed that an n-port is passive if and only if its general hybrid matrix is positive real.

Finally, we introduced the discrete-frequency concepts of passivity and activity for a class of most significant and useful signals comprised of two complex exponential signals at the complex conjugate pair of frequencies. In this way, the closed RHS can be divided into regions of passivity and activity by a given n-port network. In the passive region, the net energy delivered to the n-port is always nonnegative. Thus, the n-port behaves like a passive network and can be replaced by a passive n-port that exhibits identical behavior with respect to all real signals comprised of two complex exponential signals at the complex conjugate pair of frequencies. In the active region, the n-port network can be made to achieve power gain or to oscillate by means of an appropriate passive imbedding network. In fact, we demonstrated that there exists an excitation of the above type such that, for sufficient time, an arbitrary, large amount of energy can be extracted from the n-port network. For practical applications, we are concerned mainly with the behavior of the n-port networks on the real-frequency axis and, less frequently, on the positive σ-axis. For these purposes, we introduced the notions of maximum frequency of oscillation and fastest regenerative mode. The former determines the maximum frequency of sinusoidal oscillation and the latter imposes a limitation on the period of relaxation oscillation, such as multivibrators and blocking oscillators, that an active n-port device can attain under passive imbedding.

PROBLEMS

1.1 Show that a properly biased tunnel diode of Fig. 1.12*a* with its *vi*-characteristic as shown in Fig. 1.12*b* is active.

1.2 Use Definition 1.3 to show that a linear time-invariant inductor is passive if its inductance is nonnegative and active if its inductance is negative.

1.3 Demonstrate that the unbiased tunnel diode of Fig. 1.11 is a causal one-port under voltage excitation-current response, and a noncausal one-port under current excitation-voltage response.

1.4 Consider a linear time-varying resistor whose resistance is characterized by the equation

$$R(t) = r_0 + r_1 \cos(\omega_0 t + \theta) \tag{1.201}$$

where $r_0 \geqslant r_1 > 0$ and $\omega_0 > 0$. Determine whether this device is passive or active.

1.5 Let $\mathbf{y}(t)$ be the response vector of an n-port network resulting from the excitation vector $\mathbf{u}(t)$. If $\mathbf{v}(t)$ and $\mathbf{i}(t)$ denote the port-voltage and port-current vectors, is it generally true that

$$\mathbf{v}'(t)\mathbf{i}(t) = \mathbf{u}'(t)\mathbf{y}(t) \tag{1.202}$$

1.6 Demonstrate that the open circuit is noncausal under current excitation-voltage response.

1.7 For the given one-port network of Fig. 1.29, determine the regions of passivity and activity in the closed RHS. Choose a point in the passive region and demonstrate that this one-port has a passive equivalent one-port at the chosen point. Choose a point in the active region and show that this one-port can be employed for energy gain at the chosen complex frequency.

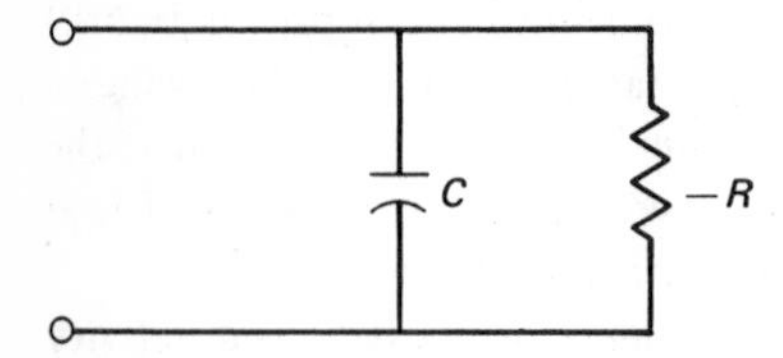

Figure 1.29 An active one-port network.

1.8 For the two-port network considered in Example 1.4, compute the hybrid parameters $h_{12}(s)$ and $h_{22}(s)$, the impedance parameters $z_{12}(s)$ and $z_{22}(s)$, the admittance parameters $y_{12}(s)$ and $y_{22}(s)$, and the transmission parameters $B(s)$ and $D(s)$.

1.9 Suppose that we have a two-port network that can admit only excitation signals of the type

$$\mathbf{u}(t) = a\begin{bmatrix}1\\1\end{bmatrix} \qquad t > 0 \tag{1.203a}$$

$$= \begin{bmatrix}0\\0\end{bmatrix} \qquad t \leqslant 0 \tag{1.203b}$$

where a is real and arbitrary. Assume that the response of the two-port caused by the above excitation (1.203) is given by

$$y(t) = a\begin{bmatrix}1\\-1\end{bmatrix} \tag{1.204}$$

Show that the two-port is linear and passive. The response is evidently a single-valued function of excitation. Demonstrate that this two-port is noncausal.

1.10 Determine the condition under which the impedance matrix (1.69) is positive real.

1.11 Confirm that the Fourier transform of the signal (1.111) is given by (1.112).

1.12 For the two-port network of Fig. 1.17, use the procedure outlined in Example 1.5 to compute the hybrid matrix and the transmission matrix from the known impedance matrix (1.69).

1.13 Repeat Prob. 1.12 if the impedance parameters are given as $z_{ij}(s)$, $i, j = 1, 2$.

1.14 Show that for $\omega_0 = 0$ Eq. (1.130) requires that Re $\mathbf{u}_0^*\mathbf{H}(s_0)\mathbf{u}_0 \geqslant 0$, and this inequality is equivalent to (1.131) for all real $\mathbf{u}_0$.

1.15 In deriving (1.131) from (1.129), we assume that $\sigma_0 \neq 0$. Show that for $\sigma = 0$, the passivity requirement implies Re $\mathbf{u}_0^*\mathbf{H}(j\omega_0)\mathbf{u}_0 \geqslant 0$. This indicates that the inequality (1.131) is valid for all s_0 in the closed RHS for a passive n-port network.

1.16 Consider a one-port admittance $y(s)$ that is passive at a point s_0 in the closed RHS. Obtain an equivalent passive replacement one-port network that exhibits identical behavior with respect to all real signals of the type (1.127).

1.17 A nonlinear capacitor is characterized by the equation

$$i(t) = 2v(t)\frac{dv(t)}{dt} \tag{1.205}$$

Determine whether this capacitor is passive or active.

1.18 Prove that Theorem 1.6 is valid for $\sigma_0 = 0$ or $\omega_0 = 0$, where $s_0 = \sigma_0 + j\omega_0$ is a point in the closed RHS where the given n-port is passive.

1.19 A nonlinear inductor is characterized by the equation

$$v(t) = 2i^2(t)\frac{di(t)}{dt} \tag{1.206}$$

Determine whether this inductor is passive or active.

1.20 Repeat Example 1.15 for a one-port impedance $z_a(s)$ that is active at a closed RHS point s_0, and show that this one-port can be employed for energy gain at s_0.

1.21 Figure 1.30 shows the small-signal high-frequency hybrid-pi equivalent network of a bipolar transistor. Compute its admittance matrix.

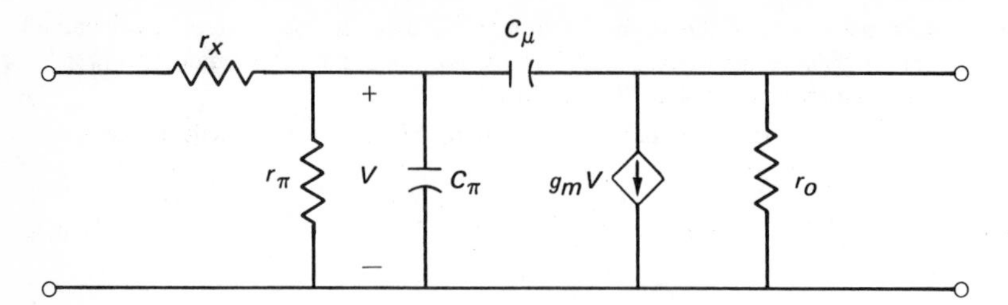

Figure 1.30 The small-signal high-frequency hybrid-pi equivalent network of a bipolar transistor.

1.22 Let $\tilde{u}(s) = [I_1(s),\ V_2(s)]'$ be the excitation vector of the two-port network of Fig. 1.30. Compute the associated hybrid matrix.

1.23 The matrix given below is known to be positive real. Confirm this claim.

$$\mathbf{H}(s) = \frac{1}{2s(s^2+1)}\begin{bmatrix} 2s^2+1 & 1 \\ 1 & 2s^2+1 \end{bmatrix} \tag{1.207}$$

1.24 Typical values of the hybrid-pi parameters for a type 2N2614 transistor of Fig. 1.30 are given as follows:

$$\begin{aligned} r_x &= 300\ \Omega & g_m &= 0.0385 \text{ mho} \\ r_\pi &= 4\text{ k}\Omega & C_\pi &= 750 \text{ pF} \\ r_o &= 167\text{ k}\Omega & C_\mu &= 9 \text{ pF} \end{aligned} \tag{1.208}$$

These values pertain at $I_C = 1$ mA and $V_{CE} = 6$ V. Using these values, find the maximum frequency of oscillation and the fastest regenerative mode of the transistor.

1.25 For the one-port network of Fig. 1.31, determine the regions of passivity and activity.

1.26 Test the following matrix to see if it is positive real:

$$\mathbf{H}(s) = \frac{1}{46s+1}\begin{bmatrix} 2s+1 & 2s-10 \\ 2s-10 & 2s+1 \end{bmatrix} \tag{1.209}$$

1.27 Assume that the one-port network of Fig. 1.32 is initially relaxed. Determine if this linear time-varying one-port is active.

1.28 A nonlinear resistor is characterized by the equation

$$v(t) = i(t) + 2i^2(t) \tag{1.210}$$

Is this resistor active? Is this one-port network causal under voltage excitation-current response? If the roles of $v(t)$ and $i(t)$ are interchanged in (1.210), can one draw the same conclusion?

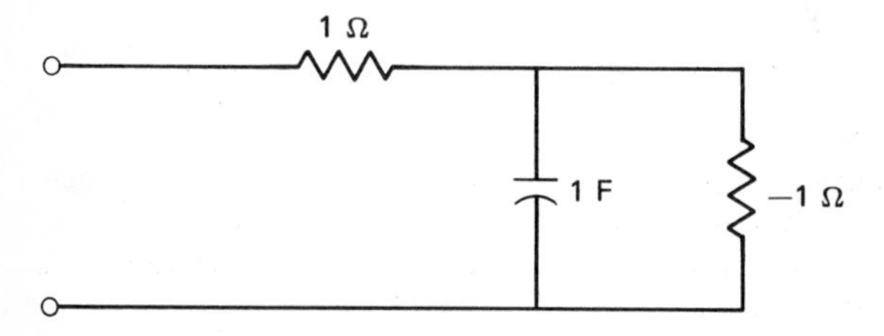

Figure 1.31 An active one-port network.

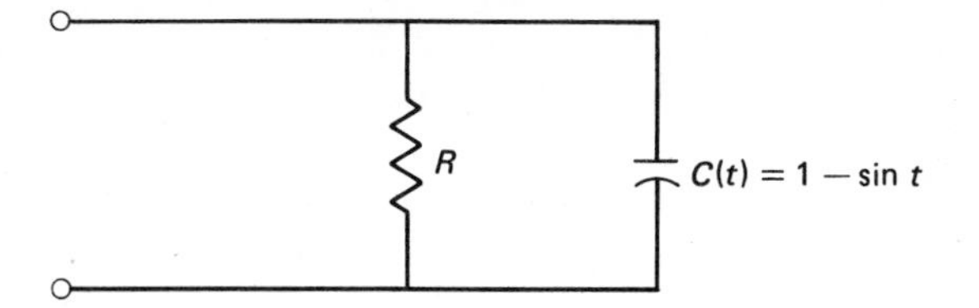

Figure 1.32 A linear time-varying one-port network.

1.29 A nonlinear resistor is characterized by the equation

$$v(t) = \tanh i(t) \tag{1.211}$$

Is this resistor active? Is this resistor causal under voltage excitation-current response?

1.30 Show that a linear time-varying inductor of inductance $L(t)$ is passive if and only if $L(t) \geqslant 0$ and $dL(t)/dt \geqslant 0$ for all t.

1.31 Show that a linear time-varying capacitor of capacitance $C(t)$ is passive if and only if $C(t) \geqslant 0$ and $dC(t)/dt \geqslant 0$ for all t.

1.32 Show that a nonlinear time-varying resistor is passive if and only if its vi-characteristic is in the first and third quadrants of the vi-plane for all time.

1.33 Consider the one-port network of Fig. 1.33. Determine the condition under which the one-port is passive.

1.34 Let a and b denote real and nonnegative constants. Determine conditons under which each of the following elements is active:

1. A resistor characterized by the resistance $R(t) = a + b \sin \omega_0 t$
2. A capacitor characterized by the capacitance $C(t) = a + b \sin \omega_0 t$
3. An inductor characterized by the inductance $L(t) = a + b \sin \omega_0 t$

1.35 Under what circumstances is the emitter-follower of Fig. 1.34 passive? Use the equivalent network of Fig. 1.17 for the transistor and ignore the biasing circuit.

1.36 Determine whether the independent sources are linear one-port networks. Are controlled sources nonlinear two-ports? If so, justify your conclusion.

1.37 Figure 1.35 shows a one-port network containing two ideal diodes. Determine the condition under which the one-port is linear.

1.38 Suppose that we have a transformer whose turns ratio is time-varying. As shown in Fig. 1.36, a capacitor of positive capacitance C is connected at the output of the transformer. Determine condition under which the resulting one-port network is not lossless.

1.39 Determine whether the ideal diode is causal or noncausal. Is it lossless?

1.40 The inductance of a linear time-varying inductor is given by

$$L(t) = t + \sin t \tag{1.212}$$

Determine whether this inductor is passive or active.

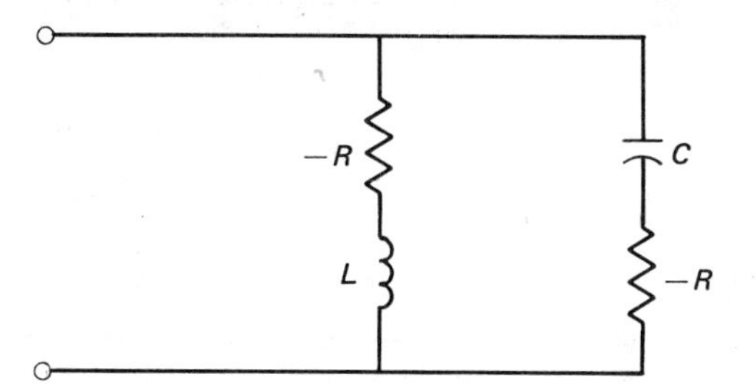

Figure 1.33 A one-port network.

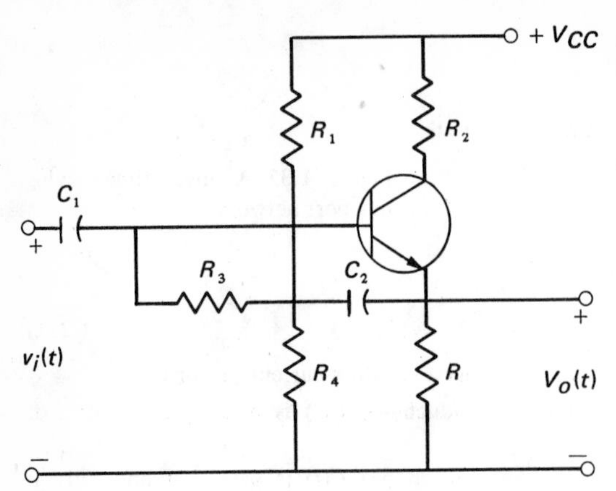

Figure 1.34 A transistor emitter-follower.

1.41 A charge-controlled capacitor is a capacitor whose terminal voltage $v(t)$ is a single-valued function of the stored electric charge $q(t)$. A charge-controlled capacitor is characterized by the equation

$$v(t) = 1 + q^2(t) \tag{1.213}$$

Determine whether this capacitor is passive or active, linear or nonlinear. Is it causal under voltage excitation-charge response?

1.42 A linear time-varying resistor of conductance

$$G(t) = g_0 + g_1 \cos \omega_p t \tag{1.214}$$

can be made active by appropriately choosing g_0 and g_1. This idea has been applied successfully to the design of *parametric amplifiers.* Figure 1.37 shows a simple parametric amplifier, where the signal to be amplified is represented by the current source $i_s(t)$. Show that power amplification at the source is possible only if $g_1 > g_0$.

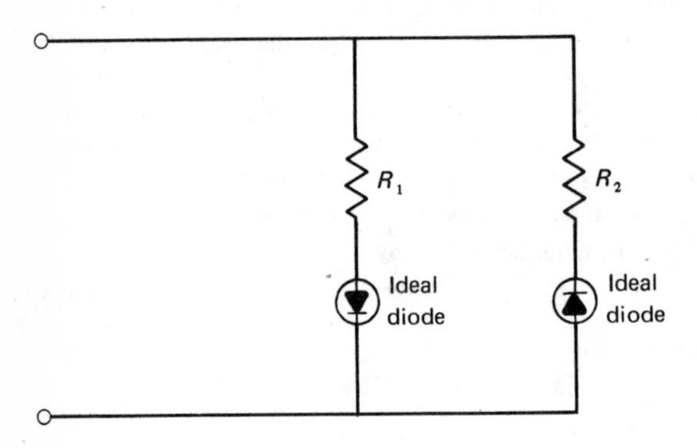

Figure 1.35 A one-port network containing two ideal diodes.

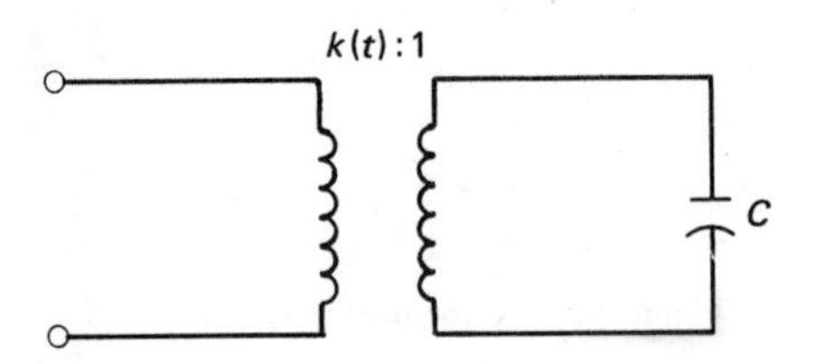

Figure 1.36 An ideal transformer whose turns ratio is time-varying.

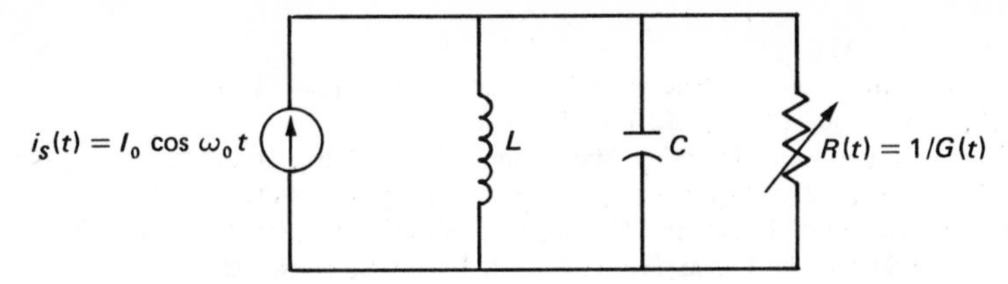

Figure 1.37 A simple parametric amplifier.

1.43 A flux-controlled inductor is an inductor whose current $i(t)$ is a single-valued function of the magnetic flux $\phi(t)$. Consider a time-invariant flux-controlled inductor with a characteristic like the one shown in Fig. 1.38. Determine the condition under which the inductor is passive. Express the condition in terms of the area under the ϕi-curve.

1.44 In the two-port network of Fig. 1.17, let

$$\mathbf{u}(t) = \mathbf{i}(t) = \begin{bmatrix} i_1(t) \\ i_2(t) \end{bmatrix} = \tfrac{1}{2}(\mathbf{I}_0 e^{s_0 t} + \bar{\mathbf{I}}_0 e^{\bar{s}_0 t}) \tag{1.215}$$

be the excitation, where $\mathbf{I}_0$ is an arbitrary complex 2-vector, and let

$$\mathbf{y}(t) = \mathbf{v}(t) = \begin{bmatrix} v_1(t) \\ v_2(t) \end{bmatrix} = \tfrac{1}{2}(\mathbf{V}_0 e^{s_0 t} + \bar{\mathbf{V}}_0 e^{\bar{s}_0 t}) \tag{1.216}$$

be the response. Show that

$$\mathbf{V}_0 = \mathbf{Z}(s_0)\mathbf{I}_0 \tag{1.217}$$

where $\mathbf{Z}(s)$ is the impedance matrix of the two-port given by (1.69). At the initial time t_0, the port voltages are required to be

$$\mathbf{v}(t_0) = \tfrac{1}{2}[\mathbf{Z}(s_0)\mathbf{I}_0 e^{s_0 t_0} + \bar{\mathbf{Z}}(s_0)\bar{\mathbf{I}}_0 e^{\bar{s}_0 t_0}] \tag{1.218}$$

Determine the initial voltages across the two capacitors at t_0. What is the amount of energy required at the initial time t_0 to set up the appropriate initial conditions so that no transient will occur?

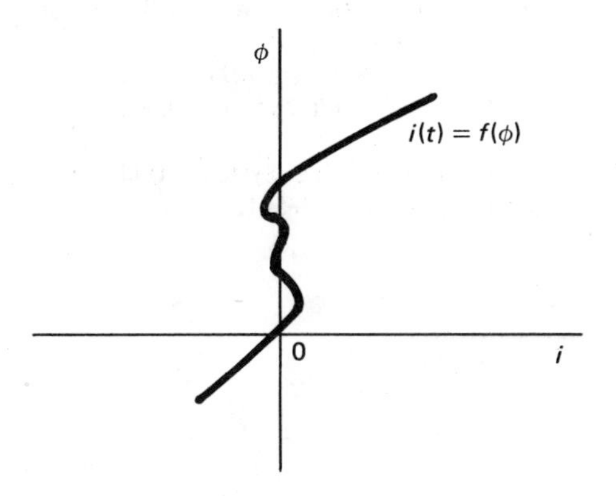

Figure 1.38 The characteristic of a time-invariant flux-controlled inductor.

BIBLIOGRAPHY

Anderson, B. D. O. and S. Vongpanitlerd: "Network Analysis and Synthesis: A Modern Systems Theory Approach," Englewood Cliffs, N.J.: Prentice-Hall, 1973.

Chen, W. K.: Relationships Between Scattering Matrix and Other Matrix Representations of Linear Two-Port Networks, *Int. J. Electronics*, vol. 38, no. 4, pp. 433–441, 1975.

Chen, W. K.: "Theory and Design of Broadband Matching Networks," Oxford: Pergamon, 1976.

Chen, W. K.: The Hybrid Matrix in Linear Multiple-Loop Feedback Networks, *IEEE Trans. Circuits and Systems*, vol. CAS-24, no. 9, pp. 469–474, 1977.

Desoer, C. A. and E. S. Kuh: Bounds on Natural Frequencies of Linear Active Networks, *Proc. Symp. on Active Networks and Feedback Systems*, Polytechnic Inst. of Brooklyn, New York, vol. 10, pp. 415–436, 1960.

Guillemin, E. A.: "The Mathematics of Circuit Analysis," Cambridge, Mass.: The M.I.T. Press, 1969.

Kuh, E. S.: Regenerative Modes of Active Networks, *IRE Trans. Circuit Theory*, vol. CT-7, no. 1, pp. 62–63, 1960.

Kuh, E. S. and R. A. Rohrer: "Theory of Linear Active Networks," San Francisco, Calif.: Holden-Day, 1967.

Kuo, Y. L.: A Note on the *n*-Port Passivity Criterion, *IEEE Trans. Circuit Theory*, vol. CT-15, no. 1, p. 74, 1968.

Lam, Y. F.: Characterization of Time-Invariant Network Elements, *J. Franklin Inst.*, vol. 298, no. 1, pp. 1–7, 1974.

Nedunuri, R.: On the Definition of Passivity, *IEEE Trans. Circuit Theory*, vol. CT-19, no. 1, p. 72, 1972.

Newcomb, R. W.: Synthesis of Passive Networks for Networks Active at p_0–I, *IRE Int. Convention Record*, part 4, pp. 162–175, 1961.

Newcomb, R. W.: On Causality, Passivity and Single-Valuedness, *IRE Trans. Circuit Theory*, vol. CT-9, no. 1, pp. 87–89, 1962.

Newcomb, R. W.: On the Definition of a Network, *Proc. IEEE*, vol. 53, no. 5, pp. 547–548, 1965.

Newcomb, R. W.: "Linear Multiport Synthesis," New York: McGraw-Hill, 1966.

Peikari, B.: "Fundamentals of Network Analysis and Synthesis," Englewood Cliffs, N.J.: Prentice-Hall, 1974.

Raisbeck, G.: A Definition of Passive Linear Networks in Terms of Time and Energy, *J. Appl. Phys.*, vol. 25, no. 12, pp. 1510–1514, 1954.

Resh, J. A.: A Note Concerning the *n*-Port Passivity Condition, *IEEE Trans. Circuit Theory*, vol. CT-13, no. 2, pp. 238–239, 1966.

Rohrer, R. A.: Lumped Network Passivity Criteria, *IEEE Trans. Circuit Theory*, vol. CT-15, no. 1, pp. 24–30, 1968.

Youla, D. C.: Physical Realizability Criteria, *IRE Trans. Circuit Theory*, vol. CT-7, Special Supplement, pp. 50–68, 1960.

Youla, D. C., L. J. Castriota, and H. J. Carlin: Bounded Real Scattering Matrices and the Foundations of Linear Passive Network Theory, *IRE Trans. Circuit Theory*, vol. CT-6, no. 1, pp. 102–124, 1959.

Zemanian, A. H.: The Passivity and Semipassivity of Time-Varying Systems Under the Admittance Formulism, *SIAM J. Appl. Math.*, vol. 21, no. 4, pp. 533–541, 1971.

CHAPTER

TWO

THE INDEFINITE–ADMITTANCE MATRIX

In the preceding chapter, networks were characterized by their port behaviors. Fundamental to the concept of a port is the assumption that the instantaneous current entering one terminal of the port is always equal to the instantaneous current leaving the other terminal of the port. However, we recognize that upon the interconnection of networks, this port constraint may be violated. Thus, it is sometimes desirable and more advantageous to consider n-terminal networks, as depicted in Fig. 2.1.

In this chapter, we discuss a useful description of the external behavior of a multiterminal network in terms of the indefinite-admittance matrix and demonstrate how it can be employed effectively for the computation of network functions. Specifically, we derive formulas expressing the network functions in terms of the first-order and the second-order cofactors of the elements of the indefinite-admittance matrix. The significance of this approach is that the indefinite-admittance matrix can usually be written down directly from the network by inspection.

Since in the remainder of this book we deal exclusively with linear, lumped, and time-invariant networks, the adjectives *linear, lumped,* and *time-invariant* are omitted in the discussion unless they are used for emphasis.

2.1 THE INDEFINITE–ADMITTANCE MATRIX

Referring to Fig. 2.1, let N be an n-terminal network comprised of an arbitrary number of active and passive network elements connected in any way whatsoever.

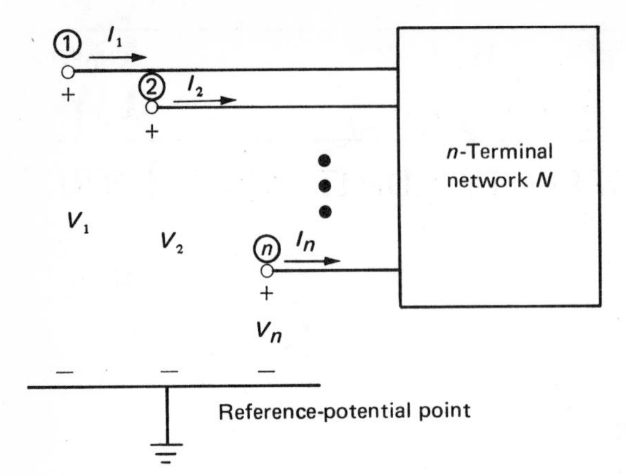

Figure 2.1 The general symbolic representation of an n-terminal network.

Let $V_1, V_2, \ldots, V_n$ be the Laplace-transformed potentials measured between terminals $1, 2, \ldots, n$ and some arbitrary but unspecified reference point, and let $I_1, I_2, \ldots, I_n$ be the Laplace-transformed currents entering the terminals $1, 2, \ldots, n$ from outside the network.

Since the network N together with its loading is linear, the terminal currents and voltages are related by the equation†

$$\begin{bmatrix} I_1 \\ I_2 \\ \vdots \\ I_n \end{bmatrix} = \begin{bmatrix} y_{11} & y_{12} & \cdots & y_{1n} \\ y_{21} & y_{22} & \cdots & y_{2n} \\ \cdots & \cdots & \cdots & \cdots \\ y_{n1} & y_{n2} & \cdots & y_{nn} \end{bmatrix} \begin{bmatrix} V_1 \\ V_2 \\ \vdots \\ V_n \end{bmatrix} + \begin{bmatrix} J_1 \\ J_2 \\ \vdots \\ J_n \end{bmatrix} \tag{2.1a}$$

or more compactly as

$$\mathbf{I}(s) = \mathbf{Y}(s)\mathbf{V}(s) + \mathbf{J}(s) \tag{2.1b}$$

where J_k $(k = 1, 2, \ldots, n)$ denotes the current flowing into the kth terminal when all terminals of N are grounded to the reference point. The coefficient matrix $\mathbf{Y}(s)$ is called the *indefinite-admittance matrix* by Shekel (1952) because the reference point for the potentials is some arbitrary but unspecified point outside the network. The short-circuit currents J_k result from the independent sources and/or initial conditions in the interior of the n-terminal network. For our purposes, we shall consider all independent sources outside the network and set all initial conditions to zero. Hence, $\mathbf{J}(s)$ is considered to be zero, and Eq. (2.1) becomes

$$\mathbf{I}(s) = \mathbf{Y}(s)\mathbf{V}(s) \tag{2.2}$$

†Instead of introducing additional symbols, $\mathbf{Y}(s)$ is used to denote either the admittance matrix or the indefinite-admittance matrix. This should not create any confusion, because the context will tell. Occasionally, we use $\mathbf{Y}_{sc}(s)$ to represent the admittance matrix in order to distinguish it from the indefinite-admittance matrix. This is similarly valid for the admittance parameters y_{ij}.

The elements of $\mathbf{Y}(s)$ are short-circuit admittances because they can be obtained by the equation

$$y_{ij} = \left.\frac{I_i}{V_j}\right|_{V_x=0,\, x\neq j} \tag{2.3}$$

showing that y_{ii} is the driving-point admittance looking into the terminal i and the reference point when all other terminals are grounded to the reference point, and that y_{ij}, $i \neq j$, is the transfer admittance from terminal j to terminal i when all terminals except the jth one are grounded to the reference point. In fact, this relation may be employed to compute the elements of $\mathbf{Y}(s)$. As will be shown in Sec. 2.2, the elements of $\mathbf{Y}(s)$ can usually be obtained directly from the network by inspection.

Example 2.1 Figure 2.2 is the network model of a transistor. Assume that each of its nodes is an accessible terminal. The suppression of inaccessible terminals will be considered later. Using Eq. (2.3), we compute the elements of its indefinite-admittance matrix $\mathbf{Y}(s)$, as follows.

For elements y_{11}, y_{21}, and y_{31}, we set $V_2 = V_3 = 0$ and apply a voltage source V_1 between the terminal 1 and the reference point as shown in Fig. 2.3. This gives

$$y_{11} = \left.\frac{I_1}{V_1}\right|_{V_2=V_3=0} = g_1 + sC_1 + sC_2 \tag{2.4a}$$

$$y_{21} = \left.\frac{I_2}{V_1}\right|_{V_2=V_3=0} = g_m - sC_2 \tag{2.4b}$$

$$y_{31} = \left.\frac{I_3}{V_1}\right|_{V_2=V_3=0} = -g_1 - sC_1 - g_m \tag{2.4c}$$

In a similar fashion, to compute y_{12}, y_{22}, and y_{32} we set $V_1 = V_3 = 0$ and apply a voltage source between terminal 2 and the reference point; and to compute y_{13}, y_{23}, and y_{33} we set $V_1 = V_2 = 0$ and apply a voltage source between terminal

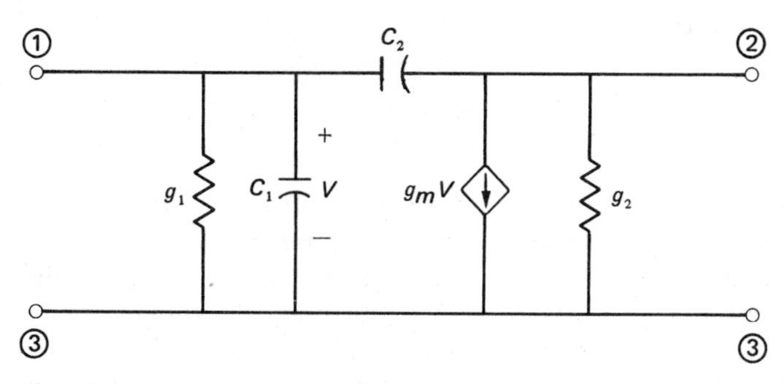

Figure 2.2 An equivalent network of a transistor.

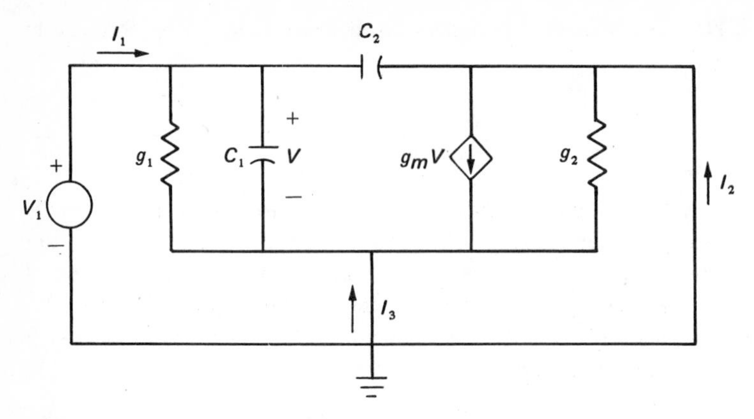

Figure 2.3 The network used to evaluate the admittance parameters y_{11}, y_{21}, and y_{31}.

3 and the reference point. The details are left as an exercise (see Prob. 2.1). The indefinite-admittance matrix of the transistor is given by

$$\mathbf{Y}(s) = \begin{bmatrix} g_1 + sC_1 + sC_2 & -sC_2 & -g_1 - sC_1 \\ g_m - sC_2 & g_2 + sC_2 & -g_2 - g_m \\ -g_1 - sC_1 - g_m & -g_2 & g_1 + g_2 + g_m + sC_1 \end{bmatrix} \tag{2.5}$$

It is significant to observe that the sum of elements of each row or column is equal to zero. That these properties are valid in general for the indefinite-admittance matrix will now be demonstrated.

First, we show that the sum of the elements in each column of $\mathbf{Y}(s)$ equals zero. To this end we add all n equations of (2.1) to yield

$$\sum_{i=1}^{n} \sum_{j=1}^{n} y_{ji} V_i = \sum_{m=1}^{n} I_m - \sum_{m=1}^{n} J_m = 0 \tag{2.6}$$

The last equation is obtained by appealing to Kirchhoff's current law for the node corresponding to the reference point. Setting all the terminal voltages to zero except the kth one, which is nonzero, gives

$$V_k \sum_{j=1}^{n} y_{jk} = 0 \tag{2.7}$$

Since $V_k \neq 0$, it follows that the sum of the elements of each column of $\mathbf{Y}(s)$ equals zero. Thus, the indefinite-admittance matrix is always singular.

To demonstrate that each row sum of $\mathbf{Y}(s)$ is also zero, we recognize that because the point of zero potential may be chosen arbitrarily, the currents J_k and I_k remain invariant when all the terminal voltages V_k are changed by the same but arbitrary constant amount. Thus, if $\mathbf{V}_0$ is an n-vector, each of its elements being $v_0 \neq 0$, then

$$\mathbf{I}(s) - \mathbf{J}(s) = \mathbf{Y}(s)[\mathbf{V}(s) + \mathbf{V}_0] = \mathbf{Y}(s)\mathbf{V}(s) + \mathbf{Y}(s)\mathbf{V}_0 \tag{2.8}$$

which after invoking (2.1*b*) shows that

$$\mathbf{Y}(s)\mathbf{V}_0 = \mathbf{0} \tag{2.9}$$

or

$$\sum_{j=1}^{n} y_{ij} = 0 \qquad i = 1, 2, \ldots, n \tag{2.10}$$

indicating that each row sum of $\mathbf{Y}(s)$ equals zero.

As a consequence of the zero-row-sum and zero-column-sum properties, all the cofactors of the elements of the indefinite-admittance matrix are equal. Before we justify this statement, we need the following definition.

Definition 2.1: Equicofactor matrix A square matrix is said to be an *equicofactor matrix* if all the cofactors of its elements are equal.

By using this definition, a fundamental property of the indefinite-admittance matrix can now be stated.

Theorem 2.1 The indefinite-admittance matrix of a linear, lumped, and time-invariant multiterminal network is an equicofactor matrix.

PROOF Let Y_{ij} be the cofactor of the element y_{ij} of $\mathbf{Y}(s)$. We first show that $Y_{iu} = Y_{iv}$ for all i, u, and v. Without loss of generality, assume that $u > v$. Since each row sum is zero, we may replace the elements y_{jv} $(v = 1, 2, \ldots, n)$ in the submatrix $\mathbf{Y}_{iu}$ obtained from $\mathbf{Y}(s)$ by deleting the ith row and uth column by

$$-\sum_{\substack{k=1 \\ k \neq v}}^{n} y_{jk} \tag{2.11}$$

without changing the value of Y_{iu}. Let the submatrix thus obtained be denoted by $\hat{\mathbf{Y}}_{iu}$. Now, adding all the columns of $\hat{\mathbf{Y}}_{iu}$ to column v and then shifting column v to the right of column $u - 1$ if $v \neq u - 1$ result in

$$\det \hat{\mathbf{Y}}_{iu} = (-1)^{u-v-1}(-1) \det \mathbf{Y}_{iv} \tag{2.12}$$

$\mathbf{Y}_{iv}$ being the submatrix obtained from $\mathbf{Y}(s)$ by deleting row i and column v. From Eq. (2.12) we have

$$Y_{iu} = (-1)^{i+u}(-1)^{u-v-1}(-1) \det \mathbf{Y}_{iv} = Y_{iv} \tag{2.13}$$

Similarly, by considering the transpose of $\mathbf{Y}(s)$, we can show that $Y_{ui} = Y_{vi}$ for all i, u, and v. Thus, we conclude that all the cofactors of the elements of the indefinite-admittance matrix $\mathbf{Y}(s)$ are equal. This completes the proof of the theorem.

As an illustration, we consider the indefinite-admittance matrix $\mathbf{Y}(s)$ of Eq. (2.5) computed in Example 2.1. It is easy to verify that all of its nine cofactors are equal to

$$s^2 C_1 C_2 + s(C_1 g_2 + C_2 g_1 + C_2 g_2 + g_m C_2) + g_1 g_2 \tag{2.14}$$

2.2 RULES FOR WRITING DOWN THE PRIMITIVE INDEFINITE-ADMITTANCE MATRIX

Suppose that every node of a network N is an accessible terminal to which external connections are to be made. The indefinite-admittance matrix associated with such a network is referred to as the *primitive indefinite-admittance matrix*. In the present section, we show that for networks comprising multiterminal devices and voltage-controlled current sources, their primitive indefinite-admittance matrices can easily be written down by inspection. We recognize that confinement to voltage-controlled current sources should not be deemed restrictive, because other types of controlled sources can usually be converted to the above type by applying Norton's theorem if necessary. In the following section, we shall describe a procedure for *suppressing* inaccessible terminals.

Consider a voltage-controlled current source, as shown in Fig. 2.4, whose indefinite-admittance matrix is clearly given by

$$\begin{array}{c} \\ \\ I_a \\ I_b \\ +I_c \\ -I_d \end{array} \begin{array}{c} \begin{array}{cccc} + & - & & \\ V_a & V_b & V_c & V_d \end{array} \\ \begin{bmatrix} 0 & 0 & 0 & 0 \\ 0 & 0 & 0 & 0 \\ y & -y & 0 & 0 \\ -y & y & 0 & 0 \end{bmatrix} \end{array} \tag{2.15}$$

where V's and I's are the terminal voltages and currents of the controlled source. Observe that the admittance y enters the indefinite-admittance matrix in a rectangular pattern that is not necessarily centered on the main diagonal. The admittance y is actuated by the voltage V_{ab} from terminal a to terminal b that corresponds to the first two columns of (2.15) and affects the currents of the terminals c and d associated with the third and fourth rows of the matrix. Thus, the

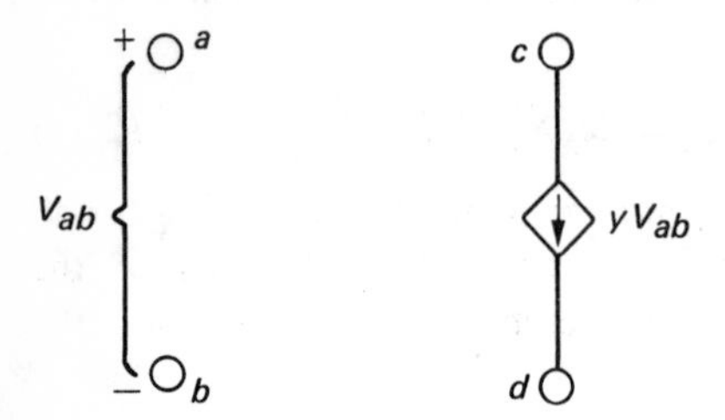

Figure 2.4 A voltage-controlled current source.

matrix (2.15) can easily be written down by inspection from the terminal labelings of the controlled source. The sign associated with the admittance y is determined according to the usual multiplication rules, as indicated in (2.15). This representation is, of course, quite general. Terminals a, b, c, and d need not be distinct. For example, if $a = c$ and $b = d$, then it represents a one-port admittance y. Thus, if these elements are imbedded in a network, its indefinite-admittance matrix is simply the sum of the indefinite-admittance matrices of the component subnetworks that form the complete network when combined in parallel by connecting together their corresponding terminals. Each of the subnetworks has n terminals, some of which may consist only of isolated nodes. The reason for this is that in parallel connection, it implies equality of voltages and addition of currents at each terminal, and the addition of the indefinite-admittance matrices is a direct result. This leads directly to the following set of rules for writing down the primitive indefinite-admittance matrix by inspection:

1. Write down that part of the indefinite-admittance matrix corresponding to the subnetwork composed of the one-port admittances according to the formulas

   $$y_{ii} = \Sigma \text{ admittances incident at the terminal } i$$

   and, for $i \neq j$,

   $$y_{ij} = -\Sigma \text{ admittances connected between terminals } i \text{ and } j$$

2. The contribution resulting from the voltage-controlled current sources of Fig. 2.4 is equivalent to adding the matrices of the type (2.15).

The primitive indefinite-admittance matrix is then obtained by adding all the component matrices discussed above after bringing them to the same order by the insertion of an appropriate number of zero rows and columns.

Example 2.2 Consider the active network N of Fig. 2.5, in which all the five nodes are assumed to be accessible terminals. The indefinite-admittance matrix corresponding to the subnetwork composed of the five one-port admittances is given by

$$\begin{bmatrix} g_1 & -g_1 & 0 & 0 & 0 \\ -g_1 & g_1 + g_2 + sC & -g_2 & 0 & -sC \\ 0 & -g_2 & g_2 + \dfrac{1}{sL} & -\dfrac{1}{sL} & 0 \\ 0 & 0 & -\dfrac{1}{sL} & g_3 + \dfrac{1}{sL} & -g_3 \\ 0 & -sC & 0 & -g_3 & g_3 + sC \end{bmatrix} \tag{2.16}$$

The indefinite-admittance matrix of the controlled source can be obtained in accordance with (2.15). The controlled source is connected between nodes 3 and 5,

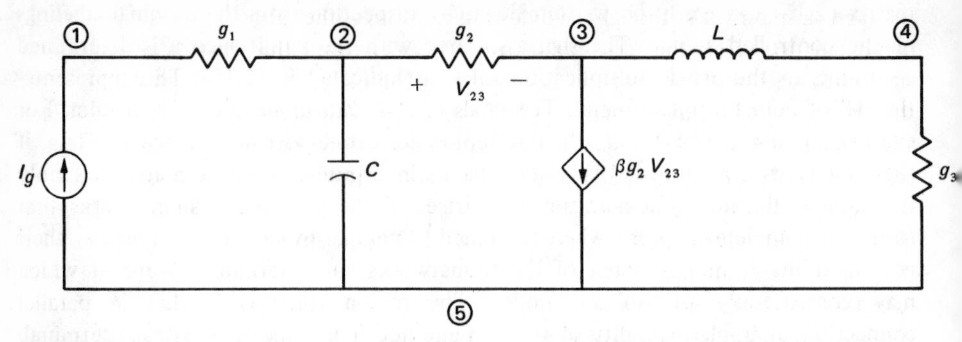

Figure 2.5 An active network used to illustrate the computation of the primitive indefinite-admittance matrix.

whose controlling voltage is the potential from node 2 to node 3. Thus, the rectangular pattern appears in rows 3 and 5 and columns 2 and 3 with the plus sign associated with column 2 and row 3. This yields

$$\begin{array}{c} \\ \\ 1 \\ 2 \\ +3 \\ 4 \\ -5 \end{array} \begin{array}{c} \begin{array}{ccccc} & + & - & & \\ 1 & 2 & 3 & 4 & 5 \end{array} \\ \left[\begin{array}{ccccc} 0 & 0 & 0 & 0 & 0 \\ 0 & 0 & 0 & 0 & 0 \\ 0 & \beta g_2 & -\beta g_2 & 0 & 0 \\ 0 & 0 & 0 & 0 & 0 \\ 0 & -\beta g_2 & \beta g_2 & 0 & 0 \end{array}\right] \end{array} \tag{2.17}$$

Observe that matrix (2.16) is symmetric because the corresponding subnetwork is reciprocal, and that in (2.17) if the sign associated with one of the terms βg_2 is known, the signs of other terms can be identified cyclically by assigning the plus and minus signs alternately as shown in (2.15). The complete primitive indefinite-admittance matrix of the active network is obtained simply by adding the two matrices (2.16) and (2.17), and the result is given by

$$\begin{bmatrix} g_1 & -g_1 & 0 & 0 & 0 \\ -g_1 & g_1 + g_2 + sC & -g_2 & 0 & -sC \\ 0 & \beta g_2 - g_2 & g_2 - \beta g_2 + \dfrac{1}{sL} & -\dfrac{1}{sL} & 0 \\ 0 & 0 & -\dfrac{1}{sL} & g_3 + \dfrac{1}{sL} & -g_3 \\ 0 & -\beta g_2 - sC & \beta g_2 & -g_3 & g_3 + sC \end{bmatrix} \tag{2.18}$$

In fact, a moment's thought would indicate that the above two steps can be combined to yield the desired matrix directly from the given network. It is significant to point out that the one-port admittances g_1, g_2, g_3, sC, and $1/sL$ enter the indefinite-admittance matrix (2.18) in a beautifully symmetric form. Each admittance appears in a square pattern of four element positions, as in (2.15), but centered about the main diagonal. The two rows and two columns associated with the square pattern have the same labels as the nodes to which the branch in question is connected in the network. The admittance enters with a positive sign on the main diagonal and with a negative sign for the off-diagonal elements. This follows directly from an earlier observation that a one-port admittance is a special type of voltage-controlled current source in which the controlling voltage is also its terminal voltage.

The rules outlined at the beginning of this section can be extended to include the situation where multiterminal devices are imbedded in the network. Like the voltage-controlled current source, all we need to derive is their associated indefinite-admittance matrices. The complete primitive indefinite-admittance matrix is then obtained by adding all the component matrices after bringing them to the same order by the insertion of an appropriate number of zero rows and columns.

An element that frequently appears in networks is the mutual inductance. Refer to Fig. 2.6; let V's and I's be the terminal voltages and currents. Then they are related by the equation

$$\begin{bmatrix} V_a - V_b \\ V_c - V_d \end{bmatrix} = \begin{bmatrix} L_1 s & Ms \\ Ms & L_2 s \end{bmatrix} \begin{bmatrix} I_a \\ I_c \end{bmatrix} \tag{2.19}$$

Assuming that the coupling is imperfect, that is, $L_1 L_2 - M^2 \neq 0$, Eq. (2.19) can be inverted, giving

$$\begin{bmatrix} I_a \\ I_c \end{bmatrix} = \frac{1}{s(L_1 L_2 - M^2)} \begin{bmatrix} L_2 & -M \\ -M & L_1 \end{bmatrix} \begin{bmatrix} V_a - V_b \\ V_c - V_d \end{bmatrix} \tag{2.20}$$

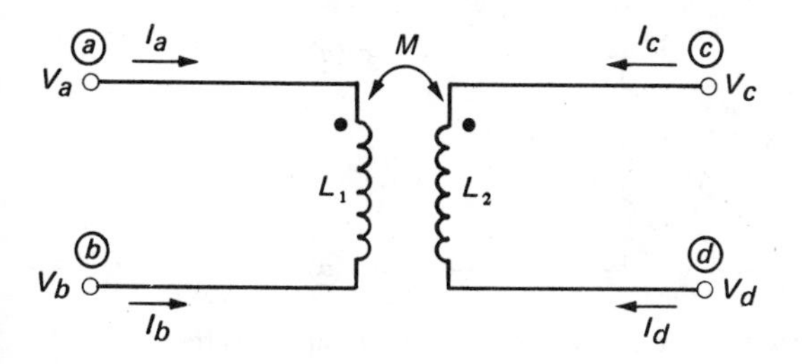

Figure 2.6 Two mutually coupled coils.

Since $I_b = -I_a$ and $I_d = -I_c$, Eq. (2.20) can be expanded and rewritten as

$$\begin{bmatrix} I_a \\ I_b \\ I_c \\ I_d \end{bmatrix} = \frac{1}{s(L_1L_2 - M^2)} \begin{bmatrix} L_2 & -L_2 & -M & M \\ -L_2 & L_2 & M & -M \\ -M & M & L_1 & -L_1 \\ M & -M & -L_1 & L_1 \end{bmatrix} \begin{bmatrix} V_a \\ V_b \\ V_c \\ V_d \end{bmatrix} \tag{2.21}$$

the coefficient matrix being the indefinite-admittance matrix of the imperfectly coupled transformer of Fig. 2.6.

The ideal transformer by itself has infinite short-circuit admittances and hence cannot be formally characterized on the nodal basis. However, in practice a transformer usually appears in series with some other elements having finite admittance as shown in Fig. 2.7. In such a case, its indefinite-admittance matrix can be determined in a similar fashion (see Prob. 2.2):

$$\begin{bmatrix} y_0 & -y_0 & -ny_0 & ny_0 \\ -y_0 & y_0 & ny_0 & -ny_0 \\ -ny_0 & ny_0 & n^2y_0 & -n^2y_0 \\ ny_0 & -ny_0 & -n^2y_0 & n^2y_0 \end{bmatrix} \tag{2.22}$$

Another useful element having two pairs of terminals is the *gyrator*, the diagrammatic symbol for which is shown in Fig. 2.8. The element is characterized by the terminal relation

$$\begin{bmatrix} I_a \\ I_c \end{bmatrix} = \begin{bmatrix} 0 & g \\ -g & 0 \end{bmatrix} \begin{bmatrix} V_a - V_b \\ V_c - V_d \end{bmatrix} \tag{2.23}$$

In the diagram, the direction of the arrow signifies the *direction of gyration.* Like the ideal transformer, the gyrator is characterized by a single parameter g, called the

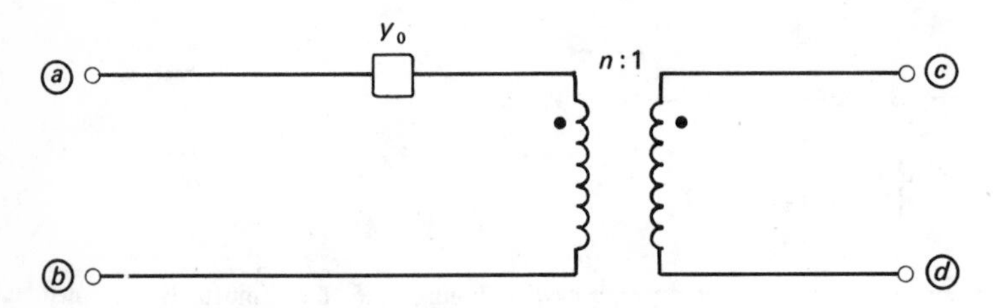

Figure 2.7 An ideal transformer in series with some other element having finite admittance.

gyration conductance. The indefinite-admittance matrix of the gyrator of Fig. 2.8 is obtained as (see Prob. 2.4)

$$\begin{bmatrix} 0 & 0 & g & -g \\ 0 & 0 & -g & g \\ -g & g & 0 & 0 \\ g & -g & 0 & 0 \end{bmatrix} \tag{2.24}$$

Example 2.3 In the active network of Fig. 2.9, assume that all five nodes are accessible terminals. Applying rules to that part of the subnetwork composed of g_1, g_2, g_3, C, and the controlled source, we can write down the indefinite-admittance matrix directly from the network by inspection:

$$\begin{array}{r} \\ 1 \\ 2 \\ +3 \\ 4 \\ -5 \end{array}\begin{array}{c} \begin{array}{ccccc} 1 & \overset{+}{2} & \overset{-}{3} & 4 & 5 \end{array} \\ \begin{bmatrix} 0 & 0 & 0 & 0 & 0 \\ 0 & g_2 + sC & -g_2 & 0 & -sC \\ 0 & \beta g_2 - g_2 & g_1 + g_2 - \beta g_2 & 0 & -g_1 \\ 0 & 0 & 0 & g_3 & -g_3 \\ 0 & -\beta g_2 - sC & \beta g_2 - g_1 & -g_3 & g_1 + g_3 + sC \end{bmatrix} \end{array} \tag{2.25}$$

The matrix of the two imperfectly coupled coils can be obtained from the coefficient matrix of (2.21) by inserting a zero row at the bottom and a zero column at the right-hand side. The sum of this matrix and (2.25) yields the desired indefinite-admittance matrix of the network:

$$\begin{bmatrix} \frac{L_2}{\delta} & -\frac{L_2}{\delta} & -\frac{M}{\delta} & \frac{M}{\delta} & 0 \\ -\frac{L_2}{\delta} & g_2 + sC + \frac{L_2}{\delta} & \frac{M}{\delta} - g_2 & -\frac{M}{\delta} & -sC \\ -\frac{M}{\delta} & \beta g_2 - g_2 + \frac{M}{\delta} & g_1 + g_2 - \beta g_2 + \frac{L_1}{\delta} & -\frac{L_1}{\delta} & -g_1 \\ \frac{M}{\delta} & -\frac{M}{\delta} & -\frac{L_1}{\delta} & g_3 + \frac{L_1}{\delta} & -g_3 \\ 0 & -\beta g_2 - sC & \beta g_2 - g_1 & -g_3 & g_1 + g_3 + sC \end{bmatrix} \tag{2.26}$$

where $\delta = s(L_1 L_2 - M^2)$

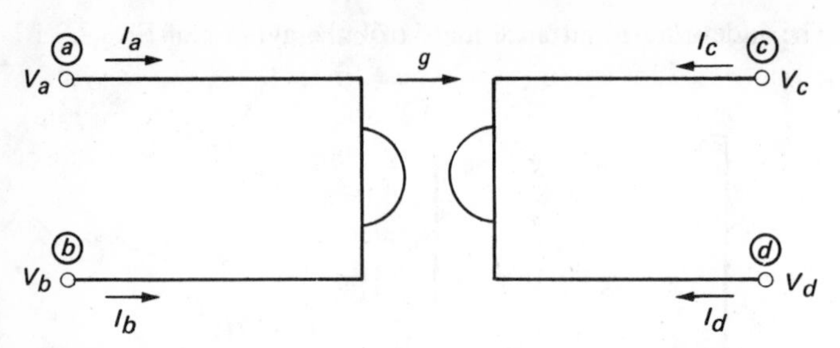

Figure 2.8 The diagrammatic symbol for a gyrator with gyration conductance g.

2.3 TERMINAL CONTRACTION AND SUPPRESSION

In the preceding section, we presented a procedure for writing down the primitive indefinite-admittance matrix of a network by inspection. Basic to the concept of the primitive indefinite-admittance matrix is the assumption that every node of the network is an accessible terminal. Also, not all the terminals of a multiterminal device are distinct; some of them may be combined together to form new terminals. Thus, in order to apply the procedure we must know a way of finding the indefinite-admittance matrix of the network in terms of those discussed in the foregoing section.

2.3.1 Terminal Contraction

By the *contraction* of a multiterminal device is meant the joining together of two or more of its terminals. We recognize that a contraction involving the joining of more than two terminals can be regarded as a composition of contractions involving only two terminals.

Observe that the joining together of two or more terminals of a multiterminal

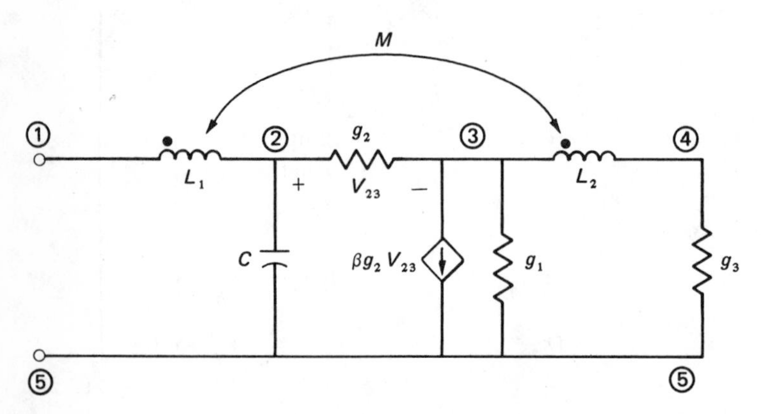

Figure 2.9 An active network used to illustrate the computation of the primitive indefinite-admittance matrix.

device implies the equality of voltages and addition of currents at the composite terminal. As a consequence, the indefinite-admittance matrix of the resulting multiterminal device can be derived from that of the original one by adding the rows and columns corresponding to the terminals being joined together to form a new row and a new column for the composition terminal (see Prob. 2.5).

As an illustration, consider the two imperfectly coupled coils shown in Fig. 2.6. On joining together terminals b and d, we get a three-terminal device. Its indefinite-admittance matrix can be derived from the coefficient matrix of (2.21) by adding row 2 to row 4 and column 2 to column 4 and then deleting row 2 and column 2 in the resulting matrix. The result is given by

$$\frac{1}{s(L_1 L_2 - M^2)}\begin{bmatrix} L_2 & -M & M-L_2 \\ -M & L_1 & M-L_1 \\ M-L_2 & M-L_1 & L_1+L_2-2M \end{bmatrix} \tag{2.27}$$

2.3.2 Terminal Suppression

A terminal is said to be *suppressed* if it is open-circuited. In other words, the associated terminal current is constrained to be zero, and the terminal is inaccessible. As in the case of contraction, the suppression of two or more terminals can be regarded as a succession of suppressions each involving a single terminal. To obtain the indefinite-admittance matrix of a network that possesses inaccessible internal terminals from its primitive indefinite-admittance matrix, we simply suppress the internal terminals to yield only the accessible ones. In terms of network equations (2.1*a*), this procedure is equivalent to eliminating the unwanted variables.

Let $\mathbf{I}_b$, $\mathbf{J}_b$, and $\mathbf{V}_b$ be the vectors of currents and voltages corresponding to the terminals to be suppressed. Partition the matrix equation (2.1*b*) according to unsuppressed and suppressed terminals as follows†:

$$\begin{bmatrix} \mathbf{I}_a \\ \mathbf{I}_b \end{bmatrix} = \begin{bmatrix} \mathbf{W}_{11} & \mathbf{W}_{12} \\ \mathbf{W}_{21} & \mathbf{W}_{22} \end{bmatrix} \begin{bmatrix} \mathbf{V}_a \\ \mathbf{V}_b \end{bmatrix} + \begin{bmatrix} \mathbf{J}_a \\ \mathbf{J}_b \end{bmatrix} \tag{2.28}$$

where the elements in $\mathbf{I}_a$, $\mathbf{J}_a$, and $\mathbf{V}_a$ correspond to unsuppressed terminals. Since suppressing a terminal is equivalent to open-circuiting that terminal, this requires that we set $\mathbf{I}_b = \mathbf{0}$. Using this and assuming that $\mathbf{W}_{22}$ is nonsingular, we can eliminate $\mathbf{V}_b$ in (2.28), yielding

$$\mathbf{I}_a = (\mathbf{W}_{11} - \mathbf{W}_{12}\mathbf{W}_{22}^{-1}\mathbf{W}_{21})\mathbf{V}_a - \mathbf{W}_{12}\mathbf{W}_{22}^{-1}\mathbf{J}_b + \mathbf{J}_a \tag{2.29}$$

Thus, the new indefinite-admittance matrix is identified as

$$\mathbf{Y}'' = \mathbf{W}_{11} - \mathbf{W}_{12}\mathbf{W}_{22}^{-1}\mathbf{W}_{21} \tag{2.30}$$

†As indicated in Eq. (2.2), if we consider all independent sources outside the network and set all initial conditions to zero, then $\mathbf{J}(s) = \mathbf{0}$. It is included here for completeness.

In particular, if a single terminal is to be suppressed, the procedure is exceedingly simple. Under this situation, to suppress the kth terminal, the ith row and jth column element y''_{ij} of $\mathbf{Y}''$ is given by

$$y''_{ij} = y_{ij} - \frac{y_{ik}y_{kj}}{y_{kk}} \tag{2.31}$$

In other words, suppressing the kth terminal is equivalent to subtracting from each of the original admittances y_{ij} the amount $y_{ik}y_{kj}/y_{kk}$ to yield the element values of the new matrix. The details of this procedure are demonstrated below:

$$\begin{array}{c} \\ i \\ \\ k \\ \\ \end{array} \left[\begin{array}{ccccc} & j & & k & \\ & \vdots & & \vdots & \\ \cdots & y_{ij} - \dfrac{y_{ik}y_{kj}}{y_{kk}} & \cdots & y_{ik} & \cdots \\ & \vdots & & \vdots & \\ \text{---} & y_{kj} & \text{---} & y_{kk} & \text{---} \\ & \vdots & & \vdots & \end{array} \right] \tag{2.32}$$

We recognize that the result is identical with that yielded by the standard method of pivotal condensation. In determinantal form, the element y''_{ij} can be expressed as

$$y''_{ij} = \frac{1}{y_{kk}} \begin{vmatrix} y_{ij} & y_{ik} \\ y_{kj} & y_{kk} \end{vmatrix} \tag{2.33}$$

In general, if more than one terminal is suppressed, it can be shown from (2.30) that $(\det \mathbf{W}_{22})y''_{ij}$ is equal to the determinant of the submatrix of $\mathbf{Y}$ formed by row i and column j and those rows and columns corresponding to the suppressed terminals.

After suppressing the unwanted terminals, an n-terminal network can be transformed into a common-terminal $(n-1)$-port network by pairing one of the terminals, say, terminal n, with $n-1$ other terminals to form $n-1$ ports, as indicated in Fig. 2.10. In doing so, we also specify terminal n to be the reference

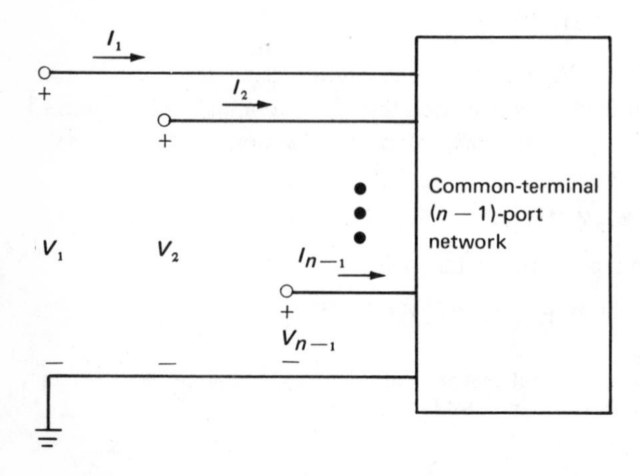

Figure 2.10 A common-terminal $(n-1)$-port network obtained from an n-terminal network by pairing one of the terminals with $n-1$ other terminals to form $n-1$ ports.

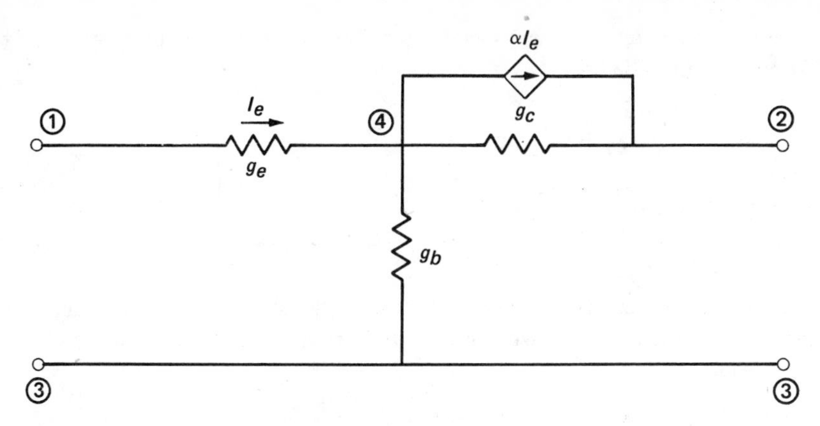

Figure 2.11 The T-equivalent network of a transistor.

potential point for $n-1$ other terminals. This is equivalent to setting V_n in (2.1*a*) to zero. Hence the last column in $\mathbf{Y}(s)$ can be removed. Since, according to Kirchhoff's current law [see Eq. (2.6)], the last equation is redundant, the last row of $\mathbf{Y}(s)$ can also be removed. This results in the admittance matrix of the common-terminal $(n-1)$-port network.

We shall illustrate the above results by the following examples.

Example 2.4 Consider the transistor network of Fig. 2.11, whose primitive indefinite-admittance matrix can easily be obtained by inspection according to the rules established in the preceding section, and is given by

$$\begin{array}{r} \\ \\ 1 \\ -\,2 \\ 3 \\ +\,4 \end{array}\begin{array}{c} \begin{array}{cccc} + & & & - \\ 1 & 2 & 3 & 4 \end{array} \\ \begin{bmatrix} g_e & 0 & 0 & -g_e \\ -\alpha g_e & g_c & 0 & \alpha g_e - g_c \\ 0 & 0 & g_b & -g_b \\ \alpha g_e - g_e & -g_c & -g_b & g_e + g_b + g_c - \alpha g_e \end{bmatrix} \end{array} \tag{2.34}$$

Suppose that terminal 4 is inaccessible and should be suppressed. By using (2.30) or (2.31), the new indefinite-admittance matrix becomes

$$\begin{bmatrix} g_e - \dfrac{g_e(1-\alpha)g_e}{w} & -\dfrac{g_e g_c}{w} & -\dfrac{g_e g_b}{w} \\ -\alpha g_e - \dfrac{(g_c - \alpha g_e)(1-\alpha)g_e}{w} & g_c - \dfrac{g_c(g_c - \alpha g_e)}{w} & -\dfrac{g_b(g_c - \alpha g_e)}{w} \\ -\dfrac{g_b(1-\alpha)g_e}{w} & -\dfrac{g_b g_c}{w} & g_b - \dfrac{g_b g_b}{w} \end{bmatrix} \tag{2.35}$$

where $w = g_e + g_c + g_b - \alpha g_e$. By expressing $r_e = 1/g_e$, $r_b = 1/g_b$, and $r_c = 1/g_c$, matrix (2.35) can be simplified to

$$\frac{1}{r_b r_c r_e w}\begin{bmatrix} r_b + r_c & -r_b & -r_c \\ -\alpha r_c - r_b & r_b + r_e & -r_e + \alpha r_c \\ -(1-\alpha)r_c & -r_e & r_e + (1-\alpha)r_c \end{bmatrix} \tag{2.36}$$

If node 3 is chosen as the common terminal, the short-circuit admittance matrix of the resulting two-port network is obtained from (2.36) by deleting row 3 and column 3, yielding

$$\mathbf{Y}_{sc} = \frac{1}{r_b r_c r_e w}\begin{bmatrix} r_b + r_c & -r_b \\ -\alpha r_c - r_b & r_b + r_e \end{bmatrix} \tag{2.37}$$

which is recognized to be the admittance matrix of the common-base transistor configuration. In fact, once the indefinite-admittance matrix (2.36) is known, the admittance matrices of the common-emitter and common-collector configurations are obtained by simply deleting row 1 and column 1, and row 2 and column 2, respectively. We shall discuss this aspect further in Sec. 2.4.

Suppose now that in (2.36) we continue our process by suppressing terminal 3. The indefinite-admittance matrix of the resulting two-terminal network becomes

$$\frac{1}{r_e + (1-\alpha)r_c}\begin{bmatrix} 1 & -1 \\ -1 & 1 \end{bmatrix} \tag{2.38}$$

the diagonal elements being its input admittance. Alternatively, the elements of (2.38) can be computed directly from (2.34), as follows: Since we know in advance that matrix (2.38) must be an equicofactor matrix, it suffices to determine only its first row and first column element y''_{11} by the procedure outlined in the paragraph following (2.33), yielding

$$y''_{11} = \frac{1}{\det \mathbf{W}_{22}} \det \begin{bmatrix} g_e & 0 & -g_e \\ 0 & g_b & -g_b \\ \alpha g_e - g_e & -g_b & g_e + g_b + g_c - \alpha g_e \end{bmatrix} = \frac{1}{r_e + (1-\alpha)r_c} \tag{2.39}$$

where

$$\mathbf{W}_{22} = \begin{bmatrix} g_b & -g_b \\ -g_b & g_e + g_b + g_c - \alpha g_e \end{bmatrix} \tag{2.40}$$

which is the submatrix formed by the rows and columns corresponding to the suppressed terminals. The third-order matrix in (2.39) is the submatrix of (2.34) formed by rows 1, 3, and 4 and columns 1, 3, and 4.

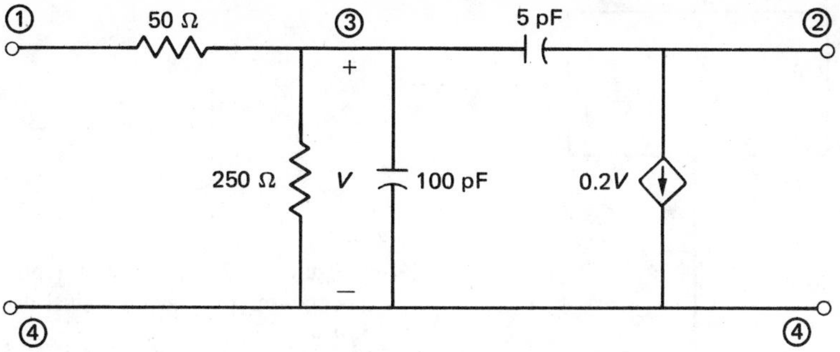

Figure 2.12 The hybrid-pi equivalent network of a transistor.

Example 2.5 Figure 2.12 is the hybrid-pi equivalent network of a transistor. By applying the procedure outlined in Sec. 2.2, the primitive indefinite-admittance matrix of the network can easily be obtained by inspection and is given by

$$\begin{array}{cc} & \begin{array}{cccc} & \phantom{-5\cdot 10^{-12}s} & + & - \end{array} \\ \begin{array}{c} \\ + \\ \\ - \end{array} & \begin{bmatrix} 0.02 & 0 & -0.02 & 0 \\ 0 & 5\cdot 10^{-12}s & 0.2-5\cdot 10^{-12}s & -0.2 \\ -0.02 & -5\cdot 10^{-12}s & 0.024+105\cdot 10^{-12}s & -0.004-10^{-10}s \\ 0 & 0 & -0.204-10^{-10}s & 0.204+10^{-10}s \end{bmatrix} \end{array} \tag{2.41}$$

Choosing node 4 as the reference potential and letting $p = s/10^9$ yield

$$\begin{bmatrix} 0.02 & 0 & -0.02 \\ 0 & 0.005p & 0.2-0.005p \\ -0.02 & -0.005p & 0.024+0.105p \end{bmatrix} \tag{2.42}$$

which is derived from (2.41) by deleting the fourth row and fourth column. The admittance matrix of the two-port network is obtained from (2.42) by suppressing the terminal corresponding to the third row or third column. Using (2.30) or (2.31) gives

$$\frac{1}{240+1050p}\begin{bmatrix} 0.8+21p & -p \\ 40-p & 11.2p+5p^2 \end{bmatrix} \tag{2.43}$$

It is significant to observe that the order of the operations of deletion and suppressing is immaterial. Of course, the suppressed terminals cannot be deleted before the operations are completed.

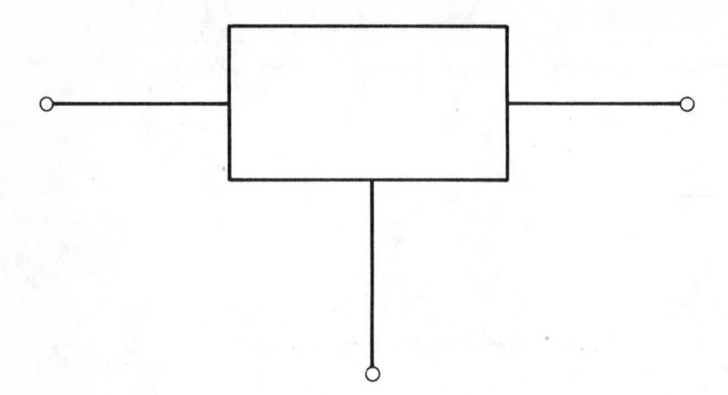

Figure 2.13 The general representation of a three-terminal device.

2.4 INTERRELATIONSHIPS OF TRANSISTOR MODELS

The transistor is a three-terminal device as shown in Fig. 2.13. If any one of the terminals is chosen as the reference-potential point, the device can be transformed into the common-terminal two-port network of Fig. 2.14, which can then be characterized by the short-circuit admittance parameters, the open-circuit impedance parameters, or the hybrid parameters. In order to distinguish the parameters associated with various terminals and configurations, the following subscripts are commonly used:

i *i*nput
o *o*utput
f *f*orward transfer
r *r*everse transfer
b common-*b*ase
e common-*e*mitter
c common-*c*ollector

The first subscript (*i, o, f,* or *r*) identifies the element in the two-port parameter matrix, and the second (*b, e,* or *c*) determines the common terminal used. In Fig. 2.15, the base terminal is chosen as the input, the collector terminal as the output, and the emitter terminal as the common terminal. Then the transistor equations of the common-emitter configuration in terms of the short-circuit admittance parameters and the hybrid parameters can be written as

$$\begin{bmatrix} I_b \\ I_c \end{bmatrix} = \begin{bmatrix} y_{ie} & y_{re} \\ y_{fe} & y_{oe} \end{bmatrix} \begin{bmatrix} V_{be} \\ V_{ce} \end{bmatrix} \tag{2.44}$$

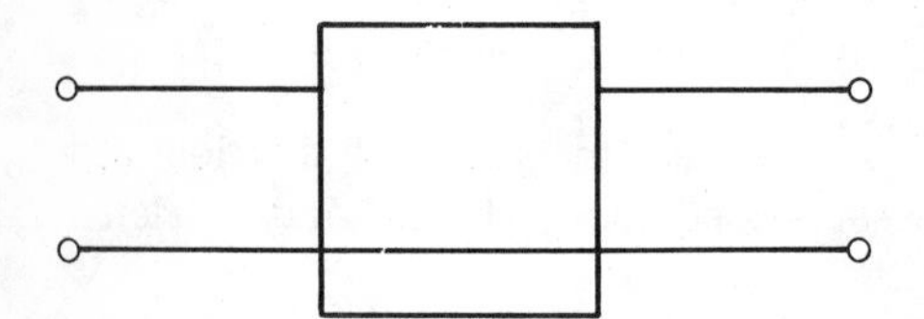

Figure 2.14 The general representation of a common-terminal two-port network.

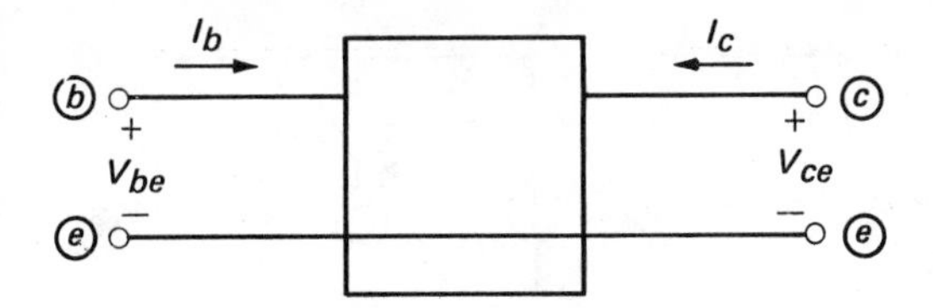

Figure 2.15 The common-emitter representation of a transistor.

$$\begin{bmatrix} V_{be} \\ I_c \end{bmatrix} = \begin{bmatrix} h_{ie} & h_{re} \\ h_{fe} & h_{oe} \end{bmatrix} \begin{bmatrix} I_b \\ V_{ce} \end{bmatrix} \tag{2.45}$$

Thus, y_{ie} denotes the input admittance and h_{fe} the forward current-transfer ratio of the common-emitter configuration when the output port is short-circuited. In general, the admittance matrix and the hybrid matrix of a common-terminal two-port device can be expressed by

$$\mathbf{Y}_x = \begin{bmatrix} y_{ix} & y_{rx} \\ y_{fx} & y_{ox} \end{bmatrix} \tag{2.46a}$$

$$\mathbf{H}_x = \begin{bmatrix} h_{ix} & h_{rx} \\ h_{fx} & h_{ox} \end{bmatrix} \tag{2.46b}$$

where $x = b$, e, or c. The elements of $\mathbf{Y}_x$ and $\mathbf{H}_x$ are related, and using the procedure outlined in Sec. 1.5, we can express the elements of $\mathbf{Y}_x$ in terms of those of $\mathbf{H}_x$ and vice versa. The results are given by

$$\mathbf{Y}_x = \begin{bmatrix} y_{ix} & y_{rx} \\ y_{fx} & y_{ox} \end{bmatrix} = \frac{1}{h_{ix}} \begin{bmatrix} 1 & -h_{rx} \\ h_{fx} & \Delta_{hx} \end{bmatrix} \tag{2.47a}$$

$$\mathbf{H}_x = \begin{bmatrix} h_{ix} & h_{rx} \\ h_{fx} & h_{ox} \end{bmatrix} = \frac{1}{y_{ix}} \begin{bmatrix} 1 & -y_{rx} \\ y_{fx} & \Delta_{yx} \end{bmatrix} \tag{2.47b}$$

where

$$\Delta_{hx} = \det \mathbf{H}_x = h_{ix}h_{ox} - h_{rx}h_{fx} \tag{2.48a}$$

$$\Delta_{yx} = \det \mathbf{Y}_x = y_{ix}y_{ox} - y_{rx}y_{fx} \tag{2.48b}$$

The most important mode of operation for transistors in practical applications is the common-emitter configuration, and the transistor parameters are usually specified by the manufacturer in this mode. In the following, we shall derive formulas expressing the parameters of other configurations in terms of those of the common-emitter connection.

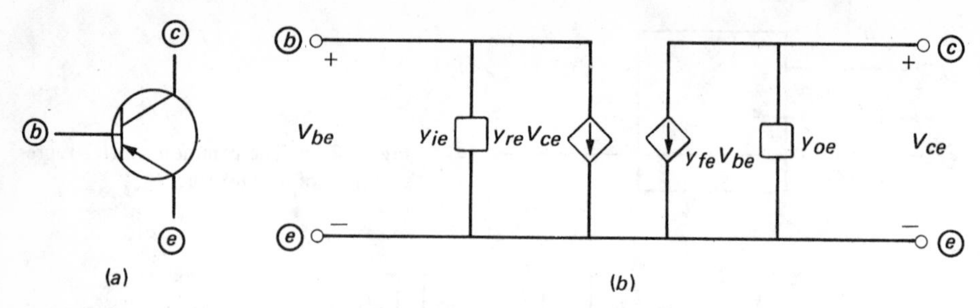

Figure 2.16 (*a*) The common-emitter configuration of a transistor and (*b*) its admittance model.

2.4.1 From the Common-Emitter Admittance Model to Other Models

The common-emitter configuration of a transistor is shown in Fig. 2.16*a* and its admittance model is presented in Fig. 2.16*b*. From Fig. 2.16*b*, the indefinite-admittance matrix is given by

$$\mathbf{Y} = \begin{matrix} b \\ e \\ c \end{matrix} \overset{\begin{matrix} b & \qquad\qquad e & \qquad\qquad c \end{matrix}}{\begin{bmatrix} y_{ie} & -y_{ie}-y_{re} & y_{re} \\ -y_{fe}-y_{ie} & y_{fe}+y_{ie}+y_{oe}+y_{re} & -y_{oe}-y_{re} \\ y_{fe} & -y_{fe}-y_{oe} & y_{oe} \end{bmatrix}} \tag{2.49}$$

To obtain the admittance matrix of the common-base configuration, we delete row 1 and column 1 from **Y**, resulting in

$$\mathbf{Y}_b = \begin{bmatrix} y_{ib} & y_{rb} \\ y_{fb} & y_{ob} \end{bmatrix} = \begin{bmatrix} y_{fe}+y_{ie}+y_{oe}+y_{re} & -y_{oe}-y_{re} \\ -y_{fe}-y_{oe} & y_{oe} \end{bmatrix} \tag{2.50}$$

whose network model is as shown in Fig. 2.17. Applying Eq. (2.47*b*), we obtain the hybrid parameters of the common-base configuration of Fig. 2.18:

$$\mathbf{H}_b = \begin{bmatrix} h_{ib} & h_{rb} \\ h_{fb} & h_{ob} \end{bmatrix} = \frac{1}{y_{fe}+y_{ie}+y_{oe}+y_{re}} \begin{bmatrix} 1 & y_{oe}+y_{re} \\ -y_{fe}-y_{oe} & y_{ie}y_{oe}-y_{fe}y_{re} \end{bmatrix} \tag{2.51}$$

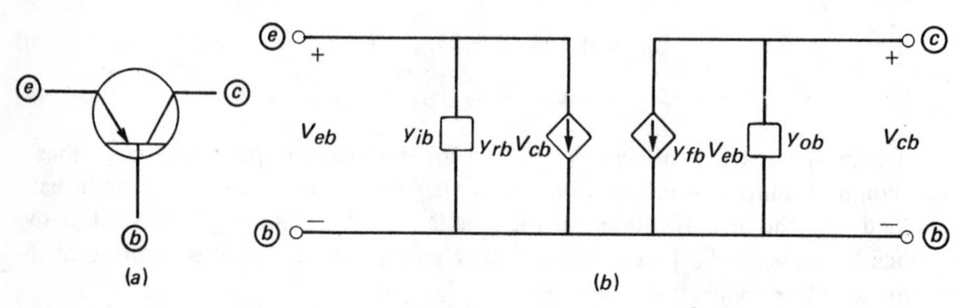

Figure 2.17 (*a*) The common-base configuration of a transistor and (*b*) its admittance model.

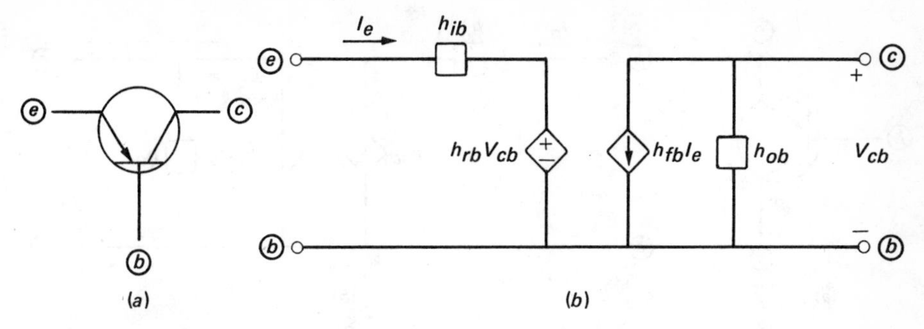

Figure 2.18 (*a*) The common-base configuration of a transistor and (*b*) its hybrid model.

For the admittance matrix of the common-collector configuration of Fig. 2.19, we delete row 3 and column 3 from **Y**, yielding

$$\mathbf{Y}_c = \begin{bmatrix} y_{ic} & y_{rc} \\ y_{fc} & y_{oc} \end{bmatrix} = \begin{bmatrix} y_{ie} & -y_{ie} - y_{re} \\ -y_{fe} - y_{ie} & y_{fe} + y_{ie} + y_{oe} + y_{re} \end{bmatrix} \tag{2.52}$$

and the corresponding hybrid matrix of Fig. 2.20 is found to be

$$\mathbf{H}_c = \begin{bmatrix} h_{ic} & h_{rc} \\ h_{fc} & h_{oc} \end{bmatrix} = \frac{1}{y_{ie}} \begin{bmatrix} 1 & y_{ie} + y_{re} \\ -y_{fe} - y_{ie} & y_{ie}y_{oe} - y_{fe}y_{re} \end{bmatrix} \tag{2.53}$$

We remark that the hybrid matrix of the common-emitter configuration of Fig. 2.21 is that of Eq. (2.47*b*) with *e* replacing *x* in all the second subscripts.

The transistor parameters vary with changes in the temperature as well as with bias-point variations. In normal operation, some of the parameters can be ignored in comparison with others. The values of the parameters listed below are typical for a variety of types of transistors for the small-signal, low-frequency, two-port model of Fig. 2.16*b*:

$$\begin{aligned} y_{ie} &= 7 \cdot 10^{-4} \text{ mho} \qquad & y_{fe} &= 0.07 \text{ mho} \\ y_{re} &= -7 \cdot 10^{-8} \text{ mho} \qquad & y_{oe} &= 8\ \mu\text{mho} \end{aligned} \tag{2.54}$$

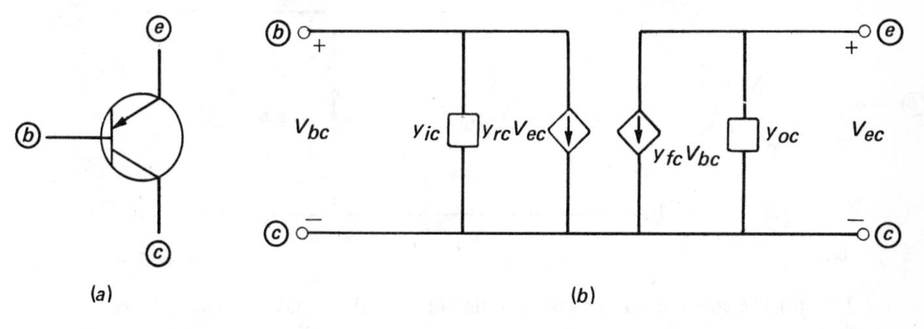

Figure 2.19 (*a*) The common-collector configuration of a transistor and (*b*) its admittance model.

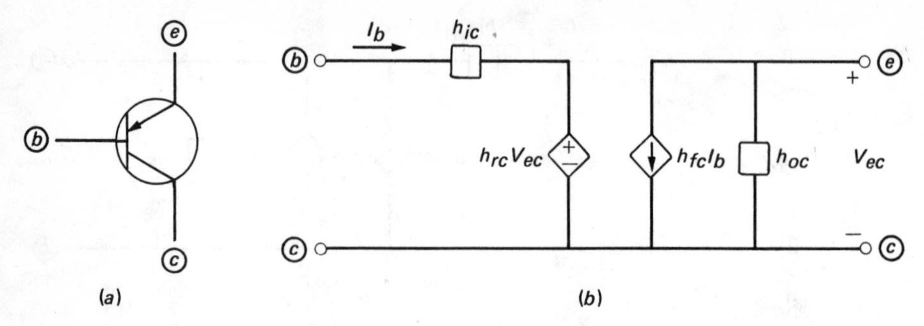

Figure 2.20 (*a*) The common-collector configuration of a transistor and (*b*) its hybrid model.

Hence, y_{re} and y_{oe} can usually be ignored in actual computations of various parameters. This approximation provides considerable simplification in design. However, at higher frequencies the admittances may be complex, and care must be taken in the simplification. For example, considering the transistor 2N697 operating at 30 MHz with $V_{CE} = 20$ V and $I_C = 20$ mA, we have†

$$\begin{aligned} y_{ie} &= (22.5 + j14.7) \cdot 10^{-3} \text{ mho} \\ y_{re} &= (-0.8 - j0.38) \cdot 10^{-3} \text{ mho} \\ y_{fe} &= (36.6 - j91.6) \cdot 10^{-3} \text{ mho} \\ y_{oe} &= (1.7 + j5.7) \cdot 10^{-3} \text{ mho} \end{aligned} \tag{2.55}$$

2.4.2 From the Common-Emitter Hybrid Model to Other Models

In this section, we derive formulas expressing parameters of various models in terms of those of the common-emitter hybrid configuration.

†Uppercase *E, C,* or *B* is used to represent the quiescent value of voltage or current, as is commonly done in electronics literature.

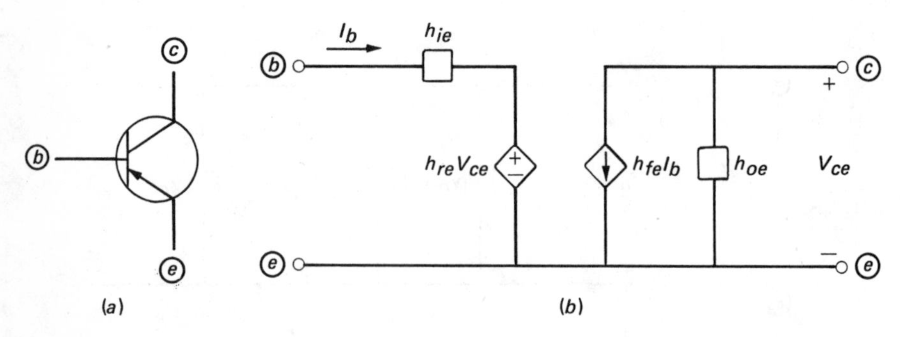

Figure 2.21 (*a*) The common-emitter configuration of a transistor and (*b*) its hybrid model.

Figure 2.21*b* is the common-emitter hybrid model of a transistor, whose admittance matrix is given by Eq. (2.47*a*) as

$$\mathbf{Y}_e = \begin{bmatrix} y_{ie} & y_{re} \\ y_{fe} & y_{oe} \end{bmatrix} = \frac{1}{h_{ie}} \begin{bmatrix} 1 & -h_{re} \\ h_{fe} & \Delta_{he} \end{bmatrix} \tag{2.56}$$

where $\Delta_{he} = h_{ie}h_{oe} - h_{re}h_{fe}$. The indefinite-admittance matrix may now be formed by adding the row and column corresponding to the emitter terminal. The elements of the new row and the new column can easily be determined from (2.56) in that each row sum and each column sum in the resulting matrix must be identically zero. For our purposes, the added row and column will occupy the second row and second column positions. We obtain

$$\mathbf{Y} = \frac{1}{h_{ie}} \begin{bmatrix} 1 & h_{re} - 1 & -h_{re} \\ -1 - h_{fe} & 1 + h_{fe} - h_{re} + \Delta_{he} & h_{re} - \Delta_{he} \\ h_{fe} & -h_{fe} - \Delta_{he} & \Delta_{he} \end{bmatrix} \begin{matrix} b \\ e \\ c \end{matrix} \quad (\text{columns } b, e, c) \tag{2.57}$$

To obtain the admittance matrix of the common-base configuration of Fig. 2.17, we delete row 1 and column 1 from **Y**, giving

$$\mathbf{Y}_b = \frac{1}{h_{ie}} \begin{bmatrix} 1 + h_{fe} - h_{re} + \Delta_{he} & h_{re} - \Delta_{he} \\ -h_{fe} - \Delta_{he} & \Delta_{he} \end{bmatrix} \tag{2.58}$$

Applying Eq. (2.47*b*), we obtain the hybrid matrix of the common-base configuration of Fig. 2.18:

$$\mathbf{H}_b = \frac{1}{1 + h_{fe} - h_{re} + \Delta_{he}} \begin{bmatrix} h_{ie} & \Delta_{he} - h_{re} \\ -h_{fe} - \Delta_{he} & h_{oe} \end{bmatrix} \tag{2.59}$$

In a similar way, if we delete row 3 and column 3 from **Y**, we obtain the admittance matrix of the common-collector configuration of Fig. 2.19 as

$$\mathbf{Y}_c = \frac{1}{h_{ie}} \begin{bmatrix} 1 & h_{re} - 1 \\ -1 - h_{fe} & 1 + h_{fe} - h_{re} + \Delta_{he} \end{bmatrix} \tag{2.60}$$

and appealing once more to Eq. (2.47*b*), we obtain the hybrid matrix of Fig. 2.20:

$$\mathbf{H}_c = \begin{bmatrix} h_{ie} & 1 - h_{re} \\ -1 - h_{fe} & h_{oe} \end{bmatrix} \tag{2.61}$$

Typical values of the hybrid parameters of a transistor operating at an emitter current $I_E = 1.3$ mA for the small-signal, low-frequency, two-port model of Fig. 2.21 are given by

$$\begin{aligned} h_{ie} &= 1.1\text{ k}\Omega \qquad & h_{fe} &= 50 \\ h_{re} &= 2.5 \cdot 10^{-4} \qquad & h_{oe} &= 25\ \mu\text{mho} \end{aligned} \tag{2.62}$$

Thus, normally, we can assume that $h_{re} \ll 1$ and $\Delta_{he} \ll h_{fe}$, which leads to considerable simplification of the formulas (2.58)-(2.61).

2.4.3 From the Common-Base T-Model to Other Models

Another common way of characterizing a transistor is by means of its equivalent T-model as shown in Fig. 2.22. This model is valid for small-signal, low-frequency operations. At higher frequencies, the model must be modified like that shown in Fig. 2.23, in which two capacitances are required. One is the junction capacitance C_c of the reverse-biased collector junction, and the other is the capacitance C_e comprised of the junction capacitance of the forward-biased emitter junction and the diffusion capacitance, which is the domain component. A typical range of values of these parameters is listed below:

$$\begin{aligned} r_e &= 5\ \Omega\ (\text{at } 25^\circ\text{C and 5 mA}) \qquad & r_c &= 1\text{ M}\Omega \\ \alpha_0 &= 0.95\text{-}0.99 \qquad & C_c &= 1\text{-}20\text{ pF} \\ r_b &= 20\text{-}100\ \Omega \qquad & C_e &= 20\text{-}5000\text{ pF} \\ \alpha &= \frac{\alpha_0}{1 + j\omega/\omega_\alpha} \qquad & \omega_\alpha &= 4 \cdot 10^9\text{ rad/s} \end{aligned} \tag{2.63}$$

In this section, we derive formulas expressing parameters of various models in terms of those of Fig. 2.22.

As shown in Example 2.4, the indefinite-admittance matrix of the T-model of Fig. 2.22, after an interchange of rows and columns, can be written as

$$\mathbf{Y} = \frac{1}{r_e(r_b + r_c) + (1-\alpha)r_b r_c} \begin{array}{c} \begin{matrix} \quad b \quad & \qquad e \qquad & \quad c \quad \end{matrix} \\ \begin{bmatrix} r_e + (1-\alpha)r_c & -(1-\alpha)r_c & -r_e \\ -r_c & r_b + r_c & -r_b \\ \alpha r_c - r_e & -\alpha r_c - r_b & r_b + r_e \end{bmatrix} \end{array} \begin{matrix} b \\ e \\ c \end{matrix} \tag{2.64}$$

For the common-base configuration of Fig. 2.17, we delete row 1 and column 1 from **Y**, resulting in

$$\mathbf{Y}_b = \frac{1}{r_e(r_b + r_c) + (1-\alpha)r_b r_c} \begin{bmatrix} r_b + r_c & -r_b \\ -\alpha r_c - r_b & r_b + r_e \end{bmatrix} \tag{2.65}$$

and from (2.47*b*) we obtain the hybrid matrix of Fig. 2.18 as

$$\mathbf{H}_b = \frac{1}{r_b + r_c} \begin{bmatrix} r_e(r_b + r_c) + (1-\alpha)r_b r_c & r_b \\ -\alpha r_c - r_b & 1 \end{bmatrix} \tag{2.66a}$$

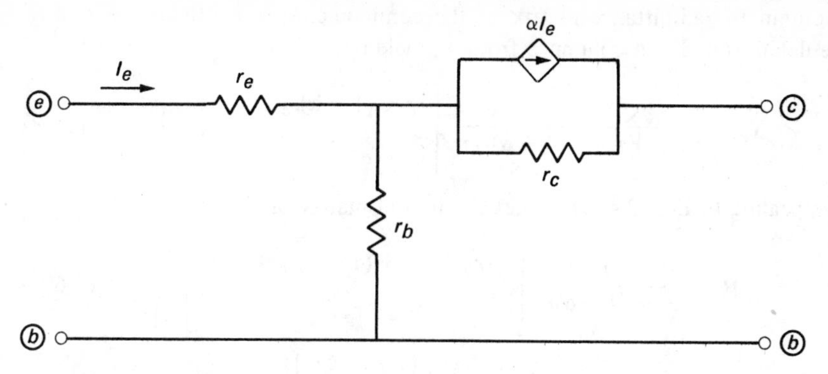

Figure 2.22 The equivalent T-model of a transistor.

$$\mathbf{H}_b \approx \begin{bmatrix} r_e + (1-\alpha)r_b & \dfrac{r_b}{r_c} \\ -\alpha & \dfrac{1}{r_c} \end{bmatrix} \tag{2.66b}$$

The second equation (2.66*b*) is an approximation and is obtained by invoking $r_b \ll r_c$ and $\alpha \approx 1$. In fact, from (2.66*b*) we can identify the elements of the T-model in terms of those of $\mathbf{H}_b$:

$$\begin{aligned} \alpha \approx -h_{fb} \qquad & r_c \approx \frac{1}{h_{ob}} \\ r_b \approx \frac{h_{rb}}{h_{ob}} \qquad & r_e \approx h_{ib} - \frac{(1+h_{fb})h_{rb}}{h_{ob}} \end{aligned} \tag{2.67}$$

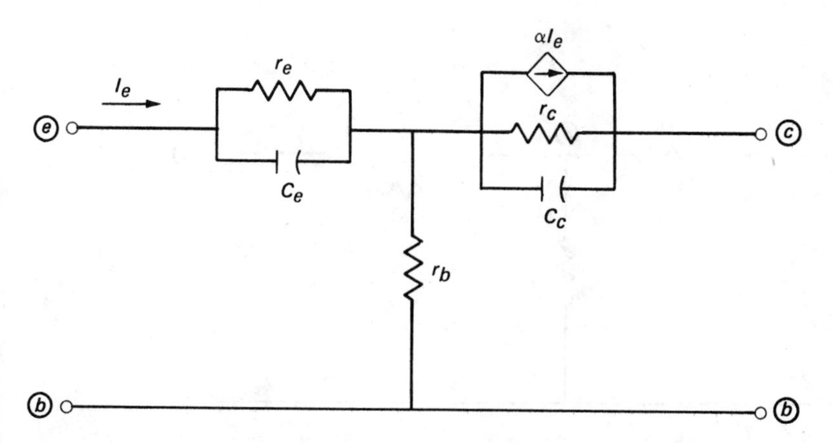

Figure 2.23 The high-frequency equivalent T-model of a transistor.

To obtain the admittance matrix of the common-emitter configuration of Fig. 2.16, we delete row 2 and column 2 from **Y**, yielding

$$\mathbf{Y}_e = \frac{1}{r_e(r_b + r_c) + (1-\alpha)r_b r_c} \begin{bmatrix} r_e + (1-\alpha)r_c & -r_e \\ \alpha r_c - r_e & r_b + r_e \end{bmatrix} \tag{2.68}$$

and by appealing to Eq. (2.47*b*) we get the hybrid matrix of Fig. 2.21 as

$$\mathbf{H}_e = \frac{1}{r_e + (1-\alpha)r_c} \begin{bmatrix} r_e(r_b + r_c) + (1-\alpha)r_b r_c & r_e \\ \alpha r_c - r_e & 1 \end{bmatrix} \tag{2.69a}$$

$$\approx \frac{1}{(1-\alpha)r_c} \begin{bmatrix} r_e r_c + (1-\alpha)r_b r_c & r_e \\ \alpha r_c & 1 \end{bmatrix} \tag{2.69b}$$

As in (2.67), the elements of Fig. 2.22 can be expressed in terms of those of Fig. 2.21 as follows:

$$\begin{aligned} \alpha &\approx \frac{h_{fe}}{1 + h_{fe}} \qquad r_c \approx \frac{1 + h_{fe}}{h_{oe}} \\ r_b &\approx h_{ie} - \frac{(1 + h_{fe})h_{re}}{h_{oe}} \qquad r_e \approx \frac{h_{re}}{h_{oe}} \end{aligned} \tag{2.70}$$

Consider the inverse $\mathbf{Z}_e$ of $\mathbf{Y}_e$ of (2.68):

$$\mathbf{Z}_e = \begin{bmatrix} r_b + r_e & r_e \\ r_e - \alpha r_c & r_e + (1-\alpha)r_c \end{bmatrix} \tag{2.71}$$

This matrix is identified as the impedance matrix of the two-port network of Fig. 2.24. Thus, Fig. 2.24 is another T-model for the transistor operating in the common-emitter mode.

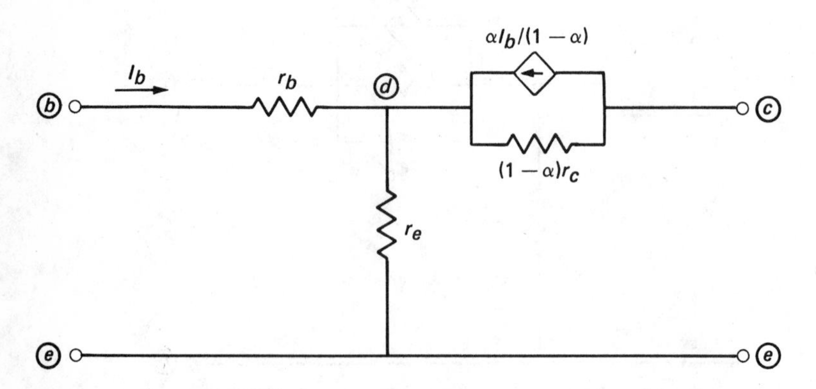

Figure 2.24 The equivalent T-model of a transistor operating in the common-emitter mode.

Finally, the admittance matrix and the hybrid matrix of the common-collector configurations of Figs. 2.19 and 2.20 are obtained from **Y** as follows:

$$\mathbf{Y}_c = \frac{1}{r_e(r_b + r_c) + (1-\alpha)r_b r_c}\begin{bmatrix} r_e + (1-\alpha)r_c & -(1-\alpha)r_c \\ -r_c & r_b + r_c \end{bmatrix} \tag{2.72}$$

$$\mathbf{H}_c = \frac{1}{r_e + (1-\alpha)r_c}\begin{bmatrix} r_e(r_b + r_c) + (1-\alpha)r_b r_c & (1-\alpha)r_c \\ -r_c & 1 \end{bmatrix} \tag{2.73a}$$

$$\approx \begin{bmatrix} r_b + \dfrac{r_e}{1-\alpha} & 1 \\ -\dfrac{1}{1-\alpha} & \dfrac{1}{(1-\alpha)r_c} \end{bmatrix} \tag{2.73b}$$

From Eq. (2.73*b*) we can identify

$$\begin{aligned} \alpha &\approx \frac{1+h_{fc}}{h_{fc}} & r_c &\approx -\frac{h_{fc}}{h_{oc}} \\ r_b &\approx h_{ic} + \frac{(1-h_{rc})h_{fc}}{h_{oc}} & r_e &\approx \frac{1-h_{rc}}{h_{oc}} \end{aligned} \tag{2.73c}$$

2.4.4 Conversion Formulas among the Hybrid Parameters

In addition to the conversion formulas derived in Sec. 2.4.2 from the common-emitter hybrid configuration to other configurations, similar relationships among the hybrid parameters of different configurations are summarized below with the approximate formulas being given in Probs. 2.31, 2.32, and 2.40:

$$\begin{bmatrix} h_{ie} & h_{re} \\ h_{fe} & h_{oe} \end{bmatrix} = \frac{1}{D_b}\begin{bmatrix} h_{ib} & \Delta_{hb} - h_{rb} \\ -\Delta_{hb} - h_{fb} & h_{ob} \end{bmatrix} = \begin{bmatrix} h_{ic} & 1 - h_{rc} \\ -(1 + h_{fc}) & h_{oc} \end{bmatrix} \tag{2.74a}$$

$$\begin{bmatrix} h_{ib} & h_{rb} \\ h_{fb} & h_{ob} \end{bmatrix} = \frac{1}{D_e}\begin{bmatrix} h_{ie} & \Delta_{he} - h_{re} \\ -\Delta_{he} - h_{fe} & h_{oe} \end{bmatrix} = \frac{1}{\Delta_{hc}}\begin{bmatrix} h_{ic} & h_{fc} + \Delta_{hc} \\ h_{rc} - \Delta_{rc} & h_{oc} \end{bmatrix} \tag{2.74b}$$

$$\begin{bmatrix} h_{ic} & h_{rc} \\ h_{fc} & h_{oc} \end{bmatrix} = \frac{1}{D_b}\begin{bmatrix} h_{ib} & 1 + h_{fb} \\ h_{rb} - 1 & h_{ob} \end{bmatrix} = \begin{bmatrix} h_{ie} & 1 - h_{re} \\ -(1 + h_{fe}) & h_{oe} \end{bmatrix} \tag{2.74c}$$

where

$$D_x = (1 + h_{fx})(1 - h_{rx}) + h_{ox}h_{ix} \tag{2.74d}$$

$$\Delta_{hx} = h_{ix}h_{ox} - h_{rx}h_{fx} \tag{2.74e}$$

2.5 THE FIRST- AND SECOND-ORDER COFACTORS

For an indefinite-admittance matrix $\mathbf{Y}(s)$ or simply $\mathbf{Y}$, denote by $\mathbf{Y}_{ij}$ the submatrix obtained from $\mathbf{Y}$ by deleting the ith row and jth column. Write $\mathbf{Y}$ explicitly as in (2.1a). Following the notation adopted in the proof of Theorem 2.1, the *(first-order) cofactor*, denoted by the symbol Y_{ij}, of the element y_{ij} of $\mathbf{Y}$, as is well known, is defined by

$$Y_{ij} = (-1)^{i+j} \det \mathbf{Y}_{ij} \tag{2.75}$$

Denote by $\mathbf{Y}_{pq,rs}$ the submatrix derived from $\mathbf{Y}$ by striking out rows p and r and columns q and s. Then the *second-order cofactor*, denoted by the symbol $Y_{pq,rs}$, of the elements y_{pq} and y_{rs} of $\mathbf{Y}$ is a scalar quantity defined by the relation

$$Y_{pq,rs} = \operatorname{sgn}(p-r) \operatorname{sgn}(q-s)(-1)^{p+q+r+s} \det \mathbf{Y}_{pq,rs} \tag{2.76}$$

where $p \neq r$ and $q \neq s$; and $\operatorname{sgn} u = +1$ if $u > 0$ and $\operatorname{sgn} u = -1$ if $u < 0$.

The symbols $\mathbf{Y}_{ij}$ and Y_{ij} or $\mathbf{Y}_{pq,rs}$ and $Y_{pq,rs}$ should create no confusion, because one is in boldface whereas the other is not. Also, for our purposes, it is convenient to define $Y_{pq,rs} = 0$ for $p = r$ or $q = s$, or $\operatorname{sgn} 0 = 0$. This convention will be followed throughout the remainder of this book.

As an illustration, consider the matrix $\mathbf{Y}$ given in (2.41). The second-order cofactors $Y_{31,42}$ and $Y_{11,34}$ of the elements of $\mathbf{Y}$ are given by

$$Y_{31,42} = \operatorname{sgn}(3-4) \operatorname{sgn}(1-2)(-1)^{3+1+4+2} \det \begin{bmatrix} -0.02 & 0 \\ 0.2 - 5 \cdot 10^{-12}s & -0.2 \end{bmatrix}$$

$$= 0.004 \tag{2.77}$$

$$Y_{11,34} = \operatorname{sgn}(1-3) \operatorname{sgn}(1-4)(-1)^{1+1+3+4} \det \begin{bmatrix} 5 \cdot 10^{-12}s & 0.2 - 5 \cdot 10^{-12}s \\ 0 & -0.204 - 10^{-10}s \end{bmatrix}$$

$$= 5 \cdot 10^{-12}s(0.204 + 10^{-10}s) \tag{2.78}$$

As indicated in Theorem 2.1, the indefinite-admittance matrix $\mathbf{Y}$ is an equicofactor matrix, meaning that all the first-order cofactors of its elements are equal. In the following, we shall demonstrate that not all of its second-order cofactors are independent; some may be obtained from the others by linear combinations. In fact, we shall show that there are only $(n-1)^2$ independent second-order cofactors for $\mathbf{Y}(s)$.

Theorem 2.2 The second-order cofactors of the elements of the indefinite-admittance matrix are related by

$$Y_{pq,rs} = Y_{pq,uv} + Y_{rs,uv} - Y_{ps,uv} - Y_{rq,uv} \tag{2.79}$$

for arbitrary u and v.

PROOF Assume that $p \neq u \neq r$ and $q \neq v \neq s$. Then in $\mathbf{Y}_{pq,rs}$, replace the column corresponding to the vth column of $\mathbf{Y}$ by the sum of all the columns of $\mathbf{Y}_{pq,rs}$. By virtue of (2.10), the element of this column in the ith row is now $-(y_{iq}+y_{is})$. If we remove the -1 from this column and let the resulting matrix be denoted by $\hat{\mathbf{Y}}_{pq,rs}$, then $\det \mathbf{Y}_{pq,rs} = -\det \hat{\mathbf{Y}}_{pq,rs}$. In $\hat{\mathbf{Y}}_{pq,rs}$, let $\hat{\mathbf{Y}}_1$ and $\hat{\mathbf{Y}}_2$ be the matrices obtained from $\hat{\mathbf{Y}}_{pq,rs}$ by setting $y_{is}=0$ and $y_{iq}=0$ $(i=1, 2, \ldots, n)$, respectively. It is obvious that we have $\det \hat{\mathbf{Y}}_{pq,rs} = \det \hat{\mathbf{Y}}_1 + \det \hat{\mathbf{Y}}_2$.

At this stage, the numerical order in which the columns q, s, and v occur is very relevant. Apart from a sign, $\det \hat{\mathbf{Y}}_1$ and $\det \hat{\mathbf{Y}}_2$ are the second-order cofactors $Y_{ps,rv}$ and $Y_{pq,rv}$, respectively.

In $\hat{\mathbf{Y}}_1$, let us shift the column containing the elements y_{iq} to the position such that the columns in the resulting matrix appear in the same relative order as those in $\mathbf{Y}$. Doing so requires $v-q-1$ adjacent transpositions for the cases $s<q<v$ and $q<v<s$; $q-v-1$ for $s<v<q$ and $v<q<s$; $v-q-2$ for $q<s<v$; and $q-v-2$ for $v<s<q$. It is not difficult to check that all the cases are clearly contained in the single equation

$$\det \hat{\mathbf{Y}}_1 = \operatorname{sgn}(q-s)\operatorname{sgn}(v-s)(-1)^{v-q-1}\det \mathbf{Y}_{ps,rv} \tag{2.80}$$

Thus, we have

$$\begin{aligned} Y_{ps,rv} &= \operatorname{sgn}(p-r)\operatorname{sgn}(s-v)(-1)^{p+s+r+v}\det \mathbf{Y}_{ps,rv} \\ &= \operatorname{sgn}(p-r)\operatorname{sgn}(q-s)(-1)^{p+q+r+s}\det \hat{\mathbf{Y}}_1 \end{aligned} \tag{2.81}$$

In a similar way, we can show that

$$Y_{pq,rv} = \operatorname{sgn}(p-r)\operatorname{sgn}(s-q)(-1)^{p+s+r+q}\det \hat{\mathbf{Y}}_2 \tag{2.82}$$

Combining these results yields

$$\begin{aligned} Y_{pq,rs} &= \operatorname{sgn}(p-r)\operatorname{sgn}(q-s)(-1)^{p+q+r+s}\det \mathbf{Y}_{pq,rs} \\ &= \operatorname{sgn}(p-r)\operatorname{sgn}(q-s)(-1)^{p+q+r+s-1}(\det \hat{Y}_1 + \det \hat{Y}_2) \\ &= Y_{pq,rv} - Y_{ps,rv} \end{aligned} \tag{2.83}$$

Observe that in deriving (2.83) we have invoked only the fact that each row sum of $\mathbf{Y}$ equals zero. Now consider the transpose $\mathbf{Y}'$ of $\mathbf{Y}$ and appeal to (2.7), which states that each column sum of $\mathbf{Y}$ or each row sum of $\mathbf{Y}'$ equals zero. By virtue of (2.83), we have

$$Y'_{pq,rs} = Y'_{pq,rv} - Y'_{ps,rv} \tag{2.84}$$

or, equivalently,

$$Y_{qp,sr} = Y_{qp,vr} - Y_{sp,vr} \tag{2.85}$$

By using (2.85), the two terms on the right-hand side of (2.83) can be expressed as

$$Y_{pq,rv} = Y_{pq,uv} - Y_{rq,uv} \tag{2.86a}$$

$$Y_{ps,rv} = Y_{ps,uv} - Y_{rs,uv} \tag{2.86b}$$

Substituting (2.86) in (2.83) yields the desired formula (2.79).

Suppose that the conditions $p \neq u \neq r$ and $q \neq v \neq s$ are not fulfilled. It is easy to check that either (2.79) is trivially satisfied or it reduces to (2.83) or (2.85), which in turn is either trivially satisfied or can be established without invoking the assumption (see Prob. 2.27). This completes the proof of the theorem.

As an example, consider the indefinite-admittance matrix **Y** given in (2.34). The following second-order cofactors are computed:

$$Y_{12,44} = \alpha g_e g_b \tag{2.87a}$$

$$Y_{33,44} = g_e g_c \tag{2.87b}$$

$$Y_{13,44} = 0 \tag{2.87c}$$

$$Y_{32,44} = 0 \tag{2.87d}$$

$$Y_{12,33} = \alpha g_e g_b + g_e g_c \tag{2.87e}$$

Thus, we have

$$Y_{12,33} = Y_{12,44} + Y_{33,44} - Y_{13,44} - Y_{32,44} \tag{2.88}$$

confirming (2.79).

It is significant to note that, by virtue of (2.79), all the second-order cofactors of the elements of **Y** can be obtained linearly in terms of the first-order cofactors of the elements of $\mathbf{Y}_{nn}$. Hence there are only $(n-1)^2$ linearly independent second-order cofactors for the nth-order indefinite-admittance matrix.

2.6 COMPUTATION OF NETWORK FUNCTIONS

The usefulness of the indefinite-admittance matrix lies in the fact that it facilitates the formulation of the driving-point or transfer functions between any pair of nodes or from any pair of nodes to any other pair of nodes in the network. In the present section, we derive explicit formulas that express the network functions in terms of the ratios of the first- and/or second-order cofactors of the elements of the indefinite-admittance matrix in an elegant and compact form.

2.6.1 Transfer Impedance and Voltage Gain

Assume that a current source is connected between any two nodes r and s so that a current I_{sr} is injected into the rth node and at the same time is extracted from the sth node. Suppose also that an ideal voltmeter is connected from node p to node q so that it indicates the potential rise from node q to node p, as depicted symbolically in Fig. 2.25. Then the *transfer impedance*, denoted by the symbol $z_{rp,sq}$, between the node pairs rs and pq of the network of Fig. 2.25 is defined by the relation

$$z_{rp,sq} = \frac{V_{pq}}{I_{sr}} \tag{2.89}$$

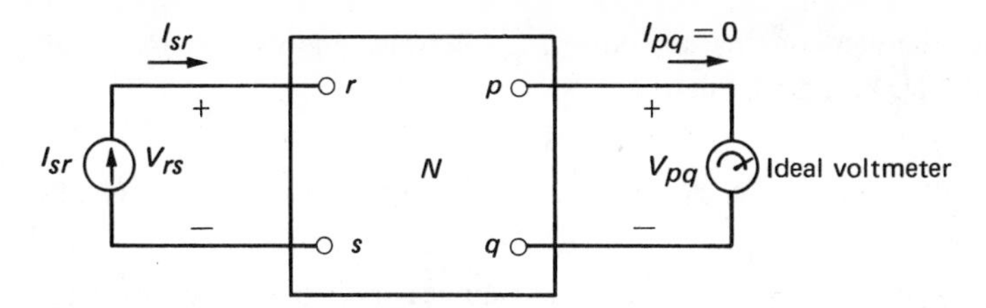

Figure 2.25 The symbolic representation for the measurement of the transfer impedance.

with all initial conditions and independent sources in the network N being set to zero. The representation is, of course, quite general. When $r = p$ and $s = q$, the transfer impedance $z_{rp,sq}$ becomes the *driving-point impedance* $z_{rr,ss}$ between the terminal pair rs of the network.

With reference to Fig. 2.25, set all initial conditions and independent sources in N to zero and choose terminal or node q to be the reference-potential point for all other terminals. In terms of the equations of (2.1), the operations are equivalent to setting $\mathbf{J} = \mathbf{0}$, $V_q = 0$, $I_x = 0$ for $x \neq r$, s and $I_r = -I_s = I_{sr}$. Since $\mathbf{Y}$ is an equicofactor matrix, the equations of (2.1) are not linearly independent and so one of them is superfluous. Let us suppress the sth equation from (2.1), which then reduces to

$$\mathbf{I}_{-s} = \mathbf{Y}_{sq}\mathbf{V}_{-q} \tag{2.90}$$

where $\mathbf{I}_{-s}$ and $\mathbf{V}_{-q}$ denote the subvectors obtained from $\mathbf{I}$ and $\mathbf{V}$ of (2.2) by deleting the sth row and qth row, respectively. Applying Cramer's rule to solve for V_p yields

$$V_p = \frac{\det \tilde{\mathbf{Y}}_{sq}}{\det \mathbf{Y}_{sq}} \tag{2.91}$$

where $\tilde{\mathbf{Y}}_{sq}$ is the matrix derived from $\mathbf{Y}_{sq}$ by replacing the column corresponding to V_p by $\mathbf{I}_{-s}$. We now expand $\det \tilde{\mathbf{Y}}_{sq}$ along the column $\mathbf{I}_{-s}$ that has $n - 2$ zeros and I_{sr}. We recognize that $\mathbf{I}_{-s}$ is in the pth column if $p < q$ but in the $(p-1)$th column if $p > q$. Furthermore, the row in which I_{sr} appears is the rth row if $r < s$ but is the $(r-1)$th row if $r > s$. With these in mind, it is not difficult to confirm that

$$(-1)^{s+q} \det \tilde{\mathbf{Y}}_{sq} = I_{sr} Y_{rp,sq} \tag{2.92}$$

In addition, we have

$$\det \mathbf{Y}_{sq} = (-1)^{s+q} Y_{sq} \tag{2.93}$$

Substituting (2.92) and (2.93) in (2.91) in conjunction with (2.89), we obtain

$$z_{rp,sq} = \frac{Y_{rp,sq}}{Y_{uv}} \tag{2.94}$$

$$z_{rr,ss} = \frac{Y_{rr,ss}}{Y_{uv}} \tag{2.95}$$

in which we have invoked the fact that $Y_{sq} = Y_{uv}$.

The *voltage gain*, denoted by the symbol $g_{rp,\,sq}$, between the node pairs rs and pq of the network of Fig. 2.25 is defined as

$$g_{rp,\,sq} = \frac{V_{pq}}{V_{rs}} \tag{2.96}$$

again with all initial conditions and independent sources in N being set to zero. Thus, from (2.94) and (2.95) we get

$$g_{rp,\,sq} = \frac{z_{rp,\,sq}}{z_{rr,\,ss}} = \frac{Y_{rp,\,sq}}{Y_{rr,\,ss}} \tag{2.97}$$

The symbols have been chosen to help us to remember. In the numerators of (2.94) and (2.95), the order of the subscripts is as follows: r, the current injecting node; p, the voltage measurement node; s, the current extracting node; and q, the voltage reference node. Nodes r and p designate the input and output transfer measurement, and nodes s and q form a sort of double datum. This is similarly valid for the numerator of $g_{rp,\,sq}$. In fact, formula (2.94) for the transfer impedance $z_{rp,\,sq}$ is all that we need to remember. The other two expressions (2.95) and (2.97) can always be deduced easily from the first. The formula (2.95) is a special case of the first in which $r=p$ and $s=q$, and (2.97) is simply the quotient of the other two. The results presented in these formulas are, of course, quite general, because the letters r, p, s, and q may refer to the labels of any four nodes or terminals in the network.

We illustrate these results by the following examples.

Example 2.6 Consider the transistor amplifier of Fig. 2.26. By using the procedure outlined in Sec. 2.2, the primitive indefinite-admittance matrix of the amplifier is found to be

$$\mathbf{Y} = \begin{array}{c} \begin{matrix} \quad 1 \quad & \quad 2 \quad & \quad 3 \quad & \quad 4 \quad \end{matrix} \\ \begin{bmatrix} g_e + sC_e & -g_e - sC_e & 0 & 0 \\ (\alpha - 1)(g_e + sC_e) & g_b + g_c + (1-\alpha)(g_e + sC_e) + sC_c & -g_c - sC_c & -g_b \\ -\alpha(g_e + sC_e) & \alpha(g_e + sC_e) - g_c - sC_c & g_c + sC_c + y_l & -y_l \\ 0 & -g_b & -y_l & g_b + y_l \end{bmatrix} \end{array} \tag{2.98}$$

To compute the transfer impedance $z_{13,44}$, we use (2.94), yielding

$$z_{13,44} = \frac{V_{34}}{I_e} = \frac{Y_{13,44}}{Y_{uv}} = \frac{Y_{13,44}}{Y_{22}} = \frac{g_c + sC_c + \alpha g_b}{(g_c + sC_c)(g_b + y_l) + g_b y_l} \tag{2.99}$$

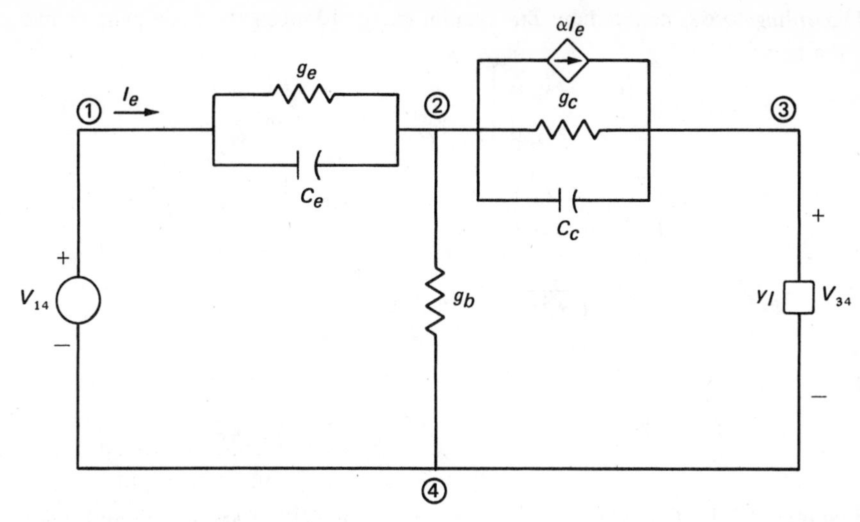

Figure 2.26 A transistor amplifier used to illustrate the computation of the transfer and driving-point impedances.

The voltage gain $g_{13,44}$ of the amplifier is determined by

$$g_{13,44}=\frac{V_{34}}{V_{14}}=\frac{Y_{13,44}}{Y_{11,44}}$$

$$=\frac{(g_e+sC_e)(g_c+sC_c+\alpha g_b)}{(g_b+g_e+sC_e+y_l)(g_c+sC_c)+(1-\alpha)(g_e+sC_e)y_l+g_by_l} \tag{2.100}$$

From (2.95), we can compute the amplifier input impedance, which is given by

$$z_{11,44}=\frac{V_{14}}{I_e}=\frac{Y_{11,44}}{Y_{uv}}=\frac{Y_{11,44}}{Y_{22}}$$

$$=\frac{(g_b+g_e+sC_e+y_l)(g_c+sC_c)+(1-\alpha)(g_e+sC_e)y_l+g_by_l}{[(g_c+sC_c)(g_b+y_l)+g_by_l](g_e+sC_e)} \tag{2.101}$$

For the amplifier output impedance with the input open-circuited, we have

$$z_{33,44}=\frac{Y_{33,44}}{Y_{uv}}=\frac{Y_{33,44}}{Y_{22}}=\frac{g_b+g_c+sC_c}{(g_c+sC_c)(g_b+y_l)+g_by_l} \tag{2.102}$$

Suppose that we wish now to compute the open-circuit impedance matrix $\mathbf{Z}(s)$ of the transistor with respect to the ports formed by the terminal pairs 1, 4 and 3, 4 after the load admittance y_l has been removed. We write

$$z^0_{rp,sq}=z_{rp,sq}|_{y_l=0} \tag{2.103}$$

Then according to our convention, $\mathbf{Z}(s)$ can be expressed as

$$\mathbf{Z}(s)=\begin{bmatrix} z^0_{11,44} & z^0_{31,44} \\ z^0_{13,44} & z^0_{33,44}\end{bmatrix}=\frac{1}{Y^0_{22}}\begin{bmatrix} Y^0_{11,44} & Y^0_{31,44} \\ Y^0_{13,44} & Y^0_{33,44}\end{bmatrix}$$

$$=\begin{bmatrix} r_b+\dfrac{r_e}{1+sr_eC_e} & r_b \\ r_b+\dfrac{\alpha r_c}{1+sr_cC_c} & r_b+\dfrac{r_c}{1+sr_cC_c}\end{bmatrix} \tag{2.104}$$

where $r_b = 1/g_b$, $r_e = 1/g_e$, $r_c = 1/g_c$, and

$$Y^0_{uv} = Y_{uv}|_{y_l=0} \quad \text{and} \quad Y^0_{rp,sq} = Y_{rp,sq}|_{y_l=0} \tag{2.105}$$

Example 2.7 In the hybrid-pi equivalent network of a transistor as shown in Fig. 2.12, suppose that we connect a 100-Ω load between nodes 2 and 4 as shown in Fig. 2.27. This is equivalent to adding 0.01 to y_{22} and y_{44} and -0.01 to y_{24} and y_{42} in (2.41). As before, let $p = s/10^9$. The indefinite-admittance matrix of the resulting amplifier is given by

$$\mathbf{Y}=\begin{bmatrix} 0.02 & 0 & -0.02 & 0 \\ 0 & 0.01+0.005p & 0.2-0.005p & -0.21 \\ -0.02 & -0.005p & 0.024+0.105p & -0.004-0.1p \\ 0 & -0.01 & -0.204-0.1p & 0.214+0.1p \end{bmatrix} \tag{2.106}$$

To compute the voltage gain $g_{12,44}$, we appeal to (2.97), which yields

$$g_{12,44}=\frac{Y_{12,44}}{Y_{11,44}}=\frac{p-40}{5p^2+21.7p+2.4} \tag{2.107}$$

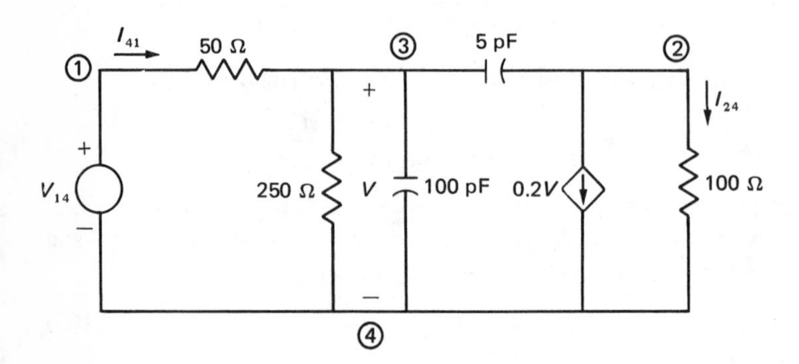

Figure 2.27 A transistor amplifier used to illustrate the computation of the voltage gain.

The input impedance looking into terminals 1 and 4 is determined by

$$z_{11,44} = \frac{Y_{11,44}}{Y_{uv}} = \frac{Y_{11,44}}{Y_{44}} = \frac{50p^2 + 217p + 24}{p^2 + 4.14p + 0.08} \tag{2.108}$$

Finally, to compute the current gain of the amplifier, which is defined as the ratio of the current I_{24} in the 100-Ω load to the current I_{41} at terminals 1 and 4 as indicated in Fig. 2.27, we apply (2.94), which shows that

$$\frac{I_{24}}{I_{41}} = 0.01z_{12,44} = 0.01\,\frac{Y_{12,44}}{Y_{44}} = \frac{0.1p - 4}{p^2 + 4.14p + 0.08} \tag{2.109}$$

The transfer admittance, which is defined as the ratio of the current I_{24} to the voltage V_{14} of Fig. 2.27, is related to the voltage gain $g_{12,44}$ by

$$\frac{I_{24}}{V_{14}} = 0.01g_{12,44} = 0.01\,\frac{Y_{12,44}}{Y_{11,44}} = \frac{p - 40}{500p^2 + 2170p + 240} \tag{2.110}$$

Thus, formulas (2.94), (2.95), and (2.97) can be employed to compute nearly all the driving-point and transfer functions except those to be discussed in the following section where the short-circuit current is involved.

2.6.2 Short-Circuit Transfer Admittance and Current Gain

Assume that a voltage source is connected between any two nodes r and s so that a current I_{sr} is injected into the rth node and at the same time is extracted from the sth node. Suppose also that an ideal ammeter is connected between nodes p and q so that it indicates a short-circuit current flow from p to q, as depicted symbolically in Fig. 2.28. Then the *short-circuit current gain*, denoted by the symbol $\alpha_{rp,sq}$, between the node pairs rs and pq of the network of Fig. 2.28 is defined by

$$\alpha_{rp,sq} = \frac{I_{pq}}{I_{sr}} \tag{2.111}$$

with all initial conditions and independent sources in the network N being set to zero. For $r = p$ and $s = q$, $\alpha_{rp,sq}$ is defined to be unity.

Likewise, the *short-circuit transfer admittance*, denoted by the symbol $y_{rp,sq}$, between the node pairs rs and pq of the network of Fig. 2.28 is defined as

$$y_{rp,sq} = \frac{I_{pq}}{V_{rs}} \tag{2.112}$$

again with all initial conditions and independent sources in N being set to zero. Note that for $r = p$ and $s = q$, $y_{rp,sq}$ cannot be meaningfully defined.

Refer to Fig. 2.28. Let $\hat{N}$ be the network obtained from N by connecting a branch of admittance y between nodes p and q. For simplicity, let us use the cap to indicate the similar quantities defined for $\hat{N}$. For example, $\hat{\mathbf{Y}}$, $\hat{z}_{rp,sq}$, and $\hat{V}_{pq}$

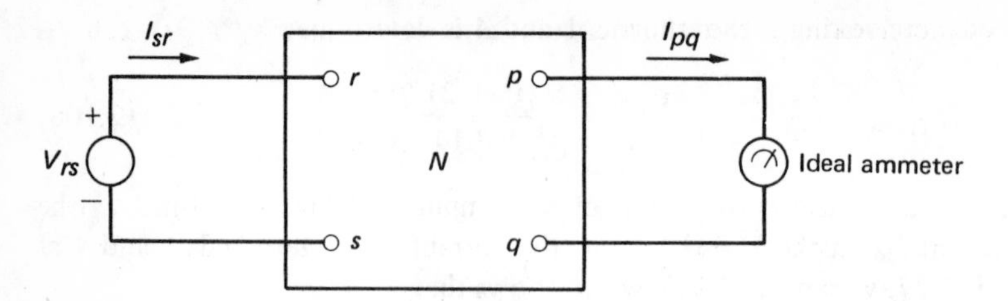

Figure 2.28 The symbolic representation for the measurement of the short-circuit current gain.

denote the indefinite-admittance matrix, the transfer impedance, and the voltage rise from q to p in $\hat{N}$, respectively. Thus, from (2.94) we have

$$y\hat{z}_{rp,sq} = y\frac{\hat{V}_{pq}}{\hat{I}_{sr}} = y\frac{\hat{Y}_{rp,sq}}{\hat{Y}_{qq}} \tag{2.113}$$

Now consider the limit by letting y approach infinity. The left-hand side of (2.113) becomes the short-circuit current gain $\alpha_{rp,sq}$ in N and the denominator $\hat{Y}_{qq}$ approaches $yY_{pp,qq}$. Since $\hat{Y}_{rp,sq} = Y_{rp,sq}$, (2.113) becomes

$$\alpha_{rp,sq} = \frac{Y_{rp,sq}}{Y_{pp,qq}} \tag{2.114}$$

Likewise, using (2.114) and the limiting process, we can show that

$$y_{rp,sq} = \frac{Y_{rp,sq}}{Y_{rr,ss,pp} + Y_{rr,ss,qq} - Y_{rr,ss,pq} - Y_{rr,ss,qp}} \tag{2.115}$$

in which the *third-order cofactor* $Y_{rr,ss,pq}$ of the elements y_{rr}, y_{ss}, and y_{pq} of $\mathbf{Y}$ is defined by the relation

$$Y_{rr,ss,pq} = \operatorname{sgn}(r-p)\operatorname{sgn}(r-q)\operatorname{sgn}(s-p)\operatorname{sgn}(s-q)(-1)^{p+q}\det \mathbf{Y}_{rr,ss,pq} \tag{2.116}$$

$\mathbf{Y}_{rr,ss,pq}$ being the submatrix obtained from $\mathbf{Y}$ by deleting the rows r, s, p and columns r, s, q. $Y_{rr,ss,pq}$ is defined to be zero if any two of the row or column indices are identical, or alternatively we can define $\operatorname{sgn} 0 = 0$. The details of the derivation of (2.115) are left as an exercise (see Prob. 2.9).

As a consequence of (2.94), (2.97), and (2.114), we conclude that for a reciprocal network, whose indefinite-admittance matrix is symmetric, the transfer impedances in opposite directions are equal, the transfer admittances in opposite directions are equal, and the open-circuit voltage gain in one direction is equal to the short-circuit current gain in the opposite direction, all being measured from one node pair to another. These are three manifestations of reciprocity, and if one of these three is known, the other two must follow. More specifically, for reciprocal networks we have $z_{rp,sq} = z_{pr,qs}$ and $g_{rp,sq} = \alpha_{pr,qs}$ (see Probs. 2.28 and 2.29).

Example 2.8 Suppose that we wish to use the above formulas to compute the short-circuit admittance parameters of the two-port network of Fig. 2.11. Then according to our convention, the short-circuit admittance matrix $\mathbf{Y}_{sc}(s)$ can be expressed as

$$\mathbf{Y}_{sc}(s) = \begin{bmatrix} \dfrac{1}{\hat{z}_{11,33}} & -y_{21,33} \\ -y_{12,33} & \dfrac{1}{\tilde{z}_{22,33}} \end{bmatrix} \tag{2.117}$$

The elements $\hat{z}_{11,33}$ and $\tilde{z}_{22,33}$ will be defined shortly.

The indefinite-admittance matrix of the network of Fig. 2.11 was computed earlier and is given by (2.34). Thus, from (2.115) we have

$$y_{12,33} = \frac{Y_{12,33}}{Y_{11,33,22}} = \frac{g_e(g_c + \alpha g_b)}{w} = \frac{\alpha r_c + r_b}{r_b r_c r_e w} \tag{2.118a}$$

$$y_{21,33} = \frac{Y_{21,33}}{Y_{22,33,11}} = \frac{g_c g_e}{w} = \frac{r_b}{r_b r_c r_e w} \tag{2.118b}$$

The input impedance $\hat{z}_{11,33}$ looking into terminals 1 and 3 when terminals 2 and 3 are joined together can be computed by (2.95) with the indefinite-admittance matrix $\hat{\mathbf{Y}}$ being derived from (2.34) by adding row 3 to row 2, column 3 to column 2, and then deleting row 3 and column 3 in the resulting matrix. The result is given by

$$\hat{\mathbf{Y}} = \begin{array}{c} \\ 1 \\ 2,3 \\ 4 \end{array} \begin{array}{c} \begin{array}{ccc} 1 & 2,3 & 4 \end{array} \\ \begin{bmatrix} g_e & 0 & -g_e \\ -\alpha g_e & g_b + g_c & \alpha g_e - g_b - g_c \\ \alpha g_e - g_e & -g_b - g_c & w \end{bmatrix} \end{array} \tag{2.119}$$

yielding

$$\hat{z}_{11,33} = \frac{\hat{Y}_{11,22}}{\hat{Y}_{33}} = \frac{w}{g_e(g_b + g_c)} = \frac{r_b r_c r_e w}{r_b + r_c} \tag{2.120}$$

Finally, $\tilde{z}_{22,33}$ is the impedance looking into terminals 2 and 3 when terminals 1 and 3 are joined together. The corresponding indefinite-admittance matrix is obtained as

$$\tilde{\mathbf{Y}} = \begin{array}{c} \\ 1,3 \\ 2 \\ 4 \end{array} \begin{array}{c} \begin{array}{ccc} 1,3 & 2 & 4 \end{array} \\ \begin{bmatrix} g_b + g_e & 0 & -g_e - g_b \\ -\alpha g_e & g_c & \alpha g_e - g_c \\ \alpha g_e - g_e - g_b & -g_c & w \end{bmatrix} \end{array} \tag{2.121}$$

giving

$$\tilde{z}_{22,33} = \frac{\tilde{Y}_{22,11}}{\tilde{Y}_{33}} = \frac{w}{g_c(g_b + g_e)} = \frac{r_b r_c r_e w}{r_b + r_e} \tag{2.122}$$

The short-circuit admittance matrix of the two-port network becomes

$$\mathbf{Y}_{sc}(s) = \frac{1}{r_b r_c r_e W} \begin{bmatrix} r_b + r_c & -r_b \\ -\alpha r_c - r_b & r_b + r_e \end{bmatrix} \tag{2.123}$$

confirming (2.37). Alternatively, the elements of $\mathbf{Y}_{sc}(s)$ can be deduced directly from (2.36) instead of (2.34), the details being left as an exercise (see Prob. 2.11).

Example 2.9 We wish to compute the short-circuit admittance matrix $\mathbf{Y}_{sc}(s)$ of the active network of Fig. 2.5 with respect to the ports formed by the terminal pairs 1, 5 and 4, 5. As before, we can write

$$\mathbf{Y}_{sc}(s) = \begin{bmatrix} \dfrac{1}{\hat{z}_{11,55}} & -y_{41,55} \\ -y_{14,55} & \dfrac{1}{\tilde{z}_{44,55}} \end{bmatrix} \tag{2.124}$$

Appealing to (2.115) and using the indefinite-admittance matrix $\mathbf{Y}$ given in (2.18) yield

$$y_{41,55} = \frac{Y_{41,55}}{Y_{44,55,11}} = \frac{g_1 g_2}{(1-\beta)g_2 LCs^2 + [(1-\beta)g_1 g_2 L + C]s + g_1 + g_2} \tag{2.125a}$$

$$y_{14,55} = \frac{Y_{14,55}}{Y_{11,55,44}} = \frac{(1-\beta)g_1 g_2}{(1-\beta)g_2 LCs^2 + [(1-\beta)g_1 g_2 L + C]s + g_1 + g_2} \tag{2.125b}$$

The input impedance $\hat{z}_{11,55}$ looking into terminals 1 and 5 with terminals 4 and 5 joined together can be determined from (2.95) by using the indefinite-admittance matrix

$$\hat{\mathbf{Y}} = \begin{bmatrix} g_1 & -g_1 & 0 & 0 \\ -g_1 & g_1 + g_2 + sC & -g_2 & -sC \\ 0 & \beta g_2 - g_2 & g_2 - \beta g_2 + \dfrac{1}{sL} & -\dfrac{1}{sL} \\ 0 & -\beta g_2 - sC & \beta g_2 - \dfrac{1}{sL} & sC + \dfrac{1}{sL} \end{bmatrix} \tag{2.126}$$

which is derived from (2.18) by adding row 5 to row 4, column 5 to column 4, and then deleting row 5 and column 5 in the resulting matrix. The result is given by

$$\hat{z}_{11,55} = \frac{\hat{Y}_{11,44}}{\hat{Y}_{44}} = \frac{(g_1 + sC)[(1-\beta)g_2 Ls + 1] + g_2}{g_1[(1-\beta)g_2 LCs^2 + sC + g_2]} \tag{2.127}$$

Finally, the input impedance $\tilde{z}_{44,55}$ is computed from the matrix

$$\tilde{\mathbf{Y}} = \begin{bmatrix} g_1 + g_3 + sC & -g_1 - \beta g_2 - sC & \beta g_2 & -g_3 \\ -g_1 - sC & g_1 + g_2 + sC & -g_2 & 0. \\ 0 & \beta g_2 - g_2 & g_2 - \beta g_2 + \dfrac{1}{sL} & -\dfrac{1}{sL} \\ -g_3 & 0 & -\dfrac{1}{sL} & g_3 + \dfrac{1}{sL} \end{bmatrix} \tag{2.128}$$

which is obtained from (2.18) by adding row 5 to row 1, column 5 to column 1, and then deleting row 5 and column 5 in the resulting matrix, giving

$$\tilde{z}_{44,55} = \frac{\tilde{Y}_{44,11}}{\tilde{Y}_{11}} = \frac{(g_1 + sC)[(1-\beta)g_2 Ls + 1] + g_2}{g_3(g_1 + g_2 + sC) + (g_1 + sC)(1-\beta)g_2(g_3 Ls + 1)} \tag{2.129}$$

The short-circuit admittance matrix of the two-port network becomes

$$\mathbf{Y}_{sc}(s) = \frac{1}{q(s)} \times \begin{bmatrix} g_1[(1-\beta)g_2 LCs^2 + Cs + g_2] & -g_1 g_2 \\ (\beta - 1)g_1 g_2 & (g_1 + sC)[g_3 + (1-\beta)g_2(g_3 Ls + 1)] + g_2 g_3 \end{bmatrix} \tag{2.130a}$$

where

$$q(s) = (g_1 + sC)[(1-\beta)g_2 Ls + 1] + g_2 \tag{2.130b}$$

2.7 ANALYSIS OF CONSTRAINED ACTIVE NETWORKS

An important network element that is used extensively to perform a wide variety of functions is the operational amplifier, abbreviated as *op-amp*. It is a direct-coupled high-gain amplifier to which feedback is added to control its overall response characteristic. The schematic diagram of an op-amp is shown in Fig. 2.29*a*, and its equivalent network in Fig. 2.29*b*. V^+ and V^- denote the voltages from terminals + and − to the ground, respectively. The + terminal is referred to as the *noninverting terminal* and the − terminal as the *inverting terminal.* Usually, one of the input terminals is grounded, and nearly all op-amps have only one output terminal. The values of the parameters of a typical op-amp are listed below:

$$A \geqslant 10{,}000 \quad \text{at frequencies } f \leqslant 10 \text{ kHz} \qquad r_i = 500 \text{ k}\Omega \quad r_o = 300\ \Omega \tag{2.131}$$

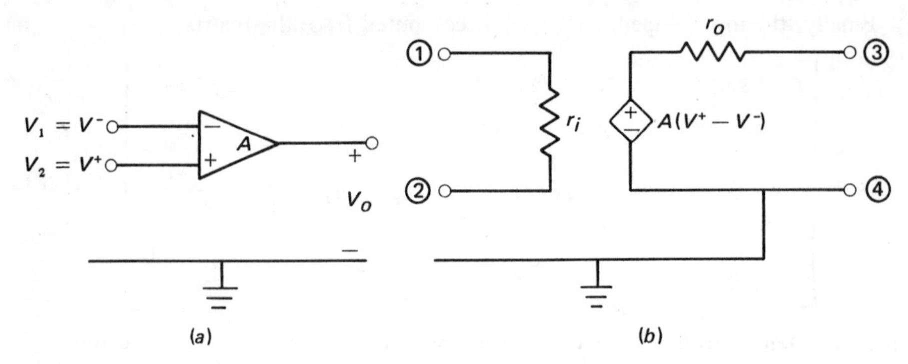

Figure 2.29 (*a*) The schematic diagram of an op-amp and (*b*) its equivalent network.

The *ideal op-amp* is the idealization of the real-world op-amp with the following characteristics:

1. Infinite input impedance, $r_i = \infty$.
2. Zero output impedance, $r_o = 0$.
3. Infinite voltage gain, $A = \infty$.
4. Infinite bandwidth.
5. Zero output signal when the input voltage is zero: $V_o = 0$ for $V^+ = V^-$. This is known as the *zero offset.*
6. Characteristics do not drift with environmental factors such as temperature.

In practice, the ideal op-amp is a good approximation to the physical op-amp of Fig. 2.29, and it can be used with considerable simplification in the design and analysis of various networks where op-amps are employed. Since the output of the ideal op-amp must be finite, $V_o = A(V^+ - V^-)$ implies that the potential difference between the input terminals of the amplifier must be zero. Moreover, because the input impedance is infinite, the input current must be identically zero, Thus, for an ideal op-amp we should apply the following rules:

Rule 1 The potential difference between the input terminals is zero.

Rule 2 The current at each of the input terminals is zero.

It is clear from the above analysis that the indefinite-admittance matrix does not exist for the ideal op-amp. For the indefinite-admittance matrix to exist, we must assume finite parasitic input and output impedances and the finite gain as shown in Fig. 2.29*b*, whose indefinite-admittance matrix is defined by the equation

$$\begin{bmatrix} I_1 \\ I_2 \\ I_3 \\ I_4 \end{bmatrix} = \begin{bmatrix} g_i & -g_i & 0 & 0 \\ -g_i & g_i & 0 & 0 \\ g_oA & -g_oA & g_o & -g_o \\ -g_oA & g_oA & -g_o & g_o \end{bmatrix} \begin{bmatrix} V_1 \\ V_2 \\ V_3 \\ V_4 \end{bmatrix} \tag{2.132}$$

where $g_i = 1/r_i$ and $g_o = 1/r_o$. To compute network functions when ideal op-amp conditions are assumed, we can let $A \to \infty$, $g_i \to 0$, and $g_o \to \infty$ after the calculations involving the indefinite-admittance matrix have been completed. This approach, although general, complicates the computation. In the following, we present a simplified procedure for computing network functions when ideal op-amps are involved. However, before we do this, we illustrate the above procedure by the following example.

Example 2.10 Consider the inverting amplifier of Fig. 2.30*a* together with its equivalent network of Fig. 2.30*b*. The indefinite-admittance matrix of the amplifier is found to be

$$\mathbf{Y} = \begin{bmatrix} y_1 + y_2 + g_i & -y_2 & -y_1 - g_i \\ -y_2 + Ag_o & g_o + y_2 & -Ag_o - g_o \\ -Ag_o - y_1 - g_i & -g_o & y_1 + g_i + (1+A)g_o \end{bmatrix} \tag{2.133}$$

Appealing to (2.94) yields

$$z_{12,33} = \frac{V_o}{y_1 V_{\text{in}}} = \frac{Y_{12,33}}{Y_{uv}} = \frac{Y_{12,33}}{Y_{33}} \tag{2.134}$$

or

$$\frac{V_o}{V_{\text{in}}} = y_1 \frac{Y_{12,33}}{Y_{33}} = \frac{y_1(y_2 - Ag_o)}{(y_1 + g_i)(y_2 + g_o) + (1+A)y_2 g_o} \tag{2.135}$$

As $g_i \to 0$ and $g_o \to \infty$, Eq. (2.135) becomes

$$\frac{V_o}{V_{\text{in}}} = -\frac{Ay_1}{y_1 + (1+A)y_2} \tag{2.136}$$

For the ideal op-amp, we let $A \to \infty$, giving

$$\frac{V_o}{V_{\text{in}}} = -\frac{y_1}{y_2} \tag{2.137}$$

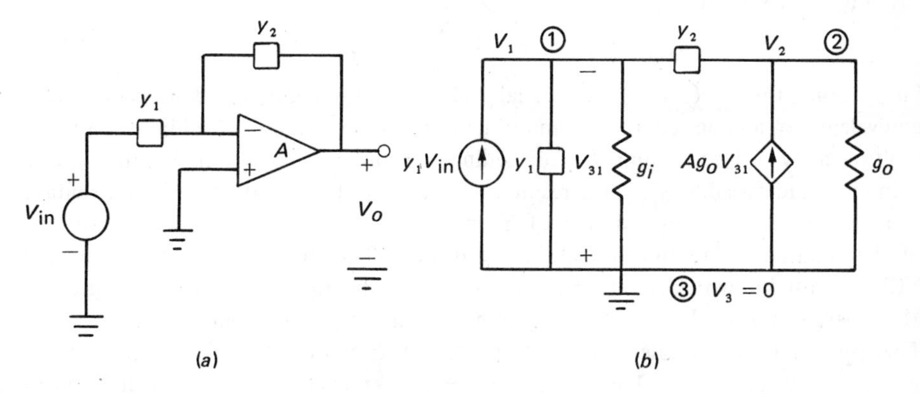

Figure 2.30 (*a*) An inverting amplifier and (*b*) its equivalent network.

For constant y_1 and y_2, (2.137) represents an ideal voltage-controlled voltage source. Thus, the inverting amplifier of Fig. 2.30*a* is a good approximation to the ideal voltage-controlled voltage source.

2.7.1 Voltage-constrained Terminals

Suppose that an ideal voltage-controlled voltage source, abbreviated as VCVS, with a finite gain A is connected between terminals c and d of an n-terminal network, as depicted in Fig. 2.31. The VCVS introduces a constraint on the network by forcing the voltage between terminals c and d to follow that between terminals a and b, so that the relation

$$V_{cd} = V_c - V_d = A V_{ab} = A(V_a - V_b) \tag{2.138}$$

holds, where as before V_x denotes the voltage of terminal x to the reference point and V_{ij} is the potential drop from terminal i to terminal j.

Consider the network equations of (2.1) with $\mathbf{J} = \mathbf{0}$. To exhibit the rows and columns corresponding to terminals *a, b, c,* and *d,* we write

$$\begin{array}{c} \\ \\ a \\ b \\ c \\ d \\ \\ \\ \end{array}
\begin{array}{c} \quad\; a \quad\;\; b \quad\;\; c \quad\;\; d \\
\begin{bmatrix}
\cdots & y_{1a} & y_{1b} & y_{1c} & y_{1d} & \cdots \\
\cdots & \cdots & \cdots & \cdots & \cdots & \cdots \\
\cdots & y_{aa} & y_{ab} & y_{ac} & y_{ad} & \cdots \\
\cdots & y_{ba} & y_{bb} & y_{bc} & y_{bd} & \cdots \\
\cdots & y_{ca} & y_{cb} & y_{cc} & y_{cd} & \cdots \\
\cdots & y_{da} & y_{db} & y_{dc} & y_{dd} & \cdots \\
\cdots & \cdots & \cdots & \cdots & \cdots & \cdots \\
\cdots & y_{na} & y_{nb} & y_{nc} & y_{nd} & \cdots
\end{bmatrix}\end{array}
\begin{bmatrix} V_1 \\ \vdots \\ V_a \\ V_b \\ V_c \\ V_d \\ \vdots \\ V_n \end{bmatrix}
=
\begin{bmatrix} I_1 \\ \vdots \\ I_a \\ I_b \\ I_c \\ I_d \\ \vdots \\ I_n \end{bmatrix} \tag{2.139}$$

From (2.138) we have

$$V_c = A V_a - A V_b + V_d \tag{2.140}$$

Substituting this in (2.139), the variable V_c can be eliminated. This operation is equivalent to adding column c multiplied by A to column a, adding column c multiplied by $-A$ to column b, and adding column c to column d. We then delete column c and variable V_c. This results in a system of n equations in $n-1$ voltage variables. Since each column sum of $\mathbf{Y}$ and the sum of elements of $\mathbf{I}$ are zero, one of the equations is redundant and can be deleted. The current I_c resulting from the VCVS is not known. For our purposes, we delete the equation corresponding to I_c. Moreover, if terminal d is chosen as the reference-potential point, we set $V_d = 0$. This operation is equivalent to deleting column d and variable V_d in Eq. (2.139), yielding a system of $n-1$ equations in $n-1$ unknowns, $n-2$ unknown terminal voltages, and the unknown terminal current $I_d = -I_c$. To compute the $n-2$

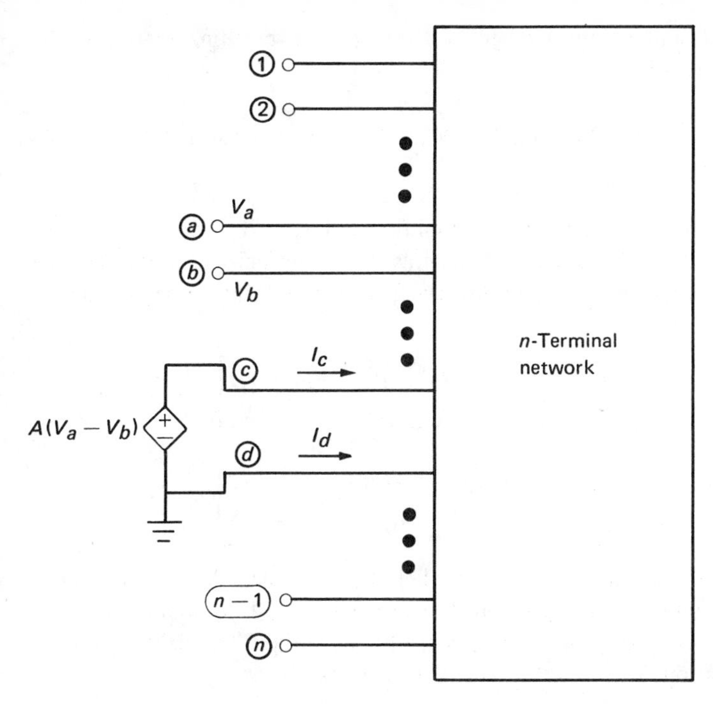

Figure 2.31 The general representation of two voltage-constrained terminals of an n-terminal network.

unknown terminal voltages V_x, $x \neq c, d$, there is no need to include the equation corresponding to I_d. Thus, it can be deleted from (2.139), and (2.139) becomes a system of $n-2$ equations in $n-2$ unknowns V_x ($x = 1, 2, \ldots, n$ and $x \neq c, d$):

$$\begin{array}{c} \\ \\ \\ a \\ b \\ \\ \\ \end{array} \begin{bmatrix} & a & b & \\ \cdots & y_{1a} + Ay_{1c} & y_{1b} - Ay_{1c} & \cdots \\ \cdots & \cdots & \cdots & \cdots \\ \cdots & y_{aa} + Ay_{ac} & y_{ab} - Ay_{ac} & \cdots \\ \cdots & y_{ba} + Ay_{bc} & y_{bb} - Ay_{bc} & \cdots \\ \cdots & \cdots & \cdots & \cdots \\ \cdots & y_{na} + Ay_{nc} & y_{nb} - Ay_{nc} & \cdots \end{bmatrix} \begin{bmatrix} V_1 \\ \vdots \\ V_a \\ V_b \\ \vdots \\ V_n \end{bmatrix} = \begin{bmatrix} I_1 \\ \vdots \\ I_a \\ I_b \\ \vdots \\ I_n \end{bmatrix} \tag{2.141}$$

The coefficient matrix of (2.141) is the admittance matrix of the constrained common-terminal $(n-2)$-port network. The rule for obtaining this matrix for finite gain A can now be stated as follows:

Rule In the unconstrained indefinite-admittance matrix $\mathbf{Y}$ of Fig. 2.31, add column c multiplied by A to column a and column c multiplied by $-A$ to column *b*. Then delete rows and columns *c* and *d* from **Y**. The resulting matrix is the coefficient matrix of (2.141).

We illustrate the above results by analyzing several active networks in which an op-amp is used as a VCVS.

Example 2.11 Consider the inverting amplifier of Fig. 2.30*a* whose equivalent network is presented in Fig. 2.32. The indefinite-admittance matrix of the unconstrained network can be obtained directly from Fig. 2.32 by inspection and is given by

$$\begin{bmatrix} y_1+y_2+g_i & -y_2 & 0 & -y_1-g_i \\ -y_2 & y_2+g_o & -g_o & 0 \\ 0 & -g_o & g_o & 0 \\ -y_1-g_i & 0 & 0 & y_1+g_i \end{bmatrix} \tag{2.142}$$

Since $a=4$, $b=1$, $c=3$, and $d=4$, according to the rule we add column 3 multiplied by A to column 4 and column 3 multiplied by $-A$ to column 1. Then we delete rows and columns 3 and 4 in the resulting matrix. This yields a desired system of nodal equations for the constrained network as

$$\begin{bmatrix} y_1+y_2+g_i & -y_2 \\ Ag_o-y_2 & y_2+g_o \end{bmatrix}\begin{bmatrix} V_1 \\ V_2 \end{bmatrix}=\begin{bmatrix} y_1 V_{\text{in}} \\ 0 \end{bmatrix} \tag{2.143}$$

Solving for V_2 yields the voltage gain

$$\frac{V_2}{V_{\text{in}}}=\frac{V_o}{V_{\text{in}}}=\frac{y_1(y_2-Ag_o)}{(y_1+g_i)(y_2+g_o)+(1+A)y_2g_o} \tag{2.144}$$

confirming (2.135)

Example 2.12 Figure 2.33*a* is a noninverting amplifier containing an op-amp in the noninverting mode to provide the VCVS characteristic. The equivalent network

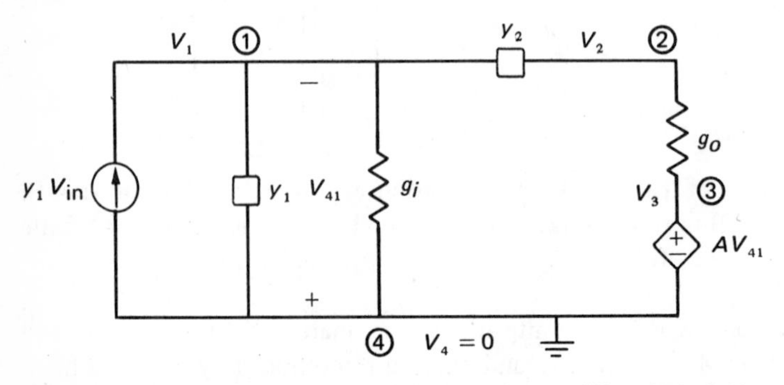

Figure 2.32 An equivalent network of the inverting amplifier of Fig. 2.30*a*.

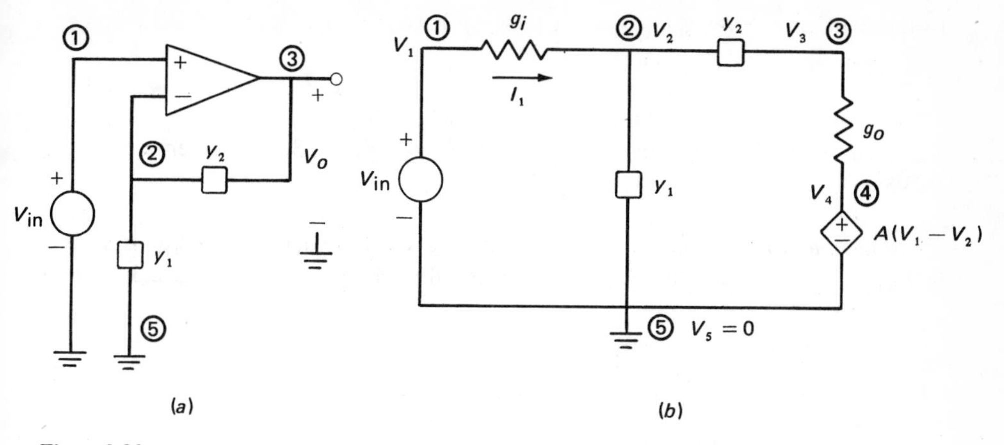

Figure 2.33 (*a*) A noninverting amplifier and (*b*) its equivalent network.

of the amplifier is shown in Fig. 2.33*b*. The indefinite-·dmittance matrix of the unconstrained network can be written down directly from Fig. 2.33*b* by inspection and is given by

$$\begin{bmatrix} g_i & -g_i & 0 & 0 & 0 \\ -g_i & y_1 + y_2 + g_i & -y_2 & 0 & -y_1 \\ 0 & -y_2 & y_2 + g_o & -g_o & 0 \\ 0 & 0 & -g_o & g_o & 0 \\ 0 & -y_1 & 0 & 0 & y_1 \end{bmatrix} \tag{2.145}$$

To obtain the constrained admittance matrix, we add column 4 multiplied by A to column 1 and column 4 multiplied by $-A$ to column 2. We then delete rows and columns 4 and 5. This yields a system of nodal equations for the constrained network as

$$\begin{bmatrix} g_i & -g_i & 0 \\ -g_i & y_1 + y_2 + g_i & -y_2 \\ -Ag_o & Ag_o - y_2 & y_2 + g_o \end{bmatrix} \begin{bmatrix} V_1 \\ V_2 \\ V_3 \end{bmatrix} = \begin{bmatrix} I_1 \\ 0 \\ 0 \end{bmatrix} \tag{2.146}$$

Since $V_1 = V_{in}$ and I_1 is unknown, the first equation need not be considered and the last two become

$$\begin{bmatrix} y_1 + y_2 + g_i & -y_2 \\ Ag_o - y_2 & y_2 + g_o \end{bmatrix} \begin{bmatrix} V_2 \\ V_3 \end{bmatrix} = \begin{bmatrix} g_i V_{in} \\ Ag_o V_{in} \end{bmatrix} \tag{2.147}$$

Solving for V_3 gives the voltage gain

$$\frac{V_3}{V_{in}} = \frac{V_o}{V_{in}} = \frac{Ag_o(y_1 + y_2) + y_2 g_i}{(y_1 + y_2 + g_i)g_o + (y_1 + g_i)y_2 + Ay_2 g_o} \tag{2.148}$$

In the limit, as $g_o \to \infty$, $g_i \to 0$, and $A \to \infty$, we have

$$\frac{V_o}{V_{\text{in}}} = 1 + \frac{y_1}{y_2} \tag{2.149}$$

Thus, for the ideal op-amp and constant y_1/y_2, Fig. 2.33*a* represents an ideal VCVS.

Example 2.13 A general configuration of a second-order active filter is shown in Fig. 2.34*a*. The equivalent network of the filter is presented in Fig. 2.34*b*. The

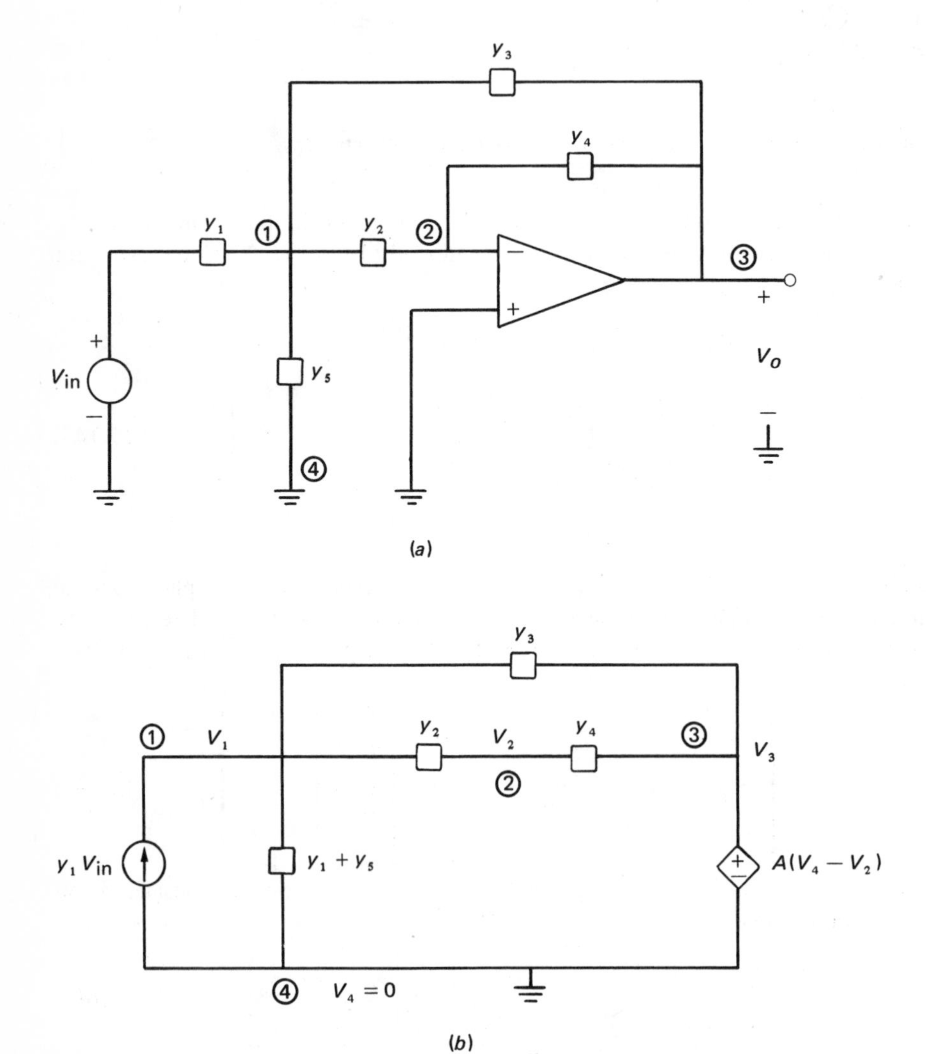

Figure 2.34 (*a*) A general configuration of a second-order active filter and (*b*) its equivalent network.

unconstrained indefinite-admittance matrix is obtained by inspection directly from Fig. 2.34*a* after a simple source transformation or from Fig. 2.34*b*:

$$\begin{bmatrix} y_1+y_2+y_3+y_5 & -y_2 & -y_3 & -y_1-y_5 \\ -y_2 & y_2+y_4 & -y_4 & 0 \\ -y_3 & -y_4 & y_3+y_4 & 0 \\ -y_1-y_5 & 0 & 0 & y_1+y_5 \end{bmatrix} \tag{2.150}$$

in which the op-amp is assumed to have infinite input impedance, zero output impedance, and a finite gain A. To obtain the admittance matrix of the constrained network, according to the rule we add column 3 multiplied by $-A$ to column 2 and then delete rows and columns 3 and 4. The resulting equation becomes

$$\begin{bmatrix} y_1+y_2+y_3+y_5 & Ay_3-y_2 \\ -y_2 & y_2+(1+A)y_4 \end{bmatrix}\begin{bmatrix} V_1 \\ V_2 \end{bmatrix}=\begin{bmatrix} y_1 V_{\text{in}} \\ 0 \end{bmatrix} \tag{2.151}$$

Solving for V_2 and substituting it in $V_3=-AV_2$ yield the voltage gain

$$\frac{V_3}{V_{\text{in}}}=\frac{V_o}{V_{\text{in}}}=-\frac{Ay_1y_2}{(y_1+y_2+y_3+y_5)(y_2+y_4+Ay_4)-y_2(y_2-Ay_3)} \tag{2.152}$$

In the limit, as $A\to\infty$, the gain reduces to

$$\frac{V_o}{V_{\text{in}}}=-\frac{y_1y_2}{y_4(y_1+y_2+y_3+y_5)+y_2y_3} \tag{2.153}$$

As an application, consider the band-pass filter of Fig. 2.35. Comparing this with Fig. 2.34*a*, we can identify that $y_1=G_1$, $y_2=sC_1$, $y_3=sC_2$, $y_4=G_2$, and $y_5=0$. Substituting these in (2.152) yields

$$\frac{V_o}{V_{\text{in}}}=-\frac{C_1G_1s}{C_1C_2s^2+(C_1+C_2)G_2s+G_1G_2+[C_1C_2s^2+(C_1G_1+C_2G_2+C_1G_2)s+G_1G_2]/A} \tag{2.154}$$

For the ideal op-amp, the gain is simplified to

$$\frac{V_o}{V_{\text{in}}}=-\frac{C_1G_1s}{C_1C_2s^2+(C_1+C_2)G_2s+G_1G_2} \tag{2.155}$$

Example 2.14 Consider the negative feedback network of Fig. 2.36*a*, which can realize the general biquadratic function of an active filter. The equivalent network of the filter is presented in Fig. 2.36*b*, assuming infinite input impedance and zero output impedance for the op-amp. The indefinite-admittance matrix of the

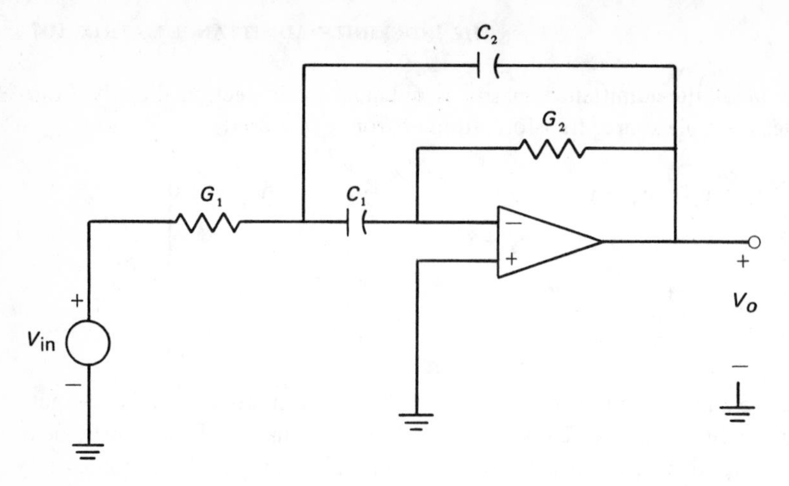

Figure 2.35 A band-pass active filter.

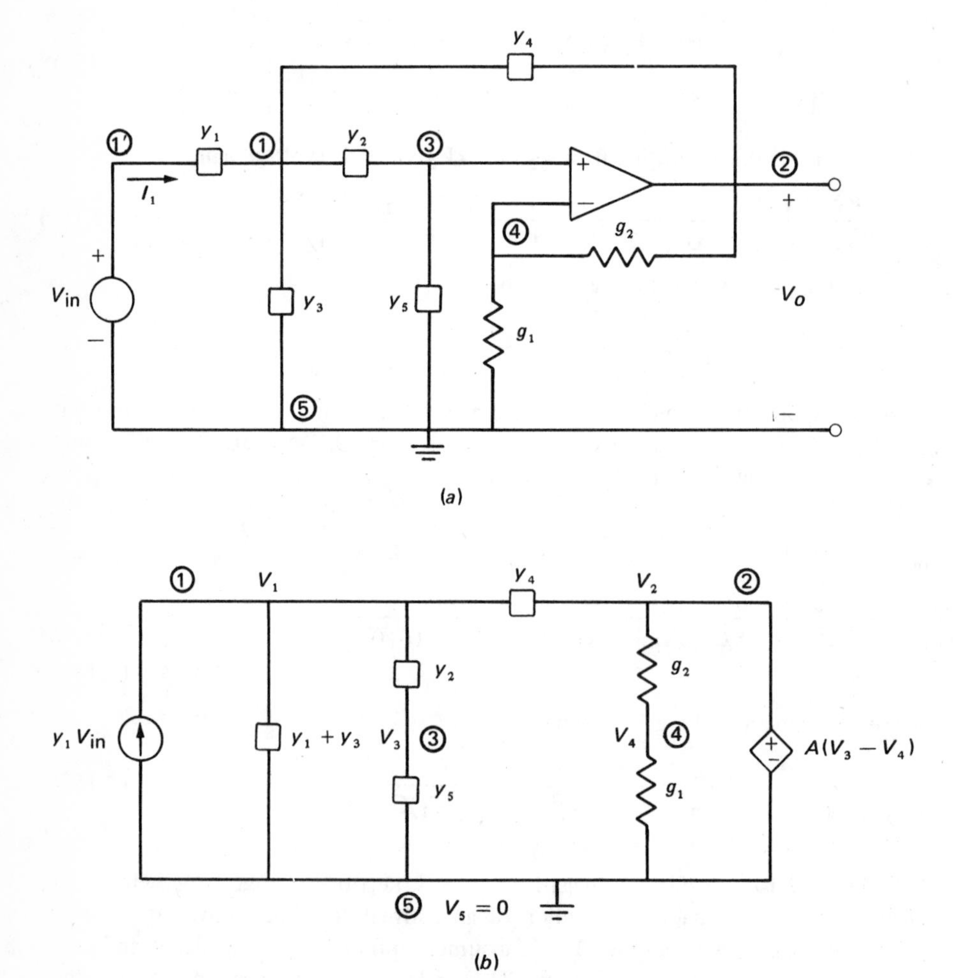

Figure 2.36 (*a*) A general biquadratic active filter and (*b*) its equivalent network.

unconstrained network is obtained from Fig. 2.36*a* after replacing the source combination V_{in} and y_1 by its Norton equivalent or from Fig. 2.36*b*:

$$\begin{bmatrix} y_1+y_2+y_3+y_4 & -y_4 & -y_2 & 0 & -y_1-y_3 \\ -y_4 & g_2+y_4 & 0 & -g_2 & 0 \\ -y_2 & 0 & y_2+y_5 & 0 & -y_5 \\ 0 & -g_2 & 0 & g_1+g_2 & -g_1 \\ -y_1-y_3 & 0 & -y_5 & -g_1 & g_1+y_1+y_3+y_5 \end{bmatrix} \tag{2.156}$$

To obtain the admittance matrix of the constrained network, according to the rule we add column 2 multiplied by A to column 3 and column 2 multiplied by $-A$ to column 4. We then delete rows and columns 2 and 5. The resulting nodal equation is given by

$$\begin{bmatrix} y_1+y_2+y_3+y_4 & -y_2-Ay_4 & Ay_4 \\ -y_2 & y_2+y_5 & 0 \\ 0 & -Ag_2 & g_1+(1+A)g_2 \end{bmatrix} \begin{bmatrix} V_1 \\ V_3 \\ V_4 \end{bmatrix} = \begin{bmatrix} y_1 V_{\text{in}} \\ 0 \\ 0 \end{bmatrix} \tag{2.157}$$

Solving for V_3 and V_4 gives

$$V_3 = \frac{y_1 y_2 (g_1 + Ag_2 + g_2) V_{\text{in}}}{q} \tag{2.158a}$$

$$V_4 = \frac{y_1 y_2 A g_2 V_{\text{in}}}{q} \tag{2.158b}$$

where

$$q = (g_1 + g_2 + Ag_2)[y_5(y_1 + y_2 + y_3 + y_4) + y_2(y_1 + y_3 + y_4) - Ay_2 y_4] + A^2 g_2 y_2 y_4 \tag{2.158c}$$

The voltage gain of the filter is obtained as

$$\frac{V_o}{V_{\text{in}}} = \frac{V_2}{V_{\text{in}}} = \frac{A(V_3 - V_4)}{V_{\text{in}}} = \frac{Ay_1 y_2 (g_1 + g_2)}{q}$$

$$= \frac{k y_1 y_2}{[(y_2 + y_5)(y_1 + y_3 + y_4) + y_2 y_5](1 + k/A) - k y_2 y_4} \tag{2.159}$$

where $k = 1 + g_1/g_2$

As an application, consider the low-pass filter of Fig. 2.37, from which we can identify $y_1 = G_1$, $y_2 = G_2$, $y_3 = 0$, $y_4 = sC_1$, and $y_5 = sC_2$. Substituting these in (2.159), we get

$$\frac{V_o}{V_{\text{in}}} = \frac{kG_1 G_2}{[(G_1 + G_2 + sC_1)(G_2 + sC_2) - G_2^2](1 + k/A) - kC_1 G_2 s} \tag{2.160}$$

which is recognized as the gain function of a low-pass characteristic.

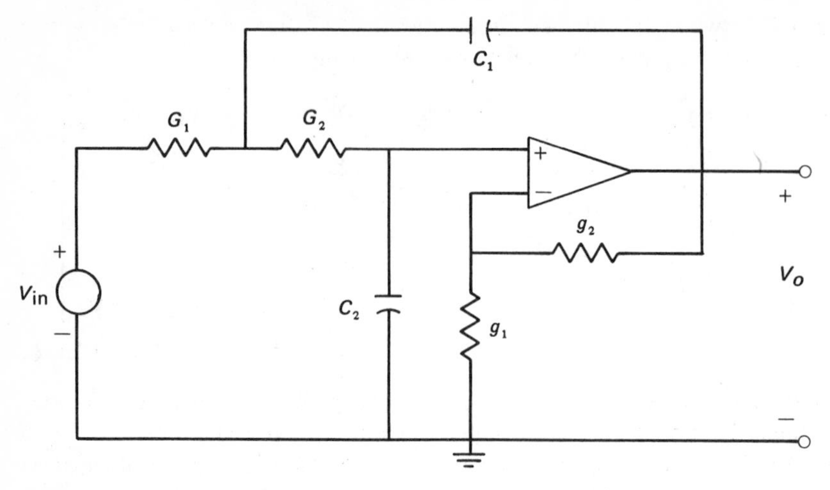

Figure 2.37 A low-pass active filter.

2.7.2 The Ideal Operational Amplifier Constraint

In the preceding section, we considered the situation where the terminal voltages are constrained by an ideal VCVS with a finite gain A, as depicted in Fig. 2.31. In this section, we present rules for writing down the admittance matrix of the constrained network when ideal op-amps are employed.

Refer again to the general configuration of Fig. 2.31, where the VCVS represents the ideal op-amp with $A \to \infty$. The noninverting terminal is terminal a, and the inverting terminal is terminal b. Since the output of the ideal op-amp must be finite, $V_{cd} = AV_{ab} = A(V_a - V_b)$ implies that the potential difference between the input terminals a and b must be zero, requiring that

$$V_a = V_b \tag{2.161}$$

By substituting (2.161) in (2.139), the variable V_b can be eliminated. This operation is equivalent to adding column b to column a and then deleting column b and V_b. This results in a system of n equations in $n-1$ voltage variables. As before, one of the equations is redundant and can be deleted. Since the current I_c is not known, we delete the equation corresponding to I_c. Moreover, since terminal d is chosen as the reference-potential point, we can delete column d and V_d and the equation corresponding to $I_d = -I_c$. The above procedure can be summarized by the following simple rule:

Rule (Ideal op-amp constraint) In the unconstrained indefinite-admittance matrix **Y** of Fig. 2.31, add column b to column a (the controlling terminals), and delete rows c and d (the controlled terminals) and columns b and d ($b \neq d$). The resulting matrix is the admittance matrix of the constrained network. For grounded terminal a or b ($a = d$ or $b = d$), the constrained admittance matrix can be obtained from **Y** simply by deleting columns a and b and rows c and d.

We illustrate the above rule by the following examples, using the same networks as in the previous examples.

Example 2.15 In Fig. 2.33*a*, assume that the op-amp is ideal. The unconstrained indefinite-admittance matrix can be obtained by inspection directly from Fig. 2.33*a*, and is given by

$$\begin{matrix} & 1 & 2 & 3 & 5 \\ 1 & 0 & 0 & 0 & 0 \\ 2 & 0 & y_1+y_2 & -y_2 & -y_1 \\ 3 & 0 & -y_2 & y_2 & 0 \\ 5 & 0 & -y_1 & 0 & y_1 \end{matrix} \tag{2.162}$$

According to the rule, to obtain the constrained admittance matrix we add column 2 to column 1 and then delete rows 3 and 4 corresponding to the constrained terminals 3 and 5 and columns 2 and 4 corresponding to terminals 2 and 5. The equation corresponding to the second row of the resulting matrix becomes

$$(y_1+y_2)V_1 - y_2 V_3 = 0 \tag{2.163}$$

which in conjunction with the facts that $V_1 = V_{\text{in}}$ and $V_3 = V_o$ gives

$$\frac{V_o}{V_{\text{in}}} = 1 + \frac{y_1}{y_2} \tag{2.164}$$

confirming (2.149).

Example 2.16 In Fig. 2.34*a*, replace the source combination of V_{in} and y_1 by its Norton equivalent network. For the ideal op-amp, the unconstrained indefinite-admittance matrix can easily be obtained by inspection and is given by (2.150). Since one of the input terminals of the op-amp is grounded, to obtain the constrained admittance matrix we delete rows 3 and 4 corresponding to the controlled terminals 3 and 4, and columns 2 and 4 corresponding to the controlling terminals 2 and 4. The resulting nodal equation is obtained as

$$\begin{matrix} & 1 & 3 \\ 1 & y_1+y_2+y_3+y_5 & -y_3 \\ 2 & -y_2 & -y_4 \end{matrix} \begin{bmatrix} V_1 \\ V_3 \end{bmatrix} = \begin{bmatrix} y_1 V_{\text{in}} \\ 0 \end{bmatrix} \tag{2.165}$$

Solving for V_3 yields

$$\frac{V_3}{V_{\text{in}}} = \frac{V_o}{V_{\text{in}}} = -\frac{y_1 y_2}{y_4(y_1+y_2+y_3+y_5)+y_2 y_3} \tag{2.166}$$

confirming (2.153).

Example 2.17 In Fig. 2.36*a*, replace the source combination of V_{in} and y_1 by its Norton equivalent network and assume that the op-amp is ideal. The unconstrained indefinite-admittance matrix can be obtained by inspection, and is given by (2.156). To obtain the constrained admittance matrix, we add column 4 to column 3 and then delete rows 2 and 5 (controlled terminals) and columns 4 and 5. The resulting matrix is the constrained admittance matrix of the network. The corresponding nodal system of equations is found to be

$$\begin{array}{c} \\ 1 \\ 3 \\ 4 \end{array}\begin{array}{c} \begin{array}{ccc} 1 & 2 & 3+4 \end{array} \\ \begin{bmatrix} y_1+y_2+y_3+y_4 & -y_4 & -y_2 \\ -y_2 & 0 & y_2+y_5 \\ 0 & -g_2 & g_1+g_2 \end{bmatrix} \end{array} \begin{bmatrix} V_1 \\ V_2 \\ V_3 \end{bmatrix} = \begin{bmatrix} y_1 V_{\text{in}} \\ 0 \\ 0 \end{bmatrix} \tag{2.167}$$

Solving for V_2 yields

$$\frac{V_2}{V_{\text{in}}} = \frac{V_o}{V_{\text{in}}} = \frac{k y_1 y_2}{(y_2+y_5)(y_1+y_3+y_4) + y_2 y_5 - k y_2 y_4} \tag{2.168}$$

where $k = 1 + g_1/g_2$, confirming (2.159) for $A = \infty$.

We remark that in writing down the unconstrained indefinite-admittance matrix, the source combination of V_{in} and y_1 need not be replaced by its Norton equivalent network. The only difference is that we have to add a row and a column corresponding to the input terminal 1′. In so doing, the constrained nodal system of equations becomes

$$\begin{array}{c} \\ 1' \\ 1 \\ 3 \\ 4 \end{array}\begin{array}{c} \begin{array}{cccc} 1' & 1 & 2 & 3+4 \end{array} \\ \begin{bmatrix} y_1 & -y_1 & 0 & 0 \\ -y_1 & y_1+y_2+y_3+y_4 & -y_4 & -y_2 \\ 0 & -y_2 & 0 & y_2+y_5 \\ 0 & 0 & -g_2 & g_1+g_2 \end{bmatrix} \end{array} \begin{bmatrix} V_{\text{in}} \\ V_1 \\ V_2 \\ V_3 \end{bmatrix} = \begin{bmatrix} I_1 \\ 0 \\ 0 \\ 0 \end{bmatrix} \tag{2.169}$$

Since the current I_1 due to the source V_{in} is unknown and since V_{in} is known, the last three equations of (2.169) can be rewritten as in (2.167), yielding the same set of equations. The details are left as an exercise (see Prob. 2.30).

In the above discussion, we have demonstrated how to obtain the constrained nodal system of equations. The procedure requires that we first write down the unconstrained indefinite-admittance matrix and then apply the rule to derive the desired constrained admittance matrix. A moment's thought would indicate that we could obtain the constrained admittance matrix directly from the network without the necessity of first writing down the unconstrained indefinite-admittance matrix. For example, the coefficient matrix of (2.165) can be written down directly from Fig. 2.34*a* as follows: The (1, 1)-element denotes the sum of admittances of all the branches incident at the terminal or node 1. The (1, 2)- and (2, 1)-elements are the negative of the admittances of the branches connecting the terminals 1 and 3, and 2 and 1, respectively. Finally, the (2, 2)-element represents the negative of the admittance of the branch connecting the terminals 2 and 3. In general, if $y(i, j)$ denotes the element of the constrained admittance matrix in the row corresponding to terminal i and in the column corresponding to terminal j, then it can be written down directly from the network by the formulas [Chen (1978)]

$$y(i, i) = \Sigma \text{ admittances incident at terminal } i \tag{2.170a}$$

and for $i \neq j$,

$$y(i, j) = -\Sigma \text{ admittances connected between terminals } i \text{ and } j \tag{2.170b}$$

We emphasize that $y(i, j)$ does not necessarily represent the ith row and jth column element. In (2.167), for example, the diagonal elements are $y(1, 1) = y_1 + y_2 + y_3 + y_4$, $y(3, 2) = 0$, and $y(4, 3+4) = y(4, 3) + y(4, 4) = 0 + g_1 + g_2 = g_1 + g_2$. The (2, 3)-element is $y(3, 3+4) = y(3, 3) + y(3, 4) = y_2 + y_5$ and the (1, 3)-element is $y(1, 3+4) = y(1, 3) + y(1, 4) = -y_2 - 0 = -y_2$. These elements are obtained directly from the network of Fig. 2.36*a*. In a similar fashion, we can obtain the other elements.

Example 2.18 One of the first three-amplifier biquads, proposed by Kerwin, Huelsman, and Newcomb (1967), is shown in Fig. 2.38. Assume that the three op-amps are ideal. To obtain the constrained admittance matrix from the unconstrained indefinite-admittance matrix, according to the rule we add column 1 to column 7 and then delete columns 1, 3, 5, and 8 and rows 2, 4, 6, and 8. The resulting nodal system of equations becomes

$$\begin{array}{c} \\ 1 \\ 3 \\ 5 \\ 7 \end{array} \begin{array}{c} \begin{array}{cccc} 2 & 4 & 6 & 1+7 \end{array} \\ \begin{bmatrix} -G_3 & 0 & -G_2 & G_2 + G_3 \\ -G_4 & -sC_1 & 0 & 0 \\ 0 & -G_5 & -sC_2 & 0 \\ 0 & -G_6 & 0 & G_1 + G_6 \end{bmatrix} \end{array} \begin{bmatrix} V_2 \\ V_4 \\ V_6 \\ V_7 \end{bmatrix} = \begin{bmatrix} 0 \\ 0 \\ 0 \\ G_1 V_{in} \end{bmatrix} \tag{2.171}$$

The (1, 1)-element, for example, is $y(1, 2) = -G_3$, and the (1, 4)-element is $y(1, 1+7) = y(1, 1) + y(1, 7) = G_2 + G_3 + 0 = G_2 + G_3$, where $G_i = 1/R_i$ $(i = 1, 2, 3, 4, 5, 6)$. Solving for V_2, V_4, and V_6 yields

$$\frac{V_2}{V_{in}} = \frac{G_1 C_1 C_2 (G_2 + G_3) s^2}{C_1 C_2 G_3 (G_1 + G_6) s^2 + C_2 G_4 G_6 (G_2 + G_3) s + G_2 G_4 G_5 (G_1 + G_6)} \tag{2.172a}$$

$$\frac{V_4}{V_{in}} = -\frac{G_1 C_2 G_4 (G_2 + G_3) s}{C_1 C_2 G_3 (G_1 + G_6) s^2 + C_2 G_4 G_6 (G_2 + G_3) s + G_2 G_4 G_5 (G_1 + G_6)} \tag{2.172b}$$

$$\frac{V_6}{V_{in}} = \frac{G_1 G_4 G_5 (G_2 + G_3)}{C_1 C_2 G_3 (G_1 + G_6) s^2 + C_2 G_4 G_6 (G_2 + G_3) s + G_2 G_4 G_5 (G_1 + G_6)} \tag{2.172c}$$

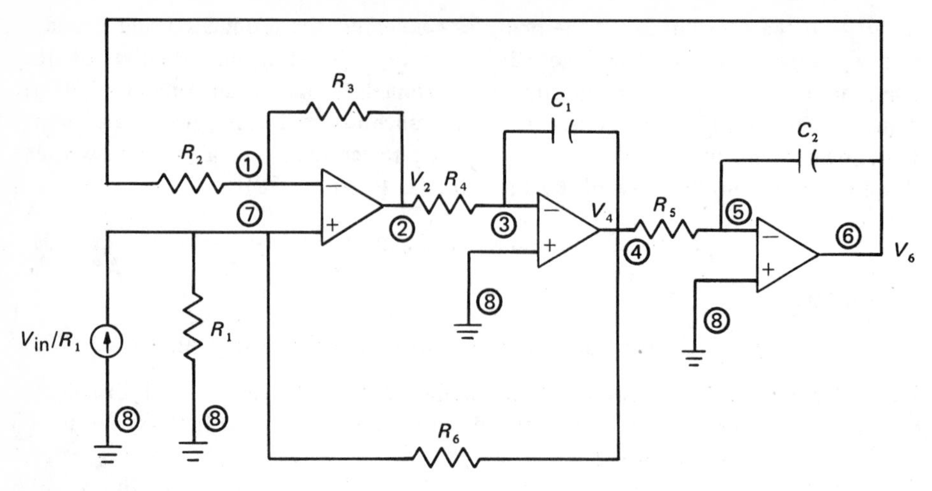

Figure 2.38 A three-amplifier biquad.

which are recognized as the functions possessing the high-pass, band-pass, and low-pass characteristics, respectively.

2.8 GENERALIZED NORTON'S THEOREM

In this section, we establish a generalized version of Norton's theorem for multiterminal active networks. Specifically, we show that any active multiterminal network can be represented as a passive multiterminal network in parallel with equivalent current sources.

Let N be an active n-terminal network connected in parallel with an arbitrary passive n-terminal network M that is initially relaxed, as shown in Fig. 2.39*a*. Suppose that additional current sources J_k are inserted between the common terminals and the reference-potential point as depicted in Fig. 2.39*b*. Assume that it is possible to choose the currents J_k, whose values will be determined shortly, so that all the terminal voltages $\hat{V}_k$ in Fig. 2.39*b* are zero. Since M is passive, the vanishing of its terminal voltages implies the vanishing of its terminal currents. Thus, M may be disconnected from the other part without affecting conditions to the right. This shows that J_k is equal to the kth terminal current of N when all of its terminal voltages vanish. In other words, J_k is the current flowing in the kth terminal of N when all of its terminals are short-circuited to the reference point. By the principle of superposition, each $\hat{V}_k$ in Fig. 2.39*b* can be regarded as the sum of two components $\hat{V}_k = \hat{V}'_k + \hat{V}''_k$, where $\hat{V}'_k$ is the kth terminal voltage when all the current sources J_k are removed, and $\hat{V}''_k$ is the kth terminal voltage when all independent current sources in N are open-circuited, all independent voltage sources are short-circuited, and all initial conditions are set to zero. Since $\hat{V}_k = 0$, we have $-\hat{V}''_k = \hat{V}'_k$, which is also equal to the kth terminal voltage V_k in Fig. 2.39*a*, or

$\hat{V}_k'' = -\hat{V}_k' = -V_k$. Thus, we have established that if we place the current sources J_k, being equal to the short-circuit terminal currents of N, in the positions and with reference directions as indicated in Fig. 2.39*b*, and open-circuit all independent current sources, short-circuit all independent voltage sources, and set all initial conditions to zero in N, then we shall obtain terminal voltages $-V_k$ in Fig. 2.39*b*. Suppose that we reverse the reference directions of J_k; then we must get V_k at the kth common terminal in Fig. 2.39*b*. Consequently, we can state the following:

> An active multiterminal network N can be represented as a parallel connection of a passive multiterminal network $\tilde{N}$ and current sources J_k as shown in Fig. 2.40. The passive network $\tilde{N}$ is derived from N by open-circuiting all of its

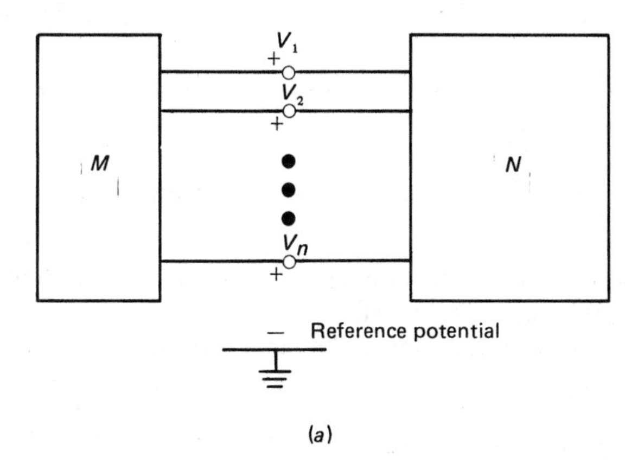

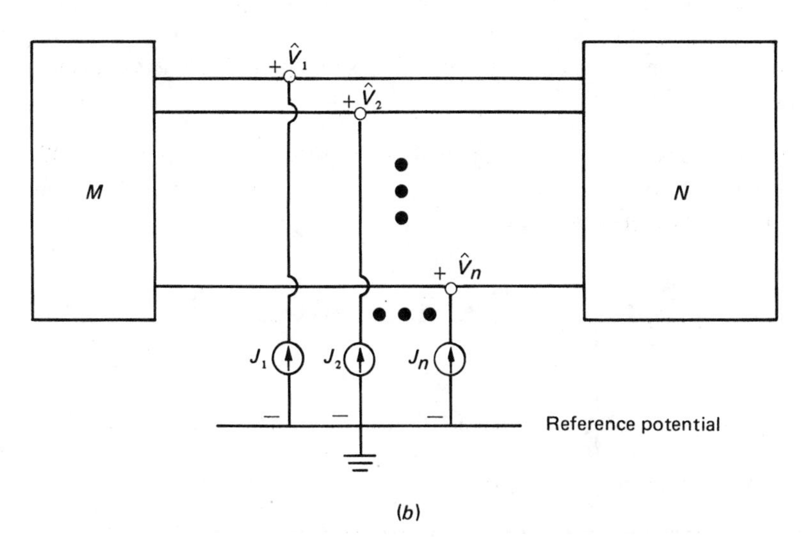

Figure 2.39 (*a*) An active n-terminal network N connected in parallel with an arbitrary passive initially relaxed n-terminal network M and (*b*) the insertion of additional current sources between the common terminals and the reference-potential point.

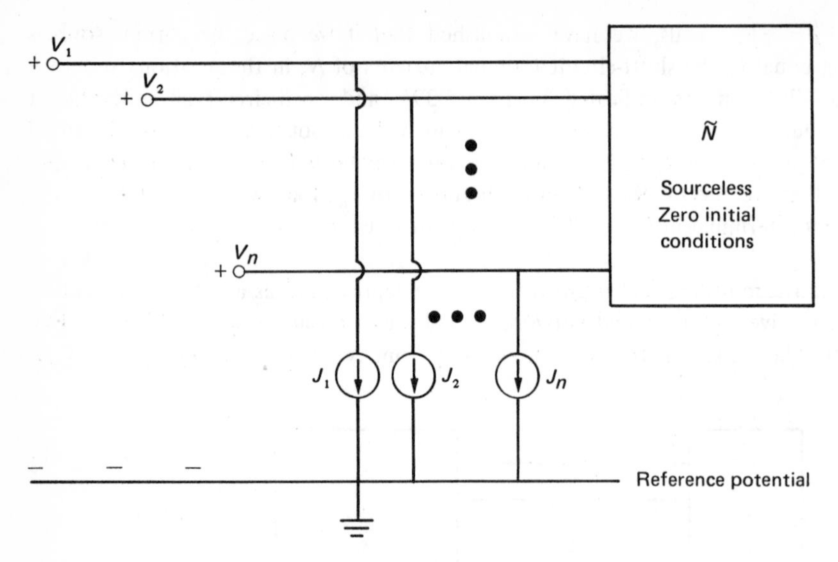

Figure 2.40 The generalized Norton's equivalent network of an n-terminal network.

independent current sources, short-circuiting all of its independent voltage sources, and reducing all of its initial conditions to zero. The value of the current source J_k is equal to the current flowing in the kth terminal of N when all of its terminals are short-circuited to the reference-potential point.

We remark that in deriving the result we assumed that the multiterminal network M was passive and initially relaxed. As a matter of fact, this assumption is not necessary, and the result applies equally well if M is active. The only restriction is that M and N not be magnetically coupled. The justification of this is too long to warrant its inclusion here.

Example 2.19 Consider the three-terminal active network N of Fig. 2.41. We wish to determine its Norton equivalent network. For convenience, we choose terminal 3 as the reference-potential point.

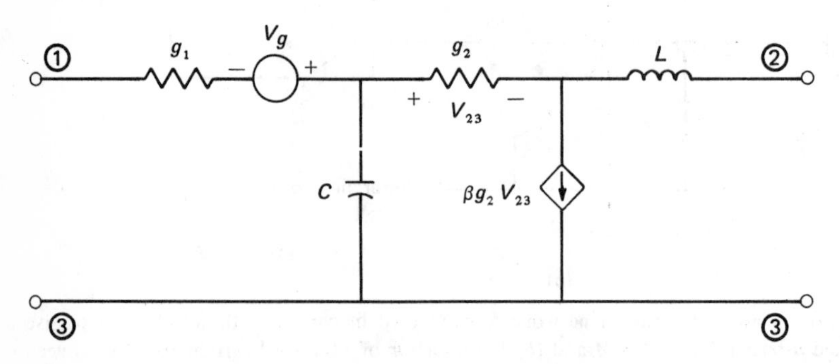

Figure 2.41 A three-terminal active network used to illustrate the determination of its Norton equivalent network.

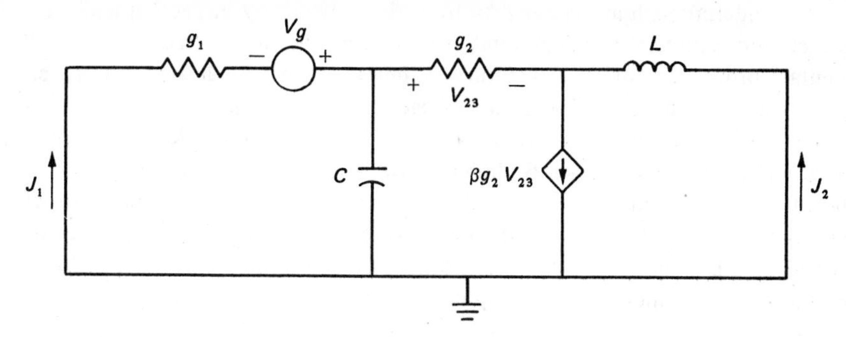

Figure 2.42 The network used to compute the equivalent current sources J_1 and J_2.

To obtain the values of the current sources J_1 and J_2, we short-circuit terminals 1 and 2 to terminal 3 as shown in Fig. 2.42. Referring to Fig. 2.5, as discussed in Example 2.9, we have

$$J_1 = \frac{V_g}{\hat{z}_{11,55}} = \frac{g_1\left[(1-\beta)g_2LCs^2 + sC + g_2\right]V_g}{q(s)} \tag{2.173a}$$

$$J_2 = -y_{14,55}V_g = \frac{(\beta-1)g_1g_2V_g}{q(s)} \tag{2.173b}$$

$q(s)$ being given in (2.130b). The corresponding passive network $\tilde{N}$ is derived from N by short-circuiting the voltage source V_g. Thus, N can be represented equivalently by a three-terminal network as indicated in Fig. 2.43.

2.9 SUMMARY

In this chapter, we described a useful characterization of the external behavior of a multiterminal network in terms of its indefinite-admittance matrix, and we presented rules for its derivation. The rules included a procedure for writing down

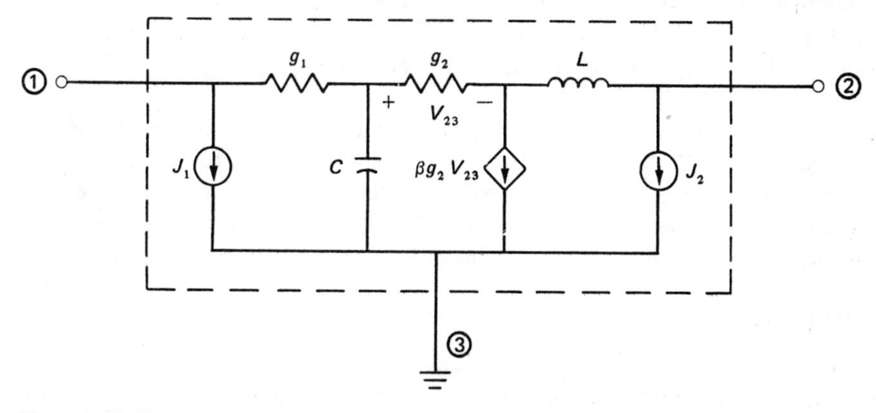

Figure 2.43 The Norton equivalent network of the three-terminal network of Fig. 2.41.

the primitive indefinite-admittance matrix from the network by inspection and ways to contract and suppress terminals and their effects on the matrix. One of the fundamental properties of the indefinite-admittance matrix is that it is an equicofactor matrix, the sum of elements of each row or column being zero. This is a direct consequence of the fact that in solving an electrical network problem, it is immaterial which terminal or node is chosen as the reference-potential point.

One of the most common and useful three-terminal devices is the transistor. If any one of its terminals is chosen as the reference-potential point, the device can be transformed into a common-terminal two-port network, yielding different configurations of the transistor model. We have demonstrated how the various configurations are interrelated.

The usefulness of the indefinite-admittance-matrix formulation of network equations lies in the fact that it facilitates the computations of the driving-point or transfer functions between any pair of terminals or nodes or from any pair of nodes to any other pair of nodes in the network. For this we introduced the concept of the second-order cofactors and derived explicit formulas for computing the network functions. We showed that these formulas can be expressed compactly and elegantly in terms of the ratios of the first- and/or second-order cofactors of the elements of the indefinite-admittance matrix.

An important network element that is used extensively to perform a wide variety of functions is the operational amplifier. In practice, the ideal op-amp is a good approximation to the physical op-amp and can be used with considerable simplification in the design and analysis of various networks where op-amps are employed. For this we presented rules for obtaining the admittance matrix of the constrained network from the indefinite-admittance matrix of the corresponding unconstrained network.

Finally, we established a generalized version of Norton's theorem by showing that any multiterminal network can be represented equivalently as a passive network in parallel with current sources that are short-circuit currents of the original multiterminal network. Clearly, an analogous extension of Thévenin's theorem can be obtained in a similar fashion by making use of the series connection of multiterminal networks. However, if we demand that series connection should correspond to an addition of the impedance parameters of the component multiterminal networks, then the desired result cannot be achieved in general without the use of ideal transformers. Thus, the result does not seem to be satisfactory. For completeness, we state the theorem as follows:

> An active multiterminal network N can be represented as a series connection of a passive network $\tilde{N}$, which has the significance as in Norton's theorem, and voltage sources whose values are equal to the open-circuit terminal potentials when all the terminals of N are open-circuited. The voltage sources are connected in series with the corresponding terminals of the passive network $\tilde{N}$.

PROBLEMS

2.1 With Eq. (2.3), confirm that the elements y_{k2} and y_{k3} $(k = 1, 2, 3)$ of the indefinite-admittance matrix of the three-terminal network of Fig. 2.2 are those given in Eq. (2.5).

2.2 Show that the indefinite-admittance matrix of the four-terminal network of Fig. 2.7 is given by Eq. (2.22).

2.3 Compute the indefinite-admittance matrix of the network of Fig. 2.9 with respect to the terminals or nodes 1, 4, and 5.

2.4 Show that the indefinite-admittance matrix of the ideal gyrator of Fig. 2.8 is that given in Eq. (2.24).

2.5 Let M be a multiterminal network derived from another one N by joining together terminals i and j. Show that the indefinite-admittance matrix of M can be deduced from that of N by adding row j to row i, column j to column i, and then deleting row j and column j in the resulting matrix.

Hint: Make use of Eq. (2.1).

2.6 Figure 2.24 is an equivalent network of the common-emitter transistor configuration.

(*a*) Write down by inspection its primitive indefinite-admittance matrix.

(*b*) Suppress the inaccessible terminal corresponding to node d. Compare your result with Eq. (2.68).

2.7 The network of a difference amplifier is presented in Fig. 2.44. Write down its primitive indefinite-admittance matrix. By using this matrix, compute the short-circuit admittance matrix of the amplifier.

2.8 Repeat Prob. 2.7 by computing the open-circuit impedance matrix.

2.9 By applying the limiting process used in the derivation of Eq. (2.114), derive the formula (2.115).

2.10 By using the indefinite-admittance matrix, compute the short-circuit admittance parameters of the two-port network of Fig. 2.45.

2.11 By using Eq. (2.36), compute $\hat{z}_{11,33}$ and $\tilde{z}_{22,33}$ as defined in Example 2.8. Compare your results with Eqs. (2.120) and (2.122).

2.12 Compute the hybrid parameters and the open-circuit impedance parameters of the two-port network of Fig. 2.45.

2.13 Determine the short-circuit admittance parameters of the two-port network of Fig. 2.46.

2.14 Determine the open-circuit impedance parameters of the two-port network of Fig. 2.46.

2.15 Refer to Fig. 2.27. A three-terminal network can be formed with respect to the nodes 1, 2, and 3. Determine its Norton equivalent three-terminal network.

2.16 Connect a voltage source V_g between nodes 1 and 5 in Fig. 2.9, so that there is a voltage rise from node 5 to node 1. Determine a three-terminal Norton equivalent network with respect to the terminals connected at nodes 2, 3, and 5.

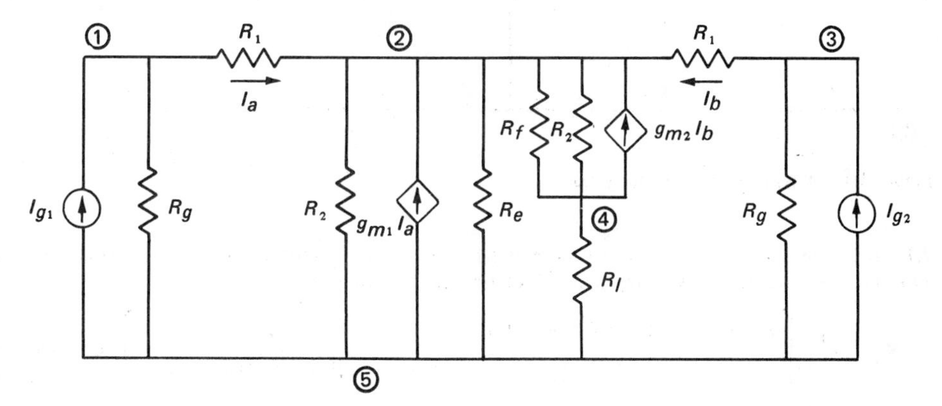

Figure 2.44 An equivalent network of a difference amplifier.

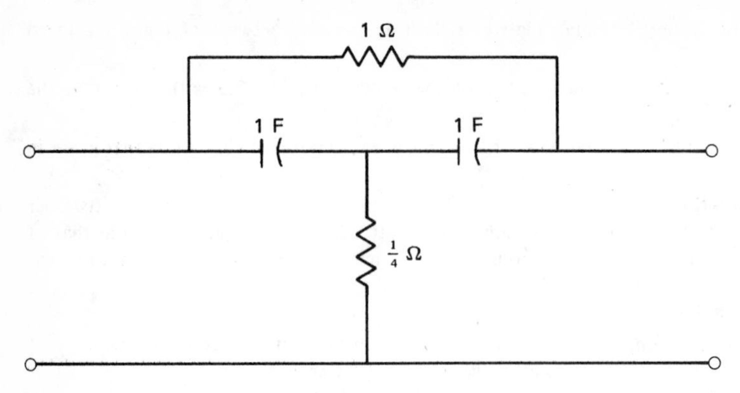

Figure 2.45 A bridged-T *RC* two-port network.

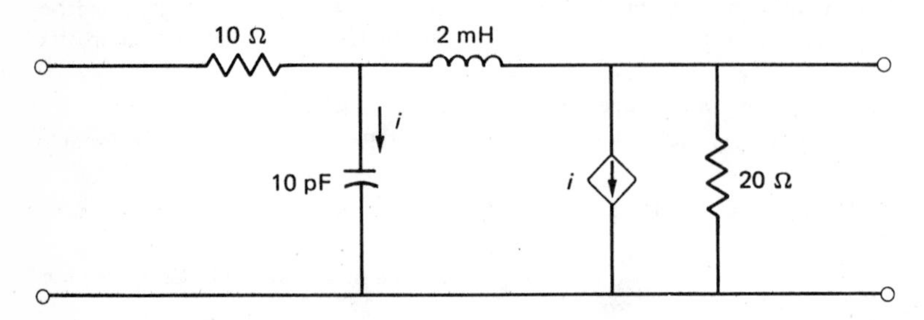

Figure 2.46 A two-port network.

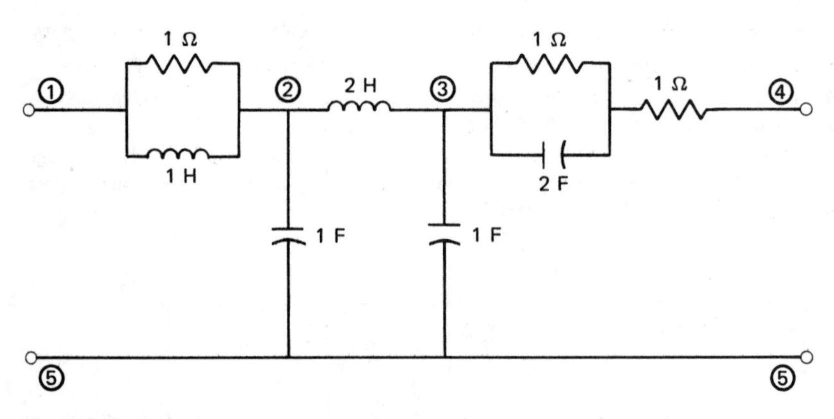

Figure 2.47 A passive two-port network.

2.17 By using the indefinite-admittance-matrix technique, confirm that the short-circuit admittance matrix of the two-port network of Fig. 2.47 is given by

$$\mathbf{Y}_{sc} = \frac{1}{p(s)} \begin{bmatrix} (s+1)(4s^4+8s^3+6s^2+6s+1) & -(s+1)(2s+1) \\ -(s+1)(2s+1) & (2s+1)(2s^4+2s^3+4s^2+s+1) \end{bmatrix} \tag{2.174}$$

where

$$p(s) = 4s^5 + 12s^4 + 18s^3 + 18s^2 + 7s + 2 \tag{2.175}$$

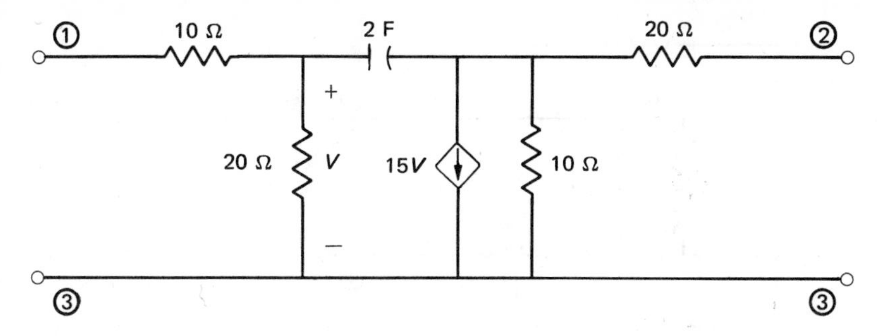

Figure 2.48 A given two-port network.

2.18 By using the indefinite-admittance-matrix technique, confirm that the open-circuit impedance matrix of the two-port network of Fig. 2.47 is given by

$$\mathbf{Z}_{oc} = \frac{1}{2s(s^2+1)}\begin{bmatrix} 2s^2+1 & 1 \\ 1 & 2s^2+1 \end{bmatrix} \tag{2.176}$$

2.19 Determine the indefinite-admittance matrix of the network of Fig. 2.48 with respect to the three terminals connected at nodes 1, 2, and 3. Obtain the short-circuit admittance matrix of the two-port network formed by the terminal pairs 1, 3 and 2, 3.

2.20 Refer to the network of Fig. 2.48. Compute the short-circuit current gain $\alpha_{12,33}$ and short-circuit transfer admittance $y_{12,33}$.

2.21 Compute the short-circuit current gain $\alpha_{14,55}$ and the short-circuit transfer admittance $y_{14,55}$ in the network of Fig. 2.47.

2.22 For Fig. 2.47, determine the impedance looking into terminals 1 and 5 when the output terminal 4 is shorted to terminal 5. Under this situation, also determine the transfer impedance $z_{13,55}$.

2.23 Figure 2.49 is an *RC* twin-T used in the design of equalizers. Compute the transfer

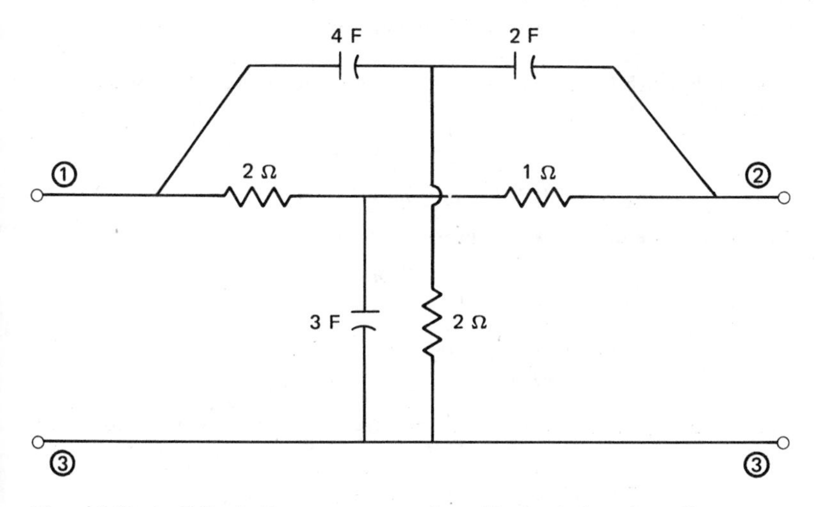

Figure 2.49 An *RC* twin-T two-port network used in the design of equalizers.

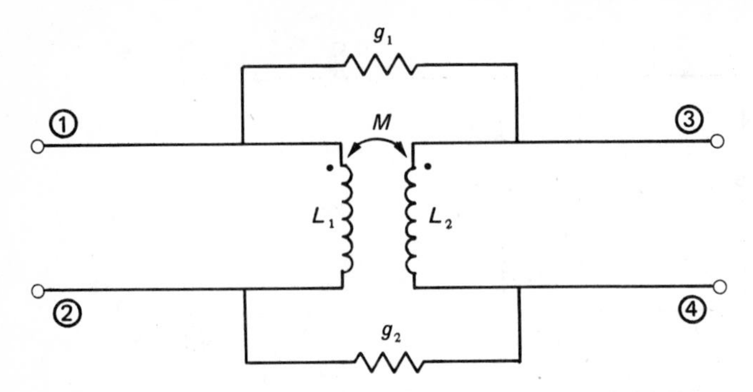

Figure 2.50 A two-port network with two mutually coupled coils.

impedance between the terminal pairs 1, 3 and 2, 3 when the output pair 2, 3 is open-circuited. Repeat the problem if the output pair 2, 3 is loaded by a 2-Ω resistor.

2.24 By using the indefinite-admittance-matrix technique, compute the short-circuit admittance parameters of the twin-T network of Fig. 2.49 with respect to the terminal pairs 1, 3 and 2, 3.

2.25 For Fig. 2.49, compute $g_{12,33}$ and $y_{12,33}$.

2.26 For the network of Fig. 2.50, find the indefinite-admittance matrix with respect to terminals 1, 2, 3, and 4. What is the resulting matrix when terminals 2 and 4 are joined together?

2.27 Show that

$$Y_{pq,rs} = -Y_{ps,rq} \tag{2.177a}$$

$$Y_{pq,rs} = Y_{rs,pq} \tag{2.177b}$$

Use these relations to confirm that (2.83) and (2.85) are valid for all p, q, r, s, and v.

2.28 Use (2.94) to show that the transfer impedances or admittances in opposite directions in a reciprocal network are equal.

2.29 Apply (2.97) and (2.114) to show that, in a reciprocal network, the voltage gain in one direction is equal to the short-circuit current gain in the opposite direction.

2.30 By applying the rule derived in Sec. 2.7.2, obtain the constrained nodal system of equations (2.169).

2.31 Typical values of the hybrid parameters of a common-base transistor for small-signal, low-frequency operation are given by

$$\begin{aligned} h_{ib} &= 20\ \Omega \qquad & h_{fb} &= -0.98 \\ h_{rb} &= 3 \cdot 10^{-4} \qquad & h_{ob} &= 0.5\ \mu\text{mho} \end{aligned} \tag{2.178}$$

With these relative values, derive the following approximate formulas:

$$\begin{aligned} h_{ie} &\approx \frac{h_{ib}}{1+h_{fb}} \qquad & h_{fe} &\approx -\frac{h_{fb}}{1+h_{fb}} \\ h_{re} &\approx \frac{h_{ib}h_{ob} - h_{rb}h_{fb} - h_{rb}}{1+h_{fb}} \qquad & h_{oe} &\approx \frac{h_{ob}}{1+h_{fb}} \end{aligned} \tag{2.179}$$

2.32 With the relative values given in Eq. (2.178), derive the following approximate formulas:

$$\begin{aligned} h_{ic} &\approx \frac{h_{ib}}{1+h_{fb}} \qquad & h_{fc} &\approx -\frac{1}{1+h_{fb}} \\ h_{rc} &\approx 1 \qquad & h_{oc} &\approx \frac{h_{ob}}{1+h_{fb}} \end{aligned} \tag{2.180}$$

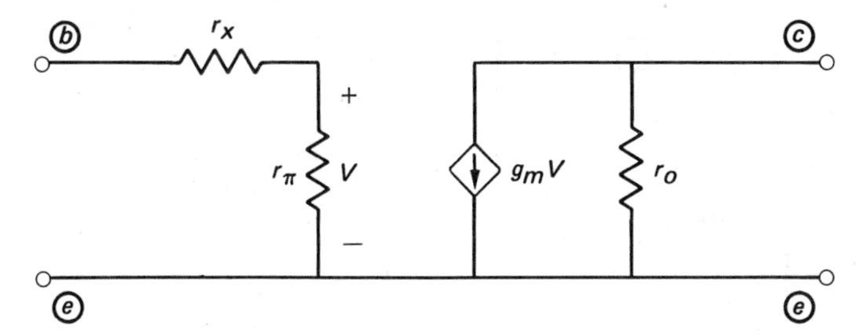

Figure 2.51 The low-frequency hybrid-pi equivalent network of a transistor.

2.33 Figure 2.51 is the hybrid-pi equivalent network of a transistor. Express the parameters of the T-model of Fig. 2.22 in terms of those of the hybrid-pi model, as follows:

$$\alpha = \frac{g_m r_\pi}{1 + g_m r_\pi} \qquad r_b = r_x + \frac{r_\pi r_\mu}{r_\mu + (1 + g_m r_\pi) r_o}$$
$$\frac{1}{r_c} = \frac{1}{r_\mu} + \frac{1}{r_o(1 + g_m r_\pi)} \qquad r_e = \frac{r_\pi r_o}{r_o + g_m r_\pi r_o + r_\mu} \tag{2.181}$$

2.34 With the relative values given in (2.62), obtain the following approximate formulas for the hybrid-pi parameters of Fig. 2.51:

$$r_\pi \approx h_{ie} - r_x \qquad r_\mu \approx \frac{h_{ie} - r_x}{h_{re}}$$
$$g_m \approx \frac{h_{fe}}{h_{ie} - r_x} \qquad \frac{1}{r_o} \approx h_{oe} - \frac{h_{fe} h_{re}}{h_{ie} - r_x} \tag{2.182}$$

2.35 Compute the voltage gain of the active filter of Fig. 2.52 under the following situations:

(*a*) The op-amp has finite input and output impedances with finite gain.

(*b*) The op-amp has infinite input impedance and zero output impedance with finite gain.

(*c*) The op-amp is ideal.

(*d*) Compare the results obtained in (*b*) and (*c*) by limiting process ($r_i \to \infty$, $r_o \to 0$, and $A \to \infty$) and by applying the rules outlined in Sec. 2.7.2.

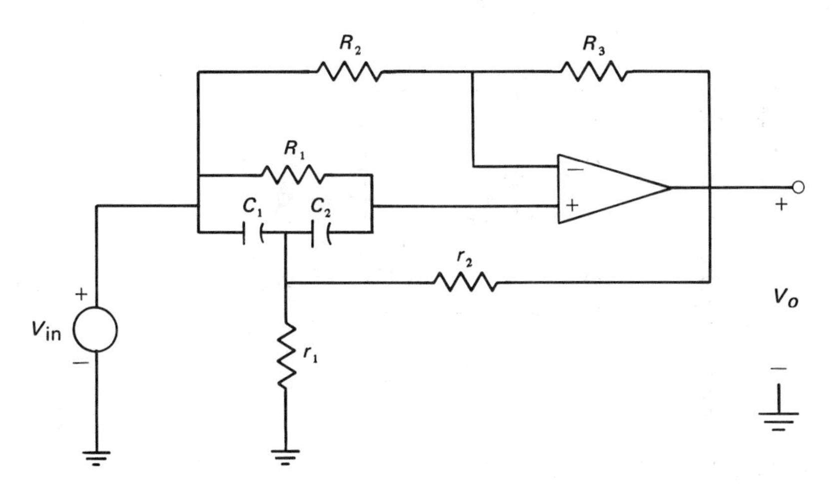

Figure 2.52 An active filter.

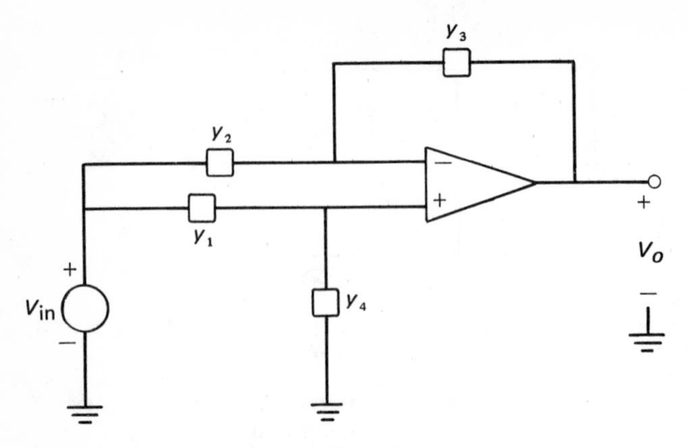

Figure 2.53 An active filter configuration.

2.36 Repeat Prob. 2.35 for the active filter configuration of Fig. 2.53.

2.37 Repeat Prob. 2.35 for the active filter configuration of Fig. 2.54.

2.38 Derive Eqs. (2.72)–(2.74) for the common-collector configurations of Figs. 2.19 and 2.20.

2.39 Compute the output impedance looking into terminals 3 and 4 of the network of Fig. 2.26 with input terminals 1 and 4 joined together.

2.40 Typical values of the hybrid parameters of a common-collector transistor for small-signal, low-frequency operation are given by

$$\begin{aligned} h_{ic} &= 1\text{ k}\Omega \qquad h_{fc} = -50 \\ h_{rc} &= 1 \qquad h_{oc} = 25\ \mu\text{mho} \end{aligned} \tag{2.183}$$

Using these relative values, derive the following approximate formulas:

$$\begin{aligned} h_{ib} &\approx -\frac{h_{ic}}{h_{fc}} \qquad h_{fb} \approx -1 - \frac{1}{h_{fc}} \\ h_{rb} &\approx h_{rc} - 1 - \frac{h_{ic}h_{oc}}{h_{fc}} \qquad h_{ob} \approx -\frac{h_{oc}}{h_{fc}} \end{aligned} \tag{2.184}$$

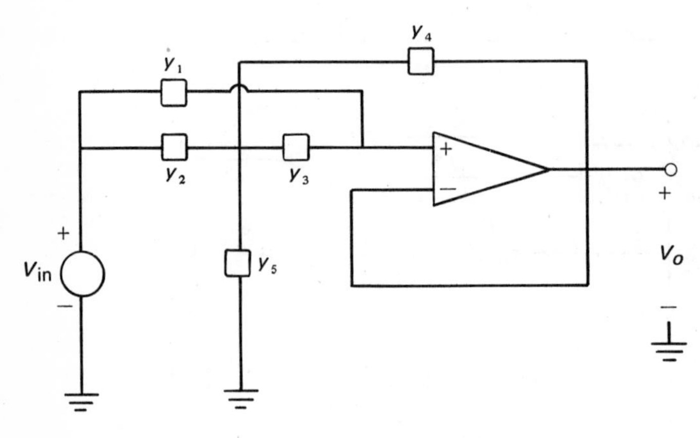

Figure 2.54 An active filter configuration.

BIBLIOGRAPHY

Alderson, G. E. and P. M. Lin: Computer Generation of Symbolic Network Functions–A New Theory and Implementation, *IEEE Trans. Circuit Theory*, vol. CT-20, no. 1, pp. 48–56, 1973.

Barabaschi, S. and E. Gatti: Moderni Metodi di Analisi delle Reti Elettriche Lineari Attive con Particolare Riguardo ai Sistemi Controreazionati, *Energia Nucleare*, vol. 2, no. 12, pp. 105–119, 1954.

Chen, W. K.: A Generalization of the Equicofactor Matrix, *IEEE Trans. Circuit Theory*, vol. CT-13, no. 4, pp. 440–442, 1966.

Chen, W. K.: On Equicofactor and Indefinite-Admittance Matrices, *Matrix and Tensor Quart.*, vol. 23, no. 1, pp. 26–28, 1972.

Chen, W. K.: "Applied Graph Theory: Graphs and Electrical Networks," 2d rev. ed., chap. 4, New York: American Elsevier, and Amsterdam: North-Holland, 1976a.

Chen, W. K.: Indefinite-Admittance Matrix Formulation of Feedback Amplifier Theory, *IEEE Trans. Circuits and Systems*, vol. CAS-23, no. 8, pp. 498–505, 1976b.

Chen, W. K.: Analysis of Constrained Active Networks, *Proc. IEEE*, vol. 66, no. 12, pp. 1655–1657, 1978.

Kerwin, W. J., L. P. Huelsman, and R. W. Newcomb: State-Variable Synthesis for Insensitive Integrated Circuit Transfer Functions, *IEEE J. Solid-State Circuits*, vol. SC-2, no. 3, pp. 87–92, 1967.

Moschytz, G. S.: "Linear Integrated Networks: Fundamentals," New York: Van Nostrand, 1974.

Nathan, A.: Matrix Analysis of Constrained Networks, *Proc. IEE (London)*, vol. 108, part C, pp. 98–106, 1954.

Nathan, A.: Matrix Analysis of Networks Having Infinite-Gain Operational Amplifiers, *Proc. IEEE*, vol. 49, no. 10, pp. 1577–1578, 1961.

Puckett, T. H.: A Note on the Admittance and Impedance Matrices of an n-Terminal Network, *IRE Trans. Circuit Theory*, vol. CT-3, no. 1, pp. 70–75, 1956.

Sharpe, G. E. and B. Spain: On the Solution of Networks by Means of the Equicofactor Matrix, *IRE Trans. Circuit Theory*, vol. CT-7, no. 3, pp. 230–239, 1960.

Sharpe, G. E. and G. P. H. Styan: Circuit Duality and the General Network Inverse, *IEEE Trans. Circuit Theory*, vol. CT-12, no. 1, pp. 22–27, 1965.

Shekel, J.: Matrix Representation of Transistor Circuits, *Proc. IRE*, vol. 40, no. 11, pp. 1493–1497, 1952.

Shekel, J.: Indefinite Admittance Representation of Linear Network Elements, *Bull. Res. Council Israel*, vol. 3, pp. 390–394, 1954a.

Shekel, J.: Matrix Analysis of Multi-terminal Transducers, *Proc. IRE*, vol. 42, no. 5, pp. 840–847, 1954b.

Skelboe, S.: A Universal Formula for Network Functions, *IEEE Trans. Circuits and Systems*, vol. CAS-22, no. 1, pp. 58–60, 1975.

Yokomoto, C. F.: An Efficient Method of Obtaining Port Parameters from the Indefinite Admittance Matrix, *IEEE Trans. Circuit Theory*, vol. CT-19, no. 5, pp. 521–524, 1972.

Zadeh, L. A.: On Passive and Active Networks and Generalized Norton's and Thevenin's Theorems, *Proc. IRE*, vol. 44, no. 3, p. 378, 1956.

Zadeh, L. A.: Multipole Analysis of Active Networks, *IRE Trans. Circuit Theory*, vol. CT-4, no. 3, pp. 97–105, 1957.

CHAPTER

THREE

ACTIVE TWO–PORT NETWORKS

In Chap. 1 we introduced many fundamental concepts related to linear, time-invariant n-port networks. Some of the results, although very general, are difficult to apply. In Chap. 2 we discussed a useful description of the external behavior of a multiterminal network in terms of the indefinite-admittance matrix, and demonstrated how it can be employed effectively for the computation of network functions.

In practical applications, the most useful class of n-port or n-terminal networks is that of two-port or three-terminal networks. Many active devices of practical importance such as transistors are naturally subsumed in this class. In this chapter, we consider the specialization of the general passivity condition for n-port networks in terms of the more immediately useful two-port parameters. We introduce various types of power gains, sensitivity, and the notion of absolute stability as opposed to potential instability. Llewellyn's conditions for absolute stability and the optimum terminations for absolutely stable two-port networks at a single frequency will be derived.

3.1 TWO–PORT PARAMETERS

In Sec. 1.5 we observed that the port behavior of an n-port network is completely characterized by giving the relationships among excitation and response signal transforms. Any n independent functions of the $2n$ port variables can be regarded as the excitation and the remaining n variables as the response. The characterization is given in the form of the general hybrid matrix whose elements are called the general hybrid parameters. For our purposes, we consider four different sets of two-port parameters, as follows:

$$\begin{bmatrix} V_1 \\ V_2 \end{bmatrix} = \begin{bmatrix} z_{11} & z_{12} \\ z_{21} & z_{22} \end{bmatrix} \begin{bmatrix} I_1 \\ I_2 \end{bmatrix} \tag{3.1a}$$

$$\begin{bmatrix} I_1 \\ I_2 \end{bmatrix} = \begin{bmatrix} y_{11} & y_{12} \\ y_{21} & y_{22} \end{bmatrix} \begin{bmatrix} V_1 \\ V_2 \end{bmatrix} \tag{3.1b}$$

$$\begin{bmatrix} V_1 \\ I_2 \end{bmatrix} = \begin{bmatrix} h_{11} & h_{12} \\ h_{21} & h_{22} \end{bmatrix} \begin{bmatrix} I_1 \\ V_2 \end{bmatrix} \tag{3.1c}$$

$$\begin{bmatrix} I_1 \\ V_2 \end{bmatrix} = \begin{bmatrix} g_{11} & g_{12} \\ g_{21} & g_{22} \end{bmatrix} \begin{bmatrix} V_1 \\ I_2 \end{bmatrix} \tag{3.1d}$$

The coefficient matrices are referred to as the *impedance matrix* (z-*matrix*), the *admittance matrix* (y-*matrix*), the *hybrid matrix* (h-*matrix*), and the *inverse hybrid matrix* (g-*matrix*), respectively, whose elements are called the z-*parameters*, y-*parameters*, h-*parameters*, and g-*parameters*. They are all special situations of the general representation of (1.60). Since h's are customarily used for (3.1c), the general two-port representation of (1.60) is written as

$$\begin{bmatrix} y_1 \\ y_2 \end{bmatrix} = \begin{bmatrix} k_{11} & k_{12} \\ k_{21} & k_{22} \end{bmatrix} \begin{bmatrix} u_1 \\ u_2 \end{bmatrix} \tag{3.2}$$

where $u_1, y_1 = V_1, I_1$ and $u_2, y_2 = V_2, I_2$. As indicated in Sec. 1.5, the coefficient matrix of (3.2) is referred to as the general hybrid matrix, whose elements are called the general hybrid parameters, which are different from the hybrid parameters h_{ij} of (3.1c).

For each parameter set of (3.1), there is a corresponding representation of the source and load terminations, as shown in Fig. 3.1. We shall demonstrate that formulas derived for any model are equally valid for the other three representations provided that we use the corresponding variables and parameter set. Thus, we can obtain the general formulas for network quantities that may be used with any consistent parameter set by simply substituting the appropriate quantities. Table 3.1 gives the corresponding quantities in the four parameter representations.

To illustrate the above procedure, we consider the impedance representation of Fig. 3.1a. Substituting $V_2 = -I_2 Z_2$ in Eq. (3.1a) and solving for I_1 yield

$$Z_{11} = \frac{V_1}{I_1} = z_{11} - \frac{z_{12} z_{21}}{z_{22} + Z_2} \tag{3.3}$$

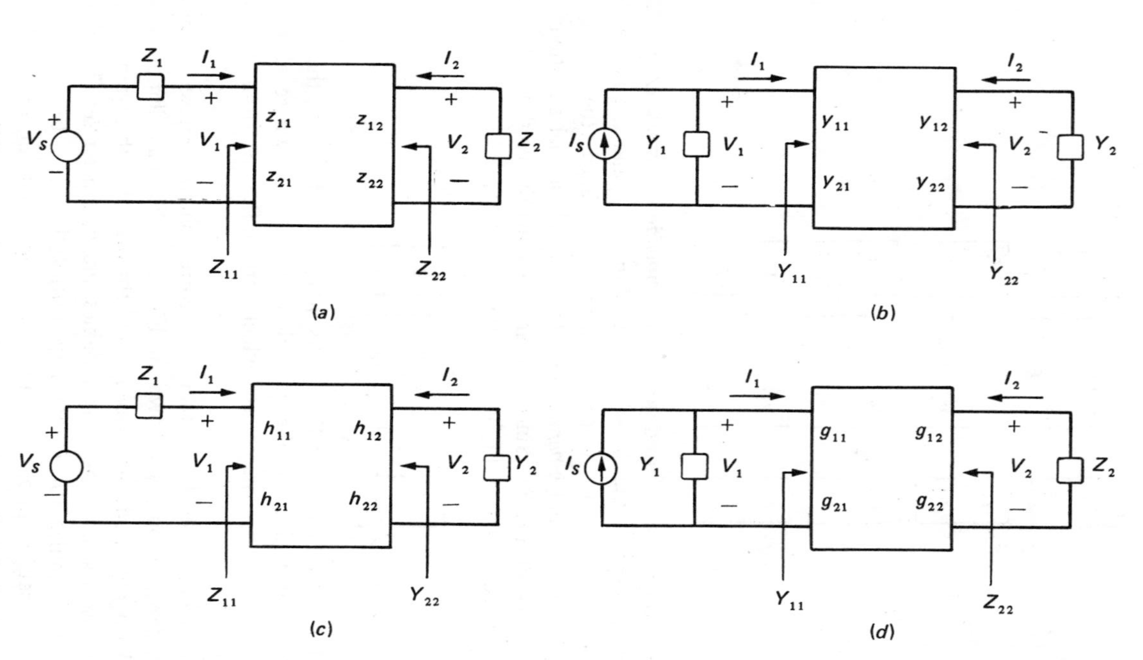

Figure 3.1 The four parameter representations of a two-port network. (*a*) The z-parameter representation. (*b*) The y-parameter representation. (*c*) The h-parameter representation. (*d*) The g-parameter representation.

Table 3.1 Corresponding quantities in four parameter representations

z's	y's	h's	g's	k's
z_{11}	y_{11}	h_{11}	g_{11}	k_{11}
z_{22}	y_{22}	h_{22}	g_{22}	k_{22}
z_{12}	y_{12}	h_{12}	g_{12}	k_{12}
z_{21}	y_{21}	h_{21}	g_{21}	k_{21}
V_S	I_S	V_S	I_S	u_S
I_1	V_1	I_1	V_1	u_1
I_2	V_2	V_2	I_2	u_2
Z_1	Y_1	Z_1	Y_1	M_1
Z_2	Y_2	Y_2	Z_2	M_2
Z_{11}	Y_{11}	Z_{11}	Y_{11}	M_{11}
Z_{22}	Y_{22}	Y_{22}	Z_{22}	M_{22}
V_1	I_1	V_1	I_1	y_1
V_2	I_2	I_2	V_2	y_2

Likewise, if we substitute $I_2 = -Y_2 V_2$ in Eq. (3.1*b*) and solve for V_1, we obtain

$$Y_{11} = \frac{I_1}{V_1} = y_{11} - \frac{y_{12}y_{21}}{y_{22} + Y_2} \tag{3.4}$$

A similar derivation using the *h*-parameters and *g*-parameters gives

$$Z_{11} = h_{11} - \frac{h_{12}h_{21}}{h_{22} + Y_2} \tag{3.5}$$

$$Y_{11} = g_{11} - \frac{g_{12}g_{21}}{g_{22} + Z_2} \tag{3.6}$$

Thus, the general formula for the input immittance can be written as

$$M_{11} = k_{11} - \frac{k_{12}k_{21}}{k_{22} + M_2} \tag{3.7}$$

The formulas (3.3)–(3.6) are the special cases of (3.7) that are obtained when appropriate substitutions are made from Table 3.1.

In a similar way, the output immittance can be expressed as

$$M_{22} = k_{22} - \frac{k_{12}k_{21}}{k_{11} + M_1} \tag{3.8}$$

3.2 POWER GAINS

Refer to the general representation of a two-port network N of Fig. 3.1. The simplest measure of power flow in N is the *power gain* Power gain is denoted by the symbol $\mathcal{G}_p$ and defined by

$$\mathcal{G}_p = \frac{\text{average power delivered to the load}}{\text{average power entering the input port}} = \frac{P_2}{P_1} \tag{3.9}$$

Clearly, it is a function of the two-port parameters and the load impedance, and does not depend on the source impedance. For a passive lossless two-port network, $\mathcal{G}_p = 1$.

The second measure of power flow is called the *available power gain*, denoted by the symbol $\mathcal{G}_a$. The available power gain is defined by

$$\mathcal{G}_a = \frac{\text{maximum available average power at the load}}{\text{maximum available average power at the source}} = \frac{P_{2a}}{P_{1a}} \tag{3.10}$$

The quantity $\mathcal{G}_a$ is a function of the two-port parameters and the source impedance, being independent of the load impedance.

The third and most useful measure of power flow is called the *transducer power gain*, denoted by the symbol $\mathcal{G}$. The transducer power gain is defined by

$$\mathcal{G} = \frac{\text{average power delivered to the load}}{\text{maximum available average power at the source}} = \frac{P_2}{P_{1a}} \tag{3.11}$$

It is a function of the two-port parameters and the source and load impedances. Its importance arises from the fact that it compares the power delivered to the load with the power that the source is capable of supplying under the optimum conditions, and therefore it measures the efficacy of using the active device and provides the most meaningful description of the power transfer capabilities of the two-port network.

To illustrate these definitions, we shall derive expressions for them in terms of the impedance parameters of Fig. 3.1*a*. Substituting $V_2 = -I_2 Z_2$ in Eq. (3.1*a*) and solving for I_1 and I_2 yield the current gain

$$\frac{I_2}{I_1} = -\frac{z_{21}}{z_{22} + Z_2} \tag{3.12}$$

The average power P_1 entering the input port and the average power P_2 delivered to the load Z_2 are given by

$$P_1 = |I_1|^2 \operatorname{Re} Z_{11} \tag{3.13a}$$

$$P_2 = |I_2|^2 \operatorname{Re} Z_2 \tag{3.13b}$$

The maximum available average power P_{1a} from the source is

$$P_{1a} = \frac{|V_s|^2}{4 \operatorname{Re} Z_1} \tag{3.14}$$

which represents the average power delivered by the given source to a conjugately matched load $Z_{11} = \bar{Z}_1$. Combining (3.3) and (3.12)–(3.14) gives

$$\mathcal{G}_p = \frac{P_2}{P_1} = \frac{|z_{21}|^2 \operatorname{Re} Z_2}{|z_{22} + Z_2|^2 \operatorname{Re} Z_{11}} \tag{3.15a}$$

$$\mathcal{G} = \frac{P_2}{P_{1a}} = \frac{4|z_{21}|^2 \operatorname{Re} Z_1 \operatorname{Re} Z_2}{|(z_{11}+Z_1)(z_{22}+Z_2) - z_{12}z_{21}|^2} \tag{3.15b}$$

The maximum available average power P_{2a} at the output port can easily be appreciated by means of the Thévenin equivalent network of Fig. 3.2 when the input is terminated in V_s in series with Z_1, giving

$$Z_{\text{eq}} = z_{22} - \frac{z_{12}z_{21}}{z_{11}+Z_1} \tag{3.16a}$$

$$V_{\text{eq}} = \frac{z_{21}V_s}{z_{11}+Z_1} \tag{3.16b}$$

The maximum available average power at the output port is obtained when $Z_2 = \bar{Z}_{\text{eq}}$, the complex conjugate of Z_{eq}, yielding

$$P_{2a} = \frac{|z_{21}|^2 |V_s|^2}{4|z_{11}+Z_1|^2 \operatorname{Re} Z_{\text{eq}}} \tag{3.17}$$

Thus, the available power gain is given by

$$\mathcal{G}_a = \frac{P_{2a}}{P_{1a}} = \frac{|z_{21}|^2 \operatorname{Re} Z_1}{|z_{11}+Z_1|^2 \operatorname{Re} Z_{\text{eq}}} \tag{3.18}$$

In a similar manner, we can evaluate the three power gains in terms of other two-port parameters of Fig. 3.1, *b*-*d*. The general formulas are obtained as (see Prob. 3.1)

$$\mathcal{G}_p = \frac{|k_{21}|^2 \operatorname{Re} M_2}{|k_{22}+M_2|^2 \operatorname{Re}\,[k_{11} - k_{12}k_{21}/(k_{22}+M_2)]} \tag{3.19a}$$

$$\mathcal{G} = \frac{4|k_{21}|^2 \operatorname{Re} M_1 \operatorname{Re} M_2}{|(k_{11}+M_1)(k_{22}+M_2) - k_{12}k_{21}|^2} \tag{3.19b}$$

$$\mathcal{G}_a = \frac{|k_{21}|^2 \operatorname{Re} M_1}{|k_{11}+M_1|^2 \operatorname{Re}\,[k_{22} - k_{12}k_{21}/(k_{11}+M_1)]} \tag{3.19c}$$

When appropriate substitutions are made from Table 3.1, we get the various gain formulas in terms of the chosen two-port parameters and the source and load immittances. We remark that the above three power gains are defined at a single real

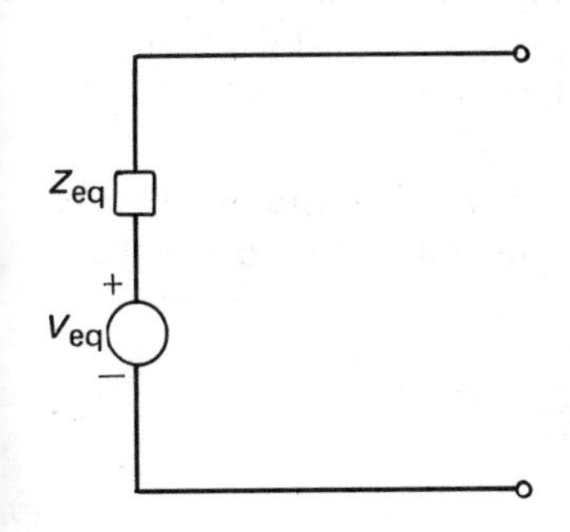

Figure 3.2 The Thévenin equivalent network looking into the output port of the network of Fig. 3.1*a*.

frequency, which may be at any point on the $j\omega$-axis. The variable $s = j\omega$ was dropped in all the expressions, for simplicity.

3.3 SENSITIVITY

Active networks are designed to perform certain functions such as signal amplification or processing. Given perfect components, there would be little difference among the many possible realizations. In practice, however, real components deviate from their nominal values because of changes in environmental conditions or aging of the components. As a result, the performance of the built networks differ from the nominal design. A practical solution to this problem is to design a circuit that has a low sensitivity to these changes. In the following, we consider the sensitivity of the network functions to the changes in the parameters. Techniques of reducing sensitivity by means of feedback will be discussed in the following chapter.

Sensitivity is a measure of the change of the overall transfer function to the change of a particular parameter in the network. It is formally defined as follows.

Definition 3.1: Sensitivity function The *sensitivity function* is defined as the ratio of the fractional change in a transfer function to the fractional change in an element for the situation when all changes concerned are differentially small.

Thus, if w is the transfer function and x is the element of interest, the sensitivity function, written as $S(x)$, is defined by the equation

$$S(x) = \lim_{\Delta x \to 0} \frac{\Delta w/w}{\Delta x/x} = \frac{x}{w}\frac{\partial w}{\partial x} \tag{3.20}$$

In terms of logarithms, Eq. (3.20) can be simplified to

$$S(x) = \frac{\partial \ln w}{\partial \ln x} = x\frac{\partial \ln w}{\partial x} \tag{3.21}$$

To compute the sensitivity function of a transfer function of Fig. 3.1 to the change of a two-port parameter, we must first express the transfer function in terms of the two-port parameters and the terminations. To this end, we shall work out the case of h-parameters in detail, leaving the other possibilities as an exercise for the reader.

For Fig. 3.1c, we have $V_1 = V_s - Z_1 I_1$ and $I_2 = -Y_2 V_2$. Substituting these in Eq. (3.1c) and solving for V_2 yield the ratio of the output voltage to source voltage as

$$\frac{V_2}{V_s} = -\frac{h_{21}}{(h_{11} + Z_1)(h_{22} + Y_2) - h_{12}h_{21}} \tag{3.22}$$

If one is interested in the ratio of output current to source voltage, (3.22) is changed to

$$\frac{I_2}{V_s} = \frac{h_{21}Y_2}{(h_{11}+Z_1)(h_{22}+Y_2)-h_{12}h_{21}} \tag{3.23}$$

For the transfer impedance V_2/I_1, we have

$$\frac{V_2}{I_1} = -\frac{h_{21}}{h_{22}+Y_2} \tag{3.24}$$

Similar results can be derived for other configurations of Fig. 3.1. The general formulas for the transfer functions become (see Prob. 3.2)

$$\frac{y_2}{u_s} = \frac{k_{21}M_2}{(k_{11}+M_1)(k_{22}+M_2)-k_{12}k_{21}} \tag{3.25a}$$

$$\frac{u_2}{u_s} = -\frac{k_{21}}{(k_{11}+M_1)(k_{22}+M_2)-k_{12}k_{21}} \tag{3.25b}$$

$$\frac{u_2}{u_1} = -\frac{k_{21}}{k_{22}+M_2} \tag{3.25c}$$

Since (3.22) and (3.23) are related only by $-Y_2$, their sensitivity functions with respect to the two-port parameters h_{ij} will be the same. To be more specific, we compute the sensitivity function of (3.22) with respect to the forward-current transfer ratio h_{21}:

$$S(h_{21}) = \frac{h_{21}}{V_2/V_s}\,\frac{\partial(V_2/V_s)}{\partial h_{21}} = \frac{(h_{11}+Z_1)(h_{22}+Y_2)}{(h_{11}+Z_1)(h_{22}+Y_2)-h_{12}h_{21}} \tag{3.26}$$

It is significant to observe that if $h_{12}h_{21}$ is small in comparison with $(h_{11}+Z_1)(h_{22}+Y_2)$, the sensitivity to h_{21} is nearly unity. This means that a 5% change in h_{21} gives a 5% change in V_2/V_s. On the other hand, if $h_{12}h_{21}$ is large in comparison with the other term in the denominator, the sensitivity decreases about inversely proportional to $h_{12}h_{21}$. This is precisely the effect that is obtained when negative feedback is employed.

In a similar way, the sensitivity functions of V_2/V_s to changes of other two-port parameters are obtained as (see Prob. 3.4)

$$S(h_{12}) = \frac{h_{12}h_{21}}{D_h} \tag{3.27a}$$

$$S(h_{11}) = -\frac{h_{11}(h_{22}+Y_2)}{D_h} \tag{3.27b}$$

$$S(h_{22}) = -\frac{h_{22}(h_{11}+Z_1)}{D_h} \tag{3.27c}$$

where

$$D_h = (h_{11}+Z_1)(h_{22}+Y_2)-h_{12}h_{21} \tag{3.28}$$

The general expressions of the sensitivity functions of the transfer functions y_2/u_s or u_2/u_s to changes of the general hybrid parameters k's are given by (see Prob. 3.5)

$$S(k_{11}) = -\frac{k_{11}(k_{22}+M_2)}{D_k} \tag{3.29a}$$

$$S(k_{12}) = \frac{k_{12}k_{21}}{D_k} \tag{3.29b}$$

$$S(k_{21}) = \frac{(k_{11}+M_1)(k_{22}+M_2)}{D_k} \tag{3.29c}$$

$$S(k_{22}) = -\frac{k_{22}(k_{11}+M_1)}{D_k} \tag{3.29d}$$

where

$$D_k = (k_{11}+M_1)(k_{22}+M_2) - k_{12}k_{21} \tag{3.30}$$

Example 3.1 Figure 3.3 shows the hybrid-pi equivalent network of a transistor. Its admittance matrix was computed earlier, as Eq. (2.43), and is repeated here:

$$\mathbf{Y}(p) = \frac{1}{240+1050p}\begin{bmatrix} 0.8+21p & -p \\ 40-p & 11.2p+5p^2 \end{bmatrix} \tag{3.31}$$

where $p = s/10^9$. Assume that the transistor is terminated at the load with 2 kΩ and the source with 100 Ω. Then $Y_1 = 0.01$ mho and $Y_2 = 5 \cdot 10^{-4}$ mho. From (3.29), the sensitivity functions of the transfer impedance V_2/I_s or the current gain I_2/I_s to the changes of the y-parameters are computed as follows:

$$S(y_{11}) = -\frac{y_{11}(y_{22}+Y_2)}{D_y} = -\frac{105p^3+250.225p^2+11.9p+0.096}{157.5p^3+384.338p^2+81.3p+0.384} \tag{3.32a}$$

$$S(y_{12}) = \frac{y_{12}y_{21}}{D_y} = \frac{p^2-40p}{157.5p^3+384.338p^2+81.3p+0.384} \tag{3.32b}$$

$$S(y_{21}) = \frac{(y_{11}+Y_1)(y_{22}+Y_2)}{D_y} = \frac{157.5p^3+385.338p^2+41.3p+0.384}{157.5p^3+384.338p^2+81.3p+0.384} \tag{3.32c}$$

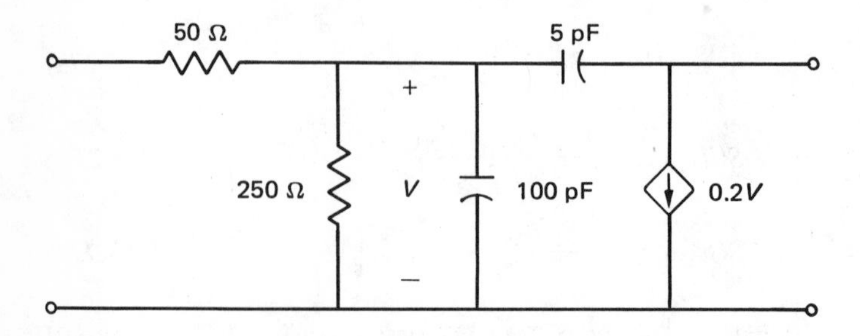

Figure 3.3 The hybrid-pi equivalent network of a transistor.

$$S(y_{22}) = -\frac{y_{22}(y_{11}+Y_1)}{D_y} = -\frac{157.5p^3 + 368.8p^2 + 35.84p}{157.5p^3 + 384.338p^2 + 81.3p + 0.384} \tag{3.32d}$$

where

$$D_y = (y_{11}+Y_1)(y_{22}+Y_2) - y_{12}y_{21} = \frac{157.5p^3 + 384.338p^2 + 81.3p + 0.384}{(1050p+240)^2} \tag{3.33}$$

3.4 PASSIVITY AND ACTIVITY

In Chap. 1 we introduced the time-energy definition of passivity and activity for an n-port network. We showed that for a linear, time-invariant n-port network to be passive, it is necessary and sufficient that its general hybrid matrix be positive real. We extended this result by introducing the discrete-frequency concepts of passivity and activity. We demonstrated that by applying the definitions of passivity and activity at a single complex frequency, a given n-port can divide the closed RHS (*r*ight *h*alf of the complex frequency s-plane) into regions of passivity and activity. In the passive region, the n-port network can be replaced by an equivalent passive n-port network. In the active region, the n-port network can be made to achieve power gain or to oscillate by means of an appropriate passive imbedding network. However, in practical applications, we are concerned mainly with the behavior of the two-port networks on the real-frequency axis and, less frequently, on the positive σ-axis rather than the n-port networks and the entire closed RHS. In this section, we consider the specialization of the general passivity condition for n-port networks in terms of the more immediately useful two-port parameters.

As indicated in Theorem 1.5, a two-port network is passive at a point s_0 on the real-frequency axis or the positive σ-axis if and only if the hermitian part of its general hybrid matrix is nonnegative definite at s_0. The general hybrid matrix referred to in the theorem is the coefficient matrix of (3.2), which is rewritten as†

$$\mathbf{H}(s) = \begin{bmatrix} k_{11} & k_{12} \\ k_{21} & k_{22} \end{bmatrix} = \begin{bmatrix} m_{11} & m_{12} \\ m_{21} & m_{22} \end{bmatrix} + j\begin{bmatrix} n_{11} & n_{12} \\ n_{21} & n_{22} \end{bmatrix} \tag{3.34}$$

where the m's and n's are real. The hermitian part of $\mathbf{H}(s)$ is found to be

$$\mathbf{H}_h(s) = \tfrac{1}{2}\,[\mathbf{H}(s) + \mathbf{H}^*(s)] = \begin{bmatrix} m_{11} & \tfrac{1}{2}(m_{12}+m_{21}+jn_{12}-jn_{21}) \\ \tfrac{1}{2}(m_{21}+m_{12}+jn_{21}-jn_{12}) & m_{22} \end{bmatrix} \tag{3.35}$$

†The letters m and n are chosen to denote the real and imaginary parts of k because M is even with respect to its midpoint and N is odd with respect to its midpoint. This helps us to remember the symbols.

The hermitian matrix $\mathbf{H}_h(s)$ is nonnegative definite if and only if

$$m_{11} \geqslant 0 \tag{3.36a}$$

$$m_{22} \geqslant 0 \tag{3.36b}$$

$$\det \mathbf{H}_h(s) \geqslant 0 \tag{3.36c}$$

The third condition (3.36*c*) is equivalent to

$$4m_{11}m_{22} - |k_{12}|^2 - |k_{21}|^2 - 2\,\mathrm{Re}\,(k_{12}k_{21}) \geqslant 0 \tag{3.37a}$$

or

$$4m_{11}m_{22} \geqslant |k_{12} + \bar{k}_{21}|^2 \tag{3.37b}$$

In the case of the reciprocal two-port network, $z_{12} = z_{21}$, $y_{12} = y_{21}$, $h_{12} = -h_{21}$, or $g_{12} = -g_{21}$, and (3.37*b*) is reduced to

$$m_{11}m_{22} \geqslant m_{12}^2 \tag{3.38a}$$

for z- or y-parameters, and

$$m_{11}m_{22} \geqslant n_{12}^2 \tag{3.38b}$$

for h- or g-parameters. Equation (3.38) together with Eqs. (3.36*a*) and (3.36*b*) is known as the *Gewertz condition for passivity* of a reciprocal two-port network, and was first given by Gewertz (1933). We summarize the above results by stating the following theorem.

Theorem 3.1 A two-port network is passive at a point s_0 on the real-frequency axis or the positive σ-axis if and only if

$$m_{11} \geqslant 0 \qquad m_{22} \geqslant 0 \tag{3.39a}$$

$$4m_{11}m_{22} \geqslant |k_{12} + \bar{k}_{21}|^2 \tag{3.39b}$$

It is strictly passive at s_0 if and only if the three conditions of (3.39) are satisfied with the strict inequality.

A two-port network is active at s_0 if it is not passive at s_0. Thus, if any one of the three conditions (3.39) is violated at s_0, the two-port is active at s_0. As stated in Sec. 1.7.2, the largest ω on the real-frequency axis at which a two-port network is active is called the maximum frequency of oscillation, and the largest σ on the nonnegative σ-axis at which a two-port is active is called the fastest regenerative mode. regenerative mode.

Example 3.2 Consider again the hybrid-pi equivalent network of a transistor of Fig. 3.3. The admittance matrix $\mathbf{Y}(p)$ of the two-port network is given by (3.31). On the real-frequency axis, the hermitian part of $\mathbf{Y}(p)$ is obtained as

$$\mathbf{Y}_h(j\omega) = \frac{1}{576 + 11{,}025\hat{\omega}^2} \begin{bmatrix} 1.92 + 220.5\hat{\omega}^2 & 48 - 10.5\hat{\omega}^2 + j210\hat{\omega} \\ 48 - 10.5\hat{\omega}^2 - j210\hat{\omega} & 105.6\hat{\omega}^2 \end{bmatrix} \tag{3.40}$$

where $p = j\hat{\omega} = j\omega/10^9$. Thus, we have

$$m_{11} = \text{Re } y_{11}(j\hat{\omega}) = \frac{1.92 + 220.5\hat{\omega}^2}{576 + 11{,}025\hat{\omega}^2} \geqslant 0 \tag{3.41a}$$

$$m_{22} = \text{Re } y_{22}(j\hat{\omega}) = \frac{105.6\hat{\omega}^2}{576 + 11{,}025\hat{\omega}^2} \geqslant 0 \tag{3.41b}$$

$$\begin{aligned} 4m_{11}m_{22} - |k_{12} + \bar{k}_{21}|^2 &= 4(\text{Re } y_{11})(\text{Re } y_{22}) - |y_{12} + \bar{y}_{21}|^2 \\ &= \frac{4(23{,}174.55\hat{\omega}^4 - 42{,}889.25\hat{\omega}^2 - 2304)}{(576 + 11{,}025\hat{\omega}^2)^2} \end{aligned} \tag{3.41c}$$

By setting the numerator of (3.41*c*) to zero, we obtain the minimum value of $|\hat{\omega}|$ for which (3.41*c*) is nonnegative:

$$23.175\hat{\omega}^4 - 42.889\hat{\omega}^2 - 2.304 = 0 \tag{3.42}$$

giving $\hat{\omega}_m^2 = 1.9$ or $\hat{\omega}_m = 1.38$. Thus, the device is passive for all $|\hat{\omega}| \geqslant \hat{\omega}_m$, and active for all $|\hat{\omega}| < \hat{\omega}_m$. The maximum frequency of oscillation of the device occurs at (see footnote on page 46)

$$\omega_m = \hat{\omega}_m \cdot 10^9 = 1.38 \cdot 10^9 \text{ rad/s} \tag{3.43}$$

which corresponds to 219.6 MHz.

On the σ-axis, conditions (3.39) become

$$m_{11} = y_{11}(\sigma) = \frac{0.8 + 21\hat{\sigma}}{240 + 1050\hat{\sigma}} \geqslant 0 \tag{3.44a}$$

$$m_{22} = y_{22}(\sigma) = \frac{11.2\hat{\sigma} + 5\hat{\sigma}^2}{240 + 1050\hat{\sigma}} \geqslant 0 \tag{3.44b}$$

for nonnegative $\hat{\sigma}$, where $\hat{\sigma} = \sigma/10^9$, and

$$\begin{aligned} 4m_{11}m_{22} - |k_{12} + \bar{k}_{21}|^2 &= 4y_{11}(\sigma)y_{22}(\sigma) - [y_{12}(\sigma) + y_{21}(\sigma)]^2 \\ &= \frac{4(105\hat{\sigma}^3 + 238.2\hat{\sigma}^2 + 48.96\hat{\sigma} - 400)}{(240 + 1050\hat{\sigma})^2} \end{aligned} \tag{3.44c}$$

Setting the numerator of (3.44*c*) to zero yields the fastest regenerative mode of the device as (see footnote on page 45)

$$\sigma_m = \hat{\sigma}_m \cdot 10^9 = 1.01 \cdot 10^9 \text{ Np/s} \tag{3.45}$$

Thus, on the nonnegative σ-axis, the device is passive for all $\sigma \geqslant \sigma_m$ and active for all $\sigma < \sigma_m$.

Example 3.3 Consider the high-frequency, small-signal equivalent network of a bipolar transistor as shown in Fig. 3.4. Its hybrid matrix was computed earlier in Example 1.4 and is given by

$$\mathbf{H}(s) = \frac{1}{G_1 + s(C_1 + C_2)} \begin{bmatrix} 1 & sC_2 \\ g_m - sC_2 & q(s) \end{bmatrix} \tag{3.46a}$$

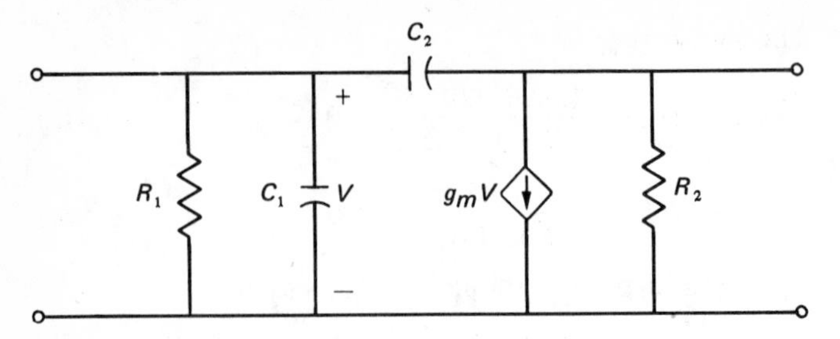

Figure 3.4 A high-frequency, small-signal equivalent network of a bipolar transistor.

where $G_1 = 1/R_1$, $G_2 = 1/R_2$, and

$$q(s) = (G_1 + sC_1)(G_2 + sC_2) + sC_2(G_2 + g_m) \tag{3.46b}$$

To test the passivity conditions (3.39) on the real-frequency axis, we compute the following: Let $a = G_1/(C_1 + C_2)$. Then we have

$$m_{11} = \text{Re } h_{11}(j\omega) = \frac{a}{(C_1 + C_2)(\omega^2 + a^2)} \geqslant 0 \tag{3.47a}$$

$$m_{22} = \text{Re } h_{22}(j\omega) = \frac{(C_1G_2 + C_2G_2 + G_1C_2 - C_1C_2a + C_2g_m)\omega^2 + G_1G_2a}{(C_1 + C_2)(\omega^2 + a^2)} \geqslant 0 \tag{3.47b}$$

$$\begin{aligned} 4m_{11}m_{22} - |k_{12} + \bar{k}_{21}|^2 &= 4(\text{Re } h_{11})(\text{Re } h_{22}) - |h_{12} + \bar{h}_{21}|^2 \\ &= \frac{4G_1G_2 - g_m^2}{(C_1 + C_2)^2(\omega^2 + a^2)} \end{aligned} \tag{3.47c}$$

The third equation is nonnegative if and only if $4G_1G_2 \geqslant g_m^2$. Thus, the device is passive for all real frequencies if $4G_1G_2 \geqslant g_m^2$ and active for all real frequencies if $4G_1G_2 < g_m^2$. This situation is evidently physically impossible. The reason is that we assume the model of Fig. 3.4 to be an adequate representation of the transistor for all frequencies. At higher frequencies, parasitic effects must be taken into consideration, yielding a finite maximum frequency of oscillation.

Now suppose that we consider the admittance matrix of the two-port network of Fig. 3.4, which is given by

$$\mathbf{Y}(s) = \begin{bmatrix} G_1 + s(C_1 + C_2) & -sC_2 \\ g_m - sC_2 & G_2 + sC_2 \end{bmatrix} \tag{3.48}$$

To test passivity on the two axes, we again appeal to (3.39). On the real-frequency axis, we have

$$m_{11} = \text{Re } y_{11}(j\omega) = G_1 \geqslant 0 \tag{3.49a}$$

$$m_{22} = \text{Re } y_{22}(j\omega) = G_2 \geqslant 0 \tag{3.49b}$$

$$4m_{11}m_{22} - |k_{12} + \bar{k}_{21}|^2 = 4(\text{Re } y_{11})(\text{Re } y_{22}) - |y_{12} + \bar{y}_{21}|^2 = 4G_1G_2 - g_m^2 \tag{3.49c}$$

Thus, the device is passive for all real frequencies if $4G_1G_2 \geqslant g_m^2$ and active for all real frequencies if $4G_1G_2 < g_m^2$, confirming an earlier assertion. On the positive σ-axis, the device is passive if and only if

$$m_{11} = y_{11}(\sigma) = G_1 + \sigma(C_1 + C_2) \geqslant 0 \tag{3.50a}$$

$$m_{22} = y_{22}(\sigma) = G_2 + \sigma C_2 \geqslant 0 \tag{3.50b}$$

$$\begin{aligned} 4m_{11}m_{22} - |k_{12} + \bar{k}_{21}|^2 &= 4y_{11}(\sigma)y_{22}(\sigma) - [y_{12}(\sigma) + y_{21}(\sigma)]^2 \\ &= 4[C_1C_2\sigma^2 + (C_1G_2 + C_2G_2 + G_1C_2 + g_mC_2)\sigma \\ &\quad + G_1G_2 - \tfrac{1}{4}g_m^2] \geqslant 0 \end{aligned} \tag{3.50c}$$

resulting in the same set of inequalities as in Eq. (1.168).

The example demonstrates that even though the same conclusion can be reached from any one of the four matrix characterizations, the amount of computation involved is quite different. Thus, to test for passivity or activity of a device, the choice of an appropriate matrix description is as important as the test itself. The reader is urged to compare the amount of labor involved in the four cases.

3.5 THE *U*-FUNCTIONS

In this section, we introduce a useful parameter associated with a two-port network N and study some of its properties first given by Mason (1954). Specifically, we show that the parameter has the physical meaning of the maximum unilateral power gain of a two-port network under a lossless reciprocal imbedding, and it is invariant under all lossless reciprocal imbedding. Before we proceed, we comment on the difference between the words *imbedding* and *terminating.* The general connection of imbedding allows external connections to be made between the ports, whereas the specialized connection of terminating does not.

Depending on the choice of the two-port parameters, the *U-function* is defined by the expression

$$U = \frac{|k_{21} - k_{12}|^2}{4(\text{Re } k_{11} \text{ Re } k_{22} - \text{Re } k_{12} \text{ Re } k_{21})} = \frac{|k_{21} - k_{12}|^2}{4(m_{11}m_{22} - m_{12}m_{21})} \tag{3.51a}$$

for z- or y-parameters, and

$$U = \frac{|k_{21} + k_{12}|^2}{4(\text{Re } k_{11} \text{ Re } k_{22} + \text{Im } k_{12} \text{ Im } k_{21})} = \frac{|k_{21} + k_{12}|^2}{4(m_{11}m_{22} + n_{12}n_{21})} \tag{3.51b}$$

for h- or g-parameters. The reason that the U-function has different expressions for the immittance and hybrid parameters is that, as will be shown shortly, it is dependent on reciprocity, and reciprocity is expressed differently for the immittance and hybrid parameters. For a reciprocal two-port network, we have

$$z_{12} = z_{21} \qquad y_{12} = y_{21} \tag{3.52a}$$

or

$$h_{12} = -h_{21} \qquad g_{12} = -g_{21} \tag{3.52b}$$

To motivate our discussion, we first demonstrate that the U-function is closely related to the determinant of the hermitian part of the general hybrid matrix $\mathbf{H}(s)$, which is not to be confused with (3.1c) although they share the same symbol. The hermitian part of $\mathbf{H}(s)$ is given by (3.35), whose determinant is found to be

$$\begin{aligned} 4 \det \mathbf{H}_h(s) &= 4m_{11}m_{22} - m_{12}^2 - m_{21}^2 - n_{12}^2 - n_{21}^2 - 2m_{12}m_{21} + 2n_{12}n_{21} \\ &= 4m_{11}m_{22} - 4m_{12}m_{21} - |k_{21} - k_{12}|^2 = 4(m_{11}m_{22} - m_{12}m_{21})(1-U) \end{aligned} \tag{3.53a}$$

for z- or y-parameters, provided that $m_{11}m_{22} - m_{12}m_{21} \neq 0$, and

$$4 \det \mathbf{H}_h(s) = 4m_{11}m_{22} + 4n_{12}n_{21} - |k_{21} + k_{12}|^2 = 4(m_{11}m_{22} + n_{12}n_{21})(1-U) \tag{3.53b}$$

for h- or g-parameters, provided that $m_{11}m_{22} + n_{12}n_{21} \neq 0$.

According to Theorem 3.1, the two-port network is passive at a point $s_0 = \sigma_0$ or $j\omega_0$ if and only if the three conditions of (3.39) are satisfied. The third condition (3.39b) is equivalent to $\det \mathbf{H}_h(s_0) \geqslant 0$. In the case where N is reciprocal, $U = 0$ because of (3.52), and (3.53) is nonnegative if and only if

$$m_{11}m_{22} \geqslant m_{12}^2 \tag{3.54a}$$

for z- or y-parameters, and

$$m_{11}m_{22} \geqslant n_{12}^2 \tag{3.54b}$$

for h- or g-parameters. On the other hand, if N is nonreciprocal, then (3.53) is nonnegative if and only if

$$0 \leqslant U \leqslant 1 \tag{3.55}$$

This follows directly from the observation that $m_{11}m_{22} - m_{12}m_{21}$ or $m_{11}m_{22} + n_{12}n_{21}$ and U must have the same sign, as dictated by (3.51). For $U = 1$, (3.53) is identically zero or (3.39b) is satisfied with the equality sign. Thus, the condition $U = 1$ serves as a convenient value for the computation of the maximum frequency of oscillation and the fastest regenerative mode of N.

Example 3.4 Consider the two-port network of Fig. 3.3, whose admittance matrix is given by (3.31). On the real-frequency axis, the admittance matrix becomes

$$\mathbf{Y}(j\omega) = \frac{1}{576 + 11{,}025\hat{\omega}^2} \begin{bmatrix} 1.92 + 220.5\hat{\omega}^2 + j42\hat{\omega} & -10.5\hat{\omega}^2 - j2.4\hat{\omega} \\ 96 - 10.5\hat{\omega}^2 - j422.4\hat{\omega} & 105.6\hat{\omega}^2 + j(26.88\hat{\omega} + 52.5\hat{\omega}^3) \end{bmatrix} \tag{3.56}$$

where $\hat{\omega} = \omega/10^9$, and the associated U-function is obtained as

$$U(j\omega) = \frac{|y_{21} - y_{12}|^2}{4(\text{Re } y_{11} \text{ Re } y_{22} - \text{Re } y_{12} \text{ Re } y_{21})} = \frac{44{,}100\hat{\omega}^2 + 2304}{23{,}174.55\hat{\omega}^4 + 1210.752\hat{\omega}^2} \tag{3.57}$$

Setting $U = 1$ yields (3.42), which determines the maximum frequency of oscillation.

On the σ-axis, the admittance matrix becomes

$$\mathbf{Y}(\sigma) = \frac{1}{240 + 1050\hat{\sigma}} \begin{bmatrix} 0.8 + 21\hat{\sigma} & -\hat{\sigma} \\ 40 - \hat{\sigma} & 11.2\hat{\sigma} + 5\hat{\sigma}^2 \end{bmatrix} \tag{3.58}$$

where $\hat{\sigma} = \sigma/10^9$, and the U-function is found to be

$$U(\sigma) = \frac{1600}{4[(0.8 + 21\hat{\sigma})(11.2\hat{\sigma} + 5\hat{\sigma}^2) - (40 - \hat{\sigma})(-\hat{\sigma})]} = \frac{400}{105\hat{\sigma}^3 + 238.2\hat{\sigma}^2 + 48.96\hat{\sigma}} \tag{3.59}$$

Setting $U = 1$ and solving for $\hat{\sigma}$, we get the fastest regenerative mode $\sigma_m = 1.01 \cdot 10^9$ Np/s, as in (3.45).

Before we turn our attention to other properties of the U-function, we remark that the significance of passivity is that it is the formal negation of activity. Thus, if $U(j\omega)$ is larger than unity, the two-port network is active. With appropriate passive reciprocal imbedding, power gain can be achieved for the device. In fact, we shall demonstrate in the succeeding section that $U(j\omega)$ has the physical meaning of the maximum unilateral power gain of a device under a lossless reciprocal imbedding, thereby making it a good measure of the inherent power-amplifying ability of the device.

3.5.1 The Invariance of the *U*-Function

In this section, we show that the U-function is invariant under the lossless reciprocal imbedding. This property was first obtained by Mason (1954) when he studied some invariant characteristics of two-port networks. For our purposes, it is sufficient to consider only the admittance parameters, leaving the other representations as being obvious.

The physical significance of this result is that if $U(j\omega)$ is used to characterize the power-amplifying capability of a device, then it is unique in that, among other reasons, it is independent of the measuring circuit, provided only that the circuit uses lossless reciprocal elements. Consequently, it measures an inherent characteristic of the device, not the device used in a particular way.

Consider a two-port network N that is imbedded in a lossless reciprocal four-port network N_0, as shown in Fig. 3.5. Let $\mathbf{Y}(s)$ and $\mathbf{Y}_0(s)$ be the admittance matrices characterizing the two-port network N and the four-port network N_0,

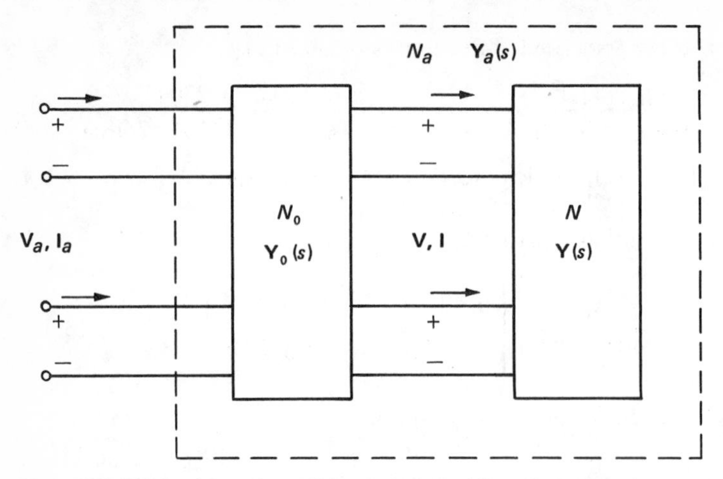

Figure 3.5 The lossless reciprocal four-port imbedding of a two-port network N.

respectively. For the time being, assume that $\mathbf{Y}_0(s)$ exists. Refer to Fig. 3.5. The matrix $\mathbf{Y}_0(s)$ is defined by the relation

$$\begin{bmatrix} \mathbf{I}_a \\ -\mathbf{I} \end{bmatrix} = \begin{bmatrix} \mathbf{Y}_{11}^0 & \mathbf{Y}_{12}^0 \\ \mathbf{Y}_{21}^0 & \mathbf{Y}_{22}^0 \end{bmatrix} \begin{bmatrix} \mathbf{V}_a \\ \mathbf{V} \end{bmatrix} \tag{3.60}$$

where the coefficient matrix is the matrix $\mathbf{Y}_0(s)$ in partitioned form. Substituting

$$\mathbf{I} = \mathbf{Y}(s)\mathbf{V} \tag{3.61}$$

in the second equation of (3.60) and solving for $\mathbf{V}$ yield

$$\mathbf{V} = -(\mathbf{Y} + \mathbf{Y}_{22}^0)^{-1}\mathbf{Y}_{21}^0\mathbf{V}_a \tag{3.62}$$

where $(\mathbf{Y} + \mathbf{Y}_{22}^0)$ is assumed to be nonsingular, to be justified shortly. Substituting (3.62) in the first equation of (3.60) gives

$$\mathbf{I}_a = [\mathbf{Y}_{11}^0 - \mathbf{Y}_{12}^0(\mathbf{Y} + \mathbf{Y}_{22}^0)^{-1}\mathbf{Y}_{21}^0]\mathbf{V}_a \tag{3.63}$$

showing that the admittance matrix $\mathbf{Y}_a(s)$ of the composite two-port network N_a of Fig. 3.5 is given by

$$\mathbf{Y}_a(s) = \mathbf{Y}_{11}^0 - \mathbf{Y}_{12}^0(\mathbf{Y} + \mathbf{Y}_{22}^0)^{-1}\mathbf{Y}_{21}^0 \tag{3.64}$$

Our objective is to show that, on the real-frequency axis, the U-functions defined by the elements of $\mathbf{Y}(s)$ and those of $\mathbf{Y}_a(s)$ are the same. To this end, let U and U_a be the U-functions associated with $\mathbf{Y}(s)$ and $\mathbf{Y}_a(s)$, respectively. Then, from the definition of the U-function, U and U_a can be manipulated into the forms

$$U = \frac{|\det [\mathbf{Y}(s) - \mathbf{Y}'(s)]|}{\det [\mathbf{Y}(s) + \bar{\mathbf{Y}}(s)]} \tag{3.65a}$$

$$U_a = \frac{|\det [\mathbf{Y}_a(s) - \mathbf{Y}_a'(s)]|}{\det [\mathbf{Y}_a(s) + \bar{\mathbf{Y}}_a(s)]} \tag{3.65b}$$

where the prime, as before, denotes the matrix transpose. In the following, we show that $U(j\omega) = U_a(j\omega)$.

Since N_0 is assumed to be lossless and reciprocal, on the $j\omega$-axis we have

$$\mathbf{Y}_{ik}^0(j\omega) = j\mathbf{B}_{ik} \qquad i, k = 1, 2 \tag{3.66a}$$

$$\mathbf{B}_{ik} = \mathbf{B}'_{ki} \qquad i, k = 1, 2 \tag{3.66b}$$

$\mathbf{B}_{ik}$ being matrices of real elements. Using (3.66) in (3.64) yields

$$\mathbf{Y}_a(j\omega) + \bar{\mathbf{Y}}_a(j\omega) = \mathbf{B}_{12}(\mathbf{W}^{-1} + \bar{\mathbf{W}}^{-1})\mathbf{B}'_{12} = \mathbf{B}_{12}\mathbf{W}^{-1}[\mathbf{Y}(j\omega) + \bar{\mathbf{Y}}(j\omega)]\bar{\mathbf{W}}^{-1}\mathbf{B}'_{12} \tag{3.67}$$

where

$$\mathbf{W} = \mathbf{Y}(j\omega) + \mathbf{Y}_{22}^0(j\omega) \tag{3.68}$$

The determinant of (3.67) is found to be

$$\det[\mathbf{Y}_a(j\omega) + \bar{\mathbf{Y}}_a(j\omega)] = \frac{(\det \mathbf{B}_{12})^2 \det[\mathbf{Y}(j\omega) + \bar{\mathbf{Y}}(j\omega)]}{|\det \mathbf{W}|^2} \tag{3.69}$$

In a similar manner, we can show that

$$\mathbf{Y}_a(j\omega) - \mathbf{Y}'_a(j\omega) = \mathbf{B}_{12}\mathbf{W}^{-1}[\mathbf{Y}'(j\omega) - \mathbf{Y}(j\omega)]\mathbf{W}'^{-1}\mathbf{B}'_{12} \tag{3.70}$$

whose determinant is given by

$$\det[\mathbf{Y}_a(j\omega) - \mathbf{Y}'_a(j\omega)] = \frac{(\det \mathbf{B}_{12})^2 \det[\mathbf{Y}'(j\omega) - \mathbf{Y}(j\omega)]}{(\det \mathbf{W})^2} \tag{3.71}$$

Substituting (3.69) and (3.71) in (3.65*b*) in conjunction with (3.65*a*) yields

$$U_a(j\omega) = U(j\omega) \tag{3.72}$$

In the case where the admittance matrix $\mathbf{Y}_0(s)$ does not exist or $\mathbf{W}$ is identically singular, we first insert some lossless reciprocal elements of finite nonzero values in N_0 so that in the resulting four-port network, $\mathbf{Y}_0(s)$ exists and $\mathbf{W}$ is nonsingular. We then take the limit in the resulting $U_a(j\omega)$ of (3.65*b*) as the values of these added elements approach zero or infinity so that the resulting network becomes N_0. Since $U_a(j\omega)$ is independent of the added elements, we arrive at the same conclusion as in (3.72).

For future reference, we summarize the above result as follows.

Theorem 3.2 On the real-frequency axis, the U-function of a linear, time-invariant two-port network is invariant under the lossless reciprocal imbedding.

We illustrate this theorem by the following examples.

Example 3.5 Figure 3.6 shows an active two-port network N_a in which a feedback admittance y_f is in shunt from output to input. Assume that the one-port admittances y_a, y_f, and y_b are lossless and reciprocal. Then N_a can be viewed as the lossless reciprocal four-port imbedding of the active two-port network N, as shown in Fig. 3.7. To compute the admittance matrix $\mathbf{Y}_a(s)$ of N_a, it is convenient to consider N_a as the parallel combination of two two-port networks, so that $\mathbf{Y}_a(s)$ can

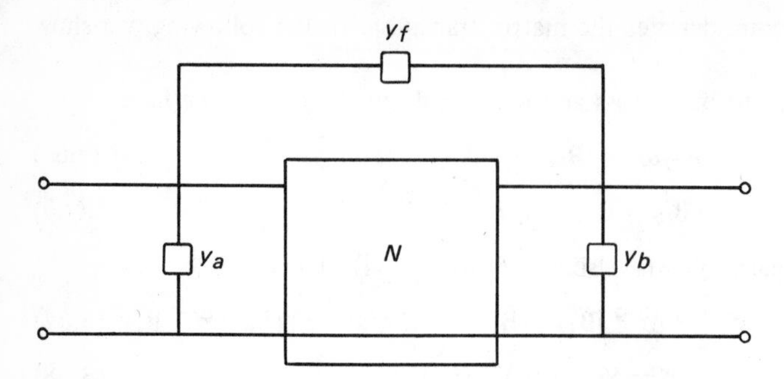

Figure 3.6 An active two-port network with a feedback admittance y_f connected in shunt from output to input.

be expressed as the sum of the admittance matrices of the component two-ports, as follows:

$$\mathbf{Y}_a(s) = \begin{bmatrix} y_{11} & y_{12} \\ y_{21} & y_{22} \end{bmatrix} + \begin{bmatrix} y_a + y_f & -y_f \\ -y_f & y_b + y_f \end{bmatrix} \tag{3.73}$$

where y_{ij} are the y-parameters of N. Since y_a, y_f, and y_b are lossless, on the real-frequency axis their real parts are zero, and the U-function of N_a is found to be

$$\begin{aligned} U_a(j\omega) &= \frac{|y_{a21} - y_{a12}|^2}{4(\text{Re } y_{a11} \text{ Re } y_{a22} - \text{Re } y_{a12} \text{ Re } y_{a21})} \\ &= \frac{|y_{21} - y_{12}|^2}{4(\text{Re } y_{11} \text{ Re } y_{22} - \text{Re } y_{12} \text{ Re } y_{21})} = U(j\omega) \end{aligned} \tag{3.74}$$

the U-function of N, where y_{aij} are the y-parameters of N_a. This confirms that the U-function is invariant under the lossless reciprocal imbedding.

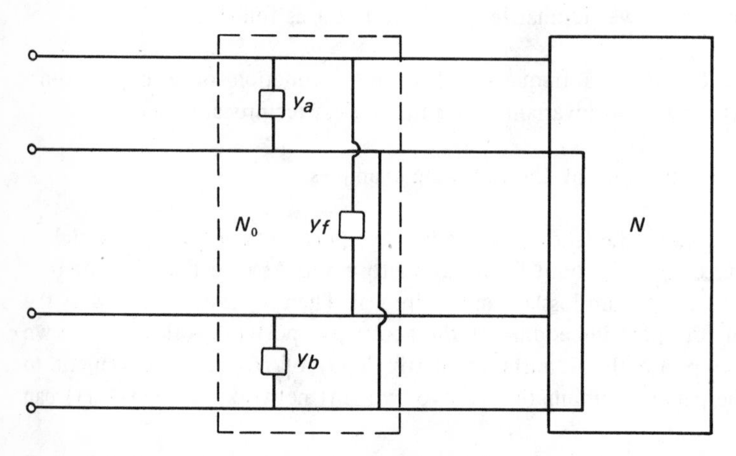

Figure 3.7 The lossless reciprocal four-port imbedding of an active two-port network N.

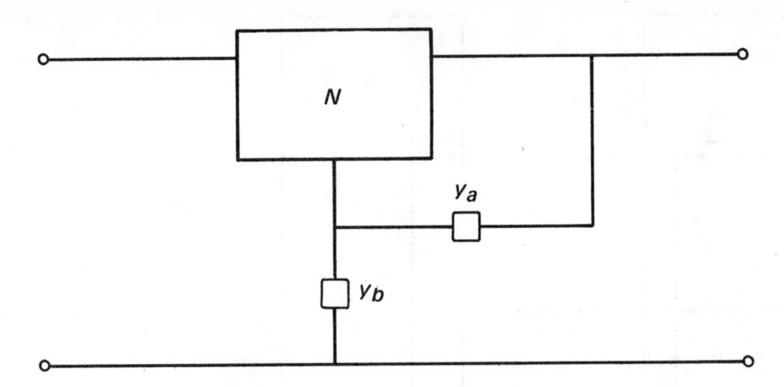

Figure 3.8 A three-terminal device N with a series feedback provided by the admittances y_a and y_b.

Example 3.6 A three-terminal device N with a series feedback is shown in Fig. 3.8. Assume that the feedback admittances y_a and y_b are lossless and reciprocal. Then the complete network N_a of Fig. 3.8 can be viewed as the lossless reciprocal four-port imbedding of the active device N, as shown in Fig. 3.9. Since in the foregoing we have concentrated on the admittance basis, in this example we shall pick the impedance matrix for a change.

Let y_{ij} and z_{aij} be the y-parameters and the z-parameters of N and N_a, respectively. To compute the impedance matrix $\mathbf{Z}_a(s)$ of N_a, it is convenient to represent N_a as the series combination of two two-ports as shown in Fig. 3.10, so that $\mathbf{Z}_a(s)$ can be written as the sum of the impedance matrices of the two component two-ports, as follows:

$$\mathbf{Z}_a(s) = \begin{bmatrix} y_{11} & y_{12} \\ y_{21} & y_{22} + y_a \end{bmatrix}^{-1} + \begin{bmatrix} \dfrac{1}{y_b} & \dfrac{1}{y_b} \\ \dfrac{1}{y_b} & \dfrac{1}{y_b} \end{bmatrix}$$

$$= \frac{1}{\Delta_y + y_{11}y_a} \begin{bmatrix} y_{22} + y_a & -y_{12} \\ -y_{21} & y_{11} \end{bmatrix} + \begin{bmatrix} \dfrac{1}{y_b} & \dfrac{1}{y_b} \\ \dfrac{1}{y_b} & \dfrac{1}{y_b} \end{bmatrix} \tag{3.75}$$

where $\Delta_y = y_{11}y_{22} - y_{12}y_{21}$. To compute the U-function $U_a(j\omega)$ of N_a, we first calculate $|z_{a21} - z_{a12}|^2$. From (3.75) we find that

$$|z_{a21} - z_{a12}|^2 = \frac{|y_{21} - y_{12}|^2}{|\Delta_y + y_{11}y_a|^2} \tag{3.76}$$

To compute the denominator of (3.51a), we use

$$4(\text{Re } z_{a11} \text{ Re } z_{a22} - \text{Re } z_{a12} \text{ Re } z_{a21}) = \det\,[\mathbf{Z}_a(j\omega) + \bar{\mathbf{Z}}_a(j\omega)]$$

$$= \det\,[\mathbf{Z}_b(j\omega) + \bar{\mathbf{Z}}_b(j\omega)] = \frac{2\text{ Re }(y_{11}\bar{y}_{22} + y_{11}\bar{y}_a - y_{12}\bar{y}_{21} + \Delta_y + y_{11}y_a)}{|\Delta_y + y_{11}y_a|^2}$$

$$= \frac{4(\text{Re } y_{11} \text{ Re } y_{22} - \text{Re } y_{12} \text{ Re } y_{21})}{|\Delta_y + y_{11}y_a|^2} \tag{3.77}$$

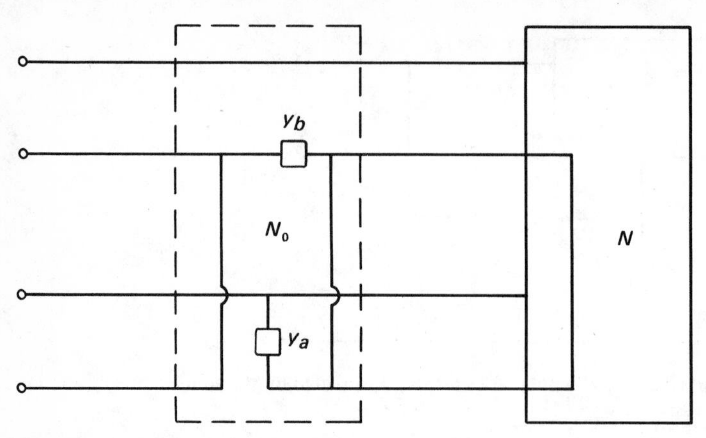

Figure 3.9 The lossless reciprocal four-port imbedding of an active device N.

in which we have invoked the relation $\bar{y}_a(j\omega) = -y_a(j\omega)$, where

$$\mathbf{Z}_b(j\omega) = \frac{1}{\Delta_y + y_{11}y_a}\begin{bmatrix} y_{22} + y_a & -y_{12} \\ -y_{21} & y_{11} \end{bmatrix} \tag{3.78}$$

Substituting (3.76) and (3.77) in (3.51a) yields

$$U_a(j\omega) = \frac{|y_{21} - y_{12}|^2}{4(\text{Re } y_{11} \text{ Re } y_{22} - \text{Re } y_{12} \text{ Re } y_{21})} \tag{3.79}$$

Next we compute the U-function of N on the impedance basis. The impedance matrix of N, being the inverse of its admittance matrix, is given by (3.78) with $y_a = 0$. Thus, if z_{ij} denote the z-parameters of N, we have

$$|z_{21} - z_{12}|^2 = \frac{|y_{21} - y_{12}|^2}{|\Delta_y|^2} \tag{3.80a}$$

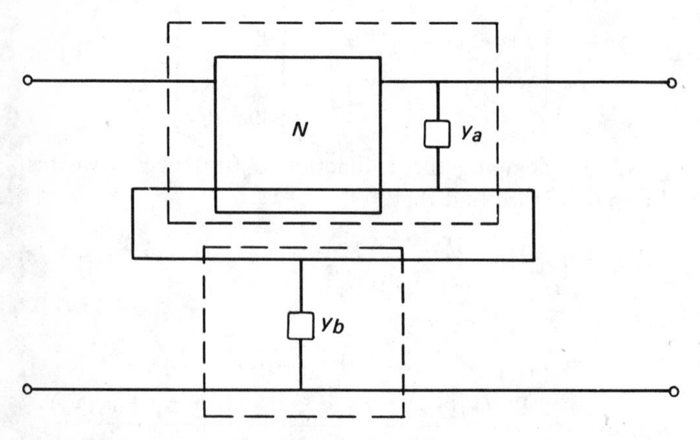

Figure 3.10 The series combination of two two-port networks.

$$4(\text{Re } z_{11} \text{ Re } z_{22} - \text{Re } z_{12} \text{ Re } z_{21}) = \det\,[\mathbf{Z}(j\omega) + \bar{\mathbf{Z}}(j\omega)]$$
$$= \frac{2 \text{ Re } (\Delta_y + y_{22}\bar{y}_{11} - \bar{y}_{12}y_{21})}{|\Delta_y|^2}$$
$$= \frac{\text{Re } y_{11} \text{ Re } y_{22} - \text{Re } y_{12} \text{ Re } y_{21}}{|\Delta_y|^2} \tag{3.80b}$$

where

$$\mathbf{Z}(j\omega) = \frac{1}{\Delta_y}\begin{bmatrix} y_{22} & -y_{12} \\ -y_{21} & y_{11} \end{bmatrix} \tag{3.81}$$

Substituting (3.80) in (3.51*a*) gives

$$U(j\omega) = \frac{|y_{21} - y_{12}|^2}{4(\text{Re } y_{11} \text{ Re } y_{22} - \text{Re } y_{12} \text{ Re } y_{21})} \tag{3.82}$$

which when compared with (3.79) shows that

$$U_a(j\omega) = U(j\omega) \tag{3.83}$$

again confirming that the U-function is invariant under the lossless reciprocal imbedding (see Prob. 3.32). In fact, (3.79) or (3.82) is also the U-function of N defined on the admittance basis. Thus, a two-port network will possess the same U-function regardless of the matrix employed in computation.

3.5.2 The Maximum Unilateral Power Gain

In the preceding section, we showed that the U-function of a two-port network is invariant under the lossless reciprocal imbedding. In the present section, we demonstrate that for a given active two-port network, its U-function is identifiable as the maximum unilateral power gain under a lossless reciprocal imbedding, the resulting structure being unilateral.

Consider a three-terminal device N that has two lossless reciprocal admittances y_α and y_f, one in series with the output and the other in shunt from output to input, as shown in Fig. 3.11. The network can be viewed as the lossless reciprocal

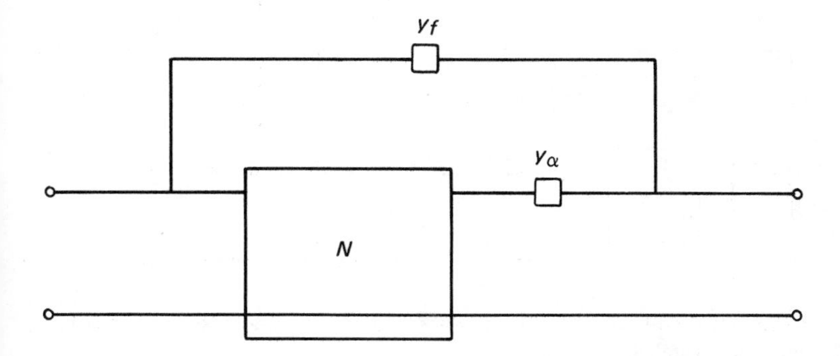

Figure 3.11 A three-terminal device N with two lossless reciprocal admittances y_α and y_f.

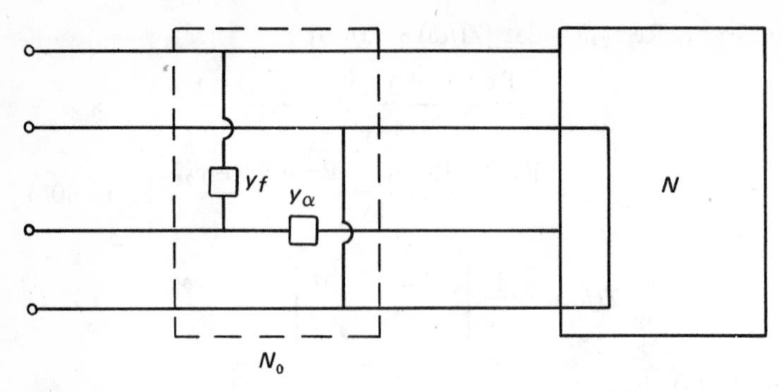

Figure 3.12 The lossless reciprocal four-port imbedding of an active device N.

four-port imbedding of the active device N as shown in Fig. 3.12. To compute the admittance matrix $\mathbf{Y}_a(s)$ of the network N_a of Fig. 3.11, it is advantageous to consider N_a as being composed of two two-ports connected in cascade and then in parallel with another, as indicated in Fig. 3.13. The transmission matrix of the two two-ports N and N_α connected in cascade is obtained as

$$\frac{1}{y_\alpha y_{21}}\begin{bmatrix} y_{22} & 1 \\ \Delta_y & y_{11} \end{bmatrix}\begin{bmatrix} -y_\alpha & -1 \\ 0 & -y_\alpha \end{bmatrix} \tag{3.84}$$

where y_{ij} are the y-parameters of N and $\Delta_y = y_{11}y_{22} - y_{12}y_{21}$. The corresponding admittance matrix becomes

$$\frac{y_\alpha}{y_\alpha + y_{22}}\begin{bmatrix} y_{11} + \dfrac{\Delta_y}{y_\alpha} & y_{12} \\ y_{21} & y_{22} \end{bmatrix} \tag{3.85}$$

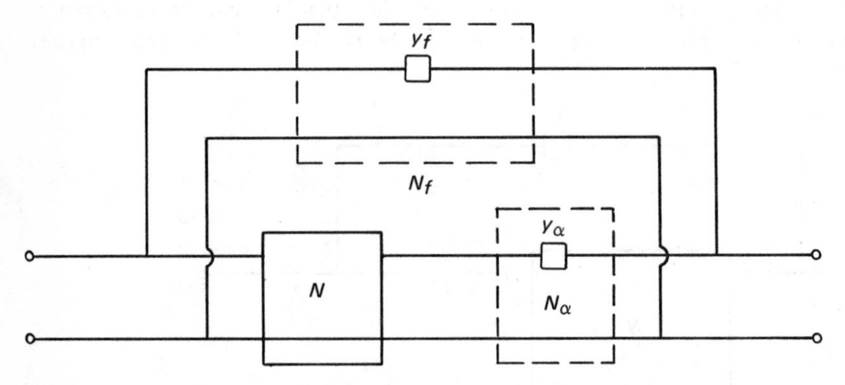

Figure 3.13 The decomposition of the two-port network of Fig. 3.11 into two two-ports connected in cascade and then in parallel with another.

The admittance matrix $\mathbf{Y}_a(s)$ of the overall network N_a of Fig. 3.11 is found to be

$$\mathbf{Y}_a(s) = \frac{y_\alpha}{y_\alpha + y_{22}} \begin{bmatrix} y_{11} + \dfrac{\Delta_y}{y_\alpha} & y_{12} \\ y_{21} & y_{22} \end{bmatrix} + \begin{bmatrix} y_f & -y_f \\ -y_f & y_f \end{bmatrix} \tag{3.86}$$

Our objective is to determine the admittances y_α and y_f so that the resulting network N_a is unilateralized at a specified frequency on the real-frequency axis. To this end, we set the reverse transfer admittance, the $(1, 2)$-element of $\mathbf{Y}_a(j\omega)$, to zero, yielding

$$y_f = \frac{y_{12} y_\alpha}{y_{22} + y_\alpha} \tag{3.87}$$

Since y_α and y_f are lossless, we can write

$$y_\alpha(j\omega) = jb_\alpha \qquad y_f(j\omega) = jb_f \tag{3.88}$$

b_α and b_f being real susceptances. Substituting (3.88) in (3.87) results in

$$b_\alpha = b_f \frac{\text{Re } y_{22}}{\text{Re } y_{12}} \tag{3.89a}$$

$$b_f = \text{Im } y_{12} - \frac{\text{Re } y_{12}}{\text{Re } y_{22}} \text{ Im } y_{22} \tag{3.89b}$$

provided that $\text{Re } y_{22} \neq 0$. By choosing these values for b_α and b_f, the two-port network N_a is unilateralized, whose admittance matrix is found to be

$$\mathbf{Y}_a(s) = \frac{j \text{ Im } (\bar{y}_{22} y_{12})}{y_{12} \text{ Re } y_{22}} \begin{bmatrix} y_{11} + y_{12} - j \dfrac{\Delta_y \text{ Re } y_{12}}{\text{Im } (\bar{y}_{22} y_{12})} & 0 \\ y_{21} - y_{12} & y_{22} + y_{12} \end{bmatrix} \tag{3.90}$$

The U-function associated with the unilateralized two-port network N_a becomes

$$U_a(j\omega) = \frac{|y_{a21}|^2}{4 \text{ Re } y_{a11} \text{ Re } y_{a22}} \tag{3.91}$$

where y_{aij} are the elements of $\mathbf{Y}_a(s)$. The equivalent network of (3.90) is presented in Fig. 3.14. Thus, any active three-terminal device can be unilateralized by the

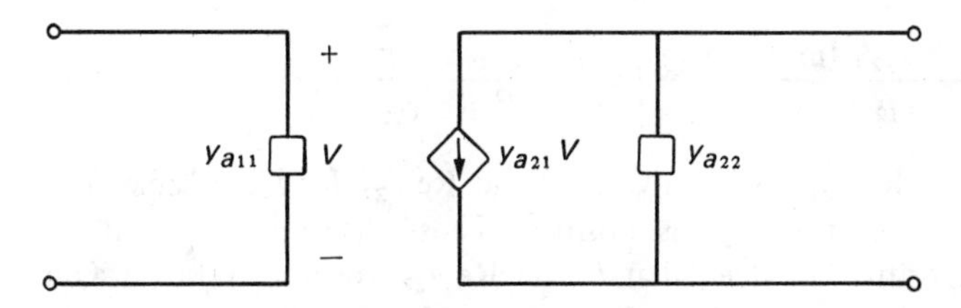

Figure 3.14 An equivalent network of the admittance matrix (3.90).

above procedure provided that its output short-circuit conductance $\text{Re}\, y_{22} \neq 0$. In the case where $\text{Re}\, y_{22} = 0$, as will be discussed in the succeeding section, the original device is potentially unstable, and an infinite power gain can be achieved, rendering the concept of maximum unilateral power gain meaningless.

Now we demonstrate that the maximum transducer power gain or the available power gain that can be achieved for the unilateralized two-port network N_a is precisely $U(j\omega)$. To this end, let us conjugate match the input and output ports as shown in Fig. 3.15. Under this situation, the maximum transducer power gain becomes equal to the available power gain and is given by

$$\mathcal{G}_a = \frac{|V_2|^2 \text{ Re } y_{a22}}{|V_1|^2 \text{ Re } y_{a11}} = \frac{|y_{a21}|^2}{4 \text{ Re } y_{a11} \text{ Re } y_{a22}} = U_a(j\omega) \tag{3.92}$$

all being evaluated on the $j\omega$-axis. The last equality follows from (3.91). Since the U-function is invariant under all lossless reciprocal imbedding, the U-function of N must be the same as $U_a(j\omega)$ (see Prob. 3.6),

$$\mathcal{G}_a = U_a(j\omega) = U(j\omega) \tag{3.93}$$

In fact, any lossless reciprocal imbedding that unilateralizes a three-terminal device will yield the same maximum transducer power gain of (3.93). Thus, the U-function of an active three-terminal device has the physical meaning of the maximum unilateral transducer power gain. Furthermore, from (3.86) we have

$$\text{Re } y_{a22}(j\omega) = \text{Re } \frac{jb_\alpha y_{22}(j\omega)}{y_{22}(j\omega) + jb_\alpha} = \frac{b_\alpha^2 \text{ Re } y_{22}}{(\text{Re } y_{22})^2 + (b_\alpha + \text{Im } y_{22})^2} \tag{3.94}$$

showing that $\text{Re}\, y_{a22}$ and $\text{Re}\, y_{22}$ have the same sign. For the unilateralized two-port network N_a, the short-circuit input and output conductances can be calculated directly from (3.90), giving

$$\begin{aligned}\text{Re } y_{a11} &= \text{Re}\left\{\frac{j \text{ Im }(\bar{y}_{22}y_{12})}{y_{12} \text{ Re } y_{22}}\left[y_{11} + y_{12} - j\frac{\Delta_y \text{ Re } y_{12}}{\text{Im }(\bar{y}_{22}y_{12})}\right]\right\}\\ &= \frac{\text{Re }\{y_{11}\bar{y}_{12}[j \text{ Im }(\bar{y}_{22}y_{12}) + y_{22} \text{ Re } y_{12}] - y_{21}|y_{12}|^2 \text{ Re } y_{12}\}}{|y_{12}|^2 \text{ Re } y_{22}}\\ &= \frac{\text{Re } y_{11} \text{ Re } y_{22} - \text{Re } y_{12} \text{ Re } y_{21}}{\text{Re } y_{22}} = \frac{|y_{21} - y_{12}|^2}{4U \text{ Re } y_{22}}\end{aligned} \tag{3.95a}$$

$$\text{Re } y_{a22} = \text{Re}\left[\frac{j(y_{22} + y_{12}) \text{ Im }(\bar{y}_{22}y_{12})}{y_{12} \text{ Re } y_{22}}\right] = \frac{|\text{Im }(\bar{y}_{22}y_{12})|^2}{|y_{12}|^2 \text{ Re } y_{22}} \tag{3.95b}$$

Observe from (3.95) that $\text{Re}\, y_{a11}$ is positive if U and $\text{Re}\, y_{22}$ have the same sign, and that $\text{Re}\, y_{a22}$ is positive if $\text{Re}\, y_{22}$ is positive. Thus, the input and output conductances are both positive provided that U and $\text{Re}\, y_{22}$ are both positive. Our conclusion is, therefore, that for positive U a unilateral network can always be chosen to give $\text{Re}\, y_{a11}$ and $\text{Re}\, y_{a22}$ the same sign as $\text{Re}\, y_{22}$. By moving the

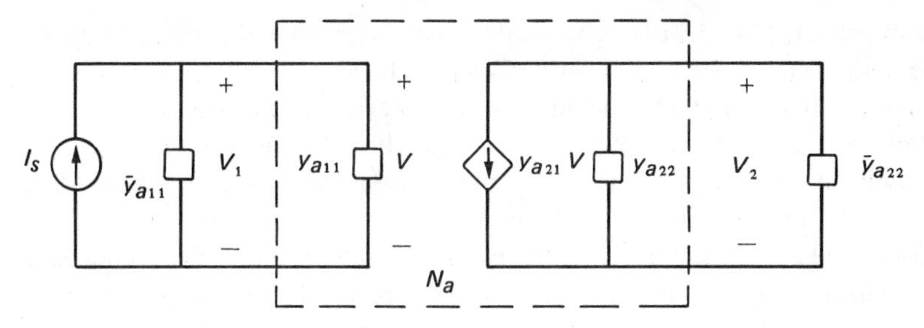

Figure 3.15 A unilateral two-port network with conjugate-matched input and output ports.

admittance y_α from the output port to the input port, the same conclusion can be reached with respect to $\text{Re}\, y_{11}$ (see Prob. 3.33). We summarize the above results by the following theorem.

Theorem 3.3 At a specified frequency on the real-frequency axis, a three-terminal device whose U-function and at least one of whose short-circuit driving-point conductances are positive can be transformed by means of a lossless reciprocal imbedding into a unilateral, common-terminal two-port network having positive short-circuit input and output conductances, its available power gain always being equal to its U-function.

We remark that, in the above theorem, the U is not the greatest attainable power gain under all passive imbedding; it is the maximum obtainable power gain under all lossless reciprocal imbedding (see Probs. 3.36 and 3.37).

The basic performance such as the power-amplifying ability of an active device can be measured meaningfully by its U-function. The measurement is unique in that it is independent of the measuring circuit, provided only that the circuit is unilateral and uses lossless reciprocal elements. As demonstrated above, such a measurement is usually physically possible and can be achieved by the network proposed above. For the transistor, for example, it is a three-terminal device. Its power-amplifying ability is usually measured in terms of the maximum power gain available in a specified network. Such a specification depends critically on the details of the particular network used. It is meaningless to compare different measurements unless the same measuring network is used. Thus, it is difficult to relate the results to basic transistor properties. Furthermore, the orientation of the transistor terminals such as the common-base, common-emitter, or common-collector configuration results in a critical difference in the measurements. The U-function, on the other hand, provides a unique measure of the degree of inherent transfer activity exhibited by a device. Since the U-function is invariant under the lossless reciprocal imbedding, it is invariant under all orientations of the device. Thus, it measures a general characteristic of the device, not merely a particular way the device is used.

As will be shown in the following section, a unilateral two-port network is "absolutely stable" at a specified frequency if and only if the real parts of its short-circuit input and output admittances are positive. It follows from Theorem 3.3

that the network of Fig. 3.11 is absolutely stable and can always be used to measure the unilateral power gain, provided that U is positive and at least one of the short-circuit driving-point conductances of the device is positive. However, if both conductances are negative and U exceeds unity, the network of Fig. 3.11 cannot be used to measure U, and a different lossless reciprocal unilateralizing network should be used (see, for example, Prob. 3.11). The situation seems unlikely to be encountered in practice. Transistors in particular obey the conditions stipulated in Theorem 3.3 in their normal operating range. The technique has been confirmed experimentally for a few transistors in the frequency range from 100 to 900 MHz by Rollett (1965).

Example 3.7 At 30 MHz, $V_{CE} = 20$ V, and $I_C = 20$ mA, the following values of the y-parameters apply to a sample type 2N697 transistor:

$$\begin{aligned} y_{11} &= y_{ie} = (22.5 + j14.7) \cdot 10^{-3} \text{ mho} \\ y_{12} &= y_{re} = (-0.8 - j0.38) \cdot 10^{-3} \text{ mho} \\ y_{21} &= y_{fe} = (36.6 - j91.6) \cdot 10^{-3} \text{ mho} \\ y_{22} &= y_{oe} = (1.7 + j5.7) \cdot 10^{-3} \text{ mho} \end{aligned} \tag{3.96}$$

We wish to design a unilateral two-port network to measure the U-value of this transistor at the frequency of 30 MHz.

By using the network of Fig. 3.11, the susceptances b_α and b_f are computed from Eq. (3.89):

$$b_f = 2.30 \cdot 10^{-3} \text{ mho} \tag{3.97a}$$

$$b_\alpha = -4.89 \cdot 10^{-3} \text{ mho} \tag{3.97b}$$

Let $y_f = j\omega C = jb_f$ and $y_\alpha = 1/j\omega L = jb_\alpha$, where $\omega = 60\pi \cdot 10^6$ rad/s. Then we have

$$C = 12.2 \text{ pF} \tag{3.98a}$$

$$L = 1.08\ \mu\text{H} \tag{3.98b}$$

The desired unilateral network is shown in Fig. 3.16. The U-value associated with the transistor at 30 MHz is found to be

$$U = \frac{|37.4 - j91.22|^2}{4[(22.5)(1.7) - (-0.8)(36.6)]} = 35.98 \tag{3.99}$$

which is the maximum transducer power gain or the available power gain of the transistor operating at 30 MHz under all lossless reciprocal imbedding.

From (3.90), the admittance matrix of the unilateral two-port network of Fig. 3.16 evaluated at $\omega = 60\pi \cdot 10^6$ rad/s is given by

$$\mathbf{Y}_a(j60\pi \cdot 10^6) = 10^{-3} \begin{bmatrix} 39.72 - j26.10 & 0 \\ -255.91 + j13.92 & 11.49 - j8.05 \end{bmatrix} \tag{3.100}$$

Appealing to (3.92) yields $U_a = 35.98$, as expected.

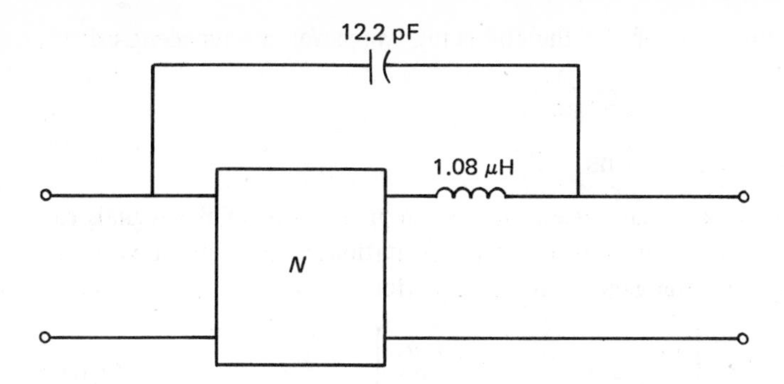

Figure 3.16 A unilateral two-port network used to measure the *U*-value of a transistor.

3.6 POTENTIAL INSTABILITY AND ABSOLUTE STABILITY

A network is said to be *stable* if all of its natural frequencies are restricted to the open LHS (*l*eft *h*alf of the complex frequency *s*-plane). The $j\omega$-axis poles of the transfer functions are not permitted for stable networks because such poles will yield an unbounded time response. If the transfer function, say, has a simple pole $j\omega_0$ on the real-frequency axis, an excitation of the type $\exp(j\omega_0 t)$ will result in an unbounded time response. For a two-port network, its stability cannot be determined by the two-port itself; it also depends on the terminations. Thus, a two-port network is said to be *potentially unstable* if it is possible to select passive one-port immittances that, when terminated at the ports, produce an unstable network. If no such passive immittances can be found, the two-port network is called *absolutely stable.* Like the discrete-frequency concepts of passivity discussed in Sec. 3.4, it is convenient to introduce single frequency concepts of stability for two-port networks.

Definition 3.2: Potential instability at $j\omega_0$ A two-port network is said to be *potentially unstable at $j\omega_0$* on the real-frequency axis if there exist two passive one-port immittances that, when terminated at the ports, produce a natural frequency at $j\omega_0$ in the overall network.

Definition 3.3: Absolute stability at $j\omega_0$ A two-port network is said to be *absolutely stable at $j\omega_0$* on the real-frequency axis if it is not potentially unstable at $j\omega_0$.

It is possible to extend the above definitions to include all the points in the closed RHS. Then a two-port network is potentially unstable if it is potentially unstable at a point in the closed RHS and is absolutely stable if it is absolutely stable at every point in the closed RHS. The restriction to the real-frequency axis is sufficient for all practical purposes.

Using the above definitions to test the stability of a two-port network is clearly difficult and tedious, because we have to examine all passive terminations, and they are not intended for this purpose. In the following, we develop equivalent criteria for these concepts.

3.6.1 Preliminary Considerations

The natural frequencies of a linear system are the frequencies for which signals can be supported by the system without external excitations. Thus, for a two-port network N, characterized by the general hybrid equation

$$\begin{bmatrix} y_1 \\ y_2 \end{bmatrix} = \begin{bmatrix} k_{11} & k_{12} \\ k_{21} & k_{22} \end{bmatrix} \begin{bmatrix} u_1 \\ u_2 \end{bmatrix} \tag{3.101}$$

and terminated in the immittances M_1 and M_2 as shown in Fig. 3.17, the natural frequencies are the values of s for which the above equation together with the boundary conditions

$$y_1 = -M_1 u_1 \tag{3.102a}$$

$$y_2 = -M_2 u_2 \tag{3.102b}$$

are satisfied simultaneously. Table 3.1 and Eq. (3.1) define the variables for various parameter sets. Combining (3.101) and (3.102) gives

$$\begin{bmatrix} k_{11} + M_1 & k_{12} \\ k_{21} & k_{22} + M_2 \end{bmatrix} \begin{bmatrix} u_1 \\ u_2 \end{bmatrix} = \mathbf{0} \tag{3.103}$$

For the above equation to possess a nontrivial solution, it is necessary and sufficient that the coefficient matrix be singular, requiring that

$$\det \begin{bmatrix} k_{11} + M_1 & k_{12} \\ k_{21} & k_{22} + M_2 \end{bmatrix} = 0 \tag{3.104a}$$

or

$$(k_{11} + M_1)(k_{22} + M_2) - k_{12}k_{21} = 0 \tag{3.104b}$$

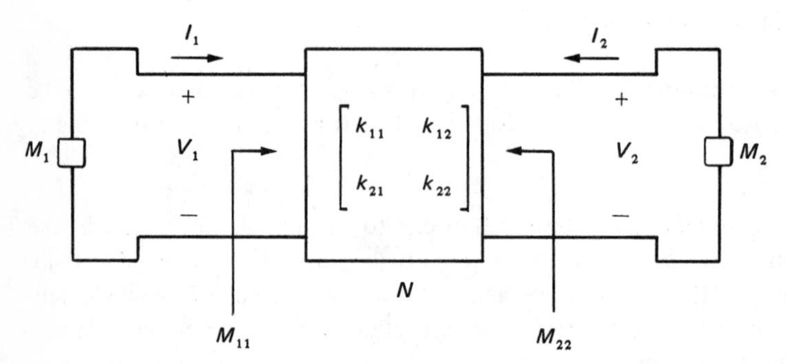

Figure 3.17 The general representation of a terminated two-port network.

which is referred to as the *characteristic equation* of the network. By using Eqs. (3.7) and (3.8), Eq. (3.104*b*) becomes

$$M_1(s) + M_{11}(s) = 0 \tag{3.105a}$$

$$M_2(s) + M_{22}(s) = 0 \tag{3.105b}$$

The left-hand sides of (3.105) are recognized as the input immittances at the input and output ports of Fig. 3.17. These provide alternative means of obtaining the characteristic equation. One may equate the impedance of the input or output loop of Fig. 3.17 to zero, or the admittance appearing across any port to zero. The zeros of these functions are the natural frequencies of the network. For example, if (3.101) is defined on the admittance basis, the left-hand sides of (3.105) denote the total admittances appearing across the input and output ports of Fig. 3.1*b*, and (3.105) can be written as

$$Y_1(s) + Y_{11}(s) = 0 \tag{3.106a}$$

$$Y_2(s) + Y_{22}(s) = 0 \tag{3.106b}$$

If (3.101) represents (3.1*c*), the corresponding equations are given by

$$Z_1(s) + Z_{11}(s) = 0 \tag{3.107a}$$

$$Y_2(s) + Y_{22}(s) = 0 \tag{3.107b}$$

The left-hand side of (3.107*a*) is the impedance looking into the voltage source V_s of Fig. 3.1*c*. The various means described above for arriving at the characteristic equation all lead to equivalent results in that the same characteristic equation is obtained.

A two-port network is therefore absolutely stable at $j\omega_0$ if and only if, for all passive terminations, its characteristic equation (3.104) does not possess a root at $j\omega_0$. In other words, at $j\omega_0$ the equation (3.103) has only a trivial solution. Thus, we can state the following.

Theorem 3.4 A linear time-invariant two-port network is absolutely stable at $j\omega_0$ if and only if the port voltages and currents are zero at $j\omega_0$ under all passive one-port terminations.

The stability of a two-port network can also be determined by its input and output conductances under all passive one-port terminations.

Theorem 3.5 A linear time-invariant two-port network is absolutely stable at $j\omega_0$ if and only if the driving-point immittance at each of its two ports has a positive-real part for all passive one-port terminations at the other port at the frequency $j\omega_0$.

PROOF It is sufficient to consider only the admittances. Referring to Fig. 3.17 and from Eq. (3.106), we conclude that to create a natural frequency at $j\omega_0$, it is necessary and sufficient that

$$\text{Re } Y_1(j\omega_0) + \text{Re } Y_{11}(j\omega_0) = 0 \tag{3.108a}$$

$$\text{Re } Y_2(j\omega_0) + \text{Re } Y_{22}(j\omega_0) = 0 \tag{3.108b}$$

Since by hypothesis $Y_1(s)$ and $Y_2(s)$ are passive, being positive-real functions, their real parts at $j\omega_0$ must be nonnegative; that is,

$$\text{Re } Y_1(j\omega_0) \geqslant 0 \tag{3.109a}$$

$$\text{Re } Y_2(j\omega_0) \geqslant 0 \tag{3.109b}$$

With the constraints (3.109), (3.108) possesses a solution if and only if Re $Y_{11}(j\omega_0) \leqslant 0$ and Re $Y_{22}(j\omega_0) \leqslant 0$ because $Y_1(s)$ and $Y_2(s)$ are arbitrary passive admittances. In other words, the two-port network is absolutely stable at $j\omega_0$ if and only if Re $Y_{11}(j\omega_0) > 0$ and Re $Y_{22}(j\omega_0) > 0$. This completes the proof of the theorem.

We illustrate the above results by the following example.

Example 3.8 Consider the two-port network N shown in Fig. 3.18, whose impedance matrix is given by

$$\mathbf{Z} = \begin{bmatrix} -1 & -2 \\ -2 & -2 \end{bmatrix} \tag{3.110}$$

In particular, if we choose the open-circuit termination, the input impedances of N become $Z_{11} = z_{11} = -1\ \Omega$ and $Z_{22} = z_{22} = -2\ \Omega$. According to Theorem 3.5, N is potentially unstable over the entire real-frequency axis. Assume that the output of N is terminated in a 1-F capacitor. By applying (3.7), the input impedance is found to be

$$Z_{11}(s) = \frac{2s+1}{2s-1} \tag{3.111}$$

Choose

$$s = j\omega_0 = j\tfrac{1}{4} \tag{3.112}$$

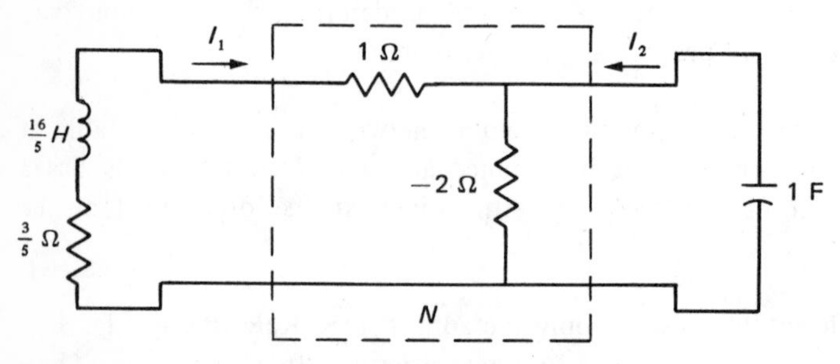

Figure 3.18 A network possessing a natural frequency at $s = j\frac{1}{4}$.

We demonstrate that a passive input termination $Z_1(s)$ can always be found, so that the resulting network possesses a natural frequency at $j\frac{1}{4}$. From (3.111) we have

$$Z_{11}(j\tfrac{1}{4}) = -\tfrac{3}{5} - j\tfrac{4}{5} \tag{3.113}$$

To create a natural frequency at $j\frac{1}{4}$, we set

$$Z_1(j\tfrac{1}{4}) = -Z_{11}(j\tfrac{1}{4}) = \tfrac{3}{5} + j\tfrac{4}{5} \tag{3.114}$$

showing that $Z_1(s)$ can be realized as the series combination of a resistor of $\frac{3}{5}$ Ω and an inductor of $\frac{16}{5}$ H. The complete network is presented in Fig. 3.18. It is straightforward to confirm that the network possesses a natural frequency at $j\frac{1}{4}$.

At $s = j\frac{1}{4}$, the two equations of (3.103) are linearly dependent, and the general solution can be written as

$$I_1 = -(1 + j2)x \tag{3.115a}$$

$$I_2 = x \tag{3.115b}$$

where x is arbitrary. Thus, for $x \neq 0$ the network can support nonzero port currents I_1 and I_2, as expected from Theorem 3.4.

3.6.2 Llewellyn's Stability Criteria

In this section, we establish Llewellyn's conditions for absolute stability of a two-port network at a specified frequency on the real-frequency axis. These conditions were first obtained by Llewellyn (1952) and are relatively simple to test. To keep our discussion general, we again consider the general hybrid matrix (3.34), whose elements are, as before, written as

$$k_{ij} = \text{Re}\ k_{ij} + j\ \text{Im}\ k_{ij} = m_{ij} + jn_{ij} \tag{3.116}$$

Definition 3.4: Stability parameter The dimensionless quantity

$$\eta = \frac{2\ \text{Re}\ k_{11}\ \text{Re}\ k_{22} - \text{Re}\ (k_{12}k_{21})}{|k_{12}k_{21}|} \tag{3.117}$$

associated with a two-port network is called the *stability parameter* of the two-port network.

An important property of this stability parameter is that it is invariant when any one set of impedance, admittance, hybrid, or inverse hybrid parameters is replaced by any other set. The operation is termed as the *immittance substitution.* The y- and h-parameters, for example, are related by the equations (see App. II)

$$y_{11} = \frac{1}{h_{11}} \qquad y_{12} = -\frac{h_{12}}{h_{11}} \tag{3.118a}$$

$$y_{21} = \frac{h_{21}}{h_{11}} \qquad y_{22} = \frac{h_{11}h_{22} - h_{12}h_{21}}{h_{11}} \tag{3.118b}$$

By substituting these in (3.117), it is straightforward to confirm that

$$\eta = \frac{2 \text{ Re } y_{11} \text{ Re } y_{22} - \text{Re } (y_{12}y_{21})}{|y_{12}y_{21}|} = \frac{2 \text{ Re } h_{11} \text{ Re } h_{22} - \text{Re } (h_{12}h_{21})}{|h_{12}h_{21}|} \tag{3.119}$$

The verification for other sets of parameters is left as an exercise (see Prob. 3.14). Since η is invariant under immittance substitution, it is immaterial as to which set is chosen in defining (3.117), and they all lead to the same result.

With these preliminaries, we now state Llewellyn's criteria for absolute stability.

Theorem 3.6 A linear time-invariant two-port network is absolutely stable at $j\omega_0$ if and only if

$$\text{Re } k_{11} = m_{11} > 0 \tag{3.120a}$$

$$\text{Re } k_{22} = m_{22} > 0 \tag{3.120b}$$

$$\eta > 1 \tag{3.120c}$$

all being evaluated at $s = j\omega_0$.

PROOF. *Necessity*: Assume that a given two-port network N is absolutely stable at $j\omega_0$. Then according to Theorem 3.5 the input immittances $M_{11}(j\omega_0)$ and $M_{22}(j\omega_0)$, as defined in Fig. 3.17, of the terminated two-port must have positive-real parts for all passive one-port terminations $M_2(j\omega_0)$ and $M_1(j\omega_0)$. In particular, if we choose the open-circuit or short-circuit termination, $M_1 = \infty$ and $M_2 = \infty$, and we have

$$\text{Re } M_{11}(j\omega_0) = m_{11}(\omega_0) > 0 \tag{3.121a}$$

$$\text{Re } M_{22}(j\omega_0) = m_{22}(\omega_0) > 0 \tag{3.121b}$$

To show that (3.120*c*) is necessary, two cases are distinguished, the second case being based on the first.

Case 1: N is reciprocal Since the stability parameter is invariant under immittance substitution, it is sufficient to consider only the admittance parameters. Thus, we write $k_{ij} = y_{ij}$, $M_{ii} = Y_{ii}$, and $M_i = Y_i$.

For a reciprocal two-port network N, $y_{12} = y_{21}$ and (3.120*c*) is reduced to

$$\text{Re } y_{11} \text{ Re } y_{22} > (\text{Re } y_{12})^2 \tag{3.122}$$

(see Prob. 3.16). By appealing to (3.7), the input admittance $Y_{11}(j\omega_0)$ is related to the y-parameters $y_{ij}(j\omega_0)$ and the terminating admittance $Y_2(j\omega_0)$ by

$$Y_{11}(j\omega_0) = y_{11} - \frac{y_{12}^2}{y_{22} + Y_2} = y_{11} + y_{12} + \frac{1}{1/(y_{22} + y_{12} + Y_2) + 1/(-y_{12})} \tag{3.123}$$

all being evaluated at $s = j\omega_0$. Choose $Y_2(j\omega_0) = jB_2$ such that

$$\frac{\text{Im } y_{22} + \text{Im } y_{12} + B_2}{\text{Re } y_{22} + \text{Re } y_{12}} = \frac{\text{Im } y_{12}}{\text{Re } y_{12}} \tag{3.124}$$

Under this situation, the admittances $(y_{22} + y_{12} + Y_2)$ and $-y_{12}$ will have the same phase at $j\omega_0$. Consequently, the real part of these two admittances in series, which corresponds to the last term on the right-hand side of (3.123), equals the series combination of their real parts (see Prob. 3.18), giving

$$\begin{aligned}\text{Re } Y_{11}(j\omega_0) &= \text{Re } y_{11} + \text{Re } y_{12} - \frac{\text{Re } y_{12} \text{ Re } (y_{22} + y_{12})}{\text{Re } (y_{22} + y_{12} - y_{12})} \\ &= \text{Re } y_{11} - \frac{(\text{Re } y_{12})^2}{\text{Re } y_{22}} \end{aligned} \tag{3.125}$$

which, according to Theorem 3.5, must be positive. Hence we have (3.122).

Case 2: N is nonreciprocal Consider the associated reciprocal two-port network N_a with admittance matrix

$$\mathbf{Y}_a(s) = \begin{bmatrix} y_{11} & \sqrt{y_{12}y_{21}} \\ \sqrt{y_{12}y_{21}} & y_{22} \end{bmatrix} \tag{3.126}$$

By applying (3.7) and (3.8), it is easy to show that the admittances looking into the corresponding ports of N and N_a are identical for identical terminations. Then, by Theorem 3.5, N is absolutely stable at $j\omega_0$ if and only if N_a is. Since N_a is reciprocal, from Case 1 it is necessary that

$$\text{Re } y_{11} \text{ Re } y_{22} > (\text{Re } \sqrt{y_{12}y_{21}})^2 \tag{3.127}$$

Now we demonstrate that (3.127) is equivalent to (3.120*c*). To this end, let

$$\sqrt{y_{12}y_{21}} = Le^{j\theta} \tag{3.128}$$

Substituting it in (3.127) yields

$$2 \text{ Re } y_{11} \text{ Re } y_{22} > 2L^2 \cos^2 \theta = L^2(1 + \cos 2\theta) = |y_{12}y_{21}| + \text{Re } y_{12}y_{21} \tag{3.129}$$

Since the stability parameter η is invariant under immittance substitution, (3.120*c*) follows.

Sufficiency: Assume that the conditions (3.120) hold at $s = j\omega_0$. We show that N is absolutely stable at $j\omega_0$ by demonstrating that the real parts of the driving-point immittances $M_{11}(j\omega_0)$ and $M_{22}(j\omega_0)$ are positive for all passive one-port terminations M_1 and M_2.

The real part of the input immittance M_{11}, as given in (3.7), can be manipulated into the form

$$\begin{aligned}\frac{|k_{22} + M_2|^2 \text{ Re } M_{11}}{\text{Re } k_{11}} &= |k_{22} + M_2|^2 - \frac{\text{Re } [k_{12}k_{21}(\bar{k}_{22} + \bar{M}_2)]}{\text{Re } k_{11}} \\ &= (A_2 + \alpha)^2 + (B_2 + \beta)^2 - \frac{|k_{12}k_{21}|^2}{(2 \text{ Re } k_{11})^2}\end{aligned} \tag{3.130}$$

where $M_2 = A_2 + jB_2$ and

$$\alpha = \text{Re } k_{22} - \frac{\text{Re } k_{12}k_{21}}{2 \text{ Re } k_{11}} \tag{3.131a}$$

$$\beta = \text{Im } k_{22} - \frac{\text{Im } k_{12}k_{21}}{2 \text{ Re } k_{11}} \tag{3.131b}$$

Since $\eta > 1$ and $\text{Re } k_{11} > 0$, we have

$$\alpha = \frac{\eta |k_{12}k_{21}|}{2 \text{ Re } k_{11}} \geqslant 0 \tag{3.131c}$$

This, together with the fact that $A_2 \geqslant 0$, indicates that the minimum of the quantity on the right-hand side of (3.130) is attained at

$$A_2 = 0 \qquad B_2 = -\beta \tag{3.132}$$

and (3.130) becomes

$$|k_{22} + M_2|^2 \text{ Re } M_{11} = \alpha^2 \text{ Re } k_{11} - \frac{|k_{12}k_{21}|^2}{4 \text{ Re } k_{11}} = \frac{|k_{12}k_{21}|^2(\eta^2 - 1)}{4 \text{ Re } k_{11}} \tag{3.133}$$

Since by hypothesis $\eta > 1$ and $\text{Re } k_{11} > 0$, $\text{Re } M_{11}$ is positive at $j\omega_0$. In a similar fashion, we can show that $\text{Re } M_{22}(j\omega_0) > 0$. Thus, by Theorem 3.5, the two-port network N is absolutely stable at $j\omega_0$. This completes the proof of the theorem.

A number of important observations can be made from the above discussions. First, as stated in Theorem 3.5, because the stability of a two-port network depends on the positiveness of the real part of the immittance looking in at either port, it must be invariant under interchange of input and output. The inequalities (3.120) are clearly invariant under the interchange of subscripts 1 and 2. Second, if the stability test is performed for one set of parameters, the same conclusions will be reached for other sets of parameters. However, the quantities m_{11} and m_{22} may be different for different sets of parameters. Finally, we mention that the stability parameter η is a measure of the degree of two-port stability. For positive m_{11} and m_{22}, the value of η lies between -1 and ∞ (see Prob. 3.19). When η is positive and large compared with unity, the degree of absolute stability is high; when η is near unity, the two-port network is close to the boundary between absolute stability and potential instability, which is defined by $\eta = 1$. When $1 \geqslant \eta \geqslant -1$, the two-port network is in the region of potential instability, meaning that we can always choose passive one-port terminations that will result in oscillations.

Two special cases are worthy of mentioning. First, for a unilateral two-port network, $k_{12} = 0$ and Eq. (3.120c) is reduced to $2m_{11}m_{22} > 0$, showing that $m_{11} > 0$ and $m_{22} > 0$ are enough to guarantee absolute stability. In other words, a unilateral two-port network is absolutely stable at $j\omega_0$ if and only if the short-circuit or open-circuit input and output immittances possess positive-real parts at $j\omega_0$. Next, for a reciprocal two-port network, (3.52) applies and (3.120c) is reduced to (see Prob. 3.16)

$$4m_{11}m_{22} > |k_{12} + \bar{k}_{21}|^2 \tag{3.134}$$

which together with $m_{11} > 0$ and $m_{22} > 0$ is precisely (3.39) with strict inequality. Thus, a reciprocal two-port network is absolutely stable at $j\omega_0$ if and only if it is strictly passive at $j\omega_0$. Mere passivity, however, will not suffice, because the passivity conditions (3.39) allow the equality to hold whereas the absolute stability conditions require the strict inequality. A reciprocal LC two-port is passive but is potentially unstable. On the other hand, an absolutely stable nonreciprocal two-port network may be active. In fact, we can show that any strictly passive two-port network must be absolutely stable. However, not all absolutely stable two-port networks are strictly passive. To see this, we consider (3.39*b*) with strict inequality, which can be written in the equivalent form

$$4m_{11}m_{22} - 2 \operatorname{Re} k_{12}k_{21} > |k_{12}|^2 + |k_{21}|^2 \tag{3.135a}$$

or

$$\eta > \frac{|k_{12}|^2 + |k_{21}|^2}{2|k_{12}k_{21}|} = 1 + \frac{(|k_{12}| - |k_{21}|)^2}{2|k_{12}k_{21}|} \tag{3.135b}$$

showing that $\eta > 1$. This together with (3.39*a*) with strict inequality guarantees the absolute stability of the two-port network.

We illustrate the above results by the following examples.

Example 3.9 Consider the two-port network N of Fig. 3.19, whose hybrid matrix is given by

$$\mathbf{H}(s) = \frac{1}{s-1}\begin{bmatrix} s & 1 \\ -1 & s-2 \end{bmatrix} \tag{3.136}$$

On the real-frequency axis, its value is found to be

$$\mathbf{H}(j\omega) = \frac{1}{\omega^2 + 1}\begin{bmatrix} \omega^2 - j\omega & -1 - j\omega \\ 1 + j\omega & \omega^2 + 2 + j\omega \end{bmatrix} \tag{3.137}$$

Thus, according to Theorem 3.6, the device is absolutely stable if and only if

$$\operatorname{Re} h_{11} = \frac{\omega^2}{\omega^2 + 1} > 0 \tag{3.138a}$$

$$\operatorname{Re} h_{22} = \frac{\omega^2 + 2}{\omega^2 + 1} > 0 \tag{3.138b}$$

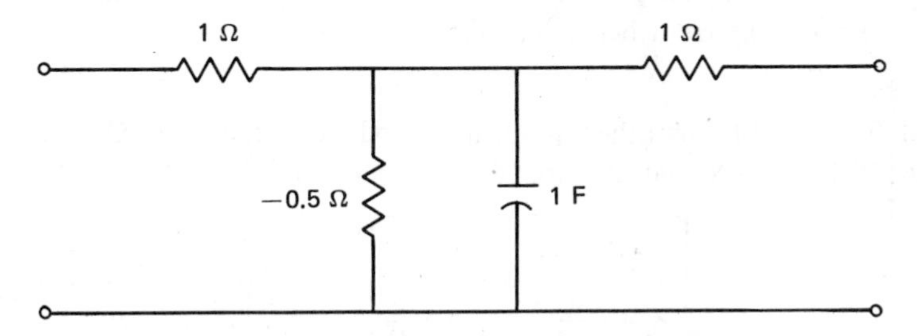

Figure 3.19 A two-port network used to illustrate absolute stability.

$$\eta = \frac{2 \text{ Re } h_{11} \text{ Re } h_{22} - \text{Re } h_{12}h_{21}}{|h_{12}h_{21}|} = 2\omega^2 + 1 > 1 \tag{3.138c}$$

Thus, the device is absolutely stable for all nonzero frequencies.

To see if the device is also passive, we compute the hermitian part of $\mathbf{H}(j\omega)$, giving

$$\mathbf{H}_h(j\omega) = \frac{1}{\omega^2 + 1}\begin{bmatrix} \omega^2 & -j\omega \\ j\omega & \omega^2 + 2 \end{bmatrix} \tag{3.139}$$

The matrix $\mathbf{H}_h(j\omega)$ is clearly nonnegative definite for all ω, indicating that the device is passive for all frequencies, zero included. As a matter of fact, $\mathbf{H}_h(j\omega)$ is positive definite for all nonzero ω. The two-port network is therefore strictly passive for all nonzero ω. This in conjunction with the fact that N is reciprocal, $\mathbf{H}(s)$ being skew symmetric, $h_{12} = -h_{21}$, implies that the device is absolutely stable and strictly passive for all nonzero real frequencies.

To demonstrate the invariance of the stability parameter under immittance substitution, we compute the impedance matrix of Fig. 3.19:

$$\mathbf{Z}(s) = \frac{1}{s-2}\begin{bmatrix} s-1 & 1 \\ 1 & s-1 \end{bmatrix} \tag{3.140}$$

which, when evaluated on the $j\omega$-axis, yields

$$\mathbf{Z}(j\omega) = \frac{1}{\omega^2 + 4}\begin{bmatrix} \omega^2 + 2 - j\omega & -2 - j\omega \\ -2 - j\omega & \omega^2 + 2 - j\omega \end{bmatrix} \tag{3.141}$$

The corresponding stability parameter is found to be

$$\eta = \frac{2 \text{ Re } z_{11} \text{ Re } z_{22} - \text{Re } z_{12}z_{21}}{|z_{12}z_{21}|} = 2\omega^2 + 1 \tag{3.142}$$

which is the same as that given in (3.138*c*). Using (3.142) in conjunction with the fact that $\text{Re } z_{11} = \text{Re } z_{22} > 0$, we arrive at the same conclusion as before. Note that the invariance of the stability parameter under immittance substitution is valid for all s, not merely for $s = j\omega$, the latter being demonstrated above.

Example 3.10 Figure 3.20 shows the equivalent T-model of a transistor. On the real-frequency axis, its impedance matrix is found to be

$$\mathbf{Z}(j\omega) = \begin{bmatrix} r_e + r_b & r_b \\ r_b - \dfrac{j\alpha}{\omega C} & r_b - \dfrac{j}{\omega C} \end{bmatrix} \tag{3.143}$$

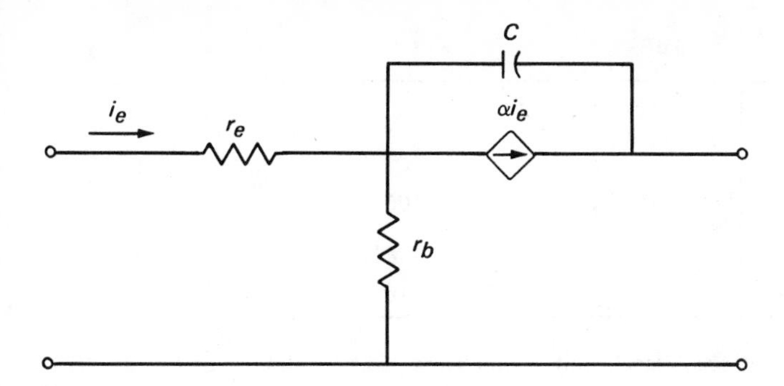

Figure 3.20 The equivalent T-model of a transistor.

Thus, according to Theorem 3.6, the transistor is absolutely stable at $j\omega_0$ if and only if

$$r_e + r_b > 0 \tag{3.144a}$$

$$r_b > 0 \tag{3.144b}$$

$$2(r_e + r_b)r_b - \text{Re}\left[r_b\left(r_b - \frac{j\alpha}{\omega_0 C}\right)\right] > \left|r_b\left(r_b - \frac{j\alpha}{\omega_0 C}\right)\right| \tag{3.144c}$$

The last inequality (3.144*c*) is simplified to

$$4\omega_0^2 C^2 r_e(r_e + r_b) > \alpha^2 \tag{3.145}$$

In Example 1.12, we showed that the transistor is active for all real frequencies

$$\omega < \frac{\alpha}{2C\sqrt{r_e r_b}} \tag{3.146}$$

Using this in conjunction with (3.145), we conclude that the transistor is both active and absolutely stable at $j\omega_0$ only if

$$\frac{\alpha^2}{4\omega_0^2 C^2} < r_e(r_e + r_b) < \frac{\alpha^2}{4\omega_0^2 C^2} + r_e^2 \tag{3.147}$$

To put it differently, for the device to be both active and absolutely stable, we must restrict the frequencies in the range

$$\frac{\alpha^2}{4C^2 r_e(r_e + r_b)} < \omega^2 < \frac{\alpha^2}{4C^2 r_b r_e} \tag{3.148}$$

Example 3.11 Figure 3.21 shows the high-frequency, small-signal equivalent network of a transistor. We wish to determine the frequency range for which the device is potentially unstable.

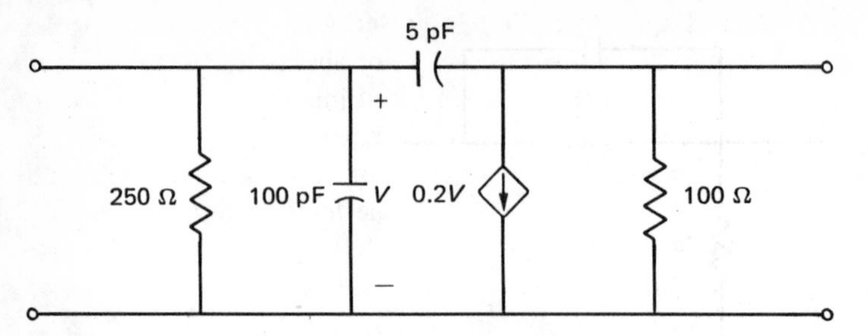

Figure 3.21 A high-frequency, small-signal equivalent network of a transistor.

First, we compute the admittance matrix, which, when evaluated on the real-frequency axis, is given by

$$\mathbf{Y}(j\omega) = \begin{bmatrix} 0.004 + j105 \cdot 10^{-12}\omega & -j5 \cdot 10^{-12}\omega \\ 0.2 - j5 \cdot 10^{-12}\omega & 0.01 + j5 \cdot 10^{-12}\omega \end{bmatrix} \tag{3.149}$$

To simplify the notation, let $\hat{\omega} = \omega/10^9$, giving

$$\mathbf{Y}(j\omega) = 10^{-3}\begin{bmatrix} 4 + j105\hat{\omega} & -j5\hat{\omega} \\ 200 - j5\hat{\omega} & 10 + j5\hat{\omega} \end{bmatrix} \tag{3.150}$$

Since

$$\operatorname{Re} y_{11} = 0.004 > 0 \tag{3.151a}$$

$$\operatorname{Re} y_{22} = 0.01 > 0 \tag{3.151b}$$

for the device to be potentially unstable, it is necessary and sufficient that $\eta \leqslant 1$ or

$$80 \leqslant -25\hat{\omega}^2 + (625\hat{\omega}^4 + 10^6\hat{\omega}^2)^{1/2} \tag{3.152}$$

yielding

$$\hat{\omega} \geqslant 80.16 \cdot 10^{-3} \tag{3.153a}$$

or

$$\omega \geqslant 80.16 \cdot 10^6 \text{ rad/s} \tag{3.153b}$$

which corresponds to 12.76 MHz. Thus, the transistor is potentially unstable for all real frequencies not less than 12.76 MHz.

3.7 OPTIMUM TERMINATIONS FOR ABSOLUTELY STABLE TWO-PORT NETWORKS

In Sec. 3.2 we introduced three types of power gains for a terminated two-port network and derived expressions for these power gains in terms of the two-port parameters and the source and load immittances. Our task here is to select the source and load terminations so that the power gain of the two-port network is maximized. We limit our discussion to absolutely stable two-port networks, because in a potentially unstable two-port network oscillation will occur for certain passive terminations. In such situations, the maximum power gain is infinite and the concept of optimum power gain has no significance. For absolutely stable two-port networks, there exists a set of *optimum* source and load immittances in relation to obtaining maximum power gain. We recognize that other performance characteristics such as the sensitivity to variations in parameters or availability of practical elements may be equally important in the selection of the terminations.

The optimum terminations to be presented in the following are optimum at a single frequency, which may be at any point on the real-frequency axis. Thus, the results are useful only for narrow-band applications.

According to the definitions of various power gains (3.9)–(3.11), the transducer power gain $\mathcal{G}$ is clearly less than or equal to the power gain $\mathcal{G}_p$:

$$\frac{P_2}{P_{1a}} = \mathcal{G} \leqslant \mathcal{G}_p = \frac{P_2}{P_1} \tag{3.154}$$

The input power P_1 to the two-port can at most equal the available power P_{1a} from the source, and this upper bound is reached when the input port is conjugate-matched. As indicated in (3.19*a*), $\mathcal{G}_p$ is a function only of the two-port parameters and the load immittance M_2. When the load $M_2 = M_{2,\text{opt}}$ is chosen to give the maximum value of the power gain $\mathcal{G}_{p,\max}$ and when the input port is conjugate-matched, the maximum value of the transducer power gain $\mathcal{G}_{\max}$ becomes equal to the maximum value of the power gain $\mathcal{G}_{p,\max}$ and we can write

$$\mathcal{G}_{\max} = \mathcal{G}_{p,\max} \tag{3.155}$$

The transducer power gain also can never exceed the available power gain $\mathcal{G}_a$, as indicated by the equation

$$\frac{P_2}{P_{1a}} = \mathcal{G} \leqslant \mathcal{G}_a = \frac{P_{2a}}{P_{1a}} \tag{3.156}$$

because the output power can at most equal the available power at the output. Equation (3.19*c*) shows that $\mathcal{G}_a$ is a function only of the two-port parameters and the source immittance M_1. When the source immittance $M_1 = M_{1,\text{opt}}$ is chosen to give the maximum value of the available power gain $\mathcal{G}_{a,\max}$ and when the output port is conjugate-matched, we can write

$$\mathcal{G}_{\max} = \mathcal{G}_{a,\max} \tag{3.157}$$

Thus, the three power gains attain a common maximum value with

$$\mathcal{G}_{\max} = \mathcal{G}_{p,\max} = \mathcal{G}_{a,\max} \tag{3.158}$$

There are many ways of determining the optimum source and load terminations of a two-port network. One way is to differentiate the transducer power gain expression (3.19*b*) with respect to the real and imaginary parts of the source and load immittances M_1 and M_2 as shown in Fig. 3.22 and to solve for the optimum M_1 and M_2 after the partial derivatives are set equal to zero. This process is rather complicated.

A second possibility is to determine the optimum terminations by simultaneous

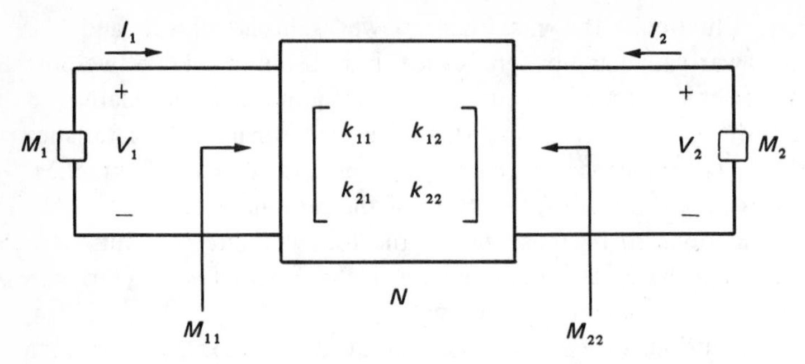

Figure 3.22 The general representation of a terminated two-port network.

conjugate match at the ports. Refer to Fig. 3.22. The maximum transducer power gain will result if the input immittance M_{11} is the conjugate of the source immittance M_1, whereas the output immittance M_{22} is the conjugate of the load immittance M_2. Under this situation and from Eqs. (3.7) and (3.8), we have

$$M_{11} = k_{11} - \frac{k_{12}k_{21}}{k_{22} + M_2} = \bar{M}_1 \tag{3.159a}$$

$$M_{22} = k_{22} - \frac{k_{12}k_{21}}{k_{11} + M_1} = \bar{M}_2 \tag{3.159b}$$

Solving for M_1 and M_2 will yield the optimum terminations of the two-port network, but the algebra involved is rather tedious.

A simpler approach is as follows. We first determine the load termination $M_{2,\mathrm{opt}}$ that will give the maximum power gain $\mathcal{G}_{p,\max}$ and then determine the source immittance by means of (3.159*a*). Alternatively, we can find the source immittance $M_{1,\mathrm{opt}}$ that will give the maximum available power gain $\mathcal{G}_{a,\max}$ and then determine the load immittance by means of (3.139*b*). In the first case, the input of the two-port is conjugate-matched by construction. We assert that the output must also be conjugate-matched, for, if not, we could get more power out by adjusting the load immittance, thus increasing $\mathcal{G}_{\max}$ above $\mathcal{G}_{p,\max}$. This is impossible because it violates Eq. (3.154). In the second case, the output port is conjugate-matched by construction. Hence, $\mathcal{G}$ is equal to $\mathcal{G}_a$. But $\mathcal{G}_a$ is already at its maximum $\mathcal{G}_{a,\max}$. Therefore, $\mathcal{G}$ is also $\mathcal{G}_{\max}$. In the following, we use this procedure to compute the optimum terminations.

From (3.19*a*), the power gain can be expressed as

$$\mathcal{G}_p = \frac{|k_{21}|^2 \operatorname{Re} M_2}{|k_{22} + M_2|^2 \operatorname{Re}\,[k_{11} - k_{12}k_{21}/(k_{22} + M_2)]} \tag{3.160}$$

To simplify our notation, as in (3.34) we write

$$k_{ik} = m_{ik} + jn_{ik} \qquad i, k = 1, 2 \tag{3.161a}$$

$$M_i = A_i + jB_i \qquad i = 1, 2 \tag{3.161b}$$

$$k_{12}k_{21} = P + jQ = Le^{j\phi} \tag{3.161c}$$

all being evaluated at the real frequency ω_0. The argument $j\omega_0$ will be dropped, as above, in all the equations, for simplicity.

By using the symbols introduced in Eq. (3.161), Eq. (3.160) becomes

$$\mathcal{G}_p = \frac{|k_{21}|^2 A_2}{m_{11}|k_{22} + M_2|^2 - P(m_{22} + A_2) - Q(n_{22} + B_2)} \tag{3.162}$$

Taking the partial derivative of $\mathcal{G}_p$ with respect to B_2 and setting it to zero yield

$$n_{22} + B_2 = \frac{Q}{2m_{11}} \tag{3.163}$$

giving the optimum load susceptance or reactance

$$\operatorname{Im} M_{2,\text{opt}} = B_2 = \frac{Q}{2m_{11}} - n_{22} \tag{3.164a}$$

Similarly, taking the partial derivative of (3.162) with respect to A_2 and setting it to zero, we obtain

$$\operatorname{Re} M_{2,\text{opt}} = A_2 = \frac{1}{2m_{11}}\sqrt{(2m_{11}m_{22} - P)^2 - L^2} \tag{3.164b}$$

Substituting $M_{2,\text{opt}}$ from (3.164) in (3.162) yields the maximum power gain of the two-port network as

$$\mathcal{G}_{p,\max} = \frac{|k_{21}|^2}{2m_{11}m_{22} - P + \sqrt{(2m_{11}m_{22} - P)^2 - L^2}} \tag{3.165}$$

Once the optimum load immittance $M_{2,\text{opt}}$ is known, the optimum source immittance $M_{1,\text{opt}}$ can easily be found from (3.159*a*). However, from the symmetry of the formulas for power gain (3.160) or (3.162) and available power gain (3.19*c*), we can write down the value of $M_{1,\text{opt}}$ directly from (3.164) as

$$\operatorname{Re} M_{1,\text{opt}} = \frac{1}{2m_{22}}\sqrt{(2m_{11}m_{22} - P)^2 - L^2} \tag{3.166a}$$

$$\operatorname{Im} M_{1,\text{opt}} = \frac{Q}{2m_{22}} - n_{11} \tag{3.166b}$$

A slight digression at this point will allow us to shine some light on our assumption that the two-port network must be absolutely stable at $j\omega_0$. This assumption is equivalent to requiring that the input immittances M_{11} and M_{22} possess positive-real parts at $j\omega_0$. From the conjugate-match conditions of (3.159), Eqs. (3.164*b*) and (3.166*a*) must be positive. This is possible if and only if the term inside the radical is positive or

$$2m_{11}m_{22} - P > L \tag{3.167}$$

showing that the stability parameter

$$\eta = \frac{2m_{11}m_{22} - P}{L} > 1 \tag{3.168}$$

Both the optimum terminations (3.164) and (3.166) and the maximum power gain (3.165) can be expressed in terms of the stability parameter as follows:

$$M_{1,\text{opt}} = \text{Re}\, M_{1,\text{opt}} + j\,\text{Im}\, M_{1,\text{opt}} = \frac{k_{12}k_{21} + |k_{12}k_{21}|(\eta + \sqrt{\eta^2 - 1})}{2\,\text{Re}\, k_{22}} - k_{11} \tag{3.169a}$$

$$M_{2,\text{opt}} = \text{Re}\, M_{2,\text{opt}} + j\,\text{Im}\, M_{2,\text{opt}} = \frac{k_{12}k_{21} + |k_{12}k_{21}|(\eta + \sqrt{\eta^2 - 1})}{2\,\text{Re}\, k_{11}} - k_{22} \tag{3.169b}$$

$$\mathcal{G}_{p,\max} = \mathcal{G}_{\max} = \mathcal{G}_{a,\max} = \left|\frac{k_{21}}{k_{12}}\right| \frac{1}{\eta + \sqrt{\eta^2 - 1}} \tag{3.170}$$

Since the two-port network is assumed to be absolutely stable at $j\omega_0$, $\eta > 1$, and the expression (3.170) is therefore meaningful.

It is significant to observe that the expression for maximum gain is made up of two factors. The one involving the stability parameter is reciprocal in that it is invariant with respect to the interchange of forward and reverse transfer parameters k_{21} and k_{12} (and also the self-immittances k_{11} and k_{22}), whereas the other is nonreciprocal. Thus, we can write

$$\mathcal{G}_{\max} = \left|\frac{k_{21}}{k_{12}}\right| \gamma \tag{3.171}$$

where

$$\gamma = (\eta + \sqrt{\eta^2 - 1})^{-1} = \eta - \sqrt{\eta^2 - 1} \tag{3.172a}$$

$$= 2(\sqrt{\eta + 1} + \sqrt{\eta - 1})^{-2} = \tfrac{1}{2}(\sqrt{\eta + 1} - \sqrt{\eta - 1})^2 \tag{3.172b}$$

The first factor $|k_{21}/k_{12}|$ is a measure of the nonreciprocity of the two-port and is called the *nonreciprocal gain* of the two-port network. The second factor γ is usually known as the *efficiency of the reciprocal part* of the two-port network. Since $\eta > 1$ for absolutely stable two-port networks, γ is allowed to vary between 0 and 1. This implies that for an absolutely stable reciprocal two-port network, its maximum gain, given by $\mathcal{G}_{\max} = \gamma$, is bounded by unity, and therefore the two-port network must be passive, a fact that was pointed out earlier.

For a unilateral two-port network, $k_{12} = 0$ and both the nonreciprocal gain and the stability parameter are infinite. The maximum unilateral power gain can be obtained directly from (3.165) and is given by

$$\mathcal{G}_{\max} = \frac{|k_{21}|^2}{4m_{11}m_{22}} \tag{3.173}$$

which agrees with (3.92).

Example 3.12 Consider the two-port network of Fig. 3.23, whose admittance matrix is given by

$$\mathbf{Y}(s) = \begin{bmatrix} g_1 + sC & -sC \\ g_m - sC & g_2 + sC \end{bmatrix} \tag{3.174}$$

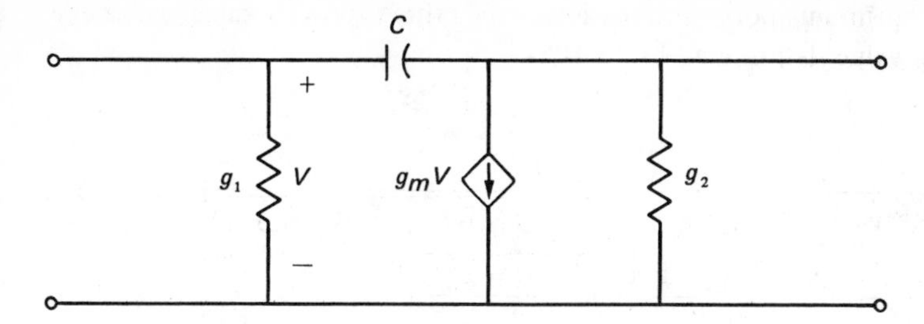

Figure 3.23 An active and absolutely stable two-port device.

It is straightforward to show that the device is both active and absolutely stable for all real frequencies ω satisfying the inequalities

$$\tfrac{1}{2}g_m^2 > 2g_1g_2 > \omega C[(g_m^2 + \omega^2C^2)^{1/2} - \omega C] \tag{3.175}$$

The details are left as an exercise (see Prob. 3.22). To determine the optimum terminating admittances

$$Y_{1,\text{opt}} = G_{1,\text{opt}} + jB_{1,\text{opt}} \tag{3.176a}$$

$$Y_{2,\text{opt}} = G_{2,\text{opt}} + jB_{2,\text{opt}} \tag{3.176b}$$

at the frequencies where the two-port is absolutely stable, we first compute

$$y_{12}y_{21} = -j\omega C(g_m - j\omega C) = -\omega^2C^2 - j\omega Cg_m \tag{3.177}$$

From (3.166) we have

$$\begin{aligned} G_{1,\text{opt}} &= \frac{1}{2\,\text{Re}\,y_{22}}\sqrt{(2\,\text{Re}\,y_{11}\,\text{Re}\,y_{22} - \text{Re}\,y_{12}y_{21})^2 - |y_{12}y_{21}|^2} \\ &= \frac{1}{2g_2}\sqrt{(2g_1g_2 + \omega^2C^2)^2 - \omega^4C^4 - \omega^2C^2g_m^2} \\ &= \frac{1}{2g_2}\sqrt{4g_1^2g_2^2 + \omega^2C^2(4g_1g_2 - g_m^2)} > 0 \end{aligned} \tag{3.178a}$$

the last inequality following from (3.175),

$$B_{1,\text{opt}} = \frac{\text{Im}\,y_{12}y_{21}}{2\,\text{Re}\,y_{22}} - \text{Im}\,y_{11} = -\frac{\omega Cg_m}{2g_2} - \omega C = -\omega C\left(1 + \frac{g_m}{2g_2}\right) \tag{3.178b}$$

and from (3.164) we get

$$\begin{aligned} G_{2,\text{opt}} &= \frac{1}{2\,\text{Re}\,y_{11}}\sqrt{(2\,\text{Re}\,y_{11}\,\text{Re}\,y_{22} - \text{Re}\,y_{12}y_{21})^2 - |y_{12}y_{21}|^2} \\ &= \frac{1}{2g_1}\sqrt{4g_1^2g_2^2 + \omega^2C^2(4g_1g_2 - g_m^2)} > 0 \end{aligned} \tag{3.179a}$$

$$B_{2,\text{opt}} = \frac{\text{Im}\,y_{12}y_{21}}{2\,\text{Re}\,y_{11}} - \text{Im}\,y_{22} = -\frac{\omega Cg_m}{2g_1} - \omega C = -\omega C\left(1 + \frac{g_m}{2g_1}\right) \tag{3.179b}$$

Under these optimum port terminations, the three power gains assume a common maximum value determined by (3.165):

$$\mathcal{G}_{\max} = \mathcal{G}_{p,\max} = \mathcal{G}_{a,\max}$$

$$= \frac{|y_{21}|^2}{2 \operatorname{Re} y_{11} \operatorname{Re} y_{22} - \operatorname{Re} y_{12}y_{21} + \sqrt{(2 \operatorname{Re} y_{11} \operatorname{Re} y_{22} - \operatorname{Re} y_{12}y_{21})^2 - |y_{12}y_{21}|^2}}$$

$$= \frac{2g_1g_2 + \omega^2C^2 - \sqrt{4g_1^2g_2^2 + \omega^2C^2(4g_1g_2 - g_m^2)}}{\omega^2C^2} \tag{3.180}$$

The nonreciprocal gain of the two-port network is found to be

$$\left|\frac{y_{21}}{y_{12}}\right| = \left|\frac{g_m - j\omega C}{-j\omega C}\right| = \sqrt{1 + \frac{g_m^2}{\omega^2C^2}} \tag{3.181}$$

and the efficiency of the reciprocal part of the two-port network is calculated from (3.172*a*) and is given by

$$\gamma = \eta - \sqrt{\eta^2 - 1} = \frac{2g_1g_2 + \omega^2C^2 - \sqrt{4g_1^2g_2^2 + \omega^2C^2(4g_1g_2 - g_m^2)}}{\omega C\sqrt{\omega^2C^2 + g_m^2}} \tag{3.182}$$

where the stability parameter is given by

$$\eta = \frac{2g_1g_2 + \omega^2C^2}{\omega C\sqrt{\omega^2C^2 + g_m^2}} \tag{3.183}$$

3.7.1 Representation of Nonreciprocal Two-Port Networks

As indicated in Eq. (3.171), the maximum power gain $\mathcal{G}_{\max}$ can be expressed as the product of the nonreciprocal gain and the efficiency of the reciprocal part of the two-port network. In this section, we demonstrate that this interpretation may be employed to decompose a general two-port network into two two-ports connected in cascade, one reciprocal and the other nonreciprocal.

Consider a general two-port network N characterized by its general hybrid equation

$$\begin{bmatrix} y_1 \\ y_2 \end{bmatrix} = \begin{bmatrix} k_{11} & k_{12} \\ k_{21} & k_{22} \end{bmatrix} \begin{bmatrix} u_1 \\ u_2 \end{bmatrix} \tag{3.184}$$

where the variables u and y for different immittance representations are tabulated in Table 3.1. Let

$$k_s = \sqrt{k_{12}k_{21}} \tag{3.185a}$$

$$\lambda = \sqrt{\frac{k_{21}}{k_{12}}} \tag{3.185b}$$

With these, Eq. (3.184) can be written as

$$\begin{bmatrix} y_1 \\ \dfrac{y_2}{\lambda} \end{bmatrix} = \begin{bmatrix} k_{11} & k_s \\ k_s & k_{22} \end{bmatrix} \begin{bmatrix} u_1 \\ \dfrac{u_2}{\lambda} \end{bmatrix} \tag{3.186a}$$

or

$$\begin{bmatrix} \lambda y_1 \\ y_2 \end{bmatrix} = \begin{bmatrix} k_{11} & k_s \\ k_s & k_{22} \end{bmatrix} \begin{bmatrix} \lambda u_1 \\ u_2 \end{bmatrix} \tag{3.186b}$$

Equation (3.186) can be represented by the cascade connection of a reciprocal two-port, whose general hybrid matrix is its coefficient matrix, and a nonreciprocal two-port on either its input or output side, as shown in Fig. 3.24. The reciprocal two-port network is the reciprocal part of the original two-port network, and the nonreciprocal two-port is characterized by the equation

$$\frac{V_\beta}{V_\alpha} = -\frac{I_\beta}{I_\alpha} = \lambda \tag{3.187}$$

where V_α and V_β are its port voltages and I_α and I_β are its port currents. Evidently, the maximum power gain for the reciprocal two-port is γ, the efficiency of the reciprocal part of the original two-port, and the power gain for the nonreciprocal two-port is $|k_{21}/k_{12}|$, the nonreciprocal gain of the original two-port network. It is

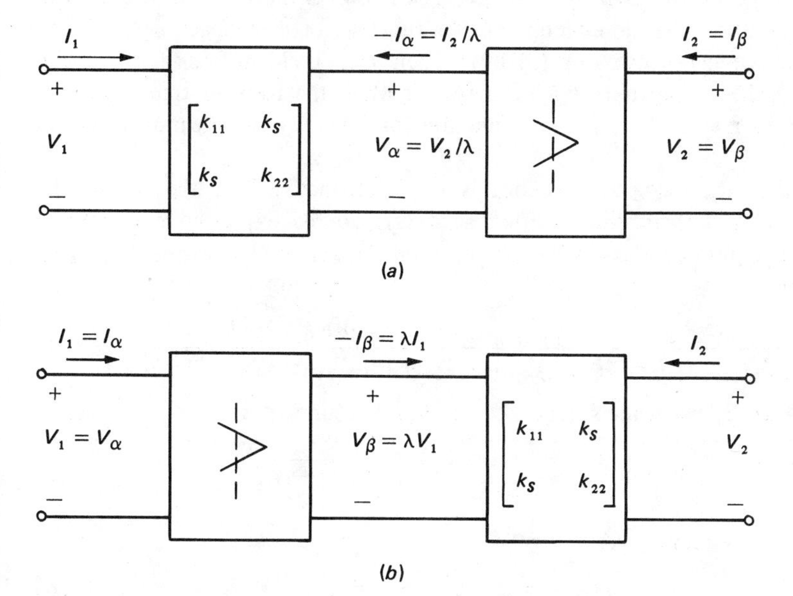

Figure 3.24 The decomposition of a general two-port network into two two-port networks connected in cascade. (*a*) A reciprocal two-port followed by a nonreciprocal one. (*b*) A nonreciprocal two-port followed by a reciprocal one.

significant to note that $|k_{21}/k_{12}|$ is invariant under immittance substitution, as it must be with the above interpretation.

3.7.2 Physical Significance of the Efficiency of the Reciprocal Part

In the foregoing, we have shown that the maximum gain $\mathcal{G}_{\max}$ of a general two-port network is made up of two factors. One is a measure of the nonreciprocity and the other the efficiency of the reciprocal part of the two-port network. The latter involves the stability parameter and is given by

$$\gamma = \eta - \sqrt{\eta^2 - 1} \tag{3.188}$$

In this section, we present a physical interpretation of γ.

Let $w(s)$ denote the transfer function of either (3.25*a*) or (3.25*b*). Then $w(s)$ can be expressed as

$$w(s) = \frac{\tilde{w}(s)}{1 - f(s)} \tag{3.189}$$

where

$$\tilde{w}(s) = w(s)|_{k_{12}=0} \tag{3.190a}$$

$$f(s) = \frac{k_{12}k_{21}}{(k_{11} + M_1)(k_{22} + M_2)} \tag{3.190b}$$

The term $f(s)$ is called the *loop transmission.* The reason is that, as will be shown in the following chapter, the closed-loop transfer function of a feedback network can also be expressed as shown in (3.189). Then $f(s)$ is identified as the loop transmission in the terminated two-port network when the forward transmission is interrupted. Observe that $f(s)$ is a function of the two-port parameters and terminations.

Assume that the two-port network is conjugate-matched at the ports. The corresponding loop transmission, written as $f_m(s)$, can be expressed explicitly in terms of the two-port parameters k_{ij} or the stability parameter η, as follows. From (3.169) we have

$$f_m(s) = \frac{k_{12}k_{21}}{(k_{11} + M_{1,\text{opt}})(k_{22} + M_{2,\text{opt}})} = \frac{4k_{12}k_{21} \operatorname{Re} k_{11} \operatorname{Re} k_{22}}{[k_{12}k_{21} + |k_{12}k_{21}|(\eta + \sqrt{\eta^2 - 1})]^2} \tag{3.191}$$

whose magnitude, by using the symbols of (3.161), is found to be

$$|f_m(s)| = \frac{1}{\eta + \sqrt{\eta^2 - 1}} = \eta - \sqrt{\eta^2 - 1} \tag{3.192a}$$

Thus, on the real-frequency axis, we get

$$|f_m(j\omega)| = \eta - \sqrt{\eta^2 - 1} = \gamma \tag{3.192b}$$

showing that the efficiency of the reciprocal part of the two-port network is equal to the real-frequency axis magnitude of its loop transmission under the situation where the two-port is conjugate-matched at its ports.

The factors $|k_{21}/k_{12}|$ and γ can also be determined experimentally. For this, let $\mathcal{G}_{\max,r}$ be the maximum power gain of the two-port network in the reverse direction, which is found by interchanging the input and the output ports:

$$\mathcal{G}_{\max,r} = \left|\frac{k_{12}}{k_{21}}\right|\gamma \tag{3.193}$$

which is obtained from (3.171) by interchanging the subscripts 1 and 2. Combining (3.171) and (3.193) yields

$$\frac{\mathcal{G}_{\max}}{\mathcal{G}_{\max,r}} = \left|\frac{k_{21}}{k_{12}}\right|^2 \tag{3.194a}$$

$$\sqrt{\mathcal{G}_{\max}\mathcal{G}_{\max,r}} = \eta - \sqrt{\eta^2 - 1} = \gamma \tag{3.194b}$$

which can then be used to determine the nonreciprocal gain $|k_{21}/k_{12}|$ and the efficiency γ of the reciprocal part of the two-port network or its stability parameter η.

3.7.3 Maximum Stable Power Gain and Nonreciprocal Gain

If a two-port network N is potentially unstable, then $\operatorname{Re} k_{11} < 0$ or $\operatorname{Re} k_{22} < 0$, or $\eta < 1$. Under these situations, lossy padding immittances $M_{1,\text{pad}}$ and $M_{2,\text{pad}}$ can always be placed at the two ports so as to make the real parts of $M_1 + M_{1,\text{pad}}$ and $M_2 + M_{2,\text{pad}}$ positive and the overall stability parameter $\tilde{\eta}$ approaching unity or, for real-positive $M_{1,\text{pad}}$ and $M_{2,\text{pad}}$,

$$M_{1,\text{pad}} + \operatorname{Re} M_1 > 0 \tag{3.195a}$$

$$M_{2,\text{pad}} + \operatorname{Re} M_2 > 0 \tag{3.195b}$$

$$\tilde{\eta} = \frac{2(M_{1,\text{pad}} + \operatorname{Re} M_1)(M_{2,\text{pad}} + \operatorname{Re} M_2) - \operatorname{Re} k_{12}k_{21}}{|k_{12}k_{21}|} \to 1 \tag{3.195c}$$

The maximum power gain then tends toward its maximum stable value,

$$\mathcal{G}_{\max} \to \mathcal{G}_0 = \left|\frac{k_{21}}{k_{12}}\right| \tag{3.196}$$

$\mathcal{G}_0$ is called the *maximum stable power gain* of the two-port network.

The above concept is defined for a potentially unstable two-port network. This definition can also be extended to include absolutely stable two-port networks by the use of negative padding resistances. The only difference is that $M_{1,\text{pad}}$ and $M_{2,\text{pad}}$ are now real and nonpositive and Eq. (3.195*c*) may be employed to determine the values of the padding resistances, so that the stability boundary $\tilde{\eta} = 1$ can again be approached. Thus, $\mathcal{G}_0$ can be defined as the maximum stable power gain for all devices.

For an absolutely stable two-port network, the maximum stable power gain is now identified as the nonreciprocal gain of the two-port, and (3.171) is reduced to

$$\mathcal{G}_{\max} = \mathcal{G}_0\gamma \tag{3.197}$$

In words, it states that the maximum power gain of an absolutely stable two-port network is equal to the maximum stable power gain times the efficiency of its reciprocal part.

Example 3.13 In the two-port network N of Fig. 3.23 considered in Example 3.12, let

$$g_1 = 4 \cdot 10^{-3} \text{ mho} \qquad g_2 = 0.01 \text{ mho}$$
$$g_m = 0.2 \text{ mho} \qquad C = 5 \text{ pF} \tag{3.198}$$

Suppose that the two-port is operating at 30 MHz. Then the second inequality of (3.175) is violated, meaning that the two-port network is active but potentially unstable at 30 MHz. One way to stabilize the two-port is to introduce sufficient loss by connecting at the input port a conductance g as shown in Fig. 3.25. To determine the value of this padding conductance so that the stability boundary $\tilde{\eta} = 1$ can be approached, we set (3.195c) to unity. From (3.183), this is equivalent to

$$\frac{2(g_1 + g)g_2 + \omega^2 C^2}{\omega C \sqrt{\omega^2 C^2 + g_m^2}} = 1 \tag{3.199}$$

giving

$$g = 5.382 \cdot 10^{-3} \text{ mho} \tag{3.200}$$

or 185.81 Ω. The maximum stable power gain is determined from (3.196) as

$$\mathcal{G}_0 = \left| \frac{y_{21}}{y_{12}} \right| = \sqrt{1 + \frac{g_m^2}{\omega^2 C^2}} = 212.21 \tag{3.201}$$

Now we lower the operating frequency from 30 MHz to 5 MHz. The inequalities (3.175) are satisfied, and the two-port network is both active and absolutely stable at 5 MHz. From (3.183) we calculate the stability parameter

$$\eta = \frac{2g_1 g_2 + \omega^2 C^2}{\omega C \sqrt{\omega^2 C^2 + g_m^2}} = 2.547 \tag{3.202}$$

The efficiency of the reciprocal part of the two-port network is found to be

$$\gamma = \eta - \sqrt{\eta^2 - 1} = 0.2045 \tag{3.203}$$

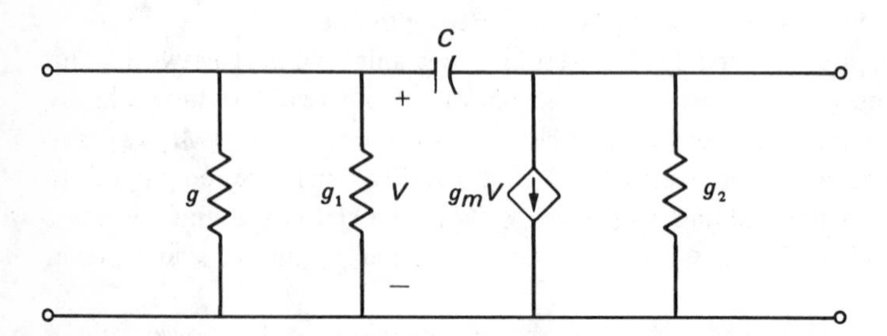

Figure 3.25 The stabilization of a two-port network by introducing at the input port a conductance g.

yielding the maximum power gain

$$\mathcal{G}_{max} = \mathcal{G}_0 \gamma = 260.38 \tag{3.204}$$

where $\mathcal{G}_0 = 1273.24$.

Suppose that we wish to determine the value of the padding conductance g connected across the input port at 5 MHz so that the stability boundary $\tilde{\eta} = 1$ is approached. Again, from (3.199) with $\omega = \pi \cdot 10^7$ rad/s, we get

$$g = -2.431 \cdot 10^{-3} \text{ mho} \tag{3.205}$$

or $-411.35\ \Omega$.

3.7.4 Maximum Power Gain and the *U*-Function

In Sec. 3.5.2 we demonstrated that for a given active two-port network, its *U*-function is identifiable as the maximum unilateral power gain under a lossless reciprocal imbedding. In the present section, we express this maximum power gain $\mathcal{G}_{max}$ in terms of the *U*-function and the complex nonreciprocal gain.

Let

$$\pm \delta = \frac{k_{21}}{k_{12}} = \mathcal{G}_0 e^{j\xi} \tag{3.206}$$

The plus sign is chosen for the z- and y-parameters, and the minus sign for the h- and g-parameters. As discussed previously, the magnitude of δ is called the nonreciprocal gain of the two-port network and can be interpreted as the maximum stable power gain. Thus, it is convenient to call δ the *complex nonreciprocal gain* of the two-port network, because it is a complex measure of nonreciprocity.

From Eqs. (3.51), (3.117), and (3.171), we have

$$\begin{aligned} |k_{12}|^2 |\delta - \mathcal{G}_{max}|^2 &= |k_{21}|^2 + |k_{12}|^2 \mathcal{G}_{max}^2 \mp 2\mathcal{G}_{max} \operatorname{Re} k_{12}\bar{k}_{21} \\ &= 2\mathcal{G}_{max} |k_{12}k_{21}| \eta \mp 2\mathcal{G}_{max} \operatorname{Re} k_{12}\bar{k}_{21} \\ &= \mathcal{G}_{max}[4 \operatorname{Re} k_{11} \operatorname{Re} k_{22} - 2 \operatorname{Re} (k_{12}k_{21} \pm k_{12}\bar{k}_{21})] \\ &= \frac{\mathcal{G}_{max}|k_{21} \mp k_{12}|^2}{U} \end{aligned} \tag{3.207}$$

which can be rewritten as

$$\frac{\mathcal{G}_{max}}{U} = \left| \frac{\delta - \mathcal{G}_{max}}{\delta - 1} \right|^2 \tag{3.208}$$

showing that $\mathcal{G}_{max}$ is a function of only two basic parameters: the *U*-function and the complex nonreciprocal gain. Since U is invariant under all lossless reciprocal imbedding, and, in particular, under all device orientations such as the common-emitter, common-base, and common-collector configurations of the transistor, a simple way to control $\mathcal{G}_{max}$ is by modifying δ with a lossless reciprocal imbedding or by changing the device orientation without changing U. Alternatively, we can alter U by lossy padding at the input and output ports, as suggested in the preceding section, without changing δ.

3.8 SUMMARY

We began this chapter by reviewing briefly the general representation of a two-port network and by defining several measures of power flow. We found that the transducer power gain is the most meaningful description of power transfer capabilities of a two-port network as it compares the power delivered to the load with the power that the source is capable of supplying under optimum conditions. We then introduced the concepts of sensitivity and derived the general expressions for various power gains, transfer functions, and the sensitivity functions in terms of the general two-port parameters and the source and load immittances. The specialization of the general passivity condition for n-port networks in terms of the more immediately useful two-port parameters was taken up next. A useful parameter associated with a two-port network called the U-function was introduced. We showed that this function is invariant under all lossless reciprocal imbedding. The physical significance of this result is that the U-function can be used to characterize the power-amplifying capability of a device. The characterization is unique in that, among other reasons, it is independent of the measuring circuit, provided only that the circuit uses lossless reciprocal elements. Furthermore, we demonstrated that for a given active two-port network, its U-function is identifiable as the maximum unilateral power gain under a lossless reciprocal imbedding, the resulting structure being unilateral.

Whenever we consider active networks, we must discuss stability. For this we introduced the single frequency concepts of stability for a two-port network: potential instability and absolute stability. Necessary and sufficient conditions for absolute stability were derived and are known as *Llewellyn's criteria*. Specifically, we showed that a two-port network is absolutely stable if and only if its open-circuit or short-circuit input and output conductances are positive and the stability parameter is greater than unity. We indicated that the stability parameter is invariant under immittance substitution, and its value remains unaltered regardless of the two-port matrix employed in computation.

For a potentially unstable two-port network, there exists a set of terminations so that the overall network possesses a natural frequency on the real-frequency axis. In such situations, the maximum power gain is infinite and the concept of optimum power gain has no significance. For an absolutely stable two-port network, we can find a set of optimum source and load immittances so that the power gain is maximized. It turns out that under this optimum matching situation, the three power gains assume a common maximum value. This maximum power gain can be written as the product of two factors. The first factor is a measure of the nonreciprocity of the two-port network and is identifiable as the maximum stable power gain. The second factor is a measure of the efficiency of the reciprocal part of the two-port network, and is bounded between 0 and 1. This product expression for the maximum power gain was employed to decompose a general two-port network into two two-ports connected in cascade, one reciprocal and the other nonreciprocal. Finally, we derived a useful relation among the maximum power gain, the *U*-function, and the complex nonreciprocal gain, a complex measure of nonreciprocity, of a two-port network.

PROBLEMS

3.1 Confirm that the three power gain expressions (3.19) are valid for (*a*) the *h*-parameters, (*b*) the *y*-parameters, and (*c*) the *g*-parameters.

3.2 Confirm that the general expressions (3.25) for the transfer functions are valid for (*a*) the *z*-parameters, (*b*) the *y*-parameters, and (*c*) the *g*-parameters.

3.3 Figure 3.26 shows the equivalent network of a field-effect transistor (FET). Determine the range of frequencies in which the device is both active and absolutely stable.

3.4 Derive the sensitivity expressions (3.27).

3.5 Confirm that the sensitivity expressions (3.29) are valid for (*a*) the *z*-parameters, (*b*) the *y*-parameters, and (*c*) the *g*-parameters.

3.6 Figure 3.11 can be viewed as the lossless reciprocal four-port imbedding of the active device *N* as shown in Fig. 3.12. Show that the *U*-function is invariant under this imbedding.

3.7 Consider the two-port network of Fig. 3.27. Determine the maximum frequency of oscillation and the fastest regenerative mode of the device from its (*a*) *y*-parameters, (*b*) *z*-parameters, (*c*) *h*-parameters, and (*d*) *g*-parameters. Compare the amount of effort involved in arriving at the same conclusion.

3.8 A transistor has the following parameter values at 3 MHz:

$$h_{11} = 50 + j1.2\ \Omega \qquad h_{12} = (1.5 + j1.6) \cdot 10^{-3}$$
$$h_{21} = -0.9 + j0.06 \qquad h_{22} = 7 + j25\ \mu\text{mho} \tag{3.209}$$

Find the optimum source and load terminations, along with the maximum power gain.

3.9 A three-terminal device *N* with a shunt and series feedback is shown in Fig. 3.28. Assume that the two feedback admittances are lossless and reciprocal. Show that the *U*-function of *N* is invariant under this lossless reciprocal imbedding.

3.10 Refer to the network of Fig. 3.28. Can appropriate feedback admittances y_f and y_α be chosen so that, at a particular frequency, the overall two-port network is unilateral?

3.11 Let *N* be a three-terminal device characterized by its impedance matrix. By using the configuration shown in Fig. 3.29, show that a lossless reciprocal four-port network can always be chosen so that, at a particular frequency, the overall two-port network is unilateral.

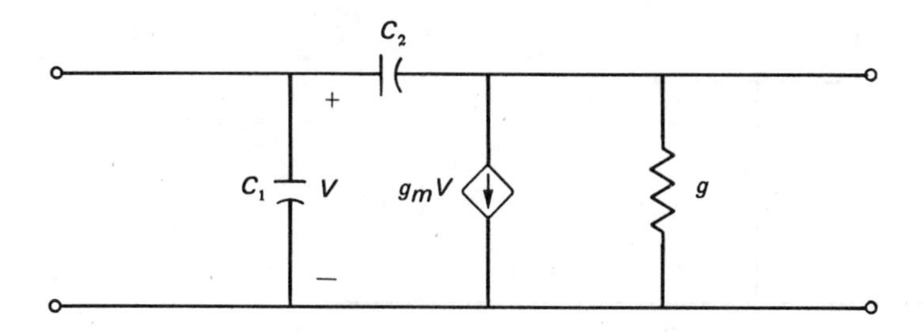

Figure 3.26 An equivalent network of an FET.

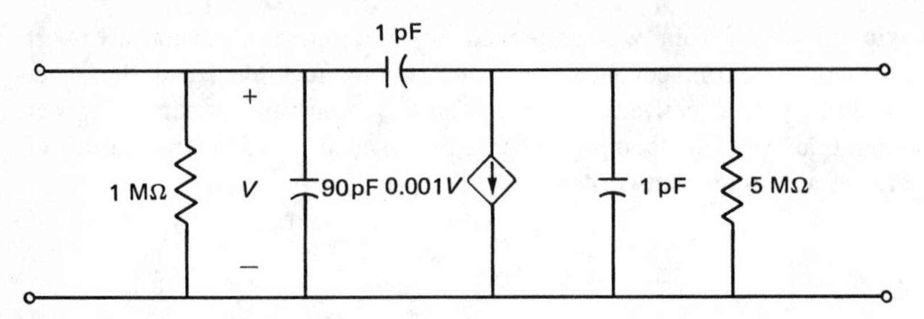

Figure 3.27 A given two-port network.

3.12 The stability parameter η and the efficiency of the reciprocal part γ of a two-port network are related by (3.172). If η lies between -1 and 1, γ is complex. Show that the locus of γ on the complex plane for values of η from -1 to 1 is a circle with the origin as center and unit radius.

3.13 Show that the efficiency of the reciprocal part γ of a two-port network can be expressed in the form

$$\gamma^2 = \frac{(1-\theta_r)^2 + \theta_x^2}{(1+\theta_r)^2 + \theta_x^2} \tag{3.210}$$

where

$$\theta_r = \frac{\text{Re } M_{1,\text{opt}}}{\text{Re } k_{11}} = \frac{\text{Re } M_{2,\text{opt}}}{\text{Re } k_{22}} \tag{3.211a}$$

$$\theta_x = \frac{\text{Im } (k_{11} + M_{1,\text{opt}})}{\text{Re } k_{11}} = \frac{\text{Im } (k_{22} + M_{2,\text{opt}})}{\text{Re } k_{22}} \tag{3.211b}$$

3.14 By substituting one set of the impedance, admittance, hybrid, or inverse hybrid parameters for any other, show that the stability parameter is invariant under this substitution.

3.15 The values of the y-parameters of a transistor operating at 30 MHz are given by (3.96). Find the optimum source and load immittances, along with the maximum power gain at this frequency.

3.16 For a reciprocal two-port network, show that (3.120c) and (3.134) are equivalent.

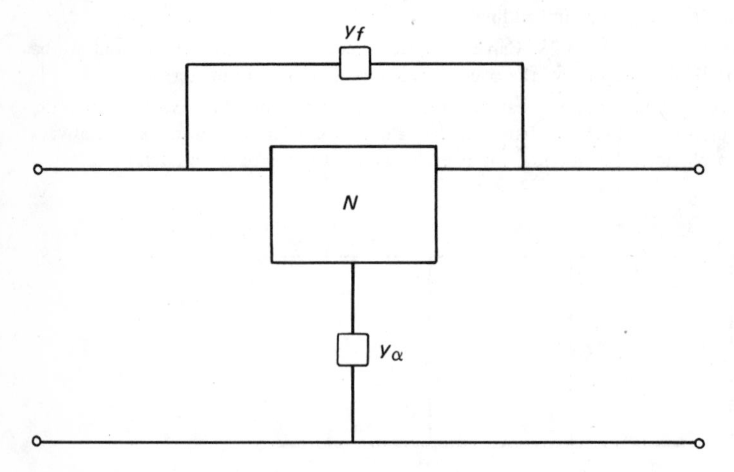

Figure 3.28 A three-terminal device N with a shunt and series feedback.

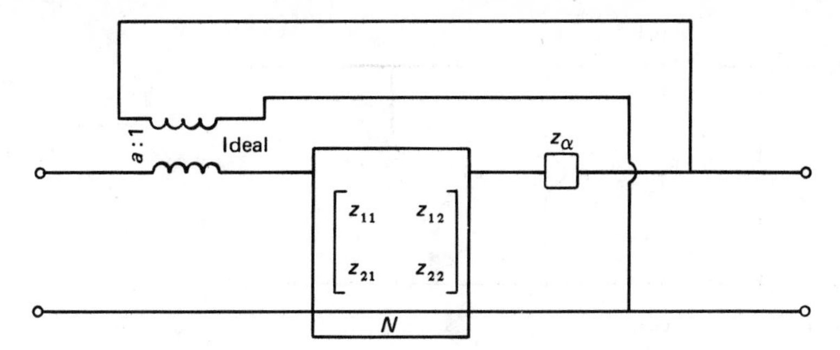

Figure 3.29 A three-terminal device N together with its feedback circuit.

3.17 Use (3.206) and show that the U-function of a two-port network can be expressed as

$$U = \frac{\frac{1}{2}(\mathcal{g}_0 + 1/\mathcal{g}_0) \pm \cos \xi}{\eta \pm \cos \xi} \tag{3.212}$$

where the plus sign applies to h- and g-parameters and the minus sign to z- and y-parameters.

3.18 Justify the statement that if two admittances have the same phase, then the real part of these two admittances in series is equal to the series combination of their real parts.

3.19 Assume that $\text{Re}\, k_{11}$ and $\text{Re}\, k_{22}$ are positive. Show that the value of the stability parameter lies between -1 and ∞.

3.20 Let

$$\epsilon = \frac{2}{\eta + 1} \tag{3.213}$$

Show that the efficiency of the reciprocal part γ of a two-port network can be written in the form

$$\gamma = \frac{1 - \sqrt{1 - \epsilon}}{1 + \sqrt{1 - \epsilon}} \tag{3.214}$$

3.21 Refer to Prob. 3.8. Determine the values of the padding resistances so that the stability boundary of the device can be approached.

3.22 Show that the two-port network of Fig. 3.23 is absolutely stable for all real frequencies satisfying the inequality

$$\omega C[(g_m^2 + \omega^2 C^2)^{1/2} - \omega C] < 2g_1 g_2 \tag{3.215}$$

3.23 The absolute stability constraint $\eta > 1$ can be expressed equivalently as

$$\zeta = \frac{2\,\text{Re}\, k_{11}\ \text{Re}\, k_{22}}{\text{Re}\, k_{12}k_{21} + |k_{12}k_{21}|} > 1 \tag{3.216}$$

Show that this quantity is *not* invariant under immittance substitution.

3.24 Consider the two-port network of Fig. 3.30. Determine the range of frequencies in which the device is both active and absolutely stable. Also, find the maximum frequency of oscillation and the fastest regenerative mode of the device.

3.25 An absolutely stable two-port network is characterized by its general hybrid parameters k_{ij}. Show that the optimum source and load immittances can be expressed as

$$\bar{M}_{1,\text{opt}} = k_{11} - \frac{k_{12}k_{21}(1 - \gamma e^{j\phi})}{2\,\text{Re}\, k_{22}} \tag{3.217a}$$

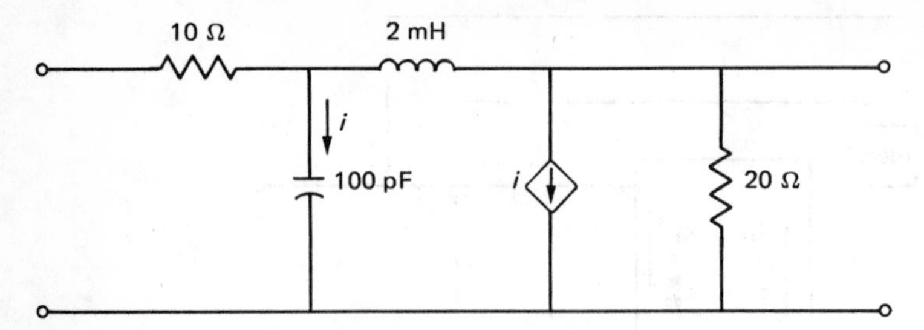

Figure 3.30 A given two-port network.

$$M_{2,\text{opt}} = -k_{22} + \frac{2 \operatorname{Re} k_{22}}{1 - \gamma e^{j\phi}} \tag{3.217b}$$

where γ is the efficiency of the reciprocal part of the two-port, and

$$-\bar{k}_{12}\bar{k}_{21} = |k_{12}k_{21}|e^{j\phi} \tag{3.217c}$$

3.26 From (3.7), the input immittance M_{11} of a two-port network when its output is terminated in M_2 can be expressed as

$$M_{11} = \frac{aM_2 + b}{M_2 + c} \tag{3.218}$$

where $a = k_{11}$, $b = k_{11}k_{22} - k_{12}k_{21}$, and $c = k_{22}$. The above relation is a simple bilinear transformation of the terminating immittance M_2. Show that for all passive M_2, that is, the closed right half of the M_2-plane, M_{11} is transformed into a circle centered at

$$p_1 = k_{11} - \frac{k_{12}k_{21}}{2 \operatorname{Re} k_{22}} \tag{3.219a}$$

with radius

$$r_1 = \frac{|k_{12}k_{21}|}{2 \operatorname{Re} k_{22}} \tag{3.219b}$$

If the two-port network is absolutely stable, then

$$\operatorname{Re} p_1 - r_1 > 0 \tag{3.220}$$

Show that (3.220) is equivalent to the familiar constraint $\eta > 1$. Instead of starting from M_{11}, demonstrate that the same conclusion can be reached if we start from the output immittance of the two-port network, as given in (3.8).

3.27 Consider the two-port network N_a of Fig. 3.28. Let y_{aij} and y_{ij} be the y-parameters of N_a and N, respectively. Show that

$$y_{a11} = y_f + \frac{y_{11}y_\alpha + y_{11}y_{22} - y_{12}y_{21}}{y_\alpha + y_{11} + y_{12} + y_{21} + y_{22}} \tag{3.221a}$$

$$y_{a22} = y_f + \frac{y_{22}y_\alpha + y_{11}y_{22} - y_{12}y_{21}}{y_\alpha + y_{11} + y_{12} + y_{21} + y_{22}} \tag{3.221b}$$

3.28 By using the parameter values of a transistor given in (3.209), determine the optimum terminations at 3 MHz that lead to the maximum power gain. If the transistor is terminated at the load with a 20-kΩ resistor and at the source with a 100-Ω resistor in shunt with a 10-pF capacitor, what is the power gain at 3 MHz? What is the transducer power gain at this frequency? Is the amplifier stable?

3.29 Repeat the second part of Prob. 3.28 if the source impedance is represented by the series connection of a 100-Ω resistor and a 10-μH inductor, everything else being the same.

3.30 For the hybrid-pi equivalent network of a transistor shown in Fig. 3.3, determine the frequency range in which the transistor is both active and absolutely stable.

3.31 Show that a lossless two-port network that is conjugate-matched at the input port also provides a conjugate match at the output port. Can the same conclusion be drawn for the lossy two-port network? If not, construct a counterexample.

3.32 Show that the *U*-function, as defined in (3.51), for a two-port network will possess the same value regardless of the matrix employed in computation. More specifically, demonstrate that the *U*-function is invariant when any one set of *z*-, *y*-, *h*-, or *g*-parameters is replaced by any other set.

3.33 Refer to Fig. 3.31. Suppose that we move the admittance y_α from the output port to the input port, as shown in Fig. 3.31. Verify that if y_{ij} are the *y*-parameters of *N*, then the admittance matrix of the two-port is

$$\frac{y_\alpha}{y_\alpha + y_{11}} \begin{bmatrix} y_{11} & y_{12} \\ y_{21} & y_{22} + \dfrac{\Delta_y}{y_\alpha} \end{bmatrix} \tag{3.222}$$

Determine the admittances y_α and y_f so that the resulting network is unilateralized at a specified frequency on the real-frequency axis. Derive expressions similar to those given by (3.95).

3.34 Assume that the two-port parameters k_{ij} are real. Show that the optimum terminations $M_{1,\text{opt}}$ and $M_{2,\text{opt}}$ and the maximum power gain $\mathcal{G}_{\max}$ can be expressed as

$$M_{1,\text{opt}} = \sqrt{\frac{\Delta_k k_{11}}{k_{22}}} \tag{3.223a}$$

$$M_{2,\text{opt}} = \sqrt{\frac{\Delta_k k_{22}}{k_{11}}} \tag{3.223b}$$

$$\mathcal{G}_{\max} = \frac{k_{21}^2}{k_{11}k_{22}[1 + (1 - k_{12}k_{21}/k_{11}k_{22})^{1/2}]^2} \tag{3.223c}$$

where $\Delta_k = k_{11}k_{22} - k_{12}k_{21}$.

3.35 An active device *N* is connected in parallel with an ideal transformer and an admittance y_α, as shown in Fig. 3.32. Demonstrate that a passive but not necessarily lossless y_α can always be chosen so that, at a particular frequency, the overall two-port network is unilateral, where *a*

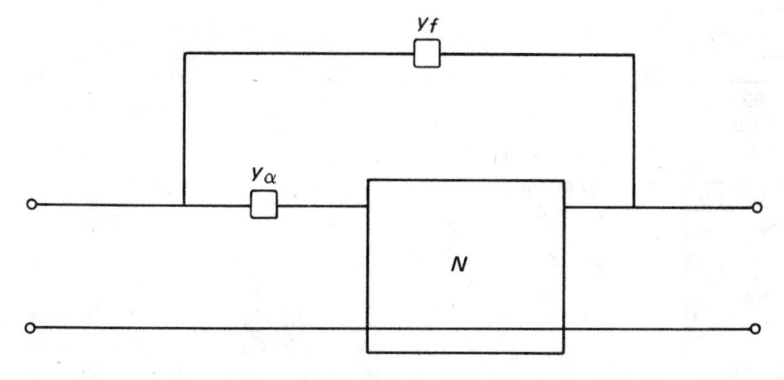

Figure 3.31 A three-terminal device with two lossless reciprocal admittances y_α and y_f.

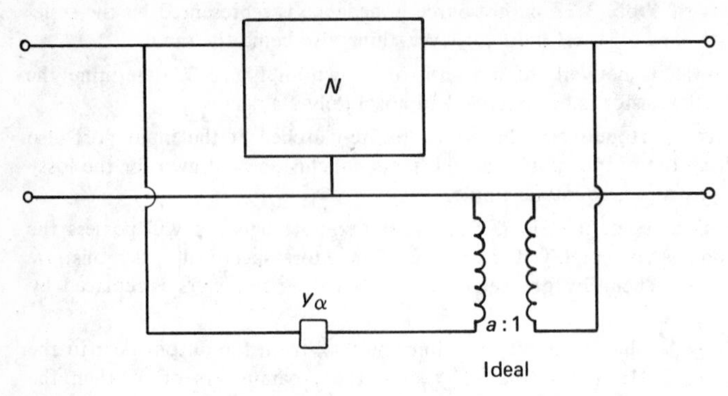

Figure 3.32 An active device N connected in parallel with an ideal transformer and an admittance y_α.

can be positive or negative. By using (3.92), the maximum unilateral transducer power gain $\mathcal{G}_a$ is obtained. Show that by optimizing $\mathcal{G}_a$ versus a, the gain expression becomes

$$\mathcal{G}_{a,\max} = \frac{|y_{21} - y_{12}|^2}{4(\sqrt{\operatorname{Re} y_{11} \operatorname{Re} y_{22}} + |\operatorname{Re} y_{12}|)^2} \tag{3.224}$$

where y_{ij} are the y-parameters of N.

3.36 Consider the unilateral two-port network N_a of Fig. 3.15. For simplicity, let $y_{a11} = y_{a22} = 1$ mho and $y_{a21} = -g_m$. Suppose that N_a is connected in parallel with an ideal gyrator N_g with the admittance matrix

$$\mathbf{Y}_g = \begin{bmatrix} 0 & -g \\ g & 0 \end{bmatrix} \tag{3.225}$$

The composite two-port network N_c can be unilateralized by a lossless reciprocal imbedding. Demonstrate that the maximum transducer power gain that can be achieved for the unilateralized N_c is precisely $U_c(j\omega)$, the U-function of N_c:

$$U_c(j\omega) = \frac{|2g - g_m|^2}{4(g^2 - g_m g + 1)} \tag{3.226}$$

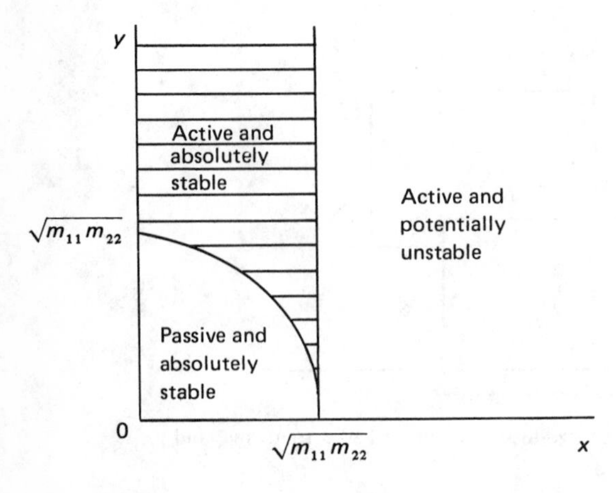

Figure 3.33 The regions of activity and stability of a two-port network in terms of its general hybrid parameters.

This indicates that for $|g_m| > 2$ there exists a real g for which $U_c(j\omega)$ is infinite. In other words, if N_a is active, then there exists a lossless *nonreciprocal* imbedding under which the maximum unilateral power gain can be made as large as one desires whereas the lossless reciprocal imbedding limits the gain to the value U, provided that the unilateralized two-port network is absolutely stable.

3.37 Consider the unilateral two-port network N_a of Fig. 3.15. Let $y_{a11} = y_{a22} = 1$ mho and $y_{a21} = -g_m$. By connecting transformers at the ports, the new admittance matrix $\mathbf{Y}_t$ can be written as

$$\mathbf{Y}_t = \begin{bmatrix} 1 & n \\ n & 1 \end{bmatrix} \begin{bmatrix} 1 & 0 \\ -g_m & 1 \end{bmatrix} \begin{bmatrix} 1 & n \\ n & 1 \end{bmatrix} = \begin{bmatrix} 1 + n(n - g_m) & n(2 - ng_m) \\ 2n - g_m & 1 + n(n - g_m) \end{bmatrix} \tag{3.227}$$

Show that by carrying out an optimum unilateralization including losses as described in Prob. 3.35, the gain expression (3.224) can be made infinite if $g_m > 2$. This result indicates that if N_a is active, then there exists a *lossy* reciprocal imbedding under which the maximum unilateral power gain can be made as large as one desires whereas the lossless reciprocal imbedding limits the gain to the value U, provided that the unilateralized two-port network is absolutely stable.

3.38 Choose $x = |\text{Re}\sqrt{k_{12}k_{21}}|$ and

$$y = \tfrac{1}{2}\left|\, |k_{12}| - |k_{21}| \,\right| = \left| \sqrt{\tfrac{1}{4}|k_{12} + \bar{k}_{21}|^2 - x^2} \right| \tag{3.228}$$

as the coordinates of a two-dimensional diagram. Show that the regions of activity and stability can be represented as in Fig. 3.33.

BIBLIOGRAPHY

Aurell, C. G.: Representation of the General Linear Four-Terminal Network and Some of Its Properties, *Ericsson Technics*, vol. 11, no. 1, pp. 155–179, 1955.

Aurell, C. G.: Some Tools for the Analysis and Representation of Linear Two-Port Networks, *IEEE Trans. Circuit Theory*, vol. CT-12, no. 1, pp. 18–21, 1965.

Bolinder, E. F.: Survey of Some Properties of Linear Networks, *IRE Trans. Circuit Theory*, vol. CT-4, no. 3, pp. 70–78, 1957 (correction in vol. CT-5, no. 2, p. 139, 1958).

Brodie, J. H.: The Stability of Linear Two-Ports, *IEEE Trans. Circuit Theory*, vol. CT-12, no. 4, pp. 608–610, 1965.

Chen, W. K.: The Scattering Matrix and the Passivity Condition, *Matrix and Tensor Quart.*, vol. 24, nos. 1 and 2, pp. 30–32 and 74–75, 1973.

Chen, W. K.: Relationships between Scattering Matrix and Other Matrix Representations of Linear Two-Port Networks, *Int. J. Electronics*, vol. 38, no. 4, pp. 433–441, 1975.

Dietl, A.: Der optimale Wirkungsgrad von Vierpolen im Dezimeter- und Zentimeterwellenbereich, *Hochfrequenz. u. Elak.*, vol. 66, pp. 25–29, 1944.

Fjällbrant, T.: Activity and Stability of Linear Networks, *IEEE Trans. Circuit Theory*, vol. CT-12, no. 1, pp. 12–17, 1965.

Gärtner, W. W.: Maximum Available Power Gain of Linear Fourpoles, *IRE Trans. Circuit Theory*, vol. CT-5, no. 4, pp. 375–376, 1958.

Gewertz, C. M.: Synthesis of a Finite, Four-Terminal Network from Its Prescribed Driving-Point and Transfer Functions, *J. Math. Phys.*, vol. 12, nos. 1 and 2, pp. 1–257, 1933.

Jørsboe, H.: Criterion of Absolute Stability of Two-Ports, *IEEE Trans. Circuit Theory*, vol. CT-17, no. 4, pp. 639–640, 1970.

Ku, W. H.: Stability of Linear Active Nonreciprocal *n*-Ports, *J. Franklin Inst.*, vol. 276, no. 3, pp. 207–224, 1963.

Ku, W. H.: A Simple Derivation for the Stability Criterion of Linear Active Two-Ports, *Proc. IEEE*, vol. 53, no. 3, pp. 310–311, 1965.

Ku, W. H.: Extension of the Stability Criterion of Linear Active Two-Ports to the Entire Complex Frequency Plane, *Proc. IEEE*, vol. 58, no. 4, pp. 591–592, 1970.

Kuh, E. S. and R. A. Rohrer: "Theory of Linear Active Networks," San Francisco, Calif.: Holden-Day, 1967.

Leine, P. O.: On the Power Gain of Unilaterized Active Networks, *IRE Trans. Circuit Theory*, vol. CT-8, no. 3, pp. 357–358, 1961.

Linvill, J. G. and L. G. Schimpf: The Design of Tetrode Transistor Amplifiers, *Bell Syst. Tech. J.*, vol. 35, no. 4, pp. 813–840, 1956.

Llewellyn, F. B.: Some Fundamental Properties of Transmission Systems, *Proc. IRE*, vol. 40, no. 3, pp. 271–283, 1952.

Mason, S. J.: Power Gain in Feedback Amplifiers, *IRE Trans. Circuit Theory*, vol. CT-1, no. 2, pp. 20–25, 1954.

Mason, S. J.: Some Properties of Three-Terminal Devices, *IRE Trans. Circuit Theory*, vol. CT-4, no. 4, pp. 330–332, 1957.

Mathis, H. F.: Experimental Procedures for Determining the Efficiency of Four-Terminal Networks, *J. Appl. Phys.*, vol. 25, no. 8, pp. 982–986, 1954.

Page, D. F. and A. R. Boothroyd: Instability in Two-Port Active Networks, *IRE Trans. Circuit Theory*, vol. CT-5, no. 2, pp. 133–139, 1958.

Rollett, J. M.: Stability and Power-Gain Invariants of Linear Twoports, *IRE Trans. Circuit Theory*, vol. CT-9, no. 1, pp. 29–32, 1962 (correction in vol. CT-10, no. 1, p. 107, 1963).

Rollett, J. M.: The Measurement of Transistor Unilateral Gain, *IEEE Trans. Circuit Theory*, vol. CT-12, no. 1, pp. 91–97, 1965.

Sathe, S. T.: A Stability Test for a Class of Linear, Active Two-Ports with Uncertain Parameters, *IEEE Trans. Commun.*, vol. COM-23, no. 11, pp. 1357–1361, 1975.

Scanlan, J. O. and J. S. Singleton: The Gain and Stability of Linear Two-Port Amplifiers, *IRE Trans. Circuit Theory*, vol. CT-9, no. 3, pp. 240–246, 1962a.

Scanlan, J. O. and J. S. Singleton: Two-Ports–Maximum Gain for a Given Stability Factor, *IRE Trans. Circuit Theory*, vol. CT-9, no. 4, pp. 428–429, 1962b.

Sharpe, G. E., J. L. Smith, and J. R. W. Smith: A Power Theorem on Absolutely Stable Two-Ports, *IRE Trans. Circuit Theory*, vol. CT-6, no. 2, pp. 159–163, 1959.

Singhakowinta, A.: On Analysis and Representation of Two-Port, *IEEE Trans. Circuit Theory*, vol. CT-13, no. 1, pp. 102–103, 1966.

Stern, A. P.: Considerations on the Stability of Active Elements and Applications to Transistors, *IRE Natl. Conv. Rec.*, part 2, pp. 46–52, 1956.

Van der Puije, P. D.: On Negative Conductance Activity in 2-Port Networks, *IEEE Trans. Circuits and Systems*, vol. CAS-21, no. 2, pp. 194–196, 1974.

Van Heuven, J. H. C. and T. E. Rozzi: The Invariance Properties of a Multivalue *n*-Port in a Linear Embedding, *IEEE Trans. Circuit Theory*, vol. CT-19, no. 2, pp. 176–183, 1972.

Venkateswaran, S.: An Invariant Stability Factor and Its Physical Significance, *Proc. IEE* (*London*), part C, Mono. 468E, pp. 98–102, 1961.

Venkateswaran, S. and A. R. Boothroyd: Power Gain and Bandwidth of Tuned Transistor Amplifier Stages, *Proc. IEE* (*London*), vol. 106B, suppl. 15, pp. 518–529, 1960.

Woods, D.: Reappraisal of the Unconditional Stability Criteria for Active 2-Port Networks in Terms of *s* Parameters, *IEEE Trans. Circuits and Systems*, vol. CAS-23, no. 2, pp. 73–81, 1976.

Youla, D.: A Stability Characterization of the Reciprocal Linear Passive *n* Port, *Proc. IRE*, vol. 47, no. 6, pp. 1150–1151, 1959.

Youla, D.: A Note on the Stability of Linear, Nonreciprocal *n*-Ports, *Proc. IRE*, vol. 48, no. 1, pp. 121–122, 1960.

Zawels, J.: Gain-Stability Relationship, *IEEE Trans. Circuit Theory*, vol. CT-10, no. 1, pp. 109–110, 1963.

CHAPTER

FOUR

THEORY OF FEEDBACK AMPLIFIERS I

In the preceding chapter, we demonstrated that by introducing physical feedback loops externally to an active device, we can produce a particular change in the performance of the network. Specifically, we showed that a three-terminal device can be unilateralized by a lossless reciprocal imbedding. In this and following chapters, we shall study the subject of feedback in detail and demonstrate that feedback may be employed to make the gain of an amplifier less sensitive to variations in the parameters of the active components, to control its transmission and driving-point properties, to reduce the effects of noise and nonlinear distortion, and to affect the stability or instability of the network.

We first discuss the conventional treatment of feedback amplifiers, which is based on the ideal feedback model, and analyze several simple feedback networks. We then present Bode's feedback theory in detail. Bode's theory is based on the concepts of return difference and null return difference and is applicable to both simple and complicated feedback amplifiers, where the analysis by conventional method for the latter breaks down. We show that return difference is a generalization of the concept of the feedback factor of the ideal feedback model and can be interpreted physically as the returned voltage. The relationships between the network functions and return difference and null return difference are derived and are employed to simplify the calculation of driving-point impedance of an active network.

4.1 IDEAL FEEDBACK MODEL

A *feedback amplifier* is a network in which some variable, either the output variable or any other controllable one, is fed back as the input variable to some part of the network in such a way that it is able to affect its own value. The simplest form of a feedback amplifier can be represented by the ideal block diagram of Fig. 4.1, in which $\mu(s)$ represents the transfer function of the *unilateral forward path* and is conventionally called the *open-loop transfer function* or *forward amplifier gain*, defined by the relation

$$\mu(s) = \left.\frac{y(s)}{u(s)}\right|_{u_f(s)=0} = \frac{y(s)}{u_i(s)} \tag{4.1}$$

where $y(s)$ and the u's denote either the current and/or voltage variables. $\beta(s)$ is the transfer function of a *unilateral feedback path* and is defined as

$$\beta(s) = \frac{u_f(s)}{y(s)} \tag{4.2}$$

Thus, in this highly idealized model, none of the input signal is transmitted through the feedback path and none of the output signal is transmitted in the reverse direction through the forward path. Also, the presence of the ideal adder in the model indicates that there is no loading effect at the input. In other words, the transmission of signal is permitted only in the direction of the arrows as shown in the block diagram of Fig. 4.1. Under these assumptions, it is straightforward to show that the overall transfer function of the amplifier can be expressed as

$$w(s) = \frac{y(s)}{u(s)} = \frac{\mu(s)}{1 - \mu(s)\beta(s)} \tag{4.3}$$

which is referred to as the *closed-loop transfer function.* In terms of logarithms, we have

$$\ln w(s) = \ln \mu(s) - \ln\,[1 - \mu(s)\beta(s)] \tag{4.4}$$

whose $j\omega$-axis real parts are related by

$$\ln |w(j\omega)| = \ln |\mu(j\omega)| - \ln |1 - \mu(j\omega)\beta(j\omega)| \tag{4.5}$$

Observe that in the application of feedback, the open-loop transfer function is modified by a factor $1 - \mu(s)\beta(s)$. The factor $1 - \mu(s)\beta(s)$ is referred to as the *feedback factor*. At a specified frequency on the real-frequency axis, if its magnitude is greater than unity, we have *negative* or *degenerative feedback*, and if it

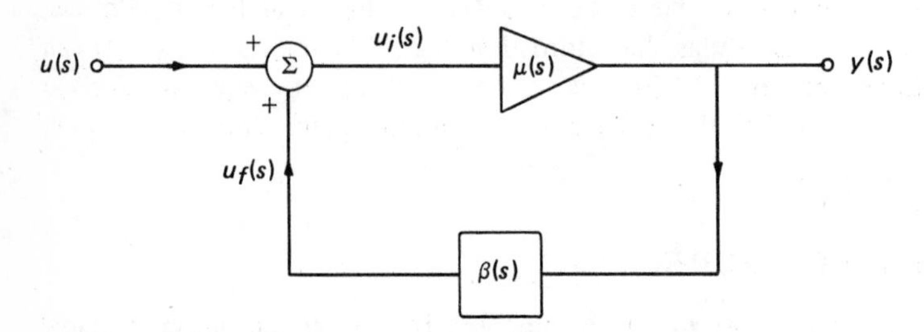

Figure 4.1 The block diagram of the ideal feedback model.

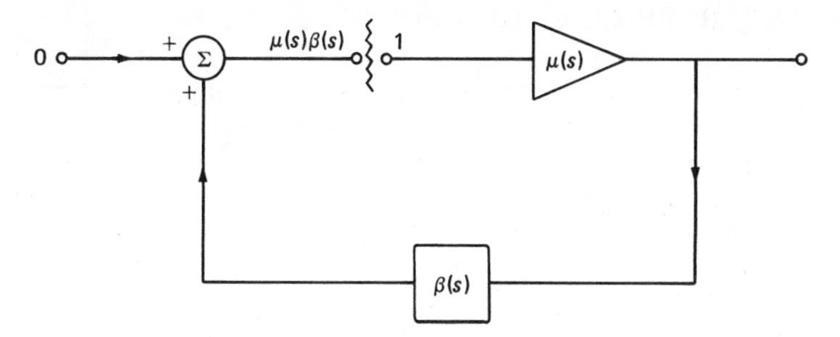

Figure 4.2 The physical interpretation of the loop transmission.

is smaller than unity, we have *positive* or *regenerative feedback*. From Eq. (4.5) we see that negative feedback reduces the forward amplifier gain by the amount of feedback $\ln |1 - \mu(j\omega)\beta(j\omega)|$ measured in nepers, or $20 \log |1 - \mu(j\omega)\beta(j\omega)|$ measured in decibels. In exchange, as we shall see, the gain becomes less sensitive to the variations of the parameters, the effects of noise and nonlinear distortion are usually reduced, and the transmission and driving-point characteristics are modified, all by the feedback factor $1 - \mu(j\omega)\beta(j\omega)$. On the other hand, the positive feedback increases the forward amplifier gain, the sensitivity, and the effects of noise and nonlinear distortion all by the factor $1 - \mu(j\omega)\beta(j\omega)$, and is therefore seldom used in amplifier design. Nevertheless, it is useful in the design of oscillator circuits. Finally, under the condition

$$1 - \mu(s)\beta(s) = 0 \tag{4.6}$$

a feedback amplifier will function as an oscillator. This condition is known as the *Barkhausen criterion*, which states that the frequency of sinusoidal oscillator is determined by the condition that the phase shift of $\mu(j\omega)\beta(j\omega)$ is zero provided that the magnitude of $\mu(j\omega)\beta(j\omega)$ equals unity. The factor $\mu(s)\beta(s)$ is often called the *loop transmission* and can be interpreted physically as follows. Suppose that no input excitation is applied to the amplifier and that the input side of the unilateral forward path is broken, as shown in Fig. 4.2, with a unit input applied to the right of the break. Then the signal appearing at the left of the break is precisely $\mu(s)\beta(s)$.

To employ the ideal feedback model to analyze a practical feedback amplifier, it becomes necessary to separate the feedback amplifier into two blocks: the basic unilateral amplifier $\mu(s)$ and the feedback network $\beta(s)$. The procedure is difficult and sometimes virtually impossible, because the forward path may not be strictly unilateral, the feedback path is usually bilateral, and the input and output coupling networks are often complicated. Thus, the ideal feedback model of Fig. 4.1 is not an adequate representation of a practical feedback amplifier. In the remainder of this chapter, we shall develop Bode's feedback theory, which is applicable to general network configuration and avoids the necessity of identifying $\mu(s)$ and $\beta(s)$. However, before we do this, we demonstrate a method for identifying $\mu(s)$ and $\beta(s)$ from given simple feedback amplifier configurations.

4.2 FEEDBACK AMPLIFIER CONFIGURATIONS

In this section, we present four simple feedback amplifier configurations based on the two-port representation, together with a general configuration.

Let an active two-port device N_a be characterized by its general hybrid matrix

$$\begin{bmatrix} y_{a1} \\ y_{a2} \end{bmatrix} = \begin{bmatrix} k_{a11} & k_{a12} \\ k_{a21} & k_{a22} \end{bmatrix} \begin{bmatrix} u_{a1} \\ u_{a2} \end{bmatrix} \tag{4.7}$$

We can consider another two-port feedback network N_f described by the general hybrid matrix

$$\begin{bmatrix} y_{f1} \\ y_{f2} \end{bmatrix} = \begin{bmatrix} k_{f11} & k_{f12} \\ k_{f21} & k_{f22} \end{bmatrix} \begin{bmatrix} u_{f1} \\ u_{f2} \end{bmatrix} \tag{4.8}$$

Upon interconnection of these two two-ports in such a way that forces u_{a1} and u_{a2} to be equal to u_{f1} and u_{f2}, respectively, the composite two-port network N is characterized by the equation

$$\begin{bmatrix} y_{1} \\ y_{2} \end{bmatrix} = \begin{bmatrix} k_{11} & k_{12} \\ k_{21} & k_{22} \end{bmatrix} \begin{bmatrix} u_{1} \\ u_{2} \end{bmatrix} \tag{4.9}$$

where

$$k_{ij} = k_{aij} + k_{fij} \qquad i, j = 1, 2 \tag{4.10}$$

More compactly, the coefficient matrix of (4.9) can be expressed as

$$\mathbf{H}(s) = \mathbf{H}_a(s) + \mathbf{H}_f(s) \tag{4.11}$$

$\mathbf{H}_a(s)$ and $\mathbf{H}_f(s)$ being the coefficient matrices of (4.7) and (4.8), respectively.

As discussed in Sec. 3.1, Eq. (4.9) can be applied to all four representations (3.1) and results in four basic feedback configurations, as shown schematically in Fig. 4.3. It is important to remember that in applying these configurations, care must be taken to ensure the validity of (4.10), which requires that the instantaneous current entering one terminal of a port must equal to the instantaneous current leaving the other terminal of the port. In Fig. 4.3*a*, for example, we must have

$$u_1 = u_{a1} = u_{f1} = I_1 \tag{4.12a}$$

$$u_2 = u_{a2} = u_{f2} = I_2 \tag{4.12b}$$

after interconnection. To this end, Brune proposes a simple procedure known as the *Brune tests.* His tests are exceedingly simple and are both necessary and sufficient. Depending on the configurations of Fig. 4.3, the tests for each case are presented in Fig. 4.4. An interconnection is permissible if the voltage marked V is zero. If this condition is not satisfied, the matrix addition, as given in (4.11), will give incorrect answers for the parameters of the composite network, unless isolating ideal transformers are introduced at one of the two ends. However, the introduction of ideal transformers is often undesirable because of network performance or cost.

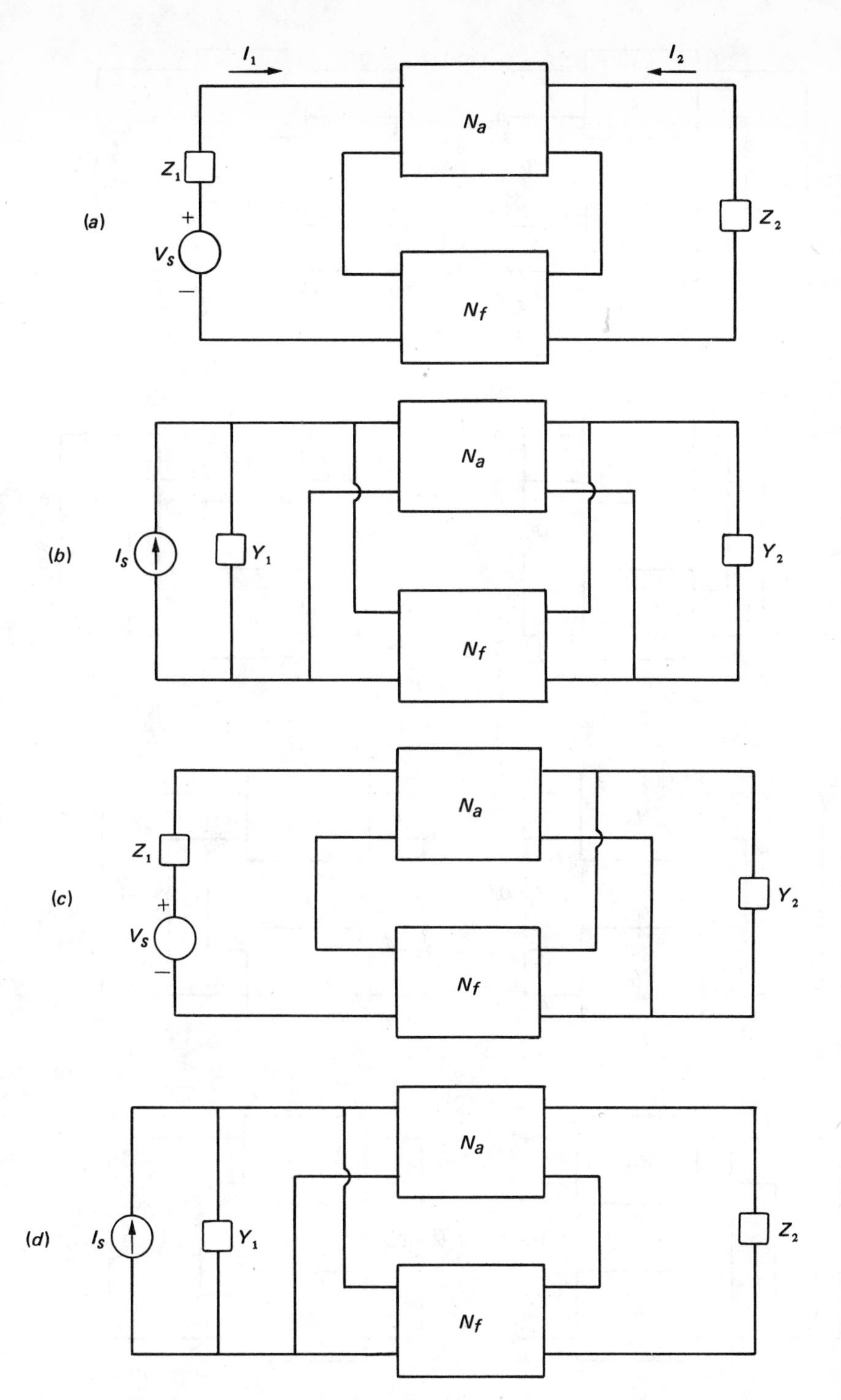

Figure 4.3 Simple feedback amplifier configurations. (*a*) The series-series or current-series feedback. (*b*) The parallel-parallel or voltage-shunt feedback. (*c*) The series-parallel or voltage-series feedback. (*d*) The parallel-series or current-shunt feedback.

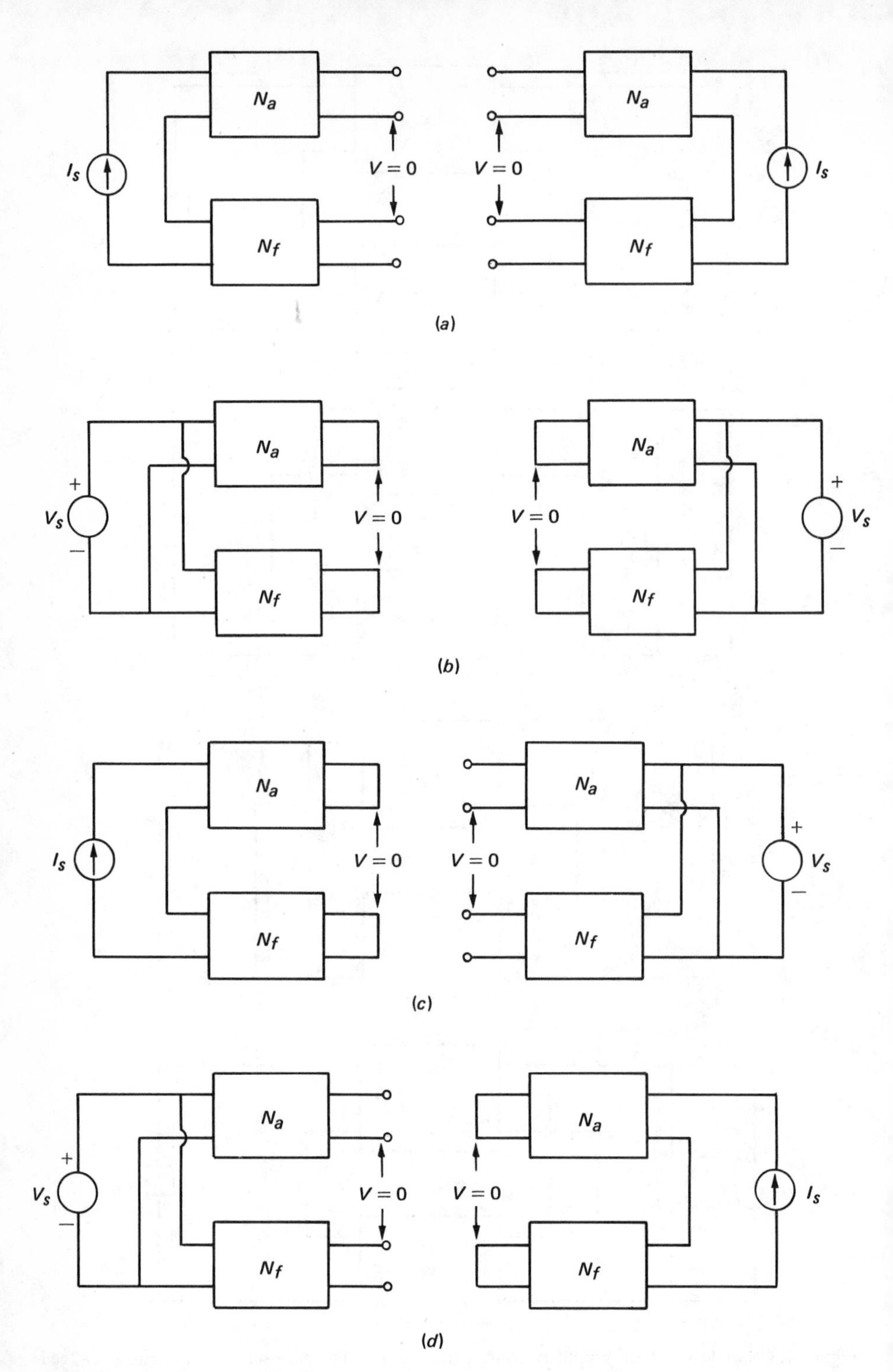

Figure 4.4 The Brune tests for (*a*) the series-series feedback, (*b*) the parallel-parallel feedback, (*c*) the series-parallel feedback, and (*d*) the parallel-series feedback.

Therefore, the most practical networks will be those that pass the Brune tests.

If the feedback parameter k_{f12} of the feedback network N_f is such that it satisfies the equation

$$k_{f12} = -k_{a12} \tag{4.13}$$

then the composite two-port network N becomes *unilateral*, and we can say that the internal feedback of the active device N_a is *neutralized.*

As indicated in Sec. 3.1, for each parameter set there is a corresponding representation of the source and load terminations, as shown in Fig. 4.3. The formulas that we shall derive are general and can be applied to all four two-port representations provided that we use the corresponding variables and parameters as listed in Table 3.1.

Refer to Fig. 4.3 and Table 3.1. The overall transfer function y_2/u_s, as given in (3.25*a*), of the composite two-port network N is

$$w(s) = \frac{y_2}{u_s} = \frac{k_{21}M_2}{(k_{11}+M_1)(k_{22}+M_2)-k_{12}k_{21}} \tag{4.14}$$

To identify the forward amplifier gain $\mu(s)$ of Fig. 4.1, as defined in Eq. (4.1), we set $k_{12}=0$ in Eq. (4.14), yielding

$$\mu(s) = \left.\frac{y_2}{u_s}\right|_{k_{12}=0} = \frac{k_{21}M_2}{(k_{11}+M_1)(k_{22}+M_2)} \tag{4.15}$$

Next, we write (4.14) in the form of (4.3) by dividing the numerator and denominator of (4.14) by $(k_{11}+M_1)(k_{22}+M_2)$, thus identifying the transfer function of the feedback path as

$$\beta(s) = \frac{k_{12}}{M_2} = \frac{k_{12}u_2}{M_2u_2} = \frac{\tau_f}{y_2} \tag{4.16a}$$

where

$$\tau_f = -k_{12}u_2 \tag{4.16b}$$

is defined as the *feedback signal*, which can either be voltage or current. Combining (4.15) and (4.16*a*), we obtain the loop transmission as

$$\mu(s)\beta(s) = \frac{k_{12}k_{21}}{(k_{11}+M_1)(k_{22}+M_2)} \tag{4.17}$$

as previously defined in Eq. (3.190*b*).

With $\mu(s)$ and $\beta(s)$ identified, the ideal feedback model can now be employed to determine the effects of feedback on network performance. One of the main disadvantages of this approach is that the quantities $\mu(s)$ and $\beta(s)$ cannot be measured experimentally, because in computing $\mu(s)$ the forward transmission of the feedback network is lumped with the active device, whereas the backward transmission of the active device is lumped with the feedback network. Nevertheless, for most practical amplifiers, the forward transmission of the active device is much larger than that of the feedback network, whereas the backward transmission of feedback network is much larger than that of the active device. Thus, we can often make the following approximation:

$$k_{21} \approx k_{a21} \tag{4.18a}$$

$$k_{12} \approx k_{f12} \tag{4.18b}$$

Because they are interconnected, the four feedback topologies of Fig. 4.3, *a–d* are usually referred to as the *series-series feedback, parallel-parallel feedback, series-parallel feedback,* and *parallel-series feedback,* respectively. The word *shunt* is sometimes used in lieu of *parallel*. In electronics, the four feedback configurations of Fig. 4.3, *a–d* are frequently called the *current-series feedback, voltage-shunt feedback, voltage-series feedback,* and *current-shunt feedback,* respectively. The word *voltage* refers to connecting the output voltage as input to the feedback network, *current* to tapping off some output current through the feedback network, *series* to connecting the feedback signal in series with the input voltage, and *shunt* to connecting the feedback signal in shunt or parallel with an input current source.

In addition to the reduction of gain by negative feedback, we shall now discuss the effects of feedback on the input and output immittances of the four feedback configurations of Fig. 4.3. From Eq. (3.7), the input immittance $M_{11}(s) + M_1(s)$ looking into the source of the composite two-port network N is found to be

$$M_{11}(s) + M_1(s) = k_{11} + M_1 - \frac{k_{12}k_{21}}{k_{22} + M_2} = (k_{11} + M_1)\left[1 - \frac{k_{12}k_{21}}{(k_{11} + M_1)(k_{22} + M_2)}\right]$$

$$= (k_{11} + M_1)[1 - \mu(s)\beta(s)] \tag{4.19a}$$

Likewise, from Eq. (3.8) we can show that the output immittance $M_{22}(s) + M_2(s)$ looking into the output port including the termination can be written as

$$M_{22}(s) + M_2(s) = (k_{22} + M_2)[1 - \mu(s)\beta(s)] \tag{4.19b}$$

Thus, we conclude that feedback affects the short-circuit or open-circuit input immittance $k_{11} + M_1$ and output immittance $k_{22} + M_2$ by the feedback factor $1 - \mu(s)\beta(s)$. Assume that $1 - \mu(0)\beta(0) > 1$. Then the effects of feedback on the low-frequency input and output immittances for the four feedback configurations are illustrated in Table 4.1.

We remark that *not* every feedback amplifier can be classified as being in one of the above four categories. A general feedback configuration may contain an input coupling network and an output coupling network as shown in Fig. 4.5. A typical

Table 4.1 Effect of feedback on impedances

Feedback configuration	Input impedance	Output impedance
Series-series	Increases	Increases
Parallel-parallel	Decreases	Decreases
Series-parallel	Increases	Decreases
Parallel-series	Decreases	Increases

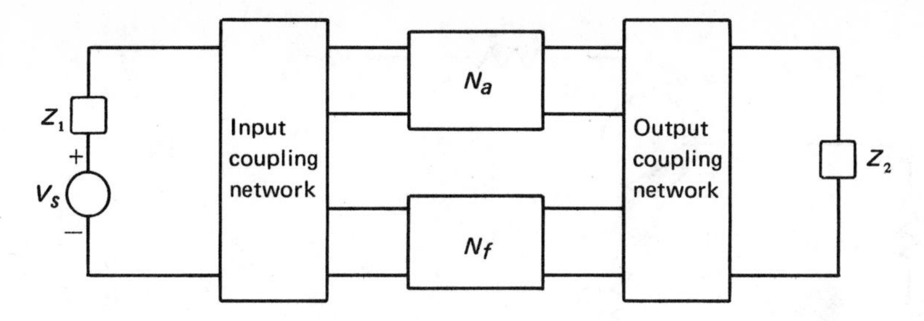

Figure 4.5 A general feedback configuration containing an input coupling network and an output coupling network.

example is the bridge feedback configuration of Fig. 4.6, which does not belong to any one of the topologies of Fig. 4.3.

We illustrate the above results by considering a few typical feedback amplifiers.

4.2.1 Series-Series Feedback

An amplifier with an unbypassed emitter resistor, as shown in Fig. 4.7, is an example of series-series or current-series feedback. Assume that the transistor is characterized by its hybrid matrix

$$\mathbf{H}_e = \begin{bmatrix} h_{ie} & h_{re} \\ h_{fe} & h_{oe} \end{bmatrix} \tag{4.20}$$

The impedance matrix of the transistor N_a becomes (see App. II)

$$\mathbf{Z}_a = \frac{1}{h_{oe}} \begin{bmatrix} h_{ie}h_{oe} - h_{fe}h_{re} & h_{re} \\ -h_{fe} & 1 \end{bmatrix} \tag{4.21}$$

The impedance matrix of the feedback network N_f is given by

$$\mathbf{Z}_f = \begin{bmatrix} R_e & R_e \\ R_e & R_e \end{bmatrix} \tag{4.22}$$

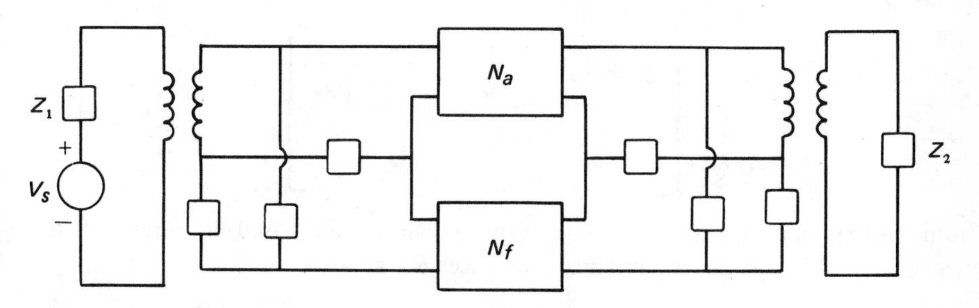

Figure 4.6 A typical example of a bridge feedback configuration.

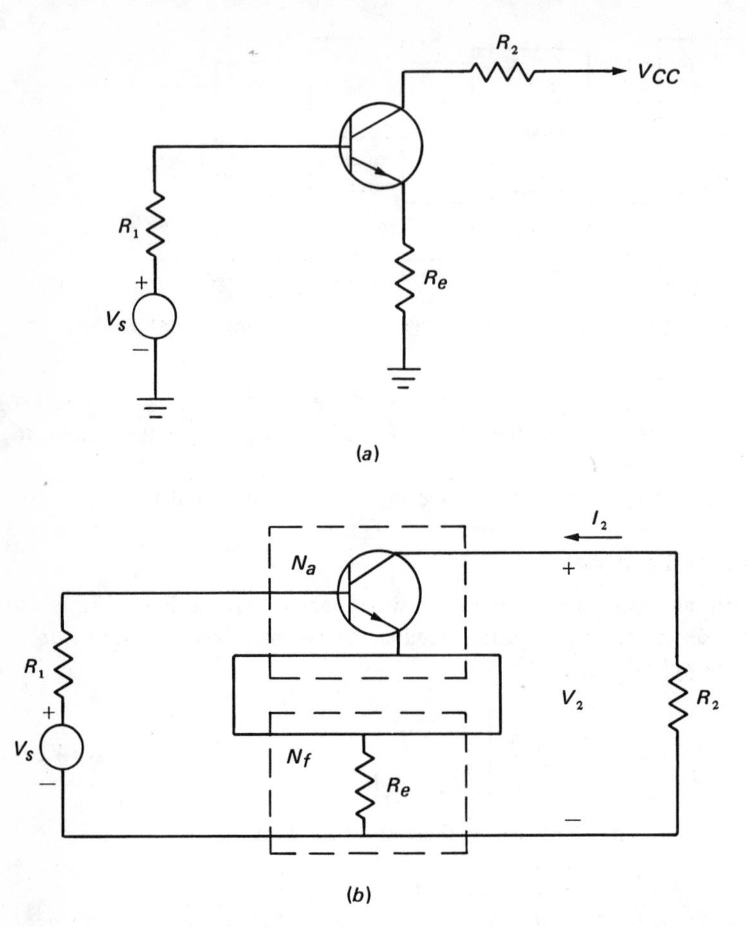

Figure 4.7 (*a*) An amplifier with an unbypassed emitter resistor and (*b*) its series-series or current-series feedback representation.

Since the Brune tests of Fig. 4.4*a* are satisfied when applied to the network configuration of Fig. 4.7*b*, the connection is permissible and the impedance matrix $\mathbf{Z}$ of the composite two-port N is simply the sum of the impedance matrices $\mathbf{Z}_a$ and $\mathbf{Z}_f$, giving

$$\mathbf{Z} = \begin{bmatrix} R_e + h_{ie} - \dfrac{h_{fe}h_{re}}{h_{oe}} & R_e + \dfrac{h_{re}}{h_{oe}} \\ R_e - \dfrac{h_{fe}}{h_{oe}} & R_e + \dfrac{1}{h_{oe}} \end{bmatrix} \tag{4.23}$$

From (4.15), the open-loop voltage gain μ can be expressed in terms of the elements z_{ij} of $\mathbf{Z}$ and the terminating resistances R_1 and R_2:

$$\mu = \left.\frac{V_2}{V_s}\right|_{z_{12}=0} = \frac{z_{21}R_2}{(z_{11}+R_1)(z_{22}+R_2)}$$

$$= \frac{(R_e - h_{fe}/h_{oe})R_2}{(R_e + h_{ie} - h_{fe}h_{re}/h_{oe} + R_1)(R_e + 1/h_{oe} + R_2)} \tag{4.24}$$

The transfer function of the unilateral feedback path is calculated from (4.16) and is given by

$$\beta = \frac{V_f}{V_2} = \frac{z_{12}}{R_2} = \frac{R_e + h_{re}/h_{oe}}{R_2} \tag{4.25a}$$

where the feedback voltage V_f is defined by

$$V_f = -z_{12}I_2 = -\left(R_e + \frac{h_{re}}{h_{oe}}\right)I_2 \tag{4.25b}$$

As indicated in (4.18), for practical amplifiers, we can usually assume that

$$\frac{h_{fe}}{h_{oe}} \gg R_e \tag{4.26a}$$

$$R_e \gg \frac{h_{re}}{h_{oe}} \tag{4.26b}$$

Also, if $1/h_{oe} \gg R_e$, then Eqs. (4.24) and (4.25*a*) can be simplified to

$$\mu \approx -\frac{h_{fe}R_2}{(h_{ie} + R_e + R_1 - h_{fe}h_{re}/h_{oe})(1 + R_2 h_{oe})} \tag{4.27a}$$

$$\beta \approx \frac{R_e}{R_2} \tag{4.27b}$$

From (4.3), the closed-loop voltage gain is obtained as

$$w \approx -\frac{h_{fe}R_2}{(h_{ie} + R_e + R_1 - h_{fe}h_{re}/h_{oe})(1 + R_2 h_{oe}) + h_{fe}R_e} \tag{4.28}$$

The above analysis can be explained by the schematic diagram of Fig. 4.8*a*. The unilateral forward path is represented by the controlled voltage source $z_{21}I_1$, whereas the unilateral feedback path is represented by the controlled voltage source $z_{12}I_2$. The calculation of μ is shown in Fig. 4.8*b*. Thus, μ can be calculated either from (4.24) or directly from the network of Fig. 4.8*b*.

Example 4.1 In the feedback amplifier of Fig. 4.7, let

$$h_{fe} = 50 \qquad h_{ie} = 1\ \text{k}\Omega$$
$$h_{re} = 2.5 \cdot 10^{-4} \qquad h_{oe} = 25\ \mu\text{mho} \tag{4.29}$$
$$R_1 = R_e = 1\ \text{k}\Omega \qquad R_2 = 10\ \text{k}\Omega$$

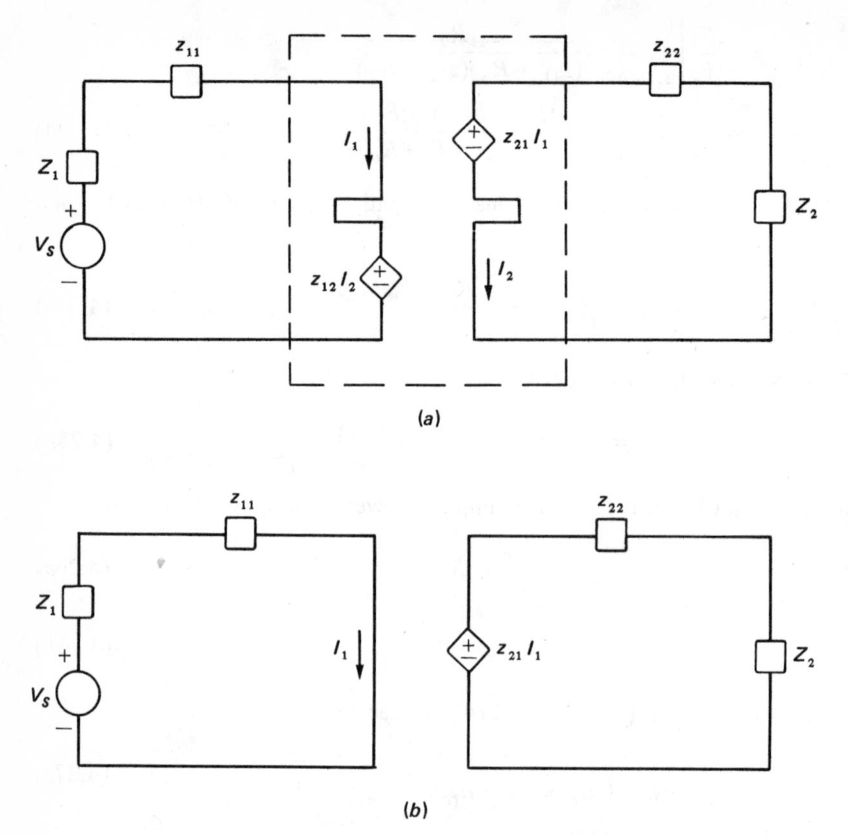

Figure 4.8 (*a*) The schematic diagram of the analysis of the series-series feedback configuration, and (*b*) the network used to calculate the voltage gain of the unilateral forward path.

Substituting these in (4.24) yields the open-loop voltage gain

$$\mu = -156.78 \tag{4.30}$$

which can also be computed directly from the network of Fig. 4.9. From (4.28) the closed-loop voltage gain is found to be

$$w = \frac{V_2}{V_s} = \frac{\mu}{1-\mu\beta} = -9.41 \tag{4.31}$$

where $\beta = 0.1$, a reduction from the gain without feedback by a factor of 16.68.

The above analysis is restricted to low-frequency applications. When transistors are used to amplify high-frequency signals, the h-parameters of (4.20) are generally complex. The most useful high-frequency model of the bipolar transistor is shown in Fig. 4.10. This model is very widely used in the literature and is called the

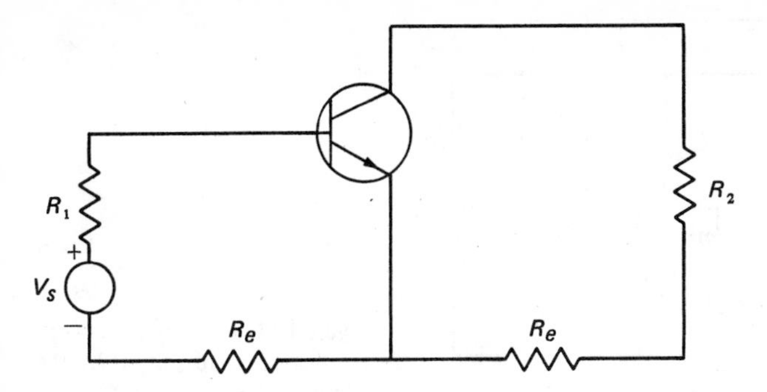

Figure 4.9 The transistor network used to calculate the open-loop voltage gain.

hybrid-pi equivalent network. The typical element values of the model are given below:

$$
\begin{gathered}
g_m \approx 40 I_e \approx 0.4 \text{ mho} \qquad \text{for } I_e = 1 \text{ mA} \\
r_x = 20\text{–}100\ \Omega \qquad r_\pi = 75\text{–}250\ \Omega \\
C_\mu = 1\text{–}20 \text{ pF} \qquad C_\pi = 10\text{–}1000 \text{ pF} \\
r_o = 10\text{–}100\ \text{k}\Omega
\end{gathered}
\tag{4.32}
$$

We may wonder at this time if we can make some simplifications of the hybrid-pi model at the outset. If the approximation can be made, we must be aware, of course, of the conditions under which it is made.

Consider the two-port network of Fig. 4.11, with the admittance matrix

$$
\mathbf{Y}(s) = \begin{bmatrix} (C_1 + C_2)s & -sC_2 \\ g_m - sC_2 & sC_2 \end{bmatrix} \tag{4.33}
$$

By appealing to Eq. (3.7), the input admittance is determined as

$$
Y_{\text{in}}(s) = s(C_1 + C_2) + \frac{sC_2(g_m - sC_2)}{Y_2 + sC_2} \tag{4.34}
$$

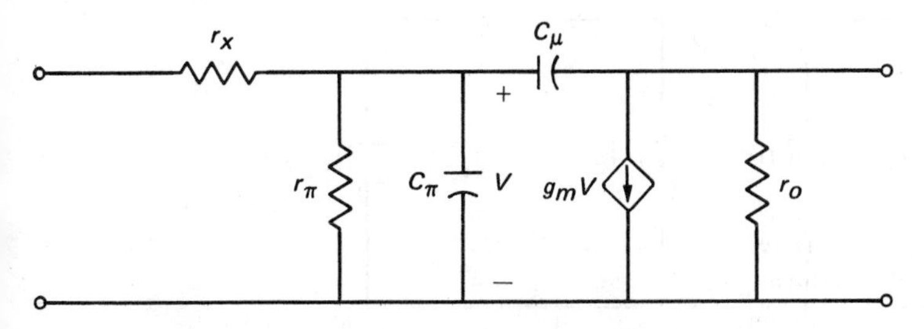

Figure 4.10 The hybrid-pi equivalent network of a transistor.

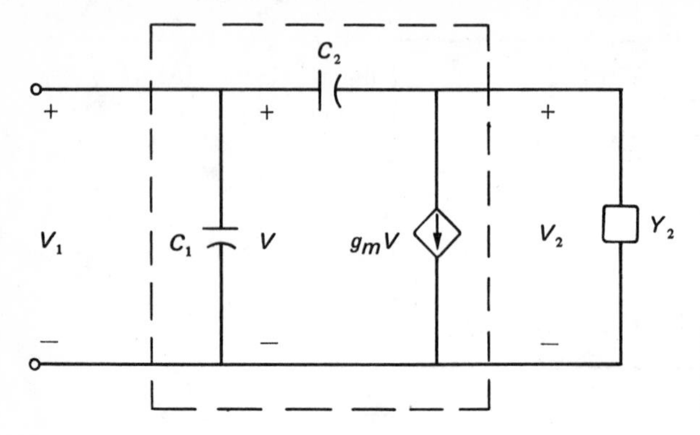

Figure 4.11 A two-port network terminated at its output by the admittance Y_2.

If we assume that in the range of frequencies of interest

$$\omega C_2 \ll \begin{cases} g_m \\ |Y_2| \end{cases} \tag{4.35}$$

then Eq. (4.34) can be approximated by

$$Y_{\text{in}}(s) \approx s\left[C_1 + C_2\left(1 + \frac{g_m}{Y_2}\right)\right] = sC_1 + Y_{\text{eq}} \tag{4.36}$$

where

$$Y_{\text{eq}}(s) = sC_2\left[1 + \frac{g_m}{Y_2(s)}\right] \tag{4.37}$$

showing that the network of Fig. 4.11 can be approximated by the unilateralized network of Fig. 4.12 as far as the input admittance is concerned. The voltage gain of Fig. 4.11 is found to be

$$\frac{V_2}{V_1} = -\frac{g_m - sC_2}{Y_2 + sC_2} \tag{4.38}$$

Under the assumption of (4.35), the above expression simplifies to

$$\frac{V_2}{V_1} \approx -\frac{g_m}{Y_2} \tag{4.39}$$

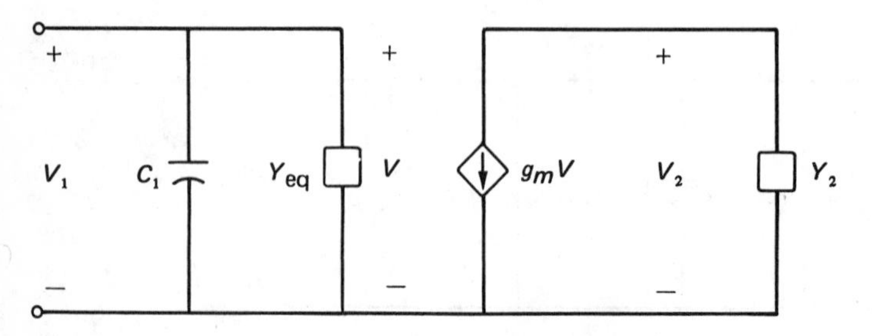

Figure 4.12 A unilateralized approximation of the network of Fig. 4.11.

which can also be computed directly from the network of Fig. 4.12. Hence, we conclude that if the approximation (4.35) is valid, the unilateralized network of Fig. 4.12 gives fairly correct results as far as the input admittance and the forward gain are concerned. However, it *does not* give the correct *output admittance* and *reverse transmission.* In situations where these two quantities are important, the simplified model should not be used.

Now we consider an important special case where the load is purely resistive with $Y_2 = 1/R_2$. Under this situation, the network of Fig. 4.12 reduces to that shown in Fig. 4.13. The input capacitance is now $C_1 + C_2(1 + g_m R_2)$. This increase results from the voltage gain of the amplifying device and is known as the *Miller effect*. The Miller effect will now be employed to unilateralize the hybrid-pi model of Fig. 4.10 when it is used for the transistor in the feedback amplifier of Fig. 4.7.

The open-loop voltage gain $\mu(s)$ of the amplifier of Fig. 4.7 can be calculated directly from the network of Fig. 4.9. By using the high-frequency model of Fig. 4.10 and applying the Miller effect, the network of Fig. 4.9 can be simplified to that of Fig. 4.14, with

$$R_1' = R_1 + R_e + r_x \tag{4.40a}$$

$$R_2' = \frac{r_o(R_2 + R_e)}{r_o + R_2 + R_e} \tag{4.40b}$$

$$C_\pi' = C_\pi + C_\mu(1 + g_m R_2') \tag{4.40c}$$

From Fig. 4.14 we obtain the open-loop voltage gain

$$\mu(s) = -\frac{A_0}{s + s_0} \tag{4.41a}$$

where

$$A_0 = \frac{g_m R_2 r_o}{R_1' C_\pi'(r_o + R_2 + R_e)} \tag{4.41b}$$

$$s_0 = \frac{1/r_\pi + 1/R_1'}{C_\pi'} \tag{4.41c}$$

Using (4.3) in conjunction with (4.27*b*) yields the closed-loop voltage gain

$$w(s) = -\frac{A_0}{s + s_0 + A_0 R_e/R_2} \tag{4.42}$$

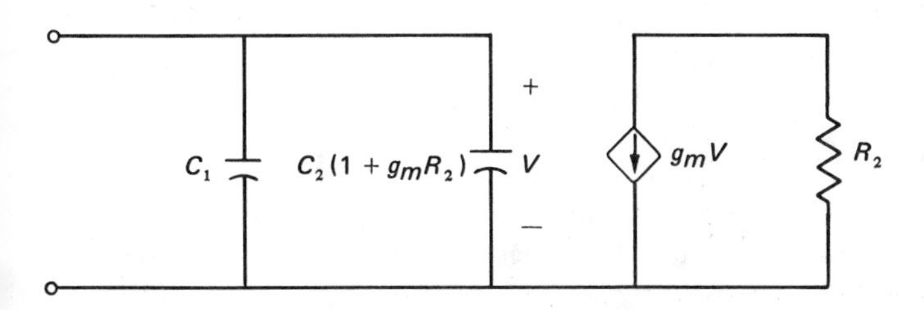

Figure 4.13 A unilateralized approximation of the network of Fig. 4.11 with resistive termination.

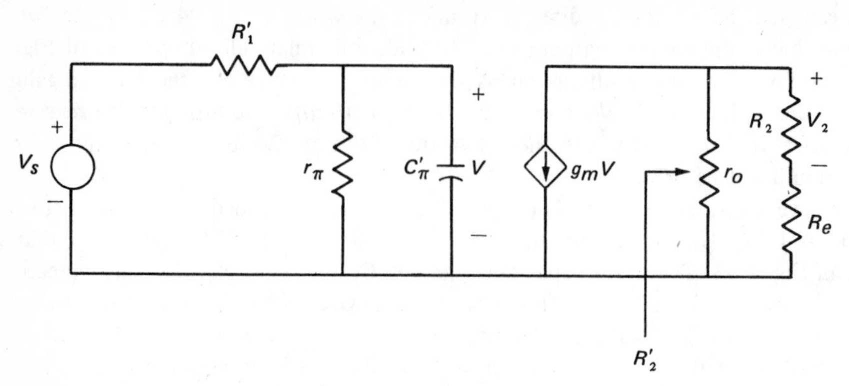

Figure 4.14 A unilateralized approximation of the network of Fig. 4.9.

Example 4.2 As a numerical example, let

$$\begin{aligned} g_m &= 0.4 \text{ mho} \qquad r_x = 50\ \Omega \\ r_\pi &= 250\ \Omega \qquad C_\pi = 195 \text{ pF} \\ C_\mu &= 5 \text{ pF} \qquad r_o = 50 \text{ k}\Omega \\ R_2 = R_e &= 1 \text{ k}\Omega \qquad R_1 = 100\ \Omega \end{aligned} \tag{4.43}$$

From (4.40) we have

$$R_1' = 1.15 \text{ k}\Omega \qquad R_2' = 1.923 \text{ k}\Omega \qquad C_\pi' = 4.05 \text{ nF} \tag{4.44}$$

giving

$$A_0 = 82.66 \cdot 10^6 \qquad s_0 = 1.20 \cdot 10^6 \text{ rad/s} \tag{4.45}$$

Thus, the closed-loop voltage gain is obtained from (4.42) as

$$w(s) = -\frac{82.66 \cdot 10^6}{s + 83.86 \cdot 10^6} \tag{4.46}$$

showing a midband gain $w(0) = -0.99$. The midband feedback factor is

$$1 - \mu(0)\beta(0) = 1 + \frac{A_0 R_e}{s_0 R_2} = 69.88 \tag{4.47}$$

where the midband open-loop gain is -68.88. From (4.19) the midband open-circuit input and output impedances are increased by a factor of 69.88 from their values $R_1' + r_\pi = 1.4$ kΩ and $R_2 + R_e + r_o = 52$ kΩ, respectively.

4.2.2 Parallel-Parallel Feedback

The network of Fig. 4.15 shows a common-emitter stage with a resistor R_f connected from the output to the input. It is clear that this configuration conforms to the parallel-parallel or voltage-shunt topology of Fig. 4.3*b*. Assume that the transistor is described by (4.20). Then, from (2.47*a*), the admittance matrix of the transistor N_a is given by

$$\mathbf{Y}_a = \frac{1}{h_{ie}} \begin{bmatrix} 1 & -h_{re} \\ h_{fe} & \Delta_{he} \end{bmatrix} \tag{4.48}$$

where $\Delta_{he} = h_{ie}h_{oe} - h_{fe}h_{re}$. The admittance matrix of the feedback network N_f with $G_f = 1/R_f$ is obtained by inspection as

$$\mathbf{Y}_f = \begin{bmatrix} G_f & -G_f \\ -G_f & G_f \end{bmatrix} \tag{4.49}$$

Since by the Brune tests of Fig. 4.4*b* the connection of N_a and N_f is permissible, the admittance matrix of the composite two-port N can be written as

$$\mathbf{Y} = \mathbf{Y}_a + \mathbf{Y}_f = \begin{bmatrix} G_f + \dfrac{1}{h_{ie}} & -G_f - \dfrac{h_{re}}{h_{ie}} \\ -G_f + \dfrac{h_{fe}}{h_{ie}} & G_f + \dfrac{\Delta_{he}}{h_{ie}} \end{bmatrix} \tag{4.50}$$

From (4.15) and (4.16), the open-loop current gain μ and the transfer function of the unilateral feedback path β, being expressible in terms of the elements y_{ij} of $\mathbf{Y}$ and the terminating conductances $G_1 = 1/R_1$ and $G_2 = 1/R_2$, are calculated as

$$\mu = \left.\frac{I_2}{I_s}\right|_{y_{12}=0} = -\frac{(G_f - h_{fe}/h_{ie})G_2}{(G_1 + G_f + 1/h_{ie})(G_2 + G_f + \Delta_{he}/h_{ie})} \tag{4.51}$$

$$\beta = \frac{I_f}{I_2} = \frac{y_{12}}{G_2} = -\frac{G_f + h_{re}/h_{ie}}{G_2} \tag{4.52a}$$

where the feedback current I_f is defined by

$$I_f = -y_{12}V_2 = \left(G_f + \frac{h_{re}}{h_{ie}}\right)V_2 \tag{4.52b}$$

As indicated in (4.18), for practical amplifiers, we can make the following approximations:

$$\frac{h_{fe}}{h_{ie}} \gg G_f \tag{4.53a}$$

$$G_f \gg \frac{h_{re}}{h_{ie}} \tag{4.53b}$$

If, in addition, $G_2 + G_f \gg \Delta_{he}/h_{ie}$, then we have

$$\mu \approx \frac{h_{fe}G_2}{(G_2 + G_f)[1 + (G_1 + G_f)h_{ie}]} \tag{4.54a}$$

$$\beta \approx -\frac{G_f}{G_2} = -\frac{R_2}{R_f} \tag{4.54b}$$

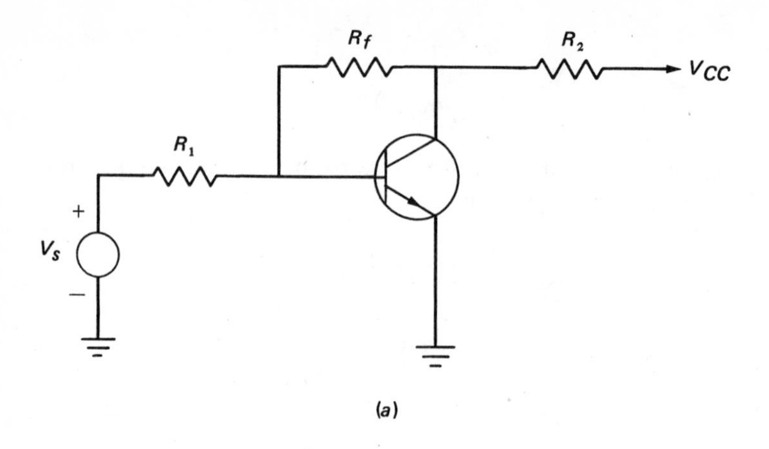

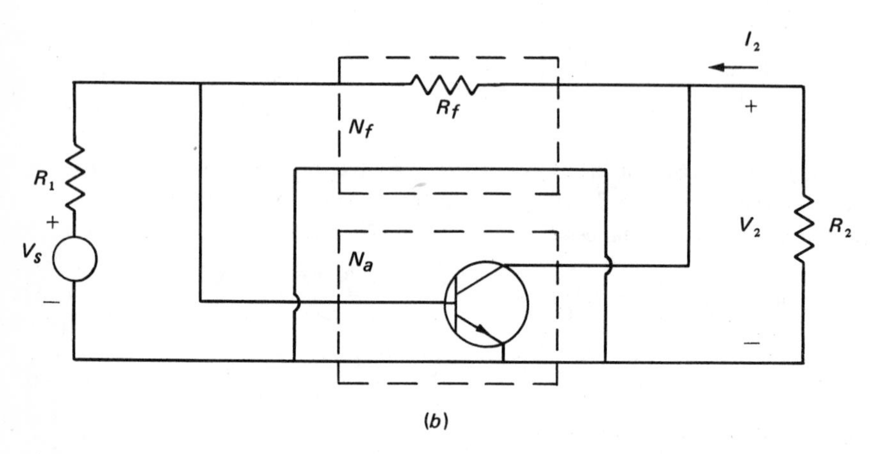

Figure 4.15 (*a*) A common-emitter stage with a resistor R_f providing the feedback, and (*b*) its parallel-parallel or voltage-shunt feedback representation.

and the closed-loop current ratio or gain is found to be

$$w = \frac{I_2}{I_s} = \frac{\mu}{1 - \mu\beta} \approx \frac{h_{fe}G_2}{(G_2 + G_f)[1 + h_{ie}(G_1 + G_f)] + h_{fe}G_f} \tag{4.55}$$

Example 4.3 In the feedback amplifier of Fig. 4.15, let

$$\begin{gathered} h_{fe} = 50 \qquad h_{ie} = 1 \text{ k}\Omega \\ h_{re} = 2.5 \cdot 10^{-4} \qquad h_{oe} = 25\ \mu\text{mho} \\ R_1 = 10 \text{ k}\Omega \qquad R_2 = 4 \text{ k}\Omega \\ R_f = 40 \text{ k}\Omega \end{gathered} \tag{4.56}$$

Then from (4.54*a*) we compute the current gain of the amplifier without feedback as

$$\mu = 40.40 \tag{4.57}$$

The amplifier current gain with feedback is found to be

$$w = \frac{I_2}{I_s} = \frac{\mu}{1 - \mu\beta} = 8.01 \tag{4.58}$$

where $\beta = -0.1$, indicating a gain reduction by a factor of 5.04. If the exact formulas (4.51) and (4.52*a*) are used, the corresponding values are found to be $\mu = 38.63$, $\beta = -0.101$, and $w = 7.88$.

At high frequencies, we use the hybrid-pi equivalent network of Fig. 4.10 for the transistor in the feedback amplifier of Fig. 4.15.

As before, let y_{aij}, y_{fij}, and y_{ij}, where $y_{ij} = y_{aij} + y_{fij}$, be the y-parameters of the transistor N_a, the feedback network N_f, and the composite two-port N of Fig. 4.15, respectively. Then, according to Eq. (4.15), the forward path is specified by the current gain

$$\mu(s) = \left.\frac{I_2}{I_s}\right|_{y_{12}=0} \tag{4.59}$$

This situation can be depicted as in Fig. 4.16. Since $y_{21} \approx y_{a21}$ and $y_{a12} \approx 0$, Fig. 4.16 can be approximated by the network of Fig. 4.17. By using the hybrid-pi model of Fig. 4.10 and applying the Miller effect, the network of Fig. 4.17 can be unilateralized as shown in Fig. 4.18. The open-loop current gain or the forward amplifier gain is found to be

$$\mu(s) = \frac{A_0}{1 + s/s_0} \tag{4.60}$$

where

$$A_0 = \frac{g_m r_\pi R_1' R_2'}{(R_1' + r_\pi + r_x)R_2} \tag{4.61a}$$

$$s_0 = \frac{1}{C}\left(\frac{1}{r_\pi} + \frac{1}{r_x + R_1'}\right) \tag{4.61b}$$

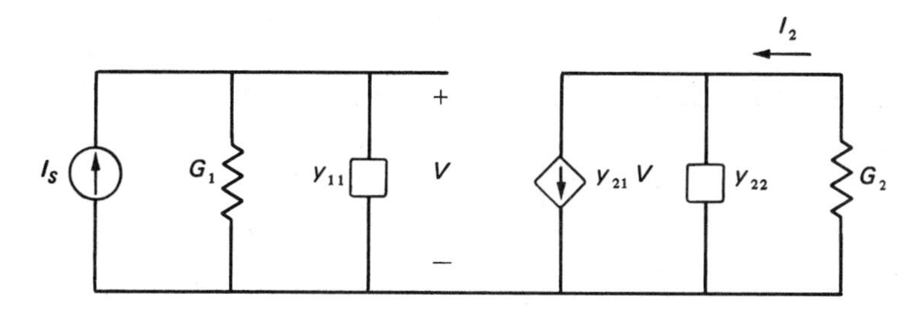

Figure 4.16 The network used to calculate the open-loop current gain of the amplifier of Fig. 4.15.

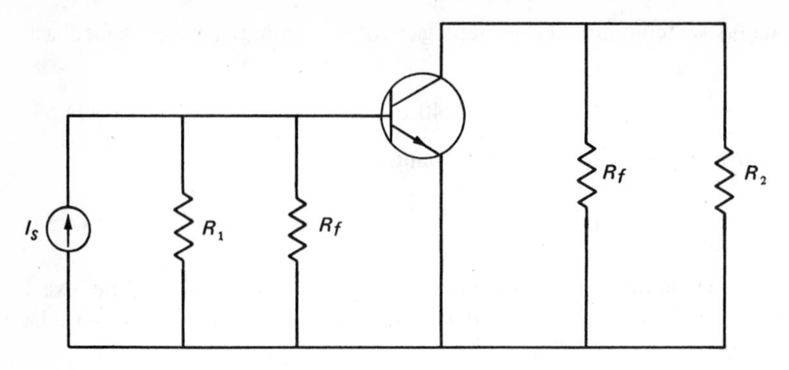

Figure 4.17 The amplifier approximating the network of Fig. 4.16.

$$C = C_\pi + C_\mu(1 + g_m R_2') \tag{4.61c}$$

$$R_1' = \frac{R_1 R_f}{R_1 + R_f} \tag{4.61d}$$

$$R_2' = \frac{r_o R_2 R_f}{r_o R_2 + r_o R_f + R_2 R_f} \tag{4.61e}$$

Since the backward transmission of a feedback network is much larger than that of the transistor, (4.53*b*) applies. The closed-loop current gain of the amplifier becomes

$$w(s) = \frac{I_2}{I_s} = \frac{A_0 s_0}{s + s_0(1 + A_0 R_2/R_f)} \tag{4.62}$$

Example 4.4 As a numerical example, let

$$\begin{aligned} g_m &= 0.4 \text{ mho} \qquad r_x = 50\ \Omega \\ r_\pi &= 250\ \Omega \qquad C_\pi = 195 \text{ pF} \\ C_\mu &= 5 \text{ pF} \qquad r_o = 50 \text{ k}\Omega \\ R_1 &= 10 \text{ k}\Omega \qquad R_f = 40 \text{ k}\Omega \\ R_2 &= 4 \text{ k}\Omega \end{aligned} \tag{4.63}$$

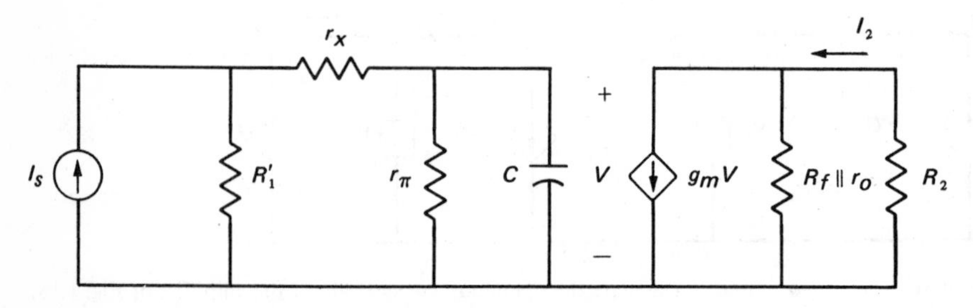

Figure 4.18 A unilateralized approximation of the network of Fig. 4.17.

From (4.61) we have

$$R_1' = 8\ \text{k}\Omega \qquad R_2' = 3.39\ \text{k}\Omega$$
$$C = 6.98\ \text{nF} \qquad s_0 = 5.9 \cdot 10^5\ \text{rad/s} \tag{4.64}$$
$$A_0 = 81.69$$

The closed-loop current gain from (4.62) is

$$w(s) = \frac{81.69}{p + 9.169} \tag{4.65}$$

where $p = s/s_0$ and $\beta = -R_2/R_f = -0.1$. The midband feedback factor is 9.169, and it reduces the midband current gain without feedback by a factor of 9.169.

From (4.19), the approximate midband input admittance facing the current source is given by

$$Y_{\text{in}} = (y_{11} + G_1)(1 - \mu\beta) = \left(\frac{1}{r_x + r_\pi} + G_f + G_1\right)(1 - \mu\beta)$$
$$= 31.71 \cdot 10^{-3}\ \text{mho or } 32\ \Omega \tag{4.66a}$$

and the approximate output admittance looking into the two terminals of R_2, including R_2, is found to be

$$Y_{\text{out}} = (y_{22} + G_2)(1 - \mu\beta) = \left(\frac{1}{r_o} + G_f + G_2\right)(1 - \mu\beta)$$
$$= 2.71 \cdot 10^{-3}\ \text{mho or } 370\ \Omega \tag{4.66b}$$

Thus, for the parallel-parallel feedback, the midband input and output impedances are decreased by a factor of 9.169 from their values without feedback. We remark that the unilateralized hybrid-pi equivalent model is not valid for the calculations of the output impedance at high frequencies, and we must use the original model of Fig. 4.10 for such calculations.

In the situation where there are several amplifier stages in cascade, as shown in Fig. 4.19, the same technique can be applied to identify $\mu(s)$ and $\beta(s)$. The open-loop current gain $\mu(s)$, for example, can be determined by the network of Fig. 4.20. However, in invoking Miller effect to decouple each amplifier stage, as shown in Fig. 4.18, only the last stage is directly applicable, because the equivalent terminations for the first two stages are not purely resistive. However, we can usually assume that Fig. 4.18 is a valid approximation for each amplifier stage. Under this assumption, it is straightforward to demonstrate that the open-loop current gain can be expressed as (see Prob. 4.2)

$$\mu(s) = \frac{I_2}{I_s} = \frac{A_1 A_2 A_3 R_f}{(1 + s/s_1)(1 + s/s_2)(1 + s/s_3)(R_2 + R_f)} \tag{4.67}$$

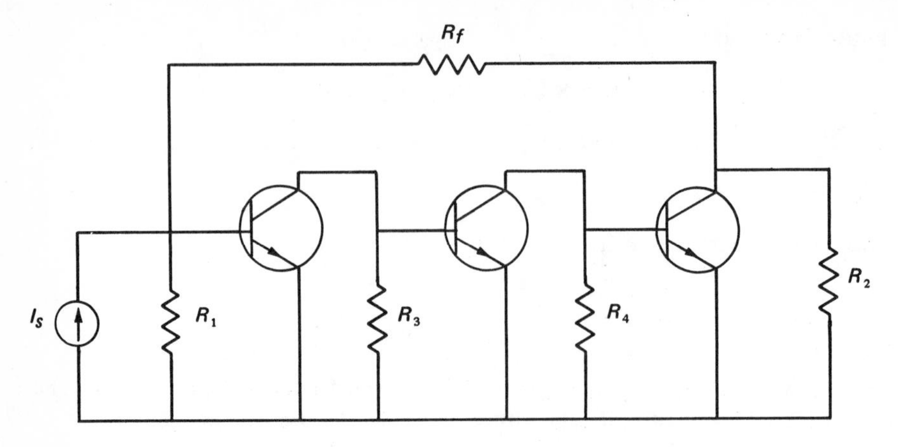

Figure 4.19 A three-transistor voltage-shunt feedback amplifier.

where A_k and s_k ($k = 1, 2, 3$) have the same significance as A_0 and s_0 in (4.61) for a single stage and are given by

$$A_k = \frac{g_{mk}\hat{R}_k r_{\pi k}}{r_{xk} + \hat{R}_k + r_{\pi k}} \tag{4.68a}$$

$$s_k = \frac{r_{xk} + \hat{R}_k + r_{\pi k}}{C_k r_{\pi k}(r_{xk} + \hat{R}_k)} \tag{4.68b}$$

in which $\hat{R}_1 = R_1 R_f/(R_1 + R_f)$, $\hat{R}_2 = R_3$, and $\hat{R}_3 = R_4$. In the above expressions, the output impedance r_o of a common-emitter stage has been ignored, because it is usually large. Otherwise, it should be included in R_3, R_4, and R_f at the output port. C_k denotes the total capacitance including the capacitance due to the Miller effect.

4.2.3 Series-Parallel Feedback

The amplifier of Fig. 4.21 makes use of series-parallel or voltage-series feedback by connecting the second collector to the first emitter through the voltage divider

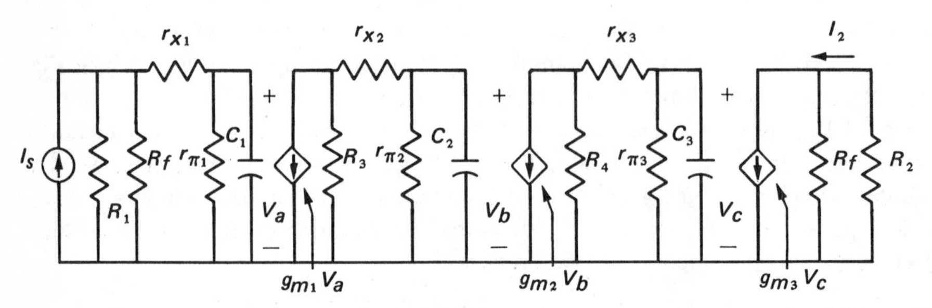

Figure 4.20 The equivalent network used to calculate the open-loop current gain of the amplifier of Fig. 4.19.

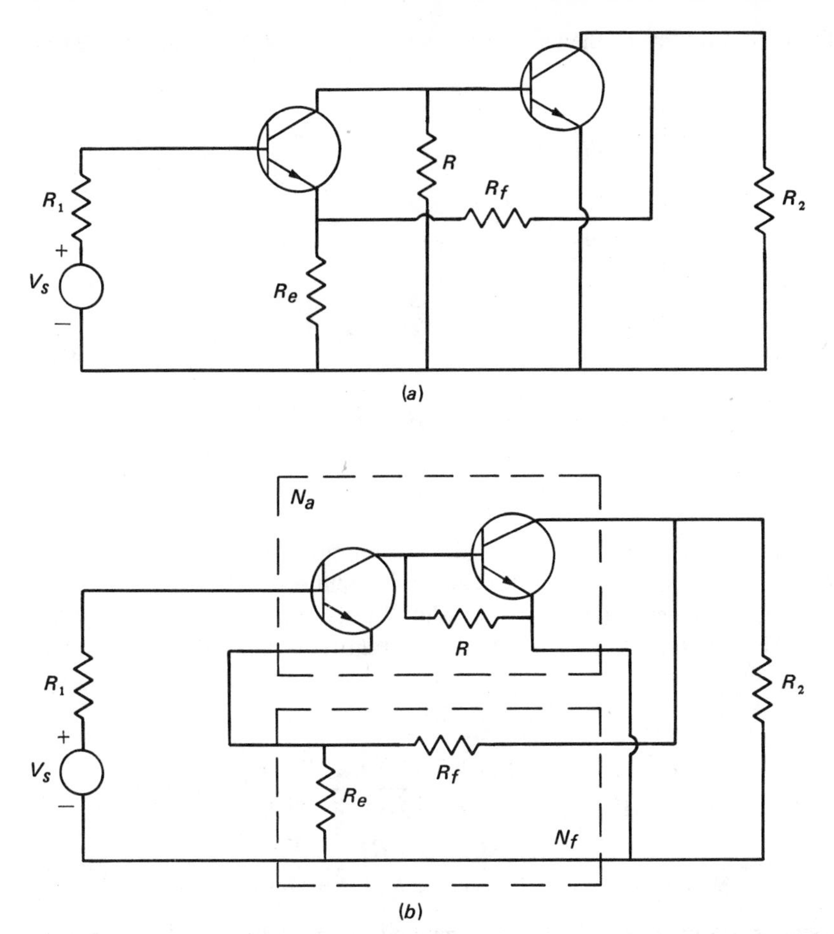

Figure 4.21 (*a*) A series-parallel or voltage-series feedback amplifier, and (*b*) its series-parallel or voltage-series feedback representation.

$R_f R_e$. As shown in Fig. 4.21*b*, the basic amplifier is represented by N_a and the feedback network by N_f, whereas the composite two-port is denoted by N. Let h_{aij}, h_{fij}, and h_{ij} be the *h*-parameters of N_a, N_f, and N, respectively. Applying the Brune tests of Fig. 4.4*c* to the network of Fig. 4.21*b* shows that the conditions are *not* satisfied, meaning that $h_{ij} \neq h_{aij} + h_{fij}$. To confirm this assertion, we compute the short-circuit input impedances h_{11}, h_{a11}, and h_{f11} by assuming identical transistors with $h_{re} = 0$ and $h_{oe} = 0$. The results are given by

$$h_{11} = h_{ie} + \frac{(1 + h_{fe}) R_e R_f}{(R_e + R_f)} \tag{4.69a}$$

$$h_{a11} = h_{ie} \tag{4.69b}$$

$$h_{f11} = \frac{R_e R_f}{R_e + R_f} \tag{4.69c}$$

In fact, the hybrid matrix of the composite two-port network N is found to be (see Prob. 4.3)

$$\mathbf{H} = \begin{bmatrix} h_{ie} + (1 + h_{fe})R'_e & \dfrac{R_e}{R_e + R_f} \\ -\dfrac{h_{fe}^2 R}{R + h_{ie}} - \dfrac{(1 + h_{fe})R'_e}{R_f} & \dfrac{1}{R_e + R_f} \end{bmatrix} \tag{4.70a}$$

$$\approx \begin{bmatrix} h_{ie} + (1 + h_{fe})R'_e & \dfrac{R_e}{R_e + R_f} \\ -\dfrac{h_{fe}^2 R}{R + h_{ie}} & \dfrac{1}{R_e + R_f} \end{bmatrix} \tag{4.70b}$$

where
$$R'_e = \frac{R_e R_f}{R_e + R_f} \tag{4.70c}$$

The approximation is obtained under the assumption that $h_{fe}^2 R/(R + h_{ie}) \gg (1 + h_{fe})R'_e/R_f$, which is usually valid for practical amplifiers.

Even though the connection failed the validity tests, formulas (4.15) and (4.16) are still valid. Thus, with $G_2 = 1/R_2$, the open-loop transfer admittance is calculated as

$$\mu = \left.\frac{I_2}{V_s}\right|_{h_{12}=0} = \frac{h_{21} G_2}{(h_{11} + R_1)(h_{22} + G_2)}$$

$$= -\frac{G_2\,[h_{fe}^2 R(R_e + R_f) + R_e(1 + h_{fe})(R + h_{ie})]}{(R + h_{ie})\,[h_{ie} + (1 + h_{fe})R'_e + R_1]\,[1 + G_2(R_e + R_f)]} \tag{4.71a}$$

$$\approx -\frac{h_{fe}^2 R G_2 (R_e + R_f)}{(R + h_{ie})\,[h_{ie} + (1 + h_{fe})R'_e + R_1]\,[1 + G_2(R_e + R_f)]} \tag{4.71b}$$

the approximate formula (4.71*b*) being computed from (4.70*b*), and the transfer impedance of the feedback network N_f is found to be

$$\beta = \frac{V_f}{I_2} = \frac{h_{12}}{G_2} = \frac{R_e R_2}{R_e + R_f} \tag{4.72a}$$

where the feedback voltage V_f is defined by

$$V_f = -h_{12} V_2 = -\frac{R_e V_2}{R_e + R_f} \tag{4.72b}$$

Finally, the closed-loop transfer admittance is determined from (4.3) and is given by (see Prob. 4.23)

$$w = \frac{I_2}{V_s} = \frac{-h_{fe}^2 R(R_e + R_f) + R_e(1 + h_{fe})(R + h_{ie})}{(R + h_{ie})\{(R_2 + R_f)[R_1 + h_{ie} + (1 + h_{fe})R_e] + (R_1 + h_{ie})R_e\} + h_{fe}^2 R R_e R_2}$$

$$\approx -\frac{h_{fe}^2 R(R_e + R_f)}{(R + h_{ie})\,[h_{ie} + (1 + h_{fe})R'_e + R_1](R_2 + R_e + R_f) + h_{fe}^2 R R_e R_2} \tag{4.73}$$

As in (4.71*b*), the approximate formula was obtained by using (4.70*b*).

Like the series-series feedback, the above analysis can also be explained as follows: The unilateral forward path is represented by the controlled current source $h_{21}I_1$, whereas the unilateral feedback path is represented by the controlled voltage source $h_{12}V_2$. Therefore, the open-loop transfer admittance μ can be computed directly from the network of Fig. 4.22 with reasonable accuracy. With $h_{re} = 0$ and $h_{oe} = 0$, the equivalent network of Fig. 4.22 is presented in Fig. 4.23, whose indefinite-admittance matrix is given by

$$\mathbf{Y} = \begin{bmatrix} G_1 + \dfrac{1}{h_{ie}} & -\dfrac{1}{h_{ie}} & 0 & 0 & -G_1 \\ -\alpha - \dfrac{1}{h_{ie}} & G'_e + \dfrac{1}{h_{ie}} + \alpha & 0 & 0 & -G'_e \\ \alpha & -\alpha & G' & 0 & -G' \\ 0 & 0 & \alpha & G'_2 & -G'_2 - \alpha \\ -G_1 & -G'_e & -G' - \alpha & -G'_2 & G_1 + G'_e + G'_2 + G' + \alpha \end{bmatrix} \tag{4.74}$$

where $G_1 = 1/R_1$, $G'_e = 1/R'_e$, $G'_2 = G_2 + 1/(R_e + R_f)$, $G' = 1/R + 1/h_{ie}$, and $\alpha = h_{fe}/h_{ie}$. By applying (2.94), the open-loop transfer admittance can be written as

$$\begin{aligned} \mu = \frac{I_2}{V_s} &= -\frac{V_2 G_2}{I_s R_1} = -G_1 G_2 z_{14,55} = -G_1 G_2 \frac{Y_{14,55}}{Y_{55}} \\ &= -\frac{G_1 G_2 G'_e \alpha^2}{G'_2(1/R + 1/h_{ie})[G'_e(G_1 + 1/h_{ie}) + G_1(\alpha + 1/h_{ie})]} \\ &= -\frac{h_{fe}^2 R G_2}{[h_{ie} + (1 + h_{fe})R'_e + R_1]\,[1/(R_f + R_e) + G_2](h_{ie} + R)} \end{aligned} \tag{4.75}$$

confirming (4.71).

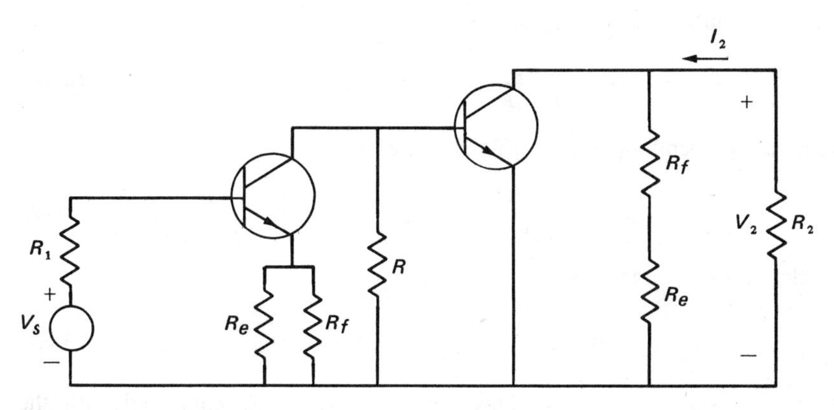

Figure 4.22 The network used to calculate the open-loop transfer admittance of the feedback amplifier of Fig. 4.21.

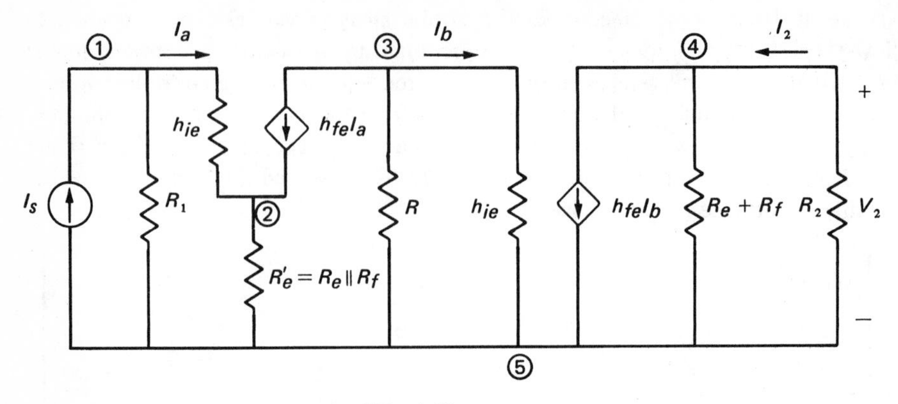

Figure 4.23 An equivalent network of Fig. 4.22.

Example 4.5 In Fig. 4.24, assume that the two transistors are identical, with

$$h_{ie} = 1.1 \text{ k}\Omega \qquad h_{fe} = 50 \qquad h_{re} = h_{oe} \doteq 0 \tag{4.76a}$$

After the biasing and coupling circuitry have been removed, the network is reduced to that of Fig. 4.21, with R denoting the parallel combination of the 10-kΩ, 47-kΩ, and 33-kΩ resistors. The corresponding element values are

$$R_f = 4.7 \text{ k}\Omega \qquad R_e = 100 \ \Omega$$
$$R_1 = 0 \qquad R_2 = 4.7 \text{ k}\Omega \tag{4.76b}$$
$$R_e' = 98 \ \Omega \qquad R = 6.6 \text{ k}\Omega$$

Substituting these in (4.71) and (4.72) yields

$$\mu = -177.55 \text{ mA/V} \tag{4.77a}$$

$$\beta = 98 \ \Omega \tag{4.77b}$$

The closed-loop transfer admittance becomes

$$w = \frac{\mu}{1 - \mu\beta} = -9.65 \text{ mA/V} \tag{4.78}$$

The corresponding open-loop voltage gain is related to μ by

$$\mu_V = -\frac{I_2 R_2}{V_s} = -\mu R_2 = 834.5 \tag{4.79}$$

giving the closed-loop voltage gain

$$w_V = \frac{\mu_V}{1 - \mu_V \beta_V} = 45.38 \tag{4.80}$$

where $\beta_V = -R_e/(R_e + R_f) = -\frac{1}{48}$. This gain value is to be compared with the approximate solution $w_V = 48$ obtained by letting $\mu_V \to \infty$.

At high frequencies, we use the hybrid-pi equivalent model for the two transistors of Fig. 4.22. We assume that the resistance r_π of the second stage is small, so that the first-stage load is approximately resistive, as shown in Fig. 4.25*a*, with

$$R' = \frac{R(r_{x2} + r_{\pi 2})}{R + r_{x2} + r_{\pi 2}} \tag{4.81}$$

where the subscript 2 denotes the transistor parameters of the second stage. The equivalent network of Fig. 4.25*a* is presented in Fig. 4.25*b*, which can be transformed equivalently to the network of Fig. 4.25*c* with the subscript 1 denoting the transistor parameters of the first stage. The input impedance of the one-port formed by the mid-four branches $r_{\pi 1}$, $C_{\pi 1}$, R_e', and $g_{m1} V_a$ is given by

$$Z' = \frac{r_{\pi 1}(1 + g_{m1}R_e')}{C_{\pi 1} r_{\pi 1} s + 1} + R_e' \approx \frac{r_{\pi 1}(1 + g_{m1}R_e')}{C_{\pi 1} r_{\pi 1} s + 1} \tag{4.82}$$

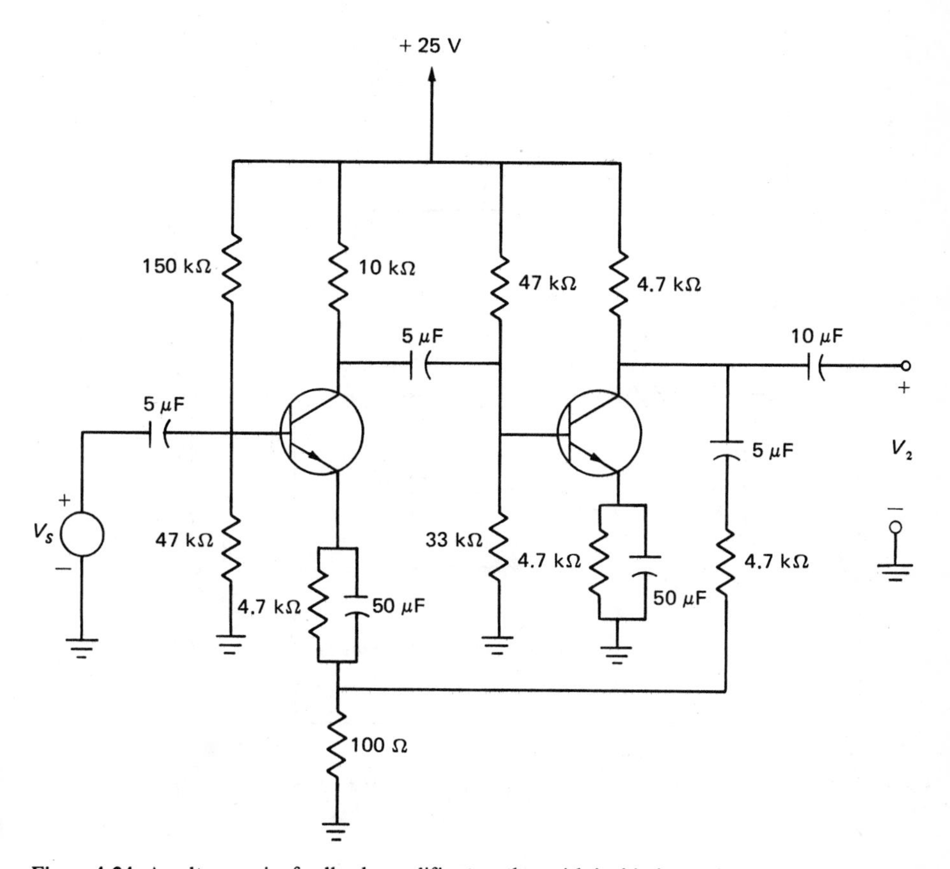

Figure 4.24 A voltage-series feedback amplifier together with its biasing and coupling circuitry.

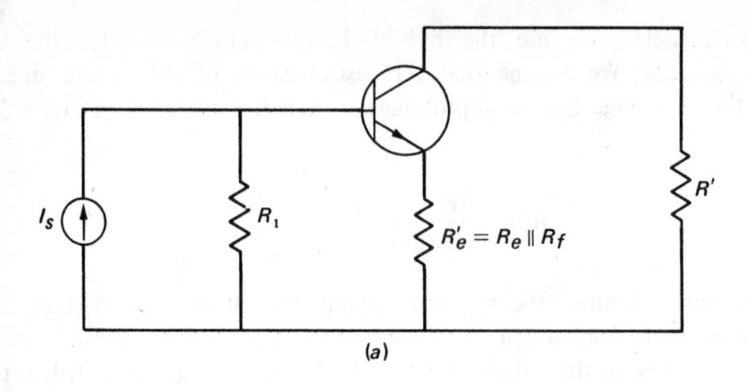

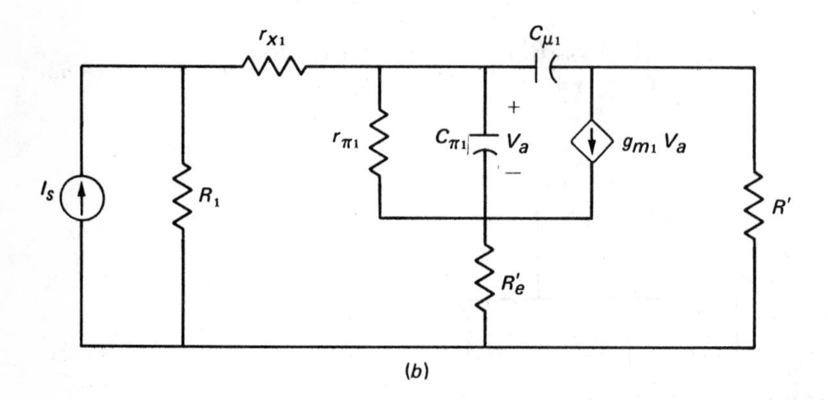

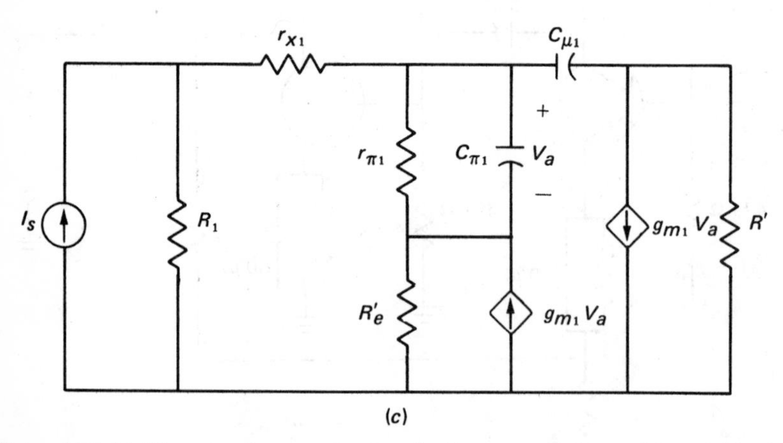

Figure 4.25 (*a*) The network approximating the first-stage of the amplifier of Fig. 4.22. (*b*) The equivalent network of (*a*). (*c*) An equivalent representation of (*b*).

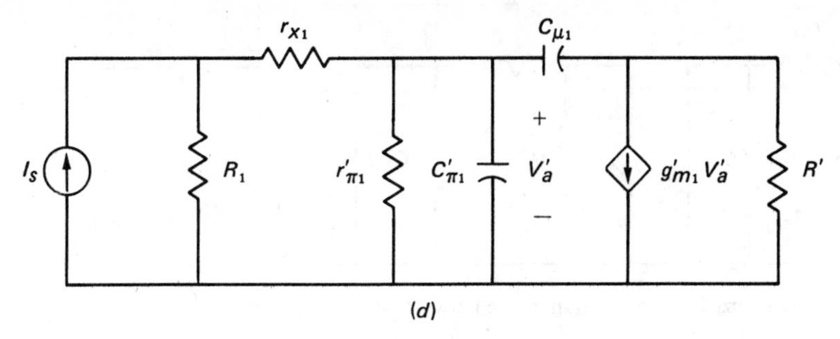

Figure 4.25 *(Continued)* (*d*) An approximate representation of (*c*).

which can be realized by the parallel combination of

$$r'_{\pi 1} = r_{\pi 1}(1 + g_{m1}R'_e) \tag{4.83a}$$

$$C'_{\pi 1} = \frac{C_{\pi 1}}{1 + g_{m1}R'_e} \tag{4.83b}$$

as indicated in Fig. 4.25*d*, where the effective transconductance becomes

$$g'_{m1} = \frac{g_{m1}}{1 + g_{m1}R'_e} \tag{4.83c}$$

By using this model for the first stage in Fig. 4.22 and applying the Miller effect, the amplifier of Fig. 4.22 can be represented by the network of Fig. 4.26 with $r_{o1} = r_{o2} = \infty$, $R'_2 = R_2(R_f + R_e)/(R_2 + R_f + R_e)$, and

$$C''_{\pi 1} = C'_{\pi 1} + C_{\mu 1}(1 + g'_{m1}R') \tag{4.84a}$$

$$C''_{\pi 2} = C_{\pi 2} + C_{\mu 2}(1 + g_{m2}R'_2) \tag{4.84b}$$

From Fig. 4.26, the open-loop transfer admittance $\mu(s)$ is determined as follows:

$$\begin{aligned}\mu(s) &= \frac{I_2}{V_s} = \frac{I_2}{I'_2}\frac{I'_2}{V}\frac{V}{I_s}\frac{I_s}{V_s}\\ &= \frac{R_f + R_e}{R_2 + R_f + R_e}\,\frac{-g'_{m1}A_2R}{1 + s/s_2}\,\frac{R_1A_1/g'_{m1}}{1 + s/s_1}\,\frac{1}{R_1}\\ &= -\frac{A_1A_2(R_e + R_f)R}{(1 + s/s_1)(1 + s/s_2)(R_f + R_2 + R_e)}\end{aligned} \tag{4.85}$$

where

$$A_1 = \frac{g'_{m1}r'_{\pi 1}}{R_1 + r_{x1} + r'_{\pi 1}} \tag{4.86a}$$

$$A_2 = \frac{g_{m2}r_{\pi 2}}{R + r_{x2} + r_{\pi 2}} \tag{4.86b}$$

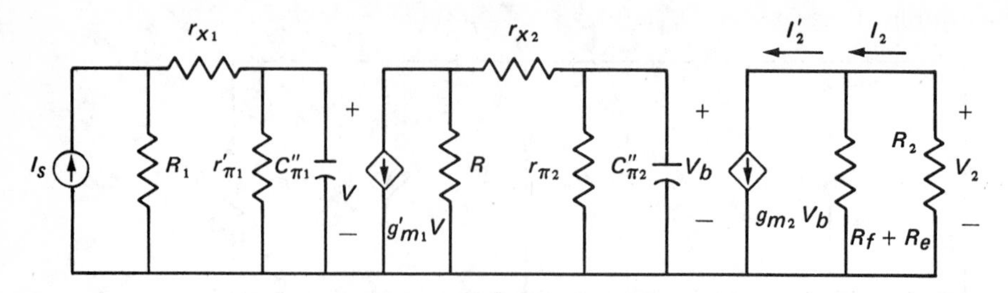

Figure 4.26 A unilateralized approximation of the network of Fig. 4.22.

$$s_1 = \frac{r'_{\pi 1} + r_{x1} + R_1}{C''_{\pi 1} r'_{\pi 1}(r_{x1} + R_1)} \tag{4.86c}$$

$$s_2 = \frac{r_{\pi 2} + r_{x2} + R}{C''_{\pi 2} r_{\pi 2}(r_{x2} + R)} \tag{4.86d}$$

The transfer impedance of the unilateral feedback path is given by (4.72*a*). Thus, the closed-loop transfer admittance is obtained as

$$w(s) = \frac{I_2}{V_s} = \frac{\mu(s)}{1 - \mu(s)\beta(s)} = -\frac{(R_e + R_f)A}{s^2 + (s_1 + s_2)s + s_1 s_2 + R_e R_2 A} \tag{4.87a}$$

where

$$A = \frac{A_1 A_2 R s_1 s_2}{R_f + R_e + R_2} \tag{4.87b}$$

As indicated in (4.19), the input impedance facing the voltage source of Fig. 4.21 is increased by a factor of $1 - \mu(s)\beta(s)$ from its value when the output port is short-circuited, whereas the output impedance looking into the two terminals of the load R_2, including R_2, is decreased by the same factor from its value when the input port is open-circuited.

Example 4.6 In Fig. 4.24, assume that the two transistors are identical, with

$$g_m = 0.4 \text{ mho} \qquad r_x = 50\ \Omega$$
$$r_\pi = 250\ \Omega \qquad C_\pi = 195 \text{ pF} \tag{4.88}$$
$$C_\mu = 5 \text{ pF} \qquad r_o = \infty$$

Then from (4.81), (4.83), and (4.84), we have

$$R' = 287\ \Omega \qquad r'_{\pi 1} = 10.05 \text{ k}\Omega$$
$$C'_{\pi 1} = 4.85 \text{ pF} \qquad g'_{m1} = 9.95 \cdot 10^{-3} \text{ mho} \tag{4.89a}$$
$$C''_{\pi 1} = 24.13 \text{ pF} \qquad C''_{\pi 2} = 4.95 \text{ nF}$$

yielding, from (4.86),

$$A_1 = 9.9 \cdot 10^{-3} \qquad A_2 = 14.49 \cdot 10^{-3}$$
$$s_1 = 83.3 \cdot 10^7 \text{ rad/s} \qquad s_2 = 8.39 \cdot 10^5 \text{ rad/s} \tag{4.89b}$$

Substituting these in (4.85) gives the open-loop transfer admittance

$$\mu(s) = -\frac{474.95}{(p + 992.85)(p + 1)} \tag{4.90}$$

where $p = s/s_2$. The closed-loop transfer admittance is computed from (4.87) as

$$w(s) = -\frac{474.95}{p^2 + 993.85p + 47{,}499} \tag{4.91}$$

where $\beta = 98\ \Omega$. The closed-loop voltage gain is related to $w(s)$ by

$$w_V(s) = -R_2 w(s) = \frac{22.33 \cdot 10^5}{p^2 + 993.85p + 47{,}499} \tag{4.92}$$

with $\beta_V = -\frac{1}{48}$, giving the midband voltage gain with feedback as $w_V(0) = 47$. This corresponds to a reduction from its midband value without feedback $\mu_V(0) = 2248$, by a factor of 47.84, the feedback factor.

As shown in (4.19), the midband input impedance with feedback is increased from its value without feedback by the feedback factor,

$$R_{\text{in}} \approx (R_1 + R_e + r_{x1} + r_{\pi 1})[1 - \mu(0)\beta(0)] = 19.14\ \text{k}\Omega \tag{4.93a}$$

whereas the output impedance with feedback is decreased from its value without feedback by the same factor:

$$R_{\text{out}} \approx \frac{R_2 R_f/(R_2 + R_f)}{1 - \mu(0)\beta(0)} = 49\ \Omega \tag{4.93b}$$

Thus, the series-parallel configuration can be used to realize approximately a voltage-controlled voltage source.

4.2.4 Parallel-Series Feedback

The amplifier of Fig. 4.27 shows two transistors in cascade with feedback from the second emitter to the first base through the feedback resistor R_f. As demonstrated in Fig. 4.27*b*, it conforms with the parallel-series or current-shunt feedback. Let g_{aij}, g_{fij}, and g_{ij} be the g-parameters of the basic amplifier N_a, the feedback network N_f, and the composite two-port network N of Fig. 4.27*b*, respectively. Like the series-parallel topology of Fig. 4.21*b*, the Brune tests of Fig. 4.4*d*, when applied to the network of Fig. 4.27*b*, are *not* satisfied, showing that $g_{ij} \neq g_{aij} + g_{fij}$. However, formulas (4.15) and (4.16) are still valid. Thus, with $G_1 = 1/R_1$, we have

$$\mu = \left.\frac{V_2}{I_s}\right|_{g_{12}=0} = \frac{g_{21}R_2}{(g_{11} + G_1)(g_{22} + R_2)} \tag{4.94}$$

$$\beta = \frac{I_f}{V_2} = \frac{g_{12}}{R_2} \tag{4.95a}$$

where the feedback current I_f is defined by

$$I_f = -g_{12} I_2 \tag{4.95b}$$

Assuming that the two transistors are identical, with $h_{re} = 0$ and

$$\frac{1}{h_{oe}} \gg R \qquad R_f \gg R_e \tag{4.96}$$

the g-parameters g_{ij} of the composite two-port network N are found to be (see Prob. 4.5)

$$\mathbf{G} = \begin{bmatrix} \dfrac{1}{h_{ie}} + \dfrac{h_{ie} + R_e + R'(1 + h_{fe}R_e/h_{ie})}{(R' + h_{ie})(R_e + R_f) + R_eR_f} & -\dfrac{R_e(R' + h_{ie})}{(R' + h_{ie})(R_e + R_f) + R_eR_f} \\ \dfrac{h_{fe}\,[h_{fe}R'(R_e + R_f)/h_{ie} + R_e]}{h_{oe}\,[(R' + h_{ie})(R_e + R_f) + R_eR_f]} & \dfrac{1}{h_{oe}}\left[1 + \dfrac{R_eR_fh_{fe}}{(R' + h_{ie})(R_e + R_f) + R_eR_f}\right] \end{bmatrix}$$

$$\approx \begin{bmatrix} \dfrac{1}{h_{ie}} + \dfrac{h_{ie} + R_e + R(1 + h_{fe}R_e/h_{ie})}{R_f(R + h_{ie} + R_e)} & -\dfrac{R_e(R + h_{ie})}{R_f(R + h_{ie} + R_e)} \\ \dfrac{h_{fe}(h_{fe}RR_f/h_{ie} + R_e)}{R_f(R + h_{ie} + R_e)h_{oe}} & \dfrac{1}{h_{oe}}\left[1 + \dfrac{R_eR_fh_{fe}}{R_f(R + h_{ie} + R_e)}\right] \end{bmatrix} \tag{4.97}$$

where

$$R' = \frac{R}{1 + Rh_{oe}} \approx R \tag{4.98a}$$

$$R'_e = \frac{R_eR_f}{R_e + R_f} \approx R_e \tag{4.98b}$$

In the limit, as h_{oe} approaches zero, the open-loop transfer impedance becomes

$$\mu = \frac{h_{fe}\,[h_{fe}R(R_e + R_f) + R_eh_{ie}]R_2q}{[(1 + G_1h_{ie})q + h_{ie}(h_{ie} + R_e + R) + h_{fe}RR_e](q + R_eR_fh_{fe})} \tag{4.99a}$$

where $q = (R_e + R_f)(R + h_{ie}) + R_eR_f \approx R_f(R + R_e + h_{ie})$. From (4.95$a$) the transfer admittance of the feedback network N_f is obtained as

$$\beta = -\frac{R_eG_2(R + h_{ie})}{q} \tag{4.99b}$$

Finally, the closed-loop transfer impedance is determined from (4.3) and is given by (see Prob. 4.24)

$$w = \frac{h_{fe}\,[h_{fe}R(R_e + R_f) + h_{ie}R_e]R_2}{(R + h_{ie})[(R_e + R_f)(1 + G_1h_{ie}) + h_{ie}] + R_e(1 + h_{fe})(R_f + h_{ie} + h_{fe}R + G_1R_fh_{ie})} \tag{4.100}$$

We remark that in deriving (4.100) there was a cancellation of the common factor q in the denominator of (4.3) or (4.14).

Example 4.7 In Fig. 4.27, assume that the two transistors are identical, with

$$\begin{aligned} h_{ie} &= 1.1\text{ k}\Omega & h_{fe} &= 50 \\ h_{re} &= h_{oe} = 0 & R &= 3\text{ k}\Omega \\ R_e &= 50\ \Omega & R_f &= 1.2\text{ k}\Omega \\ R_1 &= 1.2\text{ k}\Omega & R_2 &= 500\ \Omega \end{aligned} \tag{4.101}$$

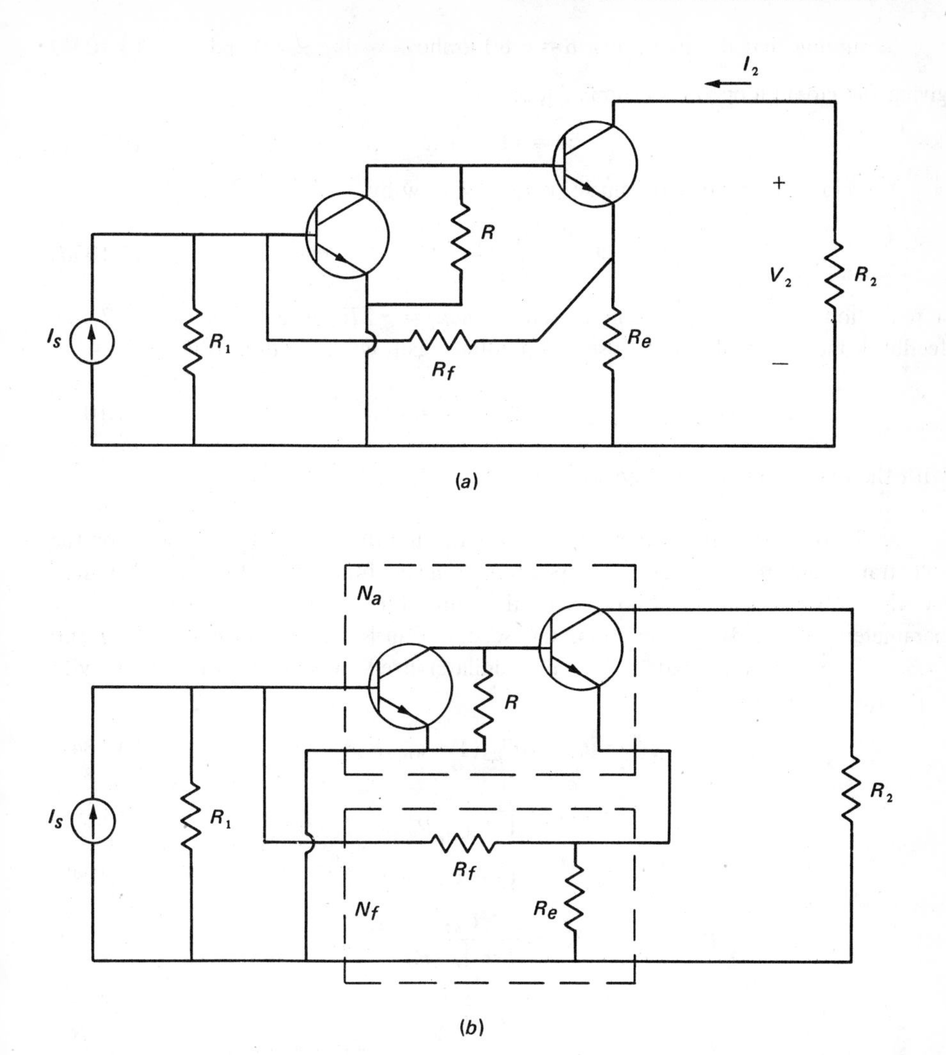

Figure 4.27 (*a*) A transistor feedback amplifier and (*b*) its parallel-series or current-shunt feedback representation.

For illustrative purposes, we first compute the inverse hybrid matrix **G** by assuming that $h_{oe} \neq 0$, since for $h_{oe} = 0$, **G** does not exist. Substituting (4.101) in (4.97) yields

$$\mathbf{G} = \begin{bmatrix} 3.03 \cdot 10^{-3} & -0.04 \\ \dfrac{1.644}{h_{oe}} & \dfrac{1.58}{h_{oe}} \end{bmatrix} \tag{4.102}$$

Thus, from (4.94) and (4.95*a*) we have

$$\mu = 135 \text{ k}\Omega \tag{4.103a}$$

$$\beta = -80 \ \mu\text{mho} \tag{4.103b}$$

giving the closed-loop transfer impedance

$$w = 11.56 \ \text{k}\Omega \tag{4.103c}$$

The closed-loop current gain w_I is related to w by

$$w_I = -\frac{w}{R_2} = -23.13 \tag{4.103d}$$

a reduction from its open-loop current gain $\mu_I = -270$ by a factor of 11.67, the feedback factor. Finally, the closed-loop voltage gain w_V is related to w by

$$w_V = \frac{w}{R_1} = 9.64 \tag{4.103e}$$

with the open-loop voltage gain $\mu_V = 112.50$.

As before, at high frequencies, we use the hybrid-pi equivalent model for the two transistors of Fig. 4.27. The open-loop transfer function $\mu(s)$ can be computed by the network of Fig. 4.28. Using the subscripts 1 and 2 to distinguish the parameters of the two transistors, the second transistor, as shown in (4.83) and (4.84), can be represented by the unilateralized model of Fig. 4.29 with $r_{o1} = r_{o2} = \infty$ and

$$C''_{\pi 2} = C'_{\pi 2} + C_{\mu 2}(1 + g'_{m2}R_2) \tag{4.104a}$$

$$g'_{m2} = \frac{g_{m2}}{1 + g_{m2}R'_e} \tag{4.104b}$$

$$r'_{\pi 2} = r_{\pi 2}(1 + g_{m2}R'_e) \tag{4.104c}$$

$$C'_{\pi 2} = \frac{C_{\pi 2}}{1 + g_{m2}R'_e} \tag{4.104d}$$

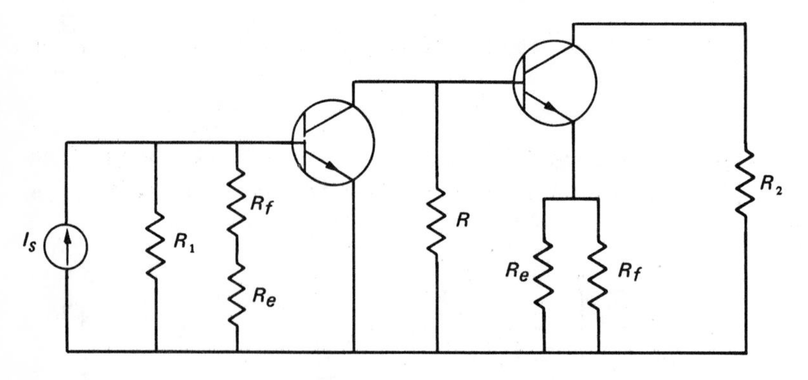

Figure 4.28 The transistor network used to calculate the open-loop transfer impedance of the amplifier of Fig. 4.27.

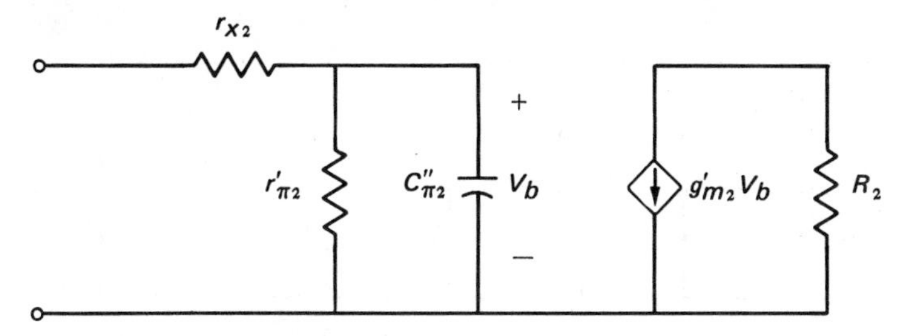

Figure 4.29 A unilateralized approximation of the second transistor of the amplifier of Fig. 4.28.

where R'_e is defined in (4.98*b*). The first stage sees approximately a parallel *RC* load, as indicated in Fig. 4.30*a*, because the effect of r_{x2} is negligible for R, $r'_{\pi 2} \gg r_{x2}$. The equivalent network of Fig. 4.30*a* is presented in Fig. 4.30*b*. By using this *RC* load for $Y_2(s)$ in (4.37), the admittance facing the capacitor $C_{\pi 1}$ can be written as

$$Y_{\text{eq}}(s) = sC_{\mu 1}\left(1 + \frac{g_{m1}}{Y_2}\right) = sC_{\mu 1} + \frac{1}{1/(R''g_{m1}C_{\mu 1}s) + C''_{\pi 2}/g_{m1}C_{\mu 1}} \tag{4.105}$$

which can be realized by the network as shown in Fig. 4.31 with

$$R_\alpha = \frac{C''_{\pi 2}}{g_{m1}C_{\mu 1}} \tag{4.106a}$$

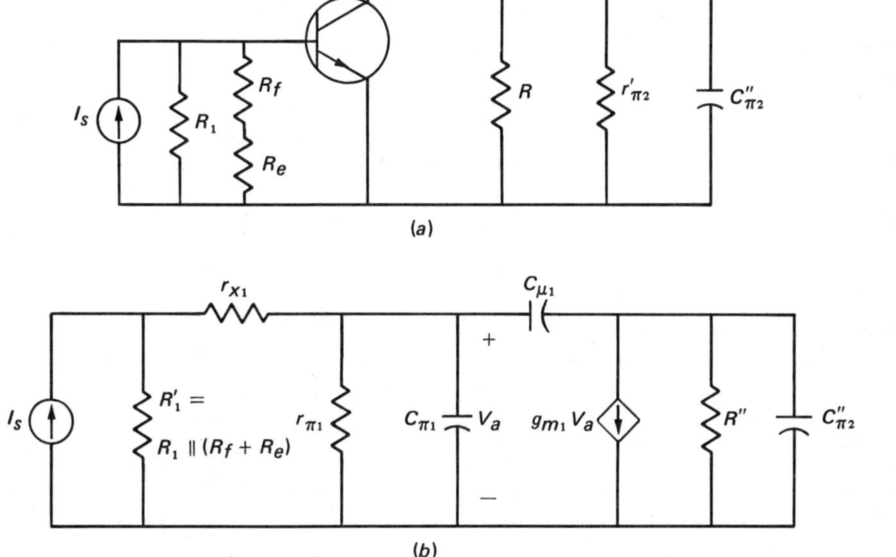

Figure 4.30 (*a*) An approximation of the first stage of the amplifier of Fig. 4.28, and (*b*) its equivalent network.

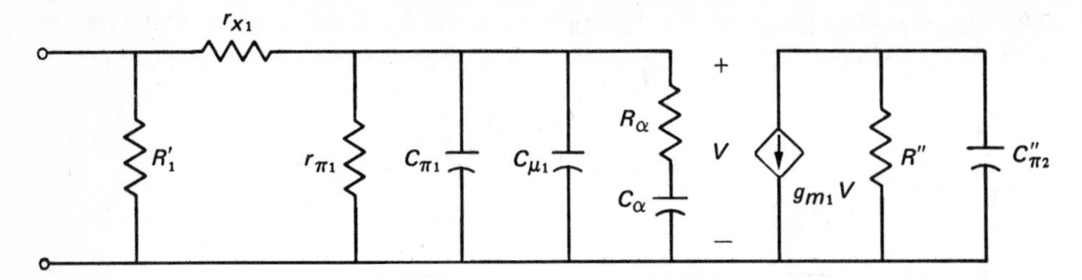

Figure 4.31 A unilateralized approximation of the network of Fig. 4.30.

$$C_\alpha = R'' g_{m1} C_{\mu 1} \tag{4.106b}$$

where

$$R'' = \frac{R r'_{\pi 2}}{R + r'_{\pi 2}} \tag{4.106c}$$

The network of Fig. 4.28 can now be represented approximately by that of Fig. 4.32 for the computation of the open-loop transfer impedance $\mu(s)$ with

$$C_1 = C_{\pi 1} + C_{\mu 1} \tag{4.107a}$$

$$R'_1 = \frac{R_1(R_e + R_f)}{R_1 + R_e + R_f} \tag{4.107b}$$

The open-loop transfer impedance is found to be

$$\mu(s) = \frac{V'_2}{I_s} = \frac{V'_2}{V_b}\frac{V_b}{V}\frac{V}{I_s} = \frac{A_0 R_2}{a_2 s^2 + a_1 s + a_0} \tag{4.108}$$

where

$$A_0 = g_{m1} g'_{m2} R'_1 R'' \tag{4.109a}$$

$$a_2 = C_1 C''_{\pi 2} R''(R'_1 + r_{x1}) \tag{4.109b}$$

$$a_1 = C''_{\pi 2} R'' + (R'_1 + r_{x1})\left(C_1 + R'' g_{m1} C_{\mu 1} + \frac{R'' C''_{\pi 2}}{r_{\pi 1}}\right) \tag{4.109c}$$

$$a_0 = 1 + \frac{R'_1 + r_{x1}}{r_{\pi 1}} \tag{4.109d}$$

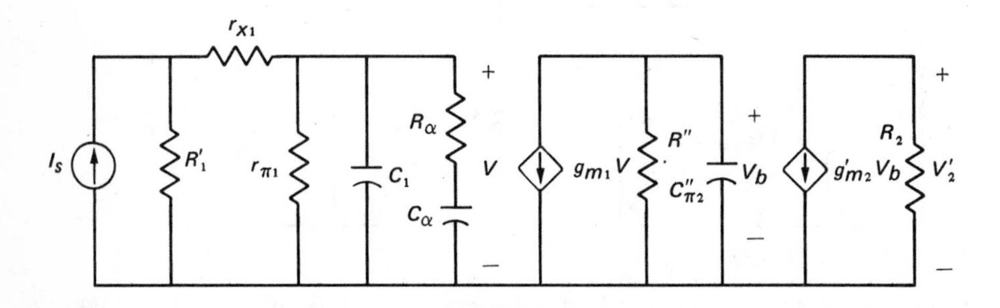

Figure 4.32 A unilateralized approximation of the amplifier of Fig. 4.28.

The transfer admittance of the feedback network N_f is the same as that given in (4.95) except that h_{ie} in g_{12}, as given in (4.97), is replaced by $r_{x2} + r_{\pi 2}$, yielding

$$\beta = -\frac{R_e/R_2}{R_e + R_f[1 + R_e/(R + r_{x2} + r_{\pi 2})]} \tag{4.110a}$$

For $R \gg R_e$, β is reduced to

$$\beta \approx -\frac{R_e}{R_2(R_e + R_f)} \tag{4.110b}$$

The closed-loop transfer impedance is found to be

$$w(s) = \frac{A_0 R_2}{a_2 s^2 + a_1 s + a_0 + A_0 R'_e/R_f} \tag{4.111}$$

We illustrate the above results by considering the following numerical example.

Example 4.8 In Fig. 4.27*a*, assume that the two transistors are identical, with

$$\begin{gathered} g_m = 0.4 \text{ mho} \qquad r_x = 50\ \Omega \\ r_\pi = 250\ \Omega \qquad C_\pi = 195 \text{ pF} \\ C_\mu = 5 \text{ pF} \qquad r_o = \infty \end{gathered} \tag{4.112a}$$

The other parameters of the amplifier are given by

$$\begin{gathered} R_1 = 1.2 \text{ k}\Omega \qquad R_2 = 500\ \Omega \\ R_f = 1.2 \text{ k}\Omega \qquad R_e = 50\ \Omega \\ R = 3 \text{ k}\Omega \end{gathered} \tag{4.112b}$$

From (4.98*b*), (4.104), (4.106*c*), and (4.107), we have

$$\begin{gathered} R'_e = 48\ \Omega \qquad C'_{\pi 2} = 9.65 \text{ pF} \\ g'_{m2} = 0.0198 \text{ mho} \qquad r'_{\pi 2} = 5.05 \text{ k}\Omega \\ C''_{\pi 2} = 64.15 \text{ pF} \qquad R'' = 1.88 \text{ k}\Omega \\ C_1 = 200 \text{ pF} \qquad R'_1 = 612.2\ \Omega \end{gathered} \tag{4.113}$$

Substituting these in (4.109) yields

$$\begin{gathered} A_0 = 9115.41 \qquad a_2 = 1.597 \cdot 10^{-14} \\ a_1 = 30.62 \cdot 10^{-7} \qquad a_0 = 3.65 \end{gathered} \tag{4.114}$$

The open-loop transfer impedance becomes

$$\mu(p) = \frac{5707.8 R_2}{(p + 0.1199)(p + 19.056)} \tag{4.115}$$

where $p = s/10^7$. The transfer admittance of the feedback network from (4.110*b*) is obtained as

$$\beta \approx -\frac{R_e}{R_2(R_e + R_f)} = -80 \ \mu\text{mho} \tag{4.116}$$

giving the closed-loop transfer impedance

$$w(s) = \frac{5707.8R_2}{p^2 + 19.18p + 230.61} = -w_I(s)R_2 \tag{4.117}$$

where $w_I(s)$ denotes the closed-loop current gain I_2/I_s (see Fig. 4.27*a*). Thus, the midband current gain with feedback is

$$w_I(0) = -24.75 \tag{4.118}$$

a reduction from the midband value without feedback $\mu_I(0) = -2498.1$ by the feedback factor $1 - \mu_I(0)\beta_I(0) = 100.9$, where $\beta_I(0) = 0.04$.

The closed-loop voltage gain $w_V(s)$ is related to $w(s)$ by $w(s)/R_1$. The midband voltage gain with feedback is

$$w_V(0) = 10.31 \tag{4.119}$$

This corresponds to a reduction from the midband value without feedback, $\mu_V(0) = 1040.9$, by the same factor $1 - \mu_V(0)\beta_V(0) = 100.9$ with $\beta_V = 0.096$.

As indicated in (4.19), the midband input admittance facing the current source of Fig. 4.27*a* is increased by the factor $1 - \mu(0)\beta(0)$ from its value without feedback, which corresponds to the open-circuiting of the output port,

$$G_{\text{in}} = \left(\frac{1}{R_1} + \frac{1}{r_{x1} + r_{\pi 1}} + \frac{1}{R_e + R_f}\right)[1 - \mu(0)\beta(0)] \approx \tfrac{1}{2} \text{ mho} \tag{4.120a}$$

whereas the output impedance is increased from its value without feedback, which corresponds to the short-circuiting of the input port, by the same factor:

$$R_{\text{out}} = (r_o + R_2 + R_e')[1 - \mu(0)\beta(0)] \tag{4.120b}$$

where r_o is the common-emitter transistor output resistance. Since r_o is on the order of 10-100 kΩ, the output resistance R_{out} is very high. The parallel-series configuration is therefore attractive for situations where low input and high output impedances are required, such as for cascading and stagger tuning to obtain multistage amplifiers with a large gain-bandwidth product.

4.3 GENERAL FEEDBACK THEORY

Up to now, we have used the ideal feedback model of Fig. 4.1 to study the properties of feedback amplifiers. The model is useful only if we can separate a feedback amplifier into two blocks: the basic amplifier $\mu(s)$ and the feedback network $\beta(s)$. For this we presented four useful feedback configurations and demonstrated in each case by a practical example how to identify $\mu(s)$ and $\beta(s)$. The four configurations are essentially the series-series, parallel-parallel, series-parallel, and parallel-series connections of two-port networks. In practice, we often run into feedback amplifiers that cannot be classified as being in one of the above four categories. As indicated earlier, a general feedback amplifier may contain an input coupling network and an output coupling network, as shown in Fig. 4.5. A typical example is the bridge feedback configuration of Fig. 4.6.

The calculation of $\mu(s)$ and $\beta(s)$, as demonstrated in the foregoing, is based on the assumption that the complete equivalent networks of the active devices and the passive two-port networks are known. Therefore they cannot be measured experimentally, because the forward transmission of the passive network is included in the basic amplifier, whereas the backward transmission of the active two-port network is lumped with the feedback network. Since the zeros of the feedback factor $1-\mu(s)\beta(s)$ are poles of the closed-loop transfer function, they are also the natural frequencies of the feedback amplifier. If $\mu(s)$ and $\beta(s)$ are known, the roots of $1-\mu(s)\beta(s)$ can readily be computed explicitly with the aid of a computer if necessary, and the stability problem can then be settled directly. However, for a physical amplifier, there remains the difficulty of getting an accurate formulation and representation of the network itself, because every equivalent network is, to a greater or lesser extent, an idealization of the physical reality. What is really needed is an equivalent measurement of the feedback factor that provides some kind of experimental verification that the system is stable and will remain so under certain prescribed conditions. Thus, the ideal feedback model is not an adequate representation of a practical feedback amplifier. In the following, we shall develop Bode's feedback theory, which is applicable to general network configuration and avoids the necessity of identifying $\mu(s)$ and $\beta(s)$.

Bode's feedback theory is based on the concept of return difference, which is defined in terms of network determinants. We shall show that the return difference is a generalization of the concept of the feedback factor of the ideal feedback model, and can be measured physically from the amplifier itself. We shall then introduce the notion of null return difference and discuss its physical significance. Among the many important properties, we shall find in the next chapter that return difference and null return difference are closely related to the sensitivity function and that they are basic to the study of the stability of the feedback amplifier. As indicated in the preceding section, feedback affects the input and output impedances. Here, we show that return difference possesses the same significance, which together with null return difference can be employed to simplify the calculation of driving-point impedance of an active network, thus observing the effects of feedback on amplifier impedance and gain.

4.3.1 The Return Difference

In the study of a feedback amplifier, we usually single out an element for particular attention. The element is generally one that is either crucial in terms of its effect on the entire system or of primary concern to the designer. The element may be the transfer function of an active device, the gain of an amplifier, or the immittance of a one-port network. For our purposes, we assume that the element x of interest is the controlling parameter of a voltage-controlled current source defined by the relation

$$I = xV \tag{4.121}$$

To focus our attention on the element x, Fig. 4.33 is the general configuration of a feedback amplifier in which the controlled source is brought out as a two-port network connected to a general four-port network, along with the input source combination of I_s and Y_1 and the load admittance Y_2.

We remark that the two-port network representation of a controlled source (4.121) is quite general. It includes as a special case a one-port element characterized by its immittance. In (4.121), if the controlling voltage V is the terminal voltage of the controlled current source I, then x represents the one-port admittance. Also, the consideration of voltage-controlled current source should not be deemed to be restrictive, because the other three types of controlled sources (current-controlled current source, voltage-controlled voltage source, and current-controlled voltage source) can usually be converted to the above type by a simple transformation and by appealing to Norton's theorem if necessary. The element x in (4.121) can either be the transconductance or more generally a transfer admittance function, depending on the type of equivalent circuit being used.

We first give a mathematical definition of return difference and then discuss its physical significance.

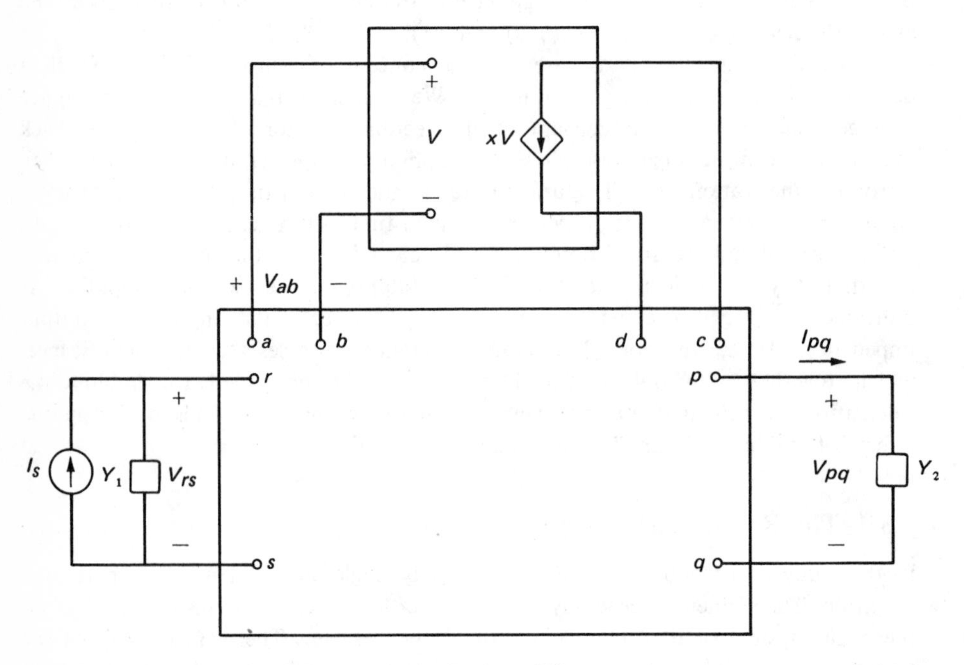

Figure 4.33 The general configuration of a feedback amplifier.

Definition 4.1: Return difference The *return difference* $F(x)$ of a feedback amplifier with respect to an element x is the ratio of the two functional values assumed by the first-order cofactor of an element of its indefinite-admittance matrix under the condition that the element x assumes its nominal value and the condition that the element x assumes the value zero.

Let $\mathbf{Y}$ be the indefinite-admittance matrix of a feedback amplifier. As shown in Theorem 2.1, $\mathbf{Y}$ is an equicofactor matrix, all of its cofactors being equal. Thus, it is immaterial which cofactor is chosen to formulate the return difference. To emphasize the importance of the feedback element x, we express $\mathbf{Y}$ as a function of x alone for the present discussion, even though it is also a function of the complex-frequency variable s, and write $\mathbf{Y} = \mathbf{Y}(x)$. Following the convention adopted in Chap. 2 for the cofactors, $Y_{ij}(x)$ denotes the cofactor of the ith row and jth column element of $\mathbf{Y}(x)$. Then the return difference $F(x)$ with respect to the element x can be expressed as

$$F(x) \equiv \frac{Y_{ij}(x)}{Y_{ij}(0)} \tag{4.122}$$

where $Y_{ij}(0) = Y_{ij}(x)|_{x=0}$

The physical significance of the return difference will now be considered. In the network of Fig. 4.33, we label the terminals of the input, the output, the controlling branch, and the controlled source as indicated. With this designation, the element x enters the indefinite-admittance matrix $\mathbf{Y}$ in a rectangular pattern as shown below:

$$\mathbf{Y} = \begin{matrix} a \\ b \\ c \\ d \end{matrix} \overset{\begin{matrix} a & b & c & d \end{matrix}}{\begin{bmatrix} & & & \\ & & & \\ x & -x & & \\ -x & x & & \end{bmatrix}} \tag{4.123}$$

If in Fig. 4.33 we replace the controlled current source xV by an independent current source of x amperes and set the excitation I_s to zero, the corresponding indefinite-admittance matrix of the resulting network is simply $\mathbf{Y}(0)$. By appealing to formula (2.94), the new voltage V'_{ab} appearing at terminals a and b of the controlling branch is given by

$$V'_{ab} = x\,\frac{Y_{da,cb}(0)}{Y_{uv}(0)} = -x\,\frac{Y_{ca,db}(0)}{Y_{uv}(0)} \tag{4.124}$$

Note that the current injecting point is terminal d, not c. Alternatively, we can reverse the direction of the current source by replacing x by $-x$ and apply the formula (2.94) directly.

The above manipulation of replacing the controlled current source by an independent current source and setting the excitation I_s to zero can be represented symbolically as shown in Fig. 4.34. The controlling branch is broken off as marked and a voltage source of 1 V is applied to the right of the breaking mark. This 1-V sinusoidal voltage of a fixed angular frequency produces a current of x amperes at the controlled current source. The voltage appearing at the left of the breaking mark caused by the 1-V excitation is then V'_{ab}, as indicated in Fig. 4.34. This returned voltage V'_{ab} has the same physical significance as the loop transmission $\mu\beta$ defined for the ideal feedback model of Fig. 4.1. To see this, we set the input excitation to the ideal feedback model to zero, break the forward path, and apply a unit input to the right of the break, as shown in Fig. 4.35. The signal appearing at the left of the break is precisely the loop transmission $\mu\beta$. For this reason, we introduce the concept of return ratio.

Definition 4.2: Return ratio The *return ratio* T with respect to a voltage-controlled current source $I = xV$ is the negative of the voltage appearing at the controlling branch when the controlled current source is replaced by an independent current source of x amperes and the input excitation is set to zero.

Thus, the return ratio T is simply the negative of the returned voltage V'_{ab}, that is, $T = -V'_{ab}$. With this in mind, we next compute the difference between the 1-V

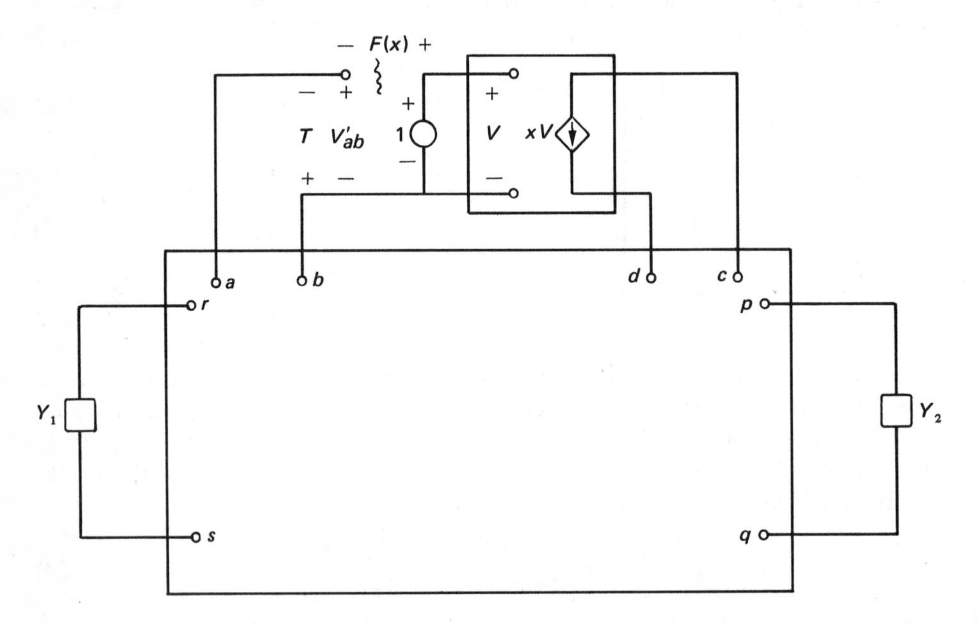

Figure 4.34 The physical interpretation of the return difference with respect to the controlling parameter of a voltage-controlled current source.

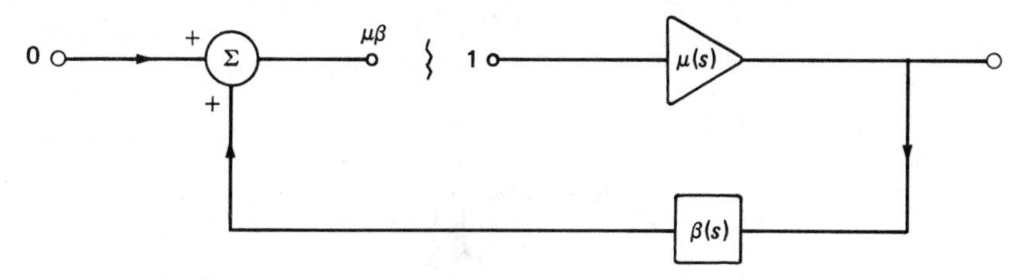

Figure 4.35 The physical interpretation of the loop transmission.

excitation and the returned voltage V'_{ab}. As before, we denote by $Y_{ab,cd}(x)$ the second-order cofactor of the elements of $\mathbf{Y}(x)$. Then we have

$$1 - V'_{ab} = 1 + x\frac{Y_{ca,db}}{Y_{uv}(0)} = \frac{Y_{uv}(0) + xY_{ca,db}}{Y_{uv}(0)} = \frac{Y_{db}(0) + xY_{ca,db}}{Y_{db}(0)}$$

$$= \frac{Y_{db}(x)}{Y_{db}(0)} = \frac{Y_{ij}(x)}{Y_{ij}(0)} = F(x) \tag{4.125}$$

in which we have invoked the identities $Y_{uv} = Y_{ij}$ and

$$Y_{db}(x) = Y_{db}(0) + xY_{ca,db} \tag{4.126}$$

$Y_{ca,db}$ being independent of x. In other words, the return difference $F(x)$ is simply the difference of the 1-V excitation and the returned voltage V'_{ab} as illustrated in Fig. 4.34, and hence its name. Since $F(x) = 1 + T = 1 - \mu\beta$, we conclude that the return difference has the same physical significance as the feedback factor of the ideal feedback model of Fig. 4.1.

The significance of the above physical interpretations is that it permits us to determine the return ratio T or $-\mu\beta$ by measurement. Once the return ratio is measured, the other quantities such as return difference and loop transmission are known. A discussion of the measurement techniques will be presented in the following chapter.

Example 4.9 Figure 4.36*a* is a common-emitter amplifier with an admittance G_f purposely added to produce an external physical feedback loop to affect specific change in the performance of the amplifier. The equivalent network of the feedback amplifier is presented in Fig. 4.36*b*, using the T-model of the transistor for a change. The amplifier is of the parallel-parallel configuration and was discussed in Sec. 4.2.2. The indefinite-admittance matrix of the amplifier can be written down by inspection as

$$\mathbf{Y} = \begin{bmatrix} G_f + g_b & -g_b & -G_f & 0 \\ -g_b & g_b + g_e + g_c - \alpha g_e & -g_c & -g_e + \alpha g_e \\ -G_f & -g_c + \alpha g_e & G_f + g_c + G_2 & -G_2 - \alpha g_e \\ 0 & -g_e & -G_2 & G_2 + g_e \end{bmatrix} \tag{4.127}$$

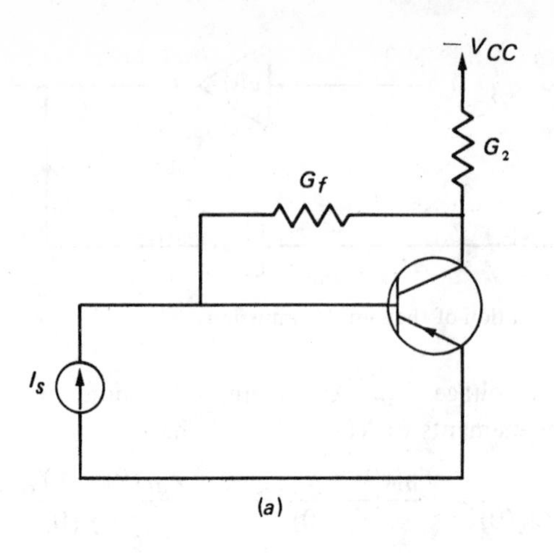

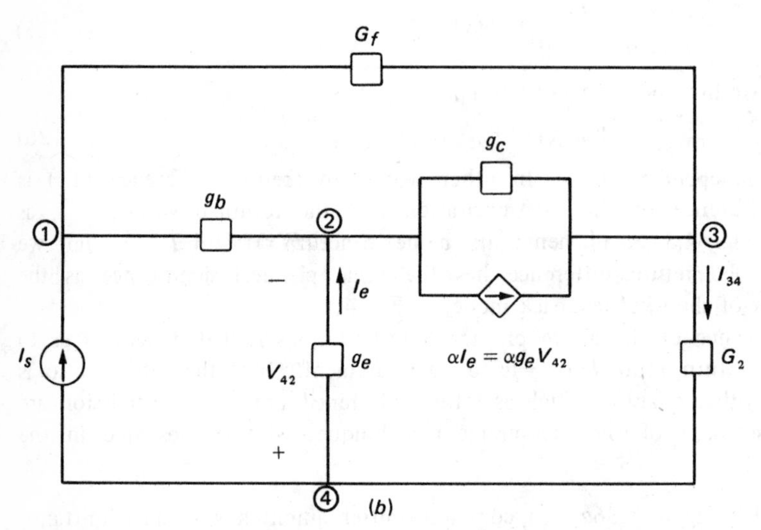

Figure 4.36 (*a*) A common-emitter feedback amplifier, and (*b*) its equivalent network.

Suppose that the controlling parameter αg_e is the element of interest. Then, according to (4.122), the return difference with respect to αg_e is given by

$$F(\alpha g_e) = \frac{Y_{ij}(\alpha g_e)}{Y_{ij}(0)} = \frac{Y_{22}(\alpha g_e)}{Y_{22}(0)}$$

$$= \frac{(G_2 + g_e)(g_c G_f + g_b g_c + g_b G_f) + (1 - \alpha) G_2 g_e (g_b + G_f)}{(G_2 + g_e)(g_c G_f + g_b g_c + g_b G_f) + G_2 g_e (g_b + G_f)} = 1 + T \quad (4.128)$$

where the return ratio

$$T = -\frac{\alpha g_e G_2 (g_b + G_f)}{(G_2 + g_e)(g_c G_f + g_b g_c + g_b G_f) + G_2 g_e (g_b + G_f)} \quad (4.129)$$

corresponds to the negative of the loop transmission $\mu\beta$ of the ideal feedback model.

For illustrative purposes, we compute the return ratio T by its definition. In the network of Fig. 4.36*b*, we remove the current source I_s and replace the controlled current source by an independent current source of αg_e amperes. The resulting network is shown in Fig. 4.37, in which the negative of the voltage V'_{42} is the return ratio T. The indefinite-admittance matrix of the network of Fig. 4.37 can be obtained from (4.127) by setting $\alpha g_e = 0$, yielding $\mathbf{Y}(0)$. Appealing again to formula (2.94) gives

$$T = -V'_{42} = -\alpha g_e \frac{Y_{34,22}(0)}{Y_{uv}(0)}$$
$$= -\frac{\alpha g_e G_2(g_b + G_f)}{(G_2 + g_e)(g_c G_f + g_b g_c + g_b G_f) + G_2 g_e(g_b + G_f)} \tag{4.130}$$

confirming (4.129).

As illustrated in this example, terminals a, b, c, and d of the controlled current source need not be distinct. In fact, Eqs. (4.124)–(4.126) remain valid even for the situation where $a = b$ or $c = d$ provided that we follow the earlier convention by defining sgn $0 = 0$.

Let us now examine the physical significance of return difference with respect to a one-port admittance x. The result can easily be deduced from (4.125) by letting $a = c$ and $b = d$ as follows:

$$F(x) = \frac{Y_{ij}(x)}{Y_{ij}(0)} = \frac{Y_{dd}(x)}{Y_{dd}(0)} = \frac{Y_{dd}(0) + xY_{cc,dd}(0)}{Y_{dd}(0)}$$
$$= 1 + x\frac{Y_{cc,dd}(0)}{Y_{dd}(0)} = 1 + \frac{x}{y} \tag{4.131}$$

where y is the admittance that x faces, as depicted in Fig. 4.38. The last equation in (4.131) follows directly from formula (2.95). The return difference can therefore be written as

$$F(x) = 1 + T = 1 - \mu\beta = 1 + \frac{x}{y} = \frac{x + y}{y} \tag{4.132}$$

In other words, the return ratio T for the one-port admittance x is equal to the ratio of admittance x to the admittance that x faces, and the return difference $F(x)$ with respect to the one-port admittance x is equal to the ratio of total admittance looking into the node pair where x is connected to the admittance that x faces.

Example 4.10 Consider the network of Fig. 4.36*b*. Suppose that we wish to compute the return difference with respect to the one-port admittance G_2. From (4.122) in conjunction with (4.127), we obtain

$$F(G_2) = \frac{Y_{ij}(G_2)}{Y_{ij}(0)} = \frac{Y_{22}(G_2)}{Y_{22}(0)}$$

$$= \frac{(G_2 + g_e)(g_c G_f + g_b g_c + g_b G_f) + (1 - \alpha) G_2 g_e (g_b + G_f)}{g_e(g_c G_f + g_b g_c + g_b G_f)} = 1 + T \quad (4.133)$$

where
$$T = G_2 \left[\frac{1}{g_e} + \frac{(1 - \alpha)(g_b + G_f)}{g_c G_f + g_b g_c + g_b G_f} \right] \quad (4.134)$$

To compute the admittance that G_2 faces, we appeal to formula (2.95) and use the matrix $\mathbf{Y}(0)$, giving

$$\frac{1}{y} = \frac{Y_{33,44}(0)}{Y_{22}(0)} = \frac{G_f(g_b + g_e + g_c - \alpha g_e) + g_b(g_e + g_c - \alpha g_e)}{g_e(g_c G_f + g_b g_c + g_b G_f)}$$

$$= \frac{1}{g_e} + \frac{(1 - \alpha)(g_b + G_f)}{g_c G_f + g_b g_c + g_b G_f} \quad (4.135)$$

This shows that

$$F(G_2) = 1 + T = 1 + \frac{G_2}{y} = \frac{G_2 + y}{y} \quad (4.136)$$

Example 4.11 Consider the series-parallel feedback amplifier of Fig. 4.24. As in Example 4.5, assume that the transistors are identical, with $h_{ie} = 1.1\ \text{k}\Omega$, $h_{fe} = 50$, and $h_{re} = h_{oe} = 0$. After the biasing and coupling circuitry have been removed, its equivalent network is presented in Fig. 4.39*a*. The effective load R' of the first transistor is composed of the parallel combination of 10-kΩ, 33-kΩ, 47-kΩ, and

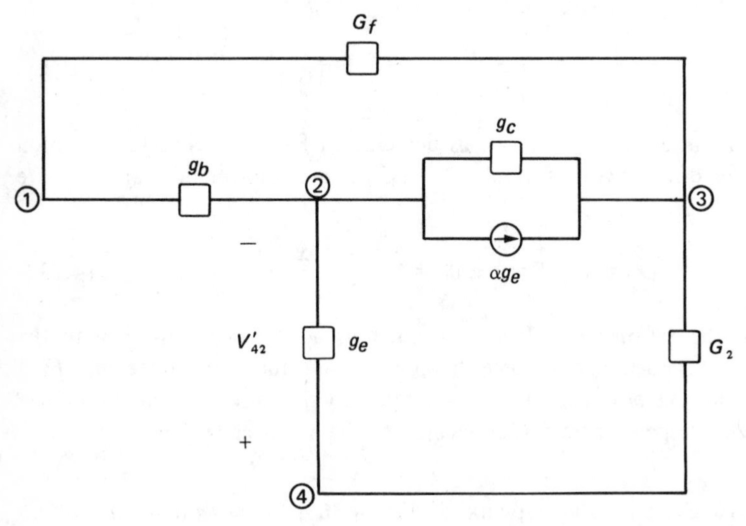

Figure 4.37 The network used to calculate the return ratio with respect to the controlling parameter αg_e of the controlled source in Fig. 4.36*b*.

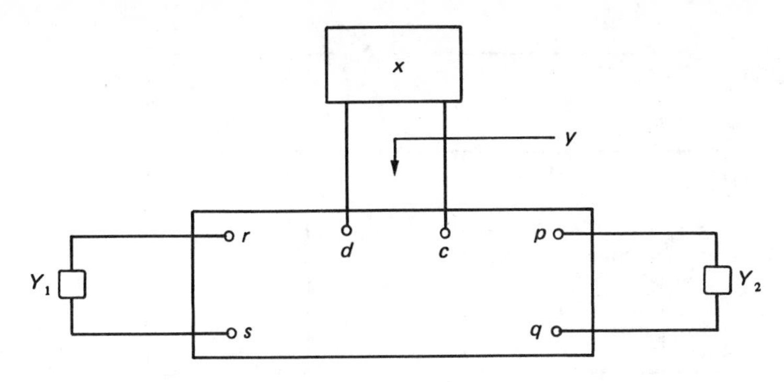

Figure 4.38 The physical interpretation of the return difference with respect to a one-port admittance.

1.1-kΩ resistors; or the parallel combination of 6.6-kΩ and 1.1-kΩ resistors, as shown in Fig. 4.39*a*. The effect of the 150-kΩ and 47-kΩ biasing resistors can be ignored; they are included in the equivalent network to show their insignificance in the computation. We remark that in some cases the effect of biasing resistors may not be entirely negligible and should therefore be included. With all the conductances denoted in mho, the network of Fig. 4.39*a* is redrawn as in Fig. 4.39*b* with

$$\tilde{\alpha}_k = \alpha_k \cdot 10^{-4} = \frac{h_{fe}}{h_{ie}} = 455 \cdot 10^{-4} \tag{4.137}$$

where $k = 1, 2$ is used to distinguish the transconductances of the first and second transistors. The indefinite-admittance matrix can now be written down by inspection and is given by

$$\mathbf{Y} = 10^{-4}\begin{bmatrix} 9.37 & 0 & -9.09 & 0 & -0.28 \\ 0 & 4.256 & -2.128 & \alpha_2 & -2.128-\alpha_2 \\ -9.09-\alpha_1 & -2.128 & 111.218+\alpha_1 & 0 & -100 \\ \alpha_1 & 0 & -\alpha_1 & 10.61 & -10.61 \\ -0.28 & -2.128 & -100 & -10.61-\alpha_2 & 113.02+\alpha_2 \end{bmatrix} \tag{4.138}$$

By applying formula (2.97), the voltage gain of the amplifier is found to be

$$w = \frac{V_{25}}{V_s} = \frac{Y_{12,55}}{Y_{11,55}} = \frac{211.54 \cdot 10^{-7}}{4.66 \cdot 10^{-7}} = 45.39 \tag{4.139}$$

confirming (4.80), where

$$Y_{12,55} = (102.13\alpha_1\alpha_2 + 22.58\alpha_1 + 205.24) \cdot 10^{-12} = 211.535 \cdot 10^{-7} \tag{4.140a}$$

$$Y_{11,55} = (2.128\alpha_1\alpha_2 + 45.156\alpha_1 + 4974.132\alpha_2) \cdot 10^{-12} = 4.661 \cdot 10^{-7} \tag{4.140b}$$

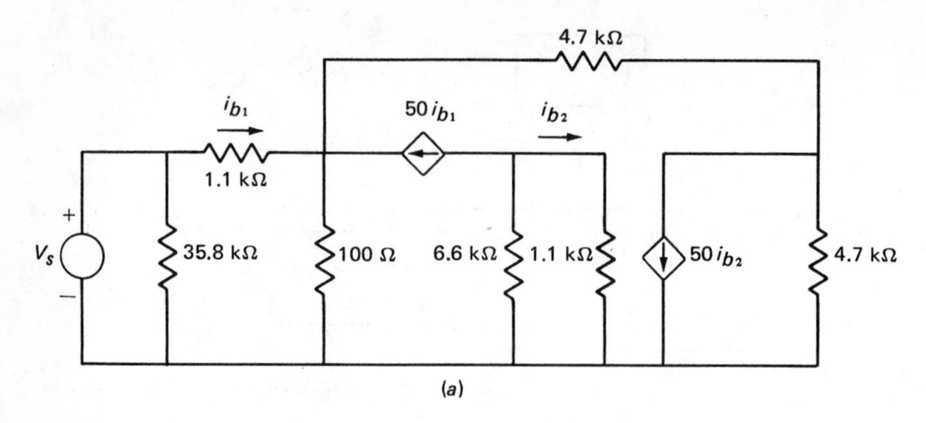

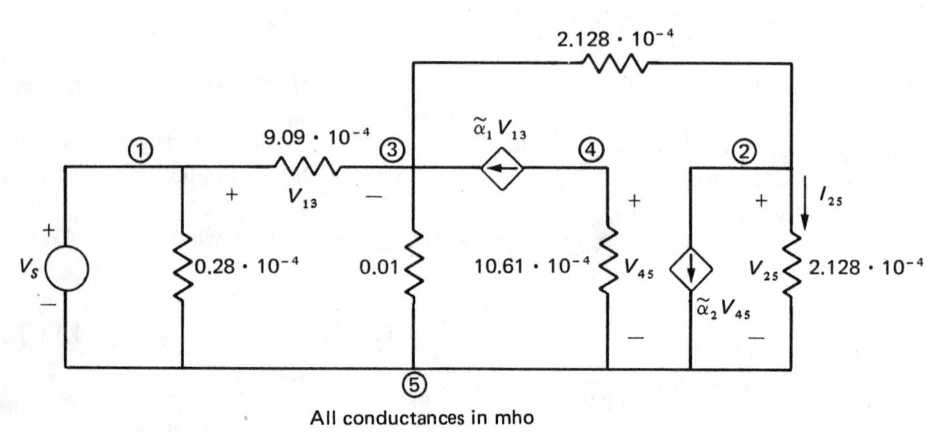

Figure 4.39 (*a*) An equivalent network of the series-parallel feedback amplifier of Fig. 4.24, and (*b*) an equivalent representation of (*a*).

To compute the return difference with respect to the transconductances $\tilde{\alpha}_k$ of the transistors, we short-circuit the voltage source V_s. The corresponding indefinite-admittance matrix is obtained by adding the first row of (4.138) to the fifth row and the first column to the fifth column and then deleting the first row and column. The first-order cofactor of the resulting matrix is simply $Y_{11,55}$. Thus, the return differences with respect to $\tilde{\alpha}_1$ and $\tilde{\alpha}_2$ are given by

$$F(\tilde{\alpha}_1) = \frac{Y_{11,55}(\tilde{\alpha}_1)}{Y_{11,55}(0)} = \frac{466.1 \cdot 10^{-9}}{4.97 \cdot 10^{-9}} = 93.70 \tag{4.141a}$$

$$F(\tilde{\alpha}_2) = \frac{Y_{11,55}(\tilde{\alpha}_2)}{Y_{11,55}(0)} = \frac{466.1 \cdot 10^{-9}}{25.52 \cdot 10^{-9}} = 18.26 \tag{4.141b}$$

For illustrative purposes, we compute the voltage gains of the first and second stages. Appealing once again to formula (2.97), we obtain

$$\frac{V_{45}}{V_{15}} = \frac{Y_{14,55}}{Y_{11,55}} = -\frac{195.7 \cdot 10^{-9}}{466.1 \cdot 10^{-9}} = -0.42 \tag{4.142a}$$

$$\frac{V_{25}}{V_{45}} = \frac{Y_{42,55}}{Y_{44,55}} = -\frac{494.53 \cdot 10^{-9}}{4.58 \cdot 10^{-9}} = -107.98 \tag{4.142b}$$

where, as in Chap. 2, V_{ij} denotes the voltage drop from terminal i to terminal j. This gives the overall amplifier voltage gain

$$w = \frac{V_{25}}{V_{15}} = \frac{V_{25}}{V_{45}} \frac{V_{45}}{V_{15}} = 45.35 \tag{4.143}$$

confirming (4.139) within computational accuracy.

4.3.2 A Useful Identity in Feedback Amplifier Theory

In this section, we establish a useful identity that is needed in interpreting physically the algebraic definition of null return difference to be presented in the following section.

Let **Y** be the indefinite-admittance matrix of the feedback amplifier of Fig. 4.33. The element x of interest enters **Y** in a rectangular pattern as shown in (4.123). Our objective is to establish the useful relation

$$Y_{uv} \dot{Y}_{rp,sq} = Y_{ra,sb} Y_{dp,cq} - Y_{da,cb} Y_{rp,sq} \tag{4.144}$$

where $\dot{Y}_{rp,sq}$ denotes the derivative of the function $Y_{rp,sq}$ with respect to x. This relation is very similar to the well-known identity [see, for example, Bocher (1938)]

$$\Delta \Delta_{ab,cd} = \Delta_{ab} \Delta_{cd} - \Delta_{ad} \Delta_{cb} \tag{4.145}$$

where $\Delta_{ab,cd}$ is the abth cofactor of Δ_{cd}, which in turn is the cdth cofactor of the determinant Δ. For our purposes, we define the *third-order cofactor* of the elements of **Y** as

$$Y_{ab,cd,nn} = \operatorname{sgn}(a-c) \operatorname{sgn}(b-d)(-1)^{a+b+c+d} \det \mathbf{Y}_{ab,cd,nn} \tag{4.146}$$

where $\mathbf{Y}_{ab,cd,nn}$ denotes the submatrix obtained from **Y**, which is of order $n \geqslant 4$, by deleting rows a, c, n and columns b, d, n for a, $c \neq n$ or b, $d \neq n$; and $Y_{ab,cd,nn} = 0$ for a, b, c, or $d = n$. Also, it is convenient to define that $Y_{11,22} = 1$ for $n = 2$ and $Y_{11,22,33} = 1$ for $n = 3$. By using this, Eq. (4.145) can be written as

$$Y_{uv} Y_{ab,cd,nn} = Y_{ab,nn} Y_{cd,nn} - Y_{ad,nn} Y_{cb,nn} \tag{4.147}$$

Also, as is known, not all the second-order cofactors are independent, and they are related by the equation [see, for example, Chen (1976a)]

$$Y_{rp,sq} = Y_{rp,uv} + Y_{sq,uv} - Y_{rq,uv} - Y_{sp,uv} \tag{4.148}$$

The other relations needed in establishing (4.144) are (see Prob. 4.18)

$$Y_{rp,sq} = -Y_{sp,rq} = -Y_{rq,sp} \tag{4.149a}$$

$$Y_{rp,sq}(x) = Y_{rp,sq}(0) + x\dot{Y}_{rp,sq} \tag{4.149b}$$

Note that $\dot{Y}_{rp,sq}$ is independent of x. With these preliminaries, we are now in a position to derive the identity (4.144).

With (4.148), the first term on the right-hand side of (4.144) can be expanded to 16 terms, as follows:

$$\begin{aligned} Y_{ra,sb}Y_{dp,cq} &= (Y_{ra,nn} + Y_{sb,nn} - Y_{rb,nn} - Y_{sa,nn})(Y_{dp,nn} + Y_{cq,nn} - Y_{dq,nn} \\ &\quad - Y_{cp,nn}) = Y_{ra,nn}Y_{dp,nn} + Y_{ra,nn}Y_{cq,nn} + Y_{sb,nn}Y_{dp,nn} \\ &\quad + Y_{sb,nn}Y_{cq,nn} + Y_{rb,nn}Y_{dq,nn} + Y_{rb,nn}Y_{cp,nn} + Y_{sa,nn}Y_{dq,nn} \\ &\quad + Y_{sa,nn}Y_{cp,nn} - Y_{ra,nn}Y_{cp,nn} - Y_{sa,nn}Y_{cq,nn} - Y_{rb,nn}Y_{dp,nn} \\ &\quad - Y_{sb,nn}Y_{dq,nn} - Y_{rb,nn}Y_{cq,nn} - Y_{sb,nn}Y_{cp,nn} - Y_{ra,nn}Y_{dq,nn} \\ &\quad - Y_{sa,nn}Y_{dp,nn} \end{aligned} \tag{4.150}$$

Likewise, the second term on the right-hand side of (4.144) can be expanded:

$$\begin{aligned} -Y_{da,cb}Y_{rp,sq} &= Y_{ca,nn}Y_{rp,nn} + Y_{ca,nn}Y_{sq,nn} + Y_{db,nn}Y_{rp,nn} + Y_{db,nn}Y_{sq,nn} \\ &\quad + Y_{cb,nn}Y_{rq,nn} + Y_{cb,nn}Y_{sp,nn} + Y_{da,nn}Y_{rq,nn} + Y_{da,nn}Y_{sp,nn} \\ &\quad - Y_{da,nn}Y_{rp,nn} - Y_{ca,nn}Y_{rq,nn} - Y_{db,nn}Y_{sp,nn} - Y_{cb,nn}Y_{sq,nn} \\ &\quad - Y_{db,nn}Y_{rq,nn} - Y_{cb,nn}Y_{rp,nn} - Y_{da,nn}Y_{sq,nn} - Y_{ca,nn}Y_{sp,nn} \end{aligned} \tag{4.151}$$

By appealing to (4.147), the first eight terms on the right-hand side of (4.150) can be combined with the corresponding last eight terms on the right-hand side of (4.151), and the first eight terms of (4.151) with the last eight terms of (4.150). The result is given by

$$\begin{aligned} Y_{ra,sb}Y_{dp,cq} - Y_{da,cb}Y_{rp,sq} = Y_{uv}(&Y_{ra,dp,nn} + Y_{ra,cq,nn} + Y_{sb,dp,nn} + Y_{sb,cq,nn} \\ &+ Y_{rb,dq,nn} + Y_{rb,cp,nn} + Y_{sa,dq,nn} + Y_{sa,cp,nn} \\ &+ Y_{ca,rp,nn} + Y_{ca,sq,nn} + Y_{db,rp,nn} + Y_{db,sq,nn} \\ &+ Y_{cb,rq,nn} + Y_{cb,sp,nn} + Y_{da,rq,nn} + Y_{da,sp,nn}) \end{aligned} \tag{4.152}$$

Referring to (4.123), we have the following expansions:

$$Y_{rp,nn}(x) = Y_{rp,nn}(0) + x(Y_{ca,rp,nn} + Y_{ra,dp,nn} + Y_{rb,cp,nn} + Y_{db,rp,nn}) \tag{4.153a}$$

$$Y_{sq,nn}(x) = Y_{sq,nn}(0) + x(Y_{ca,sq,nn} + Y_{sa,dq,nn} + Y_{sb,cq,nn} + Y_{db,sq,nn}) \tag{4.153b}$$

$$\begin{aligned} Y_{rq,nn}(x) &= Y_{rq,nn}(0) + x(Y_{ca,rq,nn} + Y_{db,rq,nn} - Y_{da,rq,nn} - Y_{cb,rq,nn}) \\ &= Y_{rq,nn}(0) - x(Y_{ra,cq,nn} + Y_{rb,dq,nn} + Y_{da,rq,nn} + Y_{cb,rq,nn}) \end{aligned} \tag{4.153c}$$

$$\begin{aligned} Y_{sp,nn}(x) &= Y_{sp,nn}(0) + x(Y_{ca,sp,nn} + Y_{db,sp,nn} - Y_{da,sp,nn} - Y_{cb,sp,nn}) \\ &= Y_{sp,nn}(0) - x(Y_{sa,cp,nn} + Y_{sb,dp,nn} + Y_{da,sp,nn} + Y_{cb,sp,nn}) \end{aligned} \tag{4.153d}$$

Appealing to (4.148) in conjunction with (4.152) and (4.153) yields

$$\begin{aligned}Y_{rp,sq}(x) &= Y_{rp,nn}(x) + Y_{sq,nn}(x) - Y_{rq,nn}(x) - Y_{sp,nn}(x)\\ &= Y_{rp,nn}(0) + Y_{sq,nn}(0) - Y_{rq,nn}(0) - Y_{sp,nn}(0)\\ &\quad + \frac{x[Y_{ra,sb}(x)Y_{dp,cq}(x) - Y_{da,cb}(x)Y_{rp,sq}(x)]}{Y_{uv}(x)}\end{aligned} \tag{4.154}$$

giving $$x(Y_{ra,sb}Y_{dp,cq} - Y_{da,cb}Y_{rp,sq}) = Y_{uv}[Y_{rp,sq}(x) - Y_{rp,sq}(0)] \tag{4.155}$$

Since from (4.149*b*) we have $[Y_{rp,sq}(x) - Y_{rp,sq}(0)] = x\dot{Y}_{rp,sq}$, Eq. (4.155) is simplified to (4.144), and we obtain the desired identity.

Example 4.12 Consider the network of Fig. 4.39*b*, whose indefinite-admittance matrix is given by Eq. (4.138). Suppose that $x = \tilde{\alpha}_2 = \alpha_2/10^4$ is the element of interest. According to the labels of Fig. 4.33, we have

$$\begin{aligned}a &= 4 \quad b = 5 \quad c = 2 \quad d = 5\\ r &= 1 \quad p = 2 \quad s = 5 \quad q = 5\end{aligned} \tag{4.156}$$

We remark that r, p, s, and q are rather arbitrary. Any two terminals can be the input terminals, and any other two the output terminals. In the present situation, we choose 1 and 5 as the input terminals and 2 and 5 as the output terminals.

With the above designations, (4.144) becomes

$$Y_{uv}\dot{Y}_{12,55} = Y_{14,55}Y_{52,25} - Y_{54,25}Y_{12,55} \tag{4.157}$$

where $\dot{Y}_{12,55}$ is the derivative of $Y_{12,55}$ with respect to $\tilde{\alpha}_2$. From (4.138), the desired quantities are calculated, as follows:

$$\begin{aligned}Y_{uv} &= Y_{55} = (271.11\alpha_2 + 48629.46) \cdot 10^{-16} = 171.983 \cdot 10^{-13}\\ Y_{12,55} &= (46{,}468.2\alpha_2 + 10{,}478.23) \cdot 10^{-12} = 211.53 \cdot 10^{-7}\\ \dot{Y}_{12,55} &= 464.682 \cdot 10^{-6}\\ Y_{14,55} &= -195.71 \cdot 10^{-9}\\ Y_{52,25} &= -Y_{22,55} = -115.32 \cdot 10^{-10}\\ Y_{54,25} &= -Y_{24,55} = -271.11 \cdot 10^{-12}\end{aligned} \tag{4.158}$$

where $\alpha_2 = 455$, giving

$$Y_{uv}\dot{Y}_{12,55} = 799.174 \cdot 10^{-17} \tag{4.159a}$$

$$Y_{14,55}Y_{52,25} - Y_{54,25}Y_{12,55} = 799.172 \cdot 10^{-17} \tag{4.159b}$$

confirming (4.157) within computational accuracy.

4.3.3 The Null Return Difference

The null return difference is found to be very useful in measurement situations and in the computation of the sensitivity for the feedback amplifiers. In this section, we introduce this concept by first giving a mathematical definition and then discussing its physical significance.

Definition 4.3: Null return difference The *null return difference* $\hat{F}(x)$ of a feedback amplifier with respect to an element x is the ratio of the two functional values assumed by the second-order cofactor $Y_{rp,sq}$ of the elements of its indefinite-admittance matrix $\mathbf{Y}$ under the condition that the element x assumes its nominal value and the condition that the element x assumes the value zero where r and s are input terminals and p and q are output terminals of the amplifier.

Thus, in computing the null return difference, we must first identify the input and output ports. Referring to the general configuration of Fig. 4.33, we have

$$\hat{F}(x) \equiv \frac{Y_{rp,sq}(x)}{Y_{rp,sq}(0)} \tag{4.160}$$

As in the case of return difference, the null return difference is simply the return difference in the network under the constraint that the input excitation I_s has been adjusted so that the output is identically zero. For this we introduce the concept of null return ratio.

Definition 4.4: Null return ratio The *null return ratio* $\hat{T}$ with respect to a voltage-controlled current source $I = xV$ is the negative of the voltage appearing at the controlling branch when the controlled current source is replaced by an independent current source of x amperes and when the input excitation is adjusted so that the output of the amplifier is identically zero.

Consider the network of Fig. 4.33. Suppose that we replace the controlled current source by an independent current source of x amperes. Then, by applying formula (2.94) and the principle of superposition, the output current I_{pq} at the load can be calculated and is found to be

$$I_{pq} = Y_2\left[I_s \frac{Y_{rp,sq}(0)}{Y_{uv}(0)} + x\frac{Y_{dp,cq}(0)}{Y_{uv}(0)}\right] \tag{4.161}$$

Setting $V_{pq} = 0$ or $I_{pq} = 0$ yields

$$I_s \equiv I_0 = -x\frac{Y_{dp,cq}(0)}{Y_{rp,sq}(0)} \tag{4.162}$$

$Y_{dp,cq}$ being independent of x. This adjustment is possible only if there is a direct transmission from the input to the output when x is set to zero. Thus, in the network of Fig. 4.34, if we connect an independent current source of this strength at its input port, the voltage V'_{ab} is the negative of the null return ratio $\hat{T}$. Using (2.94) we obtain

$$\hat{T} = -V'_{ab} = -x\frac{Y_{da,cb}(0)}{Y_{uv}(0)} - I_0\frac{Y_{ra,sb}(0)}{Y_{uv}(0)}$$

$$= -\frac{x[Y_{da,cb}(0)Y_{rp,sq}(0) - Y_{ra,sb}(0)Y_{dp,cq}(0)]}{Y_{uv}(0)Y_{rp,sq}(0)} \tag{4.163a}$$

$$= \frac{x\dot{Y}_{rp,sq}}{Y_{rp,sq}(0)} = \frac{Y_{rp,sq}(x)}{Y_{rp,sq}(0)} - 1 \tag{4.163b}$$

Equation (4.163*b*) follows directly from Eqs. (4.144) and (4.149*b*). This leads to

$$\hat{F}(x) = 1 + \hat{T} = 1 - V'_{ab} \tag{4.164}$$

Thus, like return difference, the null return difference $\hat{F}(x)$ is simply the difference of the 1-V excitation applied to the right of the breaking mark of the broken controlling branch of the controlled source and the returned voltage V'_{ab} appearing at the left of the breaking mark under the situation that the input signal I_s is adjusted so that the output is identically zero. Since, from (4.149*a*),

$$Y_{rp,sq} = -Y_{sp,rq} = -Y_{rq,sp} = Y_{sq,rp} \tag{4.165}$$

we have

$$\hat{F}(x) = \frac{Y_{rp,sq}(x)}{Y_{rp,sq}(0)} = \frac{Y_{sp,rq}(x)}{Y_{sp,rq}(0)} = \frac{Y_{rq,sp}(x)}{Y_{rq,sp}(0)} = \frac{Y_{sq,rp}(x)}{Y_{sq,rp}(0)} \tag{4.166}$$

showing that the null return difference is independent of the choice of the references for the voltages and currents at the input and output ports. In other words, in computing the null return difference $\hat{F}(x)$ it is sufficient to identify the input and the output ports.

We illustrate the above result by the following examples.

Example 4.13 In the feedback network of Fig. 4.36*b*, suppose that we wish to compute the null return difference with respect to the element αg_e. The terminals of the input port are 1 and 4 and those of the output port are 3 and 4. In terms of the labels of Fig. 4.33, we have $a = 4$, $b = 2$, $c = 2$, $d = 3$, $r = 1$, $p = 3$, $s = 4$, and $q = 4$. Thus, from (4.160) in conjunction with (4.127), we obtain

$$\hat{F}(\alpha g_e) = \frac{Y_{13,44}(\alpha g_e)}{Y_{13,44}(0)} = \frac{g_b(g_c - \alpha g_e) + G_f(g_b + g_e + g_c - \alpha g_e)}{G_f(g_b + g_e + g_c) + g_b g_c} = 1 + \hat{T} \tag{4.167}$$

where

$$\hat{T} = -\frac{\alpha g_e(g_b + G_f)}{G_f(g_b + g_e + g_c) + g_b g_c} \tag{4.168}$$

is the null return ratio.

For illustrative purposes, we compute $\hat{T}$ by (4.163*a*), yielding

$$\hat{T} = -\alpha g_e \frac{Y_{34,22}(0)Y_{13,44}(0) - Y_{14,42}(0)Y_{33,24}(0)}{Y_{22}(0)Y_{13,44}(0)}$$

$$= \frac{\alpha g_e}{Y_{22}(0)} \left\{ -G_2(G_f + g_b) - \frac{g_e(G_f + g_b)[g_b(G_f + g_c + G_2) + g_c G_f]}{G_f(g_b + g_e + g_c) + g_b g_c} \right\}$$

$$= -\frac{\alpha g_e(g_b + G_f)}{G_f(g_b + g_e + g_c) + g_b g_c} \tag{4.169}$$

confirming (4.168). Alternatively, as can be seen from Fig. 4.36*b*, if the output current I_{34} is zero, the returned voltage V_{42} is simply $-I_s/g_e$, which in conjunction with (4.162) gives

$$\hat{T} = -V_{42} = -\alpha \frac{Y_{33,24}(0)}{Y_{13,44}(0)} = -\frac{\alpha g_e(g_b + G_f)}{G_f(g_b + g_e + g_c) + g_b g_c} \tag{4.170}$$

Example 4.14 Consider the voltage-series feedback amplifier of Fig. 4.24, whose equivalent network is shown in Fig. 4.39*b*. The indefinite-admittance matrix was computed earlier in Example 4.11 and is given by Eq. (4.138). Suppose that we wish to compute the null return differences with respect to the transconductances $\tilde{\alpha}_k = \alpha_k/10^4$, as defined in (4.137). Applying definition (4.160) gives

$$\hat{F}(\tilde{\alpha}_1) = \frac{Y_{12,55}(\tilde{\alpha}_1)}{Y_{12,55}(0)} = \frac{211.54 \cdot 10^{-7}}{205.24 \cdot 10^{-12}} = 103.07 \cdot 10^3 \tag{4.171}$$

$$\hat{F}(\tilde{\alpha}_2) = \frac{Y_{12,55}(\tilde{\alpha}_2)}{Y_{12,55}(0)} = \frac{211.54 \cdot 10^{-7}}{104.79 \cdot 10^{-10}} = 2018.70 \tag{4.172}$$

where $Y_{12,55}$ is given by (4.140*a*).

Alternatively, $\hat{F}(\tilde{\alpha}_1)$ can be obtained as follows: Replace the controlled current source $\tilde{\alpha}_1 V_{13}$ in Fig. 4.39*b* by an independent current source of $\tilde{\alpha}_1$ amperes. We then adjust the voltage source V_s so that the output current I_{25} is identically zero. Let I_0 be the corresponding input current resulting from this source. The resulting network is depicted in Fig. 4.40. According to (4.164), we have in Fig. 4.40,

$$\hat{F}(\tilde{\alpha}_1) = 1 - V'_{13} = 1 - \frac{100V'_{35} + \alpha_2 V'_{45} - \alpha_1}{9.09} = 103.07 \cdot 10^3 \tag{4.173}$$

confirming (4.171), where $V'_{35} = \alpha_2 V'_{45}/2.128$ and $V'_{45} = -\alpha_1/10.61$. V'_{ij}, as before, denotes the voltage drop from terminal i to terminal j. In a similar manner we can compute $\hat{F}(\tilde{\alpha}_2)$, the details being left as an exercise (see Prob. 4.7).

4.4 THE NETWORK FUNCTIONS AND FEEDBACK

In this section, we apply the formulas derived in the foregoing directly to calculate the network functions. We show that the computation of driving-point impedance

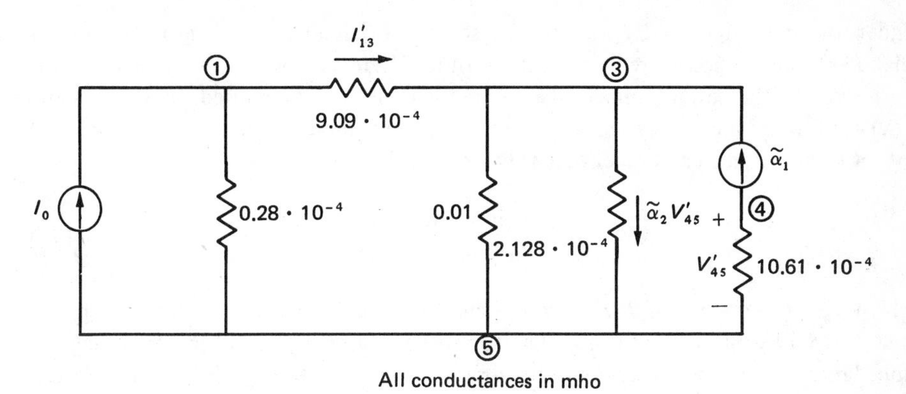

Figure 4.40 The network used to compute the null return difference $\hat{F}(\tilde{\alpha}_1)$ by its physical interpretation.

of an active network can be simplified. As a consequence, we can observe the effects of feedback on amplifier impedance and gain and obtain some useful relations among the return difference, the null return difference, and impedance functions in general.

Refer to the general feedback configuration of Fig. 4.33. Let w be a transfer function. As before, to emphasize the importance of the feedback element x, we write $w = w(x)$ even though it is also a function of the complex-frequency variable s. To be definitive, let $w(x)$ for the time being be the current gain between the input and output ports. Then from (2.94) we obtain immediately

$$w(x) = \frac{I_{pq}}{I_s} = \frac{Y_2 V_{pq}}{I_s} = \frac{Y_{rp,sq}(x)}{Y_{uv}(x)} Y_2 \tag{4.174}$$

yielding

$$\frac{w(x)}{w(0)} = \frac{Y_{rp,sq}(x)}{Y_{uv}(x)} \frac{Y_{uv}(0)}{Y_{rp,sq}(0)} = \frac{\hat{F}(x)}{F(x)} \tag{4.175}$$

provided that $w(0) \neq 0$. In fact, (4.175) remains valid if $w(x)$ represents the transfer impedance $z_{rp,sq} = V_{pq}/I_s$ (see Prob. 4.8). This gives a very useful formula for computing the current gain I_{pq}/I_s or the transfer impedance $z_{rp,sq} = V_{pq}/I_s$:

$$w(x) = w(0) \frac{\hat{F}(x)}{F(x)} \tag{4.176}$$

In the particular situation where $r = p$ and $s = q$, $w(x)$ represents the driving-point impedance $z_{rr,ss}(x)$ looking into the terminals r and s, and we have a somewhat different physical interpretation. In this case, $F(x)$ is the return difference with respect to the element x under the condition $I_s = 0$. Thus, $F(x)$ can be considered to be the return difference for the situation when the port where the input impedance is defined is left open without a source and we write $F(x) = F(\text{input open-circuited})$. Likewise, $\hat{F}(x)$ is the return difference with respect to the element x for the input excitation I_s and output response V_{rs} (Fig. 4.33) under the condition that I_s is adjusted so that V_{rs} is identically zero. This means that $\hat{F}(x)$ can be considered to be the return difference for the situation when the port where the input impedance is defined is short-circuited and we write $\hat{F}(x) = F(\text{input short-circuited})$. Consequently, the input impedance $Z(x)$ looking into a terminal pair can be conveniently expressed as

$$Z(x) = Z(0)\,\frac{F(\text{input short-circuited})}{F(\text{input open-circuited})} \tag{4.177}$$

which is the well-known *Blackman's formula* for an active impedance, which was first derived by Blackman (1943). The formula is extremely useful in computing the impedance of an active network, because the right-hand side of Eq. (4.177) can usually be determined rather easily. If x represents the controlling parameter of a controlled source in a single-loop feedback amplifier, then setting $x = 0$ opens the feedback loop and $Z(0)$ is simply a passive impedance. The return difference for x when the input port is short-circuited or open-circuited is relatively simple to compute because shorting out or opening a terminal pair frequently breaks the feedback loop. On the other hand, Blackman's formula may be used to determine the return difference by measurements. However, because it involves two return differences, only one of them can be identified; the other must be known in advance. In the case of single-loop feedback amplifiers, it is usually possible to choose a terminal pair so that either the numerator or the denominator on the right-hand side of (4.177) is unity. If $F(\text{input short-circuited}) = 1$ and $F(\text{input open-circuited}) = F(x)$, $F(x)$ being the return difference under normal operating conditions, then we have

$$F(x) = \frac{Z(0)}{Z(x)} \tag{4.178}$$

On the other hand, if $F(\text{input open-circuited}) = 1$ and $F(\text{input short-circuited}) = F(x)$, then

$$F(x) = \frac{Z(x)}{Z(0)} \tag{4.179}$$

We illustrate the above results by the following examples.

Example 4.15 Consider the network shown in Fig. 4.41, which is the equivalent network of a simplified transistor feedback amplifier. The indefinite-admittance matrix of the network is given by

$$\mathbf{Y} = \begin{bmatrix} G_1 + g_3 + sC_1 + sC_2 & -sC_2 & -G_1 - g_3 - sC_1 \\ \alpha - sC_2 & G_2 + sC_2 & -\alpha - G_2 \\ -\alpha - G_1 - g_3 - sC_1 & -G_2 & G_1 + G_2 + g_3 + sC_1 + \alpha \end{bmatrix} \tag{4.180}$$

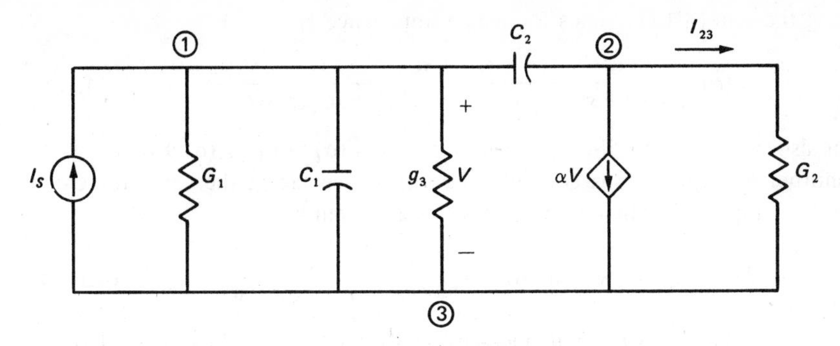

Figure 4.41 An equivalent network of a simplified transistor feedback amplifier.

Suppose that the controlling parameter α is the element of interest, I_s forms the input port, and G_2 is the output load. By using (2.94), the current gain I_{23}/I_s when α is set to zero is found to be

$$w(0) = \left.\frac{I_{23}}{I_s}\right|_{\alpha=0} = \frac{G_2 Y_{12,33}(0)}{Y_{11}(0)} = \frac{sC_2 G_2}{(G_2 + sC_2)(G_1 + g_3 + sC_1) + sC_2 G_2} \tag{4.181}$$

The return difference with respect to α is obtained as

$$F(\alpha) = \frac{Y_{11}(\alpha)}{Y_{11}(0)} = \frac{(G_2 + sC_2)(G_1 + g_3 + sC_1) + sC_2(\alpha + G_2)}{(G_2 + sC_2)(G_1 + g_3 + sC_1) + sC_2 G_2} \tag{4.182}$$

and the null return difference $\hat{F}(\alpha)$ is computed by

$$\hat{F}(\alpha) = \frac{Y_{12,33}(\alpha)}{Y_{12,33}(0)} = \frac{sC_2 - \alpha}{sC_2} = 1 - \frac{\alpha}{sC_2} \tag{4.183}$$

Substituting (4.181)–(4.183) in (4.176) yields the current gain

$$w(\alpha) = \frac{I_{23}}{I_s} = \frac{G_2(sC_2 - \alpha)}{(G_2 + sC_2)(G_1 + g_3 + sC_1) + sC_2(\alpha + G_2)} \tag{4.184}$$

which can also be computed directly by the formula $w(x) = G_2 Y_{12,33}(\alpha)/Y_{11}(\alpha)$, giving, of course, the same result.

Assume that we wish to compute the impedance $Z(\alpha)$ facing the input current I_s by Blackman's formula. F(input open-circuited) is the return difference with respect to α when the input excitation I_s is removed, yielding the return difference as shown in (4.182). F(input short-circuited) is the return difference with respect to α in the network obtained from that of Fig. 4.41 by shorting the terminals 1 and 3, resulting in F(input short-circuited) = 1. Finally, the input impedance $Z(0)$ when α is set to zero is given by

$$Z(0) = \frac{Y_{11,33}(0)}{Y_{11}(0)} = \frac{G_2 + sC_2}{(G_2 + sC_2)(G_1 + g_3 + sC_1) + sC_2 G_2} \tag{4.185}$$

Substituting these in (4.177) yields the active impedance

$$Z(\alpha) = \frac{G_2 + sC_2}{(G_2 + sC_2)(G_1 + g_3 + sC_1) + sC_2(\alpha + G_2)} \tag{4.186}$$

which can also be computed directly by the formula $Z(\alpha) = Y_{11,33}(\alpha)/Y_{11}(\alpha)$.

As another example, let us compute the impedance facing the capacitor C_2 in the network of Fig. 4.41. The desired quantities are given by

$$F(\text{input short-circuited}) = 1 + \frac{\alpha}{G_1 + G_2 + g_3 + sC_1} \tag{4.187a}$$

$$F(\text{input open-circuited}) = 1 \tag{4.187b}$$

$$Z(0) = \frac{1}{G_2} + \frac{1}{G_1 + g_3 + sC_1} \tag{4.188}$$

in which the input port is formed by removing the capacitor C_2 and the element of interest is α. The details are left as an exercise (see Prob. 4.10). Thus, using Blackman's formula, the active impedance facing the capacitor C_2 is given by

$$Z(\alpha) = Z(0)\frac{F(\text{input short-circuited})}{F(\text{input open-circuited})} = \frac{G_1 + G_2 + g_3 + sC_1 + \alpha}{G_2(G_1 + g_3 + sC_1)} \tag{4.189}$$

Example 4.16 The network of Fig. 4.42 is the equivalent network of a feedback amplifier. We use Blackman's formula to compute the driving-point impedance facing the admittance y_3.

First, we write down the indefinite-admittance matrix by the rules outlined in Sec. 2.2:

$$\mathbf{Y} = \begin{bmatrix} y_1 + y_2 & \alpha_2 - y_2 & -y_1 - \alpha_2 \\ \alpha_1 - y_2 & y_2 + y_3 & -\alpha_1 - y_3 \\ -\alpha_1 - y_1 & -y_3 - \alpha_2 & y_1 + y_3 + \alpha_1 + \alpha_2 \end{bmatrix} \tag{4.190}$$

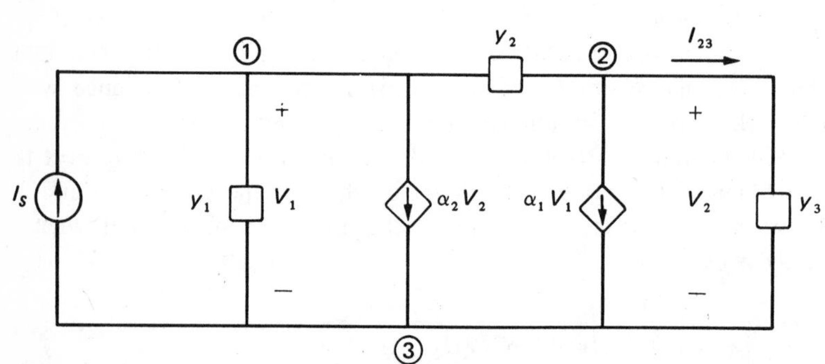

Figure 4.42 An equivalent network of a feedback amplifier.

Let α_1 be the element of interest. Then F(input open-circuited) is the return difference with respect to α_1 with the admittance y_3 removed:

$$F(\text{input open-circuited}) = F(\alpha_1)|_{y_3=0} = \left.\frac{Y_{33}(\alpha_1)}{Y_{33}(0)}\right|_{y_3=0}$$

$$= \frac{y_2(y_1 + \alpha_1 + \alpha_2) - \alpha_1\alpha_2}{y_2(y_1 + \alpha_2)} = 1 + \frac{\alpha_1(y_2 - \alpha_2)}{y_2(y_1 + \alpha_2)} \tag{4.191}$$

the last term on the right-hand side being the return ratio when $y_3 = 0$ (see Prob. 4.11). F(input short-circuited) is the return difference with respect to α_1 when the terminals of y_3 are short-circuited. Since shorting y_3 breaks the feedback loop, no voltage is returned and we have F(input short-circuited) = 1. Finally, the impedance facing y_3 when α_1 is set to zero is obtained as

$$Z(0) = \left.\frac{Y_{22,33}(0)}{Y_{33}(0)}\right|_{y_3=0} = \frac{y_1 + y_2}{y_2(y_1 + \alpha_2)} \tag{4.192}$$

Substituting these in (4.177) yields the impedance facing y_3 as

$$Z(\alpha_1) = \frac{y_1 + y_2}{y_2(y_1 + \alpha_1 + \alpha_2) - \alpha_1\alpha_2} \tag{4.193}$$

which can also be obtained directly by formula (2.95) (see Prob. 4.14).

Before we turn our attention to other transfer functions, we apply Blackman's formula to one of the four common types of feedback amplifier configurations of Fig. 4.3, as discussed in Sec. 4.2.

To determine the effect of feedback on the input and output impedances, we choose the series-parallel feedback configuration of Fig. 4.3*c*. Using Blackman's formula, we shall determine the impedance looking into the two terminals of the load admittance Y_2. Since by shorting the terminals of Y_2 we interrupt the feedback loop, formula (4.178) applies and the output impedance across the load admittance Y_2 becomes

$$Z_{\text{out}}(x) = \frac{Z_{\text{out}}(0)}{F(x)} \tag{4.194}$$

meaning that the impedance measured across the path of the feedback is reduced by the factor that is the normal value of the return difference with respect to the element x, where x is an arbitrary element of interest. For the input impedance of the amplifier looking into the voltage source V_s of Fig. 4.3*c*, we recognize that by open-circuiting or removing the voltage source V_s, we break the feedback loop, and formula (4.179) applies. Therefore, the input impedance becomes

$$Z_{\text{in}}(x) = F(x)Z_{\text{in}}(0) \tag{4.195}$$

showing that impedance measured in series lines is increased by the same factor $F(x)$. The analyses of the other three configurations are left as exercises (see Probs. 4.15-4.17).

Referring to the network of Fig. 4.33, assume that $w(x)$ represents the voltage gain V_{pq}/V_{rs} or transfer admittance I_{pq}/V_{rs}. From (2.97) we can write

$$\frac{w(x)}{w(0)} = \frac{Y_{rp,sq}(x)}{Y_{rp,sq}(0)} \frac{Y_{rr,ss}(0)}{Y_{rr,ss}(x)} \tag{4.196}$$

The first term in the product on the right-hand side is the null return difference $\hat{F}(x)$ with respect to x for the input terminals r and s and output terminals p and q. The second term is the reciprocal of the null return difference with respect to x for the same input and output port at terminals r and s. As in (4.177), this reciprocal can then be interpreted as the return difference with respect to x when the input port of the amplifier is short-circuited. Thus, the voltage gain or transfer admittance can be expressed as

$$w(x) = w(0) \frac{\hat{F}(x)}{F(\text{input short-circuited})} \tag{4.197}$$

Likewise, if $w(x)$ denotes the short-circuit current gain I_{pq}/I_s as Y_2 approaches infinity, from (2.114) we have

$$\frac{w(x)}{w(0)} = \frac{Y_{rp,sq}(x)}{Y_{rp,sq}(0)} \frac{Y_{pp,qq}(0)}{Y_{pp,qq}(x)} \tag{4.198}$$

As before, the second term in the product is the reciprocal of the return difference with respect to x when the output port of the amplifier is short-circuited, giving a formula for the short-circuit current gain

$$w(x) = w(0) \frac{\hat{F}(x)}{F(\text{output short-circuited})} \tag{4.199}$$

Example 4.17 For the network of Fig. 4.42, we compute the voltage gain V_2/V_1 and the short-circuit current gain I_{23}/I_s as $y_3 \to \infty$. For a change, we choose α_2 to be the element of interest. The null return difference $\hat{F}(\alpha_2)$ can easily be determined from the network of Fig. 4.42. First, we replace the controlled source $\alpha_2 V_2$ by an independent current source of α_2 amperes and then adjust I_s so that V_2 is identically zero, giving† $I_s = \alpha_2$ for $\alpha_1 \neq y_2$. Under this situation, the returned voltage V_2 is clearly zero and $\hat{F}(\alpha_2) = 1$. Alternatively, from the definition we have $\hat{F}(\alpha_2) = Y_{12,33}(\alpha_2)/Y_{12,33}(0) = 1$.

To compute F(input short-circuited), we short-circuit I_s and compute the return difference with respect to α_2 in the resulting network. Since short-circuiting I_s breaks the feedback loop, we have F(input short-circuited) $= 1$. Substituting the above in (4.197), we obtain the voltage gain

$$w(\alpha_2) = \frac{V_2}{V_1} = w(0) \frac{\hat{F}(\alpha_2)}{F(\text{input short-circuited})} = \left.\frac{V_2}{V_1}\right|_{\alpha_2=0} = \frac{y_2 - \alpha_1}{y_2 + y_3} \tag{4.200}$$

which can also be derived directly by formula (2.97) in conjunction with (4.190).

†For $\alpha_1 = y_2$, V_2 is identically zero for any value of I_s.

Finally, to determine F(output short-circuited), we short-circuit y_3 and compute the return difference with respect to α_2 in the resulting network. Again, the return difference is unity. Thus, from (4.199) the short-circuit current gain is obtained as

$$w(\alpha_2) = \left.\frac{I_{23}}{I_s}\right|_{y_3\to\infty} = w(0)\,\frac{\hat{F}(\alpha_2)}{F(\text{output short-circuited})}$$

$$= \left.\frac{I_{23}}{I_s}\right|_{\substack{y_3\to\infty\\ \alpha_2=0}} = \frac{y_2-\alpha_1}{y_1+y_2} \tag{4.201}$$

This gain can, of course, be computed directly by means of formula (2.114), giving

$$w(\alpha_2) = \frac{Y_{12,33}(\alpha_2)}{Y_{22,33}(\alpha_2)} = \frac{y_2-\alpha_1}{y_1+y_2} \tag{4.202}$$

Observe that in (4.200) and (4.201) we have $w(\alpha_2) = w(0)$, meaning that they are independent of the controlled current source $\alpha_2 V_2$.

Example 4.18 Consider again the voltage-series feedback amplifier of Fig. 4.24. In Examples 4.11 and 4.14, we obtained the following:

$$F(\tilde{\alpha}_1) = 93.70 \qquad F(\tilde{\alpha}_2) = 18.26 \tag{4.203a}$$

$$\hat{F}(\tilde{\alpha}_1) = 103.07\cdot 10^3 \qquad \hat{F}(\tilde{\alpha}_2) = 2018.70 \tag{4.203b}$$

$$w = \frac{V_2}{V_s} = w(\tilde{\alpha}_1) = w(\tilde{\alpha}_2) = 45.39 \tag{4.203c}$$

Appealing to (4.197) yields

$$w(\tilde{\alpha}_1) = w(0)\,\frac{\hat{F}(\tilde{\alpha}_1)}{F(\text{input short-circuited})} = 0.04126\,\frac{103.07\cdot 10^3}{93.699} = 45.39 \tag{4.204a}$$

where, from (4.138),

$$F(\text{input short-circuited}) = \frac{Y_{11,55}(\tilde{\alpha}_1)}{Y_{11,55}(0)} = \frac{466.07\cdot 10^{-9}}{4.9741\cdot 10^{-9}} = 93.699 \tag{4.204b}$$

$$w(0) = \left.\frac{Y_{12,55}(\tilde{\alpha}_1)}{Y_{11,55}(\tilde{\alpha}_1)}\right|_{\tilde{\alpha}_1=0} = \frac{205.24\cdot 10^{-12}}{497.41\cdot 10^{-11}} = 0.04126 \tag{4.204c}$$

and

$$w(\tilde{\alpha}_2) = w(0)\,\frac{\hat{F}(\tilde{\alpha}_2)}{F(\text{input short-circuited})} = 0.41058\,\frac{2018.70}{18.26} = 45.39 \tag{4.205a}$$

where

$$F(\text{input short-circuited}) = \frac{Y_{11,55}(\tilde{\alpha}_2)}{Y_{11,55}(0)} = \frac{466.07\cdot 10^{-9}}{25.52\cdot 10^{-9}} = 18.26 \tag{4.205b}$$

$$w(0) = \left.\frac{Y_{12,55}(\tilde{\alpha}_2)}{Y_{11,55}(\tilde{\alpha}_2)}\right|_{\tilde{\alpha}_2=0} = \frac{104.79\cdot 10^{-10}}{255.22\cdot 10^{-10}} = 0.41058 \tag{4.205c}$$

4.5 SUMMARY

We began this chapter by introducing the ideal feedback model. To employ the ideal feedback model to analyze a practical feedback amplifier, it becomes necessary to separate the feedback amplifier into two blocks: the basic unilateral amplifier and the feedback network. For this we considered four common types of feedback amplifier configurations. They are referred to as the series-series feedback, parallel-parallel feedback, series-parallel feedback, and parallel-series feedback, and they can be reduced to the ideal single-loop feedback model by employing z-parameters, y-parameters, h-parameters, and g-parameters, respectively. We demonstrated in each case by a practical example how to identify the transfer functions of the basic unilateral amplifier and the feedback network. We found that the gain, input, and output immittances are all affected by the same factor called the feedback factor. More specifically, for negative feedback the gain is reduced by the amount of feedback measured either in nepers or in decibels. The input and output immittances, including the terminating immittances, are increased by the feedback factor from their values when the input and output ports are either open-circuited or short-circuited. Since open-circuiting or short-circuiting the ports clearly breaks the feedback loop, these values can be considered as the input and output immittances of the amplifier without feedback.

In practice, we often run into feedback amplifiers that cannot be classified as being in one of the above four categories. A general feedback configuration may contain an input coupling network and an output coupling network in addition to the basic amplifier and the feedback network. Thus, the ideal feedback model is not an adequate representation of a practical feedback amplifier. What is needed is a more general theory that avoids the necessity of identifying the forward and backward transmission paths. To this end, we presented Bode's theory on feedback amplifiers.

Bode's feedback theory is based on the concepts of return difference and null return difference. We showed that the return difference is a generalization of the concept of the feedback factor of the ideal feedback model and can be measured physically from the amplifier itself. Also, we introduced the notion of null return difference and discussed its physical significance. We demonstrated that the ratio of the two functional values assumed by a transfer function under the condition that the element x of interest assumes its nominal value and the condition that x assumes the value zero is equal to the ratio of the return differences with respect to the element x under the condition that the input excitation I_s has been adjusted so that the output is identically zero and the condition that the input excitation I_s is removed. This result degenerates into the well-known Blackman's formula for an active impedance when the transfer impedance becomes the driving-point impedance. This relation can be employed to simplify the calculation of driving-point impedance of an active network, making it possible to observe the effects of feedback on amplifier impedance and gain.

The effect of feedback on sensitivity, stability, and other related topics will be discussed in the following chapters.

PROBLEMS

4.1 In the feedback amplifier of Fig. 4.43, assume that the two transistors are identical, with $h_{re} = h_{oe} = 0$. Compute the return differences and null return differences with respect to h_{fe}/h_{ie} of the two transistors.

4.2 Show that A_k and s_k in Eq. (4.67) are given by Eq. (4.68).

4.3 Consider the feedback amplifier of Fig. 4.21, which is of the series-parallel or voltage-series type. Assume that the two transistors are identical, with $h_{re} = h_{oe} = 0$. Confirm that the hybrid matrix of the composite two-port network is given by (4.70).

4.4 Use the equivalent network of Fig. 4.23 to compute the voltage gains of the first and second stages. Compare the results with (4.75).

4.5 The feedback amplifier of Fig. 4.27 conforms with the parallel-series or current-shunt feedback configuration. Assume that $h_{re} = 0$ and conditions (4.96) are satisfied. Confirm that the inverse hybrid matrix of the composite two-port network is given by (4.97).

4.6 Use Blackman's formula to compute the impedance facing the conductance $G_2 = 1/R_2$ in the network of Fig. 4.43.

4.7 Compute the null return difference $\hat{F}(\tilde{\alpha}_2)$ of (4.172) by its physical interpretation.

4.8 Demonstrate that (4.176) remains valid if $w(x)$ represents the transfer impedance $z_{rp,sq} = V_{pq}/I_s$ of the generai feedback configuration of Fig. 4.33. Is it still true if $w(x)$ represents the transfer admittance I_{pq}/V_{rs}? If not, how should one modify the formula?

4.9 In the feedback network of Fig. 4.36, compute the return difference and the null return difference with respect to the one-port conductance G_f. Repeat the problem by appealing to their physical significance in terms of the returned voltages.

4.10 Derive (4.187) and (4.188).

4.11 Use physical interpretation of the return difference to compute (4.191).

4.12 Use Blackman's formula to show that the impedance facing the independent voltage source V_s in the network of Fig. 4.44 is given by

$$Z(s) = \frac{y_2 y_f + (y_1 + y_g)(y_2 + y_f) + y_f g_m}{y_1 (y_2 y_f + y_g y_f + y_2 y_g + y_f g_m)} \tag{4.206}$$

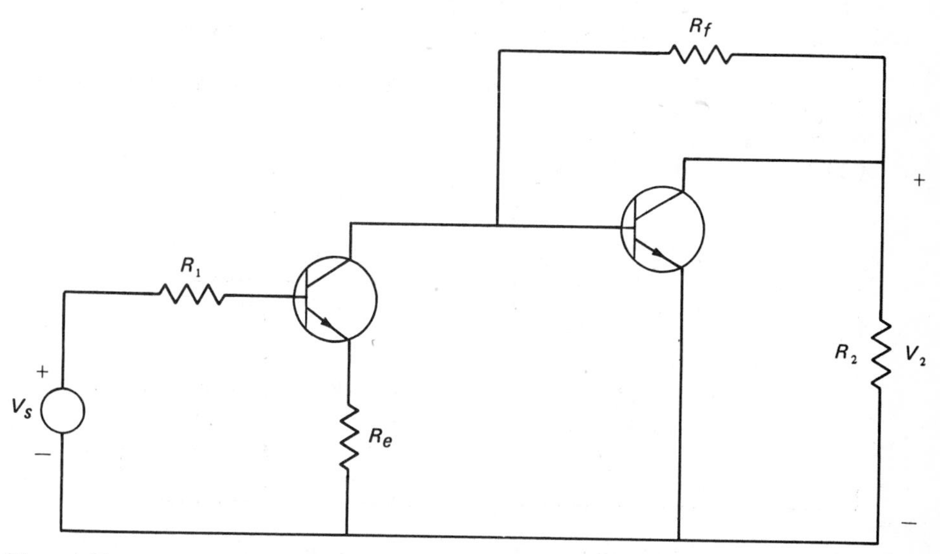

Figure 4.43 A given feedback amplifier.

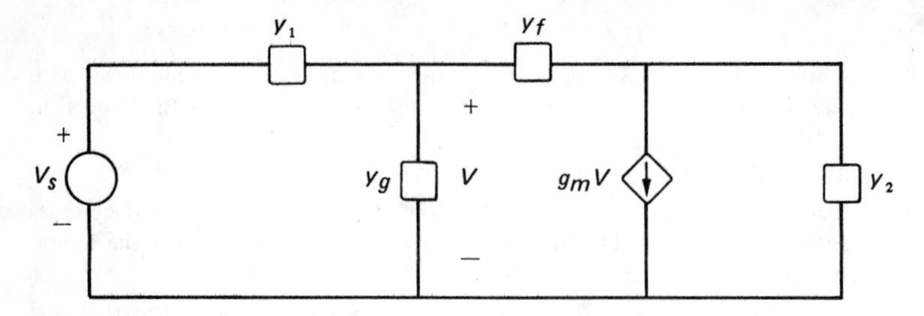

Figure 4.44 A given feedback network.

4.13 By applying Blackman's formula, confirm that the admittance facing y_2 in the feedback network of Fig. 4.44 is given by

$$Y(s) = \frac{y_f(y_1 + y_g + g_m)}{y_1 + y_f + y_g} \tag{4.207}$$

4.14 Apply formula (2.95) to confirm (4.193).

4.15 Apply Blackman's formula and determine the effect of feedback on the input and output impedances of the series-series feedback configuration of Fig. 4.3*a*.

4.16 Repeat Prob. 4.15 for the parallel-parallel feedback configuration of Fig. 4.3*b*.

4.17 Repeat Prob. 4.15 for the parallel-series feedback configuration of Fig. 4.3*d*.

4.18 Prove the identities (4.149).

Hint: Use (4.123).

4.19 For the feedback network of Fig. 4.45, compute the following:

(*a*) The return difference with respect to the element α_2
(*b*) The impedance facing the capacitor C_2
(*c*) The return difference with respect to the one-port admittance sC_2
(*d*) The null return difference with respect to the element α_1
(*e*) The return difference with respect to the element α_1

4.20 In the compound-feedback amplifier of Fig. 4.46, compute the following:

(*a*) The return difference with respect to the conductance $G_1 = 1/R_1$
(*b*) The null return difference with respect to the controlling parameter of the forward transmission of the transistor
(*c*) The voltage gain of the amplifier
(*d*) The impedance facing the resistor R_f

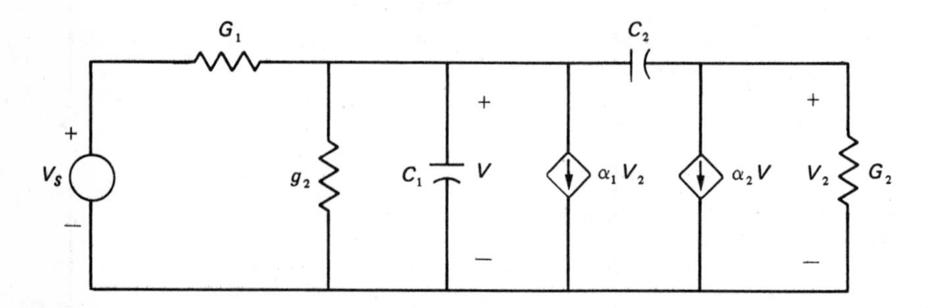

Figure 4.45 A given feedback network with two controlled sources.

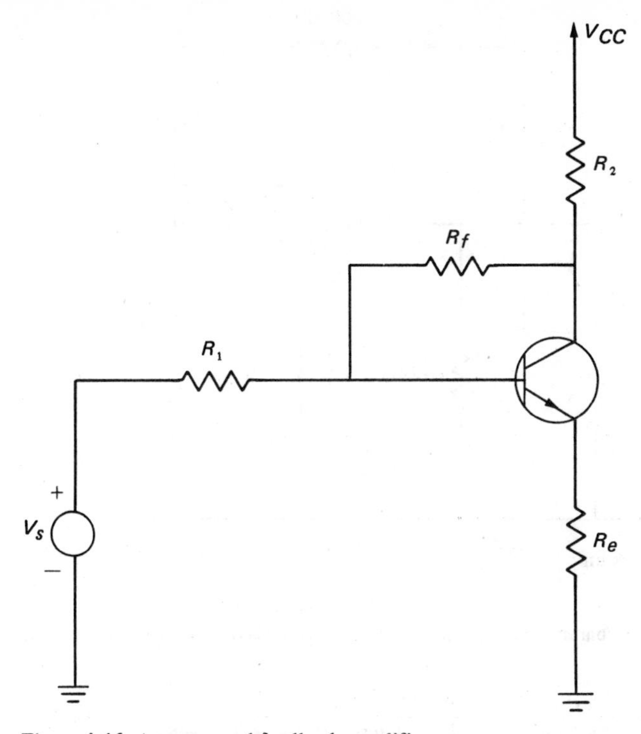

Figure 4.46 A compound-feedback amplifier.

4.21 In the two-stage feedback amplifier of Fig. 4.47, assume that the two transistors are identical, with $h_{ie} = 1$ kΩ, $h_{fe} = 45$, and $h_{re} = h_{oe} = 0$. Determine the following:

(*a*) The return difference with respect to the controlling parameter h_{fe}/h_{ie} of the second transistor

(*b*) The null return difference with respect to the controlling parameter h_{fe}/h_{ie} of the second transistor

(*c*) The input impedance of the feedback amplifier

(*d*) The output impedance of the feedback amplifier

4.22 Repeat (*a*) and (*b*) of Prob. 4.21 for the controlling parameter h_{fe}/h_{ie} of the first transistor. Also compute the current gain of the feedback amplifier.

4.23 To distinguish the transistor parameters in the voltage-series feedback amplifier of Fig. 4.21, we use subscripts 1 and 2. Assume that $R_1 = 0$, $h_{re1} = h_{re2} = 0$, and $h_{oe1} = h_{oe2} = 0$. Determine the h-parameters of the composite two-port network N. Use these parameters to show that the closed-loop voltage gain is given by

$$w_V = \frac{h_{fe1}h_{fe2}RR_2(R_e + R_f) + R_eR_2(1 + h_{fe1})(R + h_{ie2})}{(R + h_{ie2})\{h_{ie1}R_e + (R_f + R_2)[h_{ie1} + (1 + h_{fe1})R_e]\} + h_{fe1}h_{fe2}RR_eR_2} \tag{4.208}$$

For large h_{fe1} and h_{fe2}, demonstrate that the voltage gain w_V approaches the limiting value $1 + R_f/R_e$, thereby making it practically independent of the transistor parameters.

4.24 Use the subscripts 1 and 2 to distinguish the transistor parameters of Fig. 4.27. Let $h_{oe1} = h_{oe2} = 0$, $h_{re1} = h_{re2} = 0$, and $R_1 = \infty$. Determine the g-parameters of the composite

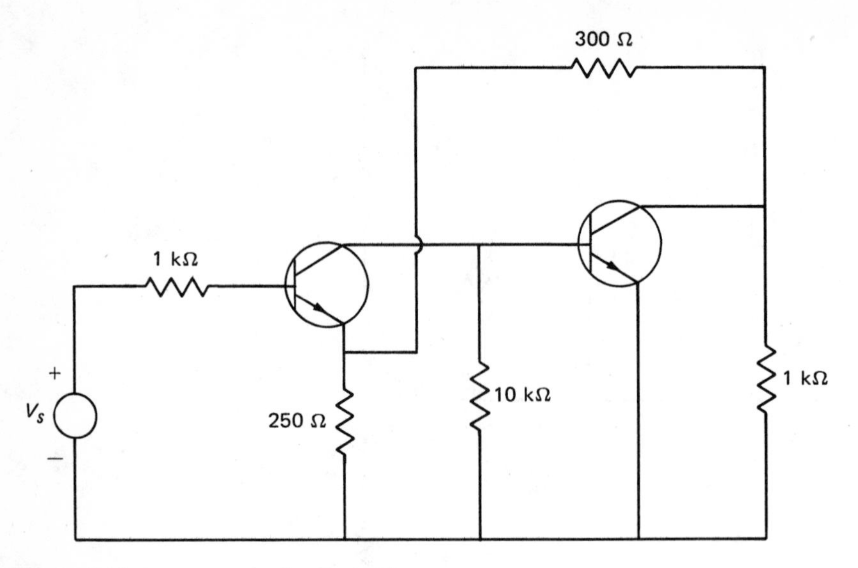

Figure 4.47 A two-stage feedback amplifier.

two-port network N. Use these parameters to show that the closed-loop current gain I_2/I_s is given by

$$w_I = \frac{-h_{fe_1}h_{fe_2}R(R_e + R_f) - h_{fe_2}h_{ie_1}R_e}{(R + h_{ie_2})(R_e + R_f + h_{ie_1}) + R_e(1 + h_{fe_2})(Rh_{fe_1} + h_{ie_1} + R_f)} \tag{4.209}$$

For large h_{fe_1} and h_{fe_2}, demonstrate that the current gain w_I approaches the limiting value $-(1 + R_f/R_e)$, thereby making it practically independent of the transistor parameters.

4.25 The emitter-follower as shown in Fig. 4.48 is an example of voltage-series feedback amplifier. Compute the following:

(*a*) The return difference with respect to the element $\alpha = h_{fe}/h_{ie}$ both by its definition and by its physical interpretation

(*b*) The impedance facing the resistor R_2

(*c*) The null return difference with respect to the element α both by its definition and by its physical interpretation

(*d*) The input and output impedances of the amplifier

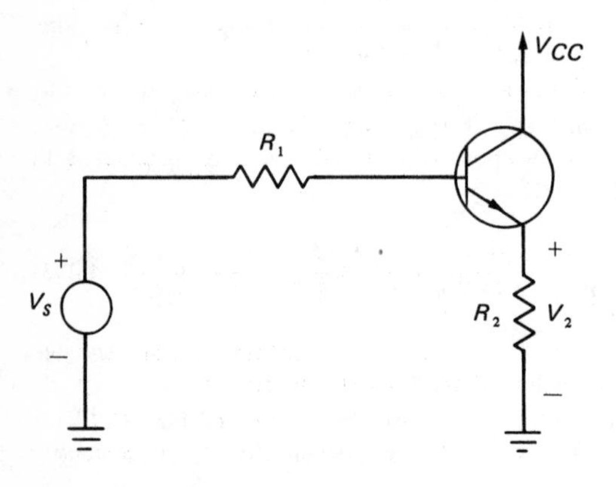

Figure 4.48 A voltage-series feedback amplifier.

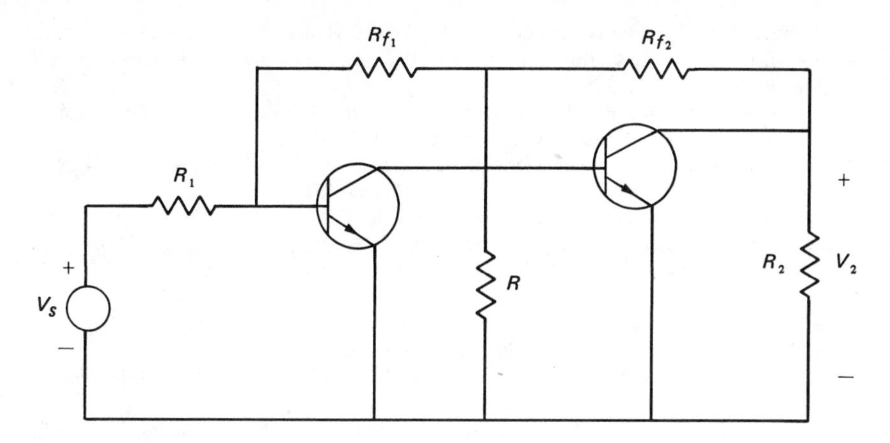

Figure 4.49 A transistor feedback amplifier.

4.26 In the series-parallel feedback amplifier of Fig. 4.21, assume that the two transistors are identical, with

$$\begin{aligned} h_{fe} &= 60 \qquad h_{ie} = 1.8 \text{ k}\Omega \\ h_{re} &= h_{oe} = 0 \qquad R_1 = 0 \\ R &= 8 \text{ k}\Omega \qquad R_f = 2 \text{ k}\Omega \\ R_e &= 200 \ \Omega \qquad R_2 = 6.8 \text{ k}\Omega \end{aligned} \tag{4.210}$$

Determine the voltage gains of the amplifier with and without feedback. Also compute the input and output immittances of the amplifier.

4.27 Consider the feedback amplifier of Fig. 4.49. To simplify the computation, assume that the transistors are identical, with $h_{re} = h_{oe} = 0$. Determine the open-loop and closed-loop voltage gains and the transfer function of the feedback network.

4.28 At high frequencies, we use the hybrid-pi equivalent model of Fig. 4.10 for the two transistors of Fig. 4.49. Determine the open-loop and closed-loop voltage gains of the amplifier.

4.29 In Fig. 4.15, replace the feedback resistor R_f by a capacitor of capacitance C_f. Show that the voltage gain V_2/V_s of the amplifier with $G_1 = 1/R_1$ and $G_2 = 1/R_2$ is given by

$$w = \frac{-h_{fe} + sC_f h_{ie}}{(G_2 + h_{oe})(R_1 + h_{ie}) - h_{re}h_{fe} + sC_f R_1 [1 + h_{ie}(h_{oe} + G_2 + G_1) + h_{fe}]} \tag{4.211}$$

BIBLIOGRAPHY

Black, H. S.: Inventing the Negative Feedback Amplifier, *IEEE Spectrum*, vol. 14, no. 12, pp. 55–60, 1977.

Blackman, R. B.: Effect of Feedback on Impedance, *Bell Syst. Tech. J.*, vol. 22, no. 4, pp. 268–277, 1943.

Bocher, M.: "Introduction to Higher Algebra," New York: Macmillan, 1938.

Bode, H. W.: "Network Analysis and Feedback Amplifier Design," Princeton, N.J.: Van Nostrand, 1945.

Chen, W. K.: Graph-Theoretic Considerations on the Invariance of Return Difference, *J. Franklin Inst.*, vol. 298, no. 2, pp. 81–100, 1974a.

Chen, W. K.: Invariance and Mutual Relations of the General Null-Return-Difference Functions, *Proc. 1974 Eur. Conf. Circuit Theory and Design*, IEE Conf. Publ. No. 116, Inst. of Electrical Engineers, London, pp. 371–376, 1974b.

Chen, W. K.: Graph Theory and Feedback Systems, *Proc. Ninth Asilomar Conf. on Circuits, Systems, and Computers*, Pacific Grove, Calif., pp. 26–30, 1975.

Chen, W. K.: "Applied Graph Theory: Graphs and Electrical Networks," 2d rev. ed., chap. 4, New York: American Elsevier, and Amsterdam: North-Holland, 1976a.

Chen, W. K.: Indefinite-Admittance Matrix Formulation of Feedback Amplifier Theory, *IEEE Trans. Circuits and Systems*, vol. CAS-23, no. 8, pp. 498–505, 1976b.

Chen, W. K.: Network Functions and Feedback, *Int. J. Electronics*, vol. 42, no. 6, pp. 617–618, 1977.

Chen, W. K.: On Second-Order Cofactors and Null Return Difference in Feedback Amplifier Theory, *Int. J. Circuit Theory and Applications*, vol. 6, no. 3, pp. 305–312, 1978.

Cheng, C. C.: Neutralization and Unilateralization, *IRE Trans. Circuit Theory*, vol. CT-2, no. 2, pp. 138–145, 1955.

Chu, G. Y.: Unilateralization of Junction-Transistor Amplifiers at High Frequencies, *Proc. IRE*, vol. 43, no. 8, pp. 1001–1006, 1955.

Ghausi, M. S.: Optimum Design of the Shunt-Series Feedback Pair with a Maximally Flat Magnitude Response, *IRE Trans. Circuit Theory*, vol. CT-8, no. 4, pp. 448–453, 1961.

Ghausi, M. S.: "Principles and Design of Linear Active Circuits," New York: McGraw-Hill, 1965.

Ghausi, M. S.: "Electronic Circuits," New York: Van Nostrand, 1971.

Hakim, S. S.: Aspects of Return-Difference Evaluation in Transistor Feedback Amplifiers, *Proc. IEE (London)*, vol. 112, no. 9, pp. 1700–1704, 1965.

Haykin, S. S.: "Active Network Theory," Reading, Mass.: Addison-Wesley, 1970.

Hoskins, R. F.: Definition of Loop Gain and Return Difference in Transistor Feedback Amplifiers, *Proc. IEE (London)*, vol. 112, no. 11, pp. 1995–2001, 1965.

Kuh, E. S. and R. A. Rohrer: "Theory of Linear Active Networks," San Francisco, Calif.: Holden-Day, 1967.

Millman, J. and C. C. Halkias: "Electronic Fundamentals and Applications: For Engineers and Scientists," New York: McGraw-Hill, 1976.

Mulligan, J. H., Jr.: Signal Transmission in Nonreciprocal Systems, *Proc. Symp. on Active Networks and Feedback Systems*, Polytechnic Inst. of Brooklyn, New York, vol. 10, pp. 125–153, 1960.

Truxal, J. G.: "Automatic Feedback Control System Synthesis," New York: McGraw-Hill, 1955.

CHAPTER

FIVE

THEORY OF FEEDBACK AMPLIFIERS II

In the preceding chapter, we studied the ideal feedback model and demonstrated by several practical examples how to calculate the transfer functions $\mu(s)$ and $\beta(s)$ of the basic amplifier and the feedback network of a given feedback configuration. We introduced Bode's feedback theory, which is based on the concepts of return difference and null return difference. We showed that the return difference is a generalization of the concept of the feedback factor of the ideal feedback model and can be interpreted physically as the difference between the 1-V excitation and the returned voltage. We demonstrated that return difference and null return difference are closely related to network functions and can therefore be employed to simplify the calculation of driving-point impedance of an active network, thereby observing the effects of feedback on amplifier impedance and gain.

In the present chapter, we continue our study of feedback amplifier theory. We show that feedback may be employed to make the gain of an amplifier less sensitive to variations in the parameters of the active components and to reduce the effects of noise and nonlinear distortion, and to affect the stability of the network. The concepts of return difference, null return difference, and sensitivity function will be generalized by introducing the general reference value, which is very useful in measurement situations. Since the zeros of the return difference are also the natural frequencies of the network, they are essential for the stability study. To this end, we present three procedures for the physical measurements of return difference. This is especially important in view of the fact that it is difficult to get an accurate formulation of the equivalent network, which, to a greater or lesser extent, is an idealization of the physical reality. The measurement of return difference provides an experimental verification that the system is stable and will remain so under certain prescribed conditions. Finally, we discuss the invariance of return difference and null return difference under different formulations of network equations.

5.1 SENSITIVITY FUNCTION AND FEEDBACK

One of the most important effects of feedback is its ability to make an amplifier less sensitive to the variations of its parameters because of aging, temperature change, or other environmental changes. In this section, we study this effect.

A useful quantitative measure for the degree of dependence of an amplifier on a particular parameter is sensitivity. Sensitivity is a measure of the change of the overall transfer function to the change of a particular parameter in the network. As indicated in Sec. 3.3, the sensitivity function, written as $S(x)$, with respect to an element x is defined as the ratio of the fractional change in a transfer function to the fractional change in x for the situation when all changes concerned are differentially small (see Definition 3.1). Thus, if $w(x)$ is the transfer function, the sensitivity function can be written as

$$S(x) = \frac{x}{w}\frac{\partial w}{\partial x} = \frac{\partial \ln w}{\partial \ln x} = x\frac{\partial \ln w}{\partial x} \tag{5.1}$$

Consider the general feedback configuratin of Fig. 5.1, whose indefinite-admittance matrix is denoted by $\mathbf{Y}$. As before, to emphasize the importance of the element x, we express w and $\mathbf{Y}$ as functions of x alone, even though they are also functions of the complex-frequency variable s, and we write $w = w(x)$ and $\mathbf{Y} = \mathbf{Y}(x)$.

To be definitive, let $w(x)$ represent either the current gain I_{pq}/I_s or the transfer impedance V_{pq}/I_s (Fig. 5.1) for the time being. Then, from Eq. (4.174) we have

$$w(x) = Y_2 \frac{Y_{rp,sq}(x)}{Y_{uv}(x)} \quad \text{or} \quad \frac{Y_{rp,sq}(x)}{Y_{uv}(x)} \tag{5.2}$$

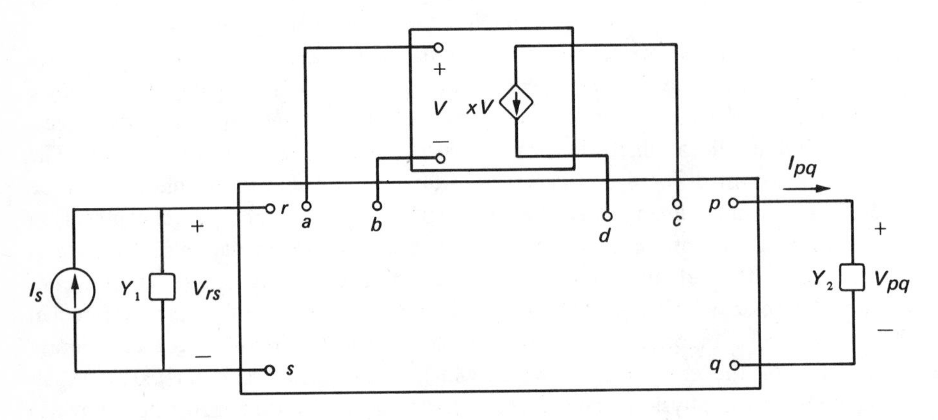

Figure 5.1 The configuration of a general feedback amplifier.

as the case may be. For simplicity, we write

$$\dot{Y}_{uv}(x) = \frac{\partial Y_{uv}(x)}{\partial x} \tag{5.3a}$$

$$\dot{Y}_{rp,sq}(x) = \frac{\partial Y_{rp,sq}(x)}{\partial x} \tag{5.3b}$$

Our objective is to establish a relation among the sensitivity function, the return difference, and the transfer functions. To this end, we need the following identities:

$$Y_{uv}(x) = Y_{uv}(0) + x\dot{Y}_{uv}(x) \tag{5.4a}$$

$$Y_{rp,sq}(x) = Y_{rp,sq}(0) + x\dot{Y}_{rp,sq}(x) \tag{5.4b}$$

as given in Eq. (4.149*b*). Substituting (5.2) in (5.1) in conjunction with (5.4), we get

$$\begin{aligned} S(x) &= x\,\frac{\dot{Y}_{rp,sq}(x)}{Y_{rp,sq}(x)} - x\,\frac{\dot{Y}_{uv}(x)}{Y_{uv}(x)} \\ &= \frac{Y_{rp,sq}(x) - Y_{rp,sq}(0)}{Y_{rp,sq}(x)} - \frac{Y_{uv}(x) - Y_{uv}(0)}{Y_{uv}(x)} \\ &= \frac{Y_{uv}(0)}{Y_{uv}(x)} - \frac{Y_{rp,sq}(0)}{Y_{rp,sq}(x)} = \frac{1}{F(x)} - \frac{1}{\hat{F}(x)} \end{aligned} \tag{5.5}$$

Combining Eqs. (4.176) and (5.5), we obtain

$$S(x) = \frac{1}{F(x)}\left[1 - \frac{w(0)}{w(x)}\right] \tag{5.6}$$

an equation relating the sensitivity function, the return difference, and the transfer functions. As pointed out in Chap. 4, the concept of null return difference is invalid if $Y_{rp,sq}(0)$ is zero. However, (5.6) remains valid, because under the stipulated condition $w(0) = 0$, showing that (5.5) and (5.6) are again equivalent. It is significant to observe that if $w(0) = 0$, then (5.6) becomes $S(x) = 1/F(x)$, meaning that the sensitivity is equal to the reciprocal of the return difference. For the ideal feedback model of Fig. 4.1, the feedback path is unilateral. Hence $w(0) = 0$ and $S = 1/F = 1/(1 + T) = 1/(1 - \mu\beta)$. For a practical amplifier, $w(0)$ is usually very much smaller than $w(x)$ in the passband, and $F \approx 1/S$ may be used as a good estimate of the reciprocal of the sensitivity in the same frequency band. A single-loop feedback amplifier composed of a cascade of common-emitter stages with a passive network providing the desired feedback fulfills this requirement. If in such a structure any one of the transistors fails, the forward transmission is nearly zero and $w(0)$ is practically zero. Thus, we conclude that the sensitivity and reciprocal of the return difference are equal for any element whose failure would interrupt the transmission through the amplifier as a whole to zero.

In the particular situation where $r = p$ and $s = q$, the transfer impedance becomes the driving-point impedance, and formula (5.5) or (5.6) is the sensitivity function of a one-port impedance with respect to an element x of interest. We recognize that in this case $w(0)$ is not usually smaller than $w(x)$.

Refer again to Fig. 5.1. Assume that $w(x)$ represents the voltage gain V_{pq}/V_{rs}. From (2.97) we have $w(x) = Y_{rp,sq}(x)/Y_{rr,ss}(x)$. Using this in (5.1) gives

$$\begin{aligned} S(x) &= x\frac{\dot{Y}_{rp,sq}(x)}{Y_{rp,sq}(x)} - x\frac{\dot{Y}_{rr,ss}(x)}{Y_{rr,ss}(x)} \\ &= \frac{Y_{rp,sq}(x) - Y_{rp,sq}(0)}{Y_{rp,sq}(x)} - \frac{Y_{rr,ss}(x) - Y_{rr,ss}(0)}{Y_{rr,ss}(x)} \\ &= \frac{Y_{rr,ss}(0)}{Y_{rr,ss}(x)} - \frac{Y_{rp,sq}(0)}{Y_{rp,sq}(x)} \\ &= \frac{1}{F(\text{input short-circuited})} - \frac{1}{\hat{F}(x)} \end{aligned} \tag{5.7}$$

By using (4.197), the sensitivity function can be expressed as

$$S(x) = \frac{1}{F(\text{input short-circuited})}\left[1 - \frac{w(0)}{w(x)}\right] \tag{5.8}$$

Likewise, if $w(x)$ represents the short-circuit current gain I_{pq}/I_s as Y_2 approaches infinity, the sensitivity function can be written as

$$S(x) = \frac{Y_{pp,qq}(0)}{Y_{pp,qq}(x)} - \frac{Y_{rp,sq}(0)}{Y_{rp,sq}(x)} = \frac{1}{F(\text{output short-circuited})} - \frac{1}{\hat{F}(x)} \tag{5.9}$$

which when combined with (4.199) gives

$$S(x) = \frac{1}{F(\text{output short-circuited})}\left[1 - \frac{w(0)}{w(x)}\right] \tag{5.10}$$

The derivation of (5.9) is left as an exercise (see Prob. 5.1). Observe that formulas (5.6), (5.8), and (5.10) are quite similar. If the return difference $F(x)$ is interpreted properly, they can all be represented by the single relation (5,6). As before, if $w(0) = 0$, the sensitivity for the voltage gain function is equal to the reciprocal of the return difference under the situation that the input port of the amplifier is short-circuited, and the sensitivity for the short-circuit current gain is equal to the reciprocal of the return difference when the output port is short-circuited.

We illustrate these by the following examples.

Example 5.1 In the feedback network of Fig. 5.2, let α_2 be the element of interest. As demonstrated in Example 4.17, the voltage gain or short-circuit current gain $w(\alpha_2)$ is independent of the controlling parameter α_2, giving $w(x) = w(0)$. From (5.8) and (5.10), we have $S(\alpha_2) = 0$, meaning that $w(\alpha_2)$ is invariant to the change of the value of α_2, as expected.

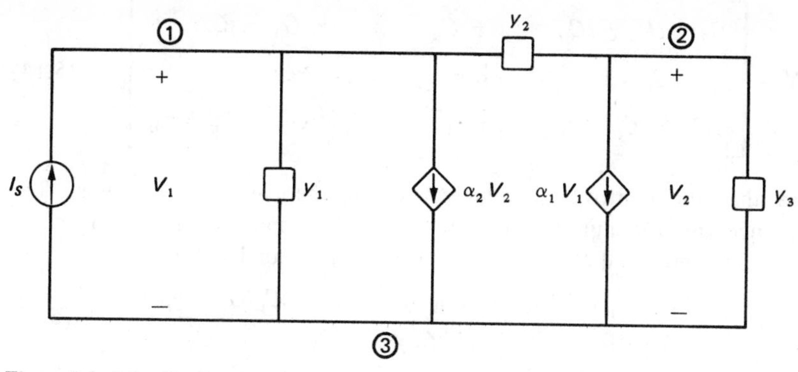

Figure 5.2 A feedback network used to illustrate the computation of the sensitivity function.

On the other hand, suppose that we pick α_1 to be the element of interest. The voltage gain was computed earlier in Example 4.17 and is given by

$$w(\alpha_1) = \frac{y_2 - \alpha_1}{y_2 + y_3} \tag{5.11}$$

F(input short-circuited) is the return difference with respect to α_1 when terminals 1 and 3 are short-circuited. Since shorting terminals 1 and 3 breaks the feedback loop, F(input short-circuited) = 1. Substituting these in (5.8) yields the sensitivity function for the voltage gain with respect to the controlling parameter α_1:

$$S(\alpha_1) = 1 - \frac{w(0)}{w(\alpha_1)} = \frac{\alpha_1}{\alpha_1 - y_2} \tag{5.12}$$

Likewise, applying (4.201) and the fact that F(output short-circuited) = 1, we can show that the sensitivity function for the short-circuit current gain with respect to α_1 is again given by Eq. (5.12) (see Prob. 5.3).

Example 5.2 Consider the common-emitter transistor amplifier of Fig. 5.3*a*. After removing the biasing circuit, its equivalent network is presented in Fig. 5.3*b* with

$$I_s' = \frac{V_s}{R_1 + r_x} \tag{5.13a}$$

$$G_1' = \frac{1}{R_1'} = \frac{1}{R_1 + r_x} + \frac{1}{r_\pi} \tag{5.13b}$$

$$G_2' = \frac{1}{R_2'} = \frac{1}{R_2} + \frac{1}{R_c} \tag{5.13c}$$

The indefinite-admittance matrix of the network of Fig. 5.3*b* can be written down directly by inspection as

$$\mathbf{Y} = \begin{bmatrix} G_1' + sC_\pi + sC_\mu & -sC_\mu & -G_1' - sC_\pi \\ -sC_\mu + g_m & G_2' + sC_\mu & -G_2' - g_m \\ -G_1' - sC_\pi - g_m & -G_2' & G_1' + G_2' + sC_\pi + g_m \end{bmatrix} \tag{5.14}$$

Suppose that the controlling parameter g_m is the element of interest. The return difference and the null return difference with respect to g_m in Fig. 5.3*b*, with I_s' as the input port and R_2' as the output port, are found to be

$$F(g_m) = \frac{Y_{33}(g_m)}{Y_{33}(0)} = \frac{(G_1' + sC_\pi)(G_2' + sC_\mu) + sC_\mu(G_2' + g_m)}{(G_1' + sC_\pi)(G_2' + sC_\mu) + sC_\mu G_2'} \tag{5.15}$$

$$\hat{F}(g_m) = \frac{Y_{12,33}(g_m)}{Y_{12,33}(0)} = \frac{sC_\mu - g_m}{sC_\mu} = 1 - \frac{g_m}{sC_\mu} \tag{5.16}$$

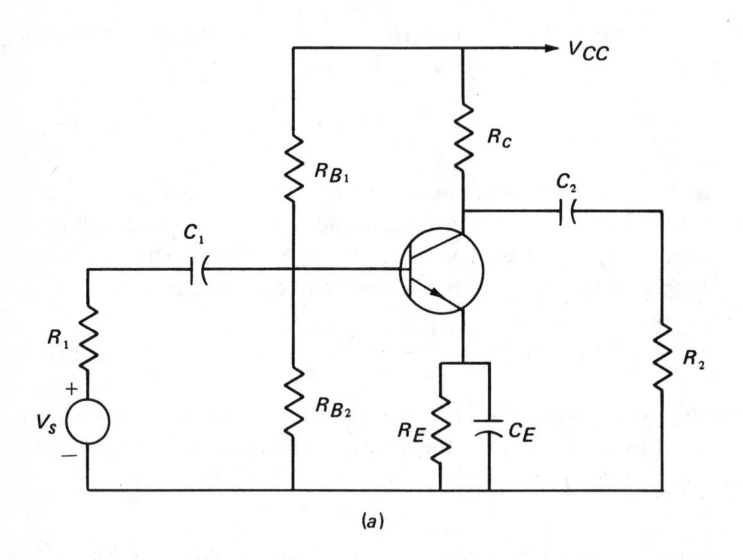

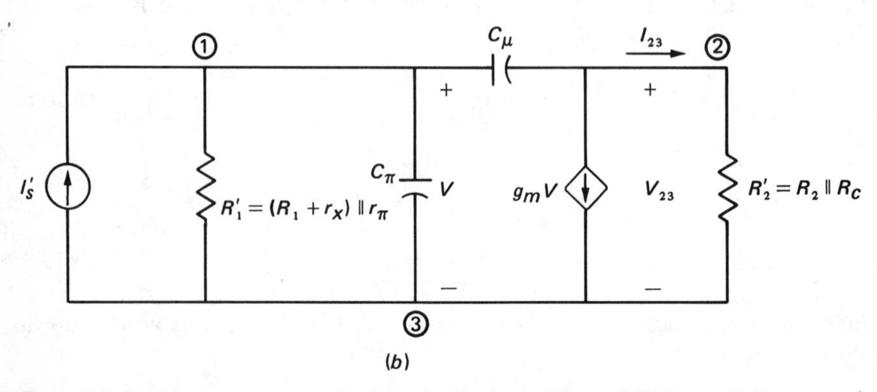

Figure 5.3 (*a*) A common-emitter transistor feedback amplifier and (*b*) its equivalent network.

The current gain I_{23}/I'_s, as defined in Fig. 5.3*b*, is determined to be

$$w(g_m) = \frac{Y_{12,33}(g_m)}{R'_2 Y_{33}(g_m)} = \frac{sC_\mu - g_m}{R'_2\,[(G'_1 + sC_\pi)(G'_2 + sC_\mu) + sC_\mu(G'_2 + g_m)]} \tag{5.17}$$

Substituting (5.15) and (5.16) in (5.5) or (5.15) and (5.17) in (5.6), we obtain the sensitivity function for the current gain I_{23}/I'_s or the transfer impedance V_{23}/I'_s with respect to the transconductance g_m:

$$S(g_m) = -\frac{g_m(G'_2 + sC_\mu)(G'_1 + sC_\pi + sC_\mu)}{(sC_\mu - g_m)[(G'_1 + sC_\pi)(G'_2 + sC_\mu) + sC_\mu(G'_2 + g_m)]} \tag{5.18}$$

Finally, we compute the sensitivity for the driving-point impedance facing the current source I'_s. From (5.6) we have

$$S(g_m) = \frac{1}{F(g_m)}\left[1 - \frac{Z(0)}{Z(g_m)}\right] = -\frac{sC_\mu g_m}{(G'_1 + sC_\pi)(G'_2 + sC_\mu) + sC_\mu(G'_2 + g_m)} \tag{5.19}$$

where
$$Z(g_m) = \frac{Y_{11,33}(g_m)}{Y_{33}(g_m)} = \frac{G'_2 + sC_\mu}{(G'_1 + sC_\pi)(G'_2 + sC_\mu) + sC_\mu(G'_2 + g_m)} \tag{5.20}$$

Example 5.3 Figure 5.4*a* represents a common-emitter stage with a resistor R_f connected from the output to the input to provide an external physical feedback. By using the common-emitter hybrid model of a transistor at low frequencies, an equivalent network of the feedback amplifier is presented in Fig. 5.4*b*, whose indefinite-admittance matrix is given by

$$\mathbf{Y} = \begin{bmatrix} G_1 + G_f + \dfrac{1}{h_{ie}} & -G_f & -G_1 - \dfrac{1}{h_{ie}} \\ \alpha - G_f & G_2 + G_f & -G_2 - \alpha \\ -G_1 - \dfrac{1}{h_{ie}} - \alpha & -G_2 & G_1 + G_2 + \dfrac{1}{h_{ie}} + \alpha \end{bmatrix} \tag{5.21}$$

where $\alpha = h_{fe}/h_{ie}$ and h_{re} and h_{oe} are assumed to be zero. The return difference with respect to the controlling parameter α is found to be

$$F(\alpha) = \frac{Y_{33}(\alpha)}{Y_{33}(0)} = \frac{(G_2 + G_f)(G_1 + 1/h_{ie}) + G_f(\alpha + G_2)}{(G_2 + G_f)(G_1 + 1/h_{ie}) + G_2 G_f} = 1 + \frac{h_{fe}R'_2 R}{R_f(h_{ie} + R)} \tag{5.22}$$

where $R'_2 = R_2R_f/(R_2 + R_f)$ and $R = R_1(R_2 + R_f)/(R_1 + R_2 + R_f)$. By applying (2.97), the voltage gain V_{23}/V_{13} is obtained as

$$w(\alpha) = \frac{V_{23}}{V_{13}} = \frac{Y_{12,33}(\alpha)}{Y_{11,33}(\alpha)} = \frac{G_f - \alpha}{G_2 + G_f} = \frac{R_2(1 - \alpha R_f)}{R_2 + R_f} \tag{5.23}$$

giving
$$w(\alpha) = (1 - \alpha R_f)w(0) \tag{5.24}$$

Since for a practical amplifier $|1 - \alpha R_f| \gg 1$, the ratio $w(0)/w(\alpha)$ is practically zero, and from (5.8) the sensitivity of the voltage gain with respect to α approximately

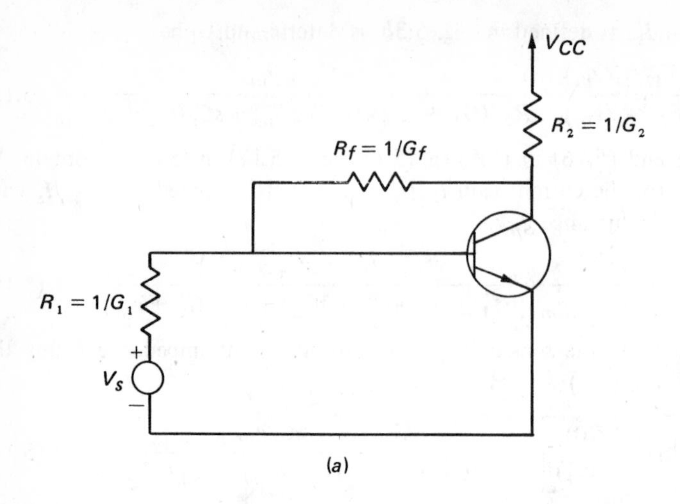

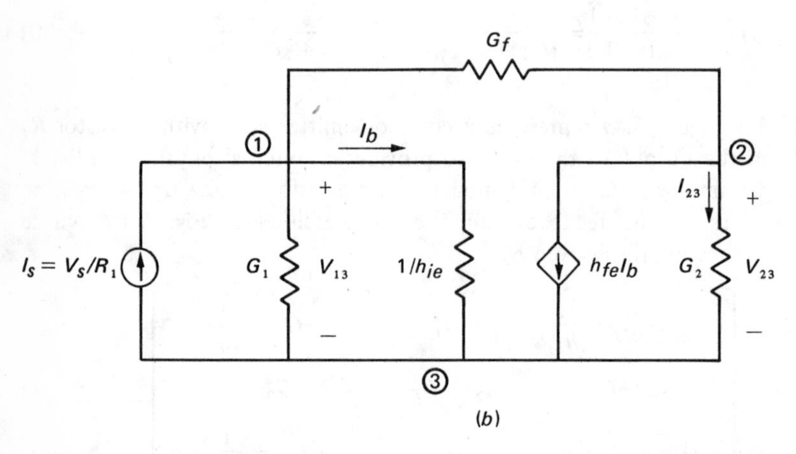

Figure 5.4 (*a*) A common-emitter feedback amplifier and (*b*) its low-frequency equivalent network.

equals the reciprocal of F(input short-circuited), which is unity. Thus, the sensitivity is about unity, being independent of α.

Suppose that we are interested in the sensitivity of the voltage gain V_{23}/V_s of the feedback amplifier with respect to α. We consider the equivalent network of Fig. 5.5, whose indefinite-admittance matrix is given by

$$\mathbf{Y} = \begin{bmatrix} G_1 & -G_1 & 0 & 0 \\ -G_1 & G_f + G_1 + \dfrac{1}{h_{ie}} & -G_f & -\dfrac{1}{h_{ie}} \\ 0 & \alpha - G_f & G_f + G_2 & -\alpha - G_2 \\ 0 & -\alpha - \dfrac{1}{h_{ie}} & -G_2 & \alpha + G_2 + \dfrac{1}{h_{ie}} \end{bmatrix} \tag{5.25}$$

By appealing once more to (2.97) and using (5.25), the amplifier gain is given by

$$w(\alpha) = \frac{V_{34}}{V_s} = \frac{Y_{13,44}(\alpha)}{Y_{11,44}(\alpha)} = \frac{G_1(G_f - \alpha)}{(G_2 + G_f)(G_1 + 1/h_{ie}) + G_f(G_2 + \alpha)} \tag{5.26}$$

which can be expressed in the form of (4.3) as

$$w(\alpha) = \frac{\mu(\alpha)}{1 - \mu(\alpha)\beta(\alpha)} = \frac{\mu(\alpha)}{F(\alpha)} \tag{5.27}$$

where $F(\alpha)$ is given in (5.22) and (see Prob. 5.4)

$$\mu(\alpha) = -\frac{(\alpha R_f - 1)h_{ie}R_2'R}{R_1 R_f(h_{ie} + R)} \tag{5.28a}$$

$$\beta(\alpha) = \frac{R_1 \alpha}{\alpha R_f - 1} \tag{5.28b}$$

$\mu(\alpha)$ may be considered as the amplifier gain without feedback. Thus, the amplifier gain with feedback is reduced by a factor equal to the return difference under normal operating conditions. Consider the ratio

$$\frac{w(0)}{w(\alpha)} = \frac{F(\alpha)\mu(0)}{F(0)\mu(\alpha)} = \frac{F(\alpha)}{F(0)}\,\frac{1}{1 - \alpha R_f} \tag{5.29}$$

which is usually very small compared with unity because $|1 - \alpha R_f| \gg 1$. Thus, from (5.8) the sensitivity of the voltage gain of the feedback amplifier with respect to the controlling parameter α is approximately equal to the reciprocal of the return difference $F(\alpha)$ when the voltage source V_s is short-circuited. This is precisely the same return difference obtained in (5.22) for the network of Fig. 5.4*b* under normal operating conditions. This leads to

$$S(\alpha) \approx \frac{R_f(h_{ie} + R)}{R_f(h_{ie} + R) + h_{fe}R_2'R} \tag{5.30}$$

We note that the matrix (5.21) can be derived from (5.25) by first adding row 1 to row 4, then column 1 to column 4, and finally deleting row 1 and column 1. The operations are equivalent to shorting terminals 1 and 4.

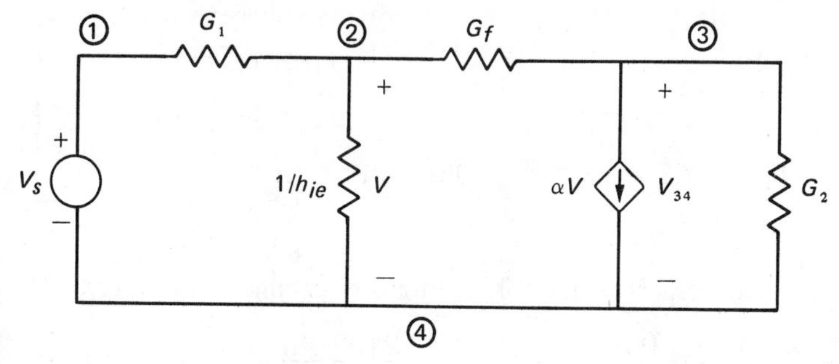

Figure 5.5 An equivalent network of the feedback amplifier of Fig. 5.4.

As an illustration, we use the following set of parameters:

$$\begin{aligned} &R_f = 40\ \text{k}\Omega \qquad R_1 = 10\ \text{k}\Omega \qquad R_2 = 4\ \text{k}\Omega \\ &h_{fe} = 50 \qquad h_{ie} = 1.1\ \text{k}\Omega \qquad h_{re} = h_{oe} = 0 \end{aligned} \tag{5.31}$$

Substituting these in (5.22) and (5.26)-(5.30), we obtain

$$\begin{aligned} F(\alpha) &= 5.01 \\ \mu(\alpha) &= -16.01 \\ w(\alpha) &= -3.20 \\ \frac{w(0)}{w(\alpha)} &= -0.0028 \\ S(\alpha) &= 0.20 \end{aligned} \tag{5.32}$$

It is instructive to compare the above results with those obtained by approximating the feedback amplifier to the ideal feedback model of Fig. 4.1. From Prob. 5.6 and using the same set of parameters as given in (5.31), we obtain the gain without feedback to be -16, the feedback factor to be 5, and the gain with feedback to be -3.2. The results are nearly the same as those given above.

5.2 THE RETURN DIFFERENCE AND TWO-PORT FUNCTIONS

In Secs. 3.1-3.3, we derived the general expressions of the network functions and power gains in terms of the general hybrid parameters. In the present section, we show the general relations between these expressions and feedback, and we demonstrate by a specific example how they can be employed to design amplifiers that achieve maximum gain with a prescribed sensitivity.

Consider an active two-port device N_a characterized by its y-parameters y_{ij}. Suppose that the two-port is terminated at its input and output ports by the admittances Y_1 and Y_2, as shown in Fig. 3.1b. The indefinite-admittance matrix is found to be

$$\mathbf{Y} = \begin{bmatrix} y_{11} + Y_1 & y_{12} & -y_{11} - y_{12} - Y_1 \\ y_{21} & y_{22} + Y_2 & -y_{21} - y_{22} - Y_2 \\ -y_{11} - y_{21} - Y_1 & -y_{12} - y_{22} - Y_2 & y_{11} + y_{12} + y_{21} + y_{22} + Y_1 + Y_2 \end{bmatrix} \tag{5.33}$$

The return difference with respect to the forward-transfer parameter y_{21} is given by

$$F(y_{21}) = \frac{Y_{33}(y_{21})}{Y_{33}(0)} = \frac{(y_{11} + Y_1)(y_{22} + Y_2) - y_{12}y_{21}}{(y_{11} + Y_1)(y_{22} + Y_2)} = 1 + T_y \tag{5.34}$$

where
$$T_y = -\frac{y_{12}y_{21}}{(y_{11} + Y_1)(y_{22} + Y_2)} \tag{5.35}$$

is the return ratio with respect to y_{21}.

From App. II, we can express the y-parameters y_{ij} in terms of the h-parameters h_{ij} of N_a, giving

$$Y_{33} = \left(\frac{1}{h_{11}} + Y_1\right)\left(h_{22} - \frac{h_{12}h_{21}}{h_{11}} + Y_2\right) + \frac{h_{12}h_{21}}{h_{11}^2} \tag{5.36}$$

The return difference with respect to h_{21} becomes

$$F(h_{21}) = \frac{Y_{33}(h_{21})}{Y_{33}(0)} = \frac{(h_{11} + Z_1)(h_{22} + Y_2) - h_{12}h_{21}}{(h_{11} + Z_1)(h_{22} + Y_2)} = 1 + T_h \tag{5.37}$$

where $Z_1 = 1/Y_1$ and

$$T_h = -\frac{h_{12}h_{21}}{(h_{11} + Z_1)(h_{22} + Y_2)} \tag{5.38}$$

is the return ratio with respect to h_{21}. Likewise, in terms of the g-parameters g_{ij} of N_a, we have

$$Y_{33} = \left(g_{11} - \frac{g_{12}g_{21}}{g_{22}} + Y_1\right)\left(\frac{1}{g_{22}} + Y_2\right) + \frac{g_{12}g_{21}}{g_{22}^2} \tag{5.39}$$

yielding the return difference with respect to g_{21} as

$$F(g_{21}) = \frac{Y_{33}(g_{21})}{Y_{33}(0)} = \frac{(g_{11} + Y_1)(g_{22} + Z_2) - g_{12}g_{21}}{(g_{11} + Y_1)(g_{22} + Z_2)} = 1 + T_g \tag{5.40}$$

where $Z_2 = 1/Y_2$ and

$$T_g = -\frac{g_{12}g_{21}}{(g_{11} + Y_1)(g_{22} + Z_2)} \tag{5.41}$$

is the return ratio with respect to g_{21}.

From the above discussion, we recognize that return difference and return ratio with respect to the forward-transfer parameter are invariant under immittance substitution for y-, h-, and g-parameters, and all can be written in terms of the general hybrid parameters k_{ij} by the single expression

$$F(k_{21}) = \frac{Y_{33}(k_{21})}{Y_{33}(0)} = \frac{(k_{11} + M_1)(k_{22} + M_2) - k_{12}k_{21}}{(k_{11} + M_1)(k_{22} + M_2)} = 1 + T_k \tag{5.42}$$

where, as before, M_1 and M_2 denote the terminating immittances, and

$$T_k = -\frac{k_{12}k_{21}}{(k_{11} + M_1)(k_{22} + M_2)} \tag{5.43}$$

We emphasize that the above expressions are valid for y-, h-, or g-parameters. For the z-parameters z_{ij}, the situation is somewhat different and will be elaborated in Sec. 5.9.2. In this case, the first-order cofactor Y_{33} becomes

$$Y_{33} = \left(\frac{z_{22}}{\Delta_z} + Y_1\right)\left(\frac{z_{11}}{\Delta_z} + Y_2\right) - \frac{z_{12}z_{21}}{\Delta_z^2} \tag{5.44}$$

where $\Delta_z = z_{11}z_{22} - z_{12}z_{21}$, yielding

$$F(z_{21}) = \frac{z_{11}z_{22}\,[(z_{11} + Z_1)(z_{22} + Z_2) - z_{12}z_{21}]}{\Delta_z(z_{11} + Z_1)(z_{22} + Z_2)} = \frac{z_{11}z_{22}}{\Delta_z}F^m(z_{21}) \tag{5.45a}$$

where $F^m(z_{21}) = 1 + T_z^m$ and

$$T_z^m = -\frac{z_{12}z_{21}}{(z_{11} + Z_1)(z_{22} + Z_2)} \tag{5.45b}$$

Even though the return difference $F(k_{21})$ and return ratio T_k are invariant under immittance substitution for the y-, h-, and g-parameters, it does not mean that their values are the same. For a given active two-port device, the return difference is not unique and depends on the matrix representation used to characterize its external port behavior. We illustrate this by the following example.

Example 5.4 The transistor amplifier of Fig. 5.6 is in the common-emitter connection. Suppose that the parameters of the amplifier are specified by

$$\begin{gathered} h_{ie} = 1.1\ \text{k}\Omega \qquad h_{re} = 2.5 \cdot 10^{-4} \\ h_{fe} = 50 \qquad h_{oe} = 25\ \mu\text{mho} \\ R_1 = 10\ \text{k}\Omega \qquad R_2 = 4\ \text{k}\Omega \end{gathered} \tag{5.46}$$

From App. II, the corresponding y-, g-, and z-parameters are found to be

$$\begin{gathered} y_{11} = 0.91 \cdot 10^{-3}\ \text{mho} \qquad y_{12} = -0.23\ \mu\text{mho} \\ y_{21} = 45.46 \cdot 10^{-3}\ \text{mho} \qquad y_{22} = 13.64\ \mu\text{mho} \\ g_{11} = 1.67 \cdot 10^{-3}\ \text{mho} \qquad g_{12} = -16.67 \cdot 10^{-3} \\ g_{21} = -3.33 \cdot 10^{3} \qquad g_{22} = 73.33\ \text{k}\Omega \\ z_{11} = 600\ \Omega \qquad z_{12} = 10\ \Omega \\ z_{21} = -2\ \text{M}\Omega \qquad z_{22} = 40\ \text{k}\Omega \end{gathered} \tag{5.47}$$

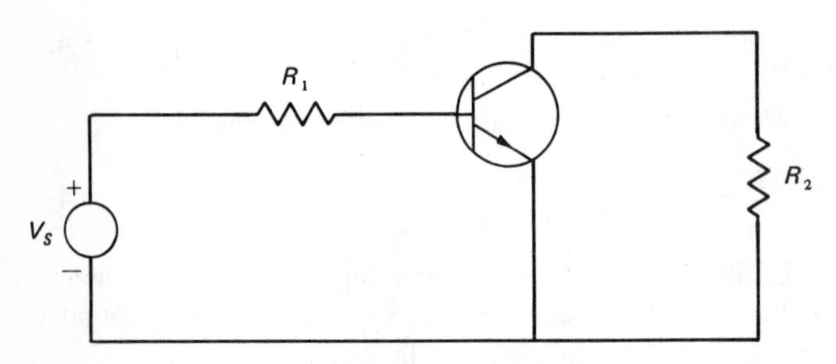

Figure 5.6 A common-emitter transistor amplifier.

From these we obtain

$$\begin{aligned} T_y &= 39.3 \cdot 10^{-3} \qquad F(y_{21}) = 1.039 \\ T_h &= -4.10 \cdot 10^{-3} \qquad F(h_{21}) = 0.996 \\ T_g &= -0.41 \qquad F(g_{21}) = 0.59 \end{aligned} \tag{5.48}$$

and $F^m(z_{21}) = 1.043$, indicating that the internal feedback for the common-emitter stage is negative for the y- and z-representations and is positive for the h- and g-representations of the transistor.

In terms of the return difference, the general formulas for the transfer functions, as defined by Table 3.1 and given by Eq. (3.25), can be written as

$$\frac{y_2}{u_s} = \frac{k_{21} M_2}{(k_{11} + M_1)(k_{22} + M_2) - k_{12} k_{21}} = -\frac{M_2 T_k / k_{12}}{F(k_{21})} \tag{5.49a}$$

$$\frac{u_2}{u_s} = -\frac{k_{21}}{(k_{11} + M_1)(k_{22} + M_2) - k_{12} k_{21}} = \frac{T_k / k_{12}}{F(k_{21})} \tag{5.49b}$$

From (3.29), the sensitivity functions of (5.49) to changes of the general hybrid parameters are given by

$$S(k_{11}) = -\frac{k_{11}/(k_{11} + M_1)}{1 + T_k} = -\frac{k_{11}/(k_{11} + M_1)}{F(k_{21})} \tag{5.50a}$$

$$S(k_{12}) = -\frac{T_k}{1 + T_k} = -\frac{T_k}{F(k_{21})} \tag{5.50b}$$

$$S(k_{21}) = \frac{1}{1 + T_k} = \frac{1}{F(k_{21})} \tag{5.50c}$$

$$S(k_{22}) = -\frac{k_{22}/(k_{22} + M_2)}{1 + T_k} = -\frac{k_{22}/(k_{22} + M_2)}{F(k_{21})} \tag{5.50d}$$

From (5.50) we observe that in making T_k or $F(k_{21})$ large and positive, the sensitivities to changes of k_{11}, k_{21}, and k_{22} are reduced, whereas the sensitivity to changes of k_{12} is close to unity. However, too much feedback may bring a system to the condition of oscillation, which is not what we want. Also, from (5.50) we have

$$S(k_{21}) - S(k_{12}) = 1 \tag{5.51}$$

In amplifier performance, the transfer functions (5.49) are less significant than the transducer power gain. Therefore, we express the transducer power gain, as given in (3.19b), in terms of the return difference, which is important in determining the sensitivities.

$$\begin{aligned} \mathcal{G} &= \frac{4|k_{21}|^2 \operatorname{Re} M_1 \operatorname{Re} M_2}{|(k_{11} + M_1)(k_{22} + M_2) - k_{12} k_{21}|^2} = \frac{4|k_{21}|^2 \operatorname{Re} M_1 \operatorname{Re} M_2}{|k_{11} + M_1|^2 |k_{22} + M_2|^2 |1 + T_k|^2} \\ &= \frac{4|k_{21}|^2 \operatorname{Re} M_1 \operatorname{Re} M_2}{|k_{11} + M_1|^2 |k_{22} + M_2|^2 |F(k_{21})|^2} = \frac{4|k_{21}|^2 \operatorname{Re} M_1 \operatorname{Re} M_2 |S(k_{21})|^2}{|k_{11} + M_1|^2 |k_{22} + M_2|^2} \end{aligned} \tag{5.52}$$

As a result, to increase feedback to reduce the sensitivity, the magnitudes of the transfer functions and transducer power gain are also reduced. In general, an improvement in sensitivity is accompanied by a reduction in gain. However, the gain reduction resulting from feedback is far outweighed by the improvement in sensitivity, because the loss of gain can easily be made up by another feedback amplifier connected in cascade if necessary. In the following, we demonstrate by a specific example the design of a feedback amplifier that attains the maximum transducer power gain with prescribed sensitivity.

Example 5.5 A transistor in common-emitter mode is characterized by its hybrid parameters:

$$\begin{aligned} h_{a11} &= h_{ie} = 1.1\ \text{k}\Omega \qquad h_{a12} = h_{re} = 2.5 \cdot 10^{-4} \\ h_{a21} &= h_{fe} = 50 \qquad h_{a22} = h_{oe} = 25\ \mu\text{mho} \end{aligned} \tag{5.53}$$

The most significant parameter that will affect amplifier performance is the forward-transfer parameter h_{fe}, which is subject to variations caused by drift. Thus, we assume that the sensitivity $S(h_{fe})$ is an appropriate measure of amplifier performance. Suppose that we wish to design a feedback amplifier whose sensitivity is 10% of that of the amplifier without feedback in which the transistor is terminated for maximum transducer power gain. To this end, we first determine the terminations that should be used with the transistor to yield maximum transducer power gain. From (3.164)-(3.166), we get

$$Z_{1,\text{opt}} = 812.41\ \Omega \tag{5.54a}$$

$$Y_{2,\text{opt}} = 18.46\ \mu\text{mho} \tag{5.54b}$$

$$\mathcal{G}_{\max} = 30077 \tag{5.54c}$$

As indicated in Sec. 4.2, there are four basic feedback configurations, as shown schematically in Fig. 4.3. For our purposes, we choose the series-parallel topology of Fig. 4.3*c* to desensitize the amplifier performance to variations of h_{fe}. To simplify our computation, the passive feedback network N_f is an ideal transformer of turns ratio $b:1$, as shown in Fig. 5.7. Our objective is to determine the turns ratio and the terminating immittances that will reduce the effect of the variations of h_{fe} on amplification by a factor of 10.

Applying the Brune tests of Fig. 4.4*c* to the network of Fig. 5.7 shows that the conditions are satisfied. The *h*-parameters h_{ij} of the composite two-port network N are therefore given by

$$h_{11} = h_{a11} + h_{f11} = h_{a11} \tag{5.55a}$$

$$h_{12} = h_{a12} + h_{f12} = h_{a12} + b \tag{5.55b}$$

$$h_{21} = h_{a21} + h_{f21} = h_{a21} - b \tag{5.55c}$$

$$h_{22} = h_{a22} + h_{f22} = h_{a22} \tag{5.55d}$$

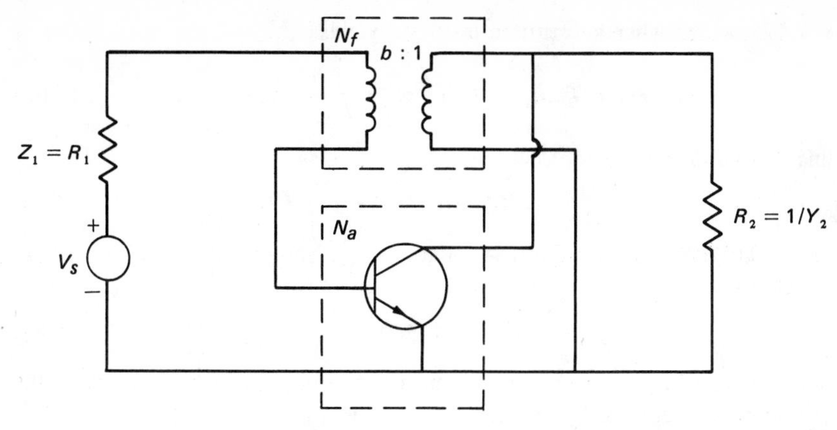

Figure 5.7 A transistor amplifier with the ideal transformer providing the desired external feedback.

Under the optimum terminations of (5.54), the return difference of the transistor N_a with respect to h_{fe} is found from (5.37) to be

$$F(h_{fe}) = F(h_{a21}) = 0.85 \tag{5.56}$$

showing a positive internal feedback. From (5.50*c*) we get

$$S(h_{fe}) = S(h_{a21}) = 1.18 \tag{5.57}$$

As indicated in (4.18*a*), the difference between h_{21} and h_{a21} is usually very small, and our requirement can be stated as one of reducing $S(h_{21})$ to 10% of $S(h_{a21}) = 1.18$ with minimum loss in gain.

From (5.52) we have the transducer power gain

$$\mathcal{G} = \frac{4h_{21}^2 R_1 S^2(h_{21})/R_2}{(h_{11} + R_1)^2 (h_{22} + 1/R_2)^2} \tag{5.58}$$

since all the quantities are real. To maximize $\mathcal{G}$ with prescribed $S(h_{21})$, we must minimize

$$\frac{(h_{11} + R_1)^2}{R_1} \quad \text{and} \quad \left(h_{22} + \frac{1}{R_2}\right)^2 R_2 \tag{5.59}$$

by choosing appropriate values of R_1 and R_2. The minimum values of (5.59) occur at

$$R_1 = h_{11} = h_{a11} = 1.1 \text{ k}\Omega \tag{5.60a}$$

$$R_2 = \frac{1}{h_{22}} = \frac{1}{h_{a22}} = 40 \text{ k}\Omega \tag{5.60b}$$

From (5.50*c*), the desired amount of feedback is found to be

$$F(h_{21}) = \frac{1}{S(h_{21})} = \frac{1}{0.1 \cdot 1.18} = 8.475 \tag{5.61}$$

or $T_h = 7.475$, which when substituted in (5.38) yields

$$-h_{12}h_{21} = T_h(h_{11} + R_1)\left(h_{22} + \frac{1}{R_2}\right) = 0.822 \tag{5.62}$$

Appealing to (5.55*b*) and (5.55*c*), we get

$$b^2 - 50b - 0.8345 = 0 \tag{5.63}$$

giving $b = -0.01669$ and $b = 50.0164$. Since $|h_{a21}|$ should be much larger than $|h_{f21}|$, we choose

$$b = -0.01669 \tag{5.64}$$

The minus sign indicates the polarity of the transformer. The h-parameters of the composite two-port network N become

$$\begin{aligned} h_{11} &= 1.1 \text{ k}\Omega \qquad h_{12} = -16.44 \cdot 10^{-3} \\ h_{21} &\approx 50 \qquad h_{22} = 25\ \mu\text{mho} \end{aligned} \tag{5.65}$$

Thus, the major effect of adding the feedback network N_f is to change the reverse-transfer parameter, and the other parameters are essentially the same as the transistor N_a. From (3.7), (3.8), and (5.58), the desensitized amplifier input and output immittances and transducer power gain are given by

$$Z_{\text{in}} = Z_{11} = 17.54 \text{ k}\Omega \tag{5.66a}$$

$$Y_{\text{out}} = Y_{22} = 3.99 \cdot 10^{-4} \text{ mho} \tag{5.66b}$$

$$\mathcal{G} = 316.46 \tag{5.66c}$$

These results are to be compared with the situation where N is optimally terminated. Again, from (3.164)–(3.166), (5.42), and (5.50*c*), we obtain

$$\begin{aligned} Z_{1,\text{opt}} &= 6.11 \text{ k}\Omega \qquad Y_{2,\text{opt}} = 139\ \mu\text{mho} \\ \mathcal{G}_{\max} &= 2114 \qquad S(h_{21}) = 0.59 \end{aligned} \tag{5.67}$$

where $F(h_{21}) = 1.695$, showing that the feedback is now negative.

Before we turn our attention to another subject, we mention the effect of feedback on input and output immittances. As shown in (4.19*a*), the input immittance $M_{11} + M_1$ looking into the source of the composite two-port network of Fig. 4.3 can be written as

$$M_{11} + M_1 = (k_{11} + M_1)(1 + T_k) = (k_{11} + M_1)F(k_{21})$$

In words, this states that the input immittance is increased by a factor that is the normal value of the return difference with respect to the forward-transfer parameter k_{21} from its value when the output port is open-circuited or short-circuited, depending on the two-port parameters used. Since by open-circuiting or short-circuiting the output port we break the feedback loop, we can say that feedback increases the input immittance by the factor $F(k_{21})$ from its value without feedback.

As an example, consider the series-parallel configuration of Fig. 4.3*c*. It is convenient to use the *h*-parameters h_{ij}. Then $k_{11} + M_1 = h_{11} + Z_1$ denotes the input impedance looking into the voltage source when the output port is short-circuited. The short-circuiting of the output port breaks the feedback loop. Consequently, the input impedance with feedback is increased by the factor $F(h_{21})$ from its value without feedback.

Likewise, from Eq. (4.19*b*), the output immittance $M_{22} + M_2$ looking into the output port, including the load M_2, can be written as

$$M_{22} + M_2 = (k_{22} + M_2)(1 + T_k) = (k_{22} + M_2)F(k_{21}) \tag{5.69}$$

Thus, the output immittance with feedback is also increased from its value without feedback by the factor $F(k_{21})$.

In Example 5.5, we use (5.61), (5.65), (5.66*a*), and (5.66*b*). It is straightforward to confirm the above assertion for the *h*-parameters.

5.3 RETURN DIFFERENCE AND NULL RETURN DIFFERENCE WITH RESPECT TO TWO ELEMENTS

In this section, we introduce return difference and null return difference with respect to two elements, which are found helpful in feedback calculations.

Suppose that x_1 and x_2 are two elements of interest in a feedback network. To express the fact that the first- and second-order cofactors are functions of both x_1 and x_2, we write them as $Y_{uv}(x_1, x_2)$ and $Y_{rp,sq}(x_1, x_2)$ so that the elements can be exhibited explicitly. The return differences with respect to x_1 and x_2 individually can be written as

$$F(x_1) = \frac{Y_{uv}(x_1, x_2)}{Y_{uv}(0, x_2)} \tag{5.70}$$

for the element x_1, and

$$F(x_2) = \frac{Y_{uv}(x_1, x_2)}{Y_{uv}(x_1, 0)} \tag{5.71}$$

for the element x_2. Their ratio is therefore

$$\frac{F(x_1)}{F(x_2)} = \frac{Y_{uv}(x_1, 0)}{Y_{uv}(0, x_2)} = \frac{Y_{uv}(x_1, 0)/Y_{uv}(0, 0)}{Y_{uv}(0, x_2)/Y_{uv}(0, 0)} \tag{5.72}$$

provided that $Y_{uv}(0, 0) \neq 0$. This is equivalent to

$$\frac{F(x_1)}{F(x_2)} = \frac{F(x_1)|_{x_2=0}}{F(x_2)|_{x_1=0}} \tag{5.73}$$

In words, this states that the ratio of the return differences with respect to the elements x_1 and x_2 individually is equal to the ratio of the return difference with respect to x_1 when x_2 is set to zero and that with respect to x_2 when x_1 is set to zero. Thus, it may be employed to compute the return difference with respect to a given element from the known return difference with respect to another element.

Likewise, the null return differences with respect to the elements x_1 and x_2 individually can be expressed as

$$\hat{F}(x_1) = \frac{Y_{rp,sq}(x_1, x_2)}{Y_{rp,sq}(0, x_2)} \tag{5.74}$$

$$\hat{F}(x_2) = \frac{Y_{rp,sq}(x_1, x_2)}{Y_{rp,sq}(x_1, 0)} \tag{5.75}$$

Their ratio becomes

$$\frac{\hat{F}(x_1)}{\hat{F}(x_2)} = \frac{Y_{rp,sq}(x_1, 0)/Y_{rp,sq}(0, 0)}{Y_{rp,sq}(0, x_2)/Y_{rp,sq}(0, 0)} \tag{5.76}$$

provided that $Y_{rp,sq}(0, 0) \neq 0$, yielding

$$\frac{\hat{F}(x_1)}{\hat{F}(x_2)} = \frac{\hat{F}(x_1)|_{x_2=0}}{\hat{F}(x_2)|_{x_1=0}} \tag{5.77}$$

Example 5.6 In the feedback network of Fig. 5.4*b*, we compute the return difference with respect to the conductance G_1 by means of (5.73). Let $x_1 = G_1$ and $x_2 = \alpha \equiv h_{fe}/h_{ie}$. Then $F(\alpha)|_{G_1=0}$ can easily be deduced from (5.22) and is given by

$$F(\alpha)|_{G_1=0} = 1 + \frac{\alpha G_f}{(G_2 + G_f)/h_{ie} + G_2 G_f} \tag{5.78}$$

To compute $F(G_1)|_{\alpha=0}$, we use the concept that the return difference with respect to a one-port admittance is equal to the ratio of the total admittance at the terminal pair where the admittance is connected to the admittance that it faces. Thus, by setting $\alpha = 0$ in Fig. 5.4*b*, we obtain

$$F(G_1)|_{\alpha=0} = 1 + \frac{G_1(G_2 + G_f)}{G_f G_2 + (G_2 + G_f)/h_{ie}} \tag{5.79}$$

Substituting (5.22), (5.78), and (5.79) in (5.73) yields

$$F(G_1) = F(\alpha)\frac{F(G_1)|_{\alpha=0}}{F(\alpha)|_{G_1=0}} = 1 + \frac{G_1(G_2 + G_f)}{(G_2 + G_f)/h_{ie} + G_f(G_2 + \alpha)} \tag{5.80}$$

As a check, we compute the active admittance that G_1 faces. Invoking (2.95) in conjunction with (5.21) gives the active admittance

$$y = \left.\frac{Y_{33}}{Y_{11,33}}\right|_{G_1=0} = \frac{(G_2 + G_f)/h_{ie} + G_f(\alpha + G_2)}{G_2 + G_f} \tag{5.81}$$

showing that

$$F(G_1) = 1 + T = 1 + \frac{G_1}{y} = \frac{y + G_1}{y} \tag{5.82}$$

5.4 EXTENSIONS TO FEEDBACK CONCEPTS

In this section, the concepts of return difference and null return difference are generalized. Their properties, physical significance, and interrelations are discussed.

5.4.1 The General Return Difference

As a direct extension to the concept of return difference as given in Definition 4.1, we define the return difference with respect to an element for a general reference value.

Definition 5.1: General return difference The *general return difference* $F_k(x)$ of a feedback amplifier with respect to an element x for a general reference value k is the ratio of the two functional values assumed by the first-order cofactor of an element of its indefinite-admittance matrix under the condition that the element x assumes its nominal value and the condition that the element x assumes the value k. For $k = 0$ we write $F_0(x) = F(x)$.

Thus, it reduces to ordinary return difference when the reference value k is zero. This gives

$$F_k(x) = \frac{Y_{ij}(x)}{Y_{ij}(k)} \tag{5.83}$$

where $Y_{ij}(k) = Y_{ij}(x)|_{x=k}$. Its relation to return differences for the zero reference value can easily be derived by a simple manipulation:

$$F_k(x) = \frac{Y_{ij}(x)}{Y_{ij}(k)} = \frac{Y_{ij}(x)/Y_{ij}(0)}{Y_{ij}(k)/Y_{ij}(0)} \tag{5.84}$$

giving

$$F_k(x) = \frac{F(x)}{F(k)} \tag{5.85}$$

In words, this states that the return difference with respect to an element x for a general reference value k is equal to the ratio of the return difference with respect to x to that with respect to k, both for the zero reference value.

Following Definition 4.2, we define the general return ratio.

Definition 5.2: General return ratio The *general return ratio* T_k with respect to a voltage-controlled current source $I = xV$ for a general reference value k is the negative of the voltage appearing at the controlling branch when the controlled current source is replaced by a parallel combination of an independent current source of $x - k$ amperes and a controlled current source $I = kV$ and when the input excitation is set to zero. For $k = 0$ we write $T_0 = T$.

Refer to the general feedback configuration of Fig. 5.8. The general return ratio T_k is the negative of the voltage V'_{ab} appearing at terminals a and b of Fig. 5.9. We

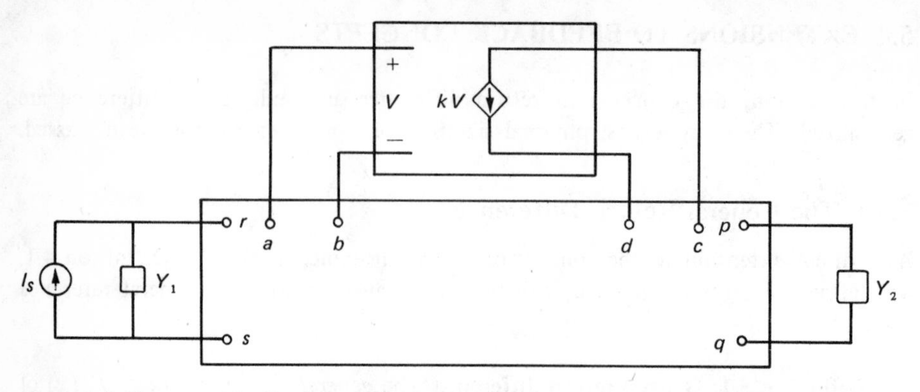

Figure 5.8 The general feedback amplifier configuration.

show that the general return difference $F_k(x)$ and general return ratio T_k are related by the equation $F_k(x) = 1 + T_k$.

As discussed in Chap. 4, the element x enters the indefinite-admittance matrix **Y** in a rectangular pattern as shown in (4.123). The cofactor $Y_{db}(x)$ can be expanded as

$$Y_{db}(x) = Y_{db}(0) + xY_{ca,db} \tag{5.86}$$

where $Y_{ca,db}$ is independent of x. Replacing x by k yields

$$Y_{db}(k) = Y_{db}(0) + kY_{ca,db} \tag{5.87}$$

which, when combined with (5.86), gives

$$Y_{db}(x) = Y_{db}(k) + x'Y_{ca,db} \tag{5.88}$$

where

$$x' = x - k \tag{5.89}$$

Observe that the indefinite-admittance matrix of the network of Fig. 5.9 can be obtained from $\mathbf{Y}(x)$ by replacing the element x by k; the resulting matrix is simply

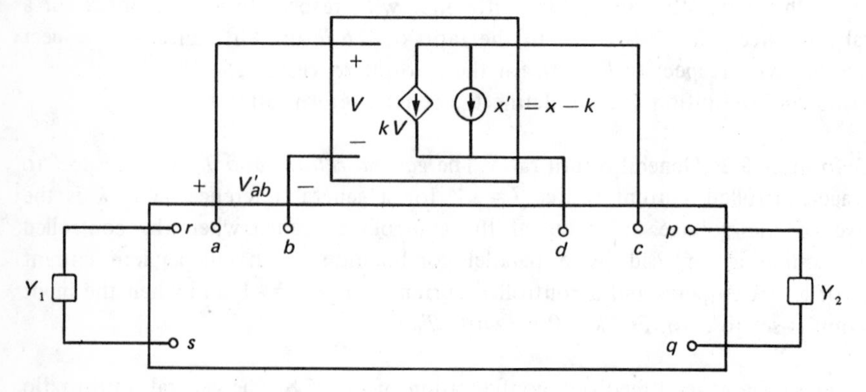

Figure 5.9 The physical interpretation of the general return ratio with respect to the controlling parameter x of a voltage-controlled current source for a general reference value k.

$\mathbf{Y}(k)$. By appealing to formula (2.94), the voltage V'_{ab} appearing at terminals a and b of Fig. 5.9 is found to be

$$V'_{ab} = -x' \frac{Y_{ca,db}(k)}{Y_{uv}(k)} = -x' \frac{Y_{ca,db}}{Y_{uv}(k)} \tag{5.90}$$

Now we are in a position to show that

$$\begin{aligned} F_k(x) &= \frac{Y_{ij}(x)}{Y_{ij}(k)} = \frac{Y_{db}(x)}{Y_{uv}(k)} = \frac{Y_{db}(k) + x' Y_{ca,db}}{Y_{db}(k)} = 1 + x' \frac{Y_{ca,db}}{Y_{uv}(k)} \\ &= 1 - V'_{ab} = 1 + T_k \end{aligned} \tag{5.91}$$

Thus, physically the general return difference $F_k(x)$ can be interpreted as follows: Replace the voltage-controlled current source $I = xV$ of Fig. 5.8 by two controlled current sources $I_1 = kV_1$ and $I_2 = x'V_2$ connected in parallel as shown in Fig. 5.10. The controlling branch of the controlled current source $I_2 = x'V_2$ is broken off as marked and a voltage source of 1 V is applied to the right of the breaking mark. The voltage appearing at the left of the breaking mark resulting from the 1-V excitation is then V'_{ab}, as indicated in Fig. 5.10. The negative of V'_{ab} is the general return ratio T_k. The general return difference $F_k(x)$ is simply the difference of the 1-V excitation and the returned voltage V'_{ab} as illustrated in the figure. The significance of this interpretation is that it is useful in the calculation of the sensitivity function and in the measurement of the return ratio for practical feedback amplifiers, because the interpretation corresponds to the situation where the feedback amplifier under study is made partially active rather than completely dead, as in the original interpretation of the return ratio and return difference for the zero reference value.

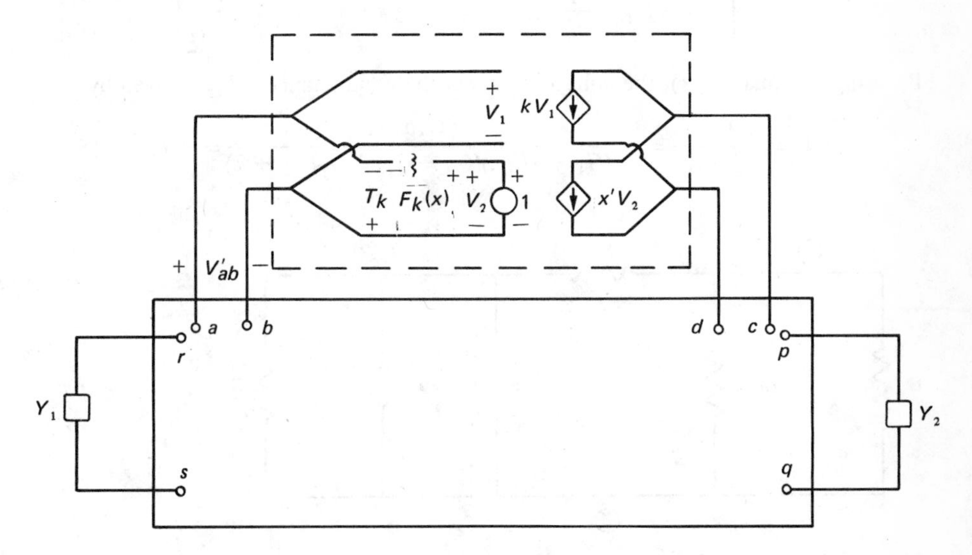

Figure 5.10 The physical interpretation of the general return difference with respect to the controlling parameter x of a voltage-controlled current source for a general reference value k.

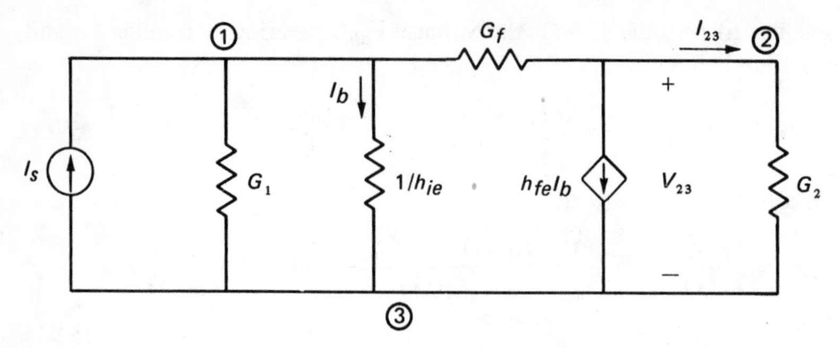

Figure 5.11 An equivalent network of a common-emitter feedback amplifier.

Example 5.7 The network of Fig. 5.11 was considered in Example 5.3. We wish to compute the general return difference with respect to the controlling parameter $\alpha \equiv h_{fe}/h_{ie}$ for a general reference value k. For this we replace the controlled current source $h_{fe}I_b$ by a parallel combination of an independent current source of α' amperes and a voltage-controlled current source $I = kV$ as illustrated in Fig. 5.12. The voltage across the conductance $1/h_{ie}$ is the negative of the general return ratio T_k. To compute T_k, we first write down the indefinite-admittance matrix of the network as

$$\mathbf{Y} = \begin{bmatrix} G_f + G_1 + \dfrac{1}{h_{ie}} & -G_f & -G_1 - \dfrac{1}{h_{ie}} \\ k - G_f & G_2 + G_f & -k - G_2 \\ -k - G_1 - \dfrac{1}{h_{ie}} & -G_2 & k + G_1 + G_2 + \dfrac{1}{h_{ie}} \end{bmatrix} \tag{5.92}$$

By using formula (2.94), the voltage V across the conductance $1/h_{ie}$ is given by

$$V = \alpha' \frac{Y_{31,23}}{Y_{33}} = -\frac{G_f\alpha'}{(G_1 + 1/h_{ie})(G_2 + G_f) + G_f(G_2 + k)} = -T_k \tag{5.93}$$

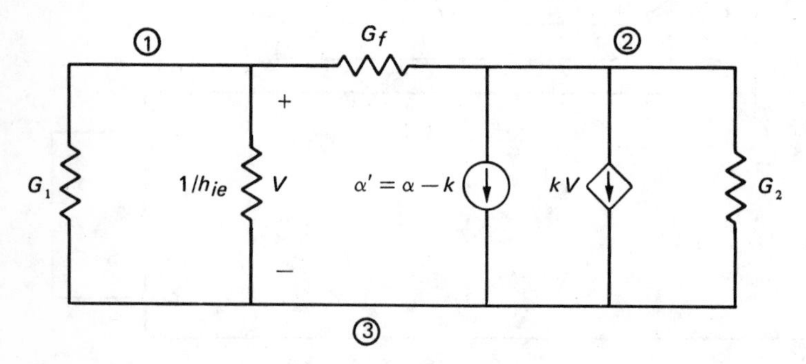

Figure 5.12 The network used to compute the general return difference $F_k(\alpha)$.

The general return difference is obtained as

$$F_k(\alpha) = 1 + T_k = \frac{(G_1 + 1/h_{ie})(G_2 + G_f) + G_f(G_2 + \alpha)}{(G_1 + 1/h_{ie})(G_2 + G_f) + G_f(G_2 + k)} \tag{5.94}$$

which is equal to the ratio of the two functional values assumed by the cofactor of an element of the matrix (5.21) under the condition that α assumes its nominal value and the condition that α assumes the value k.

5.4.2 The General Null Return Difference

The extension is straightforward, and we first state the definition.

Definition 5.3: General null return difference The *general null return difference* $\hat{F}_k(x)$ of a feedback amplifier with respect to an element x for a general reference value k is the ratio of the two functional values assumed by the second-order cofactor $Y_{rp,sq}$ of the elements of its indefinite-admittance matrix $\mathbf{Y}$ under the condition that the element x assumes its nominal value and the condition that the element x assumes the value k, where r and s are input terminals and p and q are output terminals of the amplifier. For $k = 0$ we write $\hat{F}_0(x) = \hat{F}(x)$.

Referring to the general configuration of Fig. 5.8, we have

$$\hat{F}_k(x) = \frac{Y_{rp,sq}(x)}{Y_{rp,sq}(k)} \tag{5.95}$$

provided, of course, that $Y_{rp,sq}(k) \neq 0$. Likewise, we can define the general null return ratio.

Definition 5.4: General null return ratio The *general null return ratio* $\hat{T}_k$ with respect to a voltage-controlled current source $I = xV$ for a general reference value k is the negative of the voltage appearing at the controlling branch when the controlled current source is replaced by a parallel combination of an independent current source of $x - k$ amperes and a controlled current source $I = kV$ and when the input excitation is adjusted so that the output of the amplifier is identically zero. For $k = 0$ we write $\hat{T}_0 = \hat{T}$.

As in (5.85), we can derive a relation between the general null return difference and the null return differences. Equation (5.95) can be rewritten as

$$\hat{F}_k(x) = \frac{Y_{rp,sq}(x)}{Y_{rp,sq}(k)} = \frac{Y_{rp,sq}(x)/Y_{rp,sq}(0)}{Y_{rp,sq}(k)/Y_{rp,sq}(0)} \tag{5.96}$$

provided that $Y_{rp,sq}(0) \neq 0$, giving

$$\hat{F}_k(x) = \frac{\hat{F}(x)}{\hat{F}(k)} \tag{5.97}$$

Thus, the general null return difference with respect to an element x for a general reference value k is equal to the ratio of the null return difference with respect to x to that with respect to k, both for the zero reference value.

5.5 THE NETWORK FUNCTIONS AND GENERAL RETURN DIFFERENCE AND GENERAL NULL RETURN DIFFERENCE

We demonstrate that formulas (4.176), (4.197), and (4.199) can be expressed in terms of the general return difference, the general null return difference, and the transfer function $w(k)$. We also give a simple physical interpretation of the general return difference for a one-port admittance.

Combining (4.176) with (5.85) and (5.97) yields a relation among the transfer impedance or current gain and return difference and null return difference:

$$\frac{w(x)}{w(k)} = \frac{\hat{F}(x)}{F(x)} \frac{F(k)}{\hat{F}(k)} \tag{5.98}$$

giving

$$w(x) = w(k) \frac{\hat{F}_k(x)}{F_k(x)} \tag{5.99}$$

If $w(x)$ represents the voltage gain, then as in (4.196) we can write

$$\frac{w(x)}{w(k)} = \frac{Y_{rp,sq}(x)}{Y_{rp,sq}(k)} \frac{Y_{rr,ss}(k)}{Y_{rr,ss}(x)} \tag{5.100}$$

The first term in the product on the right-hand side is the general null return difference $\hat{F}_k(x)$. The second term can be interpreted as the reciprocal of the general return difference with respect to x for a general reference value k when the input port of the amplifier is short-circuited. Thus, the voltage gain V_{pq}/V_{rs} (see Fig. 5.8) can be expressed as

$$w(x) = w(k) \frac{\hat{F}_k(x)}{F_k(\text{input short-circuited})} \tag{5.101}$$

Likewise, if $w(x)$ denotes the short-circuit current gain, then as in (4.198) we can write

$$\frac{w(x)}{w(k)} = \frac{Y_{rp,sq}(x)}{Y_{rp,sq}(k)} \frac{Y_{pp,qq}(k)}{Y_{pp,qq}(x)} \tag{5.102}$$

As before, the second term in the product can be interpreted as the reciprocal of the general return difference with respect to x for a general reference value k when the output port of the amplifier is short-circuited, giving a formula for the short-circuit current gain as

$$w(x) = w(k) \frac{\hat{F}_k(x)}{F_k(\text{output short-circuited})} \tag{5.103}$$

Finally, we give a physical interpretation of the general return difference $F_k(x)$ with respect to a one-port admittance x for a general reference value k. The result can easily be deduced from (5.91) by letting $a = c$ and $b = d$ as follows:

$$F_k(x) = \frac{Y_{dd}(x)}{Y_{dd}(k)} = \frac{Y_{dd}(k) + x' Y_{cc,dd}}{Y_{dd}(k)} = 1 + x' \frac{Y_{cc,dd}}{Y_{dd}(k)}$$
$$= 1 + \frac{x'}{y + k} = \frac{y + x}{y + k} \tag{5.104}$$

where y is the admittance that x faces. We remark that $\mathbf{Y}(k)$ is the indefinite-admittance matrix of the original feedback network when x is replaced by k. Thus, according to (2.94), $Y_{cc,dd}/Y_{dd}(k)$ is the impedance looking into terminals c and d when x is replaced by k, whose reciprocal is $y + k$. In other words, the general return difference $F_k(x)$ with respect to the one-port admittance x is equal to the ratio of total admittance looking into the node pair where x is connected to that when x is replaced by its reference value k.

Example 5.8 In the feedback network of Fig. 5.11, suppose that we wish to compute the transfer impedance V_{23}/I_s by means of formula (5.99). Let $\alpha \equiv h_{fe}/h_{ie}$ be the element of interest. To compute $\hat{F}_k(\alpha)$, we insert a current source I_s between terminals 1 and 3 in the network of Fig. 5.12. The resulting network is presented in Fig. 5.13. We then adjust I_s so that the output voltage across the conductance G_2 is zero. This yields

$$I_s = I_0 \equiv \frac{(\alpha - k)(G_1 + G_f + 1/h_{ie})}{G_f - k} \tag{5.105}$$

Under this situation, the returned voltage V is $I_0/(G_1 + G_f + 1/h_{ie})$, which is the negative of the general null return ratio $\hat{T}_k$. Thus, we obtain

$$\hat{F}_k(\alpha) = \frac{G_f - \alpha}{G_f - k} \tag{5.106}$$

The transfer impedance when α is replaced by k is given by

$$w(k) = \frac{G_f - k}{(G_1 + 1/h_{ie})(G_2 + G_f) + G_f(G_2 + k)} \tag{5.107}$$

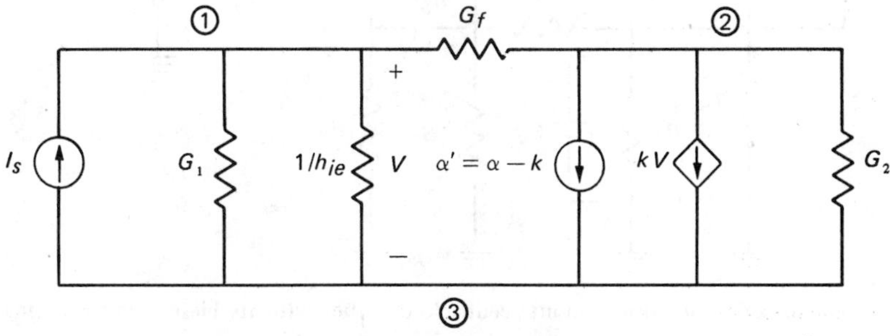

Figure 5.13 The network used to compute the general null return difference $\hat{F}_k(\alpha)$.

Substituting (5.94), (5.106), and (5.107) in (5.99) gives the desired transfer impedance

$$w(\alpha) = w(k)\,\frac{\hat{F}_k(\alpha)}{F_k(\alpha)} = \frac{G_f - \alpha}{(G_1 + 1/h_{ie})(G_2 + G_f) + G_f(G_2 + \alpha)} \tag{5.108}$$

Now suppose that the element of interest is the one-port admittance G_1. The admittance that G_1 faces is given by

$$y = \frac{(G_2 + G_f)/h_{ie} + G_f(G_2 + \alpha)}{G_2 + G_f} \tag{5.109}$$

Thus, from (5.104) the general return difference $F_k(G_1)$ with respect to G_1 for a general reference value k is obtained as

$$F_k(G_1) = \frac{y + G_1}{y + k} = 1 + \frac{(G_1 - k)(G_2 + G_f)}{(G_2 + G_f)(k + 1/h_{ie}) + G_f(G_2 + \alpha)} \tag{5.110}$$

the second term on the right-hand side being the general return ratio T_k.

Example 5.9 Consider the parallel-series or current-shunt feedback amplifier of Fig. 5.14, which was studied in Sec. 4.2.4. After the biasing and coupling circuitry

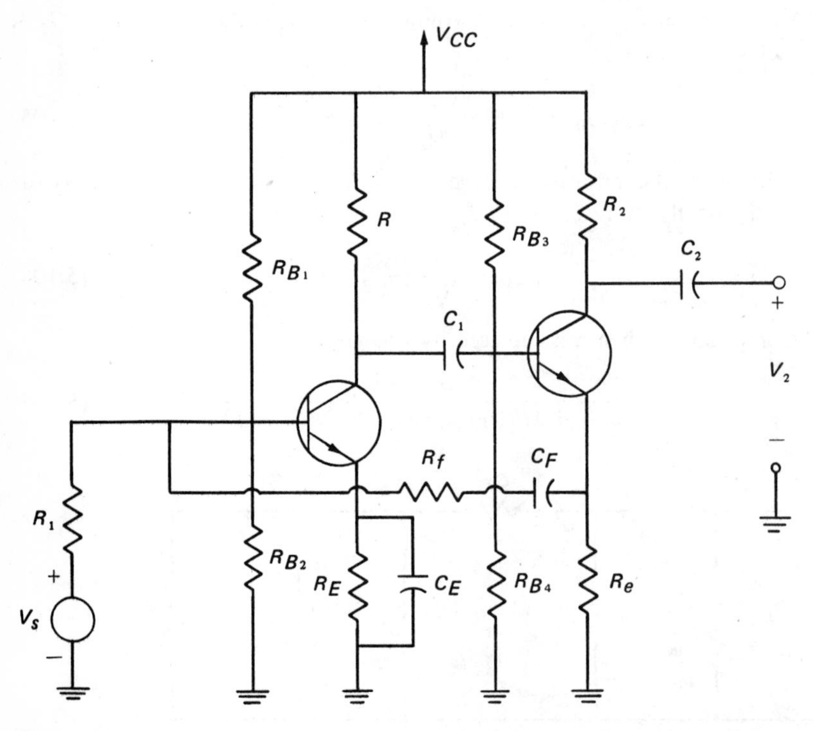

Figure 5.14 A parallel-series or current-shunt feedback amplifier with its biasing and coupling circuitry.

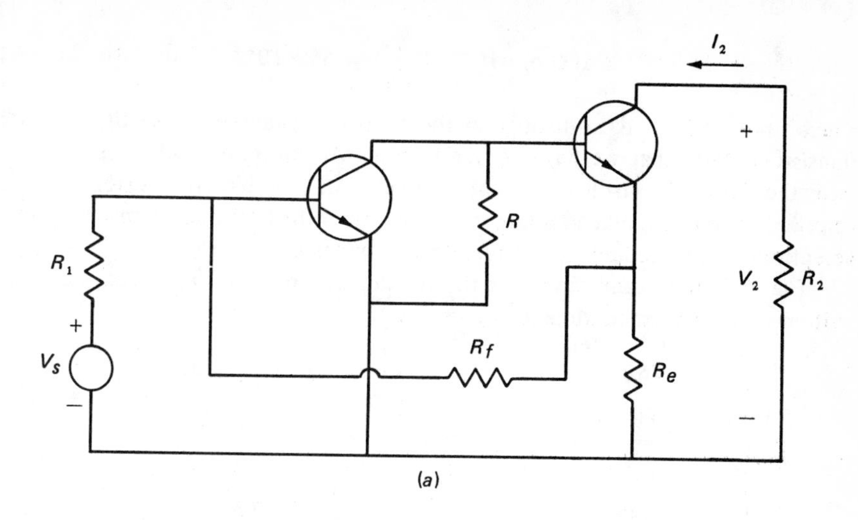

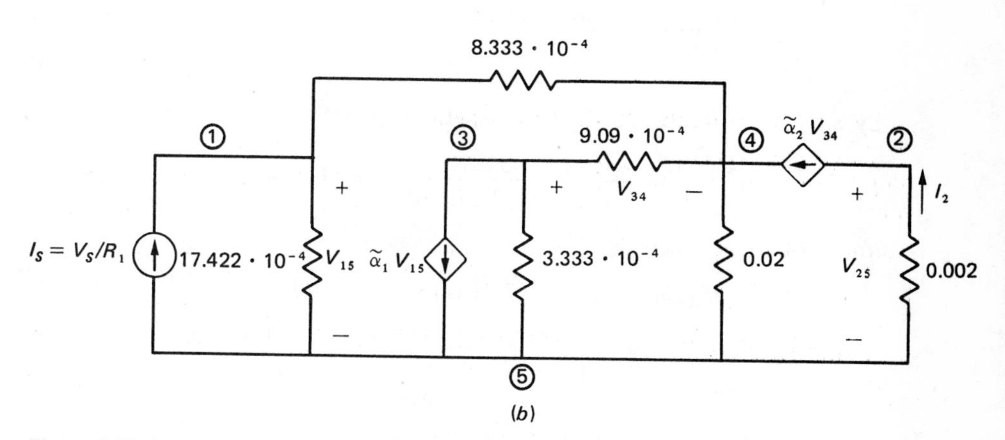

Figure 5.15 (*a*) The amplifier of Fig. 5.14 after the removal of its biasing and coupling circuitry and (*b*) its equivalent network.

have been removed, the network reduces to that of Fig. 5.15*a*. Assume that the two transistors are identical, with the following specifications:

$$\begin{aligned} h_{ie} &= 1.1\ \text{k}\Omega \qquad h_{fe} = 50 \\ h_{re} &= h_{oe} = 0 \qquad R = 3\ \text{k}\Omega \\ R_e &= 50\ \Omega \qquad R_f = 1.2\ \text{k}\Omega \\ R_1 &= 1.2\ \text{k}\Omega \qquad R_2 = 500\ \Omega \end{aligned} \tag{5.111}$$

An equivalent network of Fig. 5.15*a* is presented in Fig. 5.15*b* with all the conductances denoted in mho and

$$\tilde{\alpha}_j \equiv \alpha_j \cdot 10^{-4} \equiv \frac{h_{fe}}{h_{ie}} = 455 \cdot 10^{-4} \tag{5.112}$$

where $j = 1, 2$ is used to distinguish the transconductances of the first and second transistors. Note that we have converted the two current-controlled current sources into the voltage-controlled current sources. All the biasing resistors have been ignored. We remark that in some situations, the effect of biasing resistors may not be entirely negligible and should therefore be included.

The indefinite-admittance matrix of the network of Fig. 5.15*b* can now be written down by inspection and is given by

$$\mathbf{Y} = 10^{-4} \begin{bmatrix} 25.755 & 0 & 0 & -8.333 & -17.422 \\ 0 & 20 & \alpha_2 & -\alpha_2 & -20 \\ \alpha_1 & 0 & 12.423 & -9.09 & -3.333 - \alpha_1 \\ -8.333 & 0 & -9.09 - \alpha_2 & 217.423 + \alpha_2 & -200 \\ -17.422 - \alpha_1 & -20 & -3.333 & -200 & 240.755 + \alpha_1 \end{bmatrix} \tag{5.113}$$

By applying formula (2.94), the closed-loop transfer impedance is found to be

$$w = \frac{V_{25}}{I_s} = \frac{Y_{12,55}}{Y_{55}} = \frac{431.42 \cdot 10^{-7}}{373.05 \cdot 10^{-11}} = 11.564 \text{ k}\Omega \tag{5.114}$$

confirming (4.103*c*), where

$$Y_{12,55} = (208.33\alpha_1\alpha_2 + 27.77\alpha_2) \cdot 10^{-12} = 431.42 \cdot 10^{-7} \tag{5.115a}$$

$$\begin{aligned} Y_{uv} = Y_{55} &= (166.66\alpha_1\alpha_2 + 1514.94\alpha_1 + 1716.83\alpha_2 + 1{,}331{,}494) \cdot 10^{-16} \\ &= 373.05 \cdot 10^{-11} \end{aligned} \tag{5.115b}$$

The closed-loop voltage and current gains are obtained as

$$w_V = \frac{V_{25}}{V_s} = \frac{w}{R_1} = \frac{11.56 \cdot 10^3}{R_1} = 9.64 \tag{5.116a}$$

$$w_I = \frac{I_2}{I_s} = -\frac{V_{25}}{I_s R_2} = -\frac{w}{R_2} = -23.13 \tag{5.116b}$$

The return differences with respect to the transconductances $\tilde{\alpha}_j$ for a general reference value k are given by

$$F_k(\tilde{\alpha}_1) = \frac{Y_{55}(\tilde{\alpha}_1)}{Y_{55}(k)} = \frac{773.45\tilde{\alpha}_1 + 2.1127}{773.45k + 2.1127} = \frac{37.31}{773.45k + 2.1127} \tag{5.117a}$$

$$F_k(\tilde{\alpha}_2) = \frac{Y_{55}(\tilde{\alpha}_2)}{Y_{55}(k)} = \frac{775.47\tilde{\alpha}_2 + 2.0208}{775.47k + 2.0208} = \frac{37.31}{775.47k + 2.0208} \tag{5.117b}$$

yielding

$$F(\tilde{\alpha}_1) = F_0(\tilde{\alpha}_1) = 17.66 \tag{5.118a}$$

$$F(\tilde{\alpha}_2) = F_0(\tilde{\alpha}_2) = 18.46 \tag{5.118b}$$

The null return differences with respect to $\tilde{\alpha}_j$ for a general reference value k are determined from (5.95) with terminals 1 and 5 as the input and terminals 2 and 5 as the output:

$$\hat{F}_k(\tilde{\alpha}_1) = \frac{Y_{12,55}(\tilde{\alpha}_1)}{Y_{12,55}(k)} = \frac{9479\tilde{\alpha}_1 + 0.1263}{9479k + 0.1263} = \frac{431.42}{9479k + 0.1263} \tag{5.119a}$$

$$\hat{F}_k(\tilde{\alpha}_2) = \frac{Y_{12,55}(\tilde{\alpha}_2)}{Y_{12,55}(k)} = \frac{948.18\tilde{\alpha}_2}{948.18k} = \frac{\tilde{\alpha}_2}{k} = \frac{0.0455}{k} \tag{5.119b}$$

giving

$$\hat{F}(\tilde{\alpha}_1) = \hat{F}_0(\tilde{\alpha}_1) = 3415.84 \tag{5.120a}$$

$$\hat{F}(\tilde{\alpha}_2) = \hat{F}_0(\tilde{\alpha}_2) = \infty \tag{5.120b}$$

From (5.115), the transfer impedance w can be written as

$$w = w(\tilde{\alpha}_1, \tilde{\alpha}_2) = \frac{208.33\tilde{\alpha}_1\tilde{\alpha}_2 \cdot 10^4 + 27.76\tilde{\alpha}_2}{166.66\tilde{\alpha}_1\tilde{\alpha}_2 + (1514.94\tilde{\alpha}_1 + 1716.83\tilde{\alpha}_2 + 133.15) \cdot 10^{-4}} \tag{5.121}$$

which leads to

$$w(k, \tilde{\alpha}_2) = \frac{94{,}790k + 1.2631}{7.735k + 211.27 \cdot 10^{-4}} \tag{5.122a}$$

$$w(\tilde{\alpha}_1, k) = \frac{94{,}818k}{7.755k + 202.08 \cdot 10^{-4}} \tag{5.122b}$$

Substituting (5.117), (5.119), and (5.122) in (5.99), we get

$$w(\tilde{\alpha}_1, \tilde{\alpha}_2) = w(k, \tilde{\alpha}_2)\frac{\hat{F}_k(\tilde{\alpha}_1)}{F_k(\tilde{\alpha}_1)} = 11.563 \text{ k}\Omega \tag{5.123a}$$

$$w(\tilde{\alpha}_1, \tilde{\alpha}_2) = w(\tilde{\alpha}_1, k)\frac{\hat{F}_k(\tilde{\alpha}_2)}{F_k(\tilde{\alpha}_2)} = 11.563 \text{ k}\Omega \tag{5.123b}$$

confirming (5.114).

Assume now that $w = g_{12,55}$ denotes the voltage gain V_{25}/V_{15}, as defined in Fig. 5.15*b*. Appealing to formula (2.97) in conjunction with (5.115*a*), we obtain

$$g_{12,55} = \frac{V_{25}}{V_{15}} = \frac{Y_{12,55}}{Y_{11,55}} = \frac{431.42 \cdot 10^{-7}}{826.98 \cdot 10^{-10}} = 521.68 \tag{5.124}$$

where $$Y_{11,55} = (66.66\alpha_2 + 52{,}368.36) \cdot 10^{-12} = 826.98 \cdot 10^{-10} \tag{5.125}$$

As in (5.122), $g_{12,55}$ can also be written as

$$g_{12,55} = g_{12,55}(\tilde{\alpha}_1, \tilde{\alpha}_2) = \frac{208.33\tilde{\alpha}_1\tilde{\alpha}_2 \cdot 10^4 + 27.77\tilde{\alpha}_2}{66.66\tilde{\alpha}_2 + 5.2368} \tag{5.126}$$

which gives

$$g_{12,55}(k, \tilde{\alpha}_2) = \frac{94{,}790k + 1.2635}{8.2698} \tag{5.127a}$$

$$g_{12,55}(\tilde{\alpha}_1, k) = \frac{94818k}{66.66k + 5.2368} \tag{5.127b}$$

To compute the return difference F_k(input short-circuited), we short-circuit the input terminals 1 and 5. The corresponding indefinite-admittance matrix is obtained by adding the first row of (5.113) to the fifth row and the first column to the fifth column and then deleting the first row and column. The first-order cofactor of an element of the resulting matrix is simply $Y_{11,55}$. Thus, we have

$$F_k(\text{input short-circuited}) = \frac{Y_{11,55}(\tilde{\alpha}_1)}{Y_{11,55}(k)} = 1 \tag{5.128a}$$

with respect to $\tilde{\alpha}_1$, and

$$F_k(\text{input short-circuited}) = \frac{Y_{11,55}(\tilde{\alpha}_2)}{Y_{11,55}(k)} = \frac{8.2698}{66.66k + 5.2368} \tag{5.128b}$$

with respect to $\tilde{\alpha}_2$.

Substituting (5.119), (5.127), and (5.128) in (5.101) results in

$$g_{12,55} = g_{12,55}(k, \tilde{\alpha}_2) \frac{\hat{F}_k(\tilde{\alpha}_1)}{F_k(\text{input short-circuited})} = 521.68 \tag{5.129a}$$

$$g_{12,55} = g_{12,55}(\tilde{\alpha}_1, k) \frac{\hat{F}_k(\tilde{\alpha}_2)}{F_k(\text{input short-circuited})} = 521.68 \tag{5.129b}$$

confirming (5.124).

As in (4.177), (5.99) is also valid for the driving-point impedance. This is the particular situation where the input and output are taken to be the same port. In this case, $F_k(x)$ denotes the general return difference for the situation when the port where the driving-point impedance is defined is left open without a source, and we write $F_k(x) = F_k$(input open-circuited). Likewise, $\hat{F}_k(x)$ is the general return difference for the situation when the port where the driving-point impedance is measured is short-circuited, and we write $\hat{F}_k(x) = F_k$(input short-circuited). Consequently, the driving-point impedance $Z(x)$ looking into a terminal pair can be conveniently expressed as

$$Z(x) = Z(k) \frac{F_k(\text{input short-circuited})}{F_k(\text{input open-circuited})} \tag{5.130}$$

We emphasize that the word *input* means the terminal pair where the driving-point impedance is measured, not necessarily the input port of the feedback amplifier.

Example 5.10 Consider the same current-shunt feedback amplifier of Fig. 5.14. We compute the driving-point impedances facing the current source I_s and looking into the output terminals 2 and 5 in the network of Fig. 5.15b. By appealing to (2.95) in conjunction with (5.113), the amplifier input and output impedances are found to be

$$z_{11,55} = \frac{Y_{11,55}}{Y_{55}} = \frac{82{,}698}{3731} = 22.17\ \Omega \tag{5.131a}$$

$$z_{22,55} = \frac{Y_{22,55}}{Y_{55}} = \frac{1{,}864{,}617}{3731} = 499.76\ \Omega \tag{5.131b}$$

where

$$Y_{22,55} = (8.33\alpha_1\alpha_2 + 75.75\alpha_1 + 85.84\alpha_2 + 66{,}575)\cdot 10^{-12} = 1{,}864{,}617\cdot 10^{-12} \tag{5.132}$$

Suppose that $\tilde{\alpha}_1$ is the element of interest. Then we can write

$$z_{11,55}(k) = \frac{826.98}{773.45k + 2.1127} \tag{5.133a}$$

$$z_{22,55}(k) = \frac{386{,}590k + 1056.32}{773.45k + 2.1127} \tag{5.133b}$$

To compute F_k(input short-circuited) for $z_{11,55}$, we short-circuit terminals 1 and 5 in the network of Fig. 5.15*b*, giving F_k(input short-circuited) = 1, as in Eq. (5.128*a*). For F_k(input open-circuited), we remove the current source I_s and compute $F_k(\tilde{\alpha}_1)$, which is given by (5.117*a*). Substituting these in (5.130) yields

$$z_{11,55} = z_{11,55}(k)\frac{F_k(\text{input short-circuited})}{F_k(\text{input open-circuited})} = 22.17\ \Omega \tag{5.134}$$

confirming (5.131*a*).

To compute F_k(input short-circuited) for $z_{22,55}$, we short-circuit terminals 2 and 5 in the network of Fig. 5.15*b*. The first-order cofactor of an element of the resulting indefinite-admittance matrix is simply $Y_{22,55}$. Thus, we have

$$F_k(\text{input short-circuited}) = \frac{Y_{22,55}(\tilde{\alpha}_1)}{Y_{22,55}(k)} = \frac{18{,}646.17}{386{,}590k + 1056.32} \tag{5.135}$$

For F_k(input open-circuited), we leave terminals 2 and 5 open as shown in Fig. 5.15*b* and compute $F_k(\tilde{\alpha}_1)$, which is again given by Eq. (5.117*a*). Substituting (5.117*a*), (5.133*b*), and (5.135) in (5.130), we obtain

$$z_{22,55} = z_{22,55}(k)\frac{F_k(\text{input short-circuited})}{F_k(\text{input open-circuited})} = 499.76\ \Omega \tag{5.136}$$

confirming (5.131*b*).

5.6 THE RELATIVE SENSITIVITY FUNCTION AND FEEDBACK

As a generalization of the concept of sensitivity function discussed in Sec. 5.1, we introduce the *relative sensitivity function*, which is defined according to the equation

$$S'(x') = \frac{x'}{w}\frac{\partial w}{\partial x'} \tag{5.137}$$

where, as before, $x' = x - k$. Comparing this with (5.1) for the sensitivity function, we see that the relative sensitivity function is defined with respect to the altered element x'. From their definitions, we find that the sensitivity function and the relative sensitivity function are related by the equation

$$x'S(x) = xS'(x') \tag{5.138}$$

Since the element x enters the indefinite-admittance matrix **Y** in a rectangular pattern as shown in (4.123), we have the following expansions:

$$Y_{uv}(x) = Y_{uv}(0) + x\dot{Y}_{uv} \tag{5.139a}$$

$$Y_{rp,sq}(x) = Y_{rp,sq}(0) + x\dot{Y}_{rp,sq} \tag{5.139b}$$

as in (5.4). Note that $\dot{Y}_{uv}$ and $\dot{Y}_{rp,sq}$ are independent of x. Following (5.87) we establish the identities

$$Y_{uv}(x) = Y_{uv}(k) + x'\dot{Y}_{uv} \tag{5.140a}$$

$$Y_{rp,sq}(x) = Y_{rp,sq}(k) + x'\dot{Y}_{rp,sq} \tag{5.140b}$$

We now proceed to derive the general relative sensitivity functions for various network functions. To start, let $w(x)$ represent either the current gain I_{pq}/I_s or the transfer impedance V_{pq}/I_s, as indicated in Fig. 5.8. From Eq. (5.138) in conjunction with Eqs. (5.2) and (5.140) we obtain

$$S'(x') = x'\frac{\dot{Y}_{rp,sq}(x)}{Y_{rp,sq}(x)} - x'\frac{\dot{Y}_{uv}(x)}{Y_{uv}(x)} = \frac{Y_{rp,sq}(x) - Y_{rp,sq}(k)}{Y_{rp,sq}(x)} - \frac{Y_{uv}(x) - Y_{uv}(k)}{Y_{uv}(x)}$$

$$= \frac{Y_{uv}(k)}{Y_{uv}(x)} - \frac{Y_{rp,sq}(k)}{Y_{rp,sq}(x)} \tag{5.141}$$

giving

$$S'(x') = \frac{1}{F_k(x)} - \frac{1}{\hat{F}_k(x)} \tag{5.142}$$

Using (5.99) in (5.142) yields

$$S'(x') = \frac{1}{F_k(x)}\left[1 - \frac{w(k)}{w(x)}\right] \tag{5.143}$$

In the special situation where the reference value k is zero, the relative sensitivity function $S'(x')$ becomes the familiar sensitivity function $S(x)$, and (5.143) reduces to (5.6). Also, we see that if $w(k) = 0$, then $S'(x') = 1/F_k(x)$, the relative sensitivity being equal to the reciprocal of the general return difference $F_k(x)$. In fact, since the reference value k is arbitrary, we can choose the particular value for which $w(k) = 0$. This yields a direct generalization of the simple relation between the sensitivity function and the return difference for the ideal feedback model. The desired reference value k_0 is obtained by setting $w(k_0)$ of (5.2) to zero, giving

$$Y_{rp,sq}(k_0) = 0 \tag{5.144}$$

Likewise, following the derivations of (5.8) and (5.10) in conjunction with the interpretations of (5.101) and (5.103) for F_k(input short-circuited) and F_k(output short-circuited), we can show that the relative sensitivity function for the voltage gain V_{pq}/V_{rs} can be expressed as

$$S'(x') = \frac{1}{F_k(\text{input short-circuited})}\left[1 - \frac{w(k)}{w(x)}\right] \tag{5.145}$$

and the relative sensitivity function for the short-circuit current gain can be expressed as

$$S'(x') = \frac{1}{F_k(\text{output short-circuited})}\left[1 - \frac{w(k)}{w(x)}\right] \tag{5.146}$$

Their derivations are straightforward and are left as exercises (see Probs. 5.10 and 5.11).

Example 5.11 Again consider the feedback network of Fig. 5.11. We compute the relative sensitivity function for the transfer impedance V_{23}/I_s or the current gain I_{23}/I_s with respect to the controlling parameter α. This can easily be accomplished by simply substituting (5.94) and (5.106) in (5.142). However, if we choose the reference value $k = k_0$ satisfying the equation $Y_{12,33}(k_0) = 0$, where $Y_{12,33}(\alpha)$ is a second-order cofactor of the elements of the matrix (5.21), the reciprocal of the general null return difference $\hat{F}_k(\alpha)$ will be zero or, equivalently, $w(k_0) = 0$. The desired reference value k_0 for which $w(k_0) = 0$ occurs at $k_0 = G_f$. Under this condition, the relative sensitivity function is simply equal to the reciprocal of the general return difference with respect to α for the reference value G_f, and is given by

$$S'(\alpha') = \frac{1}{F_{G_f}(\alpha)} = \frac{(G_1 + 1/h_{ie})(G_2 + G_f) + G_f(G_2 + G_f)}{(G_1 + 1/h_{ie})(G_2 + G_f) + G_f(G_2 + \alpha)} \tag{5.147}$$

where $\alpha' = \alpha - G_f$.

Example 5.12 We compute the relative sensitivities of the network functions considered in Examples 5.10 and 5.11. To compute the relative sensitivity of the transfer impedance $w(\tilde{\alpha}_1)$ with respect to $\tilde{\alpha}_1$, we have, from (5.114), (5.117*a*), and (5.122*a*),

$$w = w(\tilde{\alpha}_1) = 11{,}564\ \Omega \tag{5.148a}$$

$$F_k(\tilde{\alpha}_1) = \frac{37.31}{773.45k + 2.1127} \tag{5.148b}$$

$$w(k) = \frac{94{,}790k + 1.2631}{7.735k + 211.27 \cdot 10^{-4}} \tag{5.148c}$$

Substituting these in (5.143) yields

$$S'(\tilde{\alpha}_1') = -1.2383k + 0.0563 \tag{5.149}$$

where $\tilde{\alpha}_1' = \tilde{\alpha}_1 - k$. Likewise, from (5.114), (5.117*b*), and (5.122*b*), we have

$$S'(\tilde{\alpha}_2') = -1.1911k + 0.0542 \tag{5.150}$$

where $\tilde{\alpha}_2' = \tilde{\alpha}_2 - k$. For $k = 0$, we obtain the sensitivities of the transfer function with respect to the transconductances $\tilde{\alpha}_1$ and $\tilde{\alpha}_2$:

$$S(\tilde{\alpha}_1) = 0.0563 \approx \frac{1}{F(\tilde{\alpha}_1)} \tag{5.151a}$$

$$S(\tilde{\alpha}_2) = 0.0542 = \frac{1}{F(\tilde{\alpha}_2)} \tag{5.151b}$$

since $w = 11{,}564 \gg w(0) = 59.786$ for $\tilde{\alpha}_1$ and $w(0) = 0$ for $\tilde{\alpha}_2$.

For the relative sensitivity of the voltage gain $g_{12,55}$, we use (5.124), (5.127*a*), and (5.128*a*) in (5.145). The result is obtained as

$$S'(\tilde{\alpha}_1') = 0.9997 - 21.9716k \tag{5.152}$$

giving the sensitivity of $g_{12,55}$ with respect to $\tilde{\alpha}_1$ as

$$S(\tilde{\alpha}_1) = 0.9997 \approx \frac{1}{F(\text{input short-circuited})} \tag{5.153}$$

Likewise, the relative sensitivity of $g_{12,55}$ with respect to $\tilde{\alpha}_2$ is found to be

$$S'(\tilde{\alpha}_2') = -13.9175k + 0.6332 \tag{5.154}$$

giving the sensitivity of $g_{12,55}$ with respect to $\tilde{\alpha}_2$ as

$$S(\tilde{\alpha}_2) = 0.6332 = \frac{1}{F(\text{input short-circuited})} \tag{5.155}$$

since $g_{12,55}(0) = 0$.

The above results confirm an earlier assertion that for practical amplifiers, the transfer function $|w(0)|$ is usually very much smaller than $|w(x)|$ in the passband, and $F \approx 1/S$ may be used as a good estimate of the reciprocal of the sensitivity in the same frequency band. However, for the one-port impedance, the conclusion is not generally valid, because $|w(0)|$ is not usually smaller than $|w(x)|$. We illustrate this by calculating the relative sensitivities of the input impedance $z_{11,55}$ and the output impedance $z_{22,55}$ of the feedback network of Fig. 5.15*b*.

From (5.117*a*), (5.131*a*), and (5.133*a*), we have

$$F_k(\tilde{\alpha}_1) = \frac{37.31}{773.45k + 2.1127} \tag{5.156a}$$

$$z_{11,55}(\tilde{\alpha}_1) = 22.17\ \Omega \tag{5.156b}$$

$$z_{11,55}(k) = \frac{826.98}{773.45k + 2.1127} \tag{5.156c}$$

Substituting these in (5.143) gives the relative sensitivity of $z_{11,55}$ with respect to the transconductance $\tilde{\alpha}_1$ as

$$S'(\tilde{\alpha}_1') = 20.7304k - 0.9432 \tag{5.157}$$

giving the sensitivity of $z_{11,55}$ to $\tilde{\alpha}_1$ as

$$S(\tilde{\alpha}_1) = -0.9432 \not\approx \frac{1}{F(\tilde{\alpha}_1)} = 0.0566 \tag{5.158}$$

where $z_{11,55}(0) = 391.43\ \Omega$, which, in fact, is larger than $z_{11,55}(\tilde{\alpha}_1) = 22.17\ \Omega$.

For the relative sensitivity of $z_{22,55}$ with respect to $\tilde{\alpha}_1$, we use (5.117*a*), (5.131*b*), and (5.133*b*). The result is given by

$$S'(\tilde{\alpha}_1') = -0.002715k - 0.000026 \tag{5.159}$$

giving the sensitivity of $z_{22,55}$ to $\tilde{\alpha}_1$ as

$$S(\tilde{\alpha}_1) = -0.000026 \not\approx \frac{1}{F(\tilde{\alpha}_1)} = 0.0566 \tag{5.160}$$

where $z_{22,55}(0) = 499.99\ \Omega$, which is about the same as $z_{22,55}(\tilde{\alpha}_1) = 499.76\ \Omega$.

Finally, we determine the reference values $k = k_0$ so that the relative sensitivities of the network functions become equal to the reciprocal of the general return difference,

$$S'(x') = \frac{1}{F_{k_0}(x)} \tag{5.161}$$

This is equivalent to setting

$$w(k_0) = 0 \tag{5.162}$$

For the transfer impedance $w(\tilde{\alpha}_1)$, we set (5.122*a*) to zero, giving

$$k_0 = -1.33 \cdot 10^{-5} \tag{5.163}$$

and for $w(\tilde{\alpha}_2)$ we set (5.122*b*) to zero, giving $k_0 = 0$, meaning that $F_{k_0}(\tilde{\alpha}_2) = F(\tilde{\alpha}_2)$. For the voltage gain, we set (5.127) to zero and obtain the reference values $-1.33 \cdot 10^{-5}$ and 0 for $F_k(\tilde{\alpha}_1)$ and $F_k(\tilde{\alpha}_2)$, respectively, the results being the same as those for the transfer impedance.

For the output impedance $z_{22,55}(\tilde{\alpha}_1)$, we set (5.133*b*) to zero, yielding

$$k_0 = -2.73 \cdot 10^{-3} \tag{5.164}$$

The reference value for the input impedance $z_{11,55}(\tilde{\alpha}_1)$ is infinite and therefore cannot be meaningfully defined.

The above results are a direct generalization of the simple relation between the sensitivity function and the feedback factor for the ideal feedback model. From (4.3) we calculate the sensitivity function of the closed-loop transfer function $w(\mu)$ with respect to the forward amplifier gain, giving

$$S(\mu) = \frac{1}{1 - \mu\beta} = \frac{1}{F(\mu)} \tag{5.165}$$

5.7 SIGNAL-FLOW GRAPH FORMULATION OF FEEDBACK AMPLIFIER THEORY

So far the feedback amplifier theory has been formulated algebraically in terms of the return difference and the null return difference, which are the ratios of the two functional values assumed by the first- and second-order cofactors of the elements of the indefinite-admittance matrix. In the present section, we demonstrate that it can be equivalently formulated in terms of a signal-flow graph representing the signal transmissions among the various ports. One of the advantages of this formulation is that it can easily be generalized to multiple-loop feedback networks to be discussed in Chap. 7.

Consider the four-port network in the general feedback configuration of Fig. 5.8, which is redrawn in Fig. 5.16. Since the network is linear, the port current I_2' and voltage V can be expressed in terms of the port currents I_s and I, as follows:

$$I_2' = t_{11}I_s + t_{12}I \tag{5.166a}$$

$$V = t_{21}I_s + t_{22}I \tag{5.166b}$$

where the transmittances t_{ij} $(i, j = 1, 2)$ are all independent of the controlling parameter x. However, the port current I and voltage V are not independent, and as indicated in Fig. 5.8, they are related by the equation

$$I = xV \tag{5.167}$$

Equations (5.166) and (5.167) may be represented by the signal-flow graph of Fig. 5.17, which is known as the *fundamental feedback-flow graph.*

Suppose that in Fig. 5.17 we break the branch with transmittance x, as illustrated in Fig. 5.18. Then the return ratio T for the element x is simply the

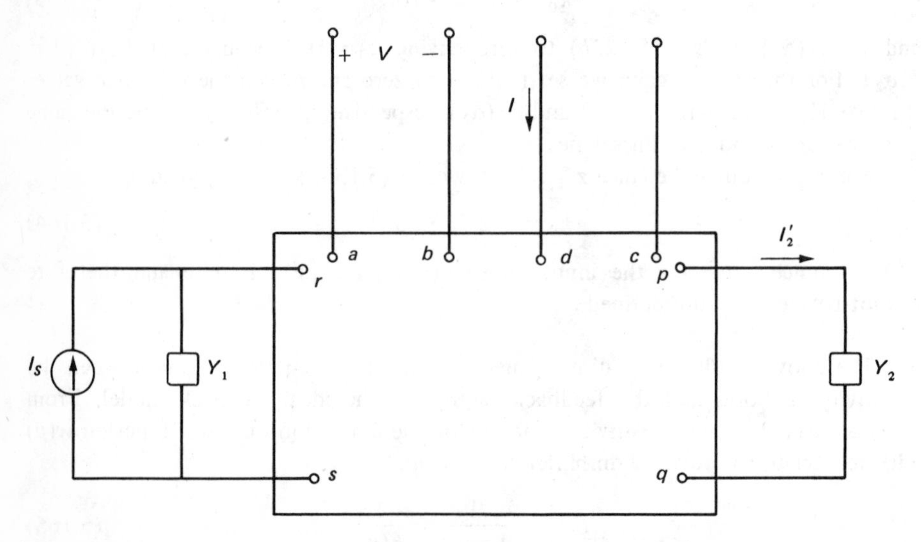

Figure 5.16 The four-port network in the general feedback configuration of Fig. 5.8.

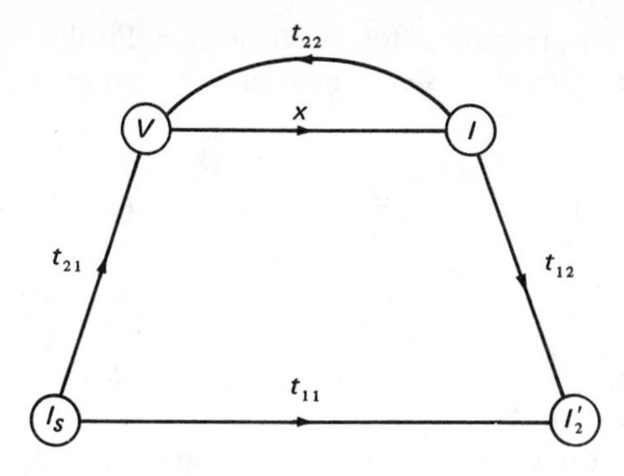

Figure 5.17 The fundamental feedback-flow graph.

negative of the signal returned to the left of the breaking mark when a unit signal is applied to the right of the breaking mark and when the input excitation is set to zero. This gives

$$T = -xt_{22} \tag{5.168}$$

If in (5.166*b*) we set $I_s = 0$ and $I = x$, we find that $V = xt_{22}$. This means that the return ratio T is equal to the negative of the voltage V appearing across the controlling branch of the controlled current source $I = xV$ of the feedback amplifier when the controlled source is replaced by an independent current source of x amperes and when the input excitation I_s is set to zero. The return difference is then the difference between the unit signal input and the returned signal xt_{22} under the situation that the input excitation to the amplifier is set to zero:

$$F(x) = 1 - xt_{22} = 1 + T \tag{5.169}$$

Refer again to Fig. 5.18. A unit signal is applied to the right of the breaking mark. Now we adjust the current source I_s so that the output current I_2' is reduced to zero. This requires that $xt_{12} + I_s t_{11} = 0$, giving the required $I_s = -xt_{12}/t_{11}$. The corresponding value of the returned signal to the left of the breaking mark becomes

$$V = xt_{22} - \frac{xt_{12}t_{21}}{t_{11}} \tag{5.170}$$

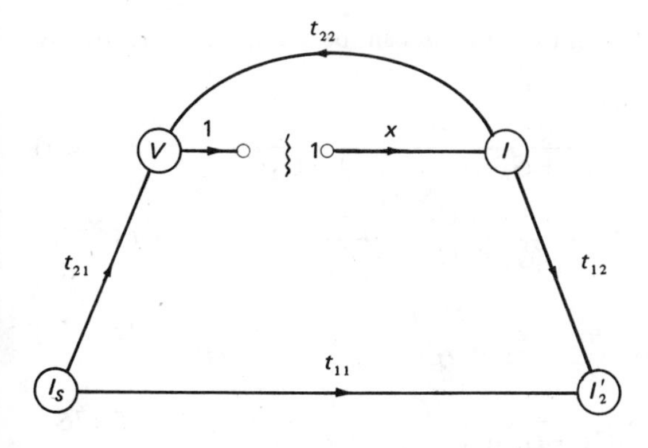

Figure 5.18 The signal-flow graph interpretation of return ratio and return difference functions.

which is equal to the negative of the null return ratio $\hat{T}$ for the element x. Finally, the null return difference is the difference of the unit signal input and the returned signal (5.170), so that

$$\hat{F}(x) = 1 - V = 1 + \hat{T} = 1 - xt_{22} + \frac{xt_{12}t_{21}}{t_{11}} \tag{5.171}$$

Combining (5.170) and (5.171) yields

$$\hat{F}(x) = F(x) + \frac{xt_{12}t_{21}}{t_{11}} \tag{5.172}$$

From Fig. 5.17, the current gain I_2'/I_s can easily be obtained by inspection and is given by

$$w(x) = \frac{I_2'}{I_s} = t_{11} + \frac{t_{21}xt_{12}}{1 - xt_{22}} \tag{5.173}$$

By applying (5.169) and (5.172) in (5.173), the current gain can be written in the familiar form:

$$w(x) = w(0)\,\frac{\hat{F}(x)}{F(x)} \tag{5.174}$$

Example 5.13 We wish to compute the feedback quantities of the amplifier of Fig. 5.11 by means of the signal-flow graph technique discussed above.

The transmittances t_{ij} must first be determined. For this we replace the controlled current source $h_{fe}I_b$ in the network of Fig. 5.11 by an independent current source I as shown in Fig. 5.19. The indefinite-admittance matrix of network of Fig. 5.19 can easily be obtained by inspection and is given by

$$\mathbf{Y} = \begin{bmatrix} G_f + G_1 + \dfrac{1}{h_{ie}} & -G_f & -G_1 - \dfrac{1}{h_{ie}} \\ -G_f & G_2 + G_f & -G_2 \\ -G_1 - \dfrac{1}{h_{ie}} & -G_2 & G_1 + G_2 + \dfrac{1}{h_{ie}} \end{bmatrix} \tag{5.175}$$

From (5.166), we see that the transmittances t_{ij} can be computed directly by formula (2.94), as follows:

$$t_{11} = \left.\frac{I_{23}}{I_s}\right|_{I=0} = G_2\,\frac{Y_{12,33}}{Y_{33}} = \frac{G_2G_f}{(G_2 + G_f)(G_1 + 1/h_{ie}) + G_2G_f} \tag{5.176a}$$

$$t_{21} = \left.\frac{V}{I_s}\right|_{I=0} = \frac{Y_{11,33}}{Y_{33}} = \frac{G_2 + G_f}{(G_2 + G_f)(G_1 + 1/h_{ie}) + G_2G_f} \tag{5.176b}$$

$$t_{12} = \left.\frac{I_{23}}{I}\right|_{I_s=0} = -G_2\,\frac{Y_{22,33}}{Y_{33}} = -\frac{G_2(G_f + G_1 + 1/h_{ie})}{(G_2 + G_f)(G_1 + 1/h_{ie}) + G_2G_f} \tag{5.176c}$$

$$t_{22} = \left.\frac{V}{I}\right|_{I_s=0} = \frac{Y_{31,23}}{Y_{33}} = -\frac{G_f}{(G_2 + G_f)(G_1 + 1/h_{ie}) + G_2G_f} \tag{5.176d}$$

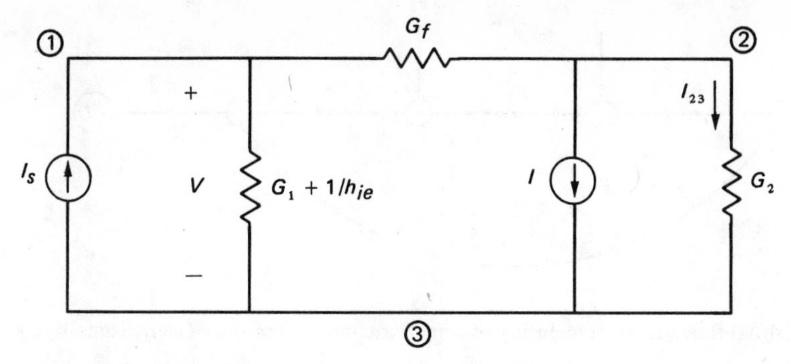

Figure 5.19 The network used to compute the transmittances t_{ij}.

Substituting (5.176*d*) in (5.169) with $x = \alpha \equiv h_{fe}/h_{ie}$ yields the return difference with respect to α:

$$F(\alpha) = 1 - \alpha t_{22} = 1 + \frac{\alpha G_f}{(G_2 + G_f)(G_1 + 1/h_{ie}) + G_2 G_f} \tag{5.177}$$

confirming (5.22). The second term on the right-hand side of (5.177) is the return ratio T. Likewise, substituting (5.176) in (5.171) results in the null return difference

$$\hat{F}(\alpha) = 1 - \alpha t_{22} + \frac{\alpha t_{12} t_{21}}{t_{11}} = 1 - \frac{\alpha}{G_f} \tag{5.178}$$

confirming (5.106) for $k = 0$. The second term on the right-hand side of (5.178) is the null return ratio $\hat{T}$.

In the foregoing, we have demonstrated that feedback can be used to control the driving-point and transmission characteristics of an amplifier and can be employed to make an amplifier less sensitive to the variation of its parameters caused by aging, temperature change, or other environmental changes. In the following, we indicate that, in some cases, negative feedback can be used to reduce the effect of noise and nonlinear distortion generated within an amplifier.

To demonstrate this effect, consider an amplifier composed of three internal amplifying stages without feedback represented by the signal-flow graph of Fig. 5.20, in which the extraneous signals x_1 and x_2 represent noise or nonlinear distortion introduced at some arbitrary points within the amplifier. The output signal x_l in terms of the input signal x_s and the disturbing signals x_1 and x_2 is

$$x_l = \mu_1 \mu_2 \mu_3 x_s + \mu_2 \mu_3 x_1 + \mu_3 x_2 \tag{5.179}$$

The output *signal-to-noise ratio* in the absence of feedback is therefore given by $\mu_1 \mu_2 \mu_3 x_s / (\mu_2 \mu_3 x_1 + \mu_3 x_2)$.

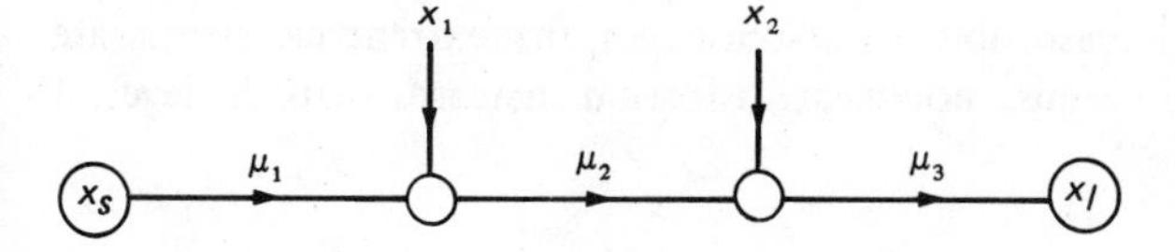

Figure 5.20 The signal-flow graph representing an amplifier composed of three internal amplifying stages without feedback.

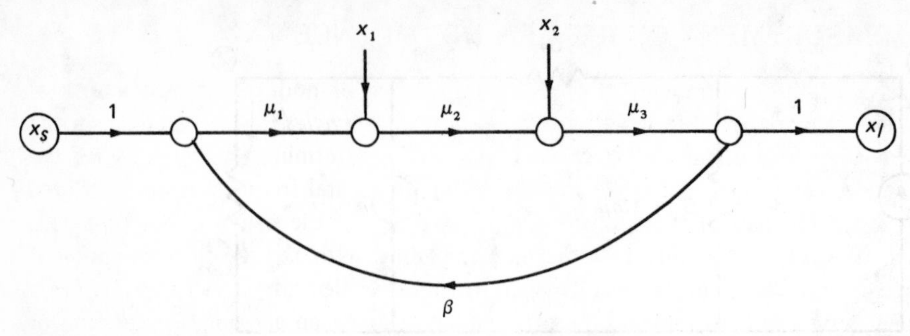

Figure 5.21 The signal-flow graph representing an amplifier composed of three internal amplifying stages with feedback.

Now, if a simple feedback loop is added as shown in Fig. 5.21, the output signal becomes

$$x_l = \frac{\mu_1 \mu_2 \mu_3 x_s + \mu_2 \mu_3 x_1 + \mu_3 x_2}{1 - \mu_1 \mu_2 \mu_3 \beta} \tag{5.180}$$

It is easy to check that the output signal-to-noise ratio with feedback is the same as that without feedback. Thus, feedback has no direct effect in improving the output signal-to-noise ratio. However, it does help indirectly. To see this, we observe from (5.180) that the effect of negative feedback on the output signal is the reduction of both signal and noise by a factor of $(1 - \mu_1 \mu_2 \mu_3 \beta)$. If the input signal x_s is increased so that the output signal remains at the same level as it had before the application of feedback, we see that the net effect at the output will be the reduction of the contribution of the extraneous signals x_1 and x_2 by the same factor $(1 - \mu_1 \mu_2 \mu_3 \beta)$. In other words, if the output of the amplifier is maintained constant with and without negative feedback, the effect of negative feedback increases the signal-to-noise ratio by the factor $(1 - \mu_1 \mu_2 \mu_3 \beta)$. Clearly, this improvement is possible only if the extraneous signals are generated within the amplifier. The increase of input signal x_s to the amplifier without feedback is impractical, because such an increase may cause excessive distortion due to nonlinearity.

For $|\mu_1 \mu_2 \mu_3 \beta| \gg 1$, Eq. (5.180) reduces to

$$x_l = -\frac{x_s}{\beta} - \frac{x_1}{\mu_1 \beta} - \frac{x_2}{\mu_1 \mu_2 \beta} \tag{5.181}$$

showing that the contribution of the extraneous signal x_2 at the output has been reduced by a factor of $\mu_1 \mu_2 \beta$, whereas that of x_1 is reduced by $\mu_1 \beta$ and that of x_s by β. Therefore, the magnitude of the reduction depends largely on the origin of the extraneous signals. It is reasonable to assume that the extraneous signals are generated in the final stage, because nonlinear distortion increases with the level of the output signal.

5.8 MEASUREMENT OF RETURN DIFFERENCE

The zeros of the determinant of the loop-impedance or node-admittance matrix of an active network are referred to as the *natural frequencies.* Their locations in the complex-frequency plane are extremely important in determining the stability of the network. A network is said to be *stable* if all of its natural frequencies are restricted to the open left half of the complex-frequency plane. If the network determinant is known, its roots can readily be computed explicitly with the aid of a computer if necessary, and the stability problem can then be settled directly. However, for a physical network there remains the difficulty of getting an accurate formulation of the network determinant itself, because every equivalent network is, to a greater or lesser extent, an idealization of the physical reality. What is really needed is some kind of experimental verification that the system is stable and will remain so under certain prescribed conditions. The measurement of return difference provides a solution to this problem.

In a feedback amplifier, the return difference with respect to a controlling parameter x is defined by $F(x) = Y_{ij}(x)/Y_{ij}(0)$. Since $Y_{ij}(x)$ also denotes the nodal determinant, we see that the zeros of the return difference are exactly the same as the zeros of the nodal determinant provided that there is no cancellation of common factors between $Y_{ij}(x)$ and $Y_{ij}(0)$. Thus, if $Y_{ij}(0)$ is known to have no zeros in the closed right half of the complex-frequency plane, which is usually the case in single-loop feedback systems, $F(x)$ gives precisely the same information about the stability of a feedback amplifier as does the nodal determinant itself. Since $F(x)$, as will be shown shortly, can be measured quite readily in practice, it can be used as the criterion of stability study for the feedback amplifiers.

The difficulty inherent in the measurement of the return difference with respect to the controlling parameter of a controlled source is that, in a physical system, the controlling branch and the controlled source both form part of a single device such as a transistor, and cannot be physically separated. Thus, we must devise a scheme that does not require the physical decomposition of a device.

In a feedback amplifier, let a device of interest be brought out as a two-port network connected to a general four-port network as shown in Fig. 5.22 along with the input and the load of the amplifier. For our purposes, assume that this device is represented by its y-parameter equivalent two-port network as indicated in Fig. 5.23, in which the parameter y_{21} controls signal transmission in the forward direction through the device whereas y_{12} gives the reverse transmission, accounting for the internal feedback within the device. Our objective is to measure the return difference with respect to the forward short-circuit transfer admittance y_{21}. In the following, we describe three procedures for measuring the return ratio and hence the return difference.

5.8.1 Blecher's Procedure [Blecher (1957) and Hakim (1965b)]

Let the device be a transistor operated in the common-emitter configuration with terminals a, $b = d$, and c representing, respectively, the base, emitter, and collector

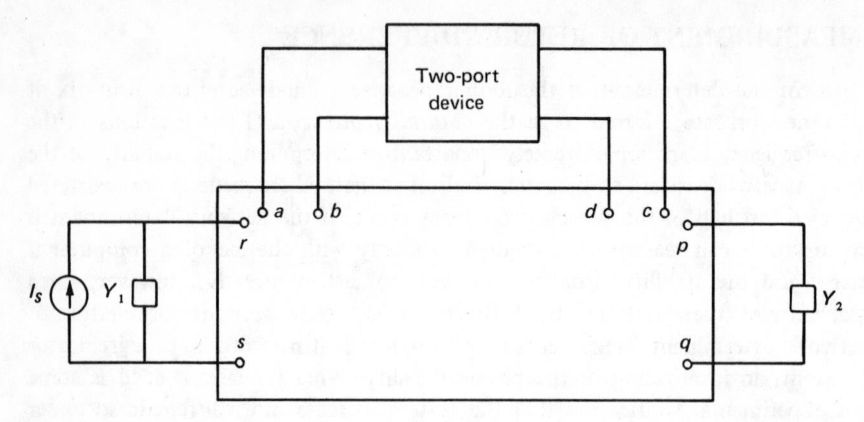

Figure 5.22 The general configuration of a feedback amplifier with a two-port device exhibited explicitly.

terminals. For convenience, let $a = 1$, $b = d = 3$, and $c = 2$ as indicated in Fig. 5.24. To measure $F(y_{21})$, we break the base terminal of the transistor and apply a 1-V excitation at its input as shown in Fig. 5.24. To ensure that the controlled current source $y_{21} V_{13}$ drives a replica of what it sees during normal operation, we connect an active one-port network composed of a parallel combination of the admittance y_{11} and a controlled current source $y_{12} V_{23}$ at terminals 1 and 3. The returned voltage V_{13} is precisely the negative of the return ratio with respect to the element y_{21}. If, in the frequency band of interest, the externally applied feedback is large compared with the internal feedback of the transistor, the controlled source $y_{12} V_{23}$

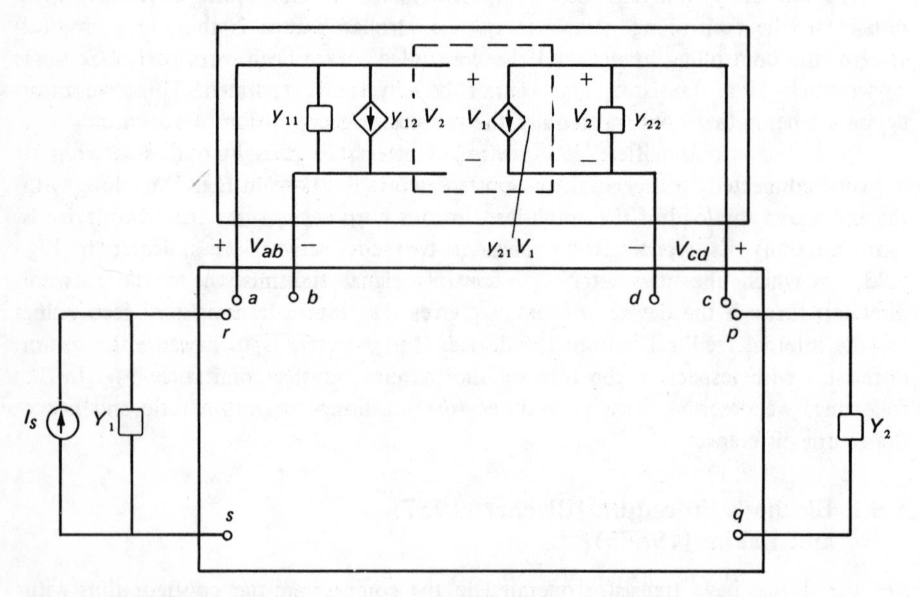

Figure 5.23 The representation of the two-port device in Fig. 5.22 by its y-parameters.

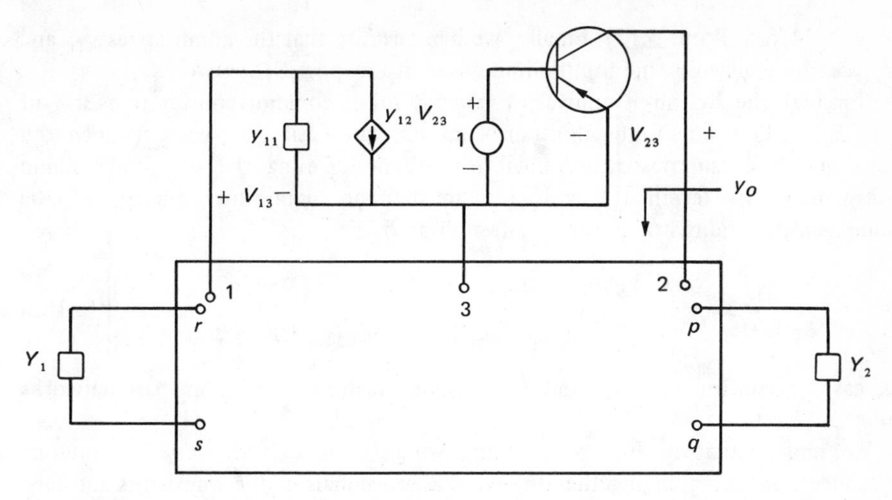

Figure 5.24 The physical interpretation of return difference $F(y_{21})$ for a transistor operated in the common-emitter configuration and represented by its y-parameters y_{ij}.

can be ignored. If, however, we find that this internal feedback cannot be ignored, we can simulate it by using an additional transistor, connected as shown in Fig. 5.25. This additional transistor must be matched as closely as possible to the one in question. The one-port admittance y_o denotes the admittance presented to the output port of the transistor under consideration as indicated in Figs. 5.24 and 5.25. For a common-emitter stage, it is perfectly reasonable to assume that $|y_o| \gg |y_{12}|$ and $|y_{11}| \gg |y_{12}|$. Under these assumptions, it is straightforward to show that the Norton equivalent network looking into the two-port network at terminals 1 and 3 of Fig. 5.25 can be approximated by the parallel combination of y_{11} and $y_{12} V_{23}$ as shown

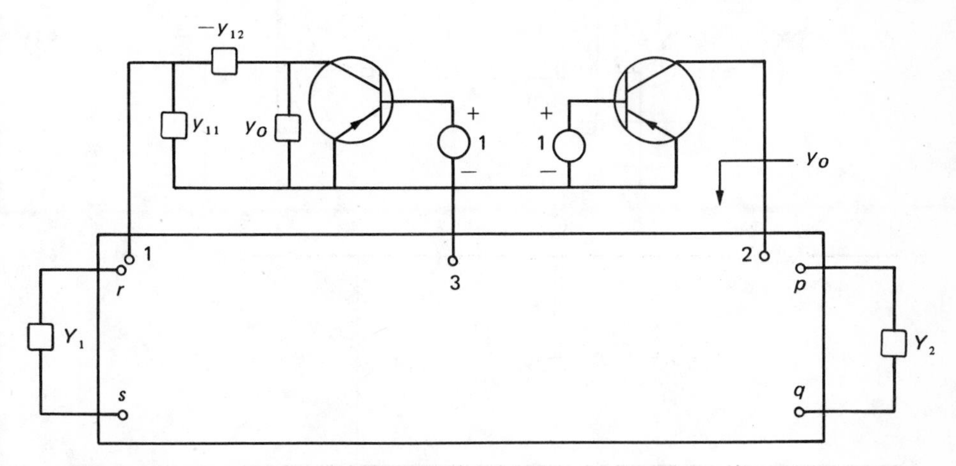

Figure 5.25 The measurement of return difference $F(y_{21})$ for a transistor operated in the common-emitter configuration and represented by its y-parameters y_{ij}.

in Fig. 5.24 (see Prob. 5.15). Finally, we demonstrate that the admittances y_{11} and $-y_{12}$ can be realized as the input admittances of one-port RC networks.

Consider the hybrid-pi equivalent network of a common-emitter transistor of Fig. 5.26, whose indefinite-admittance matrix can easily be written down by inspection. After suppressing terminal $B' = 4$ and deleting the row and column corresponding to terminal $E = 3$, the short-circuit admittance matrix of the common-emitter transistor is given by (see Prob. 5.16)

$$\mathbf{Y} = \frac{1}{g_x + g_\pi + sC_\pi + sC_\mu} \begin{bmatrix} g_x(g_\pi + sC_\pi + sC_\mu) & -g_x sC_\mu \\ g_x(g_m - sC_\mu) & sC_\mu(g_x + g_\pi + sC_\pi + g_m) \end{bmatrix} \tag{5.182}$$

It is easy to confirm that y_{11} and $-y_{12}$ can be realized by the one-part networks of Fig. 5.27.

We remark that in Fig. 5.25 if the voltage sources have very low internal impedances, we can join together the two base terminals of the transistors and feed them both from a single voltage source of very low internal impedance. In this way, we avoid the need of using two separate sources.

5.8.2 Hakim's Procedure [Hakim (1965a)]

In this section, we discuss a procedure that avoids the need to use a matched transistor to simulate the controlled source responsible for the internal feedback.

Refer to the network of Fig. 5.28. The controlling parameters y_{12} and y_{21} of the two controlled sources enter the indefinite-admittance matrix of the amplifier in the following pattern:

$$\mathbf{Y} = \begin{array}{c} \\ 1 \\ 2 \\ 3 \end{array} \begin{array}{c} \begin{array}{ccc} 1 & 2 & 3 \end{array} \\ \begin{bmatrix} & y_{12} & -y_{12} \\ y_{21} & & -y_{21} \\ -y_{21} & -y_{12} & y_{12} + y_{21} \end{bmatrix} \end{array} \tag{5.183}$$

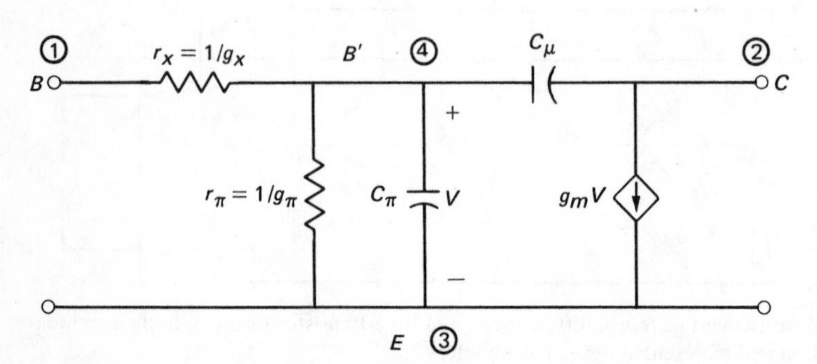

Figure 5.26 The hybrid-pi equivalent network of a common-emitter transistor.

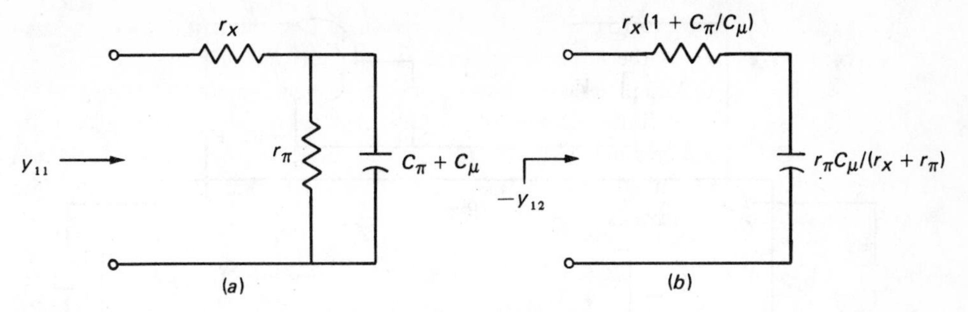

Figure 5.27 (*a*) The realization of y_{11} and (*b*) the realization of $-y_{12}$.

As before, to indicate the dependence of **Y** on y_{12} and y_{21}, we write

$$\mathbf{Y} = \mathbf{Y}(y_{12}, y_{21}) \tag{5.184}$$

In Fig. 5.28, suppose that we set $y_{12} = 0$, remove the independent current source I_s, and replace the controlled current source $y_{21} V_{13}$ by an independent current source of y_{21} amperes. The resulting indefinite-admittance matrix is simply $\mathbf{Y}(0, 0)$. Under this situation, the voltages across terminals 1 and 3 and terminals 2 and 3 can be obtained directly by appealing to formula (2.94) and are given by

$$V_{13}\Big|_{\substack{I_s=0\\ y_{12}=0}} = V'_{13} = -y_{21}\frac{Y_{21,33}(0, 0)}{Y_{33}(0, 0)} \tag{5.185a}$$

$$V_{23}\Big|_{\substack{I_s=0\\ y_{12}=0}} = V'_{23} = -y_{21}\frac{Y_{22,33}(0, 0)}{Y_{33}(0, 0)} \tag{5.185b}$$

If the device is a transistor operated in the common-emitter configuration, the voltages can be measured using the arrangement of Fig. 5.29, in which the feedback loop has been broken at the base of the transistor. The desired voltages V'_{13} and V'_{23} are as indicated in the figure.

Likewise, if we set $y_{21} = 0$, remove the independent current source I_s, and

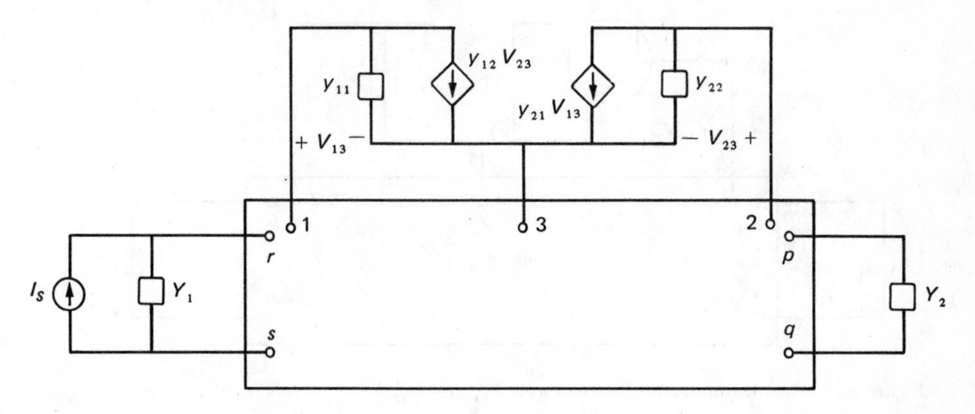

Figure 5.28 The representation of the two-port device in Fig. 5.22 by its y-parameters.

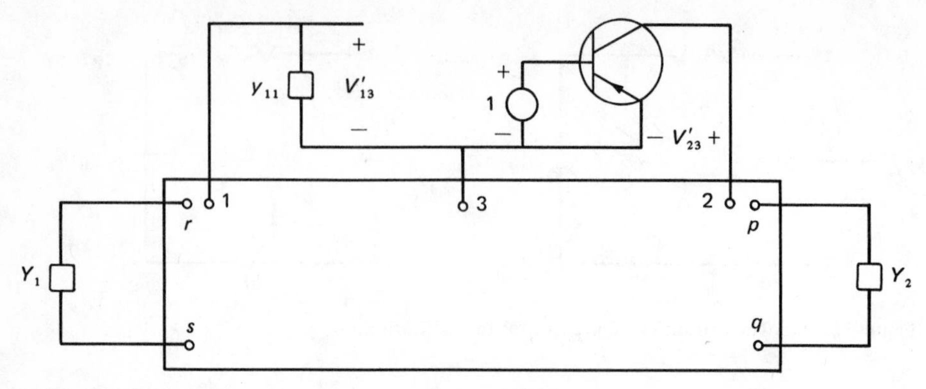

Figure 5.29 The measurement of the voltages V'_{13} and V'_{23}.

replace the controlled current source $y_{12}V_{23}$ by an independent current source of y_{12} amperes in the network of Fig. 5.28, the voltages across terminals 1 and 3 and terminals 2 and 3 are obtained as

$$V_{13}\Big|_{\substack{I_s=0\\ y_{21}=0}} = V''_{13} = -y_{12}\frac{Y_{11,33}(0,0)}{Y_{33}(0,0)} \tag{5.186a}$$

$$V_{23}\Big|_{\substack{I_s=0\\ y_{21}=0}} = V''_{23} = -y_{12}\frac{Y_{12,33}(0,0)}{Y_{33}(0,0)} \tag{5.186b}$$

If the device is a transistor operated in the common-emitter configuration, these voltages can be measured using the arrangement of Fig. 5.30, in which the feedback loop has been broken at the collector of the transistor. The desired voltages V''_{13} and V''_{23} are as indicated in the figure. In fact, since the voltage V''_{13} developed at the base is practically independent of termination y_{22}, and since V''_{23} is indistinguishable from zero for all practical purposes, we could have removed y_{22} in making the measurement.

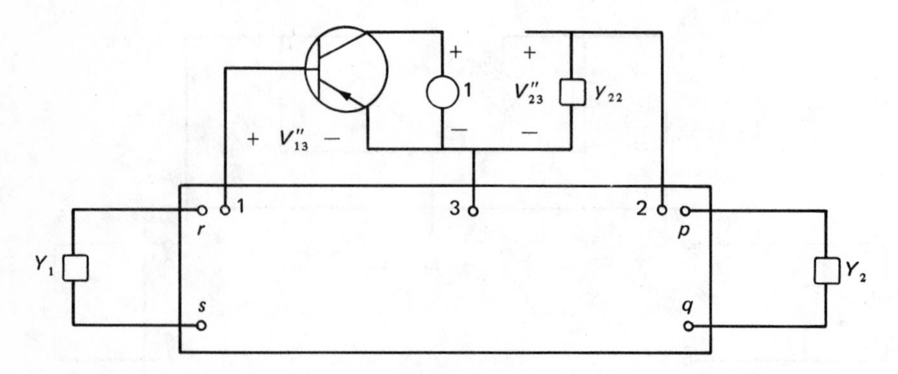

Figure 5.30 The measurement of the voltages V''_{13} and V''_{23}.

Now we show that these four measurements are sufficient to determine the return difference $F(y_{21})$ with respect to y_{21} in the original amplifier.

$$
\begin{aligned}
F(y_{21}) &= \frac{Y_{33}(y_{12}, y_{21})}{Y_{33}(y_{12}, 0)} = \frac{Y_{33}(y_{12}, 0) + y_{21} Y_{21,33}(y_{12}, 0)}{Y_{33}(y_{12}, 0)} \\
&= 1 + y_{21} \frac{Y_{21,33}(0, 0) - y_{12} Y_{11,22,33}}{Y_{33}(y_{12}, 0)} \\
&= 1 + y_{21} \frac{Y_{21,33}(0, 0) Y_{33}(y_{12}, 0) - y_{12} Y_{11,33}(0, 0) Y_{22,33}(0, 0)}{Y_{33}(0, 0) Y_{33}(y_{12}, 0)} \\
&= 1 + y_{21} \frac{Y_{21,33}(0, 0)}{Y_{33}(0, 0)} - \frac{y_{12} y_{21} Y_{11,33}(0, 0) Y_{22,33}(0, 0)}{Y_{33}(0, 0)[Y_{33}(0, 0) + y_{12} Y_{12,33}(0, 0)]} \\
&= 1 - V'_{13} - \frac{V''_{13} V'_{23}}{1 - V''_{23}}
\end{aligned}
\tag{5.187}
$$

The third line follows from (4.147) and (5.86), which can be written as

$$Y_{33}(y_{12}, 0) = Y_{33}(0, 0) + y_{12} Y_{12,33}(0, 0) \tag{5.188a}$$

$$Y_{11,22,33} Y_{33}(0, 0) = Y_{11,33}(0, 0) Y_{22,33}(0, 0) - Y_{12,33}(0, 0) Y_{21,33}(0, 0) \tag{5.188b}$$

where $Y_{11,22,33}$ denotes the determinant of the submatrix obtained from $\mathbf{Y}$ by deleting rows 1, 2, and 3 and columns 1, 2, and 3; and the fourth line follows from (5.188a). Note that $Y_{11,22,33}$ is independent of y_{12} and y_{21}. Since for all practical purposes V''_{23} is indistinguishable from zero, (5.187) can be simplified to

$$F(y_{21}) = 1 + T \approx 1 - V'_{13} - V''_{13} V'_{23} \tag{5.189}$$

where T is the return ratio with respect to the element y_{21} in the original feedback amplifier.

5.8.3 Impedance Measurements

In this section, we demonstrate that the return difference can be evaluated by measuring two driving-point impedances at a convenient port in the feedback amplifier.

Refer again to the general configuration of a feedback amplifier of Fig. 5.23. Suppose that we wish to evaluate the return difference with respect to the forward short-circuit transfer admittance y_{21}. The controlling parameters y_{12} and y_{21} enter the indefinite-admittance matrix $\mathbf{Y}$ in the rectangular patterns as shown below:

$$
\mathbf{Y} = \begin{array}{c} \\ a \\ b \\ c \\ d \end{array}
\begin{array}{c}
\begin{array}{cccc} a & b & c & d \end{array} \\
\left[\begin{array}{cccc}
 & & y_{12} & -y_{12} \\
 & & -y_{12} & y_{12} \\
y_{21} & -y_{21} & & \\
-y_{21} & y_{21} & &
\end{array}\right]
\end{array}
\tag{5.190}
$$

By appealing to formula (2.95), the impedance looking into terminals a and b of Fig. 5.23 is given by

$$z_{aa,bb}(y_{12}, y_{21}) = \frac{Y_{aa,bb}(y_{12}, y_{21})}{Y_{dd}(y_{12}, y_{21})} \tag{5.191}$$

The return difference with respect to y_{21} is defined as

$$F(y_{21}) = \frac{Y_{dd}(y_{12}, y_{21})}{Y_{dd}(y_{12}, 0)} \tag{5.192}$$

Combining (5.191) and (5.192) yields

$$\begin{aligned} F(y_{21})z_{aa,bb}(y_{12}, y_{21}) &= \frac{Y_{aa,bb}(y_{12}, y_{21})}{Y_{dd}(y_{12}, 0)} = \frac{Y_{aa,bb}(0, 0)}{Y_{dd}(y_{12}, 0)} \\ &= \frac{Y_{aa,bb}(0, 0)}{Y_{dd}(0, 0)} \frac{Y_{dd}(0, 0)}{Y_{dd}(y_{12}, 0)} = \frac{z_{aa,bb}(0, 0)}{F(y_{12})|_{y_{21}=0}} \end{aligned} \tag{5.193}$$

This gives a relation

$$F(y_{12})|_{y_{21}=0}F(y_{21}) = \frac{z_{aa,bb}(0, 0)}{z_{aa,bb}(y_{12}, y_{21})} \tag{5.194}$$

among the return differences and the driving-point impedances. $F(y_{12})|_{y_{21}=0}$ is the return difference with respect to y_{12} when y_{21} is set to zero, and it can be measured by the arrangement of Fig. 5.31. $z_{aa,bb}(y_{12}, y_{21})$ is the driving-point impedance looking into terminals a and b of the network of Fig. 5.23. $z_{aa,bb}(0, 0)$ is the impedance to which $z_{aa,bb}(y_{12}, y_{21})$ reduces when the controlling parameters y_{12} and y_{21} are both set to zero, and it can be measured by the arrangement of Fig. 5.32. Note that in all the three measurements, the independent current source I_s is removed.

As an illustration, suppose that the device is a transistor operated in the common-emitter configuration as shown in Fig. 5.28. Then, as indicated in Sec.

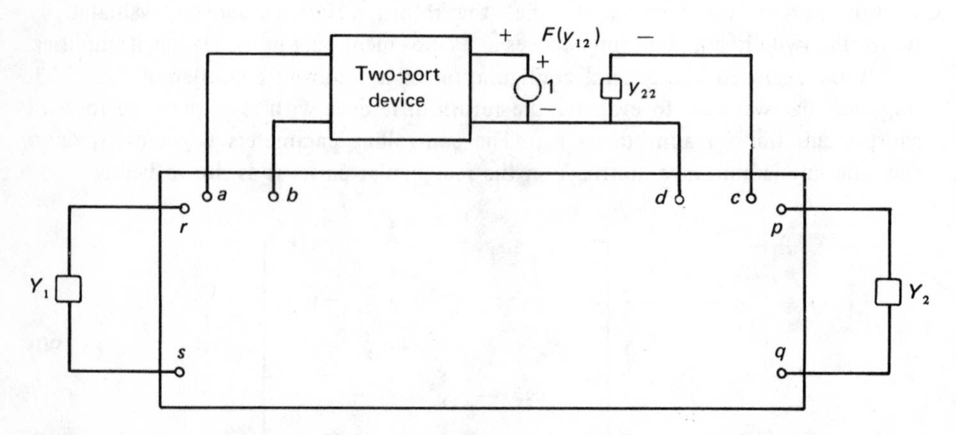

Figure 5.31 The measurement of the return difference $F(y_{12})$ with y_{21} being set to zero.

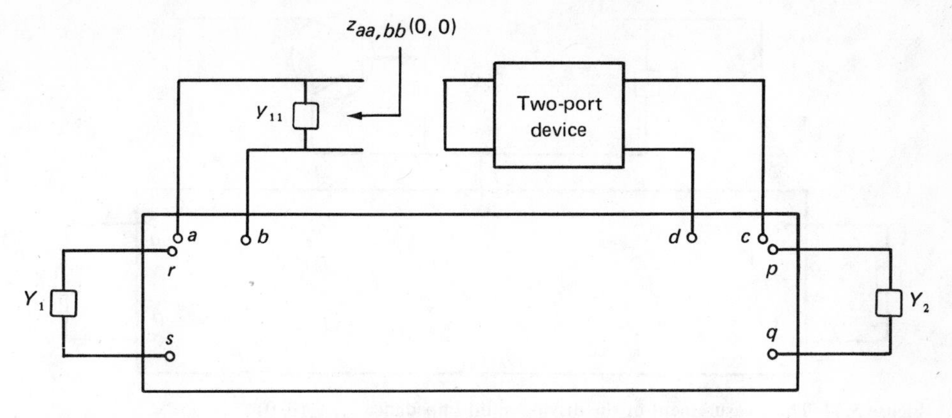

Figure 5.32 The measurement of the driving-point impedance $z_{aa,bb}(0, 0)$.

5.8.2, the return difference $F(y_{12})$ when y_{21} is set to zero is, for all practical purposes, indistinguishable from unity. Consequently, it is perfectly justifiable to reduce (5.194) to the following simpler form:

$$F(y_{21}) \approx \frac{z_{11,33}(0, 0)}{z_{11,33}(y_{12}, y_{21})} \tag{5.195}$$

showing that the return difference $F(y_{21})$ is effectively equal to the ratio of the two functional values assumed by the driving-point impedance looking into terminals 1 and 3 of Fig. 5.28 under the condition that the controlling parameters y_{12} and y_{21} are both set to zero and the condition that they assume their nominal values. These two impedances can be measured by the network arrangements of Figs. 5.33 and 5.34.

5.8.4 Measuring the Reference Value

As mentioned in reference to Eq. (5.144), if the reference value k is chosen so that $w(k) = 0$, then the relative sensitivity function is simply the reciprocal of the

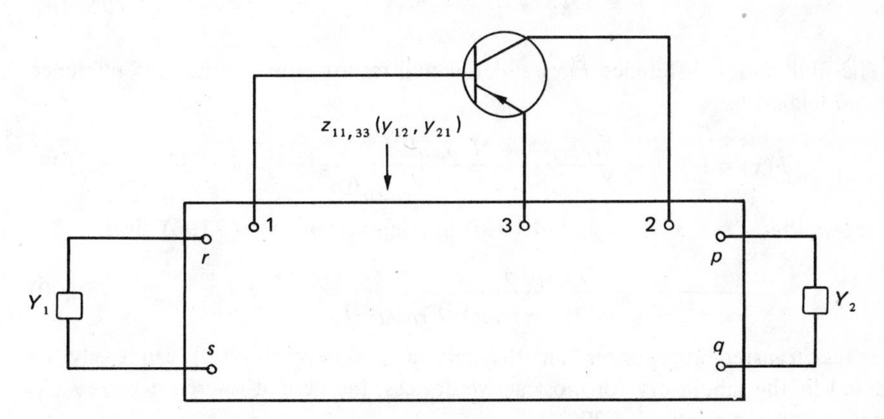

Figure 5.33 The measurement of the driving-point impedance $z_{11,33}(y_{12}, y_{21})$.

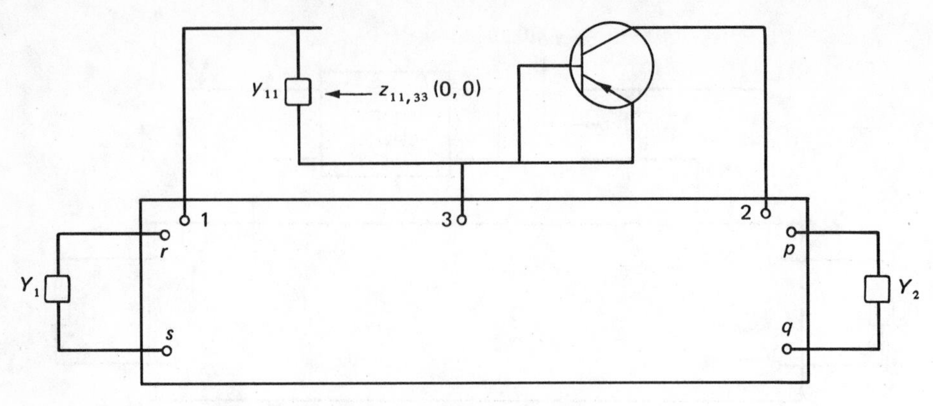

Figure 5.34 The measurement of the driving-point impedance $z_{11,33}(0, 0)$.

general return difference for this reference value. In this section, we present a technique of Bode's (1945) that permits experimental determination of this reference value k_0.

Consider the general feedback amplifier configuration of Fig. 5.8. By using the convention adopted in Chap. 2, the various transfer impedances measured under the condition $x = 0$ can be expressed as

$$z_{rp,sq}(0) = \frac{Y_{rp,sq}(0)}{Y_{uv}(0)} \qquad z_{ra,sb}(0) = \frac{Y_{ra,sb}(0)}{Y_{uv}(0)} \tag{5.196a}$$

$$z_{cp,dq}(0) = \frac{Y_{cp,dq}(0)}{Y_{uv}(0)} \qquad z_{ca,db}(0) = \frac{Y_{ca,db}(0)}{Y_{uv}(0)} \tag{5.196b}$$

From (5.144), the desired reference value k_0 is obtained by setting $Y_{rp,sq}(k_0)$ equal to zero:

$$Y_{rp,sq}(k_0) = Y_{rp,sq}(0) + k_0 \dot{Y}_{rp,sq} = 0 \tag{5.197}$$

giving

$$k_0 = -\frac{Y_{rp,sq}(0)}{\dot{Y}_{rp,sq}} \tag{5.198}$$

The null return difference $\hat{F}(x)$ and the null return ratio for the zero reference value are related by

$$\hat{F}(x) = 1 + \hat{T} = \frac{Y_{rp,sq}(x)}{Y_{rp,sq}(0)} = \frac{Y_{rp,sq}(0) + x\dot{Y}_{rp,sq}}{Y_{rp,sq}(0)} = 1 - \frac{x}{k_0} \tag{5.199}$$

This shows that $\hat{T} = -x/k_0$. Using (4.163*a*) in conjunction with (5.196) gives

$$\frac{z_{rp,sq}(0)}{z_{ra,sb}(0)z_{cp,dq}(0) - z_{ca,db}(0)z_{rp,sq}(0)} = -\frac{x}{\hat{T}} = k_0 \tag{5.200}$$

Since the transfer impedances on the left-hand side of (5.200) can easily be measured in the laboratory for most active devices, the desired reference value k_0 is determined by the relation (5.200).

5.9 CONSIDERATIONS ON THE INVARIANCE OF RETURN DIFFERENCE

The feedback amplifier theory developed so far is based on the concepts of return difference and null return difference, which are formulated compactly and elegantly in terms of the first- and the second-order cofactors of the elements of the indefinite-admittance matrix **Y**. The first-order cofactors of the elements of **Y** are, in fact, the determinant of the node-admittance matrix of the amplifier. In the present section, we demonstrate that the feedback amplifier theory can be dually and equally formulated in terms of the determinant of the loop-impedance matrix, and we discuss its relations to the admittance formulation.

5.9.1 Network Determinants and Their Interrelations

The term *network determinant* is usually referred to as the determinant of the loop-impedance matrix, the node-admittance matrix or, more generally, the cutset-admittance matrix. To simplify our notation, denote

Δ_m = the determinant of the loop-impedance matrix
Δ_n = the determinant of the node-admittance matrix
Δ_c = the determinant of the cutset-admittance matrix

They are called the *loop determinant*, the *nodal determinant*, and the *cutset determinant*, respectively. The value of the network determinant is dependent on the formulation of the network equations, even within the same framework of impedance or admittance consideration. If Δ_{m1} and Δ_{m2}, for example, denote the loop determinants corresponding to two different sets of loop equations, they are *not* generally equal. However, it can be shown that they are related by a real constant λ_m, depending only on the choices of the two sets of basis loops. Thus, we can write

$$\Delta_{m1} = \lambda_m \Delta_{m2} \tag{5.201}$$

Likewise, if Δ_{c1} and Δ_{c2} denote the cutset determinants corresponding to two different sets of cutset equations, we have

$$\Delta_{c1} = \lambda_c \Delta_{c2} \tag{5.202}$$

where λ_c is a real constant depending only on the choices of the two sets of basis cutsets. In particular, if Δ_{n1} and Δ_{n2} are the nodal determinants corresponding to two choices of the reference node, then (5.202) reduces to

$$\Delta_{n1} = \Delta_{n2} \tag{5.203}$$

showing that the nodal determinant is independent of the choice of the reference-potential point. We recognize that the nodal determinant is a special type of cutset determinant when basis cutsets around the nodes of the network are used.

With appropriate identifications of the basis loops and cutsets, the loop determinant and the cutset determinant of a network are related by the equation [Chen (1970)]

$$\Delta_m = \lambda_{mc}\Delta_b\Delta_c \tag{5.204}$$

where Δ_b denotes the determinant of the branch-impedance matrix of the network, and λ_{mc} is a real constant depending only on the choices of basis loops and cutsets.

We shall not discuss this aspect of the theory any further, since it is beyond the scope of this book. The proofs of the above assertions cannot be given here because they would take us far afield into the theory of graphs. For a detailed account of this subject and all of its variations and ramifications, the reader is referred to Chen (1976a).

We illustrate the above results by the following example.

Example 5.14 Figure 5.35 is an equivalent network of a general potential feedback amplifier. For our purposes, we consider the series combination of V_s and Z_1 as a single branch, and the parallel combination of $g_m V$ and Y_2 as another. For loop analysis, we use the equivalent network N_m of Fig. 5.36*a*, and for nodal or cutset analysis the equivalent network N_c of Fig. 5.36*b*. It is straightforward to calculate the loop-impedance matrix and the node-admittance matrix. Their determinants are given by

$$\begin{aligned}\Delta_m &= \det\begin{bmatrix} Z_1 + z_g & -z_g \\ -z_g - g_m z_g Z_2 & Z_2 + z_f + z_g + g_m z_g Z_2 \end{bmatrix} \\ &= Z_1(Z_2 + z_f + z_g + g_m z_g Z_2) + z_g(Z_2 + z_f) \end{aligned} \tag{5.205}$$

$$\begin{aligned}\Delta_n &= \det\begin{bmatrix} Y_1 + y_g + y_f & -y_f \\ g_m - y_f & y_f + Y_2 \end{bmatrix} \\ &= (Y_1 + y_g)(y_f + Y_2) + y_f(Y_2 + g_m) \end{aligned} \tag{5.206}$$

Our objective is to confirm (5.204). For this we express (5.206) in terms of the branch impedances:

$$\Delta_n = \frac{Z_1(Z_2 + z_f + z_g + g_m z_g Z_2) + z_g(Z_2 + z_f)}{z_g z_f Z_1 Z_2} \tag{5.207}$$

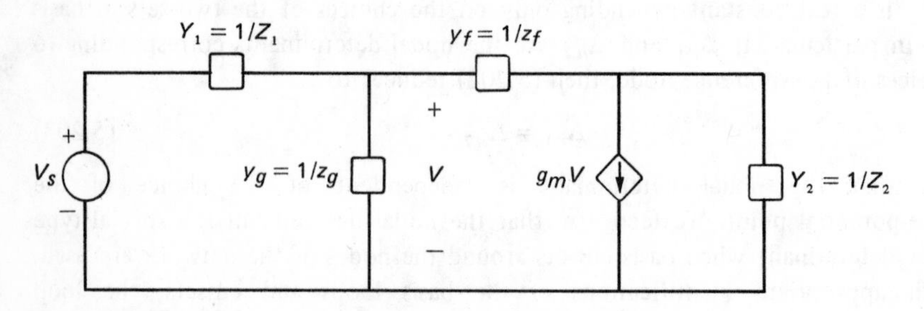

Figure 5.35 An equivalent network of a potential feedback amplifier.

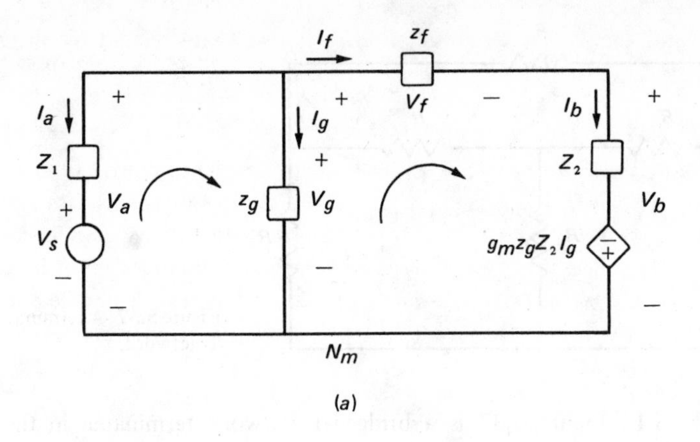

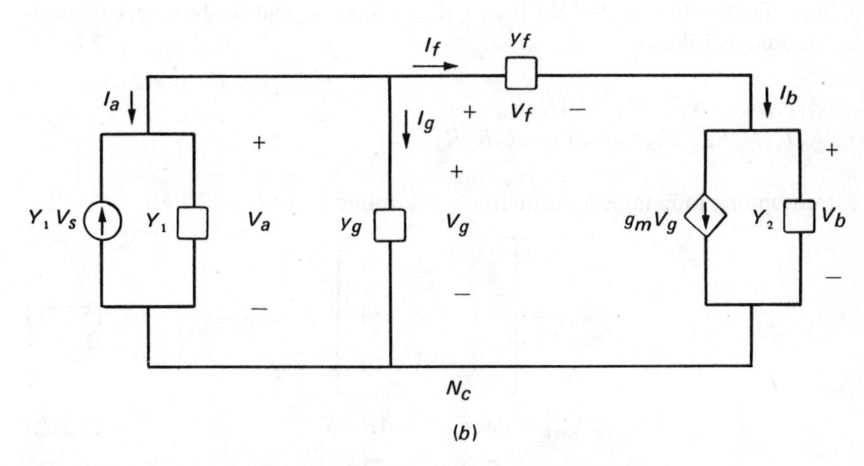

Figure 5.36 Two equivalent networks of Fig. 5.35: (*a*) N_m and (*b*) N_c.

The branch voltage-current relations are described by the equation

$$\begin{bmatrix} V_a \\ V_g \\ V_f \\ V_b \end{bmatrix} = \begin{bmatrix} Z_1 & 0 & 0 & 0 \\ 0 & z_g & 0 & 0 \\ 0 & 0 & z_f & 0 \\ 0 & -g_m z_g Z_2 & 0 & Z_2 \end{bmatrix} \begin{bmatrix} I_a \\ I_g \\ I_f \\ I_b \end{bmatrix} + \begin{bmatrix} V_s \\ 0 \\ 0 \\ 0 \end{bmatrix} \tag{5.208}$$

The coefficient matrix is the branch-impedance matrix, whose determinant is given by

$$\Delta_b = z_g z_f Z_1 Z_2 \tag{5.209}$$

From (5.205), (5.207), and (5.209), we obtain

$$\Delta_m = \Delta_b \Delta_n \tag{5.210}$$

confirming (5.204) with $\lambda_{mc} = 1$.

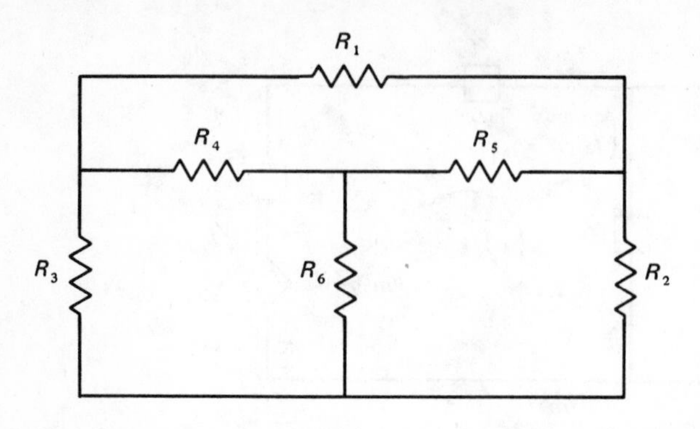

Figure 5.37 A terminated bridged-T network.

Example 5.15 Figure 5.37 is a bridged-T network terminated in the resistors R_3 and R_2. Assume that $R_i = 1\ \Omega$ for $i = 1, 2, 3, 4, 5,$ and 6. Two sets of basis loops are chosen as follows:

Set one: $R_1R_5R_4,\ \ R_2R_6R_5,\ \ R_3R_4R_6$
Set two: $R_1R_2R_6R_4,\ \ R_5R_2R_3R_4,\ \ R_3R_1R_5R_6$

Their corresponding loop-impedance matrices are found to be

$$\mathbf{Z}_{m1} = \begin{bmatrix} 3 & -1 & -1 \\ -1 & 3 & -1 \\ -1 & -1 & 3 \end{bmatrix} \tag{5.211}$$

with

$$\Delta_{m1} = \det \mathbf{Z}_{m1} = 16 \tag{5.212}$$

and

$$\mathbf{Z}_{m2} = \begin{bmatrix} 4 & 0 & 0 \\ 0 & 4 & 0 \\ 0 & 0 & 4 \end{bmatrix} \tag{5.213}$$

with

$$\Delta_{m2} = \det \mathbf{Z}_{m2} = 64 \tag{5.214}$$

This shows that

$$\Delta_{m1} = \lambda_m \Delta_{m2} \tag{5.215}$$

with $\lambda_m = 4$.

5.9.2 The Cutset and Loop Formulations

As stated in Definition 5.1, the general return difference $F_k(x)$ is defined as the ratio of the two functional values assumed by the first-order cofactor of an element of the indefinite-admittance matrix $\mathbf{Y}$ under the condition that x assumes its nominal value and the condition that x assumes the value k:

$$F_k(x) = \frac{Y_{uv}(x)}{Y_{uv}(k)} \tag{5.216}$$

Since, from (5.203),

$$Y_{uv}(x) = \Delta_n(x) \tag{5.217}$$

and since Δ_n is a special type of cutset determinant, we obtain, from (5.202),

$$F_k(x) = \frac{Y_{uv}(x)}{Y_{uv}(k)} = \frac{\Delta_n(x)}{\Delta_n(k)} = \frac{\Delta_c(x)}{\Delta_c(k)} \tag{5.218}$$

This shows that in the cutset and hence nodal formulation of the network determinant of a feedback amplifier, the general return difference is invariant under the transformations from one such system to another. Thus, it is equally valid if $F_k(x)$ is defined in terms of the cutset determinant. A natural question that arises at this point is whether or not the general return difference can also be formulated in terms of the loop determinant. This question will be answered below.

Define the general return difference based on the loop formulation by

$$F_k^m(x) = \frac{\Delta_m(x)}{\Delta_m(k)} \tag{5.219}$$

as opposed to the general return difference $F_k(x)$ based on the cutset or nodal formulation. Like $F_k(x)$, one immediate consequence of (5.201) is that $F_k^m(x)$ is invariant under the transformations from one set of basis loops to another. Now we establish a relation between $F_k^m(x)$ and $F_k(x)$. By appealing to (5.204), (5.219) can be written as

$$F_k^m(x) = \frac{\Delta_m(x)}{\Delta_m(k)} = \frac{\Delta_b(x)\Delta_c(x)}{\Delta_b(k)\Delta_c(k)} = \frac{\Delta_b(x)F_k(x)}{\Delta_b(k)} \tag{5.220}$$

giving

$$\Delta_b(k)F_k^m(x) = \Delta_b(x)F_k(x) \tag{5.221}$$

In words, this states that if the determinant of the branch-impedance matrix is nonzero under the condition that the element x assumes its nominal value and the condition that the element x assumes the value k, then the ratio of the general return difference with respect to an element x for a reference value k and based on the loop formulation to that of the same element based on the cutset or nodal formulation is equal to the ratio of the determinant of the branch-impedance matrix when the element x assumes its nominal value to that when it assumes the value k. We summarize the above results by the following theorem.

Theorem 5.1 In the cutset or nodal (loop) formulation of the network determinant of a feedback amplifier, the general return difference is invariant with respect to the transformations from one such system to another. It is invariant under the general transformations between a system of basis loops and a system of basis cutsets if, and only if, the determinant of the branch-immittance matrix remains unaltered when the element of interest is replaced by the chosen reference value.

In the theorem we implicitly assumed that $\Delta_b(x) \neq 0$ and $\Delta_b(k) \neq 0$. As a consequence, if a feedback amplifier contains only a single controlled source, then the general return difference with respect to the controlling parameter of the controlled source is invariant under the general transformations between a system of basis loops and a system of basis cutsets. Its justification is left as an exercise (see Prob. 5.18).

Consider an active two-port device characterized by its general hybrid parameters k_{ij}. Assume that the two-port device N_a is terminated at its input and output ports by the immittances M_1 and M_2, as shown in Fig. 3.1. As indicated in Eq. (5.42), the return difference

$$F(k_{21}) = 1 + T_k = 1 - \frac{k_{12}k_{21}}{(k_{11} + M_1)(k_{22} + M_2)} \tag{5.222}$$

is invariant under immittance substitution for y-, h-, and g-parameters.

For the z-parameters of Fig. 5.38, the loop-impedance matrix is given by

$$\mathbf{Z}_m = \begin{bmatrix} z_{11} + Z_1 & z_{12} \\ z_{21} & z_{22} + Z_2 \end{bmatrix} \tag{5.223}$$

whose determinant is found to be

$$\Delta_m = \det \mathbf{Z}_m = (z_{11} + Z_1)(z_{22} + Z_2) - z_{12}z_{21} \tag{5.224}$$

From (5.219) we get

$$F^m(z_{21}) \equiv F_0^m(z_{21}) = 1 + T_z^m = 1 - \frac{z_{12}z_{21}}{(z_{11} + Z_1)(z_{22} + Z_2)} \tag{5.225}$$

where T_z^m, called the *loop-based return ratio*, is identified as

$$T_z^m = -\frac{z_{12}z_{21}}{(z_{11} + Z_1)(z_{22} + Z_2)} \tag{5.226}$$

Comparing (5.45) with (5.225) yields

$$(z_{11}z_{22} - z_{12}z_{21})F(z_{21}) = z_{11}z_{22}F^m(z_{21}) \tag{5.227}$$

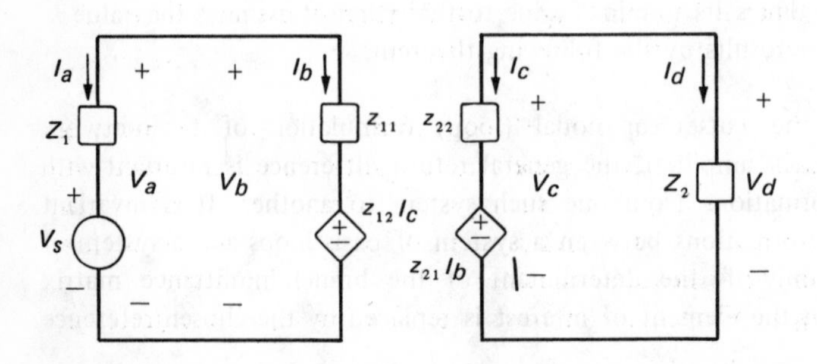

Figure 5.38 A terminated two-port network represented by its z-parameters.

From Fig. 5.38, the branch voltage-current relations are described by the equation

$$\begin{bmatrix} V_a \\ V_b \\ V_c \\ V_d \end{bmatrix} = \begin{bmatrix} Z_1 & 0 & 0 & 0 \\ 0 & z_{11} & z_{12} & 0 \\ 0 & z_{21} & z_{22} & 0 \\ 0 & 0 & 0 & Z_2 \end{bmatrix} \begin{bmatrix} I_a \\ I_b \\ I_c \\ I_d \end{bmatrix} + \begin{bmatrix} V_s \\ 0 \\ 0 \\ 0 \end{bmatrix} \tag{5.228}$$

The coefficient matrix in (5.228) is the branch-impedance matrix, whose determinant is found to be

$$\Delta_b = Z_1 Z_2 (z_{11} z_{22} - z_{12} z_{21}) \tag{5.229}$$

In terms of Δ_b, (5.227) can be written as

$$\Delta_b(z_{21}) F(z_{21}) = \Delta_b(0) F^m(z_{21}) \tag{5.230}$$

confirming (5.221).

As in (5.42), by substituting the corresponding parameters of z_{ij} we can show that

$$F^m(k_{21}) = \frac{\Delta_m(k_{21})}{\Delta_m(0)} = \frac{(k_{11} + M_1)(k_{22} + M_2) - k_{12} k_{21}}{(k_{11} + M_1)(k_{22} + M_2)} = 1 + T_k^m \tag{5.231}$$

where

$$T_k^m = -\frac{k_{12} k_{21}}{(k_{11} + M_1)(k_{22} + M_2)} \tag{5.232}$$

is invariant under immittance substitution for the z-, h-, and g-parameters. Its justification is left as an exercise (see Prob. 5.20). We remark that (5.42) and (5.231) do not imply that $F(k_{21}) = F^m(k_{21})$. The return differences $F(k_{21})$ and $F^m(k_{21})$ are related by (5.221).

Example 5.16 Consider the feedback network of Fig. 5.39*a*. Suppose that g_{m2} is the element of interest and that the reference value k is zero. Two equivalent networks of Fig. 5.39*a* are presented in Fig. 5.39, *b* and *c* with

$$\alpha_1 = -\frac{g_{m2} g_{m3} Z_2}{q} \qquad \alpha_2 = \frac{g_{m3}}{q} \tag{5.233a}$$

$$\beta_1 = \frac{g_{m2}}{q} \qquad \beta_2 = -\frac{g_{m2} g_{m3} Z_3}{q} \tag{5.233b}$$

where

$$q = Y_2 Y_3 - g_{m2} g_{m3} \tag{5.233c}$$

The branch-admittance matrix of Fig. 5.39*b* can easily be obtained by inspection and is given by

$$\mathbf{Y}_b = \begin{bmatrix} Y_1 & 0 & 0 & 0 \\ 0 & Y_3 & g_{m2} & 0 \\ 0 & g_{m3} & Y_2 & 0 \\ 0 & 0 & 0 & Y_4 \end{bmatrix} \tag{5.234}$$

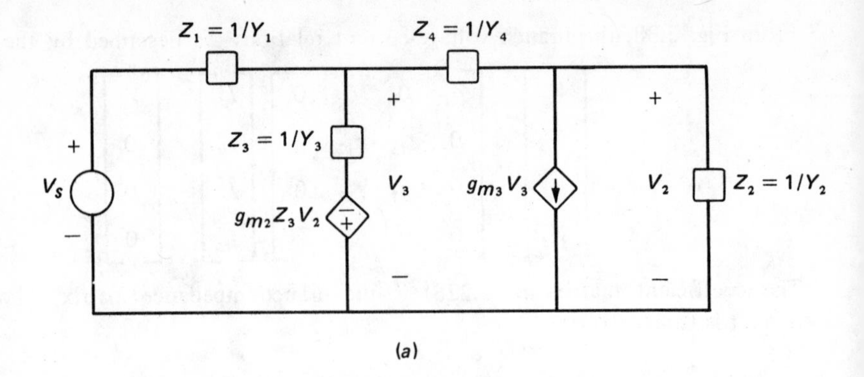

(a)

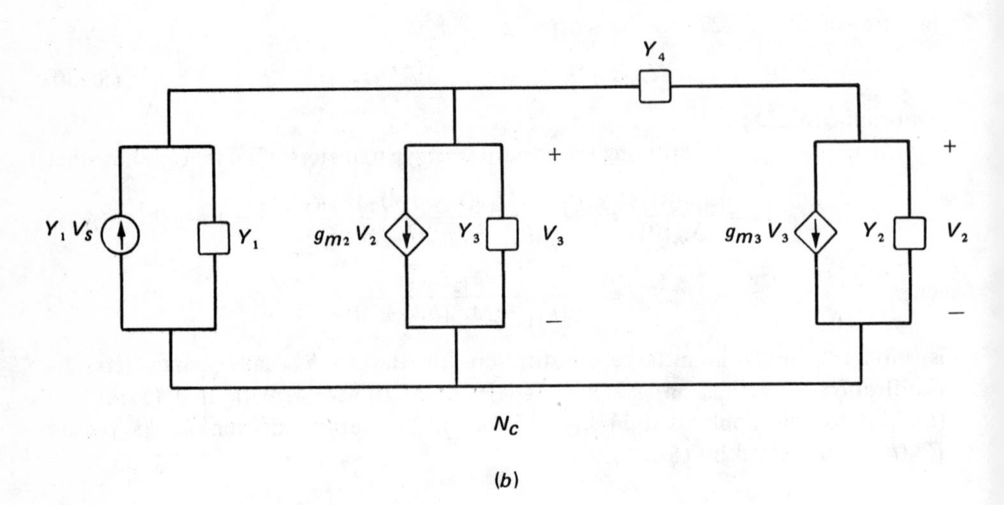

N_c

(b)

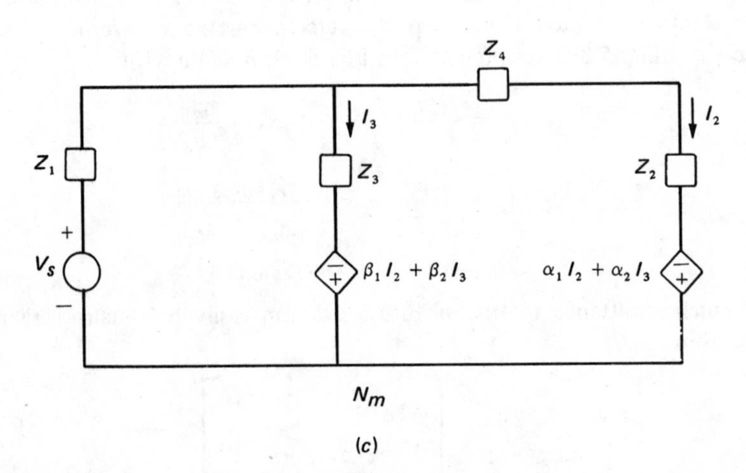

N_m

(c)

Figure 5.39 (a) A feedback network, (b) its equivalent network N_c, and (c) its equivalent network N_m.

whose determinant is found to be

$$\det \mathbf{Y}_b = Y_1 Y_4 (Y_2 Y_3 - g_{m2} g_{m3}) = \frac{1}{\Delta_b} \tag{5.235}$$

Since Δ_b does not remain unaltered when g_{m2} is replaced by 0, according to Theorem 5.1, the return differences $F(g_{m2})$ and $F^m(g_{m2})$ will be different.

Consider the node-admittance matrix of Fig. 5.39b, given as

$$\mathbf{Y}_n = \begin{bmatrix} Y_1 + Y_3 + Y_4 & g_{m2} - Y_4 \\ g_{m3} - Y_4 & Y_2 + Y_4 \end{bmatrix} \tag{5.236}$$

whose determinant is also equal to the first-order cofactor of the indefinite-admittance matrix with

$$Y_{uv} = \det \mathbf{Y}_n = (Y_1 + Y_3)(Y_2 + Y_4) + Y_4(Y_2 + g_{m2} + g_{m3}) - g_{m2} g_{m3} \tag{5.237}$$

This yields

$$F(g_{m2}) = \frac{(Y_1 + Y_3)(Y_2 + Y_4) + Y_4(Y_2 + g_{m2} + g_{m3}) - g_{m2} g_{m3}}{(Y_1 + Y_3)(Y_2 + Y_4) + Y_4(Y_2 + g_{m3})} \tag{5.238}$$

By using the equivalent network of Fig. 5.39c, the branch-impedance matrix and the loop-impedance matrix are found to be

$$\mathbf{Z}_b = \begin{bmatrix} Z_1 & 0 & 0 & 0 \\ 0 & \dfrac{Y_2}{q} & -\dfrac{g_{m2}}{q} & 0 \\ 0 & -\dfrac{g_{m3}}{q} & \dfrac{Y_3}{q} & 0 \\ 0 & 0 & 0 & Z_4 \end{bmatrix} \tag{5.239}$$

$$\mathbf{Z}_m = \begin{bmatrix} Z_1 + \dfrac{Y_2}{q} & -\dfrac{Y_2 + g_{m2}}{q} \\ -\dfrac{Y_2 + g_{m3}}{q} & Z_4 + \dfrac{Y_2 + Y_3 + g_{m2} + g_{m3}}{q} \end{bmatrix} \tag{5.240}$$

giving

$$\Delta_b = \frac{Z_1 Z_4 (Y_2 Y_3 - g_{m2} g_{m3})}{q^2} = \frac{Z_1 Z_4}{q} = \frac{1}{\det \mathbf{Y}_b} \tag{5.241}$$

$$\Delta_m = \det \mathbf{Z}_m = \frac{Z_1(Z_4 q + Y_2 + Y_3 + g_{m2} + g_{m3}) + Y_2 Z_4 + 1}{q} \tag{5.242}$$

From (5.219), we obtain

$$F^m(g_{m2}) = \frac{\Delta_m(g_{m2})}{\Delta_m(0)} = \frac{Y_1Y_2Y_3Y_4[Z_1(Z_4q + Y_2 + Y_3 + g_{m2} + g_{m3}) + Y_2Z_4 + 1]}{q[Y_2Y_3 + Y_4(Y_2 + Y_3 + g_{m3}) + Y_1Y_2 + Y_1Y_4]}$$

$$= \frac{Y_2Y_3[Y_2(Y_1 + Y_3 + Y_4) + Y_4(Y_1 + Y_3 + g_{m2} + g_{m3}) - g_{m2}g_{m3}]}{(Y_2Y_3 - g_{m2}g_{m3})[Y_2(Y_1 + Y_3 + Y_4) + Y_4(Y_1 + Y_3 + g_{m3})]} \tag{5.243}$$

Comparing (5.238) with (5.243) gives

$$\Delta_b(0)F^m(g_{m2}) = \Delta_b(g_{m2})F(g_{m2}) \tag{5.244}$$

confirming (5.221).

In the above analysis, we have used two equivalent networks N_c and N_m of a feedback amplifier N, such as the ones shown in Fig. 5.39, b and c: N_c for the cutset formulation and N_m for the loop formulation. In the equivalent network N_c, all the sources have been converted to equivalent current sources and/or voltage-controlled current sources. In N_m all the sources have been converted to equivalent voltage sources and/or current-controlled voltage sources. The n voltage-controlled current sources, characterized by the defining equations

$$I_{k_j} = x_{k_j}V_{u_j} \qquad j = 1, 2, \ldots, n \tag{5.245}$$

are said to be *cyclically coupled* if $k_{v+1} = u_v$ for $v = 1, 2, \ldots, n$, where $k_{n+1} = k_1$ and I_k and V_k denote the current and voltage of the kth branch. For example, the following three voltage-controlled current sources,

$$I_2 = x_2V_5 \qquad I_5 = x_5V_4 \qquad I_4 = x_4V_2 \tag{5.246}$$

are cyclically coupled. By interchanging the roles of I and V in (5.245), we can similarly define a set of *cyclically coupled* current-controlled voltage sources in N_m. Clearly, if every device is replaced by its equivalent network composed of the controlled sources and the passive elements, and if there are no cyclically coupled and controlled sources, we can always renumber the controlling and the controlled branches in such a way that the index of any controlled branch is not higher than the index of its corresponding controlling branch. This results in a branch-immittance matrix with all the controlling parameters appearing above the main diagonal. This means that its determinant will be independent of the controlling parameter x. Hence, from Theorem 5.1, the general return difference in such a system will be invariant under the general transformations between a system of basis loops and a system of basis cutsets.

In Example 5.16, the two voltage-controlled current sources in Fig. 5.39b are cyclically coupled. If we set $g_{m3} = 0$, the resulting network has only one controlled source, so it cannot be cyclically coupled. Thus, the general return difference with respect to g_{m2} is invariant with $F(g_{m2}) = F^m(g_{m2})$, which can be confirmed from (5.244) by noting that, for $g_{m3} = 0$, $\Delta_b(0) = \Delta_b(g_{m2})$. In the network of Fig. 5.38, the two current-controlled voltage sources are cyclically coupled. To decouple them, we set $z_{12} = 0$, resulting in the invariance of the general return difference under the general transformations of the reference frame, which can again be confirmed from (5.230) by observing that for $z_{12} = 0$, $\Delta_b(z_{21}) = \Delta_b(0)$.

As shown in (5.104), the general return difference $F_k(x)$ with respect to a one-port admittance x for a general reference value k is equal to the ratio of total admittance looking into the node pair where x is connected to that when x is replaced by its reference value k:

$$F_k(x) = \frac{y + x}{y + k} \tag{5.247}$$

A similar result can be obtained based on the loop formulation. In this case, it is more convenient to consider the one-port impedance $\alpha = 1/x$.

Since from Theorem 5.1 the return difference $F_k^m(\alpha)$ is invariant with respect to the transformations from one system of basis loops to another, for simplicity and without loss of generality, we can choose a set of loop currents such that the impedance α is contained only in one of these loops, say, loop 1. With this choice of loop currents, the impedance α will appear only in the (1, 1)-position of the corresponding loop-impedance matrix $\mathbf{Z}_m$. As before, we write $\Delta_m = \Delta_m(\alpha) = \det \mathbf{Z}_m$. Then Δ_m can be expanded as

$$\Delta_m(\alpha) = \Delta_m(k) + (\alpha - k)\Delta_{m11} \tag{5.248}$$

where Δ_{m11} denotes the cofactor of the (1, 1)-element of $\mathbf{Z}_m$ in Δ_m. Combining (5.219) with (5.248) yields

$$\begin{aligned} F_k^m(\alpha) &= \frac{\Delta_m(\alpha)}{\Delta_m(k)} = 1 + (\alpha - k)\frac{\Delta_{m11}}{\Delta_m(k)} \\ &= 1 + \frac{\alpha - k}{k + z} = \frac{z + \alpha}{z + k} \end{aligned} \tag{5.249}$$

where z is the impedance that α faces. The second line follows from the fact that $\Delta_m(k)$ is the loop determinant of the equivalent network N_m when the element α assumes the value k, that Δ_{m11} is independent of the element α, and that if, in N_m with α assuming the value k, we insert a voltage source in series with α, the ratio $\Delta_m(k)/\Delta_{m11}$ is precisely the impedance looking into the voltage source, which is $z + k$. Thus, we conclude that $F_k^m(\alpha)$ is equal to the ratio of total impedance looking into the branch where α is situated, including α, to that when α is replaced by its reference value k. Comparing (5.247) with (5.249) yields

$$kF_k(x) = xF_\zeta^m(\alpha) \tag{5.250}$$

where

$$\alpha = \frac{1}{x} \tag{5.251a}$$

$$\zeta = \frac{1}{k} \tag{5.251b}$$

Similarly, we can show that (see Prob. 5.21)

$$F_k(\alpha) = \frac{y + x}{y + \zeta} \tag{5.252a}$$

$$F_k^m(x) = \frac{z + \alpha}{z + \zeta} \tag{5.252b}$$

giving

$$\alpha F_k(\alpha) = k F_k^m(\alpha) \tag{5.253a}$$

$$k F_k(x) = x F_k^m(x) \tag{5.253b}$$

which also confirm (5.221) with $\Delta_b(\alpha)/\Delta_b(k) = \alpha/k$ and $\Delta_b(x)/\Delta_b(k) = k/x$, respectively.

Thus, we conclude that the general return difference of a feedback amplifier with respect to a one-port immittance x for a general reference value k is invariant under the general transformations between a system of basis loops and a system of basic cutsets if and only if $x = k$.

Example 5.17 Consider the same feedback network as in Example 5.16. Now we compute the return difference with respect to the one-port impedance Z_1 as shown in Fig. 5.39*c*. Applying Eq. (5.219) in conjunction with Eq. (5.242), we get

$$F_k^m(Z_1) = \frac{Z_1(Z_4 q + Y_2 + Y_3 + g_{m2} + g_{m3}) + Y_2 Z_4 + 1}{k(Z_4 q + Y_2 + Y_3 + g_{m2} + g_{m3}) + Y_2 Z_4 + 1} = \frac{z + Z_1}{z + k} \tag{5.254}$$

where z is the input impedance facing Z_1 and is given by

$$z = \frac{Y_2 Z_4 + 1}{Z_4 q + Y_2 + Y_3 + g_{m2} + g_{m3}} \tag{5.255}$$

Likewise, from (5.237) we obtain

$$F_k(Y_1) = \frac{(Y_1 + Y_3)(Y_2 + Y_4) + Y_4(Y_2 + g_{m2} + g_{m3}) - g_{m2} g_{m3}}{(k + Y_3)(Y_2 + Y_4) + Y_4(Y_2 + g_{m2} + g_{m3}) - g_{m2} g_{m3}}$$

$$= \frac{y + Y_1}{y + k} \tag{5.256}$$

where $y = 1/z$ is the admittance facing Y_1, and

$$F_k(Z_1) = \frac{(Y_1 + Y_3)(Y_2 + Y_4) + Y_4(Y_2 + g_{m2} + g_{m3}) - g_{m2} g_{m3}}{(1/k + Y_3)(Y_2 + Y_4) + Y_4(Y_2 + g_{m2} + g_{m3}) - g_{m2} g_{m3}}$$

$$= \frac{y + Y_1}{y + 1/k} \tag{5.257}$$

showing that with $\zeta = 1/k$,

$$F_k(Y_1) = F_\zeta(Z_1) \tag{5.258}$$

In a similar manner, we can demonstrate that

$$F_k^m(Z_1) = F_\zeta^m(Y_1) \tag{5.259}$$

where

$$F_k^m(Y_1) = \frac{Z_1(Z_4q + Y_2 + Y_3 + g_{m2} + g_{m3}) + Y_2Z_4 + 1}{\zeta(Z_4q + Y_2 + Y_3 + g_{m2} + g_{m3}) + Y_2Z_4 + 1} = \frac{z + Z_1}{z + \zeta} \tag{5.260}$$

Finally, comparing (5.254) with (5.257) and (5.256) with (5.260), we confirm (5.253).

5.10 SUMMARY

In this chapter, we continued our study of feedback amplifier theory begun in Chap. 4. We showed that feedback may be employed to make the amplifier gain less sensitive to variations in the parameters of the active components. If the magnitude of the ratio of the two functional values assumed by the amplifier forward transmission under the condition that an active device is deactivated and the condition that the active device is operating normally is small, then the sensitivity is approximately equal to the reciprocal of the return difference with respect to the forward-transfer parameter of the active device. In Chap. 3, we derived the general expressions of the transfer functions, driving-point functions, power gains, and the sensitivity functions of a two-port network in terms of the general two-port parameters. In this chapter, we exhibited the general relations between these expressions and feedback and demonstrated by a specific example how they can be employed to design amplifiers that achieve maximum gain with a prescribed sensitivity.

The concepts of return difference and null return difference introduced in the preceding chapter were extended and generalized by considering the general reference value. They are very useful in measurement situations in that they correspond to the situation where a feedback amplifier under study is made partially active rather than completely dead, as in the original interpretations of the return difference and null return difference for the zero reference value. Relations between network functions and the general return difference and null return difference were derived. They are generalizations of Blackman's formula introduced in Chap. 4, and they can be employed effectively to simplify the calculations of active impedances. We found that the network functions can be expressed in terms of the general return difference and the general null return difference in a manner similar to those discussed before. As an extension of the concept of sensitivity function, we defined the relative sensitivity function, which is expressed in terms of the altered parameters. The significance is that by choosing an appropriate reference value for the general return difference, we can interrupt the forward transmission of the amplifier at this reference value. Under this condition, the relative sensitivity becomes equal to the reciprocal of the general return difference. This yields a direct generalization of the simple relation between the sensitivity function and the return difference for the ideal feedback model. In general, the relative sensitivity function can be written as the difference of the reciprocals of the general return difference and the general null return difference.

In addition to the algebraic formulation of feedback amplifier theory, we indicated that the theory can also be formulated in terms of a signal-flow graph representing the signal transmissions among the various ports. One of the advantages of this formulation is that it can easily be generalized to multiple-loop feedback networks to be discussed later. One simple consequence is that, in many cases, negative feedback may be used to reduce the effect of noise and nonlinear distortion generated within an amplifier.

One useful relation found helpful in feedback calculations is the return difference and the null return difference with respect to two elements. We found that the ratio of the return differences with respect to two elements individually is equal to the ratio of the return differences with respect to one when the other one is set to zero.

Since the zeros of the return difference are the natural frequencies of the network, they are essential for the stability study. For this we presented three procedures for the physical measurements of the return difference, emphasizing the applications to transistors. The first technique is a direct one, but requires the use of a matched transistor and the simulation of three impedances. The second method overcomes the need for simulating the effect of internal feedback, but it essentially requires three measurements and the simulation of one driving-point impedance. The third method requires only two driving-point impedance measurements at a convenient port of the feedback amplifier, and is particularly attractive if the two impedances can be measured directly in their polar forms. As a consequence of these measurements, we can verify experimentally and study the stability problems of the amplifier under certain prescribed conditions. In addition to these, we discussed a procedure for experimental determination of the general reference value under which the amplifier becomes unilateral. The procedure requires four impedance measurements and can easily be performed in the laboratory for most active devices.

The feedback amplifier theory developed in this and the preceding chapters was formulated in terms of the first- and the second-order cofactors of the elements of the indefinite-admittance matrix, which is readily interpretable physically. This mathematical formulation is consistent, compact, and elegant, and it is easily applicable to most practical feedback amplifiers. It is also possible, although much more complicated, to formulate the theory in terms of the loop determinant or the cutset determinant. To this end, we outlined and derived relationships among the return differences based on various formulations. Specifically, we showed that in the cutset or nodal (loop) formulation of the network determinant, the general return difference is invariant with respect to the transformations from one such system to another. It is invariant under the general transformations of the reference frame if and only if the determinant of its branch-immittance matrix remains unaltered when the element of interest is replaced by the chosen reference value.

Unlike the indefinite-admittance-matrix formulation, the null return difference based on the cutset or loop determinant cannot generally be defined as the ratio of the two functional values assumed by one of its cofactors unless some specific details of the formulation of the network equations are also stated. Furthermore, this ratio is *variant* under the general transformations from a system of basis cutsets (loops) to another. To avoid this difficulty, we may introduce the concepts of "generalized cofactors," which are rather complicated. For a detailed account of these results and all the variations and ramifications, the reader is referred to Chen (1974a, 1974b).

PROBLEMS

5.1 Derive the identity (5.9).

5.2 Consider the series-parallel or voltage-series feedback amplifier of Fig. 4.24. Assume that the two transistors are identical, with $h_{re} = h_{oe} = 0$, $h_{ie} = 1.1$ kΩ, and $h_{fe} = 60$. Write $\tilde{\alpha}_j = \alpha_j \cdot 10^{-4} = h_{fe}/h_{ie}$ where $j = 1, 2$ is used to distinguish the transconductances of the first and second transistors. Determine the following:

(*a*) The general return differences $F_k(\tilde{\alpha}_1)$ and $F_k(\tilde{\alpha}_2)$
(*b*) The general null return differences $\hat{F}_k(\tilde{\alpha}_1)$ and $\hat{F}_k(\tilde{\alpha}_2)$
(*c*) The voltage gain in terms of $\tilde{\alpha}_1$ and $\tilde{\alpha}_2$
(*d*) The current gain in terms of $\tilde{\alpha}_1$ and $\tilde{\alpha}_2$
(*e*) The short-circuit current gain in terms of $\tilde{\alpha}_1$ and $\tilde{\alpha}_2$
(*f*) The input impedance in terms of $\tilde{\alpha}_1$ and $\tilde{\alpha}_2$
(*g*) The output impedance in terms of $\tilde{\alpha}_1$ and $\tilde{\alpha}_2$

By using the results obtained above, verify the identities (5.99), (5.101), (5.103).

5.3 In the network of Fig. 5.2, confirm that the sensitivity function for the short-circuit current gain with respect to α_1 is given by Eq. (5.12).

5.4 The open-loop current gain of the voltage-shunt feedback amplifier of Fig. 5.4 is given by Eq. (4.51). Compare this with the open-loop voltage gain of Eq. (5.28*a*). With approximation, is one deducible from the other?

5.5 In the feedback network considered in Prob. 5.2, compute the sensitivities $S(\tilde{\alpha}_1)$ and $S(\tilde{\alpha}_2)$ and the relative sensitivities $S'(\tilde{\alpha}_1')$ and $S'(\tilde{\alpha}_2')$, where $\tilde{\alpha}_1' = \tilde{\alpha}_1 - k$ and $\tilde{\alpha}_2' = \tilde{\alpha}_2 - k$.

5.6 Using the ideal feedback model, show that the open-loop voltage gain of the feedback amplifier of Fig. 5.4 with the parameters as given in Eq. (5.31) is −16 and that the closed-loop voltage gain is −3.2.

5.7 The return ratios T_h and T_g are given by Eqs. (5.38) and (5.41). Show that for $M_1 = 0$ and $M_2 = 0$, $T_h = T_g$.

5.8 Repeat Example 5.10 if the element of interest is $\tilde{\alpha}_2$.

5.9 Repeat Example 5.5 by using the parallel-parallel topology of Fig. 4.3*b*.

5.10 Derive the identity (5.145).

5.11 Derive the identity (5.146).

5.12 Repeat Example 5.5 by using the series-series topology of Fig. 4.3*a*.

5.13 Repeat Example 5.5 by using the parallel-series topology of Fig. 4.3*d*.

5.14 Consider the series-series or current-series feedback amplifier of Fig. 4.7. Determine the reference value for the general return difference with respect to the amplifier forward-transfer parameter under which the amplifier becomes unilateral.

5.15 Refer to Fig. 5.25. Assume that

$$|y_o| \gg |y_{12}| \qquad |y_{11}| \gg |y_{12}| \tag{5.261}$$

Show that the Norton equivalent network looking into the two-port network at terminals 1 and 3 can be approximated by the parallel combination of y_{11} and $y_{12}V_{23}$ as indicated in Fig. 5.24.

5.16 The hybrid-pi equivalent network of a common-emitter transistor is shown in Fig. 5.26, whose indefinite-admittance matrix can easily be written down by inspection. Show that after suppressing terminal $B' = 4$ and deleting the row and column corresponding to terminal $E = 3$, the admittance matrix is given by (5.182).

5.17 Consider the series-parallel feedback amplifier of Fig. 4.21. Determine the reference value for the general return difference with respect to the amplifier forward-transfer parameter under which the amplifier becomes unilateral.

5.18 Justify the statement that if a feedback amplifier contains only a single controlled source, then the general return difference with respect to the controlling parameter of the controlled source is invariant under the general transformations between a system of basis loops and a system of basis cutsets.

5.19 Repeat Prob. 5.14 for the parallel-series configuration of Fig. 4.27 using the hybrid parameters for the transistors.

5.20 Demonstrate that (5.232) is invariant under immittance substitution for the z-, h-, and g-parameters.

5.21 Derive the identities (5.252).

5.22 In the feedback amplifier of Fig. 4.27, apply Blecher's procedure to determine the return difference with respect to the forward-transfer parameter of the second transistor.

5.23 Repeat Prob. 5.22 using Hakim's procedure.

5.24 Repeat Prob. 5.22 by making only two driving-point impedance measurements at a convenient port.

5.25 Refer to the compound-feedback amplifier of Fig. 4.46. Compute the sensitivity and the relative sensitivity functions with respect to the amplifier forward-transfer parameter. Also determine the reference value of the general return difference with respect to this parameter under which the amplifier becomes unilateral.

5.26 In the feedback network of Fig. 5.39*b*, let $Y_1 V_s$ be the input port and Y_4 the output port. Show that

$$\hat{F}_k(g_{m2}) = 1 \tag{5.262a}$$

$$\hat{F}_k(g_{m3}) = \frac{Y_2 + g_{m3}}{Y_2 + k} \tag{5.262b}$$

Also calculate Eqs. (5.262*a*) and (5.262*b*) by their physical interpretations in terms of the returned voltages.

5.27 The transistor in the feedback amplifier of Fig. 5.4*a* can be represented by its equivalent T-model, as shown in Fig. 5.40. Show that the general null return difference with respect to the transconductance αg_e for a general reference value k is given by

$$\hat{F}_k(\alpha g_e) = \frac{g_b(g_c + G_f) + g_c G_f - \alpha g_e(g_b + G_f) + g_e G_f}{g_b(g_c + G_f) + g_c G_f - k(g_b + G_f) + g_e G_f} \tag{5.263}$$

Also show that for

$$k = k_0 = \frac{g_b(g_c + G_f) + G_f(g_c + g_e)}{g_b + G_f} \tag{5.264}$$

the relative sensitivity $S'(x')$ becomes equal to the reciprocal of the return difference $F_{k_0}(\alpha g_e)$, where $x' = \alpha g_e - k_0$.

5.28 By using Eq. (5.130), compute the impedance facing $R_f = 1/G_f$ in the network of Fig. 5.40. Also, compute the amplifier input and output impedances and determine the sensitivity function of the current gain to R_f.

5.29 Assume that the two-port network of Fig. 5.22 is a transistor and is represented by its z-parameters z_{ij}. Demonstrate that the return difference $F^m(z_{21})$ can be measured by the network of Fig. 5.41*a*. Also show that the controlled source $z_{12}I_{23}$ can be simulated using the arrangement of Fig. 5.41*b*, where z_o denotes the total impedance presented to the output port of the transistor, as shown in Fig. 5.41*a*.

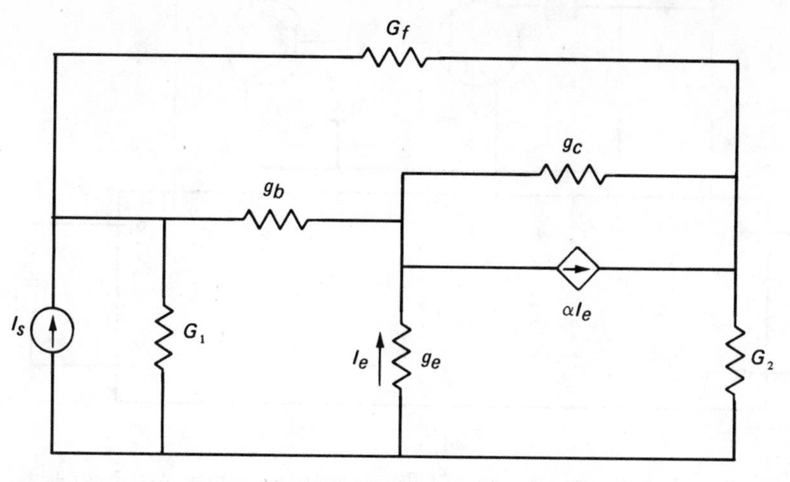

Figure 5.40 The equivalent network of the feedback amplifier of Fig. 5.4*a* with the transistor being represented by its T-model.

5.30 By using Blackman's formula, show that the input admittance Y of the network of Fig. 5.42 is given by

$$Y(s) = 1 + \frac{Z_3 - Z_4}{Z_1 - Z_2} \tag{5.265}$$

5.31 By applying Blackman's formula, show that the input impedance Z of the network of Fig. 5.43 with $\alpha = Z_3/Z_4$ can be expressed as

$$Z(s) = \frac{Z_1 Z_2 (Z_4 - Z_3)}{Z_2 Z_4 - Z_1 Z_3} \tag{5.266}$$

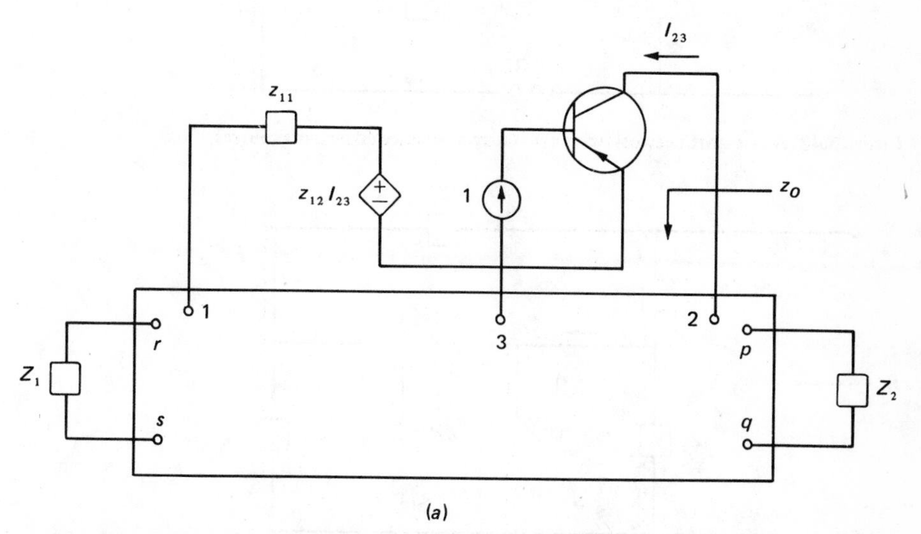

Figure 5.41 (*a*) The measurement of return difference $F^m(z_{21})$ for a transistor operated in the common-emitter configuration and represented by its z-parameters z_{ij}.

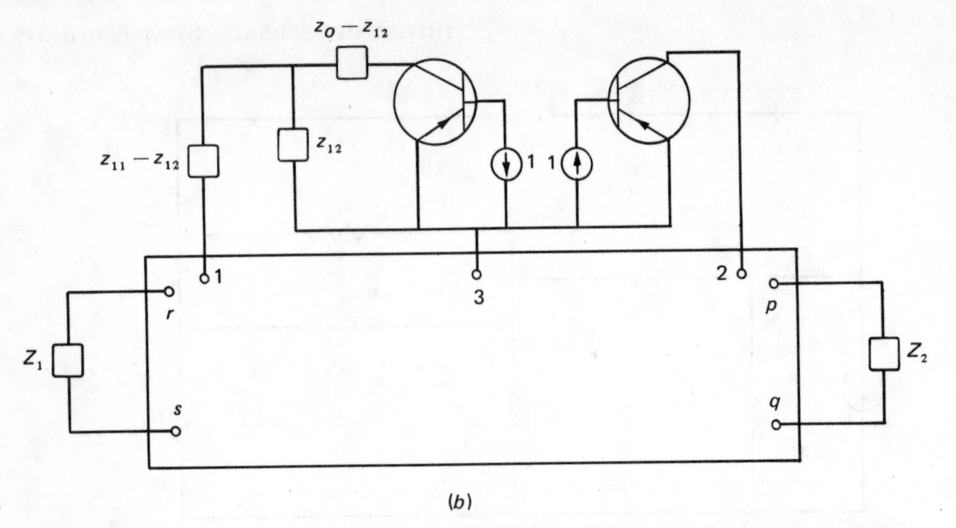

Figure 5.41 (*Continued*) (*b*) The measurement of $F^m(z_{21})$ with the simulation of the controlled source $z_{12}I_{23}$.

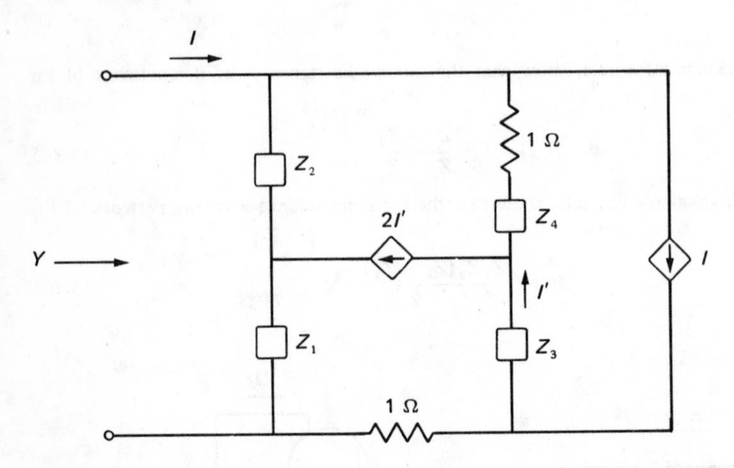

Figure 5.42 A one-port network with two current-controlled current sources.

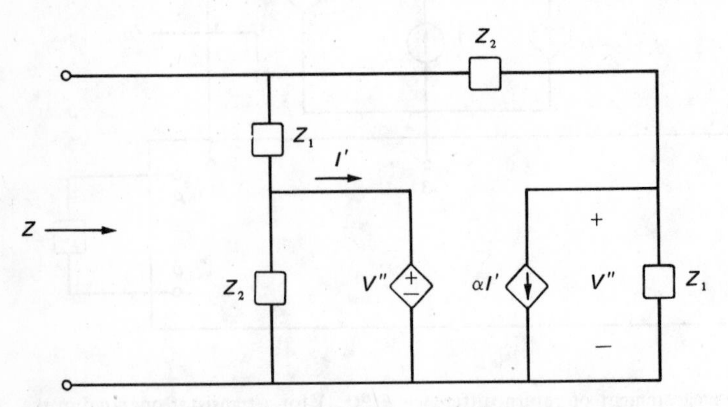

Figure 5.43 A one-port network with two controlled sources.

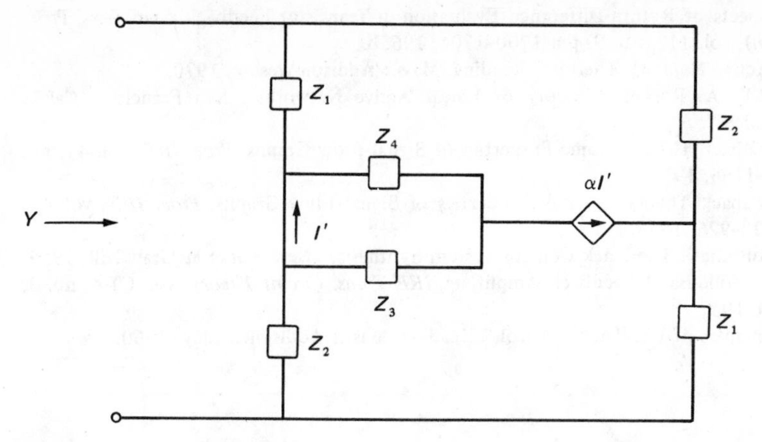

Figure 5.44 A one-port network containing a current-controlled current source.

5.32 By using Blackman's formula, demonstrate that the input admittance Y of the network of Fig. 5.44 can be written as

$$Y(s) = \frac{(\alpha + 2)Z_4 - (\alpha - 2)Z_3}{Z_4[(\alpha + 1)Z_2 + Z_1] - Z_3[(\alpha - 1)Z_1 - Z_2]} \tag{5.267}$$

5.33 In the feedback network of Fig. 5.4*a*, let $h_{fe} = 60$, $h_{ie} = 1$ kΩ, $h_{re} = 3 \cdot 10^{-4}$, $h_{oe} = 25$ μmho, $R_f = 40$ kΩ, $R_1 = 10$ kΩ, and $R_2 = 4$ kΩ. Compute the return differences with respect to the forward-transfer parameters $h_{fe} = h_{21}$, y_{21}, z_{21}, and g_{21}, respectively, of the transistor.

BIBLIOGRAPHY

Blecher, F. H.: Design Principles for Single Loop Transistor Feedback Amplifiers, *IRE Trans. Circuit Theory*, vol. CT-4, no. 3, pp. 145–156, 1957.

Bode, H. W.: "Network Analysis and Feedback Amplifier Design," Princeton, N.J.: Van Nostrand, 1945.

Chen, W. K.: Graph-Theoretic Considerations on the Invariance and Mutual Relations of the Determinants of the Generalized Network Matrices and Their Generalized Cofactors, *Quart. J. Math. (Oxford)*, 2d series, vol. 21, no. 84, pp. 459–479, 1970.

Chen, W. K.: Graph-Theoretic Considerations on the Invariance of Return Difference, *J. Franklin Inst.*, vol. 298, no. 2, pp. 81–100, 1974a.

Chen, W. K.: Invariance and Mutual Relations of the General Null-Return-Difference Functions, *Proc. 1974 Eur. Conf. Circuit Theory and Design*, IEE Conf. Publ. No. 116, pp. 371–376, 1974b.

Chen, W. K.: "Applied Graph Theory: Graphs and Electrical Networks," 2d rev. ed., chaps. 2 and 4, New York: American Elsevier, and Amsterdam: North-Holland, 1976a.

Chen, W. K.: Indefinite-Admittance Matrix Formulation of Feedback Amplifier Theory, *IEEE Trans. Circuits and Systems*, vol. CAS-23, no. 8, pp. 498–505, 1976b.

Chen, W. K.: Network Functions and Feedback, *Int. J. Electronics*, vol. 42, no. 6, pp. 617–618, 1977.

Chen, W. K.: On Second-Order Cofactors and Null Return Difference in Feedback Amplifier Theory, *Int. J. Circuit Theory and Applications*, vol. 6, no. 3, pp. 305–312, 1978.

Hakim, S. S.: Return-Difference Measurement in Transistor Feedback Amplifiers, *Proc. IEE (London)*, vol. 112, no. 5, pp. 914–915, 1965a.

Hakim, S. S.: Aspects of Return-Difference Evaluation in Transistor Feedback Amplifiers, *Proc. IEE* (*London*), vol. 112, no. 9, pp. 1700–1704, 1965b.

Haykin, S. S.: "Active Network Theory," Reading, Mass.: Addison-Wesley, 1970.

Kuh, E. S., and R. A. Rohrer: "Theory of Linear Active Networks," San Francisco, Calif.: Holden-Day, 1967.

Mason, S. J.: Feedback Theory–Some Properties of Signal Flow Graphs, *Proc. IRE*, vol. 41, no. 9, pp. 1144–1156, 1953.

Mason, S. J.: Feedback Theory–Further Properties of Signal Flow Graphs, *Proc. IRE*, vol. 44, no. 7, pp. 920–926, 1956.

Truxal, J. G.: "Automatic Feedback Control System Synthesis," New York: McGraw-Hill, 1955.

Waldhauer, F. D.: Wide-Band Feedback Amplifiers, *IRE Trans. Circuit Theory*, vol. CT-4, no. 3, pp. 178–190, 1957.

Wilts, C. H.: "Principles of Feedback Control," Reading, Mass.: Addison-Wesley, 1960.

CHAPTER
SIX
STABILITY OF FEEDBACK AMPLIFIERS

In the preceding two chapters, we demonstrated that the application of negative feedback in amplifiers tends to make the overall gain less sensitive to variations in parameters, reduce noise and nonlinear distortion, and control the input and output impedances. These improvements are all affected by the same factor, which is the normal value of the return difference. However, the price that we paid in achieving these is the net reduction of the overall gain. In addition, we are faced with the stability problem in that, for sufficient amount of feedback, at some frequency the amplifier tends to oscillate and becomes unstable. The objective of this chapter is to discuss various stability criteria and to investigate several approaches to the stabilization of feedback amplifiers.

As indicated in Sec. 5.8, the zeros of the network determinant are called the natural frequencies. A network is stable if all of its natural frequencies are restricted to the open left half of the complex frequency s-plane (LHS). If the network determinant is known, its roots can be readily computed explicitly with the aid of a computer if necessary, and the stability problem can then be settled directly. However, for a physical network, there remains the difficulty of getting an accurate formulation of the network determinant itself. Even if we have the network determinant, the roots alone do not tell us the degree of stability when the feedback amplifier is stable, nor do they provide us with any information as to how to stabilize an unstable amplifier. These limitations are overcome by applying the Nyquist criterion to the return difference, which gives precisely the same information about the stability of a feedback amplifier as does the network determinant itself. Furthermore, the return difference can be measured physically, meaning that we can include all the parasitic effects in the stability study. The discussion of this chapter is confined to single-loop feedback amplifiers. Multiple-loop feedback amplifiers are presented in the following chapter.

We first introduce the concepts of a single-loop feedback amplifier and its stability, and then review briefly the Routh-Hurwitz criterion. This is followed by a discussion of the Nyquist stability criterion and the Bode plot. The root-locus technique and the notion of root sensitivity are taken up next. The relationship between gain and phase shift are elaborated. Finally, we discuss means of stabilizing a feedback amplifier and present Bode's design theory.

6.1 THE SINGLE-LOOP FEEDBACK AMPLIFIERS

For the sake of later discussion, in this section we define several important terms.

Definition 6.1: Single-loop feedback amplifier A *single-loop feedback amplifier* is one in which the return difference with respect to the controlling parameter of any active device is equal to unity if the controlling parameter of any other active device in the network vanishes.

This definition has two important implications. The first is that the controlling parameters of the active devices can enter the network determinant only in a simple product form. Thus, for a single feedback amplifier containing n active devices, the first-order cofactor of an element of its indefinite-admittance matrix can be expressed in the form

$$Y_{uv} = g_{m1} g_{m2} \cdots g_{mn} A + B \tag{6.1}$$

where g_{mj} $(j = 1, 2, \ldots, n)$ are the controlling parameters of the active devices, and A and B are functions of s and are independent of the controlling parameters g_{mj}. As a result, the second implication is that the return differences with respect to g_{mj} of all the active devices under normal operating conditions are the same. Thus, the voltage-shunt configuration of Fig. 4.19 with its equivalent network as shown in Fig. 4.20 is a single-loop feedback amplifier. In fact, Eq. (6.1) is completely equivalent to Definition 6.1 and can be used interchangeably.

Strictly speaking, the above definition excludes amplifiers in which there is local feedback or internal feedback in one or more of the active devices. However, many of these amplifiers can be approximated by a single-loop amplifier. On the other hand, the definition includes as single-loop feedback amplifiers structures having a number of distinct feedback paths. For example, we can put another passive two-port network such as a transformer in parallel with the network of Fig. 4.19, and the resulting structure is still a single-loop feedback amplifier.

As an example of a multiple-loop feedback amplifier, we consider the parallel-series configuration of Fig. 5.14, which is represented by the equivalent network of Fig. 5.15*b*. From Example 5.9, the first-order cofactor or nodal determinant is found to be

$$Y_{uv} = (166.66\alpha_1\alpha_2 + 1514.94\alpha_1 + 1716.83\alpha_2 + 1{,}331{,}494) \cdot 10^{-16} \tag{6.2}$$

where α_1 and α_2 are the transconductances of the transistors. Since Eq. (6.2) cannot be represented in the form of Eq. (6.1), the feedback amplifier is not a single-loop one. Likewise, the series-parallel feedback amplifier of Fig. 4.24, as considered in Example 4.11, is another that is not a single-loop feedback amplifier.

The stability of a feedback amplifier is sometimes much affected by the relative rates at which the gains of the active devices decay with age, or the relative rates at which they increase as the devices warm up when power is first applied to the network. Therefore, a feedback amplifier that is stable under the designed operating conditions may become unstable as a result of variations in parameters. Since the performance of a feedback amplifier is most sensitive to the changes of the controlling parameters of the active devices, we introduce the following terms.

Definition 6.2: Absolutely stable single-loop feedback amplifier A single-loop feedback amplifier is said to be *absolutely stable* if the network is stable when the values of the controlling parameters of the active devices vary from zero to their nominal values.

Definition 6.3: Conditionally stable single-loop feedback amplifier A single-loop feedback amplifier is said to be *conditionally stable* if the network is stable when the controlling parameters of the active devices assume their nominal values, and it becomes unstable when they assume some values between zero and their nominal values.

These concepts will be elaborated below when we discuss various stability criteria.

6.2 THE ROUTH CRITERION, THE HURWITZ CRITERION, AND THE LIENARD-CHIPART CRITERION

In this section, we review briefly the Routh-Hurwitz criterion. For completeness, we also introduce the less known Liénard-Chipart criterion. As indicated at the beginning of this chapter, the criteria are of limited practical value because the network determinant or its equivalent is not usually given analytically.

During the early development of control theory, there were two sources of methods for determining the nature of the zeros of a polynomial: the Routh criterion and the Hurwitz criterion. The Routh criterion was found by E. J. Routh in 1877. Eighteen years later, in 1895, A. Hurwitz, who at the time was unaware of Routh's work, developed essentially the same result. Despite the different appearance, they have been shown to be closely related and are usually referred to as the *Routh-Hurwitz criterion.*

Consider the characteristic equation

$$P(s) = a_0 s^n + a_1 s^{n-1} + \cdots + a_{n-1} s + a_n = 0 \qquad a_0 > 0 \tag{6.3}$$

with real coefficients, which results when the network determinant is set to zero. From the coefficients of $P(s)$, we form the following array:

$$
\begin{array}{c|cccc}
s^n & a_0 & a_2 & a_4 & a_6 \quad \cdots \\
s^{n-1} & a_1 & a_3 & a_5 & a_7 \quad \cdots \\
s^{n-2} & b_1 & b_2 & b_3 & \cdots \\
s^{n-3} & c_1 & c_2 & \cdots & \\
s^{n-4} & d_1 & \cdots & & \\
\vdots & \vdots & & &
\end{array} \tag{6.4}
$$

where
$$b_i = \frac{a_1 a_{2i} - a_0 a_{2i+1}}{a_1} \tag{6.5a}$$

$$c_i = \frac{b_1 a_{2i+1} - a_1 b_{i+1}}{b_1} \tag{6.5b}$$

$$d_i = \frac{c_1 b_{i+1} - b_1 c_{i+1}}{c_1} \tag{6.5c}$$

and so on. The process is continued to the right until only zeros are obtained, filling in zeros after coefficients are exhausted, and down until there are $n + 1$ rows. This array forms the basis for the next theorem, known as the *Routh criterion* [Routh (1877)].

Theorem 6.1 The number of sign changes in the left-hand column of array (6.4) is equal to the number of zeros of $P(s)$ with positive-real parts.

Occasionally, it may occur that the first element in one of the rows vanishes but not all terms vanish. In such a case, it is immediately known that the system is unstable. However, if additional information is required regarding the characteristic roots of the system, the Routh array can be constructed by multiplying $P(s)$ by $s + \alpha$, where α is almost any positive number, and then applying the criterion to the new equation. Alternatively, we simply replace the zero with an arbitrary small quantity ϵ and continue the construction process as if no zero had been obtained. The incremental ϵ is usually taken to be positive, but a negative value is also satisfactory. It may also occur that one of the rows is entirely zero, say, the jth row. In such case, we replace the jth row by the row

$$(n + 2 - j)\alpha_0, (n - j)\alpha_1, \ldots, (n - 2i + 2 - j)\alpha_i, \ldots \tag{6.6}$$

where $\alpha_0, \alpha_1, \ldots$ are the elements of the $(j - 1)$th row. In fact, the elements of (6.6) correspond to the coefficients of the derivative of the *auxiliary polynomial* of degree $n + 2 - j$ defined by the $(j - 1)$th row of (6.4) such that the exponent of s in each term after the first, which is $n + 2 - j$, is the exponent of s in the preceding term diminished by 2.

The vanishing of a row of elements before the $(n + 2)$th row is caused by the existence of a pair of roots of equal magnitude but opposite in sign. In fact, all roots of this type of the characteristic equation are roots of the auxiliary polynomial of the row preceding the first vanishing row of the array. Thus, if the characteristic equation has roots on the real-frequency axis, they will be found among the roots of the auxiliary polynomial. The total number of vanishing rows is equal to the multiplicity of the pair of roots of the above type of the greatest multiplicity.

For the Hurwitz criterion, we define the nth-order determinant

$$\Delta_n = \det \begin{bmatrix} a_1 & a_3 & a_5 & a_7 & \cdots & & 0 & 0 \\ a_0 & a_2 & a_4 & a_6 & \cdots & & 0 & 0 \\ 0 & a_1 & a_3 & a_5 & a_7 & \cdots & 0 & 0 \\ 0 & a_0 & a_2 & a_4 & a_6 & \cdots & 0 & 0 \\ 0 & 0 & a_1 & a_3 & a_5 & \cdots & 0 & 0 \\ 0 & 0 & a_0 & a_2 & a_4 & \cdots & 0 & 0 \\ \cdots & \cdots & \cdots & \cdots & \cdots & \cdots & \cdots & \cdots \\ 0 & 0 & \cdots & \cdots & \cdots & \cdots & a_{n-1} & 0 \\ 0 & 0 & \cdots & \cdots & \cdots & \cdots & a_{n-2} & a_n \end{bmatrix} \tag{6.7}$$

Denote by Δ_{n-k} the determinant of the submatrix obtained from the matrix of (6.7) by deleting the last k rows and columns. In this way, we construct a total of n such determinants, known as the *Hurwitz determinants.* The following theorem, known as the *Hurwitz criterion*, is stated in terms of the Hurwitz determinants [Hurwitz (1895)].

Theorem 6.2 A necessary and sufficient condition that the polynomial $P(s)$ has zeros with only negative-real parts is that the values of the Hurwitz determinants Δ_j ($j = 1, 2, \ldots, n$) be all positive in sign.

A case of frequent occurrence is that in which the coefficients of $P(s)$ are functions of some parameters. In such an event, the Hurwitz criterion is reasonably easy to apply. For numerical coefficients, the Routh-Hurwitz criterion can be put into another form, as suggested by Guillemin (1949), which will now be described.

Let $P_1(s)$ and $P_2(s)$ be the polynomials derived from $P(s)$ by taking alternate terms, starting with $a_0 s^n$ and $a_1 s^{n-1}$, respectively, as follows:

$$P_1(s) = a_0 s^n + a_2 s^{n-2} + a_4 s^{n-4} + \cdots \tag{6.8a}$$

$$P_2(s) = a_1 s^{n-1} + a_3 s^{n-3} + a_5 s^{n-5} + \cdots \tag{6.8b}$$

Then form the ratio $P_1(s)/P_2(s)$ and expand it as a continued fraction, known as the *Stieltjes continued fraction*:

$$\frac{P_1(s)}{P_2(s)} = \alpha_1 s + \cfrac{1}{\alpha_2 s + \cfrac{1}{\alpha_3 s + \cfrac{1}{\ddots + \cfrac{1}{\alpha_m s}}}} \tag{6.9}$$

In general, there are n coefficients α's in the expansion with $m = n$. In the situation where $m < n$, we say that the expansion is terminated *prematurely*. With these we state the Routh-Hurwitz criterion in Guillemin form.

Theorem 6.3 A necessary and sufficient condition that the polynomial $P(s)$ has zeros with only negative-real parts is that the Stieltjes continued fraction expansion not terminate prematurely and yield only positive α_i for $i = 1, 2, \ldots, n$.

In applying Hurwitz criterion, we are required to evaluate n determinants. The number of these determinants can be reduced, and the result is known as the *Liénard-Chipart criterion* [Liénard and Chipart (1914)].

Theorem 6.4 A necessary and sufficient condition that the polynomial $P(s)$ has zeros with only negative-real parts is that all elements in any one of the following four sets are positive:

1. $a_n, a_{n-2}, a_{n-4}, \ldots$, and $\Delta_n, \Delta_{n-2}, \Delta_{n-4}, \ldots$
2. $a_n, a_{n-2}, a_{n-4}, \ldots$, and $\Delta_{n-1}, \Delta_{n-3}, \Delta_{n-5}, \ldots$
3. $a_n, a_{n-1}, a_{n-3}, \ldots$, and $\Delta_n, \Delta_{n-2}, \Delta_{n-4}, \ldots$
4. $a_n, a_{n-1}, a_{n-3}, \ldots$, and $\Delta_{n-1}, \Delta_{n-3}, \Delta_{n-5}, \ldots$

Unlike the Hurwitz criterion, the Liénard-Chipart criterion requires only that every other Hurwitz determinant be evaluated. Since the amount of work increases considerably with the size of the determinant, it is advantageous to choose either condition (2) or condition (4) of the theorem, since neither includes the largest Hurwitz determinant Δ_n.

We illustrate the above results by considering the stability of a feedback network.

Example 6.1 A practical version of a field-effect transistor (FET) Colpitts oscillator is shown in Fig. 6.1*a*. After the biasing circuitry and the radio frequency (RF) choke have been removed, the equivalent network is presented in Fig. 6.1*b*. Typical parameter values for the FETs are as follows:

$$g_m = 100\text{-}10{,}000\ \mu\text{mho} \qquad C_{gs} = 2\text{-}10\ \text{pF}$$
$$C_{gd}, C_{ds} = 0.1\text{-}2\ \text{pF} \qquad r_d = 0.05\text{-}1\ \text{M}\Omega \tag{6.10}$$

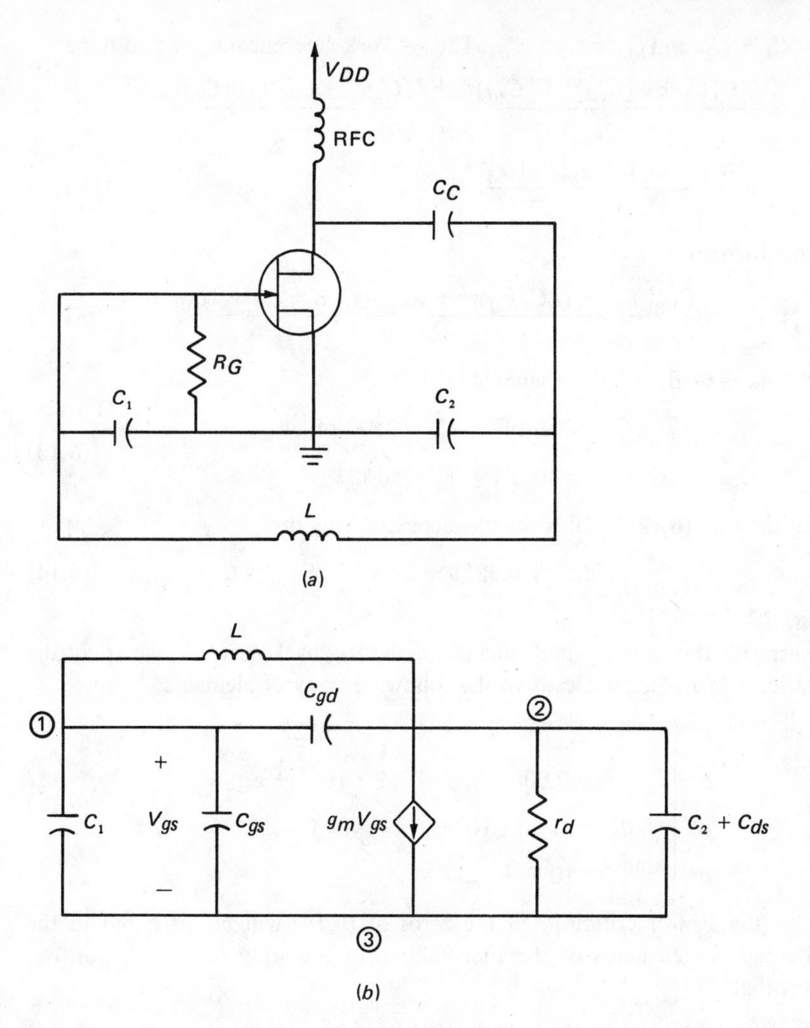

Figure 6.1 (*a*) A practical version of a FET Colpitts oscillator and (*b*) its equivalent network.

Thus, in computing the frequency of oscillation, the capacitances of the FET can usually be ignored. The indefinite-admittance matrix of the network of Fig. 6.1*b* can be written down by inspection and is given by

$$\mathbf{Y} = \begin{bmatrix} sC_1' + sC_{gd} + \dfrac{1}{sL} & -sC_{gd} - \dfrac{1}{sL} & -sC_1' \\ g_m - sC_{gd} - \dfrac{1}{sL} & sC_2' + sC_{gd} + \dfrac{1}{sL} + \dfrac{1}{r_d} & -sC_2' - g_m - \dfrac{1}{r_d} \\ -sC_1' - g_m & -sC_2' - \dfrac{1}{r_d} & sC_1' + sC_2' + g_m + \dfrac{1}{r_d} \end{bmatrix} \tag{6.11}$$

where $C_1' = C_1 + C_{gs}$ and $C_2' = C_2 + C_{ds}$. The network determinant is found to be

$$Y_{33} = \frac{L(C_1'C_2' + C_1'C_{gd} + C_2'C_{gd})s^3 + L(C_1'/r_d + C_{gd}/r_d + C_{gd}g_m)s^2}{sL} + \frac{(C_1' + C_2')s + g_m + 1/r_d}{sL} \tag{6.12a}$$

which is simplified to

$$Y_{33} \approx \frac{C_1C_2Ls^3 + (LC_1/r_d)s^2 + (C_1 + C_2)s + g_m + 1/r_d}{sL} \tag{6.12b}$$

if the capacitances of the FET are ignored. Let

$$\begin{aligned} C_1 &= 750 \text{ pF} \qquad C_2 = 2500 \text{ pF} \\ L &= 40\ \mu\text{H} \qquad r_d = 50\text{ k}\Omega \end{aligned} \tag{6.13}$$

Substituting these in (6.12b) yields the characteristic equation

$$75p^3 + 0.006p^2 + 0.325p + 2 \cdot 10^{-5} + g_m = 0 \tag{6.14}$$

where $p = s/10^8$.

To investigate the stability performance of the feedback network, we apply the Routh criterion to (6.14). This leads to the following array of elements:

$$\begin{array}{c|cc} p^3 & 75 & 0.325 \\ p^2 & 0.006 & 2 \cdot 10^{-5} + g_m \\ p^1 & 0.075 - 12.5 \cdot 10^3 g_m & 0 \\ p^0 & 2 \cdot 10^{-5} + g_m & 0 \end{array} \tag{6.15}$$

According to the Routh criterion, all the zeros of (6.14) will be restricted to the open LHS if all the elements of the first column in the array (6.15) are positive. This requires that

$$0.075 - 12.5 \cdot 10^3 g_m > 0 \tag{6.16}$$

giving

$$g_m < 6\ \mu\text{mho} \tag{6.17}$$

If we choose $g_m = 6\ \mu$mho, the third row of the array (6.15) will vanish, and the network will have a pair of natural frequencies on the real-frequency axis, which are the roots of the equation

$$0.006p^2 + 2 \cdot 10^{-5} + g_m = 0 \tag{6.18}$$

which results when the auxiliary polynomial of the second row is set to zero, giving

$$s = 10^8 p = \pm j6.583 \cdot 10^6 \tag{6.19}$$

or $f = 1.048$ MHz. Thus, for $g_m = 6\ \mu$mho, the network will oscillate with a frequency of 1.048 MHz.

Suppose that the capacitances of the FET are not ignored and let

$$C_{gd} = C_{ds} = 1 \text{ pF} \qquad C_{gs} = 5 \text{ pF} \tag{6.20}$$

The corresponding characteristic equation is determined from (6.12*a*), and is given by

$$75.66p^3 + (0.006 + 0.4g_m)p^2 + 0.326p + 2 \cdot 10^{-5} + g_m = 0 \tag{6.21}$$

To investigate stability, in this case it is convenient to apply the Liénard-Chipart criterion. The Hurwitz determinant is found to be

$$\Delta_3 = \det \begin{bmatrix} 0.006 + 0.4g_m & 2 \cdot 10^{-5} + g_m & 0 \\ 75.66 & 0.326 & 0 \\ 0 & 0.006 + 0.4g_m & 2 \cdot 10^{-5} + g_m \end{bmatrix} \tag{6.22}$$

To avoid the evaluation of Δ_3, we choose condition (2) of Theorem 6.4. For the zeros of Eq. (6.21) to be confined to the open LHS, we require that

$$a_3 = 2 \cdot 10^{-5} + g_m > 0 \tag{6.23a}$$

$$a_1 = 0.006 + 0.4g_m > 0 \tag{6.23b}$$

$$\Delta_2 = \det \begin{bmatrix} 0.006 + 0.4g_m & 2 \cdot 10^{-5} + g_m \\ 75.66 & 0.326 \end{bmatrix} = 0.443 \cdot 10^{-3} - 75.53g_m > 0 \tag{6.23c}$$

yielding

$$g_m < 5.86 \ \mu\text{mho} \tag{6.24}$$

Let $g_m = 5$ μmho; substituting it in (6.21) yields

$$75.66p^3 + 0.006p^2 + 0.326p + 25 \cdot 10^{-6} = 0 \tag{6.25}$$

To apply Theorem 6.3, we compute the Stieltjes continued fraction (6.9), which is found to be

$$\frac{P_1}{P_2} = \frac{75.66p^3 + 0.326p}{0.006p^2 + 25 \cdot 10^{-6}} = 12{,}610p + \frac{1}{0.546p + 1/440p} \tag{6.26}$$

Since all the coefficients are positive and the continued-fraction expansion does not terminate prematurely, all the roots of (6.25) are confined to the open LHS, as expected.

On the other hand, suppose that we let $g_m = 7$ μmho. We expect that the natural frequencies will move to the closed RHS and the network becomes unstable. For $g_m = 7$ μmho, (6.21) becomes

$$75.66p^3 + 0.006p^2 + 0.326p + 27 \cdot 10^{-6} = 0 \tag{6.27}$$

giving

$$\frac{P_1}{P_2} = 12{,}610p + \frac{1}{-0.429p - 1/518.52p} \tag{6.28}$$

Since two coefficients are negative, (6.27) has at least one zero in the closed RHS and the network is unstable.

6.3 THE NYQUIST CRITERION

The stability criteria described in the preceding section are useful if the network determinant is known analytically to be a function of the complex-frequency variable s. Even if we have the network determinant, the roots alone do not tell us the degree of stability when the feedback amplifier is stable, nor do they provide us with any information as to how to stabilize an unstable amplifier. Furthermore, for a physical network there remains the difficulty of getting an accurate formulation of the network determinant itself, since every equivalent network is, to a greater or lesser extent, an idealization of the physical reality. Also, the parasitic effects, which are usually ignored in the formulation of network determinant, play an important role for the stability study. These limitations are overcome by applying the Nyquist criterion to the return difference to be described below.

Let $f(s)$ be an analytic function that is regular within and on a given closed contour C in the s-plane, with the exception of a finite number of poles inside the contour. Assume that $f(s)$ has a zero or pole of order n at a point s_1 and write

$$f(s) = (s - s_1)^{\pm n} f_1(s) \tag{6.29}$$

where the plus sign denotes a zero and the minus sign a pole. Write $f'(s) = df(s)/ds$ and obtain

$$f'(s) = \pm n(s - s_1)^{\pm n - 1} f_1(s) + (s - s_1)^{\pm n} f_1'(s) \tag{6.30}$$

which when combined with (6.29) gives

$$\frac{f'(s)}{f(s)} = \frac{\pm n}{s - s_1} + \frac{f_1'(s)}{f_1(s)} \tag{6.31}$$

This result indicates that the function $f'(s)/f(s)$ has a simple pole at the point s_1, which is a zero or pole of $f(s)$, with the residue $+n$ or $-n$. The same process can now be repeated for the other zeros and poles of $f(s)$ by considering the function $f_1(s)$, which has the same zeros and poles as in $f(s)$ except for the one at s_1, multiplicity included. Thus, if n_{zi} and n_{pi} are the orders of the zeros s_{zi} and poles s_{pi} of $f(s)$, respectively, (6.31) can be expanded as

$$\frac{f'(s)}{f(s)} = \sum \frac{n_{zi}}{s - s_{zi}} - \sum \frac{n_{pi}}{s - s_{pi}} \tag{6.32}$$

where the summations are taken over all zeros and poles of $f(s)$. Let C be a closed curve in the s-plane. Consider the closed line integral

$$\oint_C \frac{f'(s)}{f(s)}\, ds = \oint_C \left(\sum \frac{n_{zi}}{s - s_{zi}} - \sum \frac{n_{pi}}{s - s_{pi}} \right) ds \tag{6.33}$$

Since the only singularities of the function $f'(s)/f(s)$ inside the closed contour C are the zeros and poles of $f(s)$ that lie inside C, by the residue theorem, the value of the closed contour integral is given by

$$\oint_C \frac{f'(s)}{f(s)}\,ds = 2\pi j(\Sigma\, n_{zi} - \Sigma\, n_{pi}) = 2\pi j(n_z - n_p) \tag{6.34}$$

where the summations are now taken only over the zeros and poles of $f(s)$ that lie within the contour C, and n_z and n_p denote, respectively, the numbers of zeros and poles of $f(s)$ within the contour C, counting each zero and pole according to its multiplicity. Since

$$\frac{f'(s)}{f(s)} = \frac{d \ln f(s)}{ds} \tag{6.35}$$

we find that with s_a and s_b symbolizing the initial and final points on C, (6.34) becomes

$$\ln |f(s)|\,\Big|_{s_a}^{s_b} + j \arg f(s)\Big|_{s_a}^{s_b} = 2\pi j(n_z - n_p) \tag{6.36}$$

As s starts at a point on the contour C, traverses along the contour in the positive direction (counterclockwise), and returns to the starting point, the locus of $f(s)$ in the f-plane will also return to its starting point. Thus, the value of $\ln|f(s)|$ will be the same at the starting point as at the end of the closed contour C; only the value of $\arg f(s)$ may be different because of the multiple-valuedness of the function $\ln f(s)$. As a result, the first contour integral on the left-hand side of (6.36) vanishes, and the value of the second integral equals j times the increase in phase angle of $f(s)$ as s traverses the contour C. If we now divide by 2π on both sides of (6.36), the result on the left-hand side should be the number of times the mapping of the contour C in the f-plane goes around its origin counterclockwise. If n_{cw} denotes the number of *clockwise* encirclements of the origin by the mapping of the closed contour C, then we have

$$-n_{cw} = n_z - n_p \tag{6.37}$$

If, instead, s traverses the contour C in the negative direction (clockwise), we introduce a minus sign on the right-hand side of (6.36), and (6.37) becomes

$$n_{cw} = n_z - n_p \tag{6.38}$$

In words, (6.38) states that the number of clockwise encirclements of the origin by the mapping of the closed contour C in the f-plane is equal to the number of zeros minus the number of poles of $f(s)$ within the contour C as s traverses the contour C in a clockwise direction under the general assumption that $f(s)$ has neither zeros nor poles on the contour C. The situation is as depicted in Fig. 6.2. As the variable s traverses a closed contour C in the s-plane in a clockwise direction, we find a closed curve representing the mapping of C in the f-plane, as illustrated in Fig. 6.2*b*. The number of clockwise encirclements of the origin by the closed curve in the f-plane is equal to the number of zeros minus the number of poles of $f(s)$ within C. By

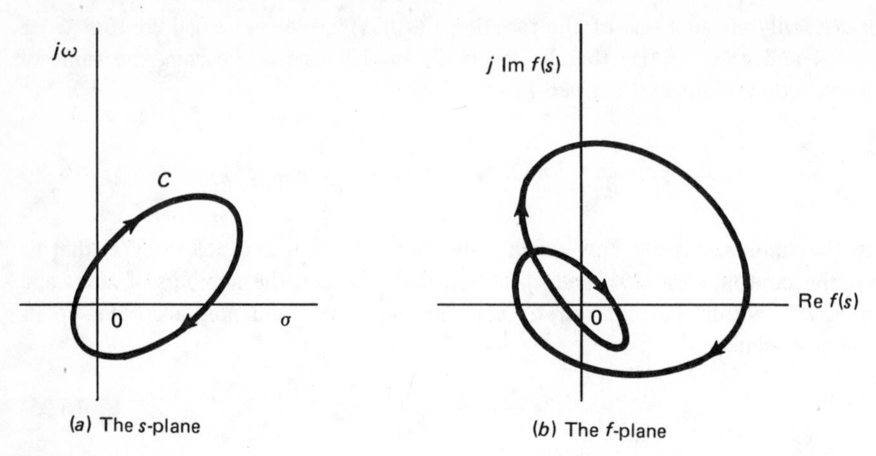

Figure 6.2 (*a*) A closed contour C in the s-plane and (*b*) the mapping of the closed contour C in the f-plane.

encirclement we mean a complete revolution of a radius vector drawn from the origin to a moving point on the closed curve in the f-plane.

The Nyquist criterion is based on (6.38). Our objective is to determine whether or not the network determinant has any poles in the closed RHS. We therefore choose the closed contour C containing the $j\omega$-axis and a semicircle of infinite radius, as indicated in Fig. 6.3, thereby enclosing the entire closed RHS. As required by (6.38), the contour C is traversed in a clockwise direction, which corresponds to increasing frequency. The contour of Fig. 6.3 is known as the *Nyquist contour.*

Suppose that the function to be tested is the return difference, written as $F(s)$ to exhibit its dependence on the complex frequency s, of a single-loop feedback amplifier with respect to the controlling parameter x of an active device:

$$F(s) = \frac{Y_{uv}(x)}{Y_{uv}(0)} \tag{6.39}$$

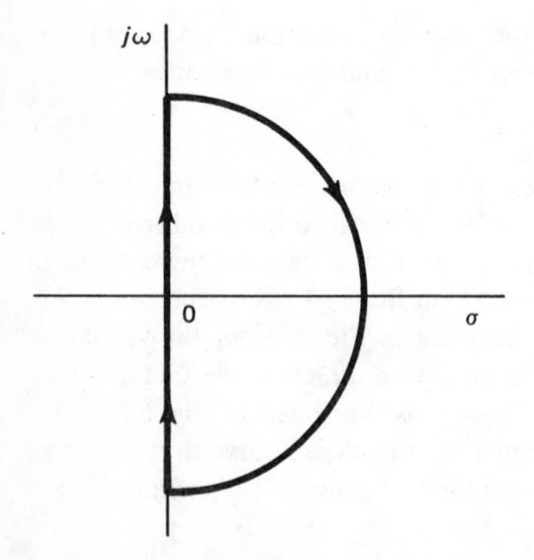

Figure 6.3 The Nyquist contour.

Then, according to (6.1), $Y_{uv}(0)$ is independent of the controlling parameters of the active devices. This implies that $Y_{uv}(0)$ is the network determinant of the amplifier when all of the active devices are made completely dead, and the resulting network is strictly passive. Therefore, the zeros of $Y_{uv}(0)$ are confined to the open LHS, and we conclude from (6.39) that the zeros of the return difference in the closed RHS are precisely those natural frequencies of the feedback amplifier that lie inside the closed RHS. To guarantee stability, the mapping of the Nyquist contour C in the F-plane should not encircle the origin. The mapping of the Nyquist contour C in the F-plane is called the *Nyquist plot.*

In practice, it is more convenient to work with the return ratio $T(s)$, which is related to $F(s)$ by $F(s) = 1 + T(s)$. Thus, the difference between a Nyquist plot of $F(s)$ and that of $T(s)$ is simply a shift in the position of the imaginary axis. To obtain a plot in the T-plane from that of the F-plane, we simply shift the imaginary axis of the F-plane to the right by a unit distance. Since the origin in the F-plane corresponds to the point $-1 + j0$ in the T-plane, we see that the amplifier will be stable if the Nyquist plot does not encircle the *critical point* $-1 + j0$ in the T-plane.

We summarize the above results by stating the following theorem, known as the *Nyquist stability criterion* [Nyquist (1932)].

Theorem 6.5 Assume that a feedback amplifier is stable when the controlling parameter of a chosen active device is set to zero. Then a necessary and sufficient condition for the feedback amplifier to be stable under normal operating conditions is that the Nyquist plot of the return ratio with respect to this controlling parameter should not encircle the critical point $-1 + j0$ in the T-plane.

Relation (6.38) was derived under the assumption that the function under test has no zeros and poles on the boundary of the contour C. For the Nyquist plot of the return ratio $T(s)$, Theorem 6.5 is valid only if $T(s)$ has no zeros and poles on the Nyquist contour of Fig. 6.3. These are not really serious restrictions, because we can tell from the Nyquist plot whether or not the assumptions hold in a given situation. If the Nyquist plot of $T(s)$ intersects the critical point $-1 + j0$ in the T-plane, or, equivalently, if the Nyquist plot of $F(s)$ intersects the origin in the F-plane, then $F(s)$ must have a zero somewhere on the Nyquist contour, either on the $j\omega$-axis or at infinity, depending on the value of s at the intersection point. This tells us that not all of the zeros of $F(s)$ are confined to the open LHS, at least one being on the real-frequency axis, and the feedback amplifier under investigation is therefore unstable. If the Nyquist plot of $T(s)$ in the T-plane becomes unbounded at some real frequency ω_0, then $F(s)$ must have a pole at $j\omega_0$. To avoid this difficulty, we can modify the Nyquist contour by indentations into the RHS with vanishingly small semicircular arcs centered at these poles, as illustrated in Fig. 6.4, because these small indentations do not affect the number of zeros of $F(s)$ computed by (6.38). This situation does not occur if all the zeros of $Y_{uv}(0)$ are confined to the open LHS, as in the case for the single-loop feedback amplifiers discussed above.

We now investigate the Nyquist plot of $T(s)$. From physical considerations, as frequency increases, the available gain of the active devices decreases. As frequency

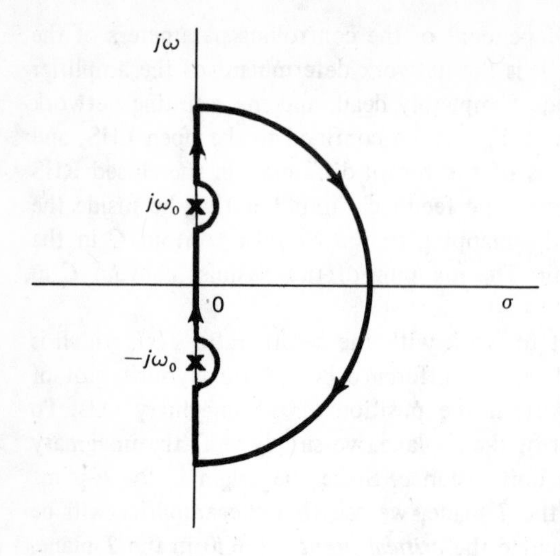

Figure 6.4 The indented Nyquist contour.

approaches infinity, the forward transmission through the active devices is reduced to zero, and the return difference becomes unity. Therefore the mapping of the points over the semicircular arc at infinity of the Nyquist contour is simply the origin in the T-plane or $F = 1$ in the F-plane. To complete the Nyquist plot, we need only check the behavior of $T(s)$ on the real-frequency axis, that is, $T(j\omega)$. Since we are concerned only with networks with real elements and since $T(s)$ is analytic on the $j\omega$-axis, we have

$$\bar{T}(j\omega) = T(\overline{j\omega}) = T(-j\omega) \tag{6.40}$$

meaning that the behavior of T on the negative $j\omega$-axis is simply the mirror image of T on the positive $j\omega$-axis with respect to the real axis of the T-plane. In other words, the Nyquist plot of the return ratio T is completely specified by its behavior for frequencies on the positive $j\omega$-axis. In fact, the Nyquist plot can be achieved experimentally, since the return ratio of an amplifier can be obtained from a frequency response measurement, which includes all parasitic effects. A typical Nyquist plot is shown in Fig. 6.5. The solid line indicates the behavior of the return ratio $T(j\omega)$ for nonnegative frequencies from $\omega = 0$ to $\omega = \infty$. The remaining plot for frequencies from $\omega = 0$ to $\omega = -\infty$ is obtained as the mirror image of the solid line with respect to the real axis of the T-plane. In the plot, since the locus does not encircle the critical point $-1 + j0$, the amplifier is stable. On the other hand, the Nyquist plot of Fig. 6.6 is unstable, because the locus encircles the critical point $-1 + j0$ twice.

6.4 APPLICATIONS OF THE NYQUIST CRITERION TO SINGLE-LOOP FEEDBACK AMPLIFIERS

In this section, we study the relations between stability and feedback for single-loop feedback amplifiers in terms of the Nyquist plot of the return ratio.

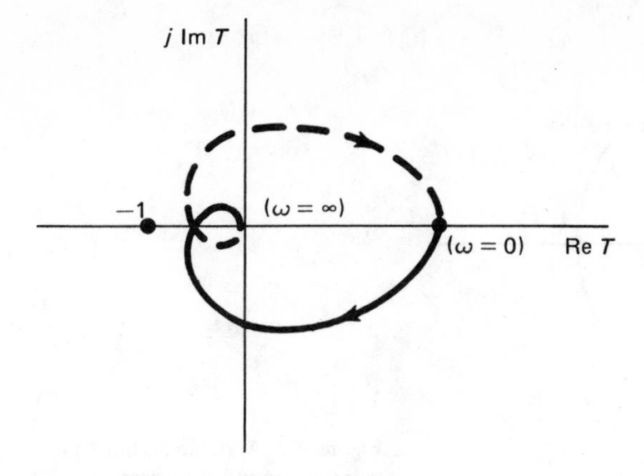

Figure 6.5 A typical Nyquist plot.

The Nyquist plot discussed in the foregoing gives a practical way for the stability study. It also provides a quantitative measure of the degree of stability of the feedback amplifier. Consider the stable plot of Fig. 6.5, which is enlarged around the origin as shown in Fig. 6.7. The feedback amplifier is stable and additional feedback is allowed until the value T_m, as indicated in Fig. 6.7, passes through the critical point $-1 + j0$, corresponding to a pair of conjugate natural frequencies on the real-frequency axis. This condition provides a measure of the maximum allowable feedback permitted to ensure amplifier stability. For this we define the following.

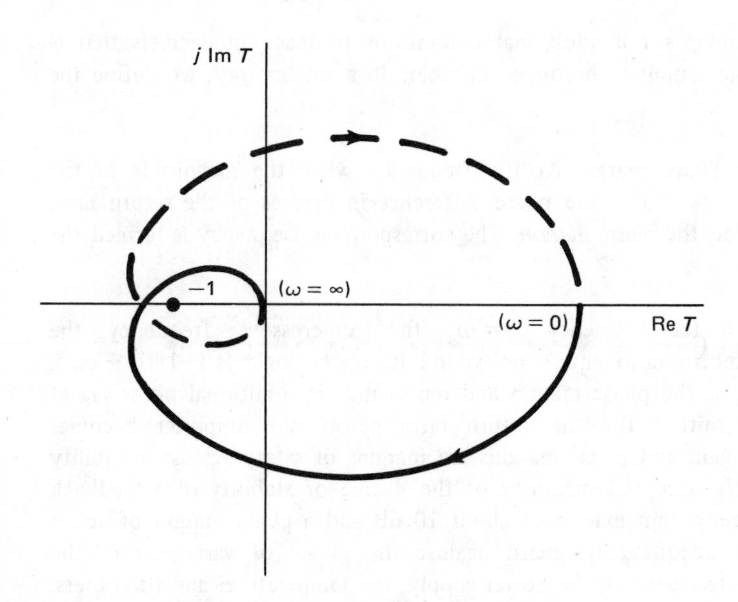

Figure 6.6 An unstable Nyquist plot.

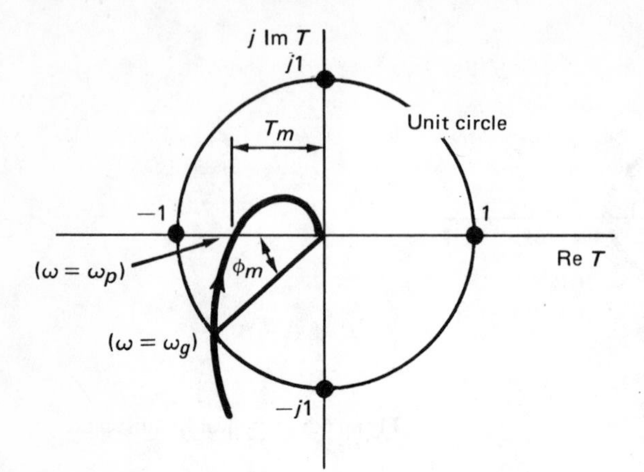

Figure 6.7 A stable Nyquist plot enlarged around the origin.

Definition 6.4: Gain margin At the frequency when the phase angle of the return ratio equals $-180°$, the magnitude in decibels of the return ratio below the 0-db level is called the *gain margin.* The corresponding frequency is termed the *phase-crossover frequency.*

In Fig. 6.7, the phase-crossover frequency is indicated by ω_p, and the gain margin is defined by the equation

$$\text{Gain margin} = -20 \log |T_m| = 20 \log \left| \frac{1}{T_m} \right| \tag{6.41}$$

Thus, the gain margin is the additional amount of feedback in decibels that is permitted before the amplifier becomes unstable. In a similar way, we define the phase margin.

Definition 6.5: Phase margin At the frequency when the magnitude of the return ratio is unity or 0 dB, the phase difference in degrees of the return ratio above $-180°$ is called the *phase margin*. The corresponding frequency is termed the *gain-crossover frequency.*

In Fig. 6.7, at the frequency $\omega = \omega_g$, the gain-crossover frequency, the magnitude of the return ratio equals unity and its phase angle is $(-180° + \phi_m)$. Thus, the angle ϕ_m is the phase margin and represents the additional phase lag at unit magnitude permitted for the return ratio before the amplifier becomes unstable. Thus, the gain and phase margins are margins of safety against instability and are the generally accepted measures of the degree of stability of a feedback amplifier. In practice, a gain margin of about 10 dB and a phase margin of about 30° are considered adequate to guard against the effect of variations of the parameters of the active devices, the power supply, the temperature, and the others.

For an absolutely stable single-loop feedback amplifier, the gain-crossover frequency ω_g is necessarily smaller than the phase-crossover frequency ω_p. In Fig.

6.5, if the curve represents the Nyquist plot of a single-loop feedback amplifier under normal operating conditions, the amplifier must be absolutely stable because any reduction of the controlling parameter will only "shrink" the curve radially about the origin. On the other hand, the Nyquist plot of Fig. 6.8*a* shows that the amplifier is stable under normal operating conditions. However, by varying the controlling parameters of some active devices between 0 and their nominal values, the Nyquist plot may shrink radially about the origin, as shown in Fig. 6.8*b*. By applying Nyquist criterion to Fig. 6.8*b*, we see that the plot encircles the critical point $-1 + j0$ twice, and the amplifier is therefore unstable. We remark that in Fig. 6.8 only the plots corresponding to positive frequencies are shown. The plots corresponding to the negative real-frequency axis, being the mirror images of those shown in Fig. 6.8, are not given, for simplicity. If the plot of Fig. 6.8*b* is shrunk further, the amplifier may become stable again. We conclude that the single-loop

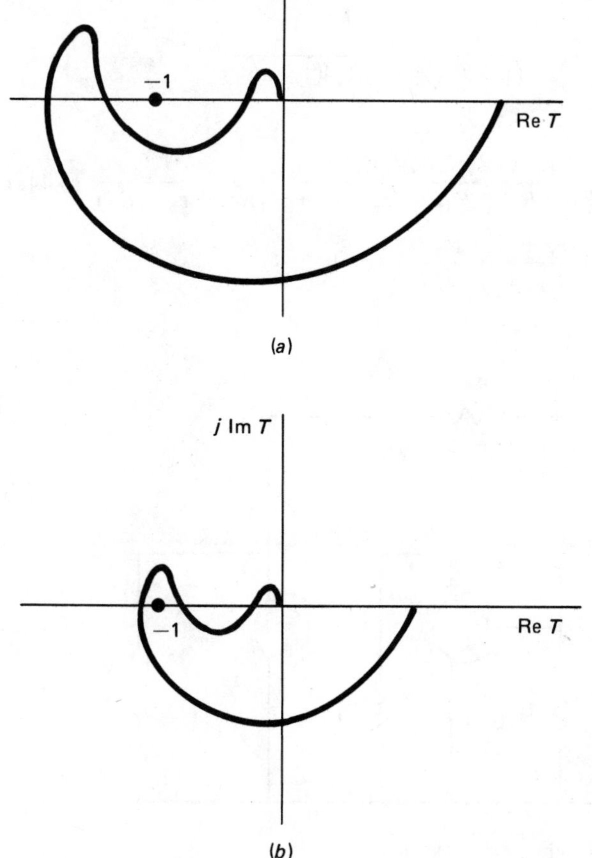

Figure 6.8 (*a*) A stable Nyquist plot under normal operating conditions and (*b*) an unstable Nyquist plot during transition.

feedback amplifier is conditionally stable. In terms of the natural frequencies, the above analysis indicates that, for a conditionally stable amplifier, as the midband value of the return ratio $T(0)$ is reduced, some of the natural frequencies cross over into the RHS. If, however, $T(0)$ is reduced further, the natural frequencies that were in the RHS return to the open LHS, thereby making the amplifier stable again.

Example 6.2 The network of Fig. 6.9 shows a three-stage common-emitter amplifier with a resistor R_f connected from the output to the input to provide a single external feedback. Assume that the three transistors are identical and are represented by their hybrid-pi equivalent network of Fig. 4.10, with

$$\begin{gathered} g_m = 0.2 \text{ mho} \qquad r_x = 50\ \Omega \qquad r_\pi = 150\ \Omega \\ r_o = \infty \qquad C_\pi = 195 \text{ pF} \qquad C_\mu = 5 \text{ pF} \\ R_1 = 100\ \Omega \qquad R_2 = 75\ \Omega \qquad R_3 = R_4 = 1\text{ k}\Omega \end{gathered} \tag{6.42}$$

As demonstrated in Sec. 4.2.2, the return ratio of the amplifier is found from Eqs. (4.54*b*) and (4.67) to be

$$T(s) = -\mu\beta = \frac{T(0)}{(1 + s/s_1)(1 + s/s_2)(1 + s/s_3)} \tag{6.43}$$

where

$$T(0) = \frac{R_1 R_2 R_3 R_4 R_f r_\pi^3 g_m^3}{(R_2 + R_f)[(r_x + r_\pi)(R_1 + R_f) + R_1 R_f](r_x + r_\pi + R_3)(r_x + r_\pi + R_4)} \tag{6.44a}$$

$$s_k = \frac{r_x + r_\pi + \hat{R}_k}{C_k r_\pi (r_x + \hat{R}_k)} \qquad k = 1, 2, 3 \tag{6.44b}$$

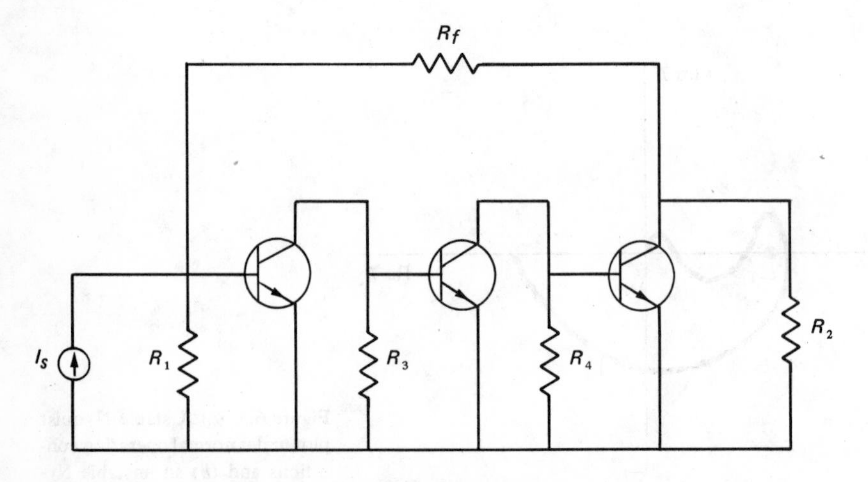

Figure 6.9 A three-stage common-emitter feedback amplifier.

in which $\hat{R}_1 = R_1R_f/(R_1 + R_f)$, $\hat{R}_2 = R_3$ and $\hat{R}_3 = R_4$, and

$$C_1 = C_\pi + C_\mu \left[1 + \frac{g_m R_3 (r_\pi + r_x)}{r_\pi + r_x + R_3}\right] \tag{6.44c}$$

$$C_2 = C_\pi + C_\mu \left[1 + \frac{g_m R_4 (r_\pi + r_x)}{r_\pi + r_x + R_4}\right] \tag{6.44d}$$

$$C_3 = C_\pi + C_\mu \left(1 + \frac{g_m R_2 R_f}{R_2 + R_f}\right) \tag{6.44e}$$

Using (6.42) in conjunction with the fact that $R_f \gg R_1, R_2$, we obtain

$$C_1 = C_2 = 367 \text{ pF} \qquad C_3 \approx 275 \text{ pF}$$

$$\begin{aligned} s_1 &\approx 3.633 \cdot 10^7 \qquad s_2 = 2.076 \cdot 10^7 \\ s_3 &= 2.770 \cdot 10^7 \qquad T(0) \approx \frac{469 \cdot 10^3}{R_f} \end{aligned} \tag{6.45}$$

giving
$$T(s) = \frac{T(0)}{(1 + p/3.633)(1 + p/2.076)(1 + p/2.770)} \tag{6.46}$$

where $p = s/10^7$. To make the Nyquist plot of the return ratio, we rewrite (6.46) as

$$T(j\lambda) = \frac{T(0)}{(1 + j\lambda/3.633)(1 + j\lambda/2.076)(1 + j\lambda/2.770)} \tag{6.47}$$

where $\lambda = \omega/10^7$. The Nyquist plots of $T(j\lambda)$ are presented in Fig. 6.10 for two different midband values of $T(0)$: $T(0) = 10$ and $T(0) = 5$. For $T(0) = 10$, the feedback resistance is found to be $R_f = 46.9$ kΩ; and for $T(0) = 5$, $R_f = 93.8$ kΩ. In Fig. 6.10, as before, only the loci corresponding to the nonnegative λ are shown, for simplicity, with λ chosen at the points

$$\lambda_k = \frac{k}{2} \qquad k = 0, 1, \ldots, 9 \tag{6.48}$$

As can be seen from the plots, for $T(0) = 10$ the amplifier is unstable, because the Nyquist locus encircles the critical point $-1 + j0$. For $T(0) = 5$ the amplifier is stable with a gain margin = 4.6 dB, because

$$T(j4.84) = 0.589 \underline{/180^\circ} \tag{6.49}$$

The phase-crossover frequency is $\omega_p = 48.4 \cdot 10^6$ rad/s or $f_p = 7.7$ MHz. The phase margin of the amplifier is about 20°, because

$$T(j3.75) = 1 \underline{/160.5^\circ} \tag{6.50}$$

The gain-crossover frequency is $\omega_g = 37.5 \cdot 10^6$ rad/s or $f_g = 5.97$ MHz.

The Nyquist plot of the return ratio $T(s)$ is nothing more than the plot of $T(j\omega)$ as a function of ω in polar form. It is sometimes more useful to plot the

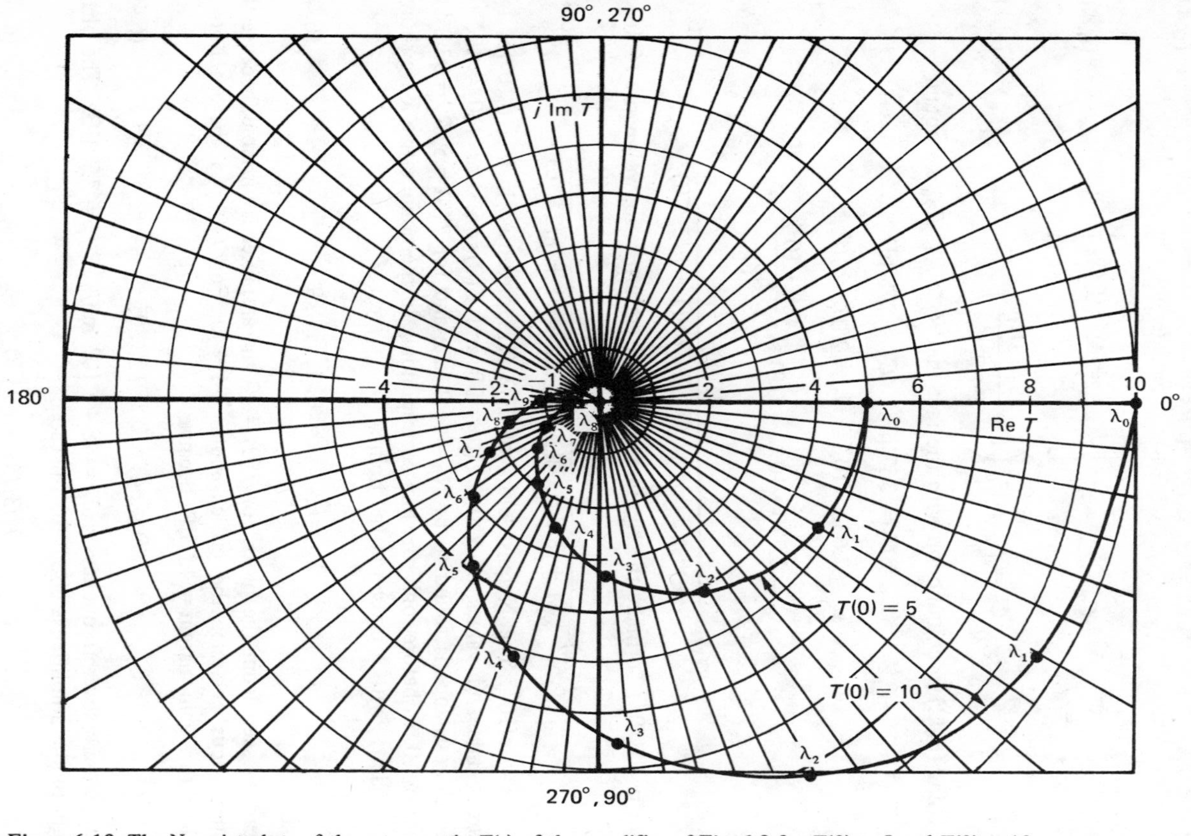

Figure 6.10 The Nyquist plots of the return ratio $T(s)$ of the amplifier of Fig. 6.9 for $T(0) = 5$ and $T(0) = 10$.

magnitude of $T(j\omega)$ in decibels and its phase in degrees each against the logarithm of frequency instead of putting them in a single curve. Thus, the critical point in the Nyquist plot corresponds to the lines of unit magnitude or zero-decibel level and $-180°$ phase in the separate frequency response plots, which are usually referred to as the *Bode plot.* The frequencies at which these lines are crossed by their respective response curves are the gain-crossover and phase-crossover frequencies.

Example 6.3 The Bode plot of Eq. (6.47) is presented in Fig. 6.11 for $T(0) = 10$ and $T(0) = 5$. For $T(0) = 10$ we see that the gain-crossover frequency ω_g,

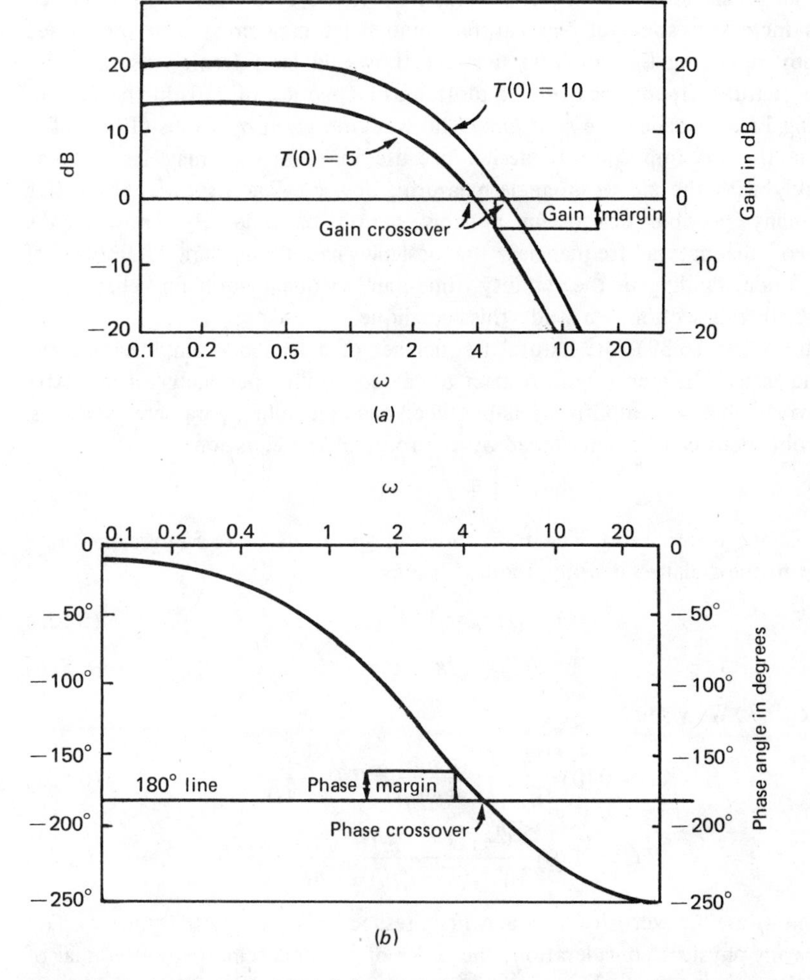

Figure 6.11 (*a*) The Bode plot of the gain of the return ratio $T(s)$ of the amplifier of Fig. 6.9 for $T(0) = 5$ and $T(0) = 10$ and (*b*) the Bode plot of the phase of the return ratio $T(s)$ of the amplifier of Fig. 6.9.

which is about $52.1 \cdot 10^6$ rad/s, is larger than the phase-crossover frequency $\omega_p = 48.4 \cdot 10^6$ rad/s, and the amplifier is unstable. For $T(0) = 5$, we have $\omega_g = 37.5 \cdot 10^6$ rad/s, which is smaller than $\omega_p = 48.4 \cdot 10^6$ rad/s, and the amplifier is therefore stable. The definitions of the gain and phase margins are indicated in the Bode plot of Fig. 6.11. From the plot, we see that for $T(0) = 5$ the gain margin is about 4.6 dB and the phase margin is about 20°, confirming Eqs. (6.49) and (6.50).

6.5 THE ROOT-LOCUS METHOD

In the foregoing, we showed that by changing the midband value of the return ratio $T(0)$, we change the locations of the natural frequencies of the feedback amplifier. As $T(0)$ is increased, some of the natural frequencies may cross into the closed RHS, thereby rendering the amplifier unstable. It would be extremely useful if the locus of the natural frequencies can be plotted as a function of $T(0)$ in the s-plane. The resulting locus is called the *root locus* and was first given by Evans (1948). The root-locus method is important in design because the root loci may be sketched rather quickly with the aid of an angle measuring device called a *spirule*, permitting study of many possible designs in a short period of time. By knowing the trajectories of the natural frequencies, the designer has a considerable amount of insight and understanding of the stability, time-, and frequency-domain behaviors of the network. In this section, we study this technique.

As indicated in (6.39), the natural frequencies of a feedback amplifier are the zeros of the return difference with respect to the controlling parameter of an active device, provided that the amplifier is stable when the controlling parameter vanishes. Thus, the root locus can be determined by the roots of the equation

$$F(s) = 1 + T(s) = 0 \tag{6.51}$$

where $T(s)$ is the return ratio. The trajectories of the roots of (6.51) are therefore the points s in the s-plane satisfying the conditions

$$|T(s)| = 1 \tag{6.52a}$$

$$\arg T(s) = (2k+1)\pi \tag{6.52b}$$

for any integer k. We write

$$T(s) = T(0)\frac{\prod_{i=1}^{m}(1 - s/z_i)}{\prod_{j=1}^{n}(1 - s/p_j)} = T(0)\frac{N(s)}{D(s)} \tag{6.53a}$$

$$F(s) = F(0)\frac{\prod_{j=1}^{n}(1 - s/s_j)}{\prod_{j=1}^{n}(1 - s/p_j)} = \frac{P(s)}{D(s)} \tag{6.53b}$$

where z_i and s_j are the zeros of $T(s)$ and $F(s)$, respectively, and p_j the poles of $T(s)$ and $F(s)$. From physical consideration, the order of the denominator polynomial of $T(s)$ is usually greater than that of the numerator polynomial, and consequently $T(s)$ has $n - m$ zeros at infinity. In the following, we study the loci of the natural frequencies s_j as a function of the midband value of the return ratio $T(0)$.

We first illustrate the root-locus method by the series-parallel and parallel-series feedback amplifiers of Secs. 4.2.3 and 4.2.4. As shown in these sections, the return ratio can generally be expressed in the form

$$T(s) = -\mu\beta = \frac{T(0)}{(1 + s/\sigma_1)(1 + s/\sigma_2)} \tag{6.54}$$

where σ_1 and σ_2 are real and positive with $\sigma_1 \leqslant \sigma_2$. The locations of the natural frequencies of the amplifier as a function of $T(0)$ are determined by setting $T(s) = -1$, yielding

$$s^2 + (\sigma_1 + \sigma_2)s + \sigma_1 \sigma_2 [1 + T(0)] = 0 \tag{6.55}$$

whose roots are given by

$$s_1, s_2 = -\tfrac{1}{2}(\sigma_1 + \sigma_2) \pm \tfrac{1}{2}\sqrt{(\sigma_1 - \sigma_2)^2 - 4\sigma_1 \sigma_2 T(0)} \tag{6.56}$$

As $T(0)$ varies from zero to infinity, the two branches of the root locus traced out by the natural frequencies s_1 and s_2 are shown in Fig. 6.12. For $T(0) \leqslant (\sigma_1 - \sigma_2)^2/4\sigma_1\sigma_2$, the two branches, starting at points $-\sigma_1$ and $-\sigma_2$, meet at $-\frac{1}{2}(\sigma_1 + \sigma_2)$. For $T(0) > (\sigma_1 - \sigma_2)^2/4\sigma_1\sigma_2$, s_1 and s_2 become complex and their loci move up and down along the vertical line $\text{Re}\,T = -\frac{1}{2}(\sigma_1 + \sigma_2)$, as depicted in Fig. 6.12. Since the locus remains in the open LHS for all nonnegative values of $T(0)$, the amplifier is therefore always stable. We remark that this conclusion is reached based on the assumption that, in the frequency band of interest, the series-parallel and parallel-series feedback amplifiers can be approximated by the unilateralized models of Figs. 4.26 and 4.32. This results in two dominant poles in $T(s)$ as shown in Eq. (6.54). However, if we use the complete hybrid-pi model for the transistors, we introduce additional nondominant poles in $T(s)$ and the root locus is modified so that some of the branches will bend toward the $j\omega$-axis, showing the possibility of instability for large amounts of feedback. The root-locus technique can be used to predict the stability of an amplifier only if we have an accurate, realistic representation of the return ratio. This is one of the

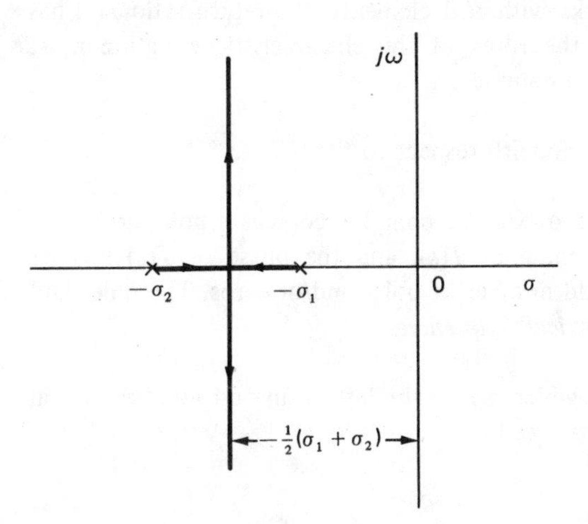

Figure 6.12 The root locus of a simplified parallel-series or series-parallel feedback amplifier.

drawbacks of testing stability using models. We shall illustrate this after presenting rules for the construction of the root locus.

6.5.1 Rules for the Construction of the Root Locus

As demonstrated in the foregoing, the construction of the root locus is relatively simple if the return ratio $T(s)$ has only two poles. The procedure becomes more complicated if $T(s)$ has three or more poles, because the solution of the characteristic equation is much more difficult. It is therefore important to develop simple rules that will enable us readily to sketch the root locus. Very often, a rough sketch will provide us sufficient information to undertake a preliminary design. The rules together with their justifications will now be presented.

By using the expression (6.53*a*), the trajectories of the roots of (6.51) are the points s in the s-plane satisfying the conditions

$$T(0) \prod_{i=1}^{m} \left| 1 - \frac{s}{z_i} \right| = \prod_{j=1}^{n} \left| 1 - \frac{s}{p_j} \right| \tag{6.57a}$$

$$\sum_{i=1}^{m} \arg \left(1 - \frac{s}{z_i} \right) - \sum_{j=1}^{n} \arg \left(1 - \frac{s}{p_j} \right) = (2k+1)\pi \tag{6.57b}$$

From (6.57*a*) we see that as $T(0)$ approaches zero, s approaches p_j, whereas as $T(0)$ approaches infinity, s approaches z_i. Thus, the branches of the locus start at the poles of $T(s)$ and terminate at the zeros of $T(s)$.

Rule 1 The root loci start from the poles and terminate at the zeros of $T(s)$.

We remark that for a rational function the number of poles is always equal to the number of its zeros if poles or zeros at the infinity are included. For $n > m$, $n - m$ branches of the locus will terminate at infinity.

Since we deal only with networks with real elements, the return ratio $T(s)$ has only real coefficients. As a result, the roots of the characteristic equation must occur in complex conjugate if they are complex.

Rule 2 The root loci are symmetric with respect to the σ-axis.

From (6.57*b*) we see that on the σ-axis, the complex conjugate poles and zeros of $T(s)$ contribute nothing to the phase of $T(s)$, and the phase of $T(s)$ will be $(2k+1)\pi$ if s is to the left of an odd number of poles and/or zeros. The poles and zeros of $T(s)$ are referred to as the *critical frequencies.*

Rule 3 The parts of the σ-axis, which are to the left of an odd number of real critical frequencies of $T(s)$, are parts of the loci.

From (6.53a) we can write, for large values of $|s|$,

$$T(s) = (-1)^{m-n} \frac{T(0)p_1p_2 \cdots p_n [s^m - (z_1 + z_2 + \cdots + z_m)s^{m-1} + \cdots]}{z_1z_2 \cdots z_m [s^n - (p_1 + p_2 + \cdots + p_n)s^{n-1} + \cdots]}$$

$$= (-1)^{m-n} T(0) \frac{p_1p_2 \cdots p_n}{z_1z_2 \cdots z_m} \left(s^m - \sum_{i=1}^{m} z_i s^{m-1} + \cdots\right)\left(s^n - \sum_{j=1}^{n} p_j s^{n-1} + \cdots\right)^{-1}$$

$$= (-1)^{m-n} T(0) \frac{p_1p_2 \cdots p_n}{z_1z_2 \cdots z_m} \left(s^m - \sum_{i=1}^{m} z_i s^{m-1} + \cdots\right)\left(s^{-n} + \sum_{j=1}^{n} p_j s^{-n-1} + \cdots\right)$$

$$= T(0) \frac{p_1p_2 \cdots p_n}{z_1z_2 \cdots z_m} (-s)^{m-n} \left(1 - \frac{\Sigma_{i=1}^m z_i - \Sigma_{j=1}^n p_j}{s} + \cdots\right) \tag{6.58}$$

The third equation is obtained after appealing to the binomial expansion. To determine the locus for large values of $|s|$, we set (6.58) to -1 or

$$T(s) = e^{j(2k+1)\pi} \tag{6.59}$$

for any integer k, yielding, for $n > m$,

$$-s\left(1 - \frac{\Sigma_{i=1}^m z_i - \Sigma_{j=1}^n p_j}{s} + \cdots\right)^{1/(m-n)} = \left[\frac{z_1z_2 \cdots z_m}{T(0)p_1p_2 \cdots p_n}\right]^{1/(m-n)} e^{-j\theta_k} \tag{6.60}$$

where

$$-\theta_k = \frac{(2k+1)\pi}{m-n} \qquad k = 0, 1, \ldots, n - m - 1 \tag{6.61}$$

Applying the binomial expansion again to the left-hand side of (6.60) gives

$$-s - \frac{\Sigma_{i=1}^m z_i - \Sigma_{j=1}^n p_j}{n-m} \approx \left[\frac{z_1z_2 \cdots z_m}{T(0)p_1p_2 \cdots p_n}\right]^{1/(m-n)} e^{-j\theta_k} \tag{6.62}$$

or, more compactly,

$$s = s_0 - M_0 e^{-j\theta_k} \qquad k = 0, 1, \ldots, n - m - 1 \tag{6.63}$$

where

$$s_0 = \frac{\Sigma_{j=1}^n p_j - \Sigma_{i=1}^m z_i}{n-m} \qquad n > m \tag{6.64}$$

$$M_0 = \left[\frac{z_1z_2 \cdots z_m}{T(0)p_1p_2 \cdots p_n}\right]^{1/(m-n)} \tag{6.65}$$

Thus, we can state the following rule.

Rule 4 For large values of $|s|$, the root loci become asymptotic to the straight lines that form angles of $-\theta_k$, as given in (6.61), with the σ-axis of the s-plane. Furthermore, the asymptotes intersect on the σ-axis at s_0 of (6.64).

We now proceed to determine the angles of departure from a pole and arrival at a zero. The *angle of departure* (*arrival*) from a point on the root locus is defined as the angle made at the point by the tangent vector for increasing (decreasing) the value of $T(0)$. Since not all of the poles and zeros of $T(s)$ will be distinct, for our purposes we express (6.53*a*) as

$$T(s) = T(0)\frac{N(s)}{D(s)} = \tilde{T}(0)\frac{\prod_{i=1}^{a}(s - z_i)^{m_i}}{\prod_{j=1}^{b}(s - p_j)^{n_j}} \tag{6.66}$$

where m_i and n_j denote the orders of the zero z_i and the pole p_j, respectively, and a and b are the numbers of distinct zeros and poles of the return ratio $T(s)$.

Let s be a point on the locus that is arbitrarily close to a pole p_j of $T(s)$ of order n_j. As $T(0)$ approaches zero, according to Rule 1 the point s approaches p_j. Substituting (6.66) in (6.52*b*) yields

$$\sum_{i=1}^{a} m_i \arg(s - z_i) - \sum_{j=1}^{b} n_j \arg(s - p_j) = (2k+1)\pi \tag{6.67}$$

for any integer k. Then we have

$$n_j \arg(s - p_j) = -(2k+1)\pi + \sum_{i=1}^{a} m_i \arg(s - z_i) - \sum_{\substack{i=1 \\ i \neq j}}^{b} n_i \arg(s - p_i) \tag{6.68}$$

As $T(0)$ approaches 0,

$$\phi_{pj} = \lim_{T(0) \to 0} \arg(s - p_j) \tag{6.69}$$

becomes the angles of departure at the pole p_j, and we obtain

$$\phi_{pj} = \frac{1}{n_j}\left[-(2k+1)\pi + \sum_{i=1}^{a} m_i \arg(p_j - z_i) - \sum_{\substack{i=1 \\ i \neq j}}^{b} n_i \arg(p_j - p_i)\right] \tag{6.70}$$

for $k = 0, 1, \ldots, n_j - 1$.

In a similar manner, by letting $T(0)$ approach infinity, the angles of arrival at a zero z_j of $T(s)$ of order m_j are found to be

$$\phi_{zj} = \frac{1}{m_j}\left[(2k+1)\pi + \sum_{i=1}^{b} n_i \arg(z_j - p_i) - \sum_{\substack{i=1 \\ i \neq j}}^{a} m_i \arg(z_j - z_i)\right] \tag{6.71}$$

for $k = 0, 1, \ldots, m_j - 1$. We summarize these results as follows.

Rule 5 The angles of departure from a pole p_j of order n_j of $T(s)$ are given by ϕ_{pj} of (6.70), and the angles of arrival at a zero z_j of order m_j are given by ϕ_{zj} of (6.71).

Finally, we determine the breakaway points on the root locus. A breakaway point s_b on the root locus is a zero of the return difference $F(s)$ of order at least 2 for some $T(0)$. Hence, at the breakaway point s_b the derivative of $F(s)$ must also be zero, showing that

$$\frac{d}{ds}F(s)\bigg|_{s=s_b} = \frac{d}{ds}T(s)\bigg|_{s=s_b} = 0 \tag{6.72}$$

Conversely, the roots of the equation $dT(s)/ds = 0$ are the zeros of $F(s)$ to the order of at least 2, and we get the following rule.

Rule 6 The breakaway points on the root locus are determined by solving the equation $dT(s)/ds = 0$ for its roots.

We remark that, barring cancellations, the roots of $dT(s)/ds = 0$ are the same as those of $dT^{-1}(s)/ds = 0$. The latter is more convenient to apply.

We illustrate the above results by the following examples.

Example 6.4 Consider the three-stage common-emitter feedback amplifier of Fig. 6.9. As computed in Example 6.2, the return ratio of the amplifier is found to be

$$T(s) = \frac{T(0)}{(1 + p/3.633)(1 + p/2.076)(1 + p/2.770)} \tag{6.73}$$

where $p = s/10^7$. We wish to construct its root locus.

The root locus will contain three branches, starting from the poles $p_1 = -3.633$, $p_2 = -2.076$, and $p_3 = -2.770$ and terminating at the three zeros z_i at infinity, all being scaled down by a factor of 10^7. According to Rule 3, the parts of the σ-axis from $\sigma = -2.076$ to -2.770 and from $\sigma = -3.633$ to $-\infty$ are parts of the loci.

For large values of $|s|$, the root loci become asymptotic to straight lines that form angles of

$$-\theta_k = -\frac{(2k+1)\pi}{3-0} \qquad k = 0, 1, 2 \tag{6.74}$$

giving $-\theta_0 = -\pi/3$, $-\theta_1 = -\pi$, and $-\theta_2 = -5\pi/3$, with the σ-axis of the s-plane. Furthermore, the asymptotes intersect on the σ-axis at the point

$$s_0 = \frac{-(3.633 + 2.076 + 2.770) + 0}{3-0} = -2.826 \tag{6.75}$$

The angles of departure at the poles p_j are obtained from (6.70) as

$$\phi_{pj} = -\pi + 0 - \sum_{\substack{i=1\\ i\neq j}}^{3} \arg(p_j - p_i) \qquad j = 1, 2, 3 \tag{6.76}$$

giving $\phi_{p1} = -3\pi$, $\phi_{p2} = -\pi$, and $\phi_{p3} = -2\pi$.

Finally, the breakaway point s_b on the locus is determined by solving for the roots of $dT^{-1}(s)/ds = 0$, which is equivalent to solving

$$\frac{d}{ds}[(p+3.633)(p+2.076)(p+2.770)] = 0 \tag{6.77a}$$

or

$$3p^2 + 16.958p + 23.356 = 0 \tag{6.77b}$$

giving $p = -2.376$ and -3.277. Since the breakaway point must lie between -2.076 and -2.770, we have $s_b = -2.376$. By summarizing all these, the root locus plot of $T(s)$ is presented in Fig. 6.13.

To determine the value of $T(0)$ for which the locus will intersect the $j\omega$-axis, we consider the numerator of $F(s) = 1 + T(s)$, yielding

$$p^3 + 8.479p^2 + 23.356p + 20.892[1 + T(0)] = 0 \tag{6.78}$$

To apply the Routh criterion, we form the array of the elements of (6.78):

$$\begin{array}{c|cc} p^3 & 1 & 23.356 \\ p^2 & 8.479 & 20.892[1+T(0)] \\ p^1 & 20.892 - 2.464T(0) & 0 \\ p^0 & 20.892[1+T(0)] & 0 \end{array} \tag{6.79}$$

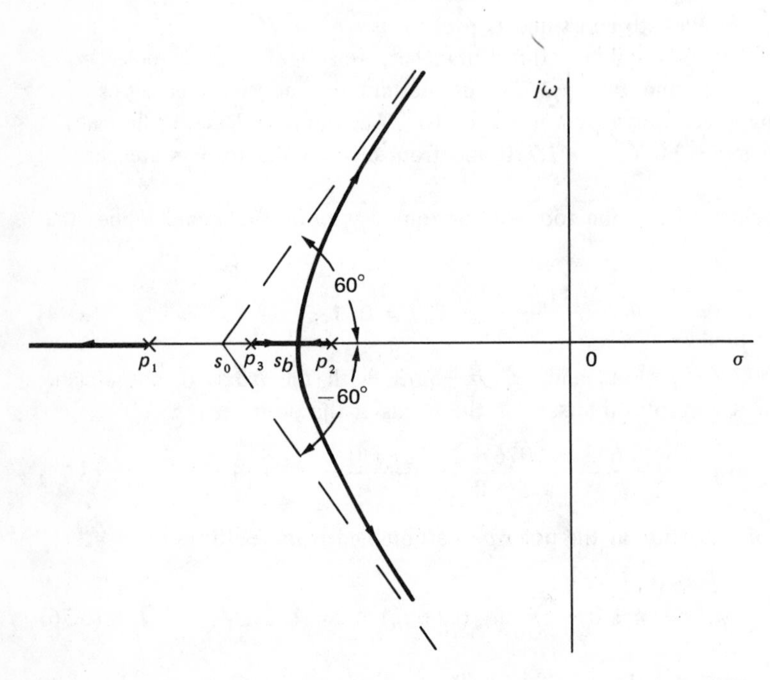

Figure 6.13 The root locus of a three-stage common-emitter feedback amplifier.

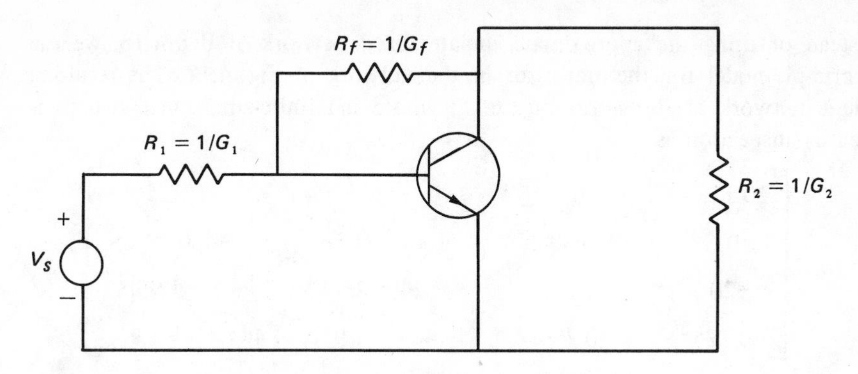

Figure 6.14 A voltage-shunt feedback amplifier.

Setting the third row to zero yields $T(0) = 8.479$. The corresponding frequencies at which the loci cross the $j\omega$-axis are obtained by setting the auxiliary polynomial of the second row to zero:

$$8.479p^2 + 198.035 = 0 \tag{6.80}$$

giving $p = \pm j4.833$.

Example 6.5 Consider the voltage-shunt feedback amplifier of Fig. 6.14. Assume that the transistor is represented by its hybrid-pi model, with

$$g_m = 0.4 \text{ mho} \qquad r_x = 50\ \Omega \qquad r_\pi = 250\ \Omega$$
$$C_\pi = 195 \text{ pF} \qquad C_\mu = 5 \text{ pF} \qquad r_o = 50 \text{ k}\Omega = \frac{1}{g_o} \tag{6.81a}$$

For the other network elements, let

$$R_1 = 10 \text{ k}\Omega \qquad R_f = 40 \text{ k}\Omega \qquad R_2 = 4 \text{ k}\Omega \tag{6.81b}$$

As demonstrated in Sec. 4.2.2, the open-loop current gain can be computed by using the network of Fig. 4.17, which after applying the Miller effect can be approximated by the unilateralized model of Fig. 4.18. In Example 4.4, the open-loop current gain was found to be

$$\mu(s) = \frac{I_2}{I_s} = \frac{81.69}{1 + p/0.059} \tag{6.82}$$

where $p = s/10^7$. The transfer current ratio of the feedback network from (4.54*b*) is given by $\beta = -R_2/R_f = -0.1$. Thus, we obtain the return ratio of the amplifier as

$$T(s) = -\mu\beta = \frac{8.169}{1 + p/0.059} \tag{6.83}$$

Since $T(s)$ has a simple pole at -0.059, the root locus is a straight line moving from the point -0.059 to $-\infty$ along the $-\sigma$-axis as $T(0)$ is increased. This implies that the amplifier is always stable.

Instead of using the approximate unilateralized network of Fig. 4.18, we use the hybrid-pi model for the transistor in the network of Fig. 4.17. The resulting equivalent network is shown in Fig. 6.15, whose indefinite-admittance matrix is obtained by inspection as

$$\mathbf{Y}(p) = 10^{-3}\begin{bmatrix} 20.125 & 0 & -20 & -0.125 \\ 0 & 0.295 + 0.05p & 400 - 0.05p & -400.295 \\ -20 & -0.05p & 24 + 2p & -4 - 1.95p \\ -0.125 & -0.295 & -404 - 1.95p & 404.42 + 1.95p \end{bmatrix} \tag{6.84}$$

By appealing to Eq. (2.94), the open-loop current gain is found to be

$$\mu(s) = \frac{I_2}{I_s} = -\frac{G_2 V_{24}}{I_s} = -\frac{G_2 Y_{12,44}}{Y_{33}}$$

$$= \frac{5(400 - 0.05p)}{1.962p^2 + 418.524p + 24.485} = \frac{2.548(400 - 0.05p)}{(p + 0.0585)(p + 213.256)} \tag{6.85}$$

giving the return ratio

$$T(s) = \frac{8.17(1 - p/8000)}{(1 + p/0.0585)(1 + p/213.256)} \tag{6.86}$$

The new return ratio now has two poles at $p_1 = -0.0585$ and $p_2 = -213.256$ and a zero at $z_1 = 8000$. If we ignore the effect of the nondominant pole p_2 and zero z_1, (6.86) reduces to (6.83). The root locus of (6.86) has two branches. The sections of the real axis from -0.0585 to 8000 and from -213.256 to $-\infty$ are parts of the loci. One of the branches will eventually terminate at $\sigma = 8000$. This indicates that for some value of $T(0)$, one of the natural frequencies will cross over the $j\omega$-axis, thereby rendering the feedback amplifier unstable. This example emphasizes the fact that the root-locus technique can be used to predict the stability of an amplifier only if we have an accurate, realistic representation of the return ratio.

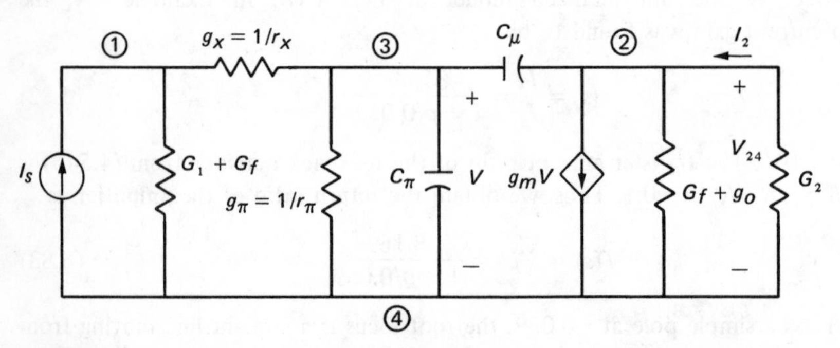

Figure 6.15 A high-frequency, small-signal equivalent network of the voltage-shunt feedback amplifier of Fig. 6.14.

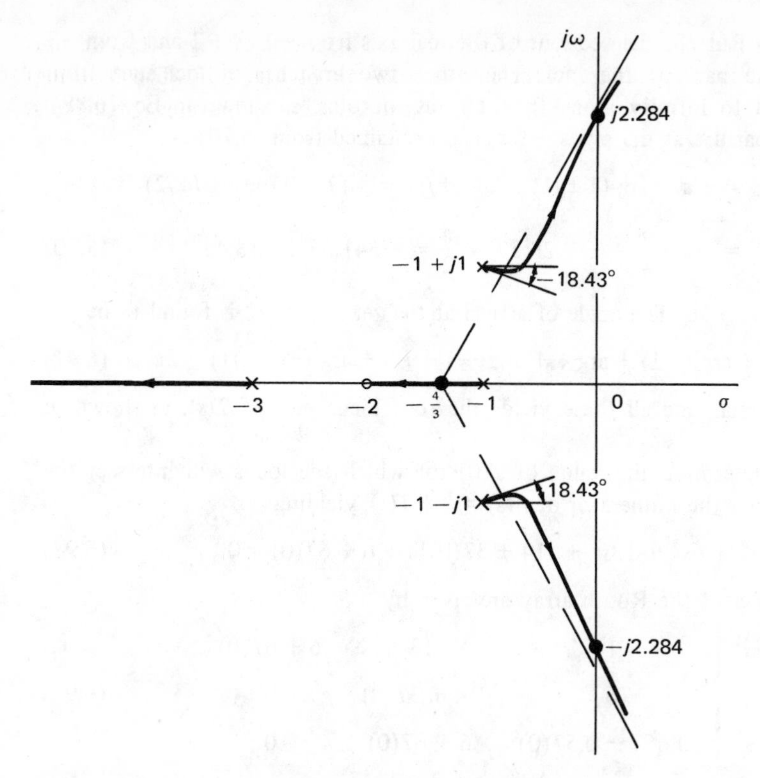

Figure 6.16 The root locus for the return ratio function (6.87).

Example 6.6 The return ratio of a feedback amplifier is described by

$$T(s) = \frac{3T(0)(s+2)}{(s+1)(s+3)(s^2+2s+2)} \tag{6.87}$$

The function has two real poles at $p_1 = -1$ and $p_2 = -3$, one pair of complex conjugate poles at $p_3 = -1 + j1$ and $p_4 = -1 - j1$, and one finite real zero at $z_1 = -2$, as indicated in Fig. 6.16. The root locus therefore consists of four branches.

As s approaches infinity, the root loci become asymptotic to straight lines that form angles of

$$-\theta_k = \frac{(2k+1)\pi}{4-1} \qquad k = 0, 1, 2 \tag{6.88}$$

giving $-\theta_0 = -\pi/3$, $-\theta_1 = -\pi$ and $-\theta_2 = -5\pi/3$, with the σ-axis of the s-plane. Furthermore, the asymptotes intersect on the σ-axis at the point

$$s_0 = \frac{-6+2}{4-1} = -\frac{4}{3} \tag{6.89}$$

According to Rule 3, the sections of the real axis from -1 to -2 and from -3 to $-\infty$ constitute part of the loci. The other two branches of loci start from $-1 \pm j1$ and tend to infinity along the $\pm 60°$ asymptotes, as shown in Eq. (6.88). The angles of departure at the poles $-1 \pm j1$ are obtained from (6.70) as

$$\phi_{p3}, \phi_{p4} = -\pi + \arg(1 \pm j1) - \arg(\pm j) - \arg(2 \pm j1) - \arg(\pm j2)$$

$$= -\pi \pm \frac{\pi}{4} \mp \frac{\pi}{2} \mp 26.57° \mp \frac{\pi}{2} = -341.57°, -18.43° \tag{6.90}$$

as indicated in Fig. 6.16. The angle of arrival at the zero $z_1 = -2$ is found to be

$$\phi_{z1} = \pi + \arg(-1) + \arg 1 + \arg(-1 - j1) + \arg(-1 + j1) = 2\pi \tag{6.91}$$

as expected. Summarizing all these yields the root locus plot of $T(s)$, as shown in Fig. 6.16.

Finally, to determine the value of $T(0)$ for which the locus will intersect the $j\omega$-axis, we consider the numerator of $F(s) = 1 + T(s)$, yielding

$$s^4 + 6s^3 + 13s^2 + [14 + 3T(0)]s + 6 + 6T(0) = 0 \tag{6.92}$$

The first three rows of the Routh array are given by

$$\begin{array}{c|ccc} s^4 & 1 & 13 & 6 + 6T(0) \\ s^3 & 6 & 14 + 3T(0) & 0 \\ s^2 & 10.667 - 0.5T(0) & 6 + 6T(0) & 0 \end{array} \tag{6.93}$$

Equation (6.92) has a pair of imaginary zeros when

$$6[6 + 6T(0)] = [10.667 - 0.5T(0)][14 + 3T(0)] \tag{6.94}$$

giving $T(0) = 5.768$. The corresponding frequencies at which the loci cross the $j\omega$-axis are obtained by setting the auxiliary polynomial of the third row to zero, yielding $s = \pm j2.284$, which are also indicated in Fig. 6.16.

6.5.2 Stabilization Techniques

To desensitize a feedback amplifier to parameter variations, we should increase the amount of negative feedback. However, as indicated above, there is a limit to the amount of feedback that is permitted before the amplifier becomes unstable. In this section, we demonstrate by using the parallel-parallel feedback configuration that the stability performance of a feedback amplifier can be improved by introducing a phase advance into the feedback network to counteract part of the phase-retarding effect of the basic amplifier.

Let the open-loop transfer function of a feedback amplifier be expressed explicitly as the ratio of two polynomials:

$$\mu(s) = \frac{N_\mu(s)}{D_\mu(s)} \tag{6.95}$$

Likewise, the transfer function of the feedback network is written as

$$\beta(s) = \frac{N_\beta(s)}{D_\beta(s)} \tag{6.96}$$

The return ratio and the closed-loop transfer function of the amplifier are obtained as

$$T(s) = -\mu(s)\beta(s) = -\frac{N_\mu(s)N_\beta(s)}{D_\mu(s)D_\beta(s)} \tag{6.97}$$

$$w(s) = \frac{\mu(s)}{1+T(s)} = \frac{N_\mu(s)D_\beta(s)}{D_\mu(s)D_\beta(s) - N_\mu(s)N_\beta(s)} \tag{6.98}$$

Observe that the zeros of $w(s)$ are either zeros of $\mu(s)$ or poles of $\beta(s)$ and that the zeros of $\beta(s)$ do not show explicitly in $w(s)$. If some of the zeros of $N_\beta(s)$ are chosen to coincide with those of $D_\mu(s)$, these cancelled zeros will not appear in $T(s)$ but they will appear as the poles of $w(s)$, as can be seen from (6.98). For this reason, the zeros of the transfer function of the feedback network are called *phantom zeros.*

Consider the three-stage common-emitter feedback amplifier of Fig. 6.9. As demonstrated in Sec. 4.2.2, the open-loop current gain can be expressed in the form

$$\mu(s) = \frac{I_2}{I_s} = \frac{T(0)}{(1-s/p_1)(1-s/p_2)(1-s/p_3)} \tag{6.99}$$

with $p_i = -s_i$ $(i = 1, 2, 3)$. The transfer current ratio of the feedback network, as defined in (4.52*a*), is given by

$$\beta(s) = y_{12}R_2 \approx y_{f12}R_2 = -\frac{R_2}{R_f} \tag{6.100}$$

where y_{12} and y_{f12} are the reverse-transfer admittance parameters of the composite two-port network and the feedback two-port network, respectively.

One way to introduce a phantom zero in $\beta(s)$ is to add a capacitor C_f across the feedback resistor R_f of Fig. 6.9. The resulting network is presented in Fig. 6.17. The amplifier will have the same open-loop current gain as in Eq. (6.99), but the current ratio of the feedback network is changed to

$$\beta(s) = -R_2C_f\left(s + \frac{1}{R_fC_f}\right) = -\frac{R_2(1-s/z_1)}{R_f} \tag{6.101}$$

where

$$z_1 = -\frac{1}{R_fC_f} \tag{6.102}$$

The corresponding return ratio becomes

$$T(s) = -\mu(s)\beta(s) = \frac{T(0)(1-s/z_1)}{(1-s/p_1)(1-s/p_2)(1-s/p_3)} \tag{6.103}$$

The location of the phantom zero z_1 is of critical importance in the nature of the root locus. In general, one branch of the root locus will terminate at the real zero

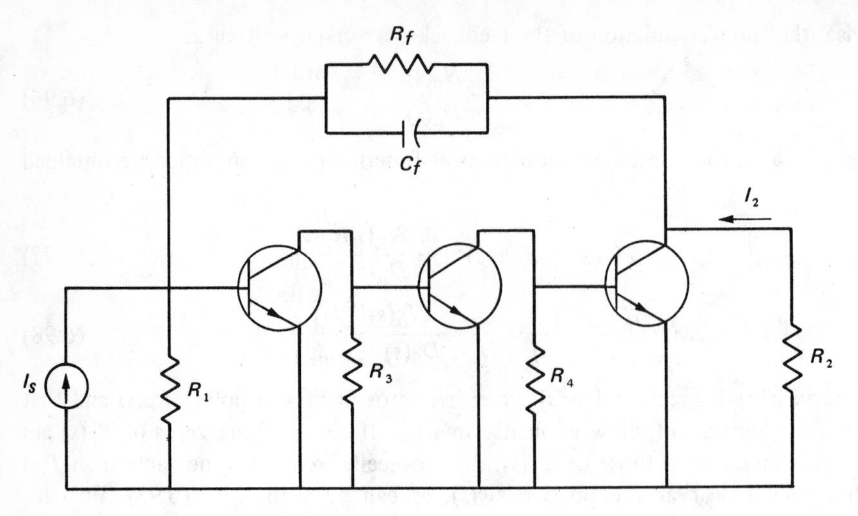

Figure 6.17 A three-stage common-emitter feedback amplifier possessing a phantom zero in its feedback circuit.

z_1, giving rise to one natural frequency in the neighborhood of the phantom zero for any appreciable amount of feedback. Suppose that we choose z_1 such that

$$z_1 = p_3 \tag{6.104}$$

The resulting return ratio will have only two real poles, whose root locus as computed in (6.54)-(6.56) is shown in Fig. 6.18. As a result, the amplifier is always

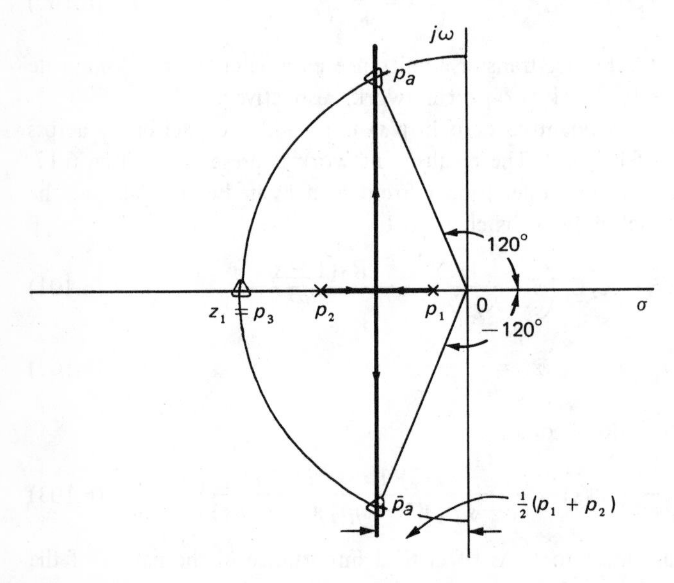

Figure 6.18 The root locus for the return ratio function (6.103) with $z_1 = p_3$.

stable. This is to be compared with the root-locus plot of (6.73), as presented in Fig. 6.13, in which one branch of the locus bends toward the $j\omega$-axis. This means that for sufficiently large $T(0)$, the amplifier will be unstable. The closed-loop current gain is found to be

$$w(s) = \frac{\mu(s)}{F(s)} = \frac{w(0)}{(1 - s/z_1)[(1 - s/p_1)(1 - s/p_2) + T(0)]} \tag{6.105}$$

Suppose that we wish to design a three-pole feedback amplifier having a maximally flat magnitude for its closed-loop current gain. This is equivalent to selecting an appropriate value of $T(0)$ so that the natural frequencies will lie on the circle of radius $|z_1|$, centered at the origin, as depicted in Fig. 6.18. The desired natural frequencies are indicated by the small triangles and are given by $z_1 = -1/R_f C_f$ and

$$p_a, \bar{p}_a = \frac{(1 \pm j\sqrt{3})z_1}{2} \tag{6.106}$$

in which we implicitly assume that $|p_3| > |p_2| \geqslant |p_1|$, giving

$$w(s) = \frac{w(0)}{(1 - s/z_1)(1 - s/p_a)(1 - s/\bar{p}_a)} \tag{6.107}$$

At $s = j1/R_f C_f$, $|w(s)| = |w(0)|/\sqrt{2}$, showing that the -3-dB bandwidth of $|w(j\omega)|$ is $|z_1|$, or

$$\omega_{3\text{-dB}} = |z_1| = |p_a| \tag{6.108}$$

The necessary value of $T(0)$ to achieve the desired pole locations is found by using the fact that $|T(s)| = 1$ on the locus and in particular at the point $s = p_a$. From (6.103) we obtain

$$T(0) = \left|\left(1 - \frac{p_a}{p_1}\right)\right| \left|\left(1 - \frac{p_a}{p_2}\right)\right| \tag{6.109}$$

We illustrate the above by considering the following example.

Example 6.7 Consider the return ratio

$$T(s) = \frac{T(0)}{(1 + p/2.076)(1 + p/2.770)(1 + p/3.633)} \tag{6.110}$$

as discussed in Example 6.2. Following Eq. (6.103), we have $p_1 = -2.076$, $p_2 = -2.770$, and $p_3 = -3.633$. Choose

$$z_1 = p_3 = -3.633 \tag{6.111}$$

giving the desired locations of the natural frequencies of the amplifier at

$$z_1 = p_3 = -\frac{1}{R_f C_f} = -3.633 \tag{6.112a}$$

$$p_a, \bar{p}_a = -1.8165 \mp j3.1463 \tag{6.112b}$$

The −3-dB bandwidth of the magnitude of the closed-loop current gain $w(j\omega)$ from (6.108) is therefore

$$\omega_{3\text{-dB}} = 36.33 \cdot 10^6 \text{ rad/s} \tag{6.113}$$

after denormalization, or $f_{3\text{-dB}} = 5.78$ MHz.

The necessary value of $T(0)$ to attain the desired pole locations is determined from (6.109):

$$T(0) = |(0.125 - j1.516)(0.344 - j1.136)| = 1.806 \tag{6.114}$$

Using a capacitor in shunt with the feedback resistor R_f, we produce a real phantom zero in the transfer function of the feedback network. It is possible to produce complex phantom zeros in $\beta(s)$ by using the bridged-T network, as shown in Fig. 6.19. In this case, the transfer current ratio of the feedback network is determined as

$$\beta(s) = y_{12}R_2 \approx y_{f12}R_2 = -R_2 C_{f1} \frac{s^2 + as + b}{s + c} \tag{6.115}$$

where
$$a = \frac{2R_{f1}C_{f1} + R_{f2}C_{f2}}{R_{f1}^2 C_{f1}C_{f2} + 2R_{f1}R_{f2}C_{f1}C_{f2}} \tag{6.116a}$$

$$b = \frac{1}{R_{f1}^2 C_{f1}C_{f2} + 2R_{f1}R_{f2}C_{f1}C_{f2}} \tag{6.116b}$$

$$c = \frac{2}{2R_{f2}C_{f2} + R_{f1}C_{f1}} \tag{6.116c}$$

With the same three-stage common-emitter basic amplifier described by Eq. (6.99), the return ratio of the feedback amplifier of Fig. 6.19 can be written as

$$T(s) = \frac{T(0)(1 - s/z_1)(1 - s/z_2)}{(1 - s/p_1)(1 - s/p_2)(1 - s/p_3)(1 - s/p_4)} \tag{6.117}$$

where z_1, $z_2 = \bar{z}_1$, and p_4 are the zeros and pole of $\beta(s)$, as given in (6.115). A typical pole-zero pattern of (6.117) together with the resulting root locus are presented in Fig. 6.20. Two branches of the locus terminate at the phantom zeros z_1 and z_2. If the other two branches of the locus are far from z_1 and z_2 for a reasonably large value of $T(0)$, the two natural frequencies near z_1 and z_2, which are poles of the closed-loop current gain $w(s)$, together with the pole p_4 of $\beta(s)$, which is the finite zero of $w(s)$, are dominant for the low-pass response. For design purposes, we need only concentrate on the loci terminating at z_1 and z_2. For example, we can choose the locations of z_1 and p_4 so that a maximally flat-magnitude or maximally flat-delay type of low-pass response is achieved. For a large value of $T(0)$, the dominant poles of $w(s)$ will be very close to z_1 and z_2. For any decrease of $T(0)$ caused by aging or environmental effect, the dominant poles move away from z_1 and z_2 only slightly. Hence, only a slight change in the dominant low-pass response is obtained. Furthermore, as will be seen in the next section, the natural frequencies in the neighborhood of phantom zeros z_1 and z_2

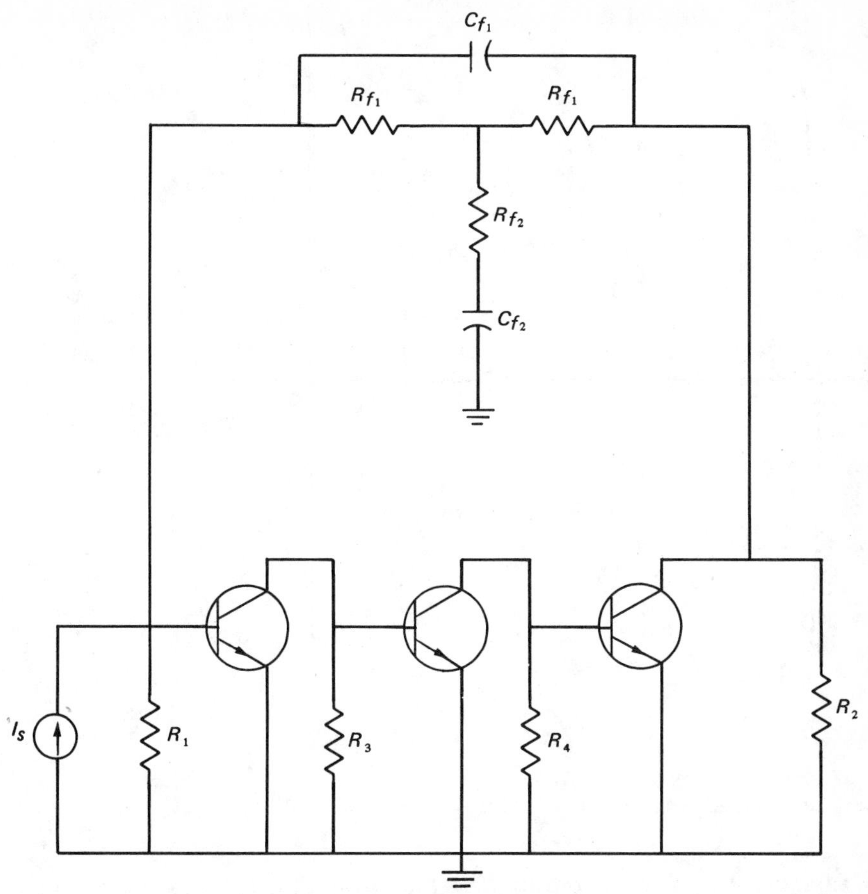

Figure 6.19 A three-stage common-emitter feedback amplifier possessing a pair of complex phantom zeros in its feedback circuit.

are relatively insensitive to the variations of the parameters. At any rate, as in the case of the real phantom zero, the technique provides a substantially large $T(0)$ without causing instability.

6.6 ROOT SENSITIVITY

In Chap. 3 we defined the sensitivity function as the ratio of the fractional change in a transfer function to the fractional change in an element (Definition 3.1). In the present situation, we are interested in knowing the change in the position of a natural frequency resulting from the variation of an element. A measure of this variation is called the root sensitivity and was first introduced by Truxal and Horowitz (1956).

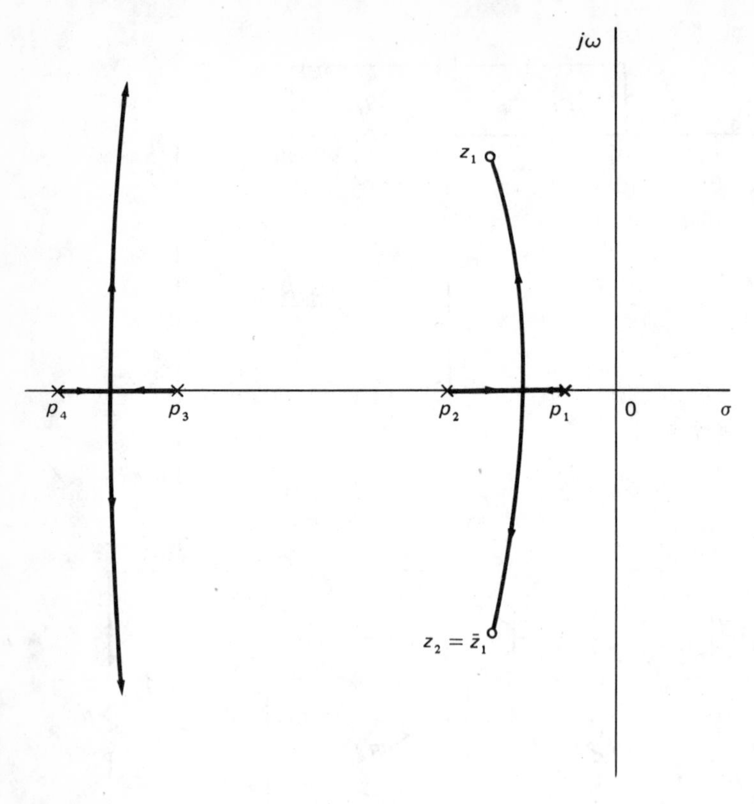

Figure 6.20 A typical root locus for the return ratio function (6.117).

Definition 6.6: Root sensitivity The *root sensitivity* is the sensitivity of a natural frequency s_k of a network with respect to an element x of interest and is defined by the relation

$$\mathcal{S}(s_k; x) = x \frac{\partial s_k}{\partial x} \tag{6.118}$$

A comparison of (6.118) with (3.21) shows that the root sensitivity is similar to the sensitivity function when ln w is under consideration. First we derive a useful relation between root sensitivity and the sensitivity function.

Refer to Eq. (6.53). The numerator polynomial of the return difference can be written as

$$P(s) = D(s) + T(0)N(s) = K \prod_{k=1}^{n} (s - s_k) \tag{6.119}$$

where s_k $(k = 1, 2, \ldots, n)$ are the natural frequencies of the feedback amplifier. On taking the logarithm, we obtain

$$\ln P(s) = \ln K + \sum_{k=1}^{n} \ln (s - s_k) \tag{6.120}$$

We compute the root sensitivity with respect to the element $x = T(0)$, the midband value of return ratio. To this end, we differentiate (6.120) with respect to x, which is $T(0)$. Assuming that s_k is simple, we get

$$\frac{\partial \ln P(s)}{\partial x} = \frac{N(s)}{P(s)} = \frac{\partial K/\partial x}{K} - \sum_{k=1}^{n} \frac{\partial s_k/\partial x}{s - s_k} \tag{6.121}$$

This shows that $-\partial s_k/\partial x$ is the residue of the function $N(s)/P(s)$ evaluated at the pole s_k. Thus, we can write

$$\frac{\partial s_k}{\partial x} = -(s - s_k)\left.\frac{N(s)}{P(s)}\right|_{s=s_k} = -\left.\frac{N(s)}{P'(s)}\right|_{s=s_k} = -\frac{N(s_k)}{P'(s_k)} \tag{6.122}$$

where the prime denotes the derivative with respect to s. Thus, from (6.119) the root sensitivity with respect to $T(0)$ can be written as

$$\mathcal{S}(s_k; T(0)) = T(0)\frac{\partial s_k}{\partial T(0)} = -\frac{T(0)N(s_k)}{P'(s_k)} = \frac{D(s_k)}{P'(s_k)} \tag{6.123}$$

From (6.53*b*), the right-hand side of (6.123) is recognized as the residue of the reciprocal of the return difference evaluated at the pole s_k. Thus, we conclude that the root sensitivity of the natural frequency s_k of a network with respect to the midband value of the return ratio is equal to the residue of the reciprocal of the return difference evaluated at its pole s_k.

As discussed in Sec. 5.1, the sensitivity function $S(x)$ with respect to an element x is related to the return difference and the transfer function $w(x)$ by

$$S(x) = \frac{1}{F(x)}\left[1 - \frac{w(0)}{w(x)}\right] \tag{6.124}$$

In the case where $w(0) = 0$, the sensitivity becomes equal to the reciprocal of the return difference. For practical amplifiers, $|w(0)|$ is usually much smaller than $|w(x)|$ in the passband, and $S(x) \approx 1/F$ may be used as a good estimate of the sensitivity in the same frequency band. For the situation where a feedback amplifier is represented by its ideal feedback model, the sensitivity function of the closed-loop transfer function $w(\mu)$ with respect to the forward amplifier gain is given by

$$S(\mu) = \frac{1}{1 - \mu(s)\beta(s)} = \frac{1}{F(\mu)} \tag{6.125}$$

Since $\beta(0)$ is considered fixed as far as the sensitivity function is concerned, we can use the variable parameter

$$T(0) = -\mu(0)\beta(0) \tag{6.126}$$

rather than $\mu(0)$, and (6.125) becomes

$$S(T(0)) = \frac{1}{F(T(0))} = \frac{D(s)}{P(s)} \tag{6.127}$$

Comparing (6.123) with (6.127) yields

$$\mathcal{S}(s_k; T(0)) = \zeta_{s_k} \tag{6.128}$$

where ζ_{s_k} denotes the residue of the sensitivity function with respect to $T(0)$ evaluated at the pole s_k. We remark that in the above analysis the return difference is written in two different ways: $F(x)$ and $F(s)$. The variable s, when shown explicitly, is used to exhibit its dependence on s, although F is dependent on both x and s. This should not create any difficulty. In the following, we express the root sensitivity in terms of the poles and zeros of the return ratio.

Consider the reciprocal of the root sensitivity, as given in (6.123), which can be manipulated into the form

$$\begin{aligned}
\frac{1}{\mathcal{S}(s_k; T(0))} &= \frac{P'(s_k)}{D(s_k)} = \frac{D'(s_k) + T(0)N'(s_k)}{D(s_k)} \\
&= \frac{D'(s_k)}{D(s_k)} + T(0)\frac{N(s_k)N'(s_k)}{D(s_k)N(s_k)} \\
&= \frac{D'(s_k)}{D(s_k)} - \frac{N'(s_k)}{N(s_k)} \\
&= \frac{d}{ds}[\ln D(s) - \ln N(s)]\Big|_{s=s_k} \\
&= \sum_{j=1}^{n} (s_k - p_j)^{-1} - \sum_{i=1}^{m} (s_k - z_i)^{-1}
\end{aligned} \tag{6.129}$$

The third line follows directly from (6.119) and the fifth line from (6.53*a*). The prime, as before, denotes the derivative with respect to s.

Since s_k is a natural frequency, it must lie on the root-locus plot of $T(s)$. In fact, from (6.129) we can determine the angle ϕ_k of the tangent to the locus at any point s_k on the locus, as follows:

$$\begin{aligned}
\phi_k &= \arg\left[\frac{\partial s_k}{\partial T(0)}\right] = \arg \mathcal{S}(s_k; T(0)) = -\arg\left[\frac{1}{\mathcal{S}(s_k; T(0))}\right] \\
&= -\arg\left[\sum_{j=1}^{n} (s_k - p_j)^{-1} - \sum_{i=1}^{m} (s_k - z_i)^{-1}\right]
\end{aligned} \tag{6.130}$$

In addition to the root sensitivity with respect to the changes of $T(0)$, in many problems we wish to evaluate the sensitivity of the natural frequency s_k to small changes in the locations of the poles and zeros of the return ratio $T(s)$. Assume first that z_i is a zero of $T(s)$. With z_i at its nominal value, by the definition of s_k we have

$$P(s_k) = D(s_k) + T(0)N(s_k) = 0 \tag{6.131}$$

Now let z_i be changed by a small amount from z_i to $z_i + \delta z_i$. The polynomial $P(s)$ will be changed from $P(s)$ to $P(s) + \delta P(s)$. Since the change is small, to a first-order approximation, the change in $P(s)$ can be written as

$$\delta P(s) \approx \frac{\partial P(s)}{\partial z_i} \delta z_i \tag{6.132}$$

With $P(s)$ changed to $P(s) + \delta P(s)$, the natural frequency s_k will correspondingly move to $s_k + \delta s_k$, so that

$$P(s_k + \delta s_k) + \delta P(s_k + \delta s_k) = 0 \tag{6.133}$$

By expanding $P(s_k + \delta s_k)$ and $\delta P(s_k + \delta s_k)$ by Taylor series, the first-order approximations are found to be

$$P(s_k + \delta s_k) \approx P(s_k) + \left.\frac{\partial P(s)}{\partial s}\right|_{s=s_k} \delta s_k = \left.\frac{\partial P(s)}{\partial s}\right|_{s=s_k} \delta s_k \tag{6.134a}$$

$$\delta P(s_k + \delta s_k) \approx \delta P(s_k) + \left.\frac{\partial \delta P(s)}{\partial s}\right|_{s=s_k} \delta s_k \tag{6.134b}$$

Substituting (6.134) in (6.133) in conjunction with (6.132) yields

$$\left.\frac{\partial P(s)}{\partial s}\right|_{s=s_k} \delta s_k + \left.\frac{\partial P(s)}{\partial z_i}\right|_{s=s_k} \delta z_i \approx 0 \tag{6.135}$$

giving

$$\mathcal{S}(s_k; z_i) = z_i \frac{\partial s_k}{\partial z_i} = -z_i \left.\frac{\partial P(s)/\partial z_i}{\partial P(s)/\partial s}\right|_{s=s_k}$$

$$= -z_i \left.\frac{T(0)\, \partial N(s)/\partial z_i}{\partial P(s)/\partial s}\right|_{s=s_k} \tag{6.136}$$

To evaluate the numerator on the right-hand side of (6.136), we write

$$N(s) = \left(1 - \frac{s}{z_i}\right) N_1(s) \tag{6.137}$$

assuming that the zero z_i is simple. Then we have

$$\frac{\partial N(s)}{\partial z_i} = \frac{sN(s)}{z_i(z_i - s)} \tag{6.138}$$

Substituting (6.138) in (6.136) and using (6.123), we obtain

$$\mathcal{S}(s_k; z_i) = -\frac{s_k T(0) N(s_k)}{(z_i - s_k)P'(s_k)} = \mathcal{S}(s_k; T(0)) \frac{s_k}{z_i - s_k} \tag{6.139}$$

In a similar manner, we can show that the root sensitivity with respect to a pole p_j of $T(s)$ is given by

$$\mathcal{S}(s_k; p_j) = \mathcal{S}(s_k; T(0)) \frac{s_k}{p_j - s_k} \tag{6.140}$$

The details are left as an exercise (see Prob. 6.11). Thus, knowing the root sensitivity with respect to the midband value of the return ratio, we can compute the root sensitivities with respect to the zeros and poles of the return ratio directly from (6.139) and (6.140).

Example 6.8 Consider the return ratio function

$$T(s) = \frac{T(0)}{s^2 + 2s + 2} = \frac{T(0)}{(s + 1 + j1)(s + 1 - j1)} \tag{6.141}$$

Then we have

$$F(s) = 1 + T(s) = \frac{s^2 + 2s + 2 + T(0)}{s^2 + 2s + 2} \tag{6.142}$$

$$S(s) = S(T(0)) = \frac{s^2 + 2s + 2}{s^2 + 2s + 2 + T(0)} \tag{6.143}$$

Assume that $T(0) = 3$. The natural frequencies are zeros of $F(s)$ or the roots of the equation

$$P(s) = s^2 + 2s + 5 = 0 \tag{6.144}$$

giving $s_1 = -1 + j2$ and $s_2 = -1 - j2$. To compute the root sensitivities of s_k with respect to $T(0)$, we apply (6.123) and obtain

$$\mathcal{S}(s_k; T(0)) = \left. \frac{s^2 + 2s + 2}{2s + 2} \right|_{s=s_k} = \pm \frac{j3}{4} \qquad k = 1, 2 \tag{6.145}$$

The partial fraction expansion of $S(s)$ is given by

$$S(s) = \frac{s^2 + 2s + 2}{s^2 + 2s + 5} = 1 + \frac{j3/4}{s + 1 - j2} - \frac{j3/4}{s + 1 + j2} \tag{6.146}$$

giving $\zeta_{s_k} = \pm j3/4$. Comparing this with (6.145) confirms (6.128). Also, $\mathcal{S}(s_k; T(0))$ can be computed directly by formula (6.129), as follows:

$$\frac{1}{\mathcal{S}(s_k; T(0))} = \frac{1}{-1 \pm j2 - (-1 + j1)} + \frac{1}{-1 \pm j2 - (-1 - j1)} = \mp \frac{j4}{3} \tag{6.147}$$

Thus, the angle of the tangent to the root locus at the point s_1 is 90°.

For illustrative purposes, we compute the root sensitivity directly from its definition. For a general value of $T(0)$, the natural frequencies are found to be

$$s_1, s_2 = -1 \pm j\sqrt{1 + T(0)} \tag{6.148}$$

Differentiating (6.148) with respect to $T(0)$ and applying the definition (6.118), we get

$$\mathcal{S}(s_k; T(0)) = T(0)\frac{ds_k}{dT(0)} = \pm j\frac{T(0)}{2\sqrt{1+T(0)}} \tag{6.149}$$

For $T(0) = 3$, $\mathcal{S}(s_k; T(0)) = \pm j3/4$, confirming (6.145).

To compute the root sensitivities of s_k with respect to the poles p_j of the return ratio (6.141), we use (6.140); the results are given by

$$\mathcal{S}(s_k; p_k) = \mathcal{S}(s_k; T(0))\frac{s_k}{p_k - s_k} = \frac{3(1 \mp j2)}{4} \qquad k = 1, 2 \tag{6.150a}$$

$$\mathcal{S}(s_1; p_2) = \bar{\mathcal{S}}(s_2; p_1) = \mathcal{S}(s_1; T(0))\frac{s_1}{p_2 - s_1} = \frac{1 - j2}{4} \tag{6.150b}$$

where $p_1 = -1 + j1$ and $p_2 = -1 - j1$. Observe that

$$\mathcal{S}(s_1; p_1) + \mathcal{S}(s_1; p_2) = \bar{\mathcal{S}}(s_2; p_1) + \bar{\mathcal{S}}(s_2; p_2) = 1 - j2 \tag{6.151}$$

This result is expected, because the natural frequencies s_k and poles p_j must move in complex conjugate pairs.

6.7 BODE FORMULAS

In Sec. 6.4, we demonstrated that the gain and phase of the return ratio are important in stability considerations. In this section, we show that they are not entirely independent; one dictates the other. Specifically, we present a number of relationships between the real and imaginary parts of a network function and its natural logarithm. These relations are well known in mathematics and are referred to as the *Hilbert transforms.* However, because Bode (1945) first applied them to network theory, we shall call them the *Bode formulas.*

Let $w(s)$ be a network function that is analytic in the entire closed RHS, and write

$$w(j\omega) = R(\omega) + jX(\omega) \tag{6.152}$$

To develop the desired relations, we shall integrate the function $w(s)/(s - j\omega_0)$ around the closed contour C as shown in Fig. 6.21, where ω_0 is any value of ω. The small indentation to the right has been introduced to avoid the pole of the function $w(s)/(s - j\omega_0)$ at the point $s = j\omega_0$, so that the integrand is analytic on the boundary and within the closed contour C. Applying Cauchy's integral theorem, we have

$$\oint_C \frac{w(s)}{s - j\omega_0}\,ds = 0 \tag{6.153}$$

The complete contour consists of three parts: the large semicircle C_1 of radius R_0, the small semicircular indentation C_2 of radius r_0 about the point $j\omega_0$, and the

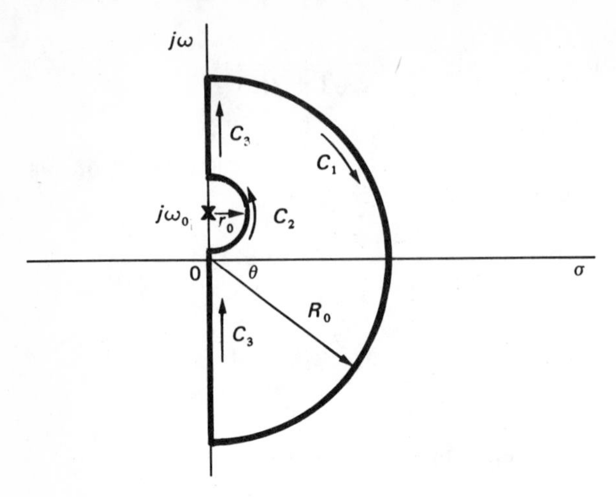

Figure 6.21 A closed contour indented around the point $j\omega_0$ in the s-plane.

imaginary axis C_3. The contour integral on the left-hand side of (6.153) can be expressed as the sum of three line integrals along the paths C_1, C_2, and C_3. For contour C_1, we have $s = R_0 e^{j\theta}$, as indicated in Fig. 6.21, and we take the limit as R_0 approaches infinity:

$$\lim_{R_0 \to \infty} \int_{C_1} \frac{w(s)}{s - j\omega_0} \, ds = \lim_{R_0 \to \infty} \int_{\pi/2}^{-\pi/2} \frac{w(R_0 e^{j\theta})}{R_0 e^{j\theta} - j\omega_0} jR_0 e^{j\theta} \, d\theta$$

$$= \int_{\pi/2}^{-\pi/2} jw(\infty) \, d\theta = -j\pi w(\infty) = -j\pi R(\infty) \qquad (6.154)$$

in which we have used the relation $w(\infty) = R(\infty)$ because the imaginary part of $w(s)$, being an odd function, must be zero at infinity.

Likewise, for the small semicircular contour C_2 we have

$$\lim_{r_0 \to 0} \int_{C_2} \frac{w(s)}{s - j\omega_0} \, ds = \lim_{r_0 \to 0} \int_{-\pi/2}^{\pi/2} \frac{w(j\omega_0 + r_0 e^{j\phi})}{r_0 e^{j\phi}} jr_0 e^{j\phi} \, d\phi$$

$$= \int_{-\pi/2}^{\pi/2} jw(j\omega_0) \, d\phi = j\pi w(j\omega_0) \qquad (6.155)$$

Finally, for the imaginary-axis contour C_3 we have

$$\int_{C_3} \frac{w(s)}{s - j\omega_0} \, ds = \lim_{\substack{R_0 \to \infty \\ r_0 \to 0}} \left[\int_{-R_0}^{\omega_0 - r_0} \frac{w(j\omega)}{\omega - \omega_0} \, d\omega + \int_{\omega_0 + r_0}^{R_0} \frac{w(j\omega)}{\omega - \omega_0} \, d\omega \right]$$

$$\equiv \int_{-\infty}^{\infty} \frac{w(j\omega)}{\omega - \omega_0} \, d\omega \qquad (6.156)$$

The integration in (6.156) must avoid the pole at $s = j\omega_0$ in a symmetric manner and will give the *principal value* of the integral on the right. In all the subsequent analysis, we must keep this point in mind. Now, substituting (6.154), (6.155), and (6.156) in (6.153) yields

$$\int_{-\infty}^{\infty} \frac{w(j\omega)}{\omega - \omega_0} \, d\omega = j\pi [R(\infty) - w(j\omega_0)] \tag{6.157}$$

If we equate the reals and imaginaries, we get

$$R(\omega_0) = R(\infty) - \frac{1}{\pi} \int_{-\infty}^{\infty} \frac{X(\omega)}{\omega - \omega_0} \, d\omega \tag{6.158a}$$

$$X(\omega_0) = \frac{1}{\pi} \int_{-\infty}^{\infty} \frac{R(\omega)}{\omega - \omega_0} \, d\omega \tag{6.158b}$$

Since ω in (6.158) is only a dummy variable of integration, and since ω_0 is an arbitrary point on the $j\omega$-axis, it is convenient to rewrite (6.158) as

$$R(\omega) = R(\infty) - \frac{1}{\pi} \int_{-\infty}^{\infty} \frac{X(u)}{u - \omega} \, du \tag{6.159a}$$

$$X(\omega) = \frac{1}{\pi} \int_{-\infty}^{\infty} \frac{R(u)}{u - \omega} \, du \tag{6.159b}$$

The above result has two important implications. It states that if a network function is devoid of poles on the $j\omega$-axis, then its imaginary part is completely determined by the behavior of its real part on the $j\omega$-axis. Conversely, if the imaginary part is specified for all ω, its real part is completely determined within an additive constant. It is significant to note that the real or imaginary part need not be a realizable rational function–its corresponding imaginary or real part can be computed from the integral. In fact, the real or imaginary part can even be specified in graphic form.

6.7.1 Reactance and Resistance Integral Theorems

We consider two special cases of the integral relations of (6.159). First, in (6.159*a*) we set $\omega = 0$. Since $X(u)/u$ is an even function, we get

$$\int_{0}^{\infty} \frac{X(u)}{u} \, du = \frac{\pi}{2} [R(\infty) - R(0)] \tag{6.160}$$

which is known as the *reactance-integral theorem.* It states that the behavior of the real part at the two extreme frequencies determines the integral of the imaginary part, weighted by the reciprocal of frequency, over all frequencies. Alternatively, if the left-hand side is plotted against the logarithmic frequency, then the area under the imaginary part is proportional to the net change in the real part between the two extreme frequencies: zero and infinite frequencies.

Next, let us multiply both sides of (6.159*b*) by ω and then take the limit as ω approaches infinity, giving

$$\lim_{\omega\to\infty} \omega X(\omega) = \frac{1}{\pi} \lim_{\omega\to\infty} \lim_{R_0\to\infty} \int_{-R_0}^{R_0} \frac{\omega R(u)}{u-\omega}\, du \tag{6.161}$$

There are two limiting operations involved on the right-hand side. The interchange of limits is permissible only if the integral is uniformly convergent for all ω. This condition is satisfied if $R(\infty)=0$. If $R(\infty)\neq 0$, we can consider $R(\omega)-R(\infty)$ as the given real part of a network function. With this manipulation, together with the fact that $R(\omega)$ is even, we get

$$\int_0^{\infty} [R(u)-R(\infty)]\, du = -\frac{\pi}{2} \lim_{\omega\to\infty} \omega X(\omega) \tag{6.162}$$

which is referred to as the *resistance-integral theorem.*

Consider a special situation where $w(s)$ has a simple zero at infinity. Then $R(\infty)=0$ and after appealing to the initial value theorem we get

$$\int_0^{\infty} R(u)\, du = \frac{\pi}{2} \lim_{s\to\infty} sw(s) = w_\delta(0^+) \tag{6.163}$$

where the limit is to be taken in the sector $|\arg s| \leqslant \pi/2 - \epsilon$, and $w_\delta(t)$ is the inverse Laplace transformation of $w(s)$ or the impulse response.

Example 6.9 Consider a one-port network that is constrained a priori by a leading shunt capacitor of value C, which may represent certain inevitable parasitic effects, as shown in Fig. 6.22. Let $Z_1(s)$ be the impedance facing the capacitor C. The total input impedance can be written as

$$Z(s) = \frac{Z_1(s)}{1 + CsZ_1(s)} \tag{6.164}$$

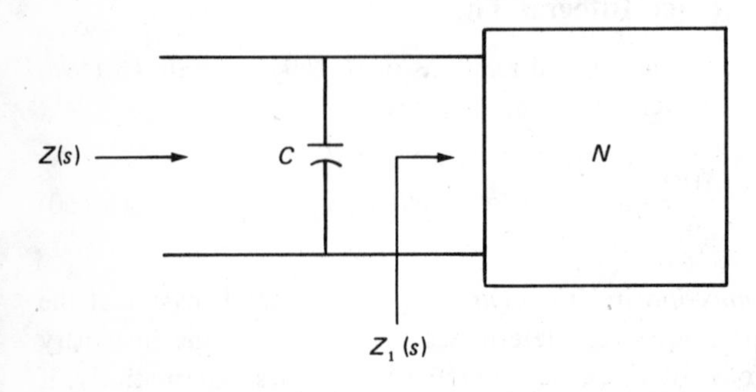

Figure 6.22 A constrained one-port network.

Assume that the one-port N is passive and that $Z_1(\infty) \neq 0$. Then $Z(s)$ must have a simple zero at infinity and (6.163) applies. Inserting (6.164) in (6.163) gives

$$\int_0^\infty R(u)\,du = \frac{\pi}{2} \lim_{s\to\infty} \frac{sZ_1(s)}{1 + CsZ_1(s)} = \frac{\pi}{2C} \tag{6.165}$$

As indicated at the beginning of this section, the resistance-integral theorem is applicable only to functions that are analytic in the closed RHS. If a function has simple poles on the $j\omega$-axis, the contour of integration must be indented to the right around these poles like the one shown in Fig. 6.21 for the pole at $s = j\omega_0$. If we go through the same procedure as in the preceding development, we find that additional terms, being proportional to the residues at these poles, must be subtracted from the right-hand side of (6.165). Since the residues are positive, for all cases, whether $Z(s)$ has simple poles on the $j\omega$-axis or not and whether $Z_1(\infty) \neq 0$ or not, the result can be written as

$$\int_0^\infty R(\omega)\,d\omega \leqslant \frac{\pi}{2C} \tag{6.166}$$

Note that the dummy variable has been changed to ω to suggest the physical meaning. Since power gain is usually associated with $R(\omega)$, the integral constraint (6.166) provides a basic limitation on the gain-bandwidth product introduced by the presence of the shunt capacitance C.

6.7.2 Phase-Integral and Gain-Integral Theorems

Consider the logarithm of a network function $w(s)$ and write

$$\ln w(j\omega) = A(\omega) + jB(\omega) \tag{6.167}$$

where $A(\omega)$ is the *gain function* and $B(\omega)$ is the *angle* or *phase function*. Assume that $w(s)$ is devoid of zeros and poles in the open RHS. We therefore permit $w(s)$ to have poles on the $j\omega$-axis, but it cannot have zero in the open RHS. Since poles and zeros of $w(s)$ are the logarithmic singularities of $\ln w(s)$, to apply Cauchy's integral theorem to $\ln w(s)$, as in (6.153), we must again make small semicircular indentations to the right around these singularities in selecting the contour of integration. If we can show that the small semicircular indentations around the logarithmic singularities will not contribute anything to the contour integral, identical results apply if we replace $w(s)$ by $\ln w(s)$, $R(\omega)$ by $A(\omega)$, and $X(\omega)$ by $B(\omega)$.

To confirm our assertion, let $j\omega_1$ be a zero or pole of order k of $w(s)$ and write

$$w(s) = (s - j\omega_1)^k w_1(s) \tag{6.168}$$

where k is either a positive or negative integer, giving

$$\ln w(s) = k \ln (s - j\omega_1) + \ln w_1(s) \tag{6.169}$$

It is sufficient to show that contour integral

$$\int_P \frac{\ln (s - j\omega_1)}{s - j\omega_0}\, ds = 0 \qquad \omega_1 \neq \omega_0 \tag{6.170}$$

where P is an infinitesimal semicircular indentation to the right around the point $s = j\omega_1$. Any point s on P is expressible as $s = j\omega_1 + re^{j\theta}$, and we have the following estimate:

$$\left| \int_P \frac{\ln (s - j\omega_1)}{s - j\omega_0}\, ds \right| = \left| \int_P \frac{(\ln r + j\theta) jre^{j\theta}}{re^{j\theta} + j(\omega_1 - \omega_0)}\, d\theta \right|$$

$$\leqslant \int_P \frac{r|\ln r|}{|\omega_1 - \omega_0|}\, d\theta + \int_P \frac{r\theta}{|\omega_1 - \omega_0|}\, d\theta = \frac{2\theta|r \ln r| + r\theta^2}{2|\omega_1 - \omega_0|} \tag{6.171}$$

Since

$$\lim_{r \to 0} r \ln r = 0 \tag{6.172}$$

we obtain

$$\lim_{r \to 0} \int_P \frac{\ln (s - j\omega_1)}{s - j\omega_0}\, ds = 0 \tag{6.173}$$

This establishes (6.170) and justifies our assertion that the logarithmic singularities lying on the path of the contour integral (6.153), with $\ln w(s)$ replacing $w(s)$, do not contribute anything to the integral.

From (6.159), (6.160), and (6.162), we can write

$$A(\omega) = A(\infty) - \frac{1}{\pi} \int_{-\infty}^{\infty} \frac{B(u)}{u - \omega}\, du \tag{6.174a}$$

$$B(\omega) = \frac{1}{\pi} \int_{-\infty}^{\infty} \frac{A(u)}{u - \omega}\, du \tag{6.174b}$$

giving

$$\int_0^{\infty} \frac{B(u)}{u}\, du = \frac{\pi}{2} [A(\infty) - A(0)] \tag{6.175}$$

which is called the *phase-integral theorem*, and

$$\int_0^{\infty} [A(u) - A(\infty)]\, du = -\frac{\pi}{2} \lim_{\omega \to \infty} \omega B(\omega) \tag{6.176}$$

which is called the *gain-integral theorem.*

6.7.3 Gain-Phase Theorem

In this section, we express the phase at a particular frequency in terms of the gain specified at all frequencies. This result is useful in the design of feedback amplifiers.

Our starting point is (6.174*b*), which, for our purposes, is rewritten as

$$B(\omega_a) = \frac{1}{\pi}\left[\int_{-\infty}^{0} \frac{A(\omega)}{\omega - \omega_a}\, d\omega + \int_{0}^{\infty} \frac{A(\omega)}{\omega - \omega_a}\, d\omega\right] \tag{6.177}$$

Notice that the dummy variable u has been changed to ω to indicate its physical significance. In the first integral on the right-hand side of (6.177), if we replace ω by $-\omega$ and change the limits accordingly, we obtain

$$\int_{-\infty}^{0} \frac{A(\omega)}{\omega - \omega_a}\, d\omega = -\int_{\infty}^{0} \frac{A(-\omega)}{-\omega - \omega_a}\, d\omega = -\int_{0}^{\infty} \frac{A(\omega)}{\omega + \omega_a}\, d\omega \tag{6.178}$$

Substituting this in (6.177) in conjunction with the fact that

$$\int_{0}^{\infty} \frac{A(\omega_a)}{\omega^2 - \omega_a^2}\, d\omega = A(\omega_a)\int_{0}^{\infty} \frac{1}{\omega^2 - \omega_a^2}\, d\omega = 0 \tag{6.179}$$

we get

$$B(\omega_a) = \frac{2\omega_a}{\pi}\int_{0}^{\infty} \frac{A(\omega) - A(\omega_a)}{\omega^2 - \omega_a^2}\, d\omega \tag{6.180}$$

A more convenient expression is obtained if a change to logarithmic frequency is made. Define

$$v = \ln\left(\frac{\omega}{\omega_a}\right) \quad \text{or} \quad \frac{\omega}{\omega_a} = e^{v} \tag{6.181}$$

By using (6.181), (6.180) can be expressed as

$$B(\omega_a) = \frac{1}{\pi}\int_{-\infty}^{\infty} \frac{A(\omega) - A(\omega_a)}{\sinh v}\, dv \tag{6.182}$$

The argument of $A(\omega)$ has been retained as ω for simplicity, although it should be written as $A(\omega_a e^{v})$. Integrating the right-hand side of (6.182) by parts, we obtain

$$\begin{aligned} B(\omega_a) &= -\frac{1}{\pi}\left\{[A(\omega) - A(\omega_a)] \ln\coth\frac{v}{2}\right\}\Bigg|_{-\infty}^{\infty} + \frac{1}{\pi}\int_{-\infty}^{\infty} \frac{dA(\omega)}{dv} \ln\coth\frac{v}{2}\, dv \\ &= \frac{1}{\pi}\int_{-\infty}^{\infty} \frac{dA(\omega)}{dv} \ln\coth\frac{|v|}{2}\, dv \end{aligned} \tag{6.183}$$

which is known as the *gain-slope theorem.* It states that the phase at any frequency depends on the slope of the gain at all frequencies when plotted against logarithmic frequency, the relative importance of any gain slope being determined by the weighting factor

$$\ln \coth \frac{|v|}{2} = \ln \left| \frac{\omega + \omega_a}{\omega - \omega_a} \right| \tag{6.184}$$

A plot of the function (6.184) is presented in Fig. 6.23. Observe that it rises sharply in the vicinity of $v = 0$ or $\omega = \omega_a$. For frequencies much higher than ω_a, the weighting factor is approximately $2\omega_a/\omega$, whereas at frequencies much lower than ω_a it is approximately $2\omega/\omega_a$. Therefore, at ω_a most of the contribution to the phase comes from the slope of the gain characteristic in the vicinity of ω_a. In terms of (6.184), (6.183) becomes

$$B(\omega_a) = \frac{1}{\pi} \int_{-\infty}^{\infty} \frac{dA(\omega)}{d\omega} \ln \left| \frac{\omega + \omega_a}{\omega - \omega_a} \right| d\omega \tag{6.185}$$

The gain slope $dA(\omega_a e^v)/dv$ in formula (6.183) is given in terms of *nepers per log radian.* This is to say that, for a unit gain slope, A will change by 1 neper between frequencies that are in the ratio $e = 2.7183$. A unit gain slope is evidently the same as a change of 6 dB per octave or 20 dB per decade.

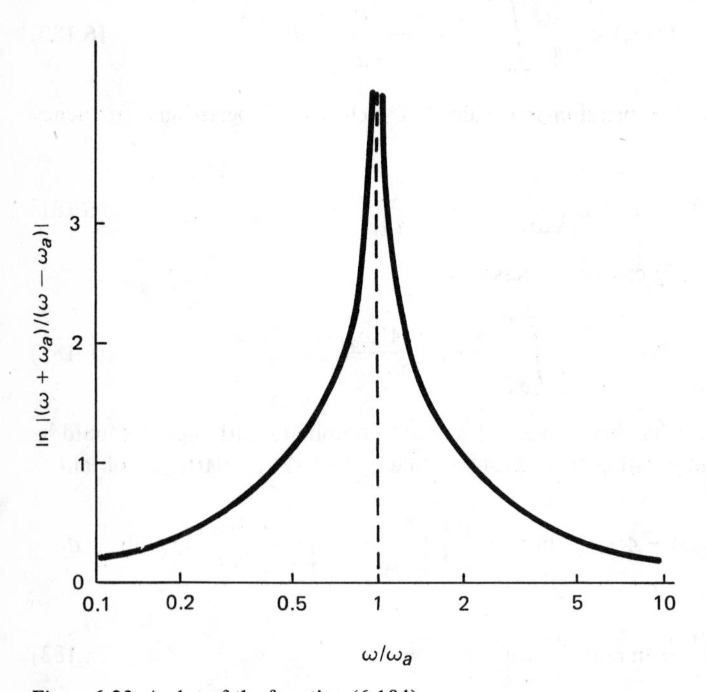

Figure 6.23 A plot of the function (6.184).

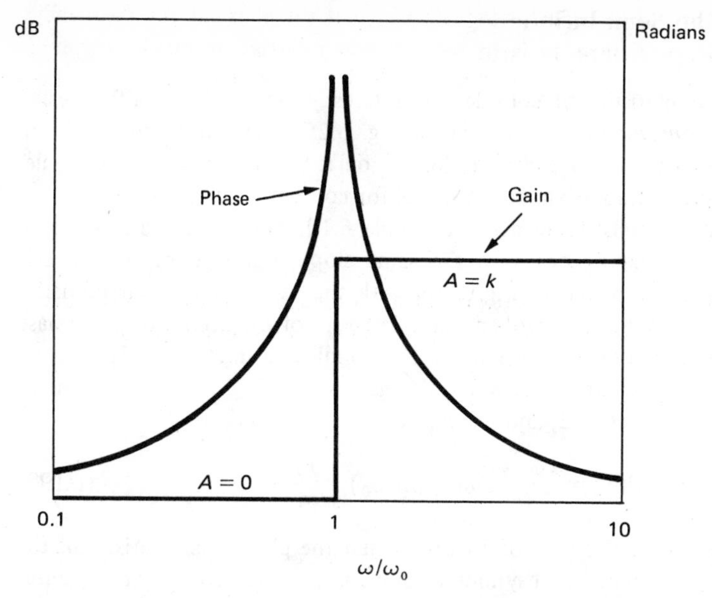

Figure 6.24 A prescribed gain characteristic and its associated phase plot.

Example 6.10 Assume that the gain characteristic has a constant slope on a logarithmic frequency scale at all frequencies. Then we have

$$A(\omega) = A(\omega_a e^{v}) = kv \tag{6.186}$$

k being the constant slope, and $dA(\omega)/dv = k$. From (6.183), the phase at any frequency ω_a is given by

$$B(\omega_a) = \frac{k}{\pi}\int_{-\infty}^{\infty} \ln \coth \frac{|v|}{2}\, dv = \frac{1}{2} k\pi \tag{6.187}$$

Thus, the phase characteristic is constant and equal to $\frac{1}{2}\pi$ times the slope of the gain characteristic.

Example 6.11 Assume that the gain characteristic is everywhere constant except for a discontinuity at frequency ω_0, as shown in Fig. 6.24. The slope is obviously an impulse function and is given by

$$\frac{dA(\omega_a e^{v})}{dv} = k\delta(\omega - \omega_0) \tag{6.188}$$

Substituting it in (6.185) yields the phase characteristic as

$$B(\omega_a) = \frac{k}{\pi} \ln \left| \frac{\omega_0 + \omega_a}{\omega_0 - \omega_a} \right| \tag{6.189}$$

A plot of (6.189) is also presented in Fig. 6.24.

6.7.4 Phase for the Semi-Infinite Constant-Slope Characteristic

One of the gain characteristics of considerable interest in feedback amplifier design is the *semi-infinite constant slope* as shown in Fig. 6.25, where the gain is zero up to $\omega = \omega_0$ and thereafter has a constant slope k on a logarithmic frequency scale. In this section, we present approximate formulas for computing its phase.

The constant slope k, as discussed in Example 6.10, can be viewed as the sum of two semi-infinite characteristics with the same slope, but running in opposite directions from ω_0, as depicted in Fig. 6.26, with the semi-infinite characteristics being identified by the solid and broken lines. The sum of the accompanying phase characteristics must therefore be equal to $k\pi/2$, as indicated in (6.187). This shows that the phase shift at ω_0 for the semi-infinite gain slope of Fig. 6.25 is equal to half of the asymptotic value $k\pi/2$, and we can write

$$B(\infty) = \frac{k\pi}{2} \quad \text{and} \quad B(\omega_0) = \frac{k\pi}{4} \tag{6.190}$$

Furthermore, we see from the plot of Fig. 6.26 that the phase characteristic of the semi-infinite gain slope exhibits odd symmetry on a logarithmic frequency scale about the point determined by $\omega = \omega_0$ and $B(\omega) = k\pi/4$. This results in the general properties of the phase characteristic as shown in Fig. 6.25, which allows us to restrict the computation of the phase to values of ω_a below ω_0.

We now proceed to develop approximate formulas for the computation of the phase at any frequency ω_a below ω_0 for the semi-infinite constant slope k of Fig. 6.25. For our purposes, we define

$$x = \frac{\omega_a}{\omega} \quad \text{and} \quad x_a = \frac{\omega_a}{\omega_0} \tag{6.191}$$

In the range of frequencies below ω_0, the gain slope is zero; for frequencies above ω_0, the gain slope is k on a logarithmic frequency scale. These are equivalent to

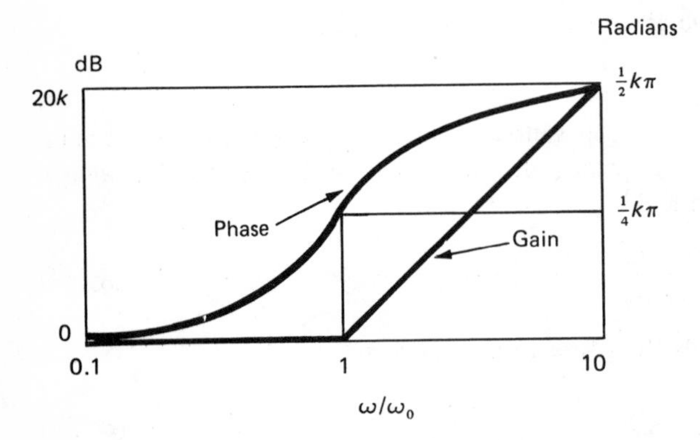

Figure 6.25 The semi-infinite constant slope and its associated phase plot.

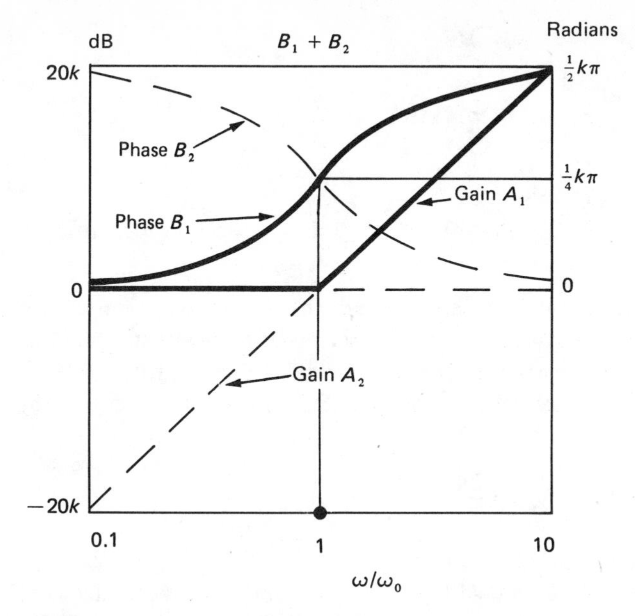

Figure 6.26 The decomposition of the constant slope into the sum of two semi-infinite characteristics with the same slope but running in the opposite directions and their associated phase plots.

$dA(\omega)/d\omega = 0$, $\omega < \omega_0$, and $dA(\omega)/d\omega = k/\omega$, $\omega \geqslant \omega_0$. Using these in (6.185), we obtain

$$B(\omega_a) = \frac{k}{\pi} \int_{\omega_0}^{\infty} \ln \left| \frac{\omega + \omega_a}{\omega - \omega_a} \right| \frac{d\omega}{\omega} = \frac{k}{\pi} \int_0^{x_a} \ln \left| \frac{1+x}{1-x} \right| \frac{dx}{x} \tag{6.192}$$

To develop the desired formulas, we set

$$y = \frac{1-x}{1+x} \quad \text{and} \quad y_a = \frac{1-x_a}{1+x_a} \tag{6.193}$$

In terms of the new variable y, (6.192) can be written as

$$B(x_a) = -\frac{k}{\pi} \int_{x=0}^{x=x_a} \ln y \, d(\ln x) \tag{6.194}$$

The above equation can be integrated by parts, and the result is given by

$$B(x_a) = -\frac{k}{\pi} (\ln x \ln y) \Big|_{x=0}^{x=x_a} + \frac{k}{\pi} \int_{x=0}^{x=x_a} \ln x \, d(\ln y) \tag{6.195}$$

which is transformed into (see Prob. 6.17)

$$B(x_a) = -\frac{k}{\pi}\ln x_a \ln y_a - \frac{k}{\pi}\int_{y=y_a}^{y=1} \ln x \, d(\ln y)$$

$$= -\frac{k}{\pi}\ln x_a \ln y_a - \frac{k}{\pi}\int_{y=0}^{y=1} \ln x \, d(\ln y) + \frac{k}{\pi}\int_{y=0}^{y=y_a} \ln x \, d(\ln y) \tag{6.196}$$

By using (6.192), it is straightforward to demonstrate that the second term on the right-hand side of (6.169) represents the phase of a semi-infinite constant slope k at the point ω_a, which according to (6.190) is equal to $k\pi/4$. Likewise, from (6.192), the third term represents the phase at the point y_a. Substituting these in (6.196) yields

$$B(x_a) + B(y_a) = \frac{k\pi}{4} - \frac{k}{\pi}\ln x_a \ln y_a \tag{6.197}$$

where x_a and y_a are related by (6.193).

If in (6.193) x_a varies from 0 to 0.414, y_a changes from 1 to 0.414. Likewise, if y_a varies from 0 to 0.414, x_a changes from 1 to 0.414. Thus, by using (6.197), the phase characteristic can be computed at all frequencies if we know it only between zero and 0.414. Within this range, we can expect a power series expansion for B to converge rapidly. Substituting the power series expansion

$$\ln\frac{1+x}{1-x} = 2\left(x + \frac{x^3}{3} + \frac{x^5}{5} + \cdots\right) \qquad x^2 < 1 \tag{6.198}$$

in (6.192) and integrating term by term, we obtain

$$B(\omega_a) = \frac{2k}{\pi}\left(x_a + \frac{x_a^3}{9} + \frac{x_a^5}{25} + \cdots\right) \tag{6.199}$$

If we use only the first term in (6.199) in conjunction with (6.197), we obtain the desired approximate formulas for the phase characteristic as

$$B(\omega_a) \approx \frac{2k\omega_a}{\pi\omega_0} \qquad 0 \leqslant \omega_a \leqslant 0.414\omega_0 \tag{6.200a}$$

$$\approx \frac{k\pi}{4} - \frac{k}{\pi}\ln\left(\frac{\omega_a}{\omega_0}\right)\ln\left(\frac{\omega_0 - \omega_a}{\omega_0 + \omega_a}\right) - \frac{2k(\omega_0 - \omega_a)}{\pi(\omega_0 + \omega_a)} \qquad 0.414\omega_0 \leqslant \omega_a \leqslant \omega_0 \tag{6.200b}$$

The maximum error in B as computed from these expressions is about 2 percent. If the first two terms in (6.199) are used, the result is almost exact. These formulas were first given by Bode (1945) and are very useful in the design of feedback amplifiers, to be discussed in the following section.

One of the major applications of the approximate formulas is the graphic evaluation of the phase shift $B(\omega)$ when the gain characteristic is known either analytically or graphically. The gain characteristic is first approximated by a series of straight-line segments. The straight-line approximation is next resolved into the sum of semi-infinite constant slopes of various steepness. With (6.200), the phase characteristics of the component semi-infinite constant slopes are then computed, and these are summed up to give the total phase characteristic of the gain response.

6.8 BODE'S DESIGN THEORY

A basic problem in the design of a feedback amplifier is the determination of the maximum amount of feedback in decibels for a given frequency band. To guard against instability, certain gain and phase margins are also specified. As discussed in this and the preceding two chapters, the return ratio T is essential for stability study and controls the amount of feedback, sensitivity, and the input and output impedance levels. In the following, we present Bode's design theory in terms of the return ratio $T(j\omega)$. We shall confine our discussion to low-pass characteristics of $T(j\omega)$, because the other types of characteristics can be obtained from the low-pass by the standard techniques of frequency transformation.

An ideal low-pass characteristic for the return ratio $T(j\omega)$ is one whose gain characteristic is completely flat from zero to cutoff frequency ω_0 and whose phase is constant beyond the cutoff frequency. In this way, the flat portion of the gain characteristic provides a constant loop transmission in the passband and the flat portion of the phase characteristic gives the desired phase margin. To this end, we introduce *Bode's ideal cutoff characteristic*, expressed by the return ratio

$$\ln T(j\omega) = A + jB = \ln\left[\frac{T(0)}{(\sqrt{1-\omega^2/\omega_0^2} + j\omega/\omega_0)^{2(1-\gamma)}}\right]$$

$$= \ln T(0) - 2(1-\gamma)\ln\left(\sqrt{1-\frac{\omega^2}{\omega_0^2}} + \frac{j\omega}{\omega_0}\right) \tag{6.201}$$

The exponent is written in the form $2(1-\gamma)$ to suit the feedback amplifier problem. A plot of A and B for the choice, $\gamma = \frac{1}{6}$, is presented in Fig. 6.27. Note that A is plotted in relation to the midband value $T(0)$ for simplicity. Thus, the 0-dB level in Fig. 6.27 corresponds to $20 \log T(0)$. Observe that the gain characteristic is completely flat from 0 to the cutoff frequency ω_0, and it drops off at the rate of $12(1-\gamma)$ dB/octave at high frequencies. The phase shift is changed from 0 to $-(1-\gamma)\pi$ when the frequency is increased from 0 to ω_0, and it remains a constant phase $-(1-\gamma)\pi$ beyond ω_0. The corresponding Nyquist plot of Fig. 6.27 is shown in Fig. 6.28. For the choice of $\gamma = \frac{1}{6}$, the corresponding phase margin is 30°. In general, the phase margin is $\gamma\pi$ in radians. In Fig. 6.28, if we set $\gamma = 0$, then the gain falls off at the rate of 12 dB/octave and the asymptotic phase reaches -180° while the gain of the return ratio is still larger than unity. The amplifier is therefore unstable.

We remark that in applying the formulas developed in the preceding sections, we implicitly assume that the transfer functions are devoid of zeros in the open RHS. Such transfer functions are termed the *minimum-phase transfer functions*. In

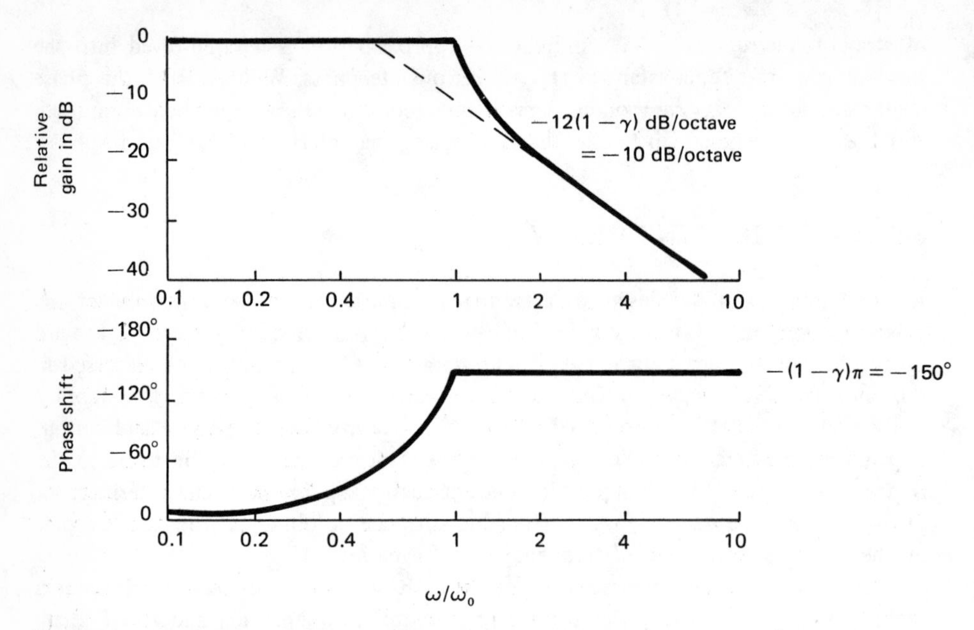

Figure 6.27 The gain and phase plots of Bode's ideal cutoff characteristic for $\gamma = \frac{1}{6}$.

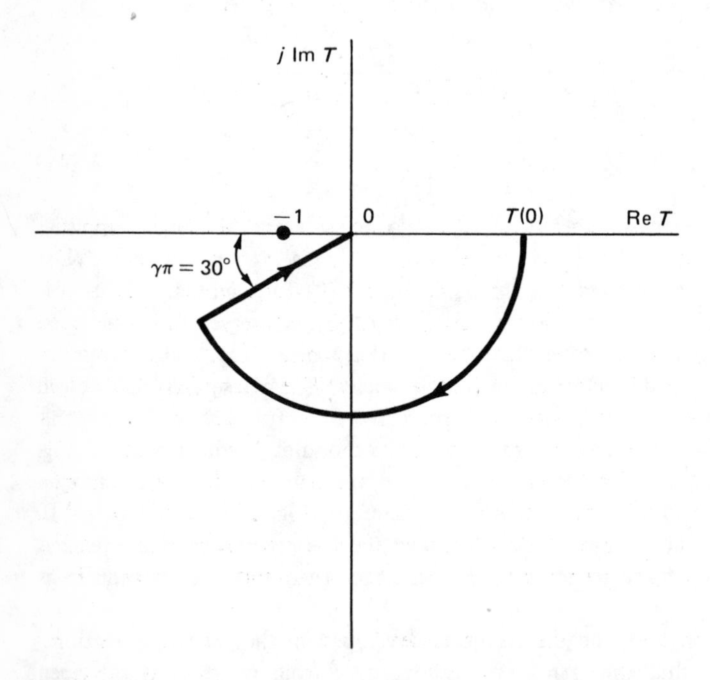

Figure 6.28 The Nyquist plot of Bode's ideal cutoff characteristic for $\gamma = \frac{1}{6}$.

feedback amplifiers, non-minimum-phase transfer functions are seldom used because excess phase shift is never desired for stability reasons. We almost always use the minimum-phase feedback and interstage networks.

6.8.1 Single-Loop Absolutely Stable Transistor Feedback Amplifiers

As demonstrated in Sec. 4.2, if a one-pole approximation is used for a common-emitter stage, its logarithmic gain falls off ultimately at the rate of at least 6 dB/octave. If n such stages are connected in cascade, the gain cuts off at a rate of $6n$ dB/octave. For broadband feedback amplifier design, we should include the nondominant pole neglected in the simplified equivalent network obtained by unilateralizing the hybrid-pi model used for the transistors. These nondominant effects are important in stability study, because they introduce an additional phase shift at low frequencies. In general, the return ratio $T(s)$ is a rational function, being the ratio of two polynomials, having m finite zeros and n finite poles. From physical considerations, $m < n$. At high frequencies, the return ratio behaves as s^{m-n} and cuts off at a rate of $-6(n-m)$ dB/octave. The asymptote with this slope in the Bode plot is called the *final asymptote.* The corresponding phase reaches $-(n-m)\pi/2$ rad asymptotically. Therefore, in practice the return ratio $T(j\omega)$ can be shaped to follow Bode's ideal characteristic only up to a certain frequency.

Consider a typical set of Bode plots as shown in Fig. 6.29, where ω_g is the gain-crossover frequency and ω_p is the phase-crossover frequency. Since $\omega_p < \omega_g$, the amplifier is unstable, and we must reduce the gain and control the phase shift by introducing the corrective networks.

Suppose that the desired shape of the return ratio $T(j\omega)$ is Bode's ideal cutoff characteristic of Fig. 6.27. This ideal cutoff characteristic is superimposed on the gain characteristic of Fig. 6.29, and the resulting curves are shown in Fig. 6.30. The shaded area is therefore the loss characteristic that should be introduced by the shaping networks. At high frequencies, the overall gain characteristic is still governed by the high-frequency asymptotic behavior of the gain curve of Fig. 6.29. Therefore, the final asymptote has a slope of $-6k$ dB/octave, as indicated in Fig. 6.30, at the frequency ω_c. To obtain the corresponding phase for this modified gain characteristic, denoted by the bottom curves in each part of Fig. 6.30, we simply add the phase contributions resulting from the ideal cutoff characteristic and a semi-infinite constant slope equal to

$$-6k + 12(1-\gamma) \text{ dB/octave} \tag{6.202}$$

which begins at the frequency ω_c. This added semi-infinite slope modifies the total phase to the shape as shown in Fig. 6.30. Observe that the much-celebrated constant-phase characteristic of the ideal cutoff characteristic has been completely destroyed. Furthermore, the plots indicate that the amplifier is unstable because of the additional phase contribution at low frequencies by the added semi-infinite slope. In the following, we present Bode's theory for obtaining a stable characteristic.

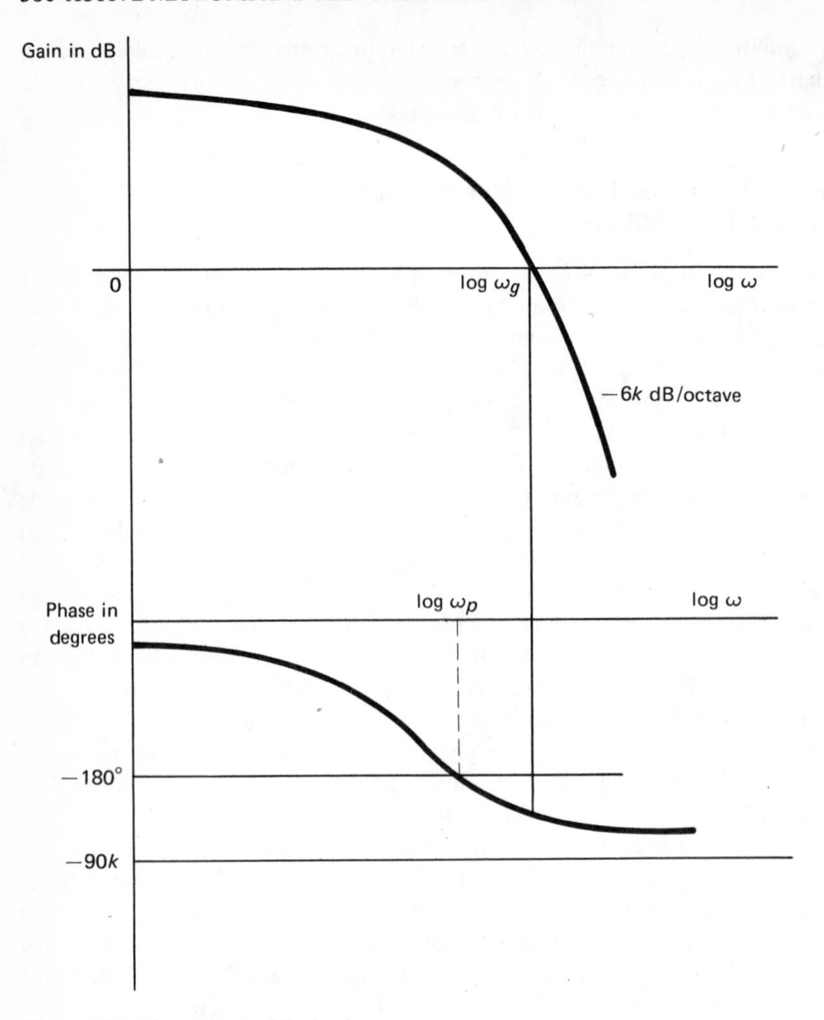

Figure 6.29 A typical set of Bode plots.

Bode's approach is by introducing a horizontal step in the gain characteristic as shown in Fig. 6.31*a*. The horizontal step is located below the 0-dB level by an amount equal to the desired gain margin G_m. The characteristic of Fig. 6.31*a* can be represented as the sum of three gain characteristics: the ideal cutoff characteristic with an asymptotic slope of $-12(1-\gamma)$ dB/octave and two semi-infinite constant slopes, one of which starts at the frequency ω_c with a slope $-6k$ dB/octave, and the other of which starts at ω_b with a slope $12(1-\gamma)$ dB/octave. The results are presented in Fig. 6.31*b*. From Eq. (6.200*a*), we see that, at low frequencies, the phase shift $B_b(\omega)$ of the semi-infinite constant slope equal to $12(1-\gamma)$ dB/octave is given by

$$B_b(\omega) \approx \frac{4(1-\gamma)\omega}{\pi\omega_b} \qquad 0 \leqslant \omega \leqslant 0.414\omega_b \tag{6.203}$$

Note that the slope used in formula (6.200*a*) is stated in terms of k nepers per log radian, which is equivalent to $6k$ dB per octave or $20k$ dB per decade. Likewise, the phase shift $B_c(\omega)$ of the semi-infinite constant slope equal to $-6k$ dB/octave is obtained as

$$B_c(\omega) \approx -\frac{2k\omega}{\pi\omega_c} \qquad 0 \leqslant \omega \leqslant 0.414\omega_c \tag{6.204}$$

To obtain the phase characteristic that follows the shape of Bode's ideal cutoff characteristic at low frequencies, we choose ω_b so that

$$B_b(\omega) + B_c(\omega) = 0 \tag{6.205}$$

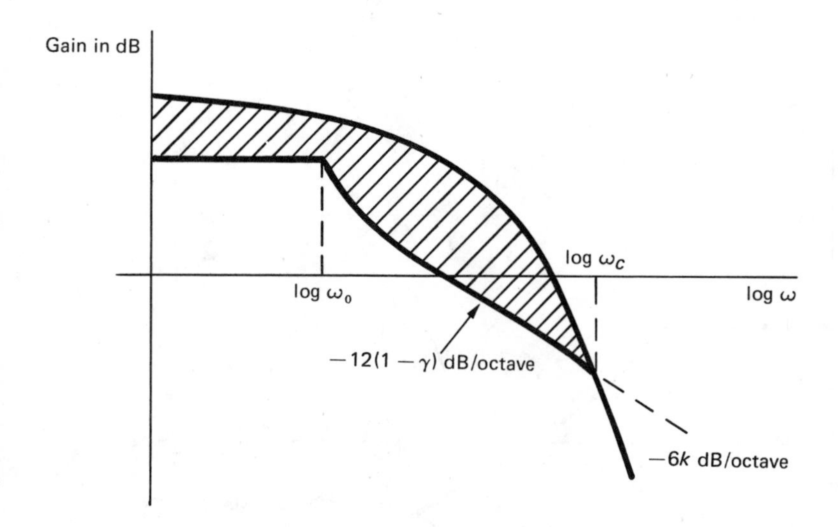

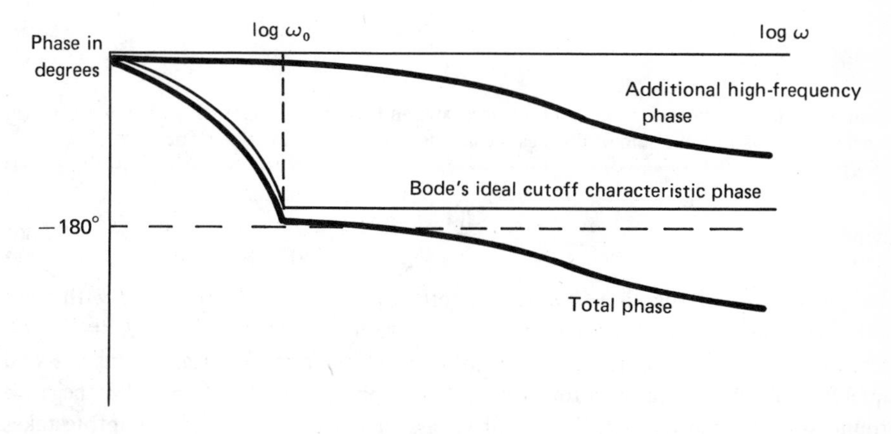

Figure 6.30 The superposition of the gain and phase plots of Bode's ideal cutoff characteristic and the Bode plots of Fig. 6.29.

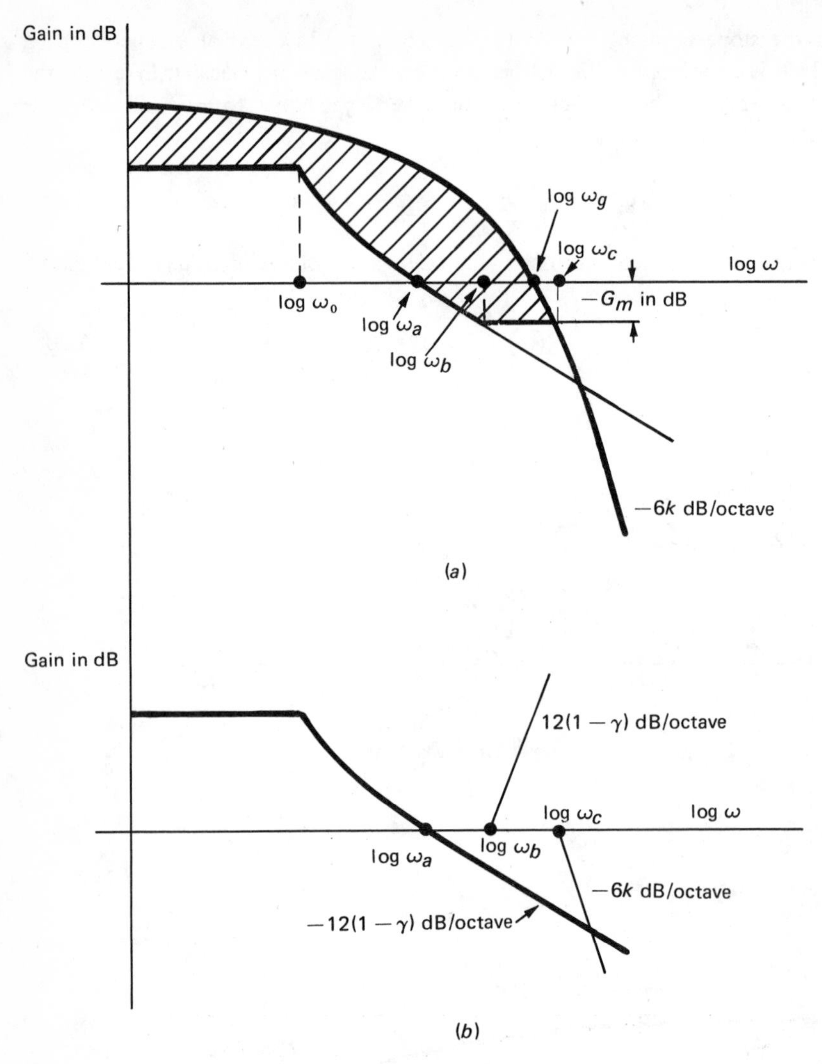

Figure 6.31 (*a*) The introduction of a horizontal step in the gain characteristic of Fig. 6.30 and (*b*) its decomposition into the sum of three gain characteristics of various asymptotic slopes.

giving

$$\omega_b = \frac{2(1-\gamma)\omega_c}{k} \tag{6.206}$$

The phase plots of the three gain characteristics of Fig. 6.31*b*, together with their sum, are presented in Fig. 6.32 for $k = 3$ and $\gamma = \frac{1}{6}$. In this case, $\omega_b = 5\omega_c/9$. Observe that at low frequencies the total phase shift is practically constant and equal to that of the ideal cutoff characteristic phase, and it becomes less negative around ω_b. At high frequencies, the phase associated with the final asymptote takes over and dominates the others. The gain characteristic of Fig. 6.31*a* is referred to as *Bode's ideal loop gain characteristic.*

To determine the maximum amount of feedback within the useful frequency band and with preassigned gain margin G_m and phase margin $\gamma\pi$, we set

$$20 \log |T(j\omega_b)| = 20 \log \left| \frac{T(0)}{(\sqrt{1 - \omega_b^2/\omega_0^2} + j\omega_b/\omega_0)^{2(1-\gamma)}} \right| = -G_m \tag{6.207}$$

giving

$$10^{G_m/20} |T(0)| = \left| \sqrt{\frac{\omega_b^2}{\omega_0^2} - 1} + \frac{\omega_b}{\omega_0} \right|^{2(1-\gamma)} \tag{6.208}$$

Since $\omega_b \gg \omega_0$, Eq. (6.208) can be simplified. In terms of decibels, we obtain the maximum amount of feedback as

$$20 \log |T(0)| \approx 40(1-\gamma) \log \left(\frac{2\omega_b}{\omega_0} \right) - G_m \tag{6.209}$$

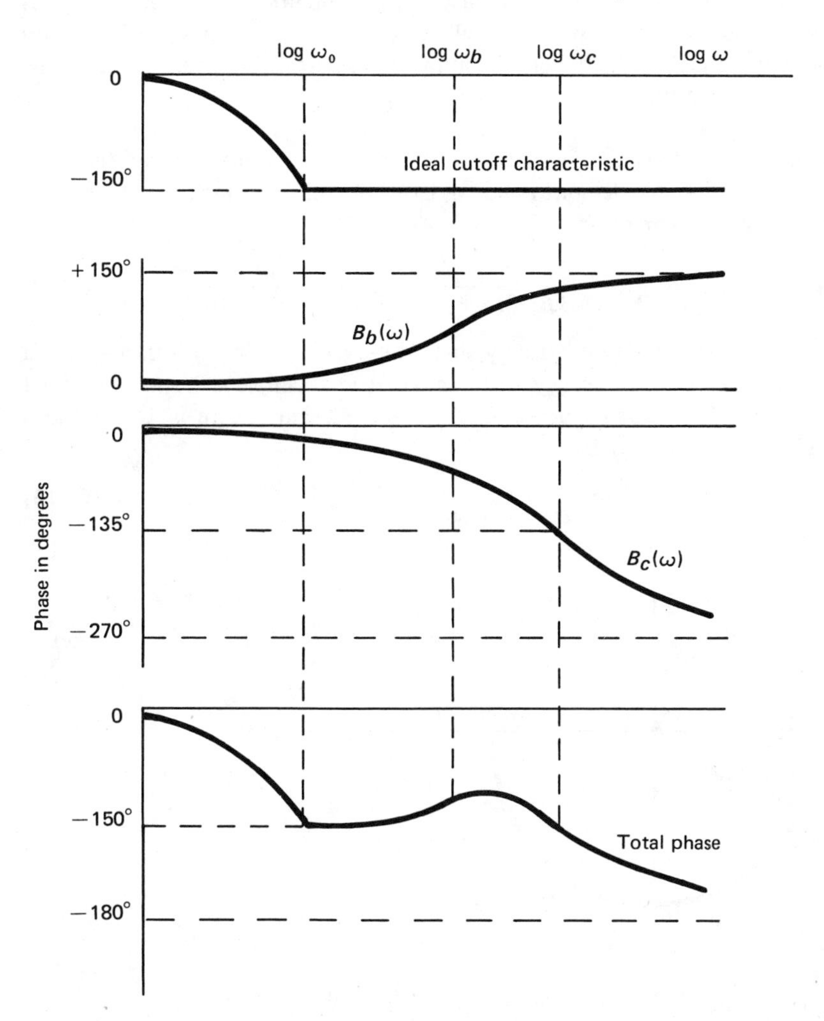

Figure 6.32 The phase plots of the three gain characteristics of Fig. 6.31*b*, together with their sum.

It is useful to express this feedback in terms of the gain-crossover frequency ω_g, the ideal cutoff band-edge frequency ω_0, and the specified gain and phase margins G_m and $\gamma\pi$. From Fig. 6.31a we find that ω_c and ω_g are related by

$$-20k(\log \omega_c - \log \omega_g) = -G_m \tag{6.210}$$

yielding

$$\omega_c = \omega_g \cdot 10^{G_m/20k} \tag{6.211}$$

By using this in conjunction with (6.206), the maximum amount of feedback is found to be (see Prob. 6.18)

$$20 \log |T(0)| = 40(1-\gamma) \log \frac{4(1-\gamma)\omega_g}{k\omega_0} + \frac{2(1-\gamma)G_m}{k} - G_m \tag{6.212}$$

As can be seen from Figs. 6.31a and 6.32, the modified gain characteristic is clearly stable. The maximum amount of allowed feedback is given by (6.212) for a prescribed gain margin G_m. For illustrative purposes, we also present the corresponding Nyquist plot in Fig. 6.33.

Example 6.12 We again consider the three-stage common-emitter feedback amplifier of Fig. 6.9. As demonstrated in Example 6.2, the return ratio can be approximated by the expression

$$T(j\lambda) = \frac{T(0)}{(1 + j\lambda/3.633)(1 + j\lambda/2.076)(1 + j\lambda/2.770)} \tag{6.213}$$

where $\lambda = \omega/10^7$. The function has only three finite poles, giving $m = 0$, $n = 3$, and $k = 3$. The gain-crossover frequency was computed in Example 6.2 and was found to be $\omega_g = 37.5 \cdot 10^6$ rad/s. Suppose that the desired gain margin is 10 dB and a phase margin is about 30°, giving

$$G_m = 10 \text{ dB} \qquad \gamma = \tfrac{30}{180} = \tfrac{1}{6} \tag{6.214}$$

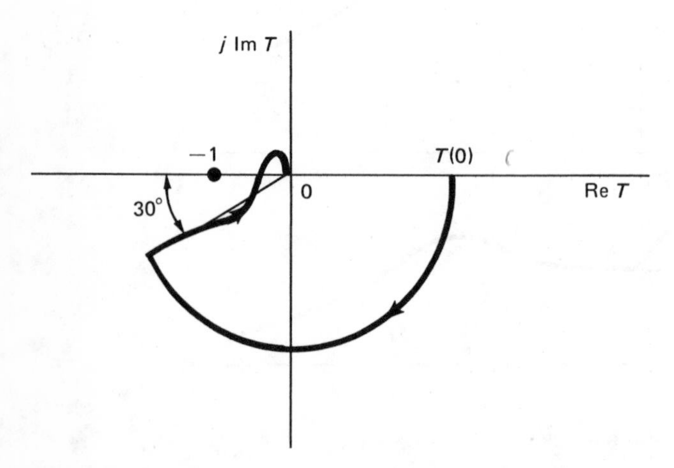

Figure 6.33 The Nyquist plot of the modified gain characteristic of Fig. 6.31a.

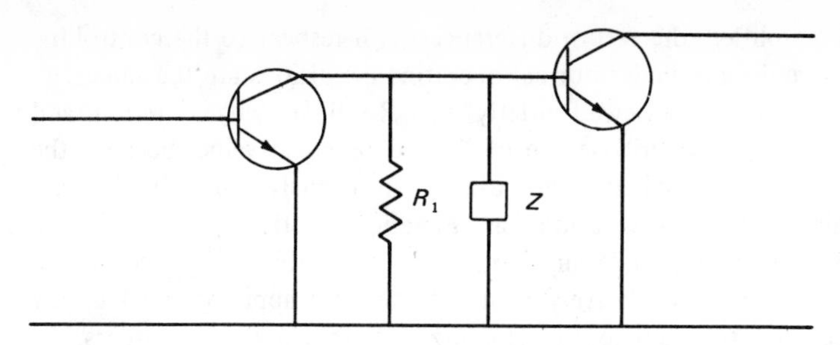

Figure 6.34 The location of a one-port corrective network.

Assume, for simplicity, that $\omega_0 = 10^6$ rad/s. Substituting these in (6.212), we obtain the maximum amount of allowed feedback:

$$20 \log |T(0)| = 54 + 5.56 - 10 = 49.56 \text{ dB} \tag{6.215}$$

6.8.2 Corrective Networks

Once the loss characteristic of the interstage shaping network is known, the desired corrective network can be synthesized. A widely used technique in practice is to employ a one-port impedance Z in one of the inter-stages, as shown in Fig. 6.34, because shaping in the overall feedback path will affect the overall transfer function. A typical one-port corrective network can be realized by a series tuned *RLC* network as illustrated in Fig. 6.35. For more complicated feedback amplifiers, more elaborate one-port networks are required. The design of these networks would, of course, differ from amplifier to amplifier. We shall terminate this discussion rather abruptly at this point, because a more detailed discussion would take us far afield into network synthesis, the insertion-loss synthesis in particular.

6.9 SUMMARY

We began this chapter by introducing the concept of a single-loop feedback amplifier and the notions of absolute stability and conditional stability. In a

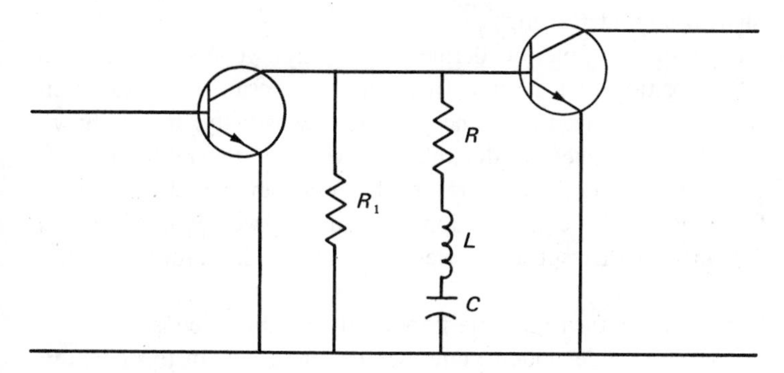

Figure 6.35 A typical one-port corrective network.

single-loop feedback amplifier, the return differences with respect to the controlling parameters of the active devices under normal operating conditions are the same.

For stability analysis, we reviewed briefly the Routh-Hurwitz criterion and Liénard-Chipart criterion. These criteria are of limited practical value, because the network determinant or its equivalent is not usually given analytically. In the case where the polynomial under consideration has numerical coefficients, the Routh-Hurwitz criterion can be put in Guillemin form in terms of the Stieltjes continued fraction. The Guillemin form is relatively simple to test. In applying the Hurwitz criterion, we are required to evaluate the Hurwitz determinants of all orders. To reduce this number, we presented the Liénard-Chipart criterion, which requires that one evaluate only about half of the Hurwitz determinants, every other one.

To study system stability when the network determinant is not known analytically, we described in some detail the Nyquist criterion. The Nyquist criterion is important in that it provides a quantitative measure of the degree of stability of a feedback amplifier. Furthermore, the parasitic effects, which play an important role in stability study, can also be included. To provide a measure of the maximum amount of feedback permitted to ensure amplifier stability, we introduced the terms of gain margin and phase margin. The gain margin is the additional amount of feedback in decibels that is permitted before the amplifier becomes unstable. Likewise, the phase margin is the additional phase lagging that is permitted before the amplifier becomes unstable. Thus, they are margins of safety against instability and are the generally accepted measures of the degree of stability of a feedback amplifier.

By changing the amount of feedback, we change the locations of the natural frequencies of a feedback amplifier. It would be extremely useful if the locus of the natural frequencies were plotted as a function of the midband value of the return ratio. For this we discussed in detail the root-locus technique. By knowing the trajectories of the natural frequencies, the designer has a considerable amount of insight and understanding of the stability, time-, and frequency-domain behaviors of the network. The construction of the root locus is relatively simple if the return ratio has only two poles. The procedure becomes more complicated if it has three or more poles. To this end, we developed simple rules that enable us readily to sketch the root locus. Very often, a rough sketch will provide sufficient information for a preliminary design to be undertaken.

In addition to the sensitivity function defined previously, we find it useful to know the change in the position of a natural frequency as a result of the variation of an element. For this we introduced the concept of root sensitivity. It was shown that the root sensitivity with respect to the midband value of the return ratio is equal to the residue of the reciprocal of the return difference evaluated at the pole that is the natural frequency under consideration. In addition, we derived formulas relating the root sensitivities with respect to poles and zeros of the return ratio and its midband value.

The gain and phase of the return ratio are important in stability considerations. However, they are not entirely independent; one dictates the other. In this chapter, we studied a number of relations between the real and imaginary parts of a network function and its natural logarithm. We showed that if a network function is devoid of poles in the closed RHS, then its imaginary part is completely determined by the behavior of its real part on the $j\omega$-axis. Conversely, if the imaginary part is specified for all ω, its real part is completely determined within an additive constant. In terms of the natural logarithm of a network function, the restrictions are that the network function be devoid of poles and zeros in the open RHS. Under these constraints, the phase is completely determined by the behavior of the gain on the $j\omega$-axis, or the gain is completely determined within an additive constant by the behavior of the phase on the $j\omega$-axis. We also derived a useful formula relating the phase at a particular frequency in terms of the gain specified at all frequencies known as the gain-phase theorem.

Finally, we discussed Bode's design theory by introducing Bode's ideal cutoff characteristic. Specifically, we demonstrated how to modify a gain characteristic to follow Bode's ideal cutoff characteristic up to certain frequencies for the preassigned gain and phase margins. A formula giving the maximum amount of feedback within the useful frequency band was derived in terms of the gain-crossover frequency, the ideal cutoff band-edge frequency, and the prescribed gain and phase margins.

PROBLEMS

6.1 Identify the polynomials whose roots are restricted to the open LHS:

(a) $s^4 + 3s^3 + 6s^2 + 3s + 2$
(b) $2s^4 + 3s^3 + 3s^2 + 3s + 3$
(c) $s^5 + 7s^4 + 14s^3 + 15s^2 + 15s + 9$
(d) $s^5 + 10s^4 + 16s^3 + 16s^2 + 14s + 9$
(e) $s^7 + 2s^6 + 6s^5 + 14s^4 + 11s^3 + 18s^2 + 6s + 6$

6.2 A practical version of a FET Hartley oscillator is shown in Fig. 6.36. Assume that the following parameters are used for the oscillator:

$$L_1 = 2 \text{ mH} \qquad L_2 = 2 \text{ mH} \qquad M = 0.5 \text{ mH}$$
$$C = 200 \text{ pF} \qquad r_d = 50 \text{ k}\Omega \tag{6.216}$$

ignoring the capacitances of the FET. Determine the frequency of oscillation and the required value of transconductance g_m.

6.3 A transistor Colpitts oscillator is shown in Fig. 6.37. Determine the frequency of oscillation and the minimum value of the transistor forward current gain h_{fe} for the following network parameters:

$$C_1 = 0.01 \ \mu\text{F} \qquad C_2 = 0.02 \ \mu\text{F}$$
$$L = 80 \ \mu\text{H} \qquad h_{ie} = 1.1 \text{ k}\Omega \tag{6.217}$$
$$h_{oe} = h_{re} = 0$$

6.4 Show that the angles of arrival at a zero z_j of the return ratio $T(s)$ on the root locus are given by (6.71).

6.5 Use the Nyquist criterion to study the stability of feedback amplifiers having the following return ratios:

(a)
$$T(s) = \frac{s+2}{(s+1)(s+3)} \tag{6.218a}$$

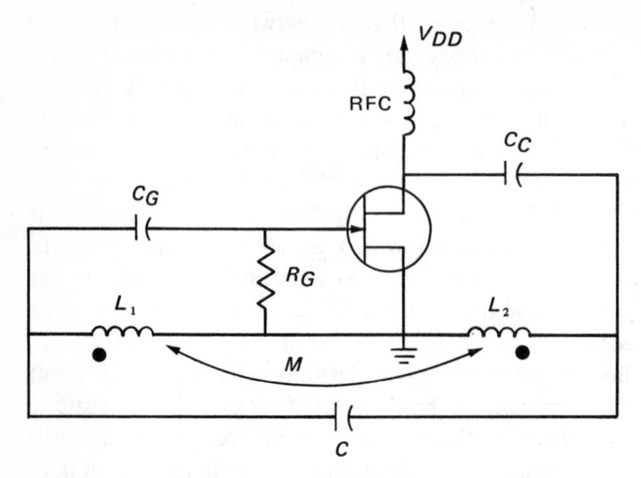

Figure 6.36 A practical version of a FET Hartley oscillator.

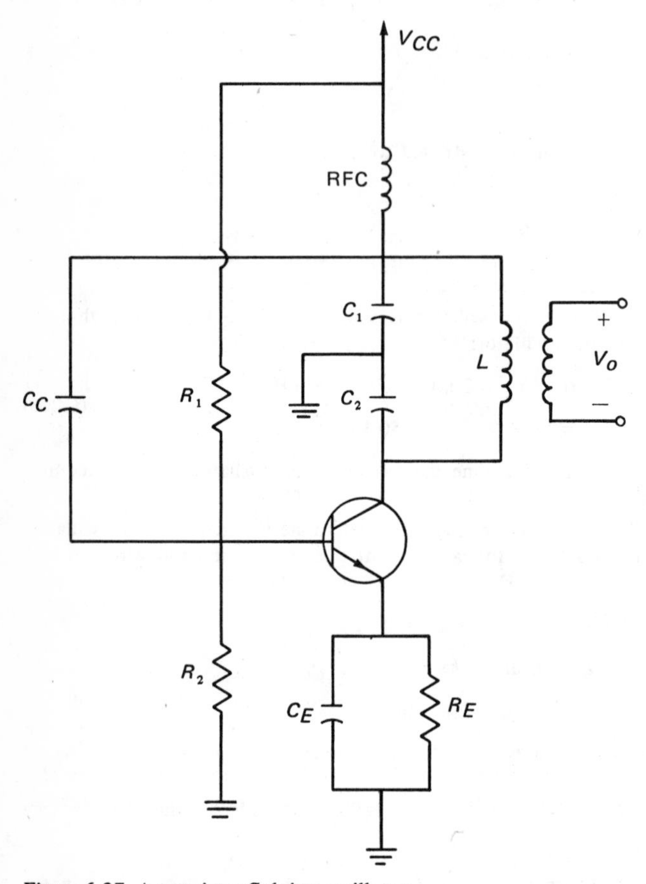

Figure 6.37 A transistor Colpitts oscillator.

(*b*)
$$T(s) = \frac{s^2 + 2s + 2}{s^2(s^2 + s + 2)} \tag{6.218b}$$

(*c*)
$$T(s) = \frac{s^2 + 3s + 2}{s^3 + 2s^2 + s + 1} \tag{6.218c}$$

6.6 Repeat Example 6.2 for the following set of network parameters:

$$g_m = 0.3 \text{ mho} \quad r_x = 50\ \Omega \quad r_\pi = 125\ \Omega$$
$$r_o = \infty \quad C_\pi = 200 \text{ pF} \quad C_\mu = 6 \text{ pF} \tag{6.219}$$
$$R_1 = R_2 = 50\ \Omega \quad R_3 = R_4 = 1\text{ k}\Omega$$

6.7 Consider the three-stage common-emitter feedback amplifier of Fig. 6.17. As indicated in Sec. 6.5.2, the return ratio has the general form shown in Eq. (6.103). Let $p_1 = 1$, $p_2 = 2$, $p_3 = 3$, and $z_1 = 2.5$. Sketch the root locus for this return ratio function.

6.8 Sketch the root locus for the following return ratio functions and determine the range of the values of the midband return ratio $T(0)$ for which the feedback amplifier is stable:

(*a*)
$$T(s) = \frac{4T(0)(s+1)}{(s+2)(s^2 + 2s + 2)} \tag{6.220a}$$

(*b*)
$$T(s) = \frac{2T(0)}{(s+1)^2(s+2)} \tag{6.220b}$$

(*c*)
$$T(s) = \frac{1.5T(0)(s^2 + 2s + 2)}{(s+1)(s^2 + s + 3)} \tag{6.220c}$$

(*d*)
$$T(s) = \frac{2T(0)(s+2)}{(s^2 + s + 1)(s^2 + 4s + 4)} \tag{6.220d}$$

6.9 Calculate the root sensitivities of the return ratio functions, as given in (6.220), with respect to the zeros and poles of $T(s)$ and also with respect to the midband value $T(0)$ of the return ratios.

6.10 Sketch the root locus for a negative feedback amplifier having the return ratio

$$T(s) = \frac{3T(0)(s+2)}{(s+1)(s+3)(s^2 + 2s + 2)} \tag{6.221}$$

6.11 Derive formula (6.140).

6.12 By integrating the function $[\ln w(s)]/(s^2 + \omega_a^2)$ around the basic contour with small semicircular indentations to the right at $s = \pm j\omega_a$, demonstrate that (6.180) can be obtained in this way.

6.13 By integrating the function $s[w(s) - R(\infty)]/(s^2 + \omega_a^2)$ around the basic contour with small semicircular indentations to the right at $s = \pm j\omega_a$, show that

$$R(\omega) = R(\infty) - \frac{2}{\pi}\int_0^\infty \frac{\omega X(\omega)}{\omega^2 - \omega_a^2}\, d\omega \tag{6.222}$$

where, as before, $w(j\omega) = R(\omega) + jX(\omega)$.

6.14 Demonstrate that the gain-integral theorem (6.176) can be derived by integrating the function $\ln [w(s)/w(\infty)]$ around the basic contour composed of the $j\omega$-axis and the infinite semicircle to the right.

6.15 By using the hybrid-pi model for the transistor in Fig. 6.38 with parameters as specified in Eq. (6.81), investigate the effect of varying the capacitance C_f on the natural frequencies of the amplifier.

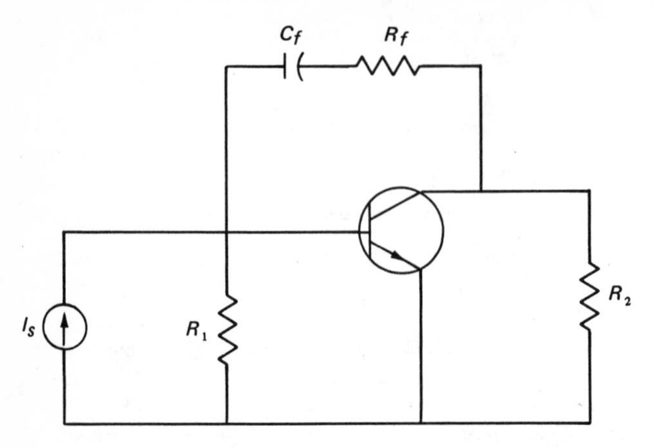

Figure 6.38 A given feedback amplifier.

6.16 A three-stage common-emitter feedback amplifier has the return ratio

$$T(s) = \frac{T(0)}{(1 + s/\sigma_1)(1 + s/\sigma_2)(1 + s/\sigma_3)} \tag{6.223}$$

where σ_i $(i = 1, 2, 3)$ are real and positive. Determine the midband value $T(0)$ of the return ratio under which the amplifier is on the verge of instability.

6.17 In (6.195), demonstrate that

$$\lim_{x \to 0} \ln x \ln y = 0 \tag{6.224}$$

where y is defined in (6.193).

6.18 Assuming that γ is small when compared with unity, we can expand $\log(1-\gamma)$ in a power series and ignore powers of γ higher than the first. Show that (6.212) can be approximated by the expression

$$A_m - A = (A_m + 17.4)\gamma + \frac{k-2}{k} G_m + \frac{2}{k}\gamma G_m \tag{6.225}$$

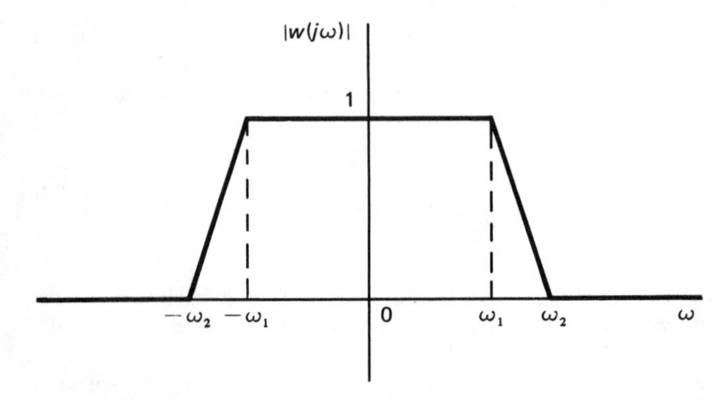

Figure 6.39 The prescribed magnitude of a minimum-phase transfer function.

where $A = 20 \log |T(0)|$ and

$$A_m = A\Big|_{\substack{G_m=0\\ \gamma=0}} \tag{6.226}$$

6.19 Demonstrate that the phase-integral theorem (6.175) can be derived by integrating the function $[\ln w(s)]/s$ around the basic contour with a small semicircular indentation to the right at the origin.

6.20 Assume that the magnitude of a transfer function is given by the nth-order maximally flat response

$$|w(j\omega)| = (1 + \omega^{2n})^{-1/2} \tag{6.227}$$

Show that the phase of the function is given by

$$B(\omega) = -\frac{n}{\pi}\int_0^{\infty} u^{-1}(1 + u^{-2n})^{-1} \ln\left|\frac{u+\omega}{u-\omega}\right| du \tag{6.228}$$

6.21 The magnitude of a minimum-phase transfer function is sketched in Fig. 6.39. By making appropriate approximations, sketch the corresponding phase as a function of ω.

6.22 As discussed in Sec. 4.2.1, Fig. 4.7 is a current-series feedback amplifier. By using the hybrid-pi model of Fig. 4.10 for the transistor, with

$$\begin{aligned} g_m &= 0.4 \text{ mho} \qquad r_x = 50\ \Omega \qquad r_\pi = 100\ \Omega \\ C_\pi &= 200 \text{ pF} \qquad C_\mu = 5 \text{ pF} \qquad r_o = 50 \text{ k}\Omega \\ R_1 &= 500\ \Omega \qquad R_2 = 5 \text{ k}\Omega \qquad R_e = 50\ \Omega \end{aligned} \tag{6.229}$$

calculate the return ratio with respect to the controlling parameter g_m over a sufficiently large range of frequencies, so that the return ratio locus can be plotted. By using the Nyquist criterion, determine the stability of this feedback amplifier.

6.23 Sketch the root locus for the return ratio considered in Prob. 6.22, assuming that its midband value can be varied independently.

6.24 The closed-loop transfer function of a feedback amplifier is written as

$$w(s) = K\frac{\prod_{i=1}^{m}(s + z_i)}{\prod_{j=1}^{n}(s + s_j)} \tag{6.230}$$

Assume that K, z_i, and s_j are functions of some parameter x. Show that

$$S_x^w = S_x^K + \sum_{i=1}^{m}\frac{\mathcal{S}(z_i; x)}{s + z_i} - \sum_{j=1}^{n}\frac{\mathcal{S}(s_j; x)}{s + s_j} \tag{6.231}$$

where S_x^w and S_x^K denote the sensitivity functions of $w(s)$ and K with respect to x, respectively.

BIBLIOGRAPHY

Barnett, S. and D. D. Šiljak: Routh's Algorithm: A Centennial Survey, *SIAM Review*, vol. 19, no. 3, pp. 472–489, 1977.

Blecher, F. H.: Design Principles for Single Loop Transistor Feedback Amplifiers, *IRE Trans. Circuit Theory*, vol. CT-4, no. 3, pp. 145–156, 1957.

Bode, H. W.: "Network Analysis and Feedback Amplifier Design," Princeton, N.J.: Van Nostrand, 1945.

Boylestad, R. and L. Nashelsky: "Electronic Devices and Circuit Theory," 2d ed., Englewood Cliffs, N.J.: Prentice-Hall, 1978.

Chen, C. F. and C. Hsu: The Determination of Root Loci Using Routh's Algorithm, *J. Franklin Inst.*, vol. 281, no. 2, pp. 114–121, 1966.

Chen, C. F. and L. S. Shieh: Continued Fraction Inversion by Routh's Algorithm, *IEEE Trans. Circuit Theory*, vol. CT-16, no. 2, pp. 197–202, 1969.

Chen, W. K.: Indefinite-Admittance Matrix Formulation of Feedback Amplifier Theory, *IEEE Trans. Circuits and Systems*, vol. CAS-23, no. 8, pp. 498–505, 1976.

Cherry, E. M.: A New Result in Negative-Feedback Theory, and Its Application to Audio Power Amplifiers, *Int. J. Circuit Theory and Applications*, vol. 6, no. 3, pp. 265–288, 1978.

Evans, W. R.: Graphical Analysis of Control Systems, *AIEE Trans.*, vol. 67, part II, pp. 547–551, 1948.

Evans, W. R.: "Control System Dynamics," New York: McGraw-Hill, 1954.

Fuller, A. T., ed.: "Stability of Motion," London: Taylor & Francis, 1975.

Gantmacher, F. R.: "The Theory of Matrices," vol. 2, New York: Chelsea, 1959.

Ghausi, M. S.: "Electronic Circuits," New York: Van Nostrand Reinhold, 1971.

Ghausi, M. S. and D. O. Pederson: A New Design Approach for Feedback Amplifiers, *IRE Trans. Circuit Theory*, vol. CT-8, no. 3, pp. 274–284, 1961.

Guillemin, E. A.: "The Mathematics of Circuit Analysis," Cambridge, Mass.: The MIT Press, 1949.

Haykin, S. S.: "Active Network Theory," Reading, Mass.: Addison-Wesley, 1970.

Horowitz, I. M.: "Synthesis of Feedback Systems," New York: Academic, 1963.

Hurwitz, A.: Über die Bedingungen, unter welchen eine Gleichung nur Wurzeln mit negativen reellen Teilen besitzt, *Math. Ann.*, vol. 46, pp. 273–284, 1895.

Krall, A. M.: An Extension and Proof of the Root-Locus Method, *J. Soc. Indust. Appl. Math.*, vol. 9, no. 4, pp. 644–653, 1961.

Kuh, E. S. and R. A. Rohrer: "Theory of Linear Active Networks," San Francisco, Calif.: Holden-Day, 1967.

Liénard, A. and M. H. Chipart: Sur la Signe de la Partie Réelle des Racines d'une Équation Algébrique, *J. Math. Pures Appl.*, vol. 10, pp. 291–346, 1914.

Mulligan, J. H.: Transient Response and the Stabilization of Feedback Amplifiers, *AIEE Trans.*, vol. 78, part II, pp. 495–503, 1960.

Nyquist, H.: Regeneration Theory, *Bell Syst. Tech. J.*, vol. 11, no. 1, pp. 126–147, 1932.

Porter, B.: "Stability Criteria for Linear Dynamical Systems," New York: Academic, 1968.

Routh, E. J.: "A Treatise on the Stability of a Given State of Motion," London: Macmillan, 1877.

Routh, E. J.: "Advanced Part of a Treatise on Advanced Rigid Dynamics," Cambridge: Cambridge University Press, 1930.

Truxal, J. G.: "Automatic Feedback Control System Synthesis," New York: McGraw-Hill, 1955.

Truxal, J. G. and I. M. Horowitz: Sensitivity Considerations in Active Network Synthesis, *Proc. Second Midwest Symp. Circuit Theory, Michigan State University, East Lansing, Mich., Dec.*, pp. 6-1–6-11, 1956.

Ur, H.: Root Locus Properties and Sensitivity Relations in Control Systems, *IRE Trans. Automatic Control*, vol. AC-5, no. 1, pp. 57–65, 1960.

CHAPTER

SEVEN

MULTIPLE–LOOP FEEDBACK AMPLIFIERS

In the preceding three chapters, we studied the theory of single-loop feedback amplifiers. The concept of feedback was introduced in terms of return difference. We found that return difference plays an important role in the study of amplifier stability, its sensitivity to the variations of the parameters, and the determination of its transfer and driving-point impedances. The fact that return difference can be measured experimentally for many practical amplifiers indicates that we can include all the parasitic effects in the stability study, and that stability problem can be reduced to a Nyquist plot.

In this chapter, we study multiple-loop feedback amplifiers, which contain a multiplicity of inputs, outputs, and feedback loops. We first review briefly the rules of the matrix signal-flow graph, and then generalize the concept of return difference for a controlled source to the notion of return difference matrix for a multiplicity of controlled sources. For measurement situations, we introduce the null return difference matrix and discuss its physical significance. In particular, we show that the determinant of the overall transfer matrix can be expressed explicitly in terms of the determinants of the return difference and the null return difference matrices, thus generalizing Blackman's formula for the input impedance. This is followed by the derivations of the generalized feedback formulas and the formulation of the multiple-loop feedback theory in terms of the hybrid matrix. The problem of multiparameter sensitivity together with its relation to the return difference matrix is discussed. Finally, we develop formulas for computing multiparameter sensitivity functions.

7.1 MATRIX SIGNAL–FLOW GRAPHS

As demonstrated in Sec. 5.7, the feedback amplifier theory can be formulated equivalently in terms of a signal-flow graph representing the signal transmissions among the various ports. To formulate the multiple-loop feedback amplifier theory, we need the matrix version of the scalar signal-flow graph. In this section, we review briefly the matrix signal-flow graph and present some of its properties.

A scalar signal-flow graph is a graphic representation of a system of linear algebraic equations, which is put into the form

$$\sum_{j=0}^{n} a_{ij}x_j = x_i \qquad (i = 1, 2, \ldots, n) \tag{7.1}$$

where x_0 is the only excitation source and x_i are the dependent variables. The scalar signal-flow graph associated with the system (7.1) is a weighted directed graph in which the nodes correspond to the dependent variables x_i and the independent variable x_0 and the weights to the *transmittances* a_{ij}. The construction of the associated scalar signal-flow graph for the ith equation is shown in Fig. 7.1. The *graph transmission* w_{0k} from x_0 to x_k is defined by the equation $x_k = w_{0k}x_0$. Application of Mason's gain formula yields the graph transmission w_{0k} by inspection of the loops and paths of the graph.

In a similar fashion, a *matrix signal-flow graph* is a directed-graph representation of a system of linear algebraic matrix equations, which is put into the form

$$\sum_{j=0}^{n} \mathbf{A}_{ij}\mathbf{X}_j = \mathbf{X}_i \qquad (i = 1, 2, \ldots, n) \tag{7.2}$$

where $\mathbf{X}_0$ is the only source vector and $\mathbf{X}_i$ are the dependent vectors. The nodes of the directed graph correspond to the independent vector $\mathbf{X}_0$ and the dependent vectors $\mathbf{X}_i$ and the weights of the directed edges to the matrices $\mathbf{A}_{ij}$. To each node $\mathbf{X}_i$, there corresponds the matrix equation (7.2). The construction of the associated matrix signal-flow graph is as illustrated in Fig. 7.2 for the ith equation in Eq. (7.2). The *graph transmission* $\mathbf{W}_{0k}$ from the source node $\mathbf{X}_0$ to the output node $\mathbf{X}_k$ is defined by the matrix equation

$$\mathbf{X}_k = \mathbf{W}_{0k}\mathbf{X}_0 \tag{7.3}$$

In general, Mason's gain formula cannot be utilized to obtain the graph transmission $\mathbf{W}_{0k}$.

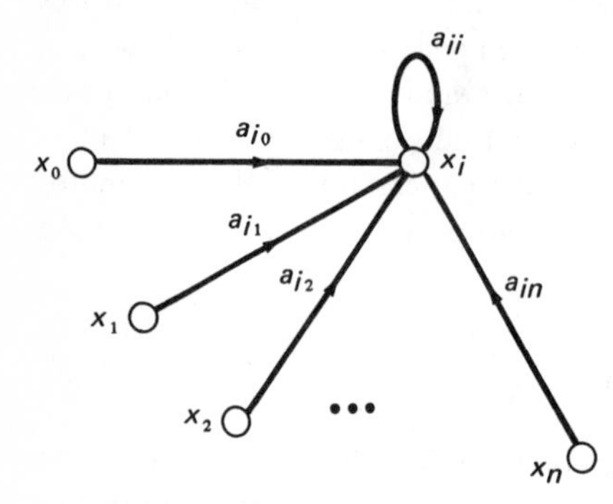

Figure 7.1 The scalar signal-flow graph for the ith equation of (7.1).

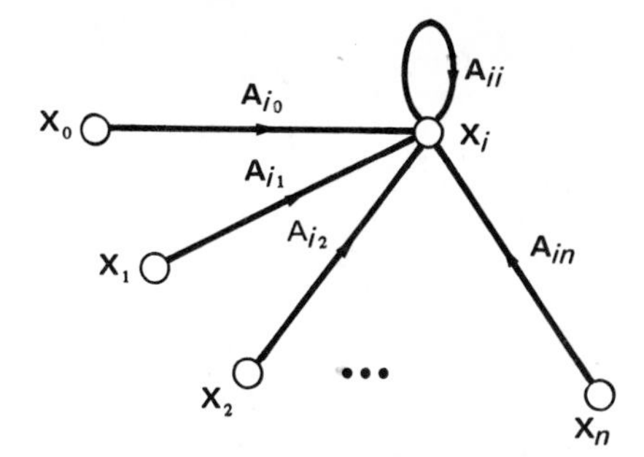

Figure 7.2 The matrix signal-flow graph for the ith equation of (7.2).

As an example of a matrix signal-flow graph, consider the following system of linear matrix equations:

$$\mathbf{X}_1 = \mathbf{A}\mathbf{X}_2 + \mathbf{B}\mathbf{X}_0 \tag{7.4a}$$

$$\mathbf{X}_3 = \mathbf{C}\mathbf{X}_2 + \mathbf{D}\mathbf{X}_0 \tag{7.4b}$$

$$\mathbf{X}_2 = \mathbf{E}\mathbf{X}_1 \tag{7.4c}$$

The associated matrix signal-flow graph is presented in Fig. 7.3.

One way to compute the graph transmission is to apply the basic reduction rules repeatedly until it can be written down directly by inspection. A direct approach is to develop a topological procedure to obtain the graph transmission without graph reduction. The best approach is probably a combination of both, so that the problem at hand can be solved in the most satisfying manner. In the following, we shall describe these two approaches.

7.1.1 Basic Reduction Rules

The basic reduction rules for a matrix signal-flow graph are given below.

1. *Series reduction.* As indicated in Fig. 7.4*a*, the directed edges with transmittances $\mathbf{A}_{ij}$ and $\mathbf{A}_{jk}$ are replaced by a single directed edge with transmittance $\mathbf{A}_{ij}\mathbf{A}_{jk}$.

2. *Parallel reduction.* As indicated in Fig. 7.4*b*, the directed edges with transmittances $\mathbf{A}_{ij}$ and $\mathbf{B}_{ij}$ are replaced by a single directed edge with transmittance $\mathbf{A}_{ij} + \mathbf{B}_{ij}$.

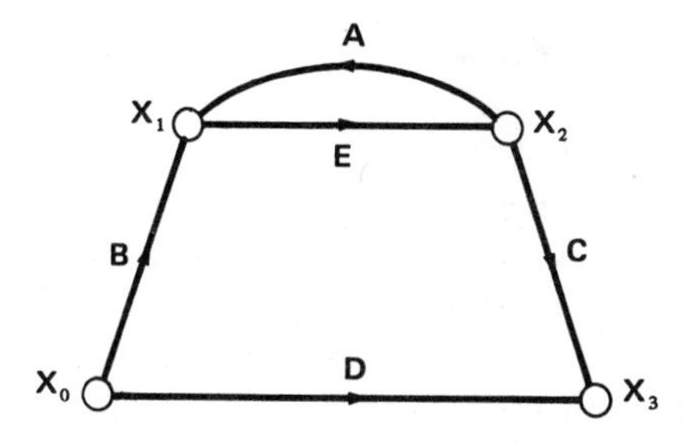

Figure 7.3 The matrix signal-flow graph associated with the system (7.4).

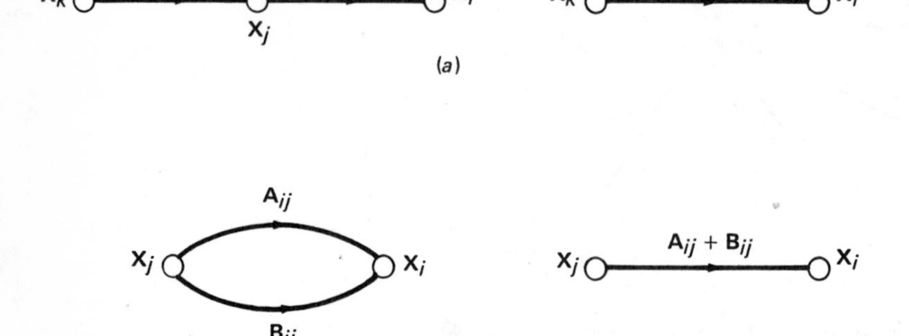

Figure 7.4 The basic matrix signal-flow graph reduction rules. (*a*) Series reduction and (*b*) parallel reduction.

3. *Removal of a self-loop.* The removal of a self-loop with transmittance $\mathbf{A}_{ii}$ at node $\mathbf{X}_i$ is equivalent to replacing the transmittances $\mathbf{A}_{ij}$ of the edges directed from nodes $\mathbf{X}_j$ (where $j \neq i$; $j = 0, 1, \ldots, n$) to node $\mathbf{X}_i$ by $(\mathbf{1} - \mathbf{A}_{ii})^{-1}\mathbf{A}_{ij}$, where $\mathbf{1}$, as before, denotes the identity matrix of appropriate order. To justify this operation, we observe that the ith equation of (7.2) can be rewritten as

$$\sum_{\substack{j=0 \\ j \neq i}}^{n} (\mathbf{1} - \mathbf{A}_{ii})^{-1}\mathbf{A}_{ij}\mathbf{X}_j = \mathbf{X}_i \tag{7.5}$$

and the result follows. The procedure is illustrated in Fig. 7.5.

4. *Absorption of a node.* To absorb a node X_k, we replace the transmittances $\mathbf{A}_{ij}$ of the edges directed from $\mathbf{X}_j$ to $\mathbf{X}_i$ by

$$\tilde{\mathbf{A}}_{ij} = \mathbf{A}_{ij} + \mathbf{A}_{ik}(\mathbf{1} - \mathbf{A}_{kk})^{-1}\mathbf{A}_{kj} \tag{7.6}$$

for all i and j. The operation is depicted in Fig. 7.6, and is equivalent to eliminating the vector $\mathbf{X}_k$ in (7.2). To see this, we replace i by k in (7.5) and substitute the resulting equation for $\mathbf{X}_k$ in (7.2), giving

$$\sum_{\substack{j=0 \\ j \neq k}}^{n} \tilde{\mathbf{A}}_{ij}\mathbf{X}_j = \mathbf{X}_i \qquad (i = 1, 2, \ldots, k-1, k+1, \ldots, n) \tag{7.7}$$

The operations described in Fig. 7.6 follow directly from Eq. (7.7).

As an illustration, we compute the graph transmission from node $\mathbf{X}_0$ to node $\mathbf{X}_3$ in the matrix flow graph of Fig. 7.3. To this end, we first remove node $\mathbf{X}_2$ by the procedure outlined in Fig. 7.6; the resulting matrix flow graph is shown in Fig. 7.7*a*. Repeating the process once more by removing node $\mathbf{X}_1$ in Fig. 7.7*a* yields the reduced matrix flow graph of Fig. 7.7*b*. The graph transmission from node $\mathbf{X}_0$ to node $\mathbf{X}_3$ is found to be

$$\mathbf{W}_{03} = \mathbf{D} + \mathbf{CE}(\mathbf{1} - \mathbf{AE})^{-1}\mathbf{B} \tag{7.8}$$

Alternatively, in Fig. 7.7*a* we can first remove the self-loop at node $\mathbf{X}_1$ by the procedure outlined in Fig. 7.5 and then apply the series and parallel reduction rules. The results are presented in Fig. 7.8, and we obtain, of course, the same graph transmission.

7.1.2 Topological Procedure

In this section, we describe a topological procedure that avoids the necessity of repeated graph reduction to obtain the desired graph transmission. To facilitate our discussion, we introduce several terms.

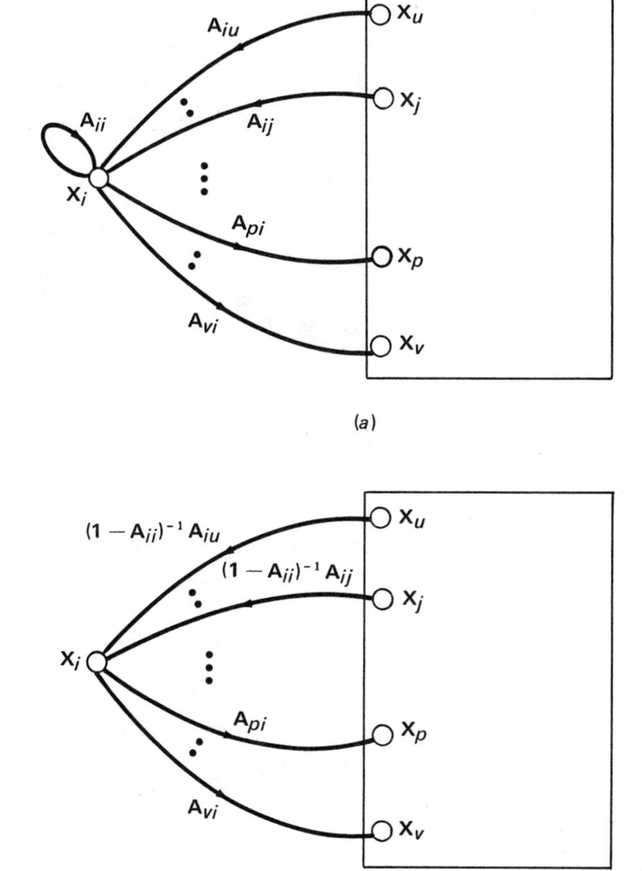

Figure 7.5 The removal of a self-loop in a matrix signal-flow graph. (*a*) Before the removal and (*b*) after the removal.

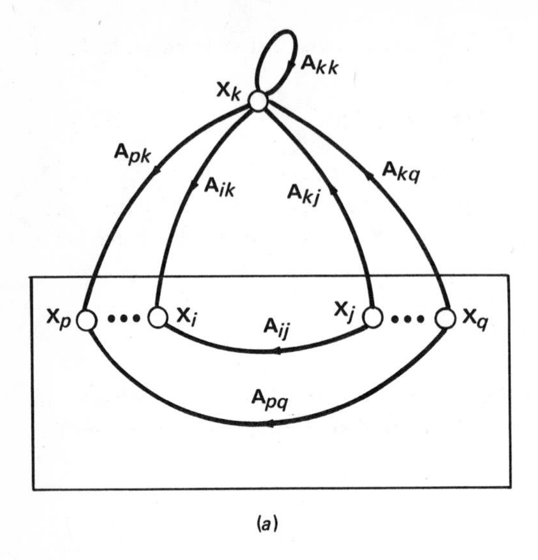

(a)

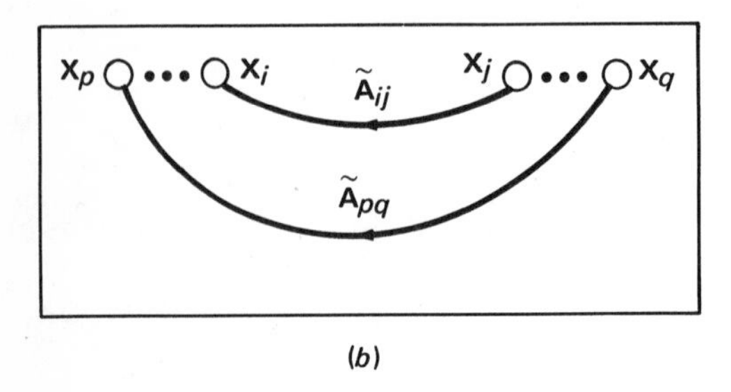

(b)

Figure 7.6 The absorption of a node in a matrix signal-flow graph. (*a*) Before the absorption and (*b*) after the absorption.

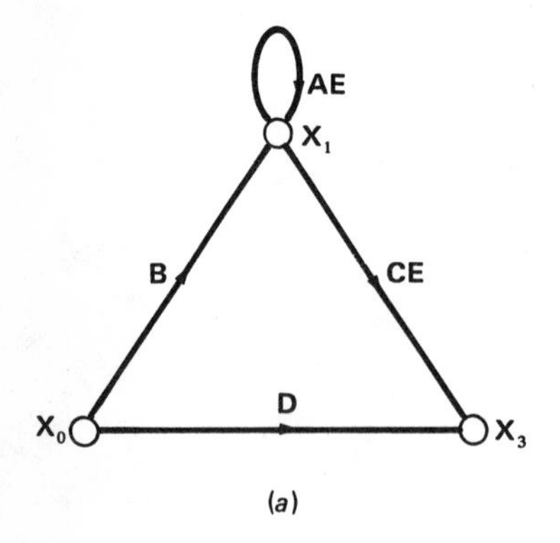

(a)

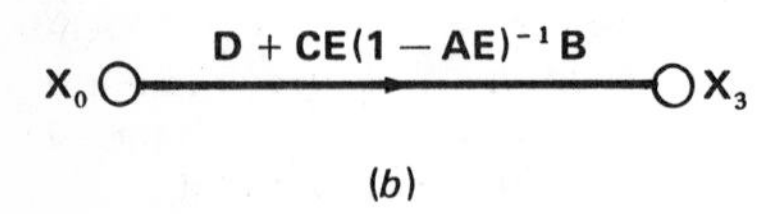

(b)

Figure 7.7 (*a*) The resulting matrix signal-flow graph after absorbing node $\mathbf{X}_2$ in that of Fig. 7.3 and (*b*) the resulting matrix signal-flow graph after absorbing node $\mathbf{X}_1$ in (*a*).

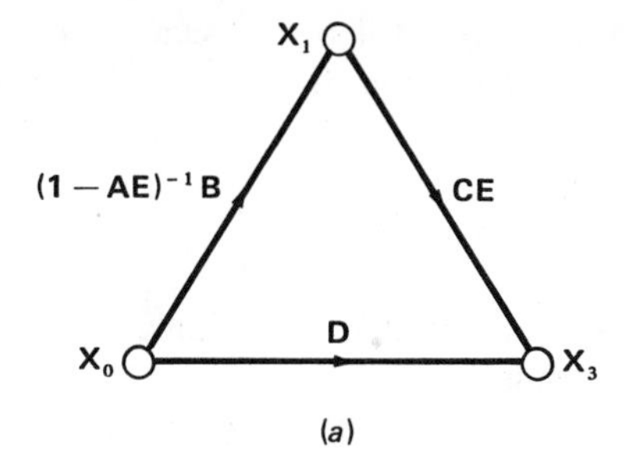

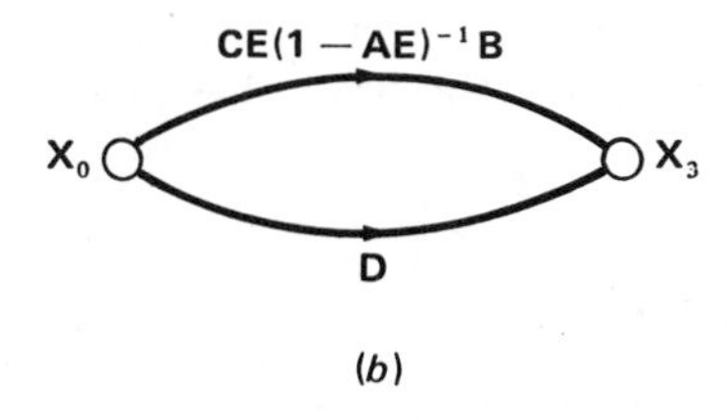

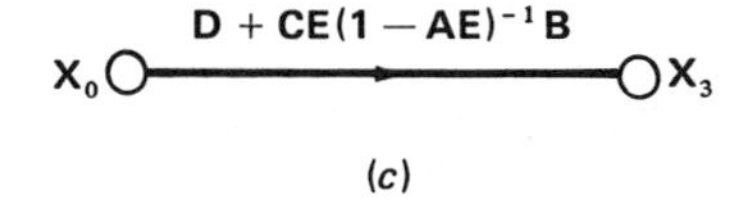

Figure 7.8 The reduction of the matrix signal-flow graph of Fig. 7.7*a*.

A node is said to be *split* when it is replaced by two nodes, a source node and a sink node, such that all edges terminating at the original node are made to terminate on the new sink node and all edges outgoing from the original node are made to originate at the new source. The *node transmission* $\mathbf{W}_i$ of a node $\mathbf{X}_i$ is the graph transmission from the source node to the sink node created by splitting node $\mathbf{X}_i$ after the original source node $\mathbf{X}_0$ is removed. The *directed path product* of a directed path from node $\mathbf{X}_i$ to node $\mathbf{X}_j$ is the product of the transmittances associated with the edges of the directed path written in the order from node $\mathbf{X}_j$ to node $\mathbf{X}_i$. Finally, the *node factor* of a node $\mathbf{X}_i$ on a directed path P is $(\mathbf{1} - \mathbf{W}_i)^{-1}$, where $\mathbf{W}_i$ is the node transmission calculated under the condition that all nodes on the directed path P between node $\mathbf{X}_i$ and the output node are split. Clearly, the node factor is dependent on the choice of the directed path.

With these preliminaries, the following steps yield the graph transmission from the source node $\mathbf{X}_0$ to any dependent node $\mathbf{X}_j$ in a matrix signal-flow graph:

1. Determine all the directed paths from node $\mathbf{X}_0$ to node $\mathbf{X}_j$.
2. For each directed path P, compute the node factors of the nodes $\mathbf{X}_i$ on P. A *modified directed path product* is formed from the directed path product of P by inserting the node factor of every node $\mathbf{X}_i$ on P between the transmittances of the edges incident at node $\mathbf{X}_i$ in P.

3. The graph transmission from node $\mathbf{X}_0$ to node $\mathbf{X}_j$ is equal to the sum of the modified directed path products of all the directed paths from node $\mathbf{X}_0$ to node $\mathbf{X}_j$.

We illustrate the above procedure by the following example.

Example 7.1 We compute the graph transmission from node $\mathbf{X}_0$ to node $\mathbf{X}_3$ in the matrix signal-flow graph of Fig. 7.3 by the procedure outlined above.

1. There are two directed paths from $\mathbf{X}_0$ to $\mathbf{X}_3$. The directed path products are given by $\mathbf{D}$ and $\mathbf{CEB}$.

2. As shown in Fig. 7.9, for the directed path $\mathbf{X}_0$, $\mathbf{X}_1$, $\mathbf{X}_2$, and $\mathbf{X}_3$ the node factors of nodes $\mathbf{X}_1$ and $\mathbf{X}_2$ are given by $(\mathbf{1}-\mathbf{0})^{-1}$ and $(\mathbf{1}-\mathbf{EA})^{-1}$, respectively. The modified directed path product becomes $\mathbf{C}(\mathbf{1}-\mathbf{EA})^{-1}\mathbf{EB}$. The other modified directed path product is simply $\mathbf{D}$.

3. The graph transmission is found to be

$$\mathbf{W}_{03} = \mathbf{D} + \mathbf{C}(\mathbf{1}-\mathbf{EA})^{-1}\mathbf{EB} \tag{7.9}$$

where $\mathbf{X}_3 = \mathbf{W}_{03}\mathbf{X}_0$. It is straightforward to demonstrate that (7.8) and (7.9) are equivalent (see Prob. 7.1). We remark that the identity matrix in (7.8) is of order

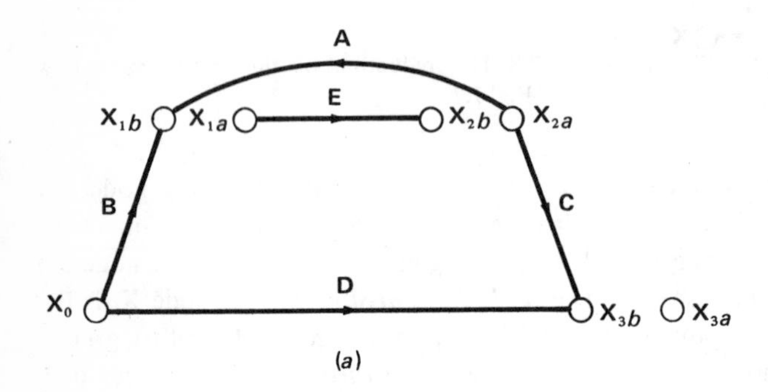

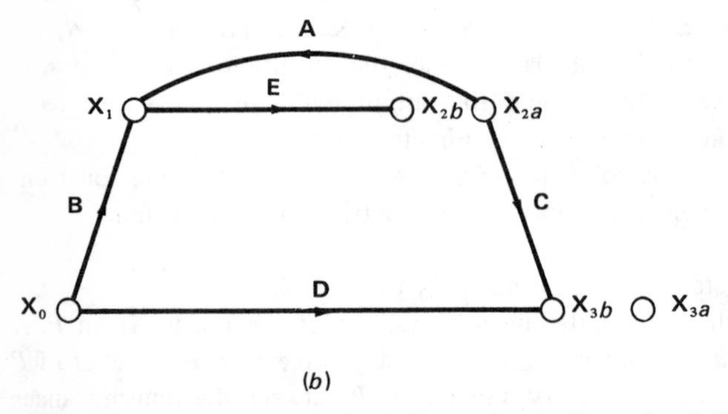

Figure 7.9 The computation of node factors for the directed path $\mathbf{X}_0, \mathbf{X}_1, \mathbf{X}_2$, and $\mathbf{X}_3$ for (*a*) node $\mathbf{X}_1$ and (*b*) node $\mathbf{X}_2$.

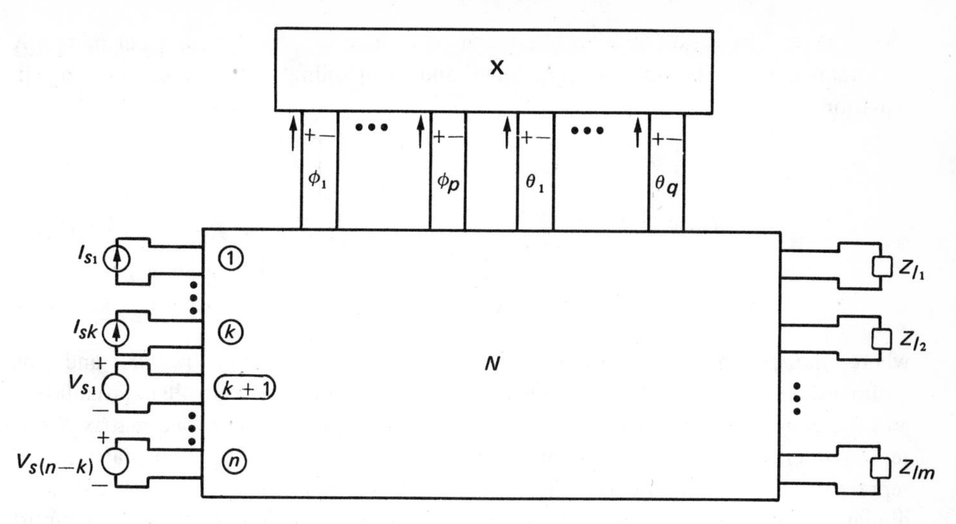

Figure 7.10 The general configuration of a multiple-input, multiple-output, and multiple-feedback loop amplifier.

equal to the number of rows of **A**, whereas the one in (7.9) is of order equal to the number of rows of **E**. Thus, their orders are generally different but the final products are the same.

7.2 THE MULTIPLE-LOOP FEEDBACK AMPLIFIER THEORY

The general configuration of a multiple-input, multiple-output, and multiple-feedback loop amplifier is shown in Fig. 7.10, in which the input, output, and feedback variables may be either currents or voltages. For the specific arrangement as indicated in Fig. 7.10, the input and output variables can be represented by an n-dimensional vector **u** and an m-dimensional vector **y** as

$$\mathbf{u}(s) = \begin{bmatrix} u_1 \\ u_2 \\ \vdots \\ u_k \\ u_{k+1} \\ u_{k+2} \\ \vdots \\ u_n \end{bmatrix} = \begin{bmatrix} I_{s1} \\ I_{s2} \\ \vdots \\ I_{sk} \\ V_{s1} \\ V_{s2} \\ \vdots \\ V_{s(n-k)} \end{bmatrix} \qquad \mathbf{y}(s) = \begin{bmatrix} y_1 \\ y_2 \\ \vdots \\ y_r \\ y_{r+1} \\ y_{r+2} \\ \vdots \\ y_m \end{bmatrix} = \begin{bmatrix} I_1 \\ I_2 \\ \vdots \\ I_r \\ V_{r+1} \\ V_{r+2} \\ \vdots \\ V_m \end{bmatrix} \tag{7.10}$$

respectively. The elements of interest can be represented by a rectangular matrix **X** of order $q \times p$ relating the controlled and controlling variables by the matrix equation

$$\boldsymbol{\theta} = \begin{bmatrix} \theta_1 \\ \theta_2 \\ \vdots \\ \theta_q \end{bmatrix} = \begin{bmatrix} x_{11} & x_{12} & \cdots & x_{1p} \\ x_{21} & x_{22} & \cdots & x_{2p} \\ \vdots & & \cdots & \\ x_{q1} & x_{q2} & \cdots & x_{qp} \end{bmatrix} \begin{bmatrix} \phi_1 \\ \phi_2 \\ \vdots \\ \phi_p \end{bmatrix} = \mathbf{X}\boldsymbol{\phi} \tag{7.11}$$

where the p-dimensional vector $\boldsymbol{\phi}$ is called the *controlling vector*, and the q-dimensional vector $\boldsymbol{\theta}$ the *controlled vector*. As before, the controlled variables θ_k and the controlling variables ϕ_k can be either currents or voltages. The matrix **X** can represent either a transfer-function matrix or a driving-point function matrix. If **X** represents a driving-point function matrix, the vectors $\boldsymbol{\theta}$ and $\boldsymbol{\phi}$ are of the same dimension ($q = p$) and their components are the currents and voltages of a p-port network.

As an illustration, consider the feedback network of Fig. 7.11, which was considered in Example 5.16. Suppose that we are interested in the effects of element Y_1 and the controlling parameters α_1, α_2, β_1, and β_2 on the whole network. We choose Y_1, α_1, α_2, β_1, and β_2 as the elements of **X**, and Eq. (7.11) becomes (see Probs. 7.10 and 7.20)

$$\boldsymbol{\theta} = \begin{bmatrix} I_1 \\ V_a \\ V_b \end{bmatrix} = \begin{bmatrix} Y_1 & 0 & 0 \\ 0 & \beta_1 & \beta_2 \\ 0 & \alpha_1 & \alpha_2 \end{bmatrix} \begin{bmatrix} V_1 \\ I_2 \\ I_3 \end{bmatrix} = \mathbf{X}\boldsymbol{\phi} \tag{7.12}$$

The general configuration of Fig. 7.10 can be represented equivalently by the block diagram of Fig. 7.12, in which N is a $(p + q + m + n)$-port network and the elements of interest are exhibited explicitly by the block **X**. Consider the

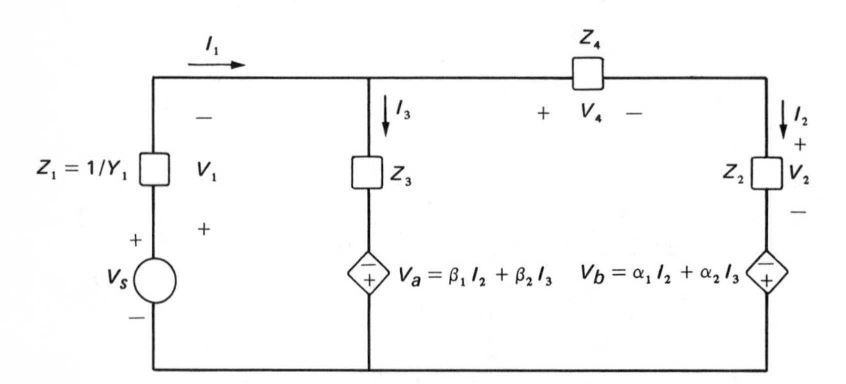

Figure 7.11 A feedback network containing two current-controlled voltage sources.

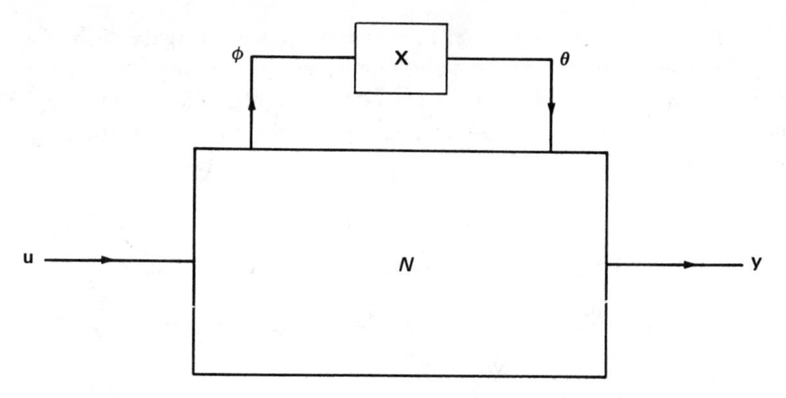

Figure 7.12 The block diagram of the general feedback configuration of Fig. 7.10.

$(p+q+m+n)$-port network N itself. The vectors **u** and θ are the inputs to N, whereas the vectors ϕ and **y** are the outputs of N. Since N is linear, the input and output vectors are related by the matrix equations

$$\phi = \mathbf{A}\theta + \mathbf{Bu} \tag{7.13a}$$

$$\mathbf{y} = \mathbf{C}\theta + \mathbf{Du} \tag{7.13b}$$

where **A**, **B**, **C**, and **D** are transfer-function matrices of orders $p \times q$, $p \times n$, $m \times q$, and $m \times n$, respectively. The vectors θ and ϕ are not independent and, from (7.11), are related by

$$\theta = \mathbf{X}\phi \tag{7.14}$$

The above system of three linear matrix equations can be represented by a matrix signal-flow graph as shown in Fig. 7.13, which is known as the *fundamental matrix feedback-flow graph*. The graph is identical to that shown in Fig. 7.3 used to illustrate the matrix signal-flow graph reduction rules. The overall transfer-function matrix of the multiple-loop feedback amplifier is defined by the equation

$$\mathbf{y} = \mathbf{W}(\mathbf{X})\mathbf{u} \tag{7.15}$$

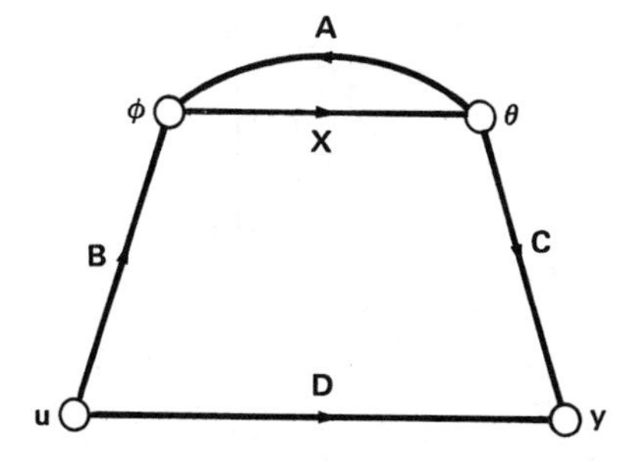

Figure 7.13 The fundamental matrix feedback-flow graph.

where $\mathbf{W(X)}$ is of order $m \times n$. As before, to emphasize the importance of $\mathbf{X}$, the matrix $\mathbf{W}$ is written as $\mathbf{W(X)}$ for the present discussion, even though it is also a function of the complex-frequency variable s. By using the matrix flow-graph reduction rules or the topological procedure discussed in the preceding section, the transfer-function matrix is found to be

$$\mathbf{W(X)} = \mathbf{D} + \mathbf{CX}(\mathbf{1}_p - \mathbf{AX})^{-1}\mathbf{B} \tag{7.16a}$$

or

$$\mathbf{W(X)} = \mathbf{D} + \mathbf{C}(\mathbf{1}_q - \mathbf{XA})^{-1}\mathbf{XB} \tag{7.16b}$$

where $\mathbf{1}_p$ denotes the identity matrix of order p. Clearly, we have

$$\mathbf{W(0)} = \mathbf{D} \tag{7.17}$$

In the particular situation where $\mathbf{X}$ is square and nonsingular, (7.16) can be written as

$$\mathbf{W(X)} = \mathbf{D} + \mathbf{C}(\mathbf{X}^{-1} - \mathbf{A})^{-1}\mathbf{B} \tag{7.18}$$

We illustrate the above procedure by the following example.

Example 7.2 Consider the voltage-series feedback amplifier of Fig. 7.14, which was discussed in Example 4.11. Assume that the two transistors are identical, with $h_{ie} = 1.1\ \text{k}\Omega$, $h_{fe} = 50$, $h_{re} = h_{oe} = 0$. The equivalent network of the amplifier is presented in Fig. 7.15. Let the controlling parameters of the two controlled current sources be the elements of interest. Then we have

$$\boldsymbol{\theta} = \begin{bmatrix} I_a \\ I_b \end{bmatrix} = 10^{-4} \begin{bmatrix} 455 & 0 \\ 0 & 455 \end{bmatrix} \begin{bmatrix} V_{13} \\ V_{45} \end{bmatrix} = \mathbf{X}\boldsymbol{\phi} \tag{7.19}$$

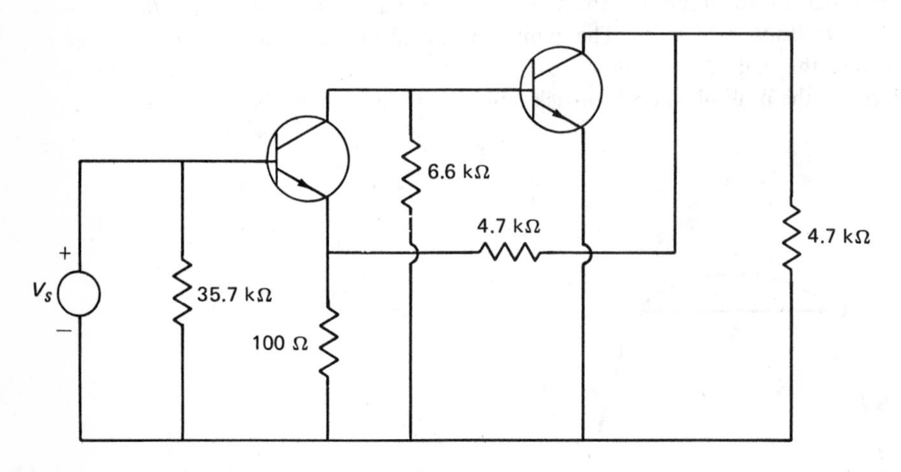

Figure 7.14 A voltage-series feedback amplifier.

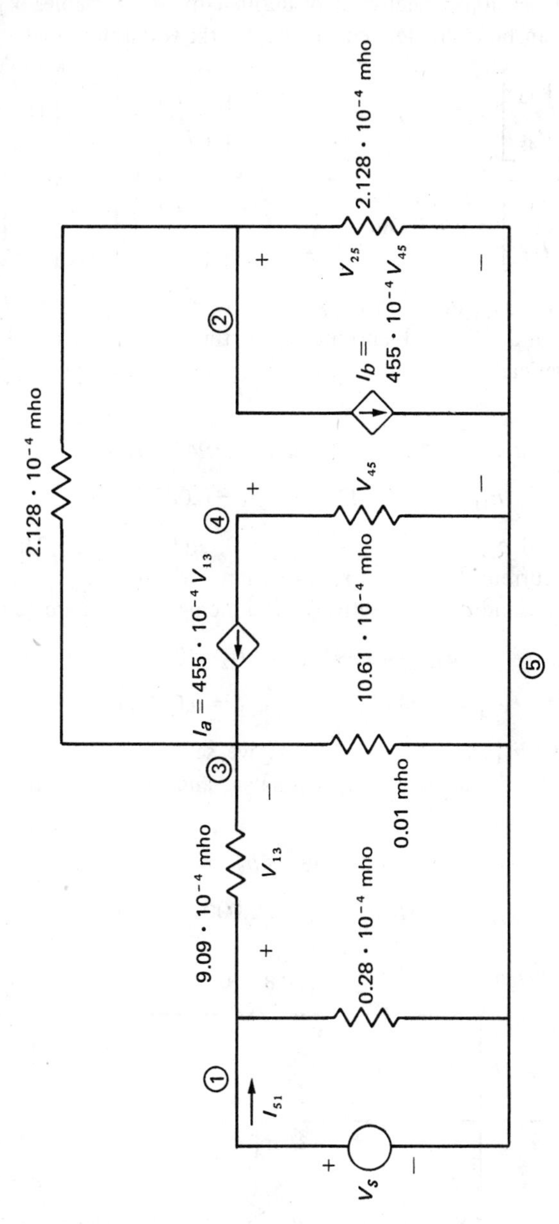

Figure 7.15 An equivalent network of the voltage-series feedback amplifier of Fig. 7.14.

Assume that the output voltage V_{25} and input current I_{51} are the output variables. Then the seven-port network N defined by the variables V_{13}, V_{45}, V_{25}, I_{51}, I_a, I_b, and V_s can be characterized by the matrix equations

$$\phi = \begin{bmatrix} V_{13} \\ V_{45} \end{bmatrix} = \mathbf{A}\theta + \mathbf{Bu} = \begin{bmatrix} a_{11} & a_{12} \\ a_{21} & a_{22} \end{bmatrix} \begin{bmatrix} I_a \\ I_b \end{bmatrix} + \begin{bmatrix} b_{11} \\ b_{21} \end{bmatrix} [V_s] \tag{7.20a}$$

$$\mathbf{y} = \begin{bmatrix} V_{25} \\ I_{51} \end{bmatrix} = \mathbf{C}\theta + \mathbf{Du} = \begin{bmatrix} c_{11} & c_{12} \\ c_{21} & c_{22} \end{bmatrix} \begin{bmatrix} I_a \\ I_b \end{bmatrix} + \begin{bmatrix} d_{11} \\ d_{21} \end{bmatrix} [V_s] \tag{7.20b}$$

To compute a_{11}, a_{21}, c_{11}, and c_{21}, we set $I_b = 0$, $V_s = 0$, and $I_a = 1$ A. The voltages V'_{13}, V'_{45}, and V'_{25} and current I'_{51} in the resulting network as shown in Fig. 7.16 are numerically equal to a_{11}, a_{21}, c_{11}, and c_{21}, respectively. The results are given by

$$\begin{aligned} a_{11} &= -90.782\ \Omega \qquad & a_{21} &= -942.507\ \Omega \\ c_{11} &= 45.391\ \Omega \qquad & c_{21} &= -0.08252 \end{aligned} \tag{7.21}$$

For a_{12}, a_{22}, c_{12}, and c_{22}, we set $I_a = 0$, $V_s = 0$, and $I_b = 1$ A. The voltages V''_{13}, V''_{45}, and V''_{25} and current I''_{51} in the resulting network of Fig. 7.17 are numerically equal to a_{12}, a_{22}, c_{12}, and c_{22}, respectively. The element values are found to be

$$\begin{aligned} a_{12} &= 45.391\ \Omega \qquad & a_{22} &= 0 \\ c_{12} &= -2372.32\ \Omega \qquad & c_{22} &= 0.04126 \end{aligned} \tag{7.22}$$

Finally, to compute b_{11}, b_{21}, d_{11}, and d_{21}, we set $I_a = I_b = 0$ and $V_s = 1$ V, as shown in Fig. 7.18, and calculate $\tilde{V}_{13}$, $\tilde{V}_{45}$, $\tilde{V}_{25}$, and $\tilde{I}_{51}$, respectively. The results are given by

$$\begin{aligned} b_{11} &= 0.91748 \qquad & b_{21} &= 0 \\ d_{11} &= 0.04126 \qquad & d_{21} &= 0.000862 \text{ mho} \end{aligned} \tag{7.23}$$

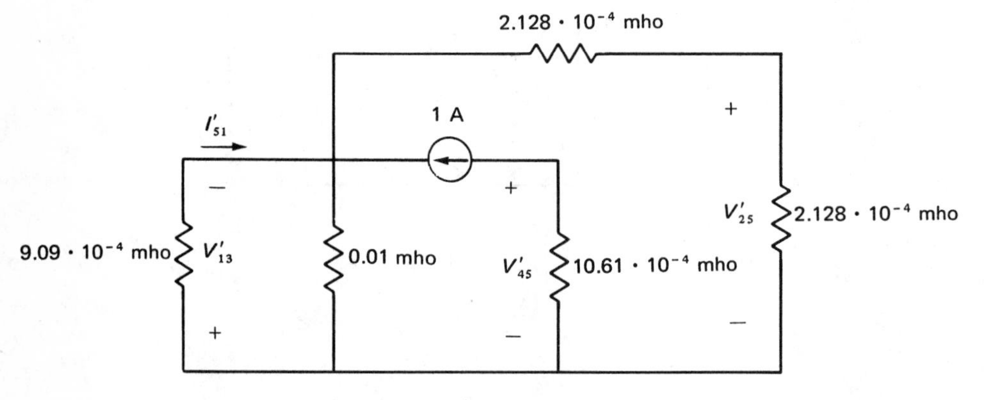

Figure 7.16 The network used for the calculation of a_{11}, a_{21}, c_{11}, and c_{21}.

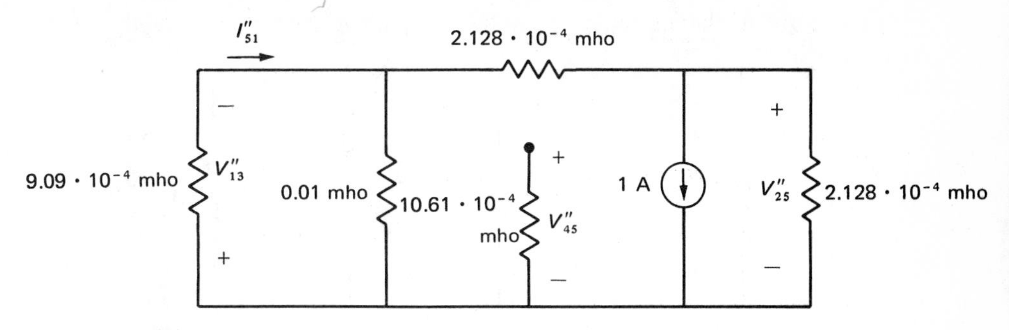

Figure 7.17 The network used for the calculation of a_{12}, a_{22}, c_{12}, and c_{22}.

Substituting these in (7.20), we obtain

$$\begin{bmatrix} V_{13} \\ V_{45} \end{bmatrix} = \begin{bmatrix} -90.782 & 45.391 \\ -942.507 & 0 \end{bmatrix} \begin{bmatrix} I_a \\ I_b \end{bmatrix} + \begin{bmatrix} 0.91748 \\ 0 \end{bmatrix} [V_s] \tag{7.24a}$$

$$\begin{bmatrix} V_{25} \\ I_{51} \end{bmatrix} = \begin{bmatrix} 45.391 & -2372.32 \\ -0.08252 & 0.04126 \end{bmatrix} \begin{bmatrix} I_a \\ I_b \end{bmatrix} + \begin{bmatrix} 0.041260 \\ 0.000862 \end{bmatrix} [V_s] \tag{7.24b}$$

According to (7.15), the transfer-function matrix of the amplifier is defined by the matrix equation

$$\mathbf{y} = \begin{bmatrix} V_{25} \\ I_{51} \end{bmatrix} = \begin{bmatrix} w_{11} \\ w_{21} \end{bmatrix} [V_s] = \mathbf{W}(\mathbf{X})\mathbf{u} \tag{7.25}$$

Since $\mathbf{X}$ is square and nonsingular, we can use (7.18) to calculate $\mathbf{W}(\mathbf{X})$, giving

$$\mathbf{W}(\mathbf{X}) = \mathbf{D} + \mathbf{C}(\mathbf{X}^{-1} - \mathbf{A})^{-1}\mathbf{B}$$

$$= \begin{bmatrix} 0.041260 \\ 0.000862 \end{bmatrix} + \begin{bmatrix} 45.391 & -2372.32 \\ -0.08252 & 0.04126 \end{bmatrix} \begin{bmatrix} 4.856 & 10.029 \\ -208.245 & 24.914 \end{bmatrix} \begin{bmatrix} 0.91748 \\ 0 \end{bmatrix}$$

$$\cdot 10^{-4} = \begin{bmatrix} 45.387 \\ 0.369 \cdot 10^{-4} \end{bmatrix} = \begin{bmatrix} w_{11} \\ w_{21} \end{bmatrix} \tag{7.26a}$$

where

$$(\mathbf{X}^{-1} - \mathbf{A})^{-1} = \begin{bmatrix} 4.856 & 10.029 \\ -208.245 & 24.914 \end{bmatrix} \cdot 10^{-4} \tag{7.26b}$$

Thus, the closed-loop voltage gain of the amplifier is given by $w_{11} = V_{25}/V_s = 45.387$, confirming Eq. (4.139). The input impedance facing the voltage source V_s, being the reciprocal of w_{21}, is given by

$$\frac{V_s}{I_{51}} = \frac{1}{w_{21}} = 27.1 \text{ k}\Omega \tag{7.27}$$

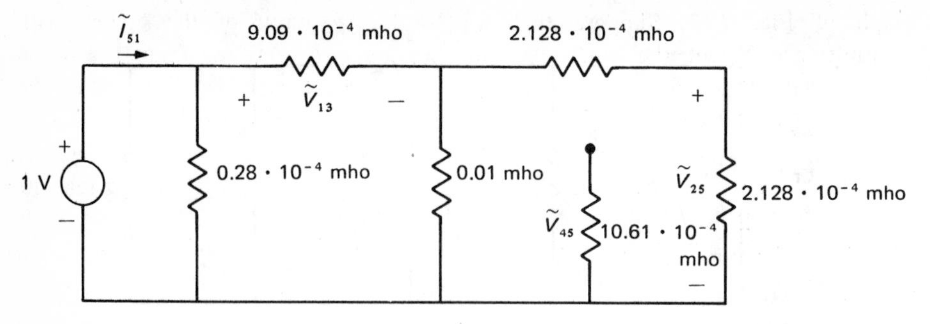

Figure 7.18 The network used for the calculation of b_{11}, b_{21}, d_{11}, and d_{21}.

To verify this result, we refer to the indefinite-admittance matrix **Y** given in Eq. (4.138). By applying formula (2.95), the input impedance facing V_s is found to be

$$z_{11,55} = \frac{V_s}{I_{51}} = \frac{Y_{11,55}}{Y_{uv}} = \frac{466{,}069 \cdot 10^{-12}}{171{,}983 \cdot 10^{-16}} = 27.1 \text{ k}\Omega \tag{7.28}$$

confirming (7.27), where

$$Y_{uv} = Y_{55} = 171{,}983 \cdot 10^{-16} \tag{7.29}$$

7.2.1 The Return Difference Matrix

In the study of a single-loop feedback amplifier, we usually single out an element for particular attention. The element is generally one that is either crucial in terms of its effect on the entire system or of main concern to the designer. For a multiple-loop feedback amplifier, instead of singling out an element, we pay particular attention to a group of elements represented by the matrix **X** and study its effects on the whole system. In this section, we generalize the concepts of return difference by introducing the return difference matrix.

As indicated in Sec. 5.7, if we break the branch with transmittance x in the fundamental feedback-flow graph of Fig. 5.17, the negative of the signal returned to the left of the breaking mark as shown in Fig. 5.18 when a unit signal is applied to the right of the breaking mark and when the input excitation is set to zero is the return ratio for the element x. Instead of applying a unit signal, we can apply any signal g to the right of the breaking mark. Then the ratio of the returned signal h to the applied signal g must be the negative of the return ratio $T(x)$ for the element x, or $h = -T(x)g$ because the network is assumed to be linear. The difference between the applied signal g and the returned signal h defines the return difference $F(x)$ with respect to the element x:

$$g - h = [1 + T(x)]g = F(x)g \tag{7.30}$$

giving $F(x) = 1 + T(x)$.

With the above interpretations, the extension of these concepts to multiple-loop feedback amplifiers is now clear. Refer to the fundamental matrix feedback-flow graph of Fig. 7.13. Suppose that we break the input of the branch with transmittance **X**, set the input excitation vector **u** to zero, and apply a signal p-vector g to the right of the breaking mark, as depicted in Fig. 7.19. Then the returned signal p-vector **h** to the left of the breaking mark is found to be

$$\mathbf{h} = \mathbf{AXg} \tag{7.31}$$

The square matrix **AX** is called the *loop-transmission matrix* and its negative is referred to as the *return ratio matrix*, denoted by the symbol **T(X)**:

$$\mathbf{T(X)} = -\mathbf{AX} \tag{7.32}$$

The difference between the applied signal vector **g** and the returned signal vector **h** is given by

$$\mathbf{g} - \mathbf{h} = (\mathbf{1}_p - \mathbf{AX})\mathbf{g} \tag{7.33}$$

The square matrix $\mathbf{1}_p - \mathbf{AX}$ is defined as the *return difference matrix* with respect to **X** and is denoted by the symbol **F(X)**:

$$\mathbf{F(X)} = \mathbf{1}_p - \mathbf{AX} \tag{7.34}$$

Combining this with (7.32) yields

$$\mathbf{F(X)} = \mathbf{1}_p + \mathbf{T(X)} \tag{7.35}$$

As an illustration, we compute the return difference matrix with respect to the matrix **X**, as given in (7.19). From (7.24*a*), the return ratio matrix is found to be

$$\mathbf{T(X)} = -\mathbf{AX} = \begin{bmatrix} 4.131 & -2.065 \\ 42.884 & 0 \end{bmatrix} \tag{7.36}$$

giving the return difference matrix as

$$\mathbf{F(X)} = \mathbf{1}_2 + \mathbf{T(X)} = \begin{bmatrix} 5.131 & -2.065 \\ 42.884 & 1 \end{bmatrix} \tag{7.37}$$

As demonstrated in Sec. 4.3.1, the scalar return ratio $T(x)$ for a one-port admittance x is equal to the ratio of the admittance x to the admittance y that x

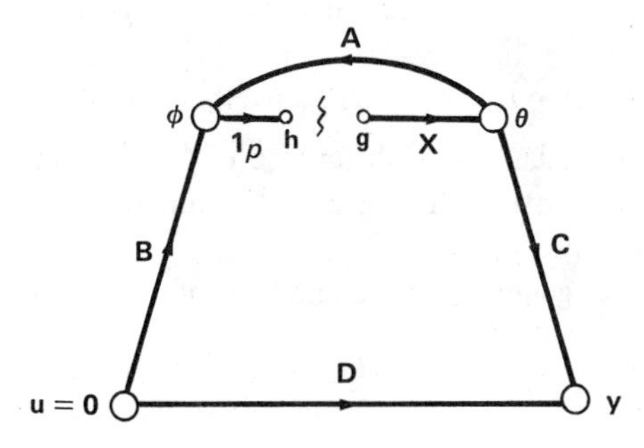

Figure 7.19 The physical interpretation of the loop-transmission matrix.

faces, and the scalar return difference $F(x)$ with respect to a one-port admittance x is equal to the ratio of the total admittance looking into the node pair where x is connected to the admittance that x faces:

$$F(x) = 1 + T(x) = 1 + \frac{x}{y} = y^{-1}(y + x) \tag{7.38}$$

A similar interpretation can be made for the return ratio matrix and the return difference matrix if $\mathbf{X}$ denotes the admittance matrix of a p-port network. In this case, the controlling vector $\boldsymbol{\phi}$ denotes the p-port voltage vector and the controlled vector $\boldsymbol{\theta}$ the p-port current vector. By writing $\boldsymbol{\phi} = \mathbf{V}$ and $\boldsymbol{\theta} = \mathbf{I}$ with $\phi_j = V_j$ and $\theta_j = I_j$ $(j = 1, 2, \ldots, p)$ denoting the port voltages and currents, (7.11) can be written as

$$I_j = x_{j1}V_1 + x_{j2}V_2 + \cdots + x_{jp}V_p \qquad (j = 1, 2, \ldots, p) \tag{7.39}$$

As discussed in Sec. 4.3.1, a one-port admittance x can be represented equivalently by a voltage-controlled current source $I = xV$, the controlling voltage V being the terminal voltage of the current source. In a similar way, a p-port admittance matrix $\mathbf{X}$ can be represented equivalently by a vector voltage-controlled current source $\mathbf{I} = \mathbf{XV}$ whose jth component is given by (7.39). After removing the p-port network, we insert at its jth port a voltage-controlled current source $I_j = x_{j1}V_1 + x_{j2}V_2 + \cdots + x_{jp}V_p$ for $j = 1, 2, \ldots, p$. Now we break the input of each of the controlling branches, as depicted in Fig. 7.20, and apply a voltage source of g_j volts just above the breaking mark, the voltage $\tilde{V}_j$ appearing at the jth port beneath the breaking mark being the returned voltage. Denoting the applied voltages g_j by the p-vector $\mathbf{g}$ and the returned voltages $\tilde{V}_j$ by $\mathbf{h}$, we have

$$\mathbf{I} = \mathbf{Xg} \tag{7.40a}$$

$$-\mathbf{I} = \mathbf{Yh} \tag{7.40b}$$

where $\mathbf{Y}$ is the admittance matrix facing the current sources I_j of Fig. 7.20. This leads to

$$\mathbf{h} = -\mathbf{Y}^{-1}\mathbf{Xg} \tag{7.41a}$$

$$\mathbf{g} - \mathbf{h} = (\mathbf{1}_p + \mathbf{Y}^{-1}\mathbf{X})\mathbf{g} \tag{7.41b}$$

provided that $\mathbf{Y}$ is nonsingular, giving

$$\mathbf{T}(\mathbf{X}) = \mathbf{Y}^{-1}\mathbf{X} \tag{7.42a}$$

$$\mathbf{F}(\mathbf{X}) = \mathbf{Y}^{-1}(\mathbf{Y} + \mathbf{X}) \tag{7.42b}$$

This is therefore a direct generalization of the scalar return difference of (7.38). The matrix $\mathbf{Y} + \mathbf{X}$ represents the total admittance matrix looking into the junctions of the p-port network and the multiport network N.

In the situation where $\mathbf{X}$ denotes the general hybrid matrix of a p-port

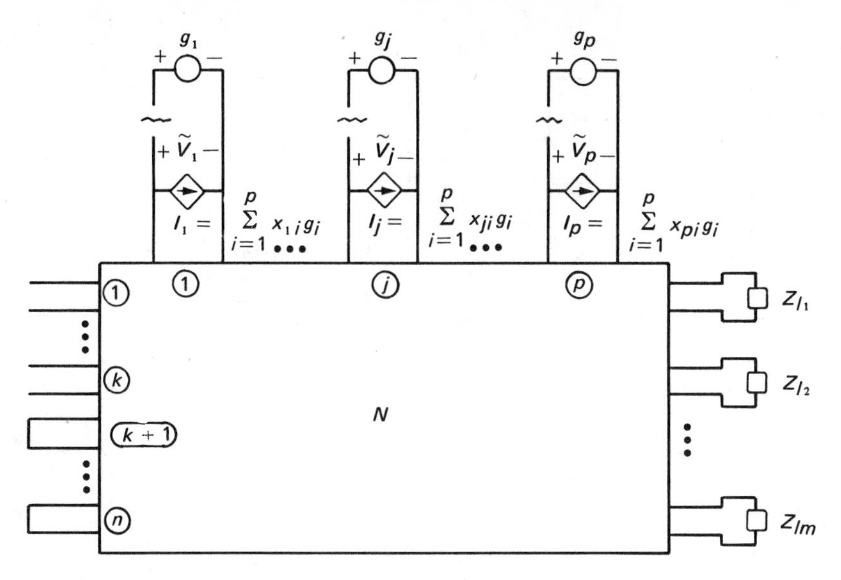

Figure 7.20 The physical interpretation of the return difference matrix with respect to a p-port admittance matrix.

network, a similar result can be obtained. Without loss of generality, we assume that $\mathbf{X}$ is characterized by the partitioned matrix equation

$$\theta = \begin{bmatrix} \mathbf{V}_\alpha \\ \mathbf{I}_\beta \end{bmatrix} = \begin{bmatrix} \mathbf{X}_{11} & \mathbf{X}_{12} \\ \mathbf{X}_{21} & \mathbf{X}_{22} \end{bmatrix} \begin{bmatrix} \mathbf{I}_\alpha \\ \mathbf{V}_\beta \end{bmatrix} = \mathbf{X}\phi \tag{7.43}$$

where $[\mathbf{I}'_\alpha, \mathbf{I}'_\beta]'$ and $[\mathbf{V}'_\alpha, \mathbf{V}'_\beta]'$ denote the port current vector and the port voltage vector of the p-port network, respectively, and the primes, as before, denote the matrix transpose. As in the admittance case, the p-port network with hybrid matrix $\mathbf{X}$ can be replaced by a controlled vector source. For each controlled variable $\theta_j = V_j$ in $\mathbf{V}_\alpha$, we insert at the jth port a controlled voltage source $V_j = x_{j1}\phi_1 + x_{j2}\phi_2 + \cdots + x_{jp}\phi_p$, and for each $\theta_j = I_j$ in $\mathbf{I}_\beta$ we insert a controlled current source $I_j = x_{j1}\phi_1 + x_{j2}\phi_2 + \cdots + x_{jp}\phi_p$. We then break the input of each of the controlling branches and apply a signal of g_j volts or amperes, as the case may be, just above the breaking mark. The current or voltage appearing at the jth port beneath the breaking mark is the returned signal $h_j = \tilde{I}_j$ or $\tilde{V}_j$, as shown in Fig. 7.21. Denote by the partitioned p-vectors

$$\mathrm{g} = \begin{bmatrix} \tilde{\mathbf{I}}_\alpha \\ \tilde{\mathbf{V}}_\beta \end{bmatrix} \qquad \mathrm{h} = \begin{bmatrix} \hat{\mathbf{I}}_\alpha \\ \hat{\mathbf{V}}_\beta \end{bmatrix} \tag{7.44}$$

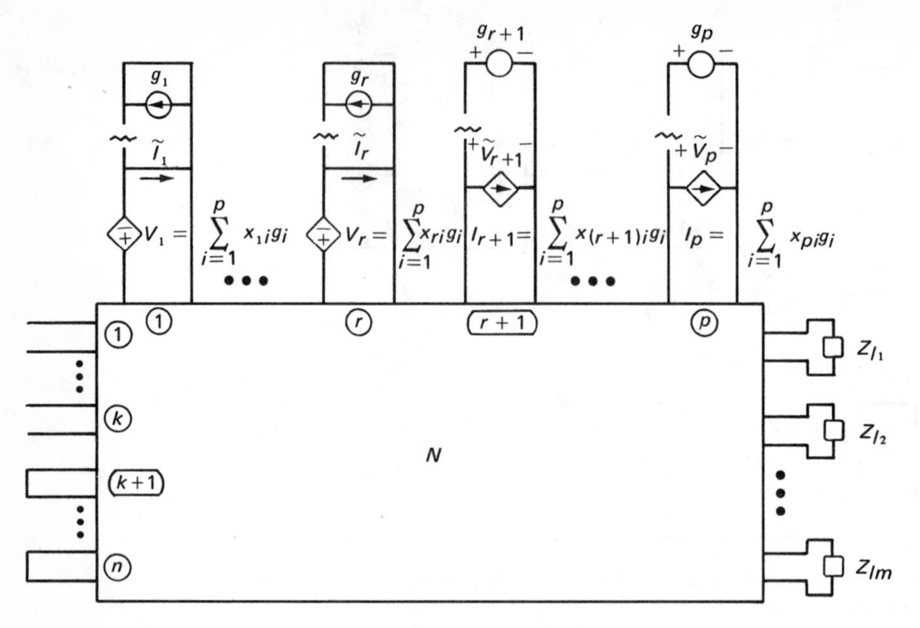

Figure 7.21 The physical interpretation of the return difference matrix with respect to a p-port general hybrid matrix.

the applied signals g_j and the returned signals h_j, respectively. Also, denote by the p-vector $\tilde{\theta}$ the signals at the p ports resulting from the controlled sources, and write

$$\tilde{\theta} = \begin{bmatrix} \tilde{\mathbf{V}}_\alpha \\ \tilde{\mathbf{I}}_\beta \end{bmatrix} \tag{7.45}$$

With these symbols, we have

$$\tilde{\theta} = \begin{bmatrix} \tilde{\mathbf{V}}_\alpha \\ \tilde{\mathbf{I}}_\beta \end{bmatrix} = \begin{bmatrix} \mathbf{X}_{11} & \mathbf{X}_{12} \\ \mathbf{X}_{21} & \mathbf{X}_{22} \end{bmatrix} \begin{bmatrix} \tilde{\mathbf{I}}_\alpha \\ \tilde{\mathbf{V}}_\beta \end{bmatrix} = \mathbf{X}\mathbf{g} \tag{7.46}$$

For the multiport network N, we open-circuit all the independent current sources and short-circuit all the independent voltage sources. The p ports facing the p-port network $\mathbf{X}$ can be characterized by its general hybrid matrix in partitioned form,

$$\begin{bmatrix} \tilde{\mathbf{V}}_\alpha \\ -\tilde{\mathbf{I}}_\beta \end{bmatrix} = \begin{bmatrix} \mathbf{H}_{\alpha\alpha} & \mathbf{H}_{\alpha\beta} \\ \mathbf{H}_{\beta\alpha} & \mathbf{H}_{\beta\beta} \end{bmatrix} \begin{bmatrix} -\hat{\mathbf{I}}_\alpha \\ \hat{\mathbf{V}}_\beta \end{bmatrix} \tag{7.47}$$

which can be rewritten as

$$\tilde{\theta} = \begin{bmatrix} \tilde{\mathbf{V}}_\alpha \\ \tilde{\mathbf{I}}_\beta \end{bmatrix} = -\begin{bmatrix} \mathbf{H}_{\alpha\alpha} & -\mathbf{H}_{\alpha\beta} \\ -\mathbf{H}_{\beta\alpha} & \mathbf{H}_{\beta\beta} \end{bmatrix} \begin{bmatrix} \hat{\mathbf{I}}_\alpha \\ \hat{\mathbf{V}}_\beta \end{bmatrix} = -\tilde{\mathbf{H}}\mathbf{h} \tag{7.48a}$$

where
$$\tilde{\mathbf{H}} = \begin{bmatrix} \mathbf{H}_{\alpha\alpha} & -\mathbf{H}_{\alpha\beta} \\ -\mathbf{H}_{\beta\alpha} & \mathbf{H}_{\beta\beta} \end{bmatrix} \tag{7.48b}$$

Combining (7.46) and (7.48*a*) yields

$$\mathbf{h} = -\tilde{\mathbf{H}}^{-1}\mathbf{X}\mathbf{g} \tag{7.49a}$$

$$\mathbf{g} - \mathbf{h} = (\mathbf{1}_p + \tilde{\mathbf{H}}^{-1}\mathbf{X})\mathbf{g} \tag{7.49b}$$

provided that $\hat{\mathbf{H}}$ is nonsingular, giving

$$\mathbf{T}(\mathbf{X}) = \tilde{\mathbf{H}}^{-1}\mathbf{X} \tag{7.50a}$$

$$\mathbf{F}(\mathbf{X}) = \mathbf{1}_p + \mathbf{T}(\mathbf{X}) = \tilde{\mathbf{H}}^{-1}(\tilde{\mathbf{H}} + \mathbf{X}) \tag{7.50b}$$

Alternatively, Eq. (7.50*b*) can be interpreted as follows. Consider an $(n+p)$-port network N_1 terminated in a p-port network N_2 as shown in Fig. 7.22. The $(n+p)$-port network N_1 can be characterized by its general hybrid matrix equation, again in partitioned form:

$$\begin{bmatrix} \mathbf{I}_1 \\ \mathbf{V}_2 \\ \mathbf{I}_3 \\ \mathbf{V}_4 \end{bmatrix} = \begin{bmatrix} \mathbf{H}_{11} & \mathbf{H}_{12} & \mathbf{H}_{13} & \mathbf{H}_{14} \\ \mathbf{H}_{21} & \mathbf{H}_{22} & \mathbf{H}_{23} & \mathbf{H}_{24} \\ \mathbf{H}_{31} & \mathbf{H}_{32} & \mathbf{H}_{33} & \mathbf{H}_{34} \\ \mathbf{H}_{41} & \mathbf{H}_{42} & \mathbf{H}_{43} & \mathbf{H}_{44} \end{bmatrix} \begin{bmatrix} \mathbf{V}_1 \\ \mathbf{I}_2 \\ \mathbf{V}_3 \\ \mathbf{I}_4 \end{bmatrix} \tag{7.51}$$

where the port current vectors $\mathbf{I}_k$ and the port voltage vectors $\mathbf{V}_k$ ($k = 1, 2, 3$, and 4) are defined in Fig. 7.22. The p-port network N_2 is described by the general hybrid matrix equation

$$\begin{bmatrix} \mathbf{V}_3 \\ -\mathbf{I}_4 \end{bmatrix} = \begin{bmatrix} \mathbf{X}_{11} & \mathbf{X}_{12} \\ \mathbf{X}_{21} & \mathbf{X}_{22} \end{bmatrix} \begin{bmatrix} -\mathbf{I}_3 \\ \mathbf{V}_4 \end{bmatrix} \tag{7.52}$$

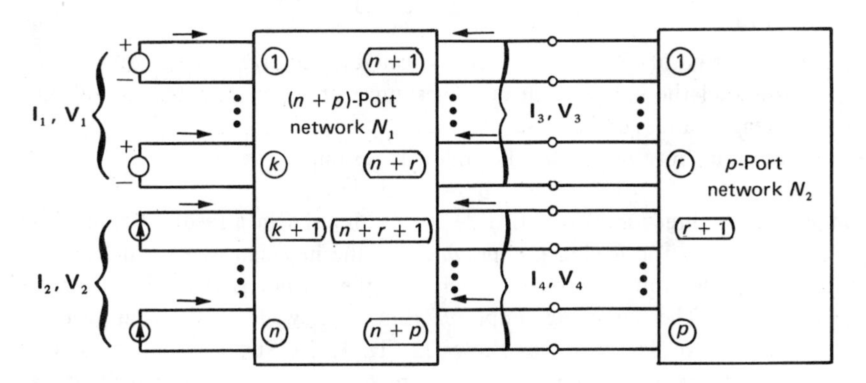

Figure 7.22 The $(n + p)$-port imbedding of a p-port network N_2.

From (7.51) and (7.52), we can solve $\mathbf{I}_1$ and $\mathbf{V}_2$ in terms of $\mathbf{V}_1$ and $\mathbf{I}_2$, and the result is given by

$$\begin{bmatrix} \mathbf{I}_1 \\ \mathbf{V}_2 \end{bmatrix} = \left(\begin{bmatrix} \mathbf{H}_{11} & \mathbf{H}_{12} \\ \mathbf{H}_{21} & \mathbf{H}_{22} \end{bmatrix} + \begin{bmatrix} \mathbf{H}_{13} & -\mathbf{H}_{14} \\ \mathbf{H}_{23} & -\mathbf{H}_{24} \end{bmatrix} \begin{bmatrix} \mathbf{M}_{11} & \mathbf{M}_{12} \\ \mathbf{M}_{21} & \mathbf{M}_{22} \end{bmatrix}^{-1} \begin{bmatrix} -\mathbf{H}_{31} & -\mathbf{H}_{32} \\ \mathbf{H}_{41} & \mathbf{H}_{42} \end{bmatrix} \right) \begin{bmatrix} \mathbf{V}_1 \\ \mathbf{I}_2 \end{bmatrix} \tag{7.53a}$$

where
$$\begin{bmatrix} \mathbf{M}_{11} & \mathbf{M}_{12} \\ \mathbf{M}_{21} & \mathbf{M}_{22} \end{bmatrix} = \begin{bmatrix} \mathbf{X}_{11} & \mathbf{X}_{12} \\ \mathbf{X}_{21} & \mathbf{X}_{22} \end{bmatrix}^{-1} + \begin{bmatrix} \mathbf{H}_{33} & -\mathbf{H}_{34} \\ -\mathbf{H}_{43} & \mathbf{H}_{44} \end{bmatrix} \tag{7.53b}$$

provided that the various inverses exist. The coefficient matrix of (7.53*a*) defines the general hybrid matrix of the n-port network as shown in Fig. 7.22. Comparing this coefficient matrix with (7.18), we can make the following identifications:

$$\mathbf{D} = \begin{bmatrix} \mathbf{H}_{11} & \mathbf{H}_{12} \\ \mathbf{H}_{21} & \mathbf{H}_{22} \end{bmatrix} \qquad \mathbf{C} = \begin{bmatrix} \mathbf{H}_{13} & -\mathbf{H}_{14} \\ \mathbf{H}_{23} & -\mathbf{H}_{24} \end{bmatrix} \tag{7.54a}$$

$$\mathbf{A} = -\begin{bmatrix} \mathbf{H}_{33} & -\mathbf{H}_{34} \\ -\mathbf{H}_{43} & \mathbf{H}_{44} \end{bmatrix} \qquad \mathbf{B} = \begin{bmatrix} -\mathbf{H}_{31} & -\mathbf{H}_{32} \\ \mathbf{H}_{41} & \mathbf{H}_{42} \end{bmatrix} \tag{7.54b}$$

From (7.34) the return difference matrix is found to be

$$\mathbf{F}(\mathbf{X}) = \mathbf{1}_p - \mathbf{A}\mathbf{X} = \tilde{\mathbf{H}}^{-1}(\tilde{\mathbf{H}} + \mathbf{X}) \tag{7.55}$$

where
$$\tilde{\mathbf{H}} = \begin{bmatrix} \mathbf{H}_{\alpha\alpha} & -\mathbf{H}_{\alpha\beta} \\ -\mathbf{H}_{\beta\alpha} & \mathbf{H}_{\beta\beta} \end{bmatrix} = \begin{bmatrix} \mathbf{H}_{33} & -\mathbf{H}_{34} \\ -\mathbf{H}_{43} & \mathbf{H}_{44} \end{bmatrix}^{-1} \tag{7.56}$$

since, as can be seen from (7.51), by setting $\mathbf{V}_1 = \mathbf{0}$ and $\mathbf{I}_2 = \mathbf{0}$ we have

$$\begin{bmatrix} \mathbf{V}_3 \\ \mathbf{I}_4 \end{bmatrix} = \begin{bmatrix} \mathbf{H}_{33} & \mathbf{H}_{34} \\ \mathbf{H}_{43} & \mathbf{H}_{44} \end{bmatrix}^{-1} \begin{bmatrix} \mathbf{I}_3 \\ \mathbf{V}_4 \end{bmatrix} \Bigg|_{\substack{\mathbf{V}_1=\mathbf{0} \\ \mathbf{I}_2=\mathbf{0}}} \tag{7.57}$$

The coefficient matrix of (7.57) is precisely the hybrid matrix looking into the p ports of N_1 when all the independent current sources are open-circuited and all the independent voltage sources are short-circuited.

We illustrate the above results by the following example.

Example 7.3 A three-terminal active device N_2 is used in a feedback amplifier as shown in Fig. 7.23*a*. The feedback is provided by the impedances Z_a and Z_b. The input excitation is the voltage source V_s, whereas the output is the voltage across the impedance Z_2. The complete amplifier can be viewed as the four-port N imbedding of the active device N_2, as shown in Fig. 7.23*b*. Suppose that the active device N_2 is the element of interest and is characterized by its hybrid parameters h_{ij}, with

$$\theta = \begin{bmatrix} V_3 \\ I_4 \end{bmatrix} = \begin{bmatrix} h_{11} & h_{12} \\ h_{21} & h_{22} \end{bmatrix} \begin{bmatrix} I_3 \\ V_4 \end{bmatrix} = \mathbf{X}\phi \tag{7.58}$$

The hybrid matrix looking into the two ports facing N_2 with V_s short-circuited is described by the hybrid matrix equation

$$\begin{bmatrix} V_3 \\ -I_4 \end{bmatrix} = \lambda_1 \begin{bmatrix} Z_2 & 1 \\ -1 & \lambda_2 \end{bmatrix} \begin{bmatrix} -I_3 \\ V_4 \end{bmatrix} \tag{7.59}$$

where

$$\lambda_1 = \frac{Z_b}{Z_b + Z_2} \tag{7.60a}$$

$$\lambda_2 = \frac{Z_a + Z_b + Z_2}{Z_a Z_b} \tag{7.60b}$$

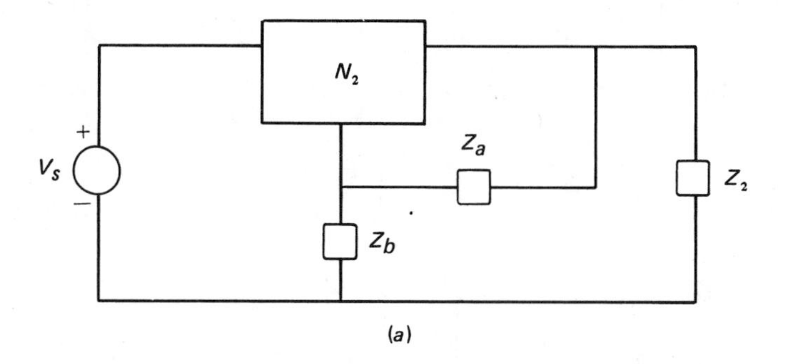

(a)

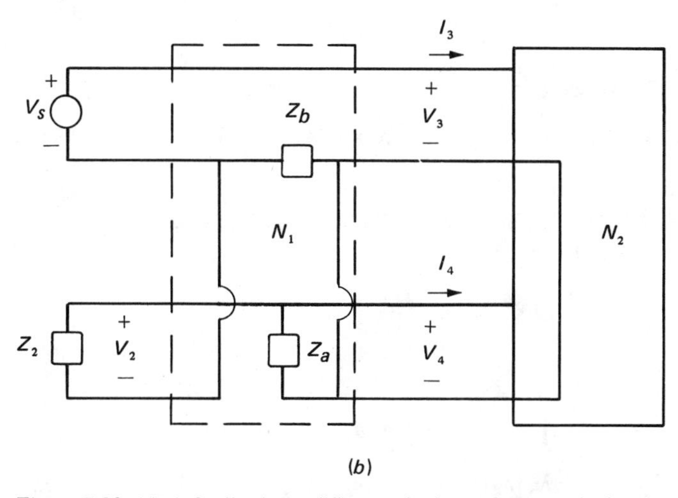

(b)

Figure 7.23 (*a*) A feedback amplifier employing a three-terminal active device and (*b*) the representation of the complete amplifier as the four-port imbedding of the active device.

giving
$$\tilde{\mathbf{H}} = \lambda_1 \begin{bmatrix} Z_2 & -1 \\ 1 & \lambda_2 \end{bmatrix} \tag{7.61}$$

whose inverse is found to be

$$\tilde{\mathbf{H}}^{-1} = \lambda_3 \begin{bmatrix} \lambda_2 & 1 \\ -1 & Z_2 \end{bmatrix} \tag{7.62a}$$

where
$$\lambda_3 = \frac{Z_a}{Z_a + Z_2} \tag{7.62b}$$

Substituting $\tilde{\mathbf{H}}^{-1}$ in (7.50*b*) yields the return difference matrix with respect to the driving-point hybrid matrix $\mathbf{X}$:

$$\mathbf{F}(\mathbf{X}) = \tilde{\mathbf{H}}^{-1}(\tilde{\mathbf{H}} + \mathbf{X}) = \begin{bmatrix} 1 + \lambda_3(h_{11}\lambda_2 + h_{21}) & \lambda_3(h_{12}\lambda_2 + h_{22}) \\ \lambda_3(h_{21}Z_2 - h_{11}) & 1 + \lambda_3(h_{22}Z_2 - h_{12}) \end{bmatrix} \tag{7.63}$$

The four-port network N_1 in Fig. 7.23*b* can be described by the matrix equations

$$\phi = \begin{bmatrix} I_3 \\ V_4 \end{bmatrix} = \lambda_3 \begin{bmatrix} -\lambda_2 & -1 \\ 1 & -Z_2 \end{bmatrix} \begin{bmatrix} V_3 \\ I_4 \end{bmatrix} + \lambda_3 \begin{bmatrix} \lambda_2 \\ -1 \end{bmatrix} [V_s] = \mathbf{A}\theta + \mathbf{B}\mathbf{u} \tag{7.64a}$$

$$\mathbf{y} = [V_2] = \lambda_3 \begin{bmatrix} -\dfrac{Z_2}{Z_a} & -Z_2 \end{bmatrix} \begin{bmatrix} V_3 \\ I_4 \end{bmatrix} + \lambda_3 \begin{bmatrix} \dfrac{Z_2}{Z_a} \end{bmatrix} [V_s] = \mathbf{C}\theta + \mathbf{D}\mathbf{u} \tag{7.64b}$$

giving
$$\mathbf{A} = \lambda_3 \begin{bmatrix} -\lambda_2 & -1 \\ 1 & -Z_2 \end{bmatrix} \tag{7.65}$$

Substituting (7.65) in (7.34), we obtain

$$\mathbf{F}(\mathbf{X}) = \mathbf{1}_2 - \mathbf{A}\mathbf{X} = \mathbf{1}_2 + \tilde{\mathbf{H}}^{-1}\mathbf{X} = \tilde{\mathbf{H}}^{-1}(\mathbf{H} + \mathbf{X}) \tag{7.66}$$

confirming (7.63).

To compute the overall voltage gain function, we appeal to (6.15) with $\mathbf{y} = [V_2]$, $\mathbf{u} = [V_s]$, and $\mathbf{W}(\mathbf{X}) = [w(\mathbf{X})]$, giving

$$\begin{aligned} w(\mathbf{X}) &= \frac{V_2}{V_s} = \mathbf{D} + \mathbf{C}(\mathbf{X}^{-1} - \mathbf{A})^{-1}\mathbf{B} \\ &= \frac{\lambda_3 Z_2}{Z_a} + \lambda_3^2 \begin{bmatrix} -\dfrac{Z_2}{Z_a} & -Z_2 \end{bmatrix} \begin{bmatrix} p_{11} & p_{12} \\ p_{21} & p_{22} \end{bmatrix} \begin{bmatrix} \lambda_2 \\ -1 \end{bmatrix} \\ &= \frac{\lambda_3 Z_2 \{1 - \lambda_3 [\lambda_2(p_{21}Z_a + p_{11}) - p_{22}Z_a - p_{12}]\}}{Z_a} \end{aligned} \tag{7.67a}$$

where
$$\begin{bmatrix} p_{11} & p_{12} \\ p_{21} & p_{22} \end{bmatrix} = \frac{1}{\Delta_\lambda} \begin{bmatrix} \lambda_3 Z_2 + \dfrac{h_{11}}{\Delta_h} & -\lambda_3 + \dfrac{h_{12}}{\Delta_h} \\ \lambda_3 + \dfrac{h_{21}}{\Delta_h} & \lambda_2\lambda_3 + \dfrac{h_{22}}{\Delta_h} \end{bmatrix} \tag{7.67b}$$

in which $\Delta_h = h_{11}h_{22} - h_{12}h_{21}$ and

$$\Delta_\lambda = \left(\lambda_2\lambda_3 + \frac{h_{22}}{\Delta_h}\right)\left(\lambda_3 Z_2 + \frac{h_{11}}{\Delta_h}\right) + \left(\lambda_3 - \frac{h_{12}}{\Delta_h}\right)\left(\lambda_3 + \frac{h_{21}}{\Delta_h}\right) \tag{7.67c}$$

Alternatively, by appealing to formula (3.25*a*) in conjunction with the impedance matrix (3.75), the voltage gain function (7.67*a*) can be obtained, the details being omitted.

7.2.2 The Null Return Difference Matrix

As a direct extension of the null return difference for the single-loop feedback amplifiers, in the present section we introduce the notion of the null return difference matrix for multiple-loop feedback networks.

Referring to the fundamental matrix feedback-flow graph of Fig. 7.13, we break, as before, the branch with transmittance **X**, as illustrated in Fig. 7.24, and apply a signal p-vector g to the right of the breaking mark. We then adjust the input excitation n-vector **u** so that the total output m-vector **y** resulting from the inputs g and **u** is zero. From Fig. 7.24, the input vectors g and **u** are related by

$$\mathbf{Du} + \mathbf{CXg} = \mathbf{0} \tag{7.68}$$

giving
$$\mathbf{u} = -\mathbf{D}^{-1}\mathbf{CXg} \tag{7.69}$$

provided that the matrix **D** is square and nonsingular. This requires that the output **y** be of the same dimension as the input **u**, that is, $m = n$. Physically, this is rather obvious, because the effects at the output caused by g can be neutralized by a unique input excitation **u** only when **u** and **y** are of the same dimension. With inputs g and **u**, as given in Eq. (7.69), the returned signal to the left of the breaking mark in Fig. 7.24 is found to be

$$\mathbf{h} = \mathbf{Bu} + \mathbf{AXg} = (-\mathbf{BD}^{-1}\mathbf{CX} + \mathbf{AX})\mathbf{g} \tag{7.70a}$$

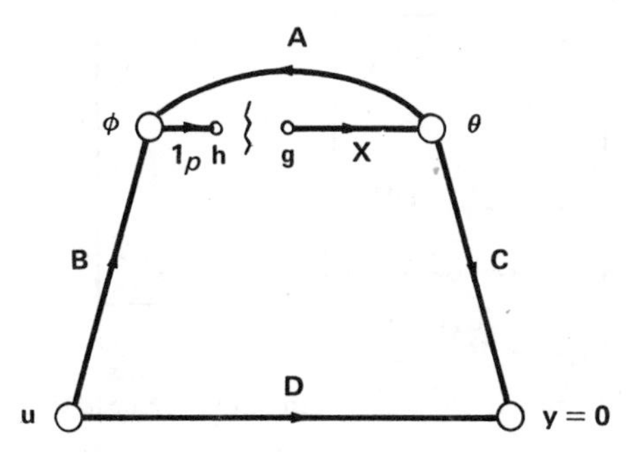

Figure 7.24 The physical interpretation of the null return difference matrix.

giving
$$\mathbf{g} - \mathbf{h} = (\mathbf{1}_p - \mathbf{AX} + \mathbf{BD}^{-1}\mathbf{CX})\mathbf{g} \tag{7.70b}$$

The square matrix

$$\hat{\mathbf{F}}(\mathbf{X}) = \mathbf{1}_p + \hat{\mathbf{T}}(\mathbf{X}) = \mathbf{1}_p - \mathbf{AX} + \mathbf{BD}^{-1}\mathbf{CX} = \mathbf{1}_p - \hat{\mathbf{A}}\mathbf{X} \tag{7.71}$$

is called the *null return difference matrix* with respect to $\mathbf{X}$, where

$$\hat{\mathbf{T}}(\mathbf{X}) = -\mathbf{AX} + \mathbf{BD}^{-1}\mathbf{CX} = -\hat{\mathbf{A}}\mathbf{X} \tag{7.72a}$$

$$\hat{\mathbf{A}} = \mathbf{A} - \mathbf{BD}^{-1}\mathbf{C} \tag{7.72b}$$

The square matrix $\hat{\mathbf{T}}(\mathbf{X})$ is referred to as the *null return ratio matrix*.

In the case where $\mathbf{X}$ denotes the admittance matrix of a p-port network, we have a simple interpretation of the null return difference matrix $\hat{\mathbf{F}}(\mathbf{X})$. Refer to the network of Fig. 7.25. An $(n+p)$-port network N_1 is terminated in a p-port network N_2. The port voltage vectors $\mathbf{V}_1$ and $\mathbf{V}_2$ and port current vectors $\mathbf{I}_s$ and $\mathbf{I}_2$ are defined in Fig. 7.25. As in the case for the return difference matrix, the p-port network N_2 can be replaced by an equivalent voltage-controlled current source vector

$$\hat{\mathbf{I}}_2 = \mathbf{Xg} \tag{7.73}$$

We then adjust the input source vector $\mathbf{I}_s$ to the n-port until the output vector $\mathbf{V}_1$ is identically zero. Under this situation, the voltage vector $\hat{\mathbf{h}}$ at the p-port network, being the returned voltage, is related to $\hat{\mathbf{I}}_2$ by

$$-\hat{\mathbf{I}}_2 = \hat{\mathbf{Y}}\hat{\mathbf{h}} \tag{7.74}$$

where $\hat{\mathbf{Y}}$ is the admittance matrix facing the p-port network N_2 when $\mathbf{V}_1$ is set to zero. In other words, $\hat{\mathbf{Y}}$ is the admittance matrix facing $\mathbf{X}$ when the input ports are short-circuited. Combining (7.73) and (7.74) yields

$$\hat{\mathbf{h}} = -\hat{\mathbf{Y}}^{-1}\mathbf{Xg} \tag{7.75}$$

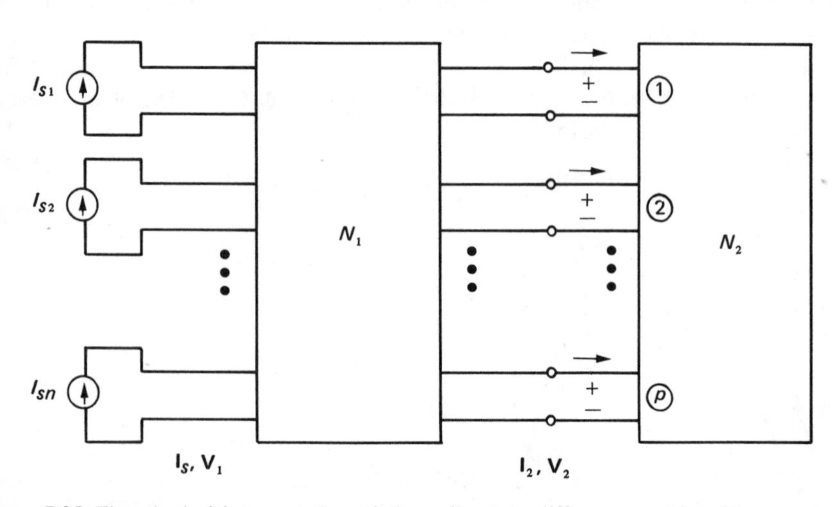

Figure 7.25 The physical interpretation of the null return difference matrix with respect to a p-port admittance matrix.

provided that $\hat{\mathbf{Y}}$ is nonsingular, giving

$$\mathbf{g} - \hat{\mathbf{h}} = (\mathbf{1}_p + \hat{\mathbf{Y}}^{-1}\mathbf{X})\mathbf{g} \tag{7.76}$$

The null return difference matrix is given by

$$\hat{\mathbf{F}}(\mathbf{X}) = \mathbf{1}_p + \hat{\mathbf{T}}(\mathbf{X}) = \mathbf{1}_p + \hat{\mathbf{Y}}^{-1}\mathbf{X} = \hat{\mathbf{Y}}^{-1}(\hat{\mathbf{Y}} + \mathbf{X}) \tag{7.77}$$

where $\hat{\mathbf{T}}(\mathbf{X}) = \hat{\mathbf{Y}}^{-1}\mathbf{X}$ is the null return ratio matrix.

We illustrate the above results by the following example.

Example 7.4 Consider the voltage-series feedback amplifier of Fig. 7.14, as discussed in Example 7.2. Assume that the voltage V_{25} is the output variable. Then, from (7.24), we have

$$\phi = \begin{bmatrix} V_{13} \\ V_{45} \end{bmatrix} = \begin{bmatrix} -90.782 & 45.391 \\ -942.507 & 0 \end{bmatrix} \begin{bmatrix} I_a \\ I_b \end{bmatrix} + \begin{bmatrix} 0.91748 \\ 0 \end{bmatrix} [V_s] = \mathbf{A}\theta + \mathbf{B}\mathbf{u} \tag{7.78a}$$

$$\mathbf{y} = [V_{25}] = [45.391 \quad -2372.32] \begin{bmatrix} I_a \\ I_b \end{bmatrix} + [0.04126]\,[V_s] = \mathbf{C}\theta + \mathbf{D}\mathbf{u} \tag{7.78b}$$

Substituting the coefficient matrices in (7.72b), we obtain

$$\hat{\mathbf{A}} = \mathbf{A} - \mathbf{B}\mathbf{D}^{-1}\mathbf{C} = \begin{bmatrix} -1100.12 & 52{,}797.6 \\ -942.507 & 0 \end{bmatrix} \tag{7.79}$$

giving the null return difference matrix with respect to $\mathbf{X}$ as

$$\hat{\mathbf{F}}(\mathbf{X}) = \mathbf{1}_2 - \hat{\mathbf{A}}\mathbf{X} = \begin{bmatrix} 51.055 & -2402.29 \\ 42.884 & 1 \end{bmatrix} \tag{7.80}$$

Suppose that the input current I_{51} is chosen as the output variable. From (7.24b) we have

$$\mathbf{y} = [I_{51}] = [-0.08252 \quad 0.04126] \begin{bmatrix} I_a \\ I_b \end{bmatrix} + [0.000862]\,[V_s] \tag{7.81}$$

and the corresponding null return difference matrix becomes

$$\hat{\mathbf{F}}(\mathbf{X}) = \mathbf{1}_2 - \hat{\mathbf{A}}\mathbf{X} = \begin{bmatrix} 1.13426 & -0.06713 \\ 42.8841 & 1 \end{bmatrix} \tag{7.82a}$$

where

$$\hat{\mathbf{A}} = \begin{bmatrix} -2.95085 & 1.47543 \\ -942.507 & 0 \end{bmatrix} \tag{7.82b}$$

7.2.3 The Transfer-Function Matrix and Feedback

In this section, we apply the results developed in the foregoing to show the effect of feedback on the transfer-function matrix $\mathbf{W}(\mathbf{X})$. Specifically, we express det $\mathbf{W}(\mathbf{X})$ in terms of det $\mathbf{X}(\mathbf{0})$ and the determinants of the return difference and null return difference matrices, thereby generalizing Blackman's impedance formula for a single input to a multiplicity of inputs.

Before we proceed to develop the desired relation, we first establish the following determinantal identity:

$$\det(\mathbf{1}_m + \mathbf{MN}) = \det(\mathbf{1}_n + \mathbf{NM}) \tag{7.83}$$

where $\mathbf{M}$ and $\mathbf{N}$ are two arbitrary matrices of orders $m \times n$ and $n \times m$, respectively. To verify this reault, we first show that (7.83) is valid when $\mathbf{M}$ and $\mathbf{N}$ are square matrices of order n. Then if $\mathbf{M}$ is nonsingular, we have

$$\begin{aligned}\det(\mathbf{1}_n + \mathbf{MN}) &= \det[\mathbf{M}(\mathbf{1}_n + \mathbf{NM})\mathbf{M}^{-1}] = \det \mathbf{M} \det(\mathbf{1}_n + \mathbf{NM}) \det \mathbf{M}^{-1} \\ &= \det(\mathbf{1}_n + \mathbf{NM})\end{aligned} \tag{7.84}$$

If $\mathbf{M}$ is singular, it has a zero eigenvalue, and hence for some λ the matrix $\lambda\mathbf{1}_n + \mathbf{M}$ is nonsingular. Thus, from (7.84), we have

$$\det[\mathbf{1}_n + (\lambda\mathbf{1}_n + \mathbf{M})\mathbf{N}] = \det[\mathbf{1}_n + \mathbf{N}(\lambda\mathbf{1}_n + \mathbf{M})] \tag{7.85}$$

Observe that both sides of (7.85) are polynomials in λ of at most degree n. They must be identical for all λ for which $\lambda\mathbf{1}_n + \mathbf{M}$ is nonsingular. Since $\lambda = 0$ is a zero of $\det(\lambda\mathbf{1}_n + \mathbf{M})$, there exists a positive number λ_0 such that $\lambda\mathbf{1}_n + \mathbf{M}$ is nonsingular for all real λ satisfying $0 < \lambda < \lambda_0$. Therefore, (7.85) is valid when $\lambda = 0$.

Consider now the situation where $\mathbf{M}$ and $\mathbf{N}$ are not square. Let

$$\tilde{\mathbf{M}} = \begin{bmatrix} \mathbf{M} & \mathbf{0} \\ \mathbf{0} & \mathbf{0} \end{bmatrix} \qquad \tilde{\mathbf{N}} = \begin{bmatrix} \mathbf{N} & \mathbf{0} \\ \mathbf{0} & \mathbf{0} \end{bmatrix} \tag{7.86}$$

be the augmented square matrices of order $n + m$. Then we have

$$\det(\mathbf{1}_{n+m} + \tilde{\mathbf{M}}\tilde{\mathbf{N}}) = \det \begin{bmatrix} \mathbf{1}_m + \mathbf{MN} & \mathbf{0} \\ \mathbf{0} & \mathbf{1}_n \end{bmatrix} = \det(\mathbf{1}_m + \mathbf{MN}) \tag{7.87a}$$

$$\det(\mathbf{1}_{n+m} + \tilde{\mathbf{N}}\tilde{\mathbf{M}}) = \det \begin{bmatrix} \mathbf{1}_n + \mathbf{NM} & 0 \\ \mathbf{0} & \mathbf{1}_m \end{bmatrix} = \det(\mathbf{1}_n + \mathbf{NM}) \tag{7.87b}$$

But $\tilde{\mathbf{M}}$ and $\tilde{\mathbf{N}}$ are square, and from the above discussion for square matrices we conclude that $\det(\mathbf{1}_{n+m} + \tilde{\mathbf{M}}\tilde{\mathbf{N}}) = \det(\mathbf{1}_{n+m} + \tilde{\mathbf{N}}\tilde{\mathbf{M}})$. The identity (7.83) follows. This completes the derivation of the identity.

Using (7.83), we next establish the following generalization of Blackman's formula for input impedance.

Theorem 7.1 In a multiple-loop feedback amplifier, if $\mathbf{W}(\mathbf{0}) = \mathbf{D}$ is nonsingular, then the determinant of the transfer-function matrix $\mathbf{W}(\mathbf{X})$ is related to the determinants of the return difference matrix $\mathbf{F}(\mathbf{X})$ and the null return difference matrix $\hat{\mathbf{F}}(\mathbf{X})$ by

$$\det \mathbf{W}(\mathbf{X}) = \det \mathbf{W}(\mathbf{0}) \frac{\det \hat{\mathbf{F}}(\mathbf{X})}{\det \mathbf{F}(\mathbf{X})} \tag{7.88}$$

PROOF From (7.16*a*) we have

$$\mathbf{W}(\mathbf{X}) = \mathbf{D}[\mathbf{1}_n + \mathbf{D}^{-1}\mathbf{C}\mathbf{X}(\mathbf{1}_p - \mathbf{A}\mathbf{X})^{-1}\mathbf{B}] \tag{7.89}$$

giving

$$\begin{aligned}\det \mathbf{W}(\mathbf{X}) &= \det \mathbf{W}(\mathbf{0}) \det [\mathbf{1}_n + \mathbf{D}^{-1}\mathbf{C}\mathbf{X}(\mathbf{1}_p - \mathbf{A}\mathbf{X})^{-1}\mathbf{B}] \\ &= \det \mathbf{W}(\mathbf{0}) \det [\mathbf{1}_p + \mathbf{B}\mathbf{D}^{-1}\mathbf{C}\mathbf{X}(\mathbf{1}_p - \mathbf{A}\mathbf{X})^{-1}] \\ &= \det \mathbf{W}(\mathbf{0}) \det (\mathbf{1}_p - \mathbf{A}\mathbf{X} + \mathbf{B}\mathbf{D}^{-1}\mathbf{C}\mathbf{X}) \det (\mathbf{1}_p - \mathbf{A}\mathbf{X})^{-1} \\ &= \frac{\det \mathbf{W}(\mathbf{0}) \det \hat{\mathbf{F}}(\mathbf{X})}{\det \mathbf{F}(\mathbf{X})}\end{aligned} \tag{7.90}$$

The second line follows directly from the identity (7.83).

The relation (7.88) is a direct extension of (4.176) for the scalar transfer function $w(x)$. As indicated in (4.177), the input impedance $Z(x)$ looking into a terminal pair can be conveniently expressed as

$$Z(x) = Z(0) \frac{F(\text{input short-circuited})}{F(\text{input open-circuited})} \tag{7.91}$$

A similar expression can be derived from (7.88) if $\mathbf{W}(\mathbf{X})$ denotes the impedance matrix of an n-port network as shown in Fig. 7.10. In this case, $\mathbf{F}(\mathbf{X})$ is the return difference matrix with respect to $\mathbf{X}$ for the situation when the n ports where the impedance matrix is defined are left open without any sources, and we write $\mathbf{F}(\mathbf{X}) = \mathbf{F}(\text{input open-circuited})$. Likewise, $\hat{\mathbf{F}}(\mathbf{X})$ is the return difference matrix with respect to $\mathbf{X}$ for the input port-current vector $\mathbf{I}_s$ and the output port-voltage vector $\mathbf{V}$ under the condition that $\mathbf{I}_s$ is adjusted so that the port-voltage vector $\mathbf{V}$ is identically zero. In other words, $\hat{\mathbf{F}}(\mathbf{X})$ is the return difference matrix for the situation when the n ports where the impedance matrix is defined are short-circuited, and we write $\hat{\mathbf{F}}(\mathbf{X}) = \mathbf{F}(\text{input short-circuited})$. Consequently, the determinant of the impedance matrix $\mathbf{Z}(\mathbf{X})$ of an n-port network can be expressed, from (7.88), as

$$\det \mathbf{Z}(\mathbf{X}) = \det \mathbf{Z}(\mathbf{0}) \frac{\det \mathbf{F}(\text{input short-circuited})}{\det \mathbf{F}(\text{input open-circuited})} \tag{7.92}$$

As an application of Eq. (7.92), let $\mathbf{X}$ be the admittance matrix of a p-port network N_2 that is imbedded in an $(n + p)$-port network N_1, as shown in Fig. 7.25. From (7.42*b*) and (7.77), we have

$$\mathbf{F}(\mathbf{X}) = \mathbf{Y}^{-1}(\mathbf{Y} + \mathbf{X}) = \mathbf{1}_p + \mathbf{Y}^{-1}\mathbf{X} \tag{7.93a}$$

$$\hat{\mathbf{F}}(\mathbf{X}) = \hat{\mathbf{Y}}^{-1}(\hat{\mathbf{Y}} + \mathbf{X}) = \mathbf{1}_p + \hat{\mathbf{Y}}^{-1}\mathbf{X} \tag{7.93b}$$

where $\mathbf{Y}$ and $\hat{\mathbf{Y}}$ denote the admittance matrices facing the p-port network N_2 when the input n ports of N_1 are open-circuited and short-circuited, respectively. The determinant of the impedance matrix $\mathbf{Z}(\mathbf{X})$ of the n-port network of Fig. 7.25 is given by

$$\det \mathbf{Z}(\mathbf{X}) = \det \mathbf{Z}(\mathbf{0}) \frac{\det (\mathbf{1}_p + \hat{\mathbf{Y}}^{-1}\mathbf{X})}{\det (\mathbf{1}_p + \mathbf{Y}^{-1}\mathbf{X})} \tag{7.94}$$

In particular, for $n = 1$ and $p = 1$, Eq. (7.94) reduces to

$$Z_{\text{in}} = z_{11} \frac{1 + x/y_{22}}{1 + xz_{22}} = z_{11} - \frac{xz_{12}z_{21}}{1 + xz_{22}} \tag{7.95}$$

where Z_{in} is the input impedance of the two-port network N_1 when its output is terminated in the admittance x, and z_{ij} and y_{ij} are the open-circuit impedance and the short-circuit admittance parameters of N_1.

Example 7.5 Refer again to the voltage-series feedback amplifier of Fig. 7.14. As computed in (7.37), the return difference matrix with respect to the two controlling parameters is given by

$$\mathbf{F}(\mathbf{X}) = \begin{bmatrix} 5.131 & -2.065 \\ 42.884 & 1 \end{bmatrix} \tag{7.96}$$

whose determinant is found to be

$$\det \mathbf{F}(\mathbf{X}) = 93.68646 \tag{7.97}$$

If V_{25} of Fig. 7.15 is chosen as the output and V_s as the input, the null return difference matrix is, from (7.80),

$$\hat{\mathbf{F}}(\mathbf{X}) = \begin{bmatrix} 51.055 & -2402.29 \\ 42.884 & 1 \end{bmatrix} \tag{7.98}$$

giving

$$\det \hat{\mathbf{F}}(\mathbf{X}) = 103{,}071 \tag{7.99}$$

By appealing to (7.88), the voltage gain V_{25}/V_s of the amplifier can be written as

$$w(\mathbf{X}) = \frac{V_{25}}{V_s} = w(\mathbf{0}) \frac{\det \hat{\mathbf{F}}(\mathbf{X})}{\det \mathbf{F}(\mathbf{X})} = 0.04126 \frac{103{,}071}{93.68646} = 45.39 \tag{7.100}$$

confirming (4.139), where $w(\mathbf{0}) = 0.04126$, as given in (7.78b).

Suppose, instead, that the input current I_{51} is chosen as the output and V_s as the input. From (7.82a) the null return difference matrix becomes

$$\hat{\mathbf{F}}(\mathbf{X}) = \begin{bmatrix} 1.13426 & -0.06713 \\ 42.8841 & 1 \end{bmatrix} \tag{7.101}$$

with
$$\det \hat{\mathbf{F}}(\mathbf{X}) = 4.01307 \tag{7.102}$$

By invoking (7.88), the amplifier input admittance is obtained as

$$w(\mathbf{X}) = \frac{I_{51}}{V_s} = w(\mathbf{0}) \frac{\det \hat{\mathbf{F}}(\mathbf{X})}{\det \mathbf{F}(\mathbf{X})} = 8.62 \cdot 10^{-4} \frac{4.01307}{93.6865} = 36.92 \ \mu\text{mho} \tag{7.103}$$

or 27.1 kΩ, confirming (7.27) or (7.28), where $w(\mathbf{0}) = 862$ μmho is found from (7.81).

Another useful application of the generalized Blackman's formula (7.88) is that it provides the basis of a procedure for the indirect measurement of return difference. Refer to Fig. 5.23. Suppose that we wish to measure the return difference $F(y_{21})$ with respect to the forward short-circuit transfer admittance y_{21} of a two-port device, which is represented by its admittance parameters y_{ij}. Choose the controlling parameters y_{21} and y_{12} as the elements of interest; then, from Fig. 5.23, we have

$$\theta = \begin{bmatrix} I_a \\ I_b \end{bmatrix} = \begin{bmatrix} y_{21} & 0 \\ 0 & y_{12} \end{bmatrix} \begin{bmatrix} V_1 \\ V_2 \end{bmatrix} = \mathbf{X}\phi \tag{7.104}$$

where I_a and I_b are the currents of the voltage-controlled current sources. By appealing to formula (7.92), the impedance looking into terminals a and b of Fig. 5.23 can be written as

$$z_{aa,bb}(y_{12}, y_{21}) = z_{aa,bb}(0, 0) \frac{\det \mathbf{F}(\text{input short-circuited})}{\det \mathbf{F}(\text{input open-circuited})} \tag{7.105}$$

When the input terminals a and b are open-circuited, the resulting return difference matrix is exactly the same as that found under normal operating conditions, and we have

$$\mathbf{F}(\text{input open-circuited}) = \mathbf{F}(\mathbf{X}) = \begin{bmatrix} F_{11} & F_{12} \\ F_{21} & F_{22} \end{bmatrix} \tag{7.106}$$

Since $\mathbf{F}(\mathbf{X}) = \mathbf{1}_2 - \mathbf{AX}$, the elements F_{11} and F_{21} are calculated with $y_{12} = 0$, whereas F_{12} and F_{22} are evaluated with $y_{21} = 0$. When the input terminals a and b are short-circuited, the feedback loop is interrupted and only the (2, 1)-element of the matrix $\mathbf{A}$ is nonzero, giving

$$\det \mathbf{F}(\text{input short-circuited}) = 1 \tag{7.107}$$

As will be shown in Eq. (7.206a), for diagonal $\mathbf{X}$ the return difference function $F(y_{21})$ can be expressed in terms of det $\mathbf{F}(\mathbf{X})$ and its (1, 1)-cofactor:

$$F(y_{21}) = \frac{\det \mathbf{F}(\mathbf{X})}{F_{22}} \tag{7.108}$$

Substituting these in (7.105) yields

$$F(y_{12})|_{y_{21}=0}F(y_{21}) = \frac{z_{aa,bb}(0, 0)}{z_{aa,bb}(y_{12}, y_{21})} \tag{7.109}$$

where $$F_{22} = 1 - a_{22}y_{12}|_{y_{21}=0} = F(y_{12})|_{y_{21}=0} \tag{7.110}$$

and a_{22} is the (2, 2)-element of **A**. Relation (7.109) was derived earlier in Chap. 5 and is given by Eq. (5.194). As demonstrated in Chap. 5, the elements $F(y_{12})|_{y_{21}=0}$ and $z_{aa,bb}(0, 0)$ can be measured by the arrangements of Figs. 5.31 and 5.32, respectively.

7.2.4 The Complementary Return Difference Matrix

In deriving the null return difference matrix $\hat{\mathbf{F}}(\mathbf{X})$ of (7.71), we assume that the matrix **D** is square and nonsingular. In the case that **D** is square and singular, the null return difference matrix is not defined. To facilitate our discussion, we introduce the concept of the complementary return difference matrix.

Refer to the matrix feedback-flow graph of Fig. 7.24. As before, we apply a signal p-vector **g** to the right of the breaking mark. We then adjust the input excitation n-vector **u** so that

$$\mathbf{u} + \check{\mathbf{y}} = \mathbf{0} \tag{7.111}$$

assuming that **u** and $\check{\mathbf{y}}$ are of the same dimension ($m = n$), where $\check{\mathbf{y}}$ denotes the output m-vector resulting from the input **g** alone with $\mathbf{u} = \mathbf{0}$. The required input vector **u** becomes

$$\mathbf{u} = -\check{\mathbf{y}} = -\mathbf{CXg} \tag{7.112}$$

and the returned signal to the left of the breaking mark is found to be

$$\mathbf{h} = \mathbf{Bu} + \mathbf{AXg} = (-\mathbf{BCX} + \mathbf{AX})\mathbf{g} \tag{7.113}$$

giving $$\mathbf{g} - \mathbf{h} = (\mathbf{1}_p - \mathbf{AX} + \mathbf{BCX})\mathbf{g} = (\mathbf{1}_p - \check{\mathbf{A}}\mathbf{X})\mathbf{g} \tag{7.114}$$

where $$\check{\mathbf{A}} = \mathbf{A} - \mathbf{BC} \tag{7.115}$$

The square matrix (see Prob. 7.29)

$$\check{\mathbf{F}}(\mathbf{X}) = \mathbf{1}_p - \check{\mathbf{A}}\mathbf{X} \tag{7.116}$$

relating the difference between the applied and returned signals to the applied signal under the condition (7.111) is called the *complementary return difference matrix*. Comparing this with (7.71), we see that $\check{\mathbf{F}}(\mathbf{X})$ can be obtained from $\hat{\mathbf{F}}(\mathbf{X})$ by setting $\mathbf{D} = \mathbf{1}_n$ in (7.71).

Theorem 7.1 does not apply when $\mathbf{W}(\mathbf{0}) = \mathbf{D}$ is singular, because in all such cases the null return difference matrix does not exist. As an application of (7.116), consider the situation where $m = n = 1$ with $w(\mathbf{0}) = D = 0$. Suppose that in Fig. 7.13 we replace the branch **D** with one having unit transmittance, as shown in Fig. 7.26. The graph transmission from u to y in Fig. 7.26 becomes $1 + w(\mathbf{X})$ and the

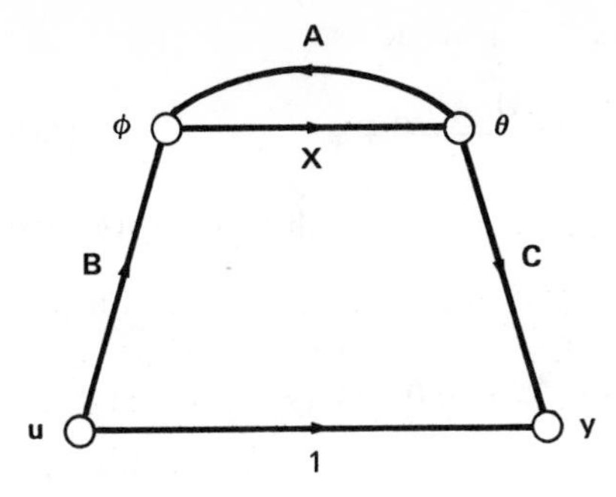

Figure 7.26 The fundamental matrix feedback-flow graph with $\mathbf{D} = 1$.

null return difference matrix with respect to $\mathbf{X}$ is obtained from (7.71) by setting $\mathbf{D} = 1$, the latter being recognized as the complementary return difference matrix with respect to $\mathbf{X}$ in Fig. 7.13. Substituting these in (7.88) gives

$$1 + w(\mathbf{X}) = [1 + w(\mathbf{0})] \frac{\det \check{\mathbf{F}}(\mathbf{X})}{\det \mathbf{F}(\mathbf{X})} \tag{7.117}$$

or (see Prob. 7.31)

$$w(\mathbf{X}) = \frac{\det \check{\mathbf{F}}(\mathbf{X}) - \det \mathbf{F}(\mathbf{X})}{\det \mathbf{F}(\mathbf{X})} \tag{7.118}$$

Example 7.6 Refer again to the voltage-series feedback amplifier of Fig. 7.14. Assume that the voltage V_{45} is the output variable and V_s is the input excitation. The required equations can be obtained from (7.24a) and are given by

$$\boldsymbol{\phi} = \begin{bmatrix} V_{13} \\ V_{45} \end{bmatrix} = \begin{bmatrix} -90.782 & 45.391 \\ -942.507 & 0 \end{bmatrix} \begin{bmatrix} I_a \\ I_b \end{bmatrix} + \begin{bmatrix} 0.91748 \\ 0 \end{bmatrix} [V_s] = \mathbf{A}\boldsymbol{\theta} + \mathbf{B}\mathbf{u} \tag{7.119a}$$

$$\mathbf{y} = [V_{45}] = [-942.507 \quad 0] \begin{bmatrix} I_a \\ I_b \end{bmatrix} + [0]\,[V_s] = \mathbf{C}\boldsymbol{\theta} + \mathbf{D}\mathbf{u} \tag{7.119b}$$

Since $\mathbf{D} = 0$, the null return difference matrix $\hat{\mathbf{F}}(\mathbf{X})$ is not defined. However, the complementary return difference matrix from (7.116) is found to be

$$\check{\mathbf{F}}(\mathbf{X}) = \mathbf{1}_2 - \check{\mathbf{A}}\mathbf{X} = \begin{bmatrix} -34.21468 & -2.06529 \\ 42.88407 & 1 \end{bmatrix} \tag{7.120}$$

where

$$\check{\mathbf{A}} = \mathbf{A} - \mathbf{B}\mathbf{C} = \begin{bmatrix} 773.949 & 45.391 \\ -942.507 & 0 \end{bmatrix} \tag{7.121}$$

giving

$$\det \check{\mathbf{F}}(\mathbf{X}) = 54.35336 \tag{7.122}$$

By appealing to (7.118), the voltage gain V_{45}/V_s is calculated as

$$w(\mathbf{X}) = \frac{V_{45}}{V_s} = \frac{\det \check{\mathbf{F}}(\mathbf{X}) - \det \mathbf{F}(\mathbf{X})}{\det \mathbf{F}(\mathbf{X})} = -0.42 \tag{7.123}$$

where $\det \mathbf{F}(\mathbf{X}) = 93.68646$ is given in (7.97). Comparing this with (4.142*a*) verifies our computation.

7.2.5 Physical Significance of the Matrices A, Â, and Ǎ

As indicated in (7.34), (7.71), and (7.116), the return difference matrix, the null return difference matrix, and the complementary return difference matrix are defined by the square matrices (see Prob. 7.30)

$$\mathbf{F}(\mathbf{X}) = \mathbf{1}_p - \mathbf{A}\mathbf{X} \tag{7.124a}$$

$$\hat{\mathbf{F}}(\mathbf{X}) = \mathbf{1}_p - \hat{\mathbf{A}}\mathbf{X} \tag{7.124b}$$

$$\check{\mathbf{F}}(\mathbf{X}) = \mathbf{1}_p - \check{\mathbf{A}}\mathbf{X} \tag{7.124c}$$

respectively, where

$$\hat{\mathbf{A}} = \mathbf{A} - \mathbf{B}\mathbf{D}^{-1}\mathbf{C} \tag{7.125a}$$

$$\check{\mathbf{A}} = \mathbf{A} - \mathbf{B}\mathbf{C} \tag{7.125b}$$

In this section, we present physical interpretations of the matrices $\mathbf{A}$, $\hat{\mathbf{A}}$, and $\check{\mathbf{A}}$ and demonstrate how they can be evaluated directly from the network.

From (7.13*a*) we see that if the controlled sources of θ are treated as the independent sources, the p-vector ϕ of the controlling variables is equal to $\mathbf{A}\theta$ when the input n-vector $\mathbf{u}$ is set to zero. In fact, we used this interpretation to calculate the elements of $\mathbf{A}$ in Example 7.2. As discussed in Sec. 7.2.2, the null return difference matrix $\hat{\mathbf{F}}(\mathbf{X})$ is the return difference matrix with respect to $\mathbf{X}$ when the input $\mathbf{u}$ is adjusted so that the output $\mathbf{y}$ is identically zero. Thus, $\hat{\mathbf{A}}$ is the matrix relating ϕ to θ when $\mathbf{u}$ is adjusted so that $\mathbf{y} = \mathbf{0}$. Let this $\mathbf{u}$ be designated as $\hat{\mathbf{u}}$, and we have

$$\phi = \hat{\mathbf{A}}\theta|_{\mathbf{u}=\hat{\mathbf{u}}} \tag{7.126}$$

Likewise, the complementary return difference matrix $\check{\mathbf{F}}(\mathbf{X})$ is the return difference matrix with respect to $\mathbf{X}$ when the input $\mathbf{u}$ is adjusted so that $\mathbf{u} + \check{\mathbf{y}} = \mathbf{0}$, where $\check{\mathbf{y}}$ is defined in (7.111). Therefore, $\check{\mathbf{A}}$ is the matrix relating ϕ to θ when $\mathbf{u}$ is adjusted so that $\mathbf{u} + \check{\mathbf{y}} = \mathbf{0}$. Let this $\mathbf{u}$ be designated as $\check{\mathbf{u}}$, and we get

$$\phi = \check{\mathbf{A}}\theta|_{\mathbf{u}=\check{\mathbf{u}}} \tag{7.127}$$

The evaluation of these matrices is simplified considerably by the fact that the controlled sources are treated as independent sources. We illustrate the above procedure by the following example.

Example 7.7 We use the equivalent network of Fig. 7.15 of the feedback amplifier of Fig. 7.14 to illustrate the evaluation of the matrices $\hat{\mathbf{A}}$ and $\check{\mathbf{A}}$. The evaluation of $\mathbf{A}$ was outlined in Example 7.2.

We first replace the two voltage-controlled current sources by two independent current sources as shown in Fig. 7.27. Assume that V_{25} is the output variable. To compute $\hat{\mathbf{A}}$, we must adjust the input voltage V_s so that the output $V_{25} = 0$. The indefinite-admittance matrix of Fig. 7.27 can be written down by inspection and is given by

$$\mathbf{Y} = 10^{-4}\begin{bmatrix} 9.37 & 0 & -9.09 & 0 & -0.28 \\ 0 & 4.256 & -2.128 & 0 & -2.128 \\ -9.09 & -2.128 & 111.218 & 0 & -100 \\ 0 & 0 & 0 & 10.61 & -10.61 \\ -0.28 & -2.128 & -100 & -10.61 & 113.02 \end{bmatrix} \tag{7.128}$$

By appealing to formulas (2.94) and (2.97), the output voltage V_{25} resulting from the three sources V_s, I_a, and I_b is found to be

$$\begin{aligned} V_{25} &= \frac{Y_{12,55}}{Y_{11,55}} V_s + \frac{Y_{32,45,11}}{Y_{11,55}} I_a + \frac{Y_{52,25,11}}{Y_{11,55}} I_b \\ &= 0.04126 V_s + 45.391 I_a - 2372.33 I_b \end{aligned} \tag{7.129}$$

confirming (7.78*b*). We remark that in applying (2.94) for the contributions of I_a and I_b, we must first short-circuit terminals 1 and 5. This is equivalent to the operations indicated in (7.129). Setting V_{25} to zero yields the desired input excitation

$$\hat{u} = \hat{V}_s = -1100.121 I_a + 57{,}497.1 I_b \tag{7.130}$$

By using this value for V_s, the voltages V_{13} and V_{45} and currents I_a and I_b in Fig. 7.27 are related by the equation (see Prob. 7.5)

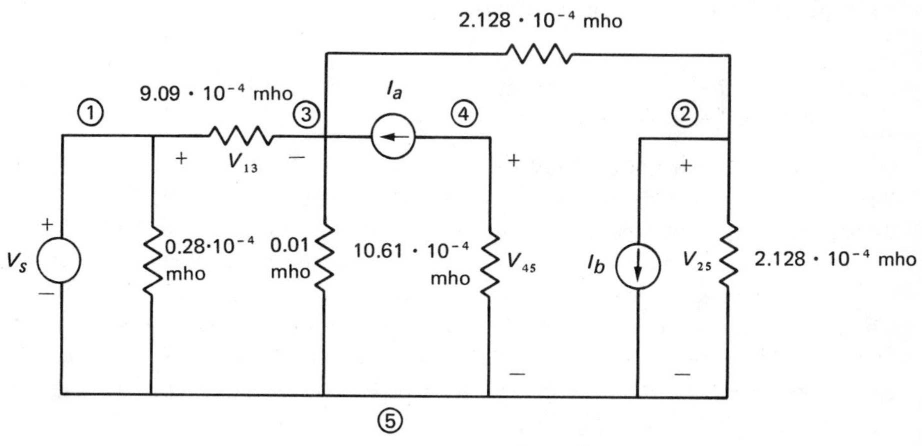

Figure 7.27 The network used to calculate the matrices $\hat{\mathbf{A}}$ and $\check{\mathbf{A}}$.

$$\boldsymbol{\phi} = \begin{bmatrix} V_{13} \\ V_{45} \end{bmatrix} = \begin{bmatrix} -1100.126 & 52,797.8 \\ -942.5071 & 0 \end{bmatrix} \begin{bmatrix} I_a \\ I_b \end{bmatrix} = \hat{\mathbf{A}}\boldsymbol{\theta} \tag{7.131}$$

The coefficient matrix is the same as that given in (7.79).

To compute $\check{\mathbf{A}}$, we must adjust the input V_s so that $V_s + \check{V}_{25} = 0$, where $\check{V}_{25}$ denotes the contributions to V_{25} by the current sources I_a and I_b. From (7.129) we have

$$\check{V}_{25} = 45.391 I_a - 2372.33 I_b \tag{7.132}$$

requiring that

$$\check{u} = \check{V}_s = -\check{V}_{25} = -45.391 I_a + 2372.33 I_b \tag{7.133}$$

Setting $V_s = \check{V}_s$ in Fig. 7.27, we obtain (see Prob. 7.6)

$$\boldsymbol{\phi} = \begin{bmatrix} V_{13} \\ V_{45} \end{bmatrix} = \begin{bmatrix} -132.427 & 2221.96 \\ -942.507 & 0 \end{bmatrix} \begin{bmatrix} I_a \\ I_b \end{bmatrix} = \check{\mathbf{A}}\boldsymbol{\theta} \tag{7.134}$$

For illustrative purposes, we also compute $\check{\mathbf{A}}$ by (7.125*b*). From (7.78) we get

$$\check{\mathbf{A}} = \mathbf{A} - \mathbf{BC} = \begin{bmatrix} -132.427 & 2221.95 \\ -942.507 & 0 \end{bmatrix} \tag{7.135}$$

giving

$$\check{\mathbf{F}}(\mathbf{X}) = \mathbf{1}_2 - \check{\mathbf{A}}\mathbf{X} = \begin{bmatrix} 7.0254 & -101.099 \\ 42.884 & 1 \end{bmatrix} \tag{7.136}$$

with det $\check{\mathbf{F}}(\mathbf{X}) = 4342.555$. By invoking (7.118), the voltage gain is found to be

$$w(\mathbf{X}) = \frac{V_{25}}{V_s} = \frac{\det \check{\mathbf{F}}(\mathbf{X}) - \det \mathbf{F}(\mathbf{X})}{\det \mathbf{F}(\mathbf{X})} = 45.35 \tag{7.137}$$

confirming (7.100) within computational accuracy.

7.2.6 Invariance of the Determinants of Feedback Matrices

So far we have introduced three feedback matrices $\mathbf{F}(\mathbf{X})$, $\hat{\mathbf{F}}(\mathbf{X})$, and $\check{\mathbf{F}}(\mathbf{X})$ and studied some of their properties. These matrices generally depend on the choice of the elements of interest $\mathbf{X}$. In this section, we demonstrate that their determinants are invariant with respect to the ordering of the components of $\boldsymbol{\theta}$ and $\boldsymbol{\phi}$ in $\boldsymbol{\theta} = \mathbf{X}\boldsymbol{\phi}$.

Let $\boldsymbol{\theta}_\gamma$ and $\boldsymbol{\phi}_\gamma$ respectively denote the vectors obtained from $\boldsymbol{\theta}$ and $\boldsymbol{\phi}$ by any reordering of the components. Then there exist nonsingular matrices $\mathbf{Q}$ and $\mathbf{P}$ of orders q and p such that $\boldsymbol{\theta}_\gamma = \mathbf{Q}\boldsymbol{\theta}$ and $\boldsymbol{\phi}_\gamma = \mathbf{P}\boldsymbol{\phi}$. In terms of $\boldsymbol{\theta}_\gamma$ and $\boldsymbol{\phi}_\gamma$, the controlled-source constraints become $\boldsymbol{\theta}_\gamma = \mathbf{X}_\gamma\boldsymbol{\phi}_\gamma$, where $\mathbf{X}_\gamma = \mathbf{QXP}^{-1}$, and the equations corresponding to (7.13) can be written as

$$\phi_\gamma = \mathbf{A}_\gamma \theta_\gamma + \mathbf{B}_\gamma \mathbf{u} = \mathbf{PAQ}^{-1}\theta_\gamma + \mathbf{PBu} \tag{7.138a}$$

$$\mathbf{y} = \mathbf{C}_\gamma \theta_\gamma + \mathbf{D}_\gamma \mathbf{u} = \mathbf{CQ}^{-1}\theta_\gamma + \mathbf{Du} \tag{7.138b}$$

where
$$\mathbf{A}_\gamma = \mathbf{PAQ}^{-1} \qquad \mathbf{B}_\gamma = \mathbf{PB} \qquad \mathbf{C}_\gamma = \mathbf{CQ}^{-1} \qquad \mathbf{D}_\gamma = \mathbf{D} \tag{7.139}$$

By applying (7.83), the determinant of the return difference matrix $\mathbf{F}(\mathbf{X}_\gamma)$ can be manipulated into the form

$$\begin{aligned} \det \mathbf{F}(\mathbf{X}_\gamma) &= \det(\mathbf{1}_p - \mathbf{A}_\gamma \mathbf{X}_\gamma) = \det(\mathbf{1}_p - \mathbf{PAQ}^{-1}\mathbf{QXP}^{-1}) \\ &= \det(\mathbf{1}_p - \mathbf{PAXP}^{-1}) = \det(\mathbf{1}_p - \mathbf{P}^{-1}\mathbf{PAX}) \\ &= \det(\mathbf{1}_p - \mathbf{AX}) = \det \mathbf{F}(\mathbf{X}) \end{aligned} \tag{7.140}$$

Likewise, the determinant of the null return difference matrix $\hat{\mathbf{F}}(\mathbf{X}_\gamma)$ can be written as

$$\begin{aligned} \det \hat{\mathbf{F}}(\mathbf{X}_\gamma) &= \det[\mathbf{1}_p - (\mathbf{A}_\gamma - \mathbf{B}_\gamma \mathbf{D}_\gamma^{-1}\mathbf{C}_\gamma)\mathbf{X}_\gamma] = \det[\mathbf{1}_q - \mathbf{X}_\gamma(\mathbf{A}_\gamma - \mathbf{B}_\gamma \mathbf{D}_\gamma^{-1}\mathbf{C}_\gamma)] \\ &= \det[\mathbf{1}_q - \mathbf{QXP}^{-1}(\mathbf{PAQ}^{-1} - \mathbf{PBD}^{-1}\mathbf{CQ}^{-1})] \\ &= \det[\mathbf{1}_q - \mathbf{QX}(\mathbf{AQ}^{-1} - \mathbf{BD}^{-1}\mathbf{CQ}^{-1})] \\ &= \det[\mathbf{1}_p - (\mathbf{AQ}^{-1} - \mathbf{BD}^{-1}\mathbf{CQ}^{-1})\mathbf{QX}] \\ &= \det[\mathbf{1}_p - (\mathbf{A} - \mathbf{BD}^{-1}\mathbf{C})\mathbf{X}] = \det \hat{\mathbf{F}}(\mathbf{X}) \end{aligned} \tag{7.141}$$

The above property is evidently shared by $\det \check{\mathbf{F}}(\mathbf{X}_\gamma)$, because $\check{\mathbf{F}}(\mathbf{X}_\gamma)$ can be obtained from $\hat{\mathbf{F}}(\mathbf{X}_\gamma)$ by setting $\mathbf{D} = \mathbf{1}_n$. Thus, $\det \mathbf{F}(\mathbf{X})$, $\det \hat{\mathbf{F}}(\mathbf{X})$, and $\det \check{\mathbf{F}}(\mathbf{X})$ are invariant with respect to the ordering of the components of the controlling and controlled variables.

Example 7.8 Suppose that we reorder the elements of ϕ in (7.19) as given below:

$$\theta_\gamma = \begin{bmatrix} I_a \\ I_b \end{bmatrix} = 10^{-4} \begin{bmatrix} 0 & 455 \\ 455 & 0 \end{bmatrix} \begin{bmatrix} V_{45} \\ V_{13} \end{bmatrix} = \mathbf{X}_\gamma \phi_\gamma \tag{7.142}$$

The equations corresponding to (7.24) are given by

$$\begin{bmatrix} V_{45} \\ V_{13} \end{bmatrix} = \begin{bmatrix} -942.507 & 0 \\ -90.782 & 45.391 \end{bmatrix} \begin{bmatrix} I_a \\ I_b \end{bmatrix} + \begin{bmatrix} 0 \\ 0.91748 \end{bmatrix} [V_s] \tag{7.143a}$$

$$[V_{25}] = [45.391 \quad -2372.32] \begin{bmatrix} I_a \\ I_b \end{bmatrix} + [0.04126]\,[V_s] \tag{7.143b}$$

From (7.124) the feedback matrices are calculated as follows:

$$\mathbf{F}(\mathbf{X}_\gamma) = \mathbf{1}_2 - \mathbf{A}_\gamma \mathbf{X}_\gamma = \begin{bmatrix} 1 & 42.8841 \\ -2.06529 & 5.13058 \end{bmatrix} \tag{7.144a}$$

$$\hat{\mathbf{F}}(\mathbf{X}_\gamma) = \mathbf{1}_2 - (\mathbf{A}_\gamma - \mathbf{B}_\gamma \mathbf{D}_\gamma^{-1} \mathbf{C}_\gamma)\mathbf{X}_\gamma = \begin{bmatrix} 1 & 42.8841 \\ -2402.29 & 51.0555 \end{bmatrix} \tag{7.144b}$$

$$\check{\mathbf{F}}(\mathbf{X}_\gamma) = \mathbf{1}_2 - (\mathbf{A}_\gamma - \mathbf{B}_\gamma \mathbf{C}_\gamma)\mathbf{X}_\gamma = \begin{bmatrix} 1 & 42.8841 \\ -101.0986 & 7.02544 \end{bmatrix} \tag{7.144c}$$

giving

$$\det \mathbf{F}(\mathbf{X}_\gamma) = 93.69868 \tag{7.145a}$$

$$\det \hat{\mathbf{F}}(\mathbf{X}_\gamma) = 103{,}071.1 \tag{7.145b}$$

$$\det \check{\mathbf{F}}(\mathbf{X}_\gamma) = 4342.548 \tag{7.145c}$$

The matrices (7.144) are different from those given in (7.96), (7.98), and (7.136), but their determinants are the same.

7.3 EXTENSIONS TO FEEDBACK MATRICES

In this section, the concepts of return difference and null return difference matrices are generalized. Their properties, physical significance, and interrelations are presented.

As discussed in Sec. 5.4, the general return difference $F_k(x)$ with respect to the controlling parameter x of a voltage-controlled current source $I = xV$ can be interpreted as follows: Replace the controlled current source $I = xV$ by the parallel combination of two controlled sources $I_1 = kV_1$ and $I_2 = x'V_2$, with $x' = x - k$, as shown in Fig. 5.10. Break the controlling branch of the controlled current source $I_2 = x'V_2$ and apply a voltage source of 1 V to the right of the breaking mark. The difference of the 1-V excitation and the returned voltage appearing at the left of the breaking mark under the condition that the input excitation to the amplifier is set to zero is the general return difference $F_k(x)$. The significance of this interpretation is that it corresponds to the situation where the feedback amplifier under study is made partially active rather than completely dead, as in the original interpretation of $F(x)$. A similar procedure can now be employed to define the general return difference and null return difference matrices.

Refer to the fundamental matrix feedback-flow graph of Fig. 7.13. The controlled source vector $\boldsymbol{\theta} = \mathbf{X}\boldsymbol{\phi}$ corresponding to the branch $\mathbf{X}$ is replaced by two controlled source p-vectors:

$$\boldsymbol{\theta} = \mathbf{X}\boldsymbol{\phi} = (\mathbf{X} - \mathbf{K})\boldsymbol{\phi} + \mathbf{K}\boldsymbol{\phi} = \boldsymbol{\theta}_1 + \mathbf{K}\boldsymbol{\phi} \tag{7.146}$$

where $\mathbf{K}$ is an arbitrary reference matrix of order $q \times p$, and

$$\boldsymbol{\theta}_1 = \mathbf{X}_1 \boldsymbol{\phi}_1 \quad \text{and} \quad \mathbf{X}_1 = \mathbf{X} - \mathbf{K} \tag{7.147}$$

with $\boldsymbol{\phi}_1 = \boldsymbol{\phi}$. By using these in conjunction with (7.13), the associated matrix signal-flow graph is presented in Fig. 7.28. To emphasize the effect of $\mathbf{X}_1$, we apply the procedure outlined in Sec. 7.1 to absorb the nodes $\boldsymbol{\phi}$ and $\boldsymbol{\theta}$ in Fig. 7.28. After

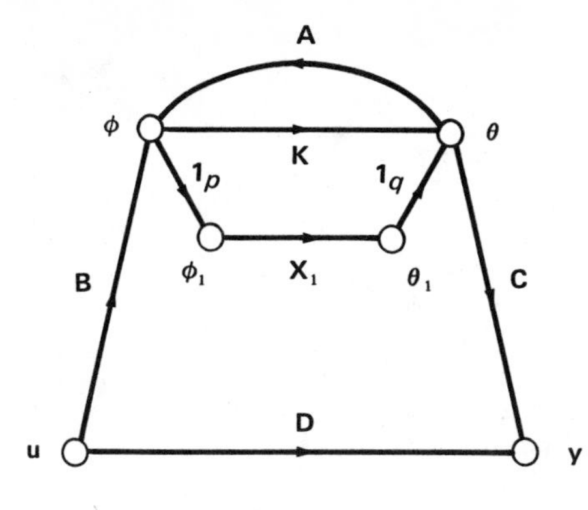

Figure 7.28 The matrix signal-flow graph associated with the systems (7.13), (7.146), and (7.147).

removing node ϕ, the resulting graph is shown in Fig. 7.29*a*. Repeat the process once more by removing node θ in Fig. 7.29*a*. The final graph is presented in Fig. 7.29*b* with

$$\mathbf{A}_1 = \mathbf{A}(\mathbf{1}_q - \mathbf{KA})^{-1} = (\mathbf{1}_p - \mathbf{AK})^{-1}\mathbf{A} \tag{7.148a}$$

$$\mathbf{B}_1 = \mathbf{B} + \mathbf{A}(\mathbf{1}_q - \mathbf{KA})^{-1}\mathbf{KB} = (\mathbf{1}_p - \mathbf{AK})^{-1}\mathbf{B} \tag{7.148b}$$

$$\mathbf{C}_1 = \mathbf{C}(\mathbf{1}_q - \mathbf{KA})^{-1} \tag{7.148c}$$

$$\mathbf{D}_1 = \mathbf{D} + \mathbf{C}(\mathbf{1}_q - \mathbf{KA})^{-1}\mathbf{KB} = \mathbf{D} + \mathbf{CK}(\mathbf{1}_p - \mathbf{AK})^{-1}\mathbf{B} \tag{7.148d}$$

In Fig. 7.29*b*, we break the input of the branch with transmittance $\mathbf{X}_1$, set the input excitation vector $\mathbf{u} = \mathbf{0}$, and apply a signal p-vector $\mathbf{g}$ to the right of the breaking mark, as shown in Fig. 7.30. The returned signal p-vector $\mathbf{h}$ to the left of

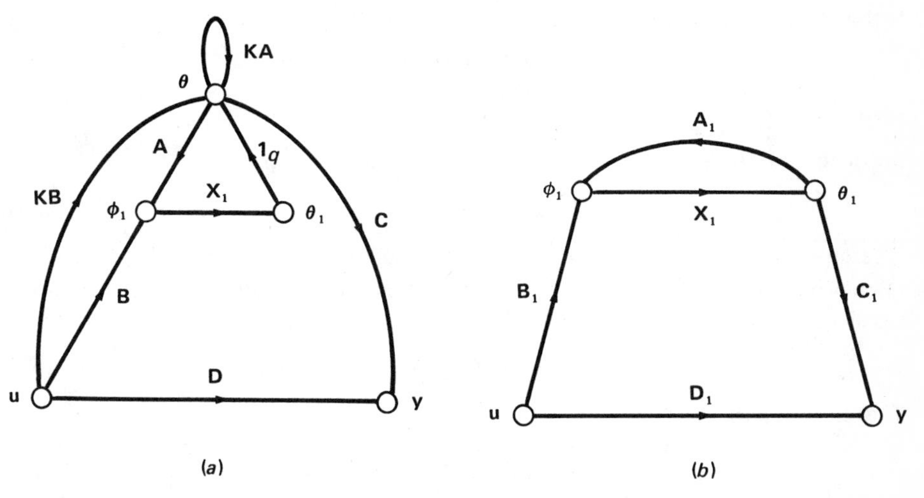

Figure 7.29 (*a*) The resulting matrix signal-flow graph after absorbing the node ϕ in that of Fig. 7.28 and (*b*) the resulting matrix signal-flow graph after absorbing the node θ in (*a*).

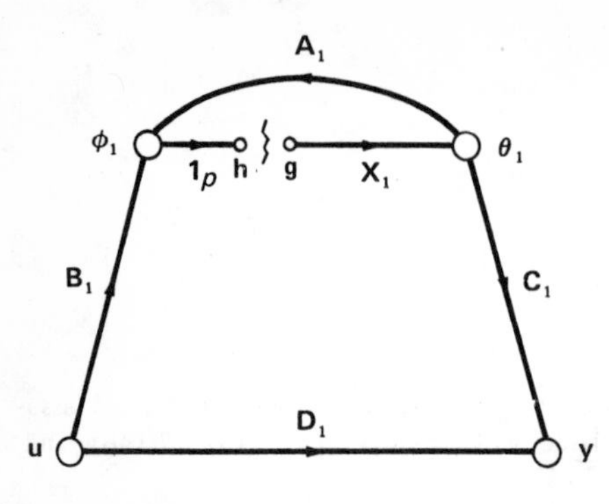

Figure 7.30 The physical interpretation of the general return difference matrix $\mathbf{F_K(X)}$.

the breaking mark is found to be $\mathbf{h} = \mathbf{A}_1\mathbf{X}_1\mathbf{g}$. The square matrix $-\mathbf{A}_1\mathbf{X}_1$ is called the *general return ratio matrix* with respect to $\mathbf{X}$ for a general reference $\mathbf{K}$ and is denoted by

$$\mathbf{T_K(X)} = -\mathbf{A}_1\mathbf{X}_1 \tag{7.149}$$

The difference between the applied signal vector $\mathbf{g}$ and the returned signal $\mathbf{h}$ is given by

$$\mathbf{g} - \mathbf{h} = (\mathbf{1}_p - \mathbf{A}_1\mathbf{X}_1)\mathbf{g} \tag{7.150}$$

The square matrix

$$\mathbf{F_K(X)} = \mathbf{1}_p - \mathbf{A}_1\mathbf{X}_1 \tag{7.151}$$

is defined as the *general return difference matrix* with respect to $\mathbf{X}$ for a reference $\mathbf{K}$. For convenience, we write $\mathbf{T_0(X)} = \mathbf{T(X)}$ and $\mathbf{F_0(X)} = \mathbf{F(X)}$.

Likewise, if, in Fig. 7.30, we adjust the input vector $\mathbf{u}$ so that the total output vector $\mathbf{y}$ resulting from $\mathbf{u}$ and $\mathbf{g}$ is zero, the returned signal becomes

$$\mathbf{h} = (\mathbf{A}_1\mathbf{X}_1 - \mathbf{B}_1\mathbf{D}_1^{-1}\mathbf{C}_1\mathbf{X}_1)\mathbf{g} = \hat{\mathbf{A}}_1\mathbf{X}_1\mathbf{g} \tag{7.152}$$

where

$$\hat{\mathbf{A}}_1 = \mathbf{A}_1 - \mathbf{B}_1\mathbf{D}_1^{-1}\mathbf{C}_1 \tag{7.153}$$

provided that $\mathbf{D}_1$ is square and nonsingular. The square matrix

$$\hat{\mathbf{T}}_\mathbf{K}(\mathbf{X}) = -\hat{\mathbf{A}}_1\mathbf{X}_1 \tag{7.154}$$

is referred to as the *general null return ratio matrix* with respect to $\mathbf{X}$ for a general reference $\mathbf{K}$. The difference between the applied signal vector $\mathbf{g}$ and the returned signal vector $\mathbf{h}$ is given by

$$\mathbf{g} - \mathbf{h} = (\mathbf{1}_p - \hat{\mathbf{A}}_1\mathbf{X}_1)\mathbf{g} = [\mathbf{1}_p + \hat{\mathbf{T}}_\mathbf{K}(\mathbf{X})]\,\mathbf{g} \tag{7.155}$$

The square matrix

$$\hat{\mathbf{F}}_\mathbf{K}(\mathbf{X}) = \mathbf{1}_p - \hat{\mathbf{A}}_1\mathbf{X}_1 = \mathbf{1}_p + \hat{\mathbf{T}}_\mathbf{K}(\mathbf{X}) \tag{7.156}$$

is the *general null return difference matrix* with respect to $\mathbf{X}$ for a general reference $\mathbf{K}$. For simplicity, we write $\hat{\mathbf{T}}_\mathbf{0}(\mathbf{X}) = \hat{\mathbf{T}}(\mathbf{X})$ and $\hat{\mathbf{F}}_\mathbf{0}(\mathbf{X}) = \hat{\mathbf{F}}(\mathbf{X})$.

Alternatively, we can break the branch $\mathbf{X}_1$ of Fig. 7.28 and apply a signal p-vector $\mathbf{g}$ to the right of the breaking mark, as shown in Fig. 7.31. We then adjust the input vector $\mathbf{u}$ so that the total output $\mathbf{y}$ resulting from $\mathbf{u}$ and $\mathbf{g}$ is zero, giving

$$\boldsymbol{\theta} = \mathbf{X}_1\mathbf{g} + \mathbf{K}\boldsymbol{\phi} \tag{7.157a}$$

$$\boldsymbol{\phi} = \mathbf{A}\boldsymbol{\theta} + \mathbf{B}\mathbf{u} \tag{7.157b}$$

$$\mathbf{y} = \mathbf{C}\boldsymbol{\theta} + \mathbf{D}\mathbf{u} = \mathbf{0} \tag{7.157c}$$

Substituting (7.157a) in (7.157b) and (7.157c) and then eliminating $\mathbf{u}$ yield

$$(\mathbf{1}_p - \mathbf{AK} + \mathbf{BD}^{-1}\mathbf{CK})\boldsymbol{\phi} = (\mathbf{A} - \mathbf{BD}^{-1}\mathbf{C})\mathbf{X}_1\mathbf{g} \tag{7.158}$$

Since $\mathbf{h} = \boldsymbol{\phi}$, the above equation can be rewritten as

$$\mathbf{h} = (\mathbf{1}_p - \hat{\mathbf{A}}\mathbf{K})^{-1}\hat{\mathbf{A}}\mathbf{X}_1\mathbf{g} \tag{7.159}$$

where $\hat{\mathbf{A}} = \mathbf{A} - \mathbf{BD}^{-1}\mathbf{C}$, as given in (7.72$b$). Comparing (7.159) with (7.152), we obtain

$$\hat{\mathbf{A}}_1 = (\mathbf{1}_p - \hat{\mathbf{A}}\mathbf{K})^{-1}\hat{\mathbf{A}} \tag{7.160}$$

We now proceed to establish relations similar to those given in (5.85) and (5.97). To this end, we substitute (7.147) and (7.148a) in (7.151), yielding

$$\begin{aligned}\mathbf{F_K}(\mathbf{X}) &= \mathbf{1}_p - \mathbf{A}_1\mathbf{X}_1 = \mathbf{1}_p - (\mathbf{1}_p - \mathbf{AK})^{-1}\mathbf{A}(\mathbf{X} - \mathbf{K})\\ &= (\mathbf{1}_p - \mathbf{AK})^{-1}(\mathbf{1}_p - \mathbf{AK} - \mathbf{AX} + \mathbf{AK})\\ &= (\mathbf{1}_p - \mathbf{AK})^{-1}(\mathbf{1}_p - \mathbf{AX})\end{aligned} \tag{7.161}$$

or

$$\mathbf{F_K}(\mathbf{X}) = \mathbf{F}^{-1}(\mathbf{K})\mathbf{F}(\mathbf{X}) \tag{7.162}$$

In words, this states that the general return difference matrix with respect to $\mathbf{X}$ for a general reference $\mathbf{K}$ is equal to the product of the inverse of the return difference matrix with respect to $\mathbf{K}$ and the return difference matrix with respect to $\mathbf{X}$, both being defined for the zero reference.

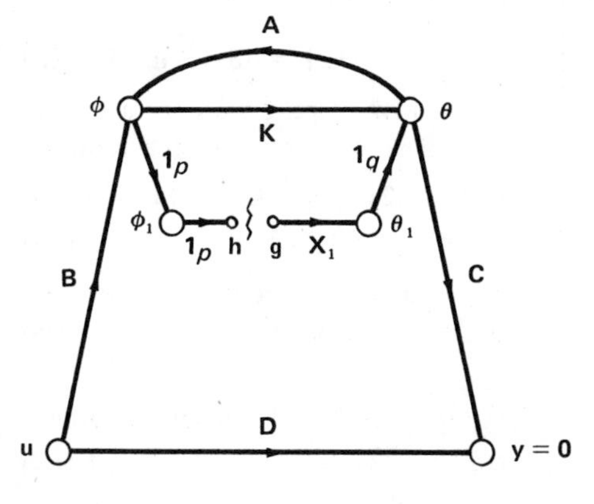

Figure 7.31 An alternative physical interpretation of the general return difference matrix $\mathbf{F_K}(\mathbf{X})$.

For the null return difference matrix, we substitute (7.160) in (7.156) and obtain

$$\begin{aligned}\hat{\mathbf{F}}_{\mathbf{K}}(\mathbf{X}) &= \mathbf{1}_p - \hat{\mathbf{A}}_1\mathbf{X}_1 = \mathbf{1}_p - (\mathbf{1}_p - \hat{\mathbf{A}}\mathbf{K})^{-1}\hat{\mathbf{A}}\mathbf{X}_1 \\ &= (\mathbf{1}_p - \hat{\mathbf{A}}\mathbf{K})^{-1}(\mathbf{1}_p - \hat{\mathbf{A}}\mathbf{K} - \hat{\mathbf{A}}\mathbf{X}_1) \\ &= (\mathbf{1}_p - \hat{\mathbf{A}}\mathbf{K})^{-1}(\mathbf{1}_p - \hat{\mathbf{A}}\mathbf{X})\end{aligned} \tag{7.163}$$

or

$$\hat{\mathbf{F}}_{\mathbf{K}}(\mathbf{X}) = \hat{\mathbf{F}}^{-1}(\mathbf{K})\hat{\mathbf{F}}(\mathbf{X}) \tag{7.164}$$

In fact, by repeated applications of (7.162) and (7.164), the return difference and null return difference matrices can be factored in finite product forms. To see this, let $\mathbf{K}_1$ denote the reference condition for $\mathbf{X}$, where $\mathbf{K}_1$ is obtained from $\mathbf{X}$ by setting some of its elements to particular values. Then from (7.162) we have

$$\mathbf{F}(\mathbf{X}) = \mathbf{F}(\mathbf{K}_1)\mathbf{F}_{\mathbf{K}_1}(\mathbf{X}) \tag{7.165a}$$

We next consider $\mathbf{K}_1$ and a reference matrix $\mathbf{K}_2$, where $\mathbf{K}_2$ is obtained from $\mathbf{K}_1$ by setting some of the elements of $\mathbf{K}_1$ to particular values. This gives

$$\mathbf{F}(\mathbf{K}_1) = \mathbf{F}(\mathbf{K}_2)\mathbf{F}_{\mathbf{K}_2}(\mathbf{K}_1) \tag{7.165b}$$

Continuing this process, $\mathbf{F}(\mathbf{X})$ can be factored into the form

$$\mathbf{F}(\mathbf{X}) = \mathbf{F}(\mathbf{K}_n) \prod_{i=n}^{1} \mathbf{F}_{\mathbf{K}_i}(\mathbf{K}_{i-1}) \tag{7.166}$$

where $\mathbf{K}_i$ is obtained from $\mathbf{K}_{i-1}$ by setting some of the elements of $\mathbf{K}_{i-1}$ to special values for $i = 1, 2, \ldots, n$ with $\mathbf{K}_0 = \mathbf{X}$. Likewise, we can show that

$$\hat{\mathbf{F}}(\mathbf{X}) = \hat{\mathbf{F}}(\mathbf{K}_n) \prod_{i=n}^{1} \hat{\mathbf{F}}_{\mathbf{K}_i}(\mathbf{K}_{i-1}) \tag{7.167}$$

Example 7.9 From Example 7.2, the equations for the voltage-series feedback amplifier of Fig. 7.15 are given by

$$\begin{bmatrix} V_{13} \\ V_{45} \end{bmatrix} = \begin{bmatrix} -90.782 & 45.391 \\ -942.507 & 0 \end{bmatrix} \begin{bmatrix} I_a \\ I_b \end{bmatrix} + \begin{bmatrix} 0.91748 \\ 0 \end{bmatrix} [V_s] = \mathbf{A}\theta + \mathbf{B}\mathbf{u} \tag{7.168a}$$

$$[I_{51}] = [-0.08252 \quad 0.04126] \begin{bmatrix} I_a \\ I_b \end{bmatrix} + [0.000862]\,[V_s] = \mathbf{C}\theta + \mathbf{D}\mathbf{u} \tag{7.168b}$$

The elements of interest are the transconductances of the two controlled sources, each having the value 0.0455 mho. Assume that their reference values are 0.01 mho and 0.02 mho. Then we have

$$\mathbf{X} = \begin{bmatrix} 0.0455 & 0 \\ 0 & 0.0455 \end{bmatrix} \qquad \mathbf{K} = \begin{bmatrix} 0.01 & 0 \\ 0 & 0.02 \end{bmatrix} \tag{7.169}$$

giving
$$\mathbf{X}_1 = \mathbf{X} - \mathbf{K} = \begin{bmatrix} 0.0355 & 0 \\ 0 & 0.0255 \end{bmatrix} \tag{7.170}$$

The corresponding matrices of (7.148) are computed as follows:

$$\mathbf{A}_1 = (\mathbf{1}_2 - \mathbf{AK})^{-1}\mathbf{A} = \begin{bmatrix} -90.44353 & 4.33779 \\ -90.06962 & -40.88412 \end{bmatrix} \tag{7.171a}$$

$$\mathbf{B}_1 = (\mathbf{1}_2 - \mathbf{AK})^{-1}\mathbf{B} = \begin{bmatrix} 0.08768 \\ -0.82638 \end{bmatrix} \tag{7.171b}$$

$$\mathbf{C}_1 = \mathbf{C}(\mathbf{1}_2 - \mathbf{KA})^{-1} = [-0.08221 \quad 0.003943] \tag{7.171c}$$

$$\mathbf{D}_1 = \mathbf{D} + \mathbf{CK}(\mathbf{1}_2 - \mathbf{AK})^{-1}\mathbf{B} = [0.0001077] \tag{7.171d}$$

The general return difference matrix with respect to **X** for the reference **K** is found to be

$$\mathbf{F}_{\mathbf{K}}(\mathbf{X}) = \mathbf{1}_2 - \mathbf{A}_1\mathbf{X}_1 = \begin{bmatrix} 4.21075 & -0.11061 \\ 3.19747 & 2.04254 \end{bmatrix} \tag{7.172}$$

Alternatively, using (7.162) in conjunction with (7.37), we get

$$\mathbf{F}_{\mathbf{K}}(\mathbf{X}) = \mathbf{F}^{-1}(\mathbf{K})\mathbf{F}(\mathbf{X}) = \begin{bmatrix} 4.21078 & -0.11059 \\ 3.19707 & 2.04229 \end{bmatrix} \tag{7.173}$$

confirming (7.172) within computational error, where

$$\mathbf{F}(\mathbf{K}) = \begin{bmatrix} 1.90782 & -0.90782 \\ 9.42507 & 1 \end{bmatrix} \tag{7.174}$$

For the null return difference matrix with respect to **X** for the reference value **K**, we first compute

$$\hat{\mathbf{A}}_1 = \mathbf{A}_1 - \mathbf{B}_1\mathbf{D}_1^{-1}\mathbf{C}_1 = \begin{bmatrix} -23.51528 & 1.12774 \\ -720.8654 & -10.62956 \end{bmatrix} \tag{7.175}$$

This matrix can also be determined from (7.160) and is found to be

$$\hat{\mathbf{A}}_1 = (\mathbf{1}_2 - \hat{\mathbf{A}}\mathbf{K})^{-1}\hat{\mathbf{A}} = \begin{bmatrix} -23.52616 & 1.12831 \\ -720.7764 & -10.63441 \end{bmatrix} \tag{7.176}$$

where $\hat{\mathbf{A}}$ is given in (7.82*b*). From (7.156) we obtain

$$\hat{\mathbf{F}}_{\mathbf{K}}(\mathbf{X}) = \mathbf{1}_2 - \hat{\mathbf{A}}_1\mathbf{X}_1 = \begin{bmatrix} 1.83518 & -0.02877 \\ 25.58756 & 1.27118 \end{bmatrix} \tag{7.177}$$

or from (7.164) in conjunction with (7.82*a*) we have

$$\hat{\mathbf{F}}_{\mathbf{K}}(\mathbf{X}) = \hat{\mathbf{F}}^{-1}(\mathbf{K})\hat{\mathbf{F}}(\mathbf{X}) = \begin{bmatrix} 1.83517 & -0.02877 \\ 25.58771 & 1.27116 \end{bmatrix} \tag{7.178}$$

where

$$\hat{\mathbf{F}}(\mathbf{K}) = \mathbf{1}_2 - \hat{\mathbf{A}}\mathbf{K} = \begin{bmatrix} 1.02951 & -0.02951 \\ 9.42507 & 1 \end{bmatrix} \tag{7.179}$$

7.3.1 The Transfer-Function Matrix and the General Feedback Matrices

As a direct extension of (7.88), we demonstrate that the determinant of the transfer-function matrix can be expressed in terms of the determinants of the general return difference matrix and the general null return difference matrix. We also give a simple physical interpretation of the general return difference and null return difference matrices for the admittance matrix of a *p*-port network.

Let $\mathbf{W}(\mathbf{X})$ be the transfer-function matrix of a multiple-loop feedback amplifier. Assume that $\mathbf{W}(\mathbf{K})$ is nonsingular, where $\mathbf{K}$ is the reference matrix. Then from (7.88) we have

$$\det \mathbf{W}(\mathbf{X}) = \det \mathbf{W}(\mathbf{0}) \frac{\det \hat{\mathbf{F}}(\mathbf{X})}{\det \mathbf{F}(\mathbf{X})} \tag{7.180a}$$

$$\det \mathbf{W}(\mathbf{K}) = \det \mathbf{W}(\mathbf{0}) \frac{\det \hat{\mathbf{F}}(\mathbf{K})}{\det \mathbf{F}(\mathbf{K})} \tag{7.180b}$$

Dividing (7.180*a*) by (7.180*b*) yields

$$\det \mathbf{W}(\mathbf{X}) = \det \mathbf{W}(\mathbf{K}) \frac{\det \hat{\mathbf{F}}(\mathbf{X}) \det \mathbf{F}(\mathbf{K})}{\det \hat{\mathbf{F}}(\mathbf{K}) \det \mathbf{F}(\mathbf{X})} \tag{7.181}$$

By appealing to (7.162) and (7.164), the above equation can be simplified and is given by

$$\det \mathbf{W}(\mathbf{X}) = \det \mathbf{W}(\mathbf{K}) \frac{\det \hat{\mathbf{F}}_{\mathbf{K}}(\mathbf{X})}{\det \mathbf{F}_{\mathbf{K}}(\mathbf{X})} \tag{7.182}$$

Consider the situation where $\mathbf{W}(\mathbf{X})$ denotes the impedance matrix of an *n*-port network of Fig. 7.10. In this case, $\mathbf{F}_{\mathbf{K}}(\mathbf{X})$ is the general return difference matrix for the situation when the *n* ports where the impedance matrix is defined are left open without any sources, and we write $\mathbf{F}_{\mathbf{K}}(\mathbf{X}) = \mathbf{F}_{\mathbf{K}}(\text{input open-circuited})$. Likewise, $\hat{\mathbf{F}}_{\mathbf{K}}(\mathbf{X})$ is the general return difference matrix for the situation when the *n* ports where the impedance matrix is defined are short-circuited, and we write $\hat{\mathbf{F}}_{\mathbf{K}}(\mathbf{X}) = \mathbf{F}_{\mathbf{K}}(\text{input short-circuited})$. As a result, the determinant of the impedance matrix $\mathbf{Z}(\mathbf{X})$ of an *n*-port network can be expressed, from (7.182), as

$$\det \mathbf{Z}(\mathbf{X}) = \det \mathbf{Z}(\mathbf{K}) \frac{\det \mathbf{F}_{\mathbf{K}}(\text{input short-circuited})}{\det \mathbf{F}_{\mathbf{K}}(\text{input open-circuited})} \tag{7.183}$$

As an application of (7.182), let $\mathbf{X}$ be the admittance matrix of a p-port network N_2 imbedded in an $(n+p)$-port network N_1, as shown in Fig. 7.25. From (7.93a) we have

$$\mathbf{F}(\mathbf{X}) = \mathbf{Y}^{-1}(\mathbf{Y}+\mathbf{X}) \tag{7.184a}$$

$$\mathbf{F}(\mathbf{K}) = \mathbf{Y}^{-1}(\mathbf{Y}+\mathbf{K}) \tag{7.184b}$$

where $\mathbf{Y}$ is the admittance matrix facing the p-port network N_2 when the input n ports of N_1 are open-circuited, giving

$$\mathbf{F}_{\mathbf{K}}(\mathbf{X}) = \mathbf{F}^{-1}(\mathbf{K})\mathbf{F}(\mathbf{X}) = (\mathbf{Y}+\mathbf{K})^{-1}(\mathbf{Y}+\mathbf{X}) \tag{7.185}$$

This is a direct extension of Eq. (5.104). In a similar manner, from (7.93b) we can show that

$$\hat{\mathbf{F}}_{\mathbf{K}}(\mathbf{X}) = \hat{\mathbf{F}}^{-1}(\mathbf{K})\hat{\mathbf{F}}(\mathbf{X}) = (\hat{\mathbf{Y}}+\mathbf{K})^{-1}(\hat{\mathbf{Y}}+\mathbf{X}) \tag{7.186}$$

where $\hat{\mathbf{Y}}$ is the admittance matrix facing the p-port N_2 when the input n ports of N_1 are short-circuited. Substituting these in (7.183) gives

$$\det \mathbf{Z}(\mathbf{X}) = \det \mathbf{Z}(\mathbf{K}) \frac{\det(\mathbf{Y}+\mathbf{K}) \det(\hat{\mathbf{Y}}+\mathbf{X})}{\det(\hat{\mathbf{Y}}+\mathbf{K}) \det(\mathbf{Y}+\mathbf{X})} \tag{7.187}$$

Example 7.10 Consider the same problem as discussed in Example 7.9. From (7.172) and (7.178) we have

$$\det \mathbf{F}_{\mathbf{K}}(\mathbf{X}) = 8.95430 \tag{7.188a}$$

$$\det \hat{\mathbf{F}}_{\mathbf{K}}(\mathbf{X}) = 3.06895 \tag{7.188b}$$

The network function of interest is the driving-point admittance facing the voltage source V_s in Fig. 7.15, that is, $w(\mathbf{X}) = I_{51}/V_s$. To apply (7.182), we first compute $w(\mathbf{K})$, which is the input admittance when the controlling parameters of the two controlled sources in Fig. 7.15 are replaced by their reference values: $I_a = 0.01V_{13}$ and $I_b = 0.02V_{45}$. The corresponding indefinite-admittance matrix can be written down by inspection and is given by

$$\mathbf{Y} = 10^{-4} \begin{bmatrix} 9.37 & 0 & -9.09 & 0 & -0.28 \\ 0 & 4.256 & -2.128 & 200 & -202.128 \\ -109.09 & -2.128 & 211.218 & 0 & -100 \\ 100 & 0 & -100 & 10.61 & -10.61 \\ -0.28 & -2.128 & -100 & -210.61 & 313.02 \end{bmatrix} \tag{7.189}$$

By appealing to formula (2.94), the input admittance $w(\mathbf{K})$ is found to be

$$w(\mathbf{K}) = \frac{Y_{uv}(\mathbf{K})}{Y_{11,55}(\mathbf{K})} = \frac{56{,}057.62 \cdot 10^{-16}}{52{,}049.75 \cdot 10^{-12}} = 107.70\ \mu\text{mho} \tag{7.190}$$

Substituting these in (7.182), we obtain the input admittance

$$w(\mathbf{X}) = w(\mathbf{K}) \frac{\det \hat{\mathbf{F}}_{\mathbf{K}}(\mathbf{X})}{\det \mathbf{F}_{\mathbf{K}}(\mathbf{X})} = 36.91 \ \mu\text{mho} \tag{7.191}$$

or the input impedance is 27.09 kΩ, confirming (7.27) or (7.28).

7.3.2 Relations to Scalar Return Difference and Null Return Difference

In Chaps. 4 and 5, we defined the return difference and the null return difference with respect to the controlling parameter x of a voltage-controlled current source $I = xV$. These concepts have been extended to the return difference and null return difference matrices with respect to a matrix $\mathbf{X}$. In the present section, we develop relations between the scalar return difference and null return difference with respect to an element of $\mathbf{X}$ and the return difference matrix and the null return difference matrix with respect to $\mathbf{X}$. The relations are useful in that we can express the sensitivity function in terms of the determinants of the return difference and null return difference matrices.

Let $\theta_i'' = x_{ij}\phi_j$ be the controlled source of interest, where x_{ij} is the (i, j)-element of $\mathbf{X}$, as defined in (7.11). The return difference $F(x_{ij})$ with respect to the controlling parameter x_{ij} can be defined as follows: Replace the controlled source $\theta_i'' = x_{ij}\phi_j$ by an independent source of strength $x_{ij}g_j$, and set all other independent sources to zero. The signal h_j appearing at the controlling branch ϕ_j is the returned signal. The return difference $F(x_{ij})$ is defined by the equation

$$g_j - h_j = F(x_{ij})g_j \tag{7.192}$$

Let $\mathbf{K}$ be the matrix obtained from $\mathbf{X}$ by setting $x_{ij} = 0$, that is,

$$\mathbf{K} = \mathbf{X}|_{x_{ij}=0} \tag{7.193}$$

Then $\mathbf{X}_1 = \mathbf{X} - \mathbf{K}$ has the form that all of its elements are zero except the one in the ith row and jth column, which is x_{ij}. Let $\mathbf{g}$ be a p-vector, all of its elements being zero except the jth row, which is g_j. The operation of replacing the controlled source $\theta_i'' = x_{ij}\phi_j$ by an independent source of strength $x_{ij}g_j$ is equivalent to replacing the controlled source vector $\theta = \mathbf{X}\phi$ by the sum of an independent source vector $\mathbf{X}_1\mathbf{g}$ and a controlled source vector $\theta'' = \mathbf{K}\phi$. In terms of the fundamental matrix feedback-flow graph of Fig. 7.28, the operation is the same as that depicted in Fig. 7.30 or Fig. 7.31. The returned signal vector $\mathbf{h}$ is therefore found to be $\mathbf{h} = \mathbf{A}_1\mathbf{X}_1\mathbf{g}$. The signal h_j, being the jth component of $\mathbf{h}$, is given by

$$h_j = a_{ji}^1 x_{ij} g_j \tag{7.194}$$

where a_{ij}^1 are the elements of $\mathbf{A}_1$. Combining this with (7.192) yields

$$F(x_{ij}) = 1 - a_{ji}^1 x_{ij} \tag{7.195}$$

The return difference matrix with respect to $\mathbf{X}$ for the reference $\mathbf{K}$ is, from (7.151),

$$\mathbf{F_K(X)} = \mathbf{1}_p - \mathbf{A}_1\mathbf{X}_1 \tag{7.196}$$

whose determinant is found to be

$$\det \mathbf{F_K(X)} = 1 - a^1_{ji}x_{ij} \tag{7.197}$$

showing that

$$F(x_{ij}) = \det \mathbf{F_K(X)} \tag{7.198}$$

or, after appealing to (7.162), we have

$$F(x_{ij}) = \frac{\det \mathbf{F(X)}}{\det \mathbf{F(X)}|_{x_{ij}=0}} \tag{7.199}$$

In words, this states that the scalar return difference of a multiple-loop feedback amplifier with respect to an element x_{ij} of a matrix **X** is equal to the ratio of the determinants of the return difference matrix with respect to **X** and that with respect to **K**, where **K** is obtained from **X** by setting x_{ij} to zero. In a similar manner, we can show that the scalar null return difference with respect to x_{ij} is equal to the ratio of the determinants of the null return difference matrix with respect to **X** and that with respect to **K**:

$$\hat{\mathbf{F}}(x_{ij}) = \frac{\det \hat{\mathbf{F}}(\mathbf{X})}{\det \hat{\mathbf{F}}(\mathbf{X})|_{x_{ij}=0}} \tag{7.200}$$

The justification of (7.200) is left as an exercise (see Prob. 7.8).

The above relations can be generalized to a general reference value. Let k_{ij} be a general reference value of x_{ij}. Then, from (7.199), we have

$$F(k_{ij}) = \frac{\det \mathbf{F(X)}|_{x_{ij}=k_{ij}}}{\det \mathbf{F(X)}|_{x_{ij}=0}} \tag{7.201}$$

Invoking (5.85) in conjunction with (7.199) and (7.201), we obtain the scalar return difference with respect to x_{ij} for a general reference value k_{ij} as

$$F_{k_{ij}}(x_{ij}) = \frac{F(x_{ij})}{F(k_{ij})} = \frac{\det \mathbf{F(X)}}{\det \mathbf{F(X)}|_{x_{ij}=k_{ij}}} \tag{7.202}$$

Likewise, we can show that the scalar null return difference with respect to x_{ij} for a general reference value k_{ij} can be written as (see Prob. 7.9)

$$\hat{F}_{k_{ij}}(x_{ij}) = \frac{\det \hat{\mathbf{F}}(\mathbf{X})}{\det \hat{\mathbf{F}}(\mathbf{X})|_{x_{ij}=k_{ij}}} \tag{7.203}$$

In the special situation where **X** is diagonal, the denominators of (7.199) and (7.200) can be simplified. In this case, let x_{ii} of **X** be the element of interest, and write

$$\Delta = \det \mathbf{F(X)} \tag{7.204a}$$

$$\hat{\Delta} = \det \hat{\mathbf{F}}(\mathbf{X}) \tag{7.204b}$$

Then we have

$$\det \mathbf{F}(\mathbf{X})|_{x_{ii}=0} = \det(\mathbf{1}_p - \mathbf{AX})|_{x_{ii}=0} = \Delta_{ii} \tag{7.205a}$$

$$\det \hat{\mathbf{F}}(\mathbf{X})|_{x_{ii}=0} = \det(\mathbf{1}_p - \hat{\mathbf{A}}\mathbf{X})|_{x_{ii}=0} = \hat{\Delta}_{ii} \tag{7.205b}$$

where Δ_{ii} and $\hat{\Delta}_{ii}$ are the cofactors of the (i, i)-element of $\mathbf{F}(\mathbf{X})$ and $\hat{\mathbf{F}}(\mathbf{X})$, respectively. This leads to

$$F(x_{ii}) = \frac{\Delta}{\Delta_{ii}} \tag{7.206a}$$

$$\hat{F}(x_{ii}) = \frac{\hat{\Delta}}{\hat{\Delta}_{ii}} \tag{7.206b}$$

In Sec. 5.6, we derived the expression for the relative sensitivity function in terms of the general return difference and null return difference. For a scalar transfer function w with respect to the element x_{ij}, the expression is given by

$$S'(x'_{ij}) = \frac{1}{F_{k_{ij}}(x_{ij})} - \frac{1}{\hat{F}_{k_{ij}}(x_{ij})} \tag{7.207}$$

where $x'_{ij} = x_{ij} - k_{ij}$. From (7.202) and (7.203), we obtain

$$S'(x'_{ij}) = \frac{\det \mathbf{F}(\mathbf{X})|_{x_{ij}=k_{ij}}}{\det \mathbf{F}(\mathbf{X})} - \frac{\det \hat{\mathbf{F}}(\mathbf{X})|_{x_{ij}=k_{ij}}}{\det \hat{\mathbf{F}}(\mathbf{X})} \tag{7.208}$$

Thus, the relative sensitivity function can be computed directly from the general return difference and null return difference matrices. For $k_{ij} = 0$, the relative sensitivity function becomes the ordinary sensitivity function:

$$S(x_{ij}) = \frac{\det \mathbf{F}(\mathbf{X})|_{x_{ij}=0}}{\det \mathbf{F}(\mathbf{X})} - \frac{\det \hat{\mathbf{F}}(\mathbf{X})|_{x_{ij}=0}}{\det \hat{\mathbf{F}}(\mathbf{X})} \tag{7.209}$$

We illustrate the above results by the following example.

Example 7.11 We wish to compute the return differences and the null return differences with respect to the controlling parameters $\tilde{\alpha}_j = 0.0455$ $(j = 1, 2)$ of the two controlled sources of Fig. 7.15. For $F(\tilde{\alpha}_1)$ and $\hat{F}(\tilde{\alpha}_1)$, we choose

$$\mathbf{K}_1 = \begin{bmatrix} 0 & 0 \\ 0 & 0.0455 \end{bmatrix} \tag{7.210}$$

The return difference with respect to $\mathbf{K}_1$ is found to be

$$\mathbf{F}(\mathbf{K}_1) = \mathbf{1}_2 - \mathbf{AK}_1 = \begin{bmatrix} 1 & -2.06529 \\ 0 & 1 \end{bmatrix} \tag{7.211}$$

where $\mathbf{A}$ is given in (7.168a), giving $\det \mathbf{F}(\mathbf{K}_1) = 1$. Using this in conjunction with (7.97), we obtain from (7.199)

$$F(\tilde{\alpha}_1) = \frac{\det \mathbf{F}(\mathbf{X})}{\det \mathbf{F}(\mathbf{K}_1)} = 93.68646 \tag{7.212}$$

confirming (4.141*a*). Alternatively, since **X** is diagonal, (7.206*a*) applies and from (7.96) we have $\Delta = 93.68646$ and $\Delta_{11} = 1$, which when substituted in (7.206*a*) yields (7.212). By using (7.79), the null return difference with respect to $\mathbf{K}_1$ is given by

$$\hat{\mathbf{F}}(\mathbf{K}_1) = \mathbf{1}_2 - \hat{\mathbf{A}}\mathbf{K}_1 = \begin{bmatrix} 1 & 2402.29 \\ 0 & 1 \end{bmatrix} \tag{7.213}$$

with $\det \hat{\mathbf{F}}(\mathbf{K}_1) = 1$. From (7.99) we get

$$\hat{F}(\tilde{\alpha}_1) = \frac{\det \hat{\mathbf{F}}(\mathbf{X})}{\det \hat{\mathbf{F}}(\mathbf{K}_1)} = 103.07 \cdot 10^3 \tag{7.214}$$

confirming (4.171). We can also apply (7.206*b*) to compute $\hat{F}(\tilde{\alpha}_1)$ with $\hat{\Delta} = 103.07 \cdot 10^3$ and $\hat{\Delta}_{11} = 1$, as given by (7.80).

In a similar manner, to compute $F(\tilde{\alpha}_2)$ and $\hat{F}(\tilde{\alpha}_2)$ we choose

$$\mathbf{K}_2 = \begin{bmatrix} 0.0455 & 0 \\ 0 & 0 \end{bmatrix} \tag{7.215}$$

From (7.168*a*) and (7.79), $\mathbf{F}(\mathbf{K}_2)$ and $\hat{\mathbf{F}}(\mathbf{K}_2)$ are determined as

$$\mathbf{F}(\mathbf{K}_2) = \begin{bmatrix} 5.13058 & 0 \\ 42.8841 & 1 \end{bmatrix} \tag{7.216a}$$

$$\hat{\mathbf{F}}(\mathbf{K}_2) = \begin{bmatrix} 51.0555 & 0 \\ 42.8841 & 1 \end{bmatrix} \tag{7.216b}$$

which in conjunction with (7.97) and (7.99) yield

$$F(\tilde{\alpha}_2) = \frac{\det \mathbf{F}(\mathbf{X})}{\det \mathbf{F}(\mathbf{K}_2)} = 18.26 \tag{7.217a}$$

$$\hat{F}(\tilde{\alpha}_2) = \frac{\det \hat{\mathbf{F}}(\mathbf{X})}{\det \hat{\mathbf{F}}(\mathbf{K}_2)} = 2018.80 \tag{7.217b}$$

confirming (4.141*b*) and (4.172), respectively. As before, we can also apply (7.206) directly to compute $F(\tilde{\alpha}_2)$ and $\hat{F}(\tilde{\alpha}_2)$ with $\Delta_{22} = 5.131$ and $\hat{\Delta}_{22} = 51.055$, as computed from (7.96) and (7.98), yielding, of course, the same results. For the sensitivity functions, we use (7.209) and obtain

$$S(\tilde{\alpha}_1) = \frac{\det \mathbf{F}(\mathbf{K}_1)}{\det \mathbf{F}(\mathbf{X})} - \frac{\det \hat{\mathbf{F}}(\mathbf{K}_1)}{\det \hat{\mathbf{F}}(\mathbf{X})} = 0.01066 \tag{7.218a}$$

$$S(\tilde{\alpha}_2) = \frac{\det \mathbf{F}(\mathbf{K}_2)}{\det \mathbf{F}(\mathbf{X})} - \frac{\det \hat{\mathbf{F}}(\mathbf{K}_2)}{\det \hat{\mathbf{F}}(\mathbf{X})} = 0.05427 \tag{7.218b}$$

which are approximately equal to the inverses of $F(\tilde{\alpha}_1)$ and $F(\tilde{\alpha}_2)$, respectively.

7.4 THE HYBRID-MATRIX FORMULATION OF MULTIPLE-LOOP FEEDBACK THEORY

In this section, we first explore the relation between the determinant of the hybrid matrix and that of the conventional cutset-admittance, node-admittance, or loop-impedance matrix, and then demonstrate how the determinant of the return difference matrix can be expressed elegantly and compactly in terms of the ratio of the two functional values assumed by the determinant of the hybrid matrix under the condition that the elements of interest assume their nominal and reference values.

7.4.1 Relations between the Hybrid Determinant and Other Network Determinants

In Sec. 5.9.1, we presented relations among the nodal, cutset, and loop determinants. In the following, we introduce the hybrid matrix and indicate how its determinant is related to the other network determinants. We assume that the reader is familiar with some of the basic terms of graph theory. For a detailed account of this subject, the reader is referred to Chen (1976a).

In a feedback network N, a branch may be taken to be either a simple resistor, inductor, capacitor, or independent or dependent generator, or a series and/or parallel connection of these elements, as we choose. For our purposes, we choose the most general branch representation of Fig. 7.32, containing both an independent and/or dependent voltage generator in series with a network branch and an independent and/or dependent current generator in parallel with this combination. The independent sources V_{sk} and I_{sk} in Fig. 7.32 may be replaced by the controlled voltage and current sources, respectively.

The starting point is the three primary systems of network equations. Let t be a tree in the directed graph G_d representing the feedback network N. Let $\mathbf{Q}_f$ and $\mathbf{B}_f$ be the fundamental cutset and fundamental circuit matrices of G_d with respect to the tree t. Partition $\mathbf{Q}_f$ and $\mathbf{B}_f$ according to the branches and chords of the tree t, and partition the branch-current vector $\mathbf{I}$ and branch-voltage vector $\mathbf{V}$ accordingly. Then we have

$$[\mathbf{1}_r \quad \mathbf{Q}_{\bar{t}}] \begin{bmatrix} \mathbf{I}_t \\ \mathbf{I}_{\bar{t}} \end{bmatrix} = \mathbf{Q}_f \mathbf{I}_s \tag{7.219a}$$

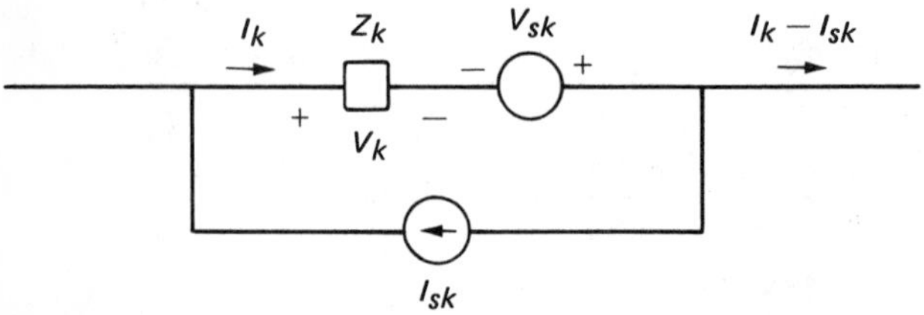

Figure 7.32 The most general branch representation.

$$[-\mathbf{Q}'_{\bar{t}} \;\; \mathbf{1}_m] \begin{bmatrix} \mathbf{V}_t \\ \mathbf{V}_{\bar{t}} \end{bmatrix} = \mathbf{B}_f \mathbf{V}_s \tag{7.219b}$$

where $\mathbf{I}_s$ and $\mathbf{V}_s$ are vectors of generator currents and voltages, and r and m denote the rank and nullity of G_d, respectively. The branch relations are characterized by the hybrid equations in the following form:

$$\begin{bmatrix} \mathbf{V}_{\bar{t}} \\ \mathbf{I}_t \end{bmatrix} = \begin{bmatrix} \mathbf{H}_{\bar{t}\bar{t}} & \mathbf{H}_{\bar{t}t} \\ \mathbf{H}_{t\bar{t}} & \mathbf{H}_{tt} \end{bmatrix} \begin{bmatrix} \mathbf{I}_{\bar{t}} \\ \mathbf{V}_t \end{bmatrix} \tag{7.220}$$

The subscript t signifies that the quantity is defined in the tree t, and $\bar{t}$ is the complement of t. Thus, the elements of $\mathbf{H}_{tt}$ have the dimension of an admittance and are defined by the tree branches, and those of $\mathbf{H}_{\bar{t}\bar{t}}$ have the dimension of an impedance and are defined by the chords of t. The elements of $\mathbf{H}_{\bar{t}t}$ and $\mathbf{H}_{t\bar{t}}$ are dimensionless and represent the coupling between the tree branches and the branches in their complement. The three systems of equations (7.219) and (7.220) can be written as a single matrix equation by considering the branch-current and branch-voltage vectors $\mathbf{I}$ and $\mathbf{V}$ as the unknowns:

$$\begin{bmatrix} -\mathbf{Q}'_{\bar{t}} & \mathbf{1}_m & \mathbf{0} & \mathbf{0} \\ \mathbf{0} & \mathbf{0} & \mathbf{1}_r & \mathbf{Q}_{\bar{t}} \\ \mathbf{H}_{tt} & \mathbf{0} & -\mathbf{1}_r & \mathbf{H}_{t\bar{t}} \\ \mathbf{H}_{\bar{t}t} & -\mathbf{1}_m & \mathbf{0} & \mathbf{H}_{\bar{t}\bar{t}} \end{bmatrix} \begin{bmatrix} \mathbf{V}_t \\ \mathbf{V}_{\bar{t}} \\ \mathbf{I}_t \\ \mathbf{I}_{\bar{t}} \end{bmatrix} = \begin{bmatrix} \mathbf{B}_f \mathbf{V}_s \\ \mathbf{Q}_f \mathbf{I}_s \\ \mathbf{0} \\ \mathbf{0} \end{bmatrix} \tag{7.221}$$

Using the elementary row operations to eliminate the variables $\mathbf{V}_{\bar{t}}$ and $\mathbf{I}_t$, we obtain a hybrid system of network equations

$$\begin{bmatrix} \mathbf{H}_{tt} & \mathbf{H}_{t\bar{t}} + \mathbf{Q}_{\bar{t}} \\ \mathbf{H}_{\bar{t}t} - \mathbf{Q}'_{\bar{t}} & \mathbf{H}_{\bar{t}\bar{t}} \end{bmatrix} \begin{bmatrix} \mathbf{V}_t \\ \mathbf{I}_{\bar{t}} \end{bmatrix} = \begin{bmatrix} \mathbf{Q}_f \mathbf{I}_s \\ \mathbf{B}_f \mathbf{V}_s \end{bmatrix} \tag{7.222}$$

or, more compactly,

$$\mathbf{H} \begin{bmatrix} \mathbf{V}_t \\ \mathbf{I}_{\bar{t}} \end{bmatrix} = \begin{bmatrix} \mathbf{Q}_f \mathbf{I}_s \\ \mathbf{B}_f \mathbf{V}_s \end{bmatrix} \tag{7.223}$$

The coefficient matrix $\mathbf{H}$ is called the *hybrid matrix* of the feedback network N, and its determinant is known as the *hybrid determinant*.

As in Sec. 5.9.1, let Δ_c, Δ_n, and Δ_m denote the cutset, nodal, and loop determinants of N, respectively. It can be shown that the hybrid determinant is related to Δ_c and Δ_m by the equations (see App. III)

$$\det \mathbf{H} = \lambda_{hc} (\det \mathbf{H}_{\bar{t}\bar{t}}) \Delta_c \tag{7.224a}$$

$$\det \mathbf{H} = \lambda_{hm} (\det \mathbf{H}_{tt}) \Delta_m \tag{7.224b}$$

where λ_{hc} and λ_{hm} are real constants, depending only on the choices of the cutsets and loops. In deriving (7.224), we implicitly assumed that the feedback amplifier N possesses the admittance and impedance representations. A formal proof of these relations can be found in Chen (1977a) and is outlined in App. III. In (7.220), if we express the branch-voltage vector $\mathbf{V}$ in terms of the branch-current vector $\mathbf{I}$, the coefficient matrix is the branch-impedance matrix and is given by

$$\mathbf{Z}_b(s) = \begin{bmatrix} \mathbf{H}_{tt}^{-1} & -\mathbf{H}_{tt}^{-1}\mathbf{H}_{t\bar{t}} \\ \mathbf{H}_{\bar{t}t}\mathbf{H}_{tt}^{-1} & \mathbf{H}_{\bar{t}\bar{t}} - \mathbf{H}_{\bar{t}t}\mathbf{H}_{tt}^{-1}\mathbf{H}_{t\bar{t}} \end{bmatrix} \tag{7.225}$$

On the other hand, if we express $\mathbf{I}$ in terms of $\mathbf{V}$, we obtain the branch-admittance matrix as

$$\mathbf{Y}_b(s) = \begin{bmatrix} \mathbf{H}_{tt} - \mathbf{H}_{t\bar{t}}\mathbf{H}_{t\bar{t}}\mathbf{H}_{\bar{t}\bar{t}}^{-1} & \mathbf{H}_{t\bar{t}}\mathbf{H}_{\bar{t}\bar{t}}^{-1} \\ -\mathbf{H}_{\bar{t}\bar{t}}^{-1}\mathbf{H}_{\bar{t}t} & \mathbf{H}_{\bar{t}\bar{t}}^{-1} \end{bmatrix} \tag{7.226}$$

We illustrate the above results by the following example.

Example 7.12 Consider the feedback network of Fig. 7.33, whose associated directed graph G_d is presented in Fig. 7.34. Choose the tree $t = e_1e_3$. The corresponding fundamental cutset matrix is found to be

$$\mathbf{Q}_f = [\mathbf{1}_2 \quad \mathbf{Q}_{\bar{t}}] = \begin{array}{c} \begin{matrix} e_1 & e_3 & e_2 & e_4 \end{matrix} \\ \begin{bmatrix} 1 & 0 & 1 & 1 \\ 0 & 1 & 0 & -1 \end{bmatrix} \end{array} \tag{7.227}$$

The branch vi-relations are described by the hybrid equation

$$\begin{bmatrix} V_2 \\ V_4 \\ \hline I_1 \\ I_3 \end{bmatrix} = \left[\begin{array}{cc|cc} z_2 & 0 & 0 & -z_2 g_{m2} \\ 0 & z_4 & 0 & 0 \\ \hline 0 & 0 & y_1 & 0 \\ g_{m3}z_2 & 0 & 0 & y_3 - z_2 g_{m2} g_{m3} \end{array}\right] \begin{bmatrix} I_2 \\ I_4 \\ \hline V_1 \\ V_3 \end{bmatrix} \tag{7.228}$$

where $z_i = 1/y_i$ $(i = 1, 2, 3, 4)$. Substituting the appropriate quantities in (7.222) yields the hybrid system of equations

$$\begin{bmatrix} y_1 & 0 & 1 & 1 \\ 0 & y_3 - g_{m2}g_{m3}z_2 & g_{m3}z_2 & -1 \\ -1 & -g_{m2}z_2 & z_2 & 0 \\ -1 & 1 & 0 & z_4 \end{bmatrix} \begin{bmatrix} V_1 \\ V_3 \\ I_2 \\ I_4 \end{bmatrix} = \begin{bmatrix} 0 \\ 0 \\ V_{s1} \\ V_{s1} \end{bmatrix} \tag{7.229}$$

The hybrid determinant is found to be

$$\det \mathbf{H} = z_2 z_4 [(y_1 + y_2)(y_3 + y_4) + y_4(y_3 + g_{m2} + g_{m3}) - g_{m2}g_{m3}] \tag{7.230}$$

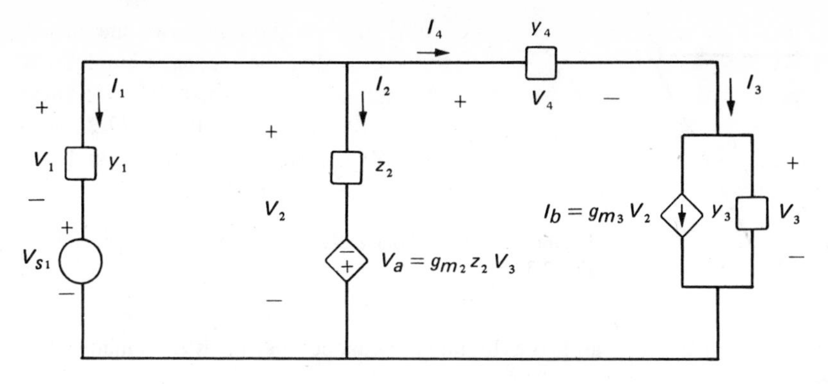

Figure 7.33 A feedback network containing two controlled sources.

The cutset-admittance matrix $\mathbf{Y}_c(s)$ is defined by the matrix triple product $\mathbf{Q}_f \mathbf{Y}_b \mathbf{Q}_f'$ and is given by

$$\mathbf{Y}_c(s) = \mathbf{Q}_f \mathbf{Y}_b \mathbf{Q}_f' = \begin{bmatrix} y_1 + y_2 + y_4 & g_{m2} - y_4 \\ g_{m3} - y_4 & y_3 + y_4 \end{bmatrix} \tag{7.231}$$

where the branch-admittance matrix is computed by means of (7.226), giving

$$\mathbf{Y}_b(s) = \begin{bmatrix} y_1 & 0 & 0 & 0 \\ 0 & y_3 & g_{m3} & 0 \\ 0 & g_{m2} & y_2 & 0 \\ 0 & 0 & 0 & y_4 \end{bmatrix} \tag{7.232}$$

It is straightforward to verify that

$$\det \mathbf{H} = \det \mathbf{H}_{\bar{t}\bar{t}} \det \mathbf{Y}_c = (\det \mathbf{H}_{\bar{t}\bar{t}})\, \Delta_c \tag{7.233}$$

where $\det \mathbf{H}_{\bar{t}\bar{t}} = z_2 z_4$. This confirms (7.224*a*) with $\lambda_{hc} = 1$. To demonstrate (7.224*b*), we compute the fundamental circuit matrix

$$\mathbf{B}_f = [-\mathbf{Q}_{\bar{t}}' \;\; \mathbf{1}_2] = \begin{array}{c} \begin{matrix} e_1 & e_3 & e_2 & e_4 \end{matrix} \\ \begin{bmatrix} -1 & 0 & 1 & 0 \\ -1 & 1 & 0 & 1 \end{bmatrix} \end{array} \tag{7.234}$$

yielding the loop determinant

$$\Delta_m = \det \mathbf{B}_f \mathbf{Z}_b \mathbf{B}_f' = \det \begin{bmatrix} z_1 + \dfrac{y_3}{P} & z_1 - \dfrac{g_{m2}}{P} \\ z_1 - \dfrac{g_{m3}}{P} & z_1 + z_4 + \dfrac{y_2}{P} \end{bmatrix}$$

$$= \frac{z_1 z_4}{P} [(y_1 + y_2)(y_3 + y_4) + y_4(y_3 + g_{m2} + g_{m3}) - g_{m2} g_{m3}] \tag{7.235}$$

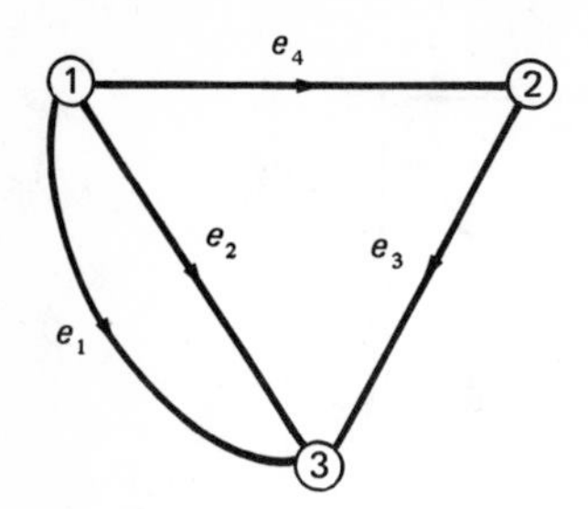

Figure 7.34 The associated directed graph of the network of Fig. 7.33.

where $P = y_2 y_3 - g_{m2} g_{m3}$, and the branch-impedance matrix is obtained from (7.225) as

$$\mathbf{Z}_b(s) = \begin{bmatrix} z_1 & 0 & 0 & 0 \\ 0 & \dfrac{y_2}{P} & -\dfrac{g_{m3}}{P} & 0 \\ 0 & -\dfrac{g_{m2}}{P} & \dfrac{y_3}{P} & 0 \\ 0 & 0 & 0 & z_4 \end{bmatrix} \tag{7.236}$$

Comparing (7.235) with (7.230) gives

$$\det \mathbf{H} = (\det \mathbf{H}_{tt})\, \Delta_m \tag{7.237}$$

where $\det \mathbf{H}_{tt} = y_1(y_3 - g_{m2} g_{m3} z_2) = y_1 z_2 P$.

7.4.2 Hybrid Formulation of the Determinant of the Return Difference Matrix

In this section, we show that, like the scalar return difference, the determinant of the return difference matrix can be expressed elegantly and compactly as the ratio of the two functional values assumed by the hybrid determinant under the condition that the elements of interest assume their nominal values and the condition that they assume their reference values.

Let the element x_{ij} of $\mathbf{X}$ be the controlling parameter of a voltage- or current-controlled current source. Then the return difference with respect to x_{ij} for a reference value k_{ij} can be expressed as the ratio of the two functional values assumed by the cutset determinant, as given in (5.218). Using this in conjunction with (7.202), we obtain

$$F_{k_{ij}}(x_{ij}) = \frac{\Delta_c(\mathbf{X})}{\Delta_c(\mathbf{X})|_{x_{ij}=k_{ij}}} = \frac{\det \mathbf{F}(\mathbf{X})}{\det \mathbf{F}(\mathbf{X})|_{x_{ij}=k_{ij}}} \tag{7.238}$$

Appealing to (7.224*a*) and assuming that $\det \mathbf{H}_{\bar{t}\bar{t}}$ is independent of x_{ij} yield

$$F_{k_{ij}}(x_{ij}) = \frac{\det \mathbf{F}(\mathbf{X})}{\det \mathbf{F}(\mathbf{X})|_{x_{ij}=k_{ij}}} = \frac{\det \mathbf{H}(\mathbf{X})}{\det \mathbf{H}(\mathbf{X})|_{x_{ij}=k_{ij}}} \tag{7.239}$$

Likewise, if x_{ij} of $\mathbf{X}$ is the controlling parameter of a voltage- or current-controlled voltage source, then the return difference with respect to x_{ij} for a reference value k_{ij} can be expressed from (5.219) and (7.202) as

$$F_{k_{ij}}^{m}(x_{ij}) = \frac{\Delta_m(\mathbf{X})}{\Delta_m(\mathbf{X})|_{x_{ij}=k_{ij}}} = \frac{\det \mathbf{F}(\mathbf{X})}{\det \mathbf{F}(\mathbf{X})|_{x_{ij}=k_{ij}}} \tag{7.240}$$

Applying (7.224*b*) and assuming that $\det \mathbf{H}_{tt}$ is independent of x_{ij}, we have

$$F_{k_{ij}}^{m}(x_{ij}) = \frac{\det \mathbf{F}(\mathbf{X})}{\det \mathbf{F}(\mathbf{X})|_{x_{ij}=k_{ij}}} = \frac{\det \mathbf{H}(\mathbf{X})}{\det \mathbf{H}(\mathbf{X})|_{x_{ij}=k_{ij}}} \tag{7.241}$$

Observe that (7.239) and (7.241) are identical in form, meaning that either one is valid for any one of the four types of controlled sources if the hybrid matrix $\mathbf{H}$ of (7.220) can be partitioned so that $\det \mathbf{H}_{\bar{t}\bar{t}}$ is independent of the controlling parameters of the controlled current sources appearing in $\mathbf{X}$ and $\det \mathbf{H}_{tt}$ is independent of the controlling parameters of the controlled voltage sources appearing in $\mathbf{X}$. These two conditions can easily be fulfilled by simple manipulation in the formulation of the hybrid matrix $\mathbf{H}$.

Suppose that $\mathbf{X}_i$ is obtained from $\mathbf{X}_{i-1}$ by setting a certain element to its reference value for $i = 1, 2, \ldots, pq$ with $\mathbf{X}_0 = \mathbf{X}$, where $\mathbf{X}$ is of order $q \times p$. Then from (7.239) and (7.241) we have the equality

$$\frac{\det \mathbf{F}(\mathbf{X}_{i-1})}{\det \mathbf{F}(\mathbf{X}_i)} = \frac{\det \mathbf{H}(\mathbf{X}_{i-1})}{\det \mathbf{H}(\mathbf{X}_i)} \tag{7.242}$$

for all i. Forming the product of all the terms of the above type yields

$$\frac{\det \mathbf{F}(\mathbf{X})}{\det \mathbf{F}(\mathbf{K})} = \frac{\det \mathbf{H}(\mathbf{X})}{\det \mathbf{H}(\mathbf{K})} \tag{7.243}$$

where $\mathbf{K}$ is the reference matrix of $\mathbf{X}$. Invoking (7.162), we get the desired relation:

$$\det \mathbf{F}_{\mathbf{K}}(\mathbf{X}) = \frac{\det \mathbf{H}(\mathbf{X})}{\det \mathbf{H}(\mathbf{K})} \tag{7.244}$$

Example 7.13 In the feedback network of Fig. 7.33, let the controlling parameters $g_{m2}z_2$ and g_{m3} be the elements of interest. Then we have

$$\boldsymbol{\theta} = \begin{bmatrix} I_b \\ V_a \end{bmatrix} = \begin{bmatrix} g_{m3} & 0 \\ 0 & g_{m2}z_2 \end{bmatrix} \begin{bmatrix} V_2 \\ V_3 \end{bmatrix} = \mathbf{X}\boldsymbol{\phi} \tag{7.245}$$

$$\boldsymbol{\phi} = \begin{bmatrix} V_2 \\ V_3 \end{bmatrix} = -\frac{1}{\lambda} \begin{bmatrix} y_4 & y_2(y_3 + y_4) \\ y_1 + y_2 + y_4 & y_2 y_4 \end{bmatrix} \begin{bmatrix} I_a \\ V_b \end{bmatrix} + \frac{1}{\lambda} \begin{bmatrix} y_1(y_3 + y_4) \\ y_1 y_4 \end{bmatrix} [V_{s1}] = \mathbf{A}\boldsymbol{\theta} + \mathbf{B}\mathbf{u} \tag{7.246}$$

where $\lambda = (y_1 + y_2)(y_3 + y_4) + y_3 y_4$. Substituting **X** and **A** from (7.245) and (7.246) in (7.34) gives the return difference matrix

$$\mathbf{F}(\mathbf{X}) = \frac{1}{\lambda}\begin{bmatrix} \lambda + g_{m3} y_4 & g_{m2}(y_3 + y_4) \\ g_{m3}(y_1 + y_2 + y_4) & \lambda + g_{m2} y_4 \end{bmatrix} \tag{7.247}$$

whose determinant is found to be

$$\det \mathbf{F}(\mathbf{X}) = \frac{(y_1 + y_2)(y_3 + y_4) + y_4(y_3 + g_{m2} + g_{m3}) - g_{m2} g_{m3}}{(y_1 + y_2)(y_3 + y_4) + y_3 y_4} \tag{7.248}$$

To compute $\det \mathbf{F}(\mathbf{X})$ by means of (7.244), the coefficient matrix **H** of (7.228) is inadequate, since $\det \mathbf{H}_{tt}$ is not independent of the controlling parameter $g_{m2} z_2$ of the controlled voltage source V_a. With a simple modification, the *vi*-relations of the network of Fig. 7.33 can be described by the hybrid equation

$$\begin{bmatrix} V_2 \\ V_4 \\ I_1 \\ I_3 \end{bmatrix} = \begin{bmatrix} z_2 & 0 & 0 & -g_{m2} z_2 \\ 0 & z_4 & 0 & 0 \\ 0 & 0 & y_1 & 0 \\ 0 & g_{m3} z_4 & 0 & y_3 + g_{m3} \end{bmatrix} \begin{bmatrix} I_2 \\ I_4 \\ V_1 \\ V_3 \end{bmatrix} \tag{7.249}$$

The determinant of the coefficient matrix **H** is obtained as

$$\det \mathbf{H} = z_2 z_4 \left[(y_1 + y_2)(y_3 + y_4) + y_4(y_3 + g_{m2} + g_{m3}) - g_{m2} g_{m3}\right] \tag{7.250}$$

which when substituted in (7.244) yields

$$\det \mathbf{F}(\mathbf{X}) = \frac{(y_1 + y_2)(y_3 + y_4) + y_4(y_3 + g_{m2} + g_{m3}) - g_{m2} g_{m3}}{(y_1 + y_2)(y_3 + y_4) + y_3 y_4} \tag{7.251}$$

confirming (7.248). We remark that $\det \mathbf{H}$ of (7.250) is the same as that given in (7.230). Suppose that k_2 and k_3 are the reference values for $g_{m2} z_2$ and g_{m3}. Then, from (7.162) and (7.244) we have

$$\det \mathbf{F}_{\mathbf{K}}(\mathbf{X}) = \frac{\det \mathbf{F}(\mathbf{X})}{\det \mathbf{F}(\mathbf{K})} = \frac{\det \mathbf{H}(\mathbf{X})}{\det \mathbf{H}(\mathbf{K})}$$

$$= \frac{(y_1 + y_2)(y_3 + y_4) + y_4(y_3 + g_{m2} + g_{m3}) - g_{m2} g_{m3}}{(y_1 + y_2)(y_3 + y_4) + y_4(y_3 + k_2 y_2 + k_3) - k_2 k_3 y_2} \tag{7.252}$$

in which

$$\mathbf{K} = \begin{bmatrix} k_3 & 0 \\ 0 & k_2 \end{bmatrix} \tag{7.253}$$

With the coefficient matrix **H** as given in (7.249), the corresponding branch-admittance matrix $\mathbf{Y}_b$ and branch-impedance matrix $\mathbf{Z}_b$ are determined from (7.226) and (7.225), and are given by

$$\mathbf{Y}_b(s) = \begin{bmatrix} y_1 & 0 & 0 & 0 \\ 0 & y_3 + g_{m3} & 0 & g_{m3} \\ 0 & g_{m2} & y_2 & 0 \\ 0 & 0 & 0 & y_4 \end{bmatrix} \tag{7.254}$$

$$\mathbf{Z}_b(s) = \begin{bmatrix} z_1 & 0 & 0 & 0 \\ 0 & \dfrac{1}{\lambda_1} & 0 & -\dfrac{g_{m3}z_4}{\lambda_1} \\ 0 & -\dfrac{g_{m2}z_2}{\lambda_1} & z_2 & \dfrac{g_{m2}g_{m3}z_2z_4}{\lambda_1} \\ 0 & 0 & 0 & z_4 \end{bmatrix} \tag{7.255}$$

where $\lambda_1 = y_3 + g_{m3}$. Using $\mathbf{Q}_f$ and $\mathbf{B}_f$ as given in (7.227) and (7.234), we can compute the cutset and loop determinants. It is straightforward to confirm that they satisfy the identities (7.224) with $\lambda_{hc} = \lambda_{hm} = 1$, $\det \mathbf{H}_{\bar{t}\bar{t}} = z_2 z_4$, and $\det \mathbf{H}_{tt} = y_1(y_3 + g_{m3})$. The details are left as an exercise (see Prob. 7.11).

7.5 THE SENSITIVITY MATRIX AND MULTIPARAMETER SENSITIVITY

So far we have studied the sensitivity of a transfer function with respect to the change of a particular element in the network. In a multiple-loop feedback network, we are usually interested in the sensitivity of a transfer function with respect to the variation of a set of elements in the network. This set may include either elements that are inherently sensitive to variation or elements whose effect on the overall amplifier performance is of paramount importance to the designers. In this section, we introduce a sensitivity matrix for a multiple-loop feedback amplifier and develop formulas for computing multiparameter sensitivity functions.

7.5.1 The Sensitivity Matrix

Figure 7.35*a* is the block diagram of a multivariable open-loop control system with n inputs and m outputs, whereas Fig. 7.35*b* shows the general feedback structure. If all feedback signals are obtainable from the output and if the controllers are linear, there is no loss of generality by assuming the controller to be of the form shown in Fig. 7.36. Denote the set of Laplace transformed input signals by the n-vector $\mathbf{u}$, the set of inputs to the network $\mathbf{X}$ in the open-loop configuration of Fig. 7.35*a* by the p-vector $\boldsymbol{\phi}_o$, and the set of outputs of the network $\mathbf{X}$ in Fig. 7.35*a* by the m-vector $\mathbf{y}_o$. Let the corresponding signals for the closed-loop configuration of Fig.

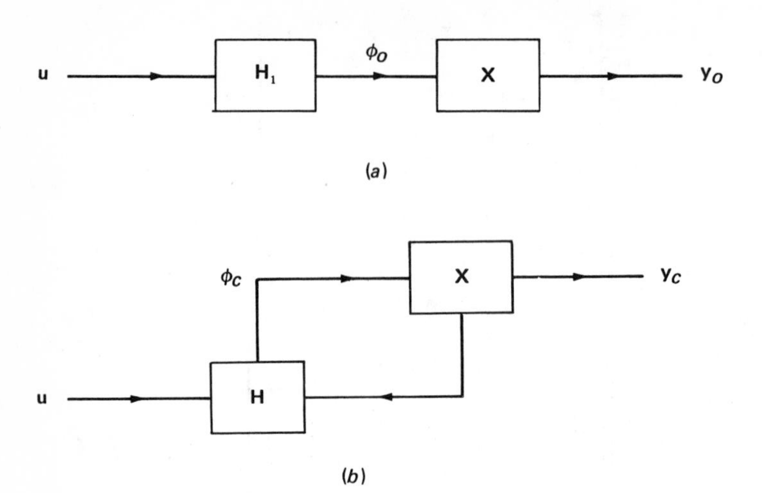

Figure 7.35 (*a*) The block diagram of a multivariable open-loop control system and (*b*) the general feedback structure.

7.36 be denoted by the n-vector $\mathbf{u}$, the p-vector ϕ_c, and the m-vector $\mathbf{y}_c$, respectively. Then from Figs. 7.35*a* and 7.36, we obtain the following relations:

$$\mathbf{y}_o = \mathbf{X}\phi_o \tag{7.256a}$$

$$\phi_o = \mathbf{H}_1 \mathbf{u} \tag{7.256b}$$

$$\mathbf{y}_c = \mathbf{X}\phi_c \tag{7.256c}$$

$$\phi_c = \mathbf{H}_2(\mathbf{u} + \mathbf{H}_3 \mathbf{y}_c) \tag{7.256d}$$

where the transfer function matrices $\mathbf{X}$, $\mathbf{H}_1$, $\mathbf{H}_2$, and $\mathbf{H}_3$ are of orders $m \times p$, $p \times n$, $p \times n$, and $n \times m$, respectively. Combining (7.256*c*) and (7.256*d*) yields

$$(\mathbf{1}_m - \mathbf{X}\mathbf{H}_2\mathbf{H}_3)\mathbf{y}_c = \mathbf{X}\mathbf{H}_2 \mathbf{u} \tag{7.257}$$

or

$$\mathbf{y}_c = (\mathbf{1}_m - \mathbf{X}\mathbf{H}_2\mathbf{H}_3)^{-1}\mathbf{X}\mathbf{H}_2 \mathbf{u} \tag{7.258}$$

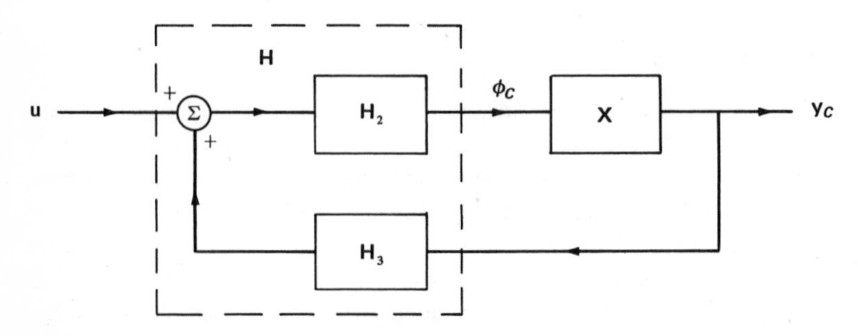

Figure 7.36 A general feedback configuration.

The closed-loop transfer function matrix that relates the input and output is defined by the equation

$$\mathbf{y}_c = \mathbf{W}(\mathbf{X})\mathbf{u} \tag{7.259}$$

where $\mathbf{W}(\mathbf{X})$ is of order $m \times n$, giving

$$\mathbf{W}(\mathbf{X}) = (\mathbf{1}_m - \mathbf{X}\mathbf{H}_2\mathbf{H}_3)^{-1}\mathbf{X}\mathbf{H}_2 \tag{7.260}$$

Now suppose that $\mathbf{X}$ is perturbed to $\mathbf{X}^+ = \mathbf{X} + \delta\mathbf{X}$. The outputs of the open-loop and closed-loop systems of Figs. 7.35*a* and 7.36 will no longer be the same as before. Distinguishing the new from the old variables by the superscript +, we have

$$\mathbf{y}_o^+ = \mathbf{X}^+\boldsymbol{\phi}_o \tag{7.261a}$$

$$\mathbf{y}_c^+ = \mathbf{X}^+\boldsymbol{\phi}_c^+ \tag{7.261b}$$

$$\boldsymbol{\phi}_c^+ = \mathbf{H}_2(\mathbf{u} + \mathbf{H}_3\mathbf{y}_c^+) \tag{7.261c}$$

where $\boldsymbol{\phi}_o$ remains the same.

We next proceed to compare the relative effects of the variations of $\mathbf{X}$ on the performance of the open-loop and the closed-loop systems. For a meaningful comparison, we assume that $\mathbf{H}_1$, $\mathbf{H}_2$, and $\mathbf{H}_3$ are such that when there is no variation of $\mathbf{X}$, $\mathbf{y}_o = \mathbf{y}_c$. Define the error vectors resulting from perturbation of $\mathbf{X}$ as

$$\mathbf{E}_o = \mathbf{y}_o - \mathbf{y}_o^+ \tag{7.262a}$$

$$\mathbf{E}_c = \mathbf{y}_c - \mathbf{y}_c^+ \tag{7.262b}$$

A square matrix relating $\mathbf{E}_c$ to $\mathbf{E}_o$ is defined as the *sensitivity matrix* for the transfer function matrix $\mathbf{W}(\mathbf{X})$ with respect to the variations of $\mathbf{X}$:

$$\mathbf{E}_c = \mathbf{S}(\mathbf{X})\mathbf{E}_o \tag{7.263}$$

In the following, we express the sensitivity matrix $\mathbf{S}(\mathbf{X})$ in terms of the system matrices $\mathbf{X}$, $\mathbf{H}_2$, and $\mathbf{H}_3$.

The output and input relation similar to that given in (7.258) for the perturbed system can be written as

$$\mathbf{y}_c^+ = (\mathbf{1}_m - \mathbf{X}^+\mathbf{H}_2\mathbf{H}_3)^{-1}\mathbf{X}^+\mathbf{H}_2\mathbf{u} \tag{7.264}$$

Substituting (7.258) and (7.264) in (7.262*b*) gives

$$\begin{aligned}\mathbf{E}_c = \mathbf{y}_c - \mathbf{y}_c^+ &= [(\mathbf{1}_m - \mathbf{X}\mathbf{H}_2\mathbf{H}_3)^{-1}\mathbf{X}\mathbf{H}_2 - (\mathbf{1}_m - \mathbf{X}^+\mathbf{H}_2\mathbf{H}_3)^{-1}\mathbf{X}^+\mathbf{H}_2]\mathbf{u} \\ &= (\mathbf{1}_m - \mathbf{X}^+\mathbf{H}_2\mathbf{H}_3)^{-1}\{[\mathbf{1}_m - (\mathbf{X}+\delta\mathbf{X})\mathbf{H}_2\mathbf{H}_3](\mathbf{1}_m - \mathbf{X}\mathbf{H}_2\mathbf{H}_3)^{-1}\mathbf{X}\mathbf{H}_2 - (\mathbf{X}+\delta\mathbf{X})\mathbf{H}_2\}\mathbf{u} \\ &= (\mathbf{1}_m - \mathbf{X}^+\mathbf{H}_2\mathbf{H}_3)^{-1}[\mathbf{X}\mathbf{H}_2 - \delta\mathbf{X}\mathbf{H}_2\mathbf{H}_3(\mathbf{1}_m - \mathbf{X}\mathbf{H}_2\mathbf{H}_3)^{-1}\mathbf{X}\mathbf{H}_2 - \mathbf{X}\mathbf{H}_2 - \delta\mathbf{X}\mathbf{H}_2]\mathbf{u} \\ &= -(\mathbf{1}_m - \mathbf{X}^+\mathbf{H}_2\mathbf{H}_3)^{-1}\delta\mathbf{X}\mathbf{H}_2[\mathbf{1}_n + \mathbf{H}_3\mathbf{W}(\mathbf{X})]\mathbf{u}\end{aligned} \tag{7.265}$$

From (7.256*d*) in conjunction with (7.259), we have

$$\boldsymbol{\phi}_c = \mathbf{H}_2[\mathbf{1}_n + \mathbf{H}_3\mathbf{W}(\mathbf{X})]\mathbf{u} \tag{7.266}$$

Since by assumption $\mathbf{y}_o = \mathbf{y}_c$,

$$\boldsymbol{\phi}_o = \boldsymbol{\phi}_c = \mathbf{H}_2 [\mathbf{1}_n + \mathbf{H}_3 \mathbf{W}(\mathbf{X})]\mathbf{u} \tag{7.267}$$

yielding $\mathbf{E}_o = \mathbf{y}_o - \mathbf{y}_o^+ = (\mathbf{X} - \mathbf{X}^+)\boldsymbol{\phi}_o = -\delta\mathbf{X}\boldsymbol{\phi}_o = -\delta\mathbf{X}\mathbf{H}_2 [\mathbf{1}_n + \mathbf{H}_3 \mathbf{W}(\mathbf{X})]\mathbf{u}$ (7.268)

Combining (7.265) and (7.268) yields the desired expression relating the error vectors $\mathbf{E}_c$ and $\mathbf{E}_o$ of the closed-loop and open-loop systems by

$$\mathbf{E}_c = (\mathbf{1}_m - \mathbf{X}^+\mathbf{H}_2\mathbf{H}_3)^{-1}\mathbf{E}_o \tag{7.269}$$

giving the sensitivity matrix first introduced by Cruz and Perkins (1964) as

$$\mathbf{S}(\mathbf{X}) = (\mathbf{1}_m - \mathbf{X}^+\mathbf{H}_2\mathbf{H}_3)^{-1} \tag{7.270}$$

For small variation of $\mathbf{X}$, $\mathbf{X}^+$ is approximately equal to $\mathbf{X}$. Refer to Fig. 7.36. The matrix triple product $\mathbf{X}\mathbf{H}_2\mathbf{H}_3$ may be regarded as the loop-transmission matrix, and $-\mathbf{X}\mathbf{H}_2\mathbf{H}_3$ as the return ratio matrix. The difference between the unit matrix and the loop-transmission matrix, $\mathbf{1}_m - \mathbf{X}\mathbf{H}_2\mathbf{H}_3$, can be defined as the return difference matrix (see Prob. 7.27). Therefore, Eq. (7.270) is a direct extension of the sensitivity function defined for a single-input, single-output system and for a single parameter, as shown in Eq. (5.165), where the sensitivity function of the closed-loop transfer function with respect to the forward amplifier gain is equal to the reciprocal of the return difference with respect to the forward amplifier gain. It is straightforward to demonstrate that (7.270) reduces to (5.165) for a single-input, single-output system and for a small perturbation of $\mathbf{X}$.

In the particular situation where $\mathbf{W}(\mathbf{X})$, $\delta\mathbf{X}$, and $\mathbf{X}$ are square and nonsingular, (7.270) can be put into the form similar to the scalar sensitivity function of (3.20). To this end, we use (7.262) in conjunction with (7.256) and (7.259) and obtain

$$\mathbf{E}_c = \mathbf{y}_c - \mathbf{y}_c^+ = [\mathbf{W}(\mathbf{X}) - \mathbf{W}^+(\mathbf{X})]\mathbf{u} = -\delta\mathbf{W}(\mathbf{X})\mathbf{u} \tag{7.271a}$$

$$\mathbf{E}_o = \mathbf{y}_o - \mathbf{y}_o^+ = (\mathbf{X}\mathbf{H}_1 - \mathbf{X}^+\mathbf{H}_1)\mathbf{u} = -\delta\mathbf{X}\mathbf{H}_1\mathbf{u} \tag{7.271b}$$

If $\mathbf{H}_1$ is nonsingular, $\mathbf{u}$ in (7.271*b*) can be solved for and substituted in the expression of (7.271*a*) to give

$$\mathbf{E}_c = \delta\mathbf{W}(\mathbf{X})\mathbf{H}_1^{-1}(\delta\mathbf{X})^{-1}\mathbf{E}_o \tag{7.272}$$

As before, for meaningful comparisons, we require that $\mathbf{y}_o = \mathbf{y}_c$ or

$$\mathbf{X}\mathbf{H}_1 = \mathbf{W}(\mathbf{X}) \tag{7.273}$$

Using this in (7.272), we get

$$\mathbf{E}_c = \delta\mathbf{W}(\mathbf{X})\mathbf{W}^{-1}(\mathbf{X})\mathbf{X}(\delta\mathbf{X})^{-1}\mathbf{E}_o \tag{7.274}$$

showing that (see Prob. 7.28)

$$\mathbf{S}(\mathbf{X}) = \delta\mathbf{W}(\mathbf{X})\mathbf{W}^{-1}(\mathbf{X})\mathbf{X}(\delta\mathbf{X})^{-1} \tag{7.275}$$

This result is to be compared with the scalar sensitivity function of Eq. (3.20), which can be rewritten in the form

$$S(x) = (\delta w)w^{-1}x(\delta x)^{-1} \tag{7.276}$$

We emphasize that the expression (7.275) is valid only when $\mathbf{W(X)}$, $\mathbf{X}$, and $\delta\mathbf{X}$ are square and nonsingular. However, (7.270) is well defined as long as the square matrix $\mathbf{1}_m - \mathbf{X}^+\mathbf{H}_2\mathbf{H}_3$ is nonsingular.

To demonstrate the use of (7.275), we consider the fundamental matrix feedback-flow graph of Fig. 7.13. As shown in Eq. (7.16*a*), the overall transfer function matrix $\mathbf{W(X)}$ of the multiple-loop feedback amplifier of Fig. 7.12 is given by

$$\mathbf{W(X)} = \mathbf{D} + \mathbf{CX}(\mathbf{1}_p - \mathbf{AX})^{-1}\mathbf{B} \tag{7.277}$$

We wish to compute the sensitivity matrix $\mathbf{S(X)}$ for the situation when all changes concerned are differentially small:

$$\mathbf{S(X)} = \delta\mathbf{W(X)W}^{-1}\mathbf{(X)X}(\delta\mathbf{X})^{-1}\big|_{\delta\mathbf{X}\to 0} \tag{7.278}$$

provided that $\mathbf{W(X)}$ and $\mathbf{X}$ are square and nonsingular. Then from (7.277) we have an expression for $\mathbf{W(X)}$ when $\mathbf{X}$ is perturbed to $\mathbf{X}^+ = \mathbf{X} + \delta\mathbf{X}$:

$$\mathbf{W(X)} + \delta\mathbf{W(X)} = \mathbf{D} + \mathbf{C}(\mathbf{X} + \delta\mathbf{X})(\mathbf{1}_p - \mathbf{AX} - \mathbf{A}\delta\mathbf{X})^{-1}\mathbf{B} \tag{7.279}$$

or

$$\delta\mathbf{W(X)} = \mathbf{C}[(\mathbf{X} + \delta\mathbf{X})(\mathbf{1}_p - \mathbf{AX} - \mathbf{A}\delta\mathbf{X})^{-1} - \mathbf{X}(\mathbf{1}_p - \mathbf{AX})^{-1}]\mathbf{B} \tag{7.280}$$

As $\delta\mathbf{X}$ approaches zero,

$$\begin{aligned}
\delta\mathbf{W(X)} &= \mathbf{C}[(\mathbf{X} + \delta\mathbf{X}) - \mathbf{X}(\mathbf{1}_p - \mathbf{AX})^{-1}(\mathbf{1}_p - \mathbf{AX} - \mathbf{A}\delta\mathbf{X})](\mathbf{1}_p - \mathbf{AX} - \mathbf{A}\delta\mathbf{X})^{-1}\mathbf{B} \\
&= \mathbf{C}[\delta\mathbf{X} + \mathbf{X}(\mathbf{1}_p - \mathbf{AX})^{-1}\mathbf{A}\delta\mathbf{X}](\mathbf{1}_p - \mathbf{AX} - \mathbf{A}\delta\mathbf{X})^{-1}\mathbf{B} \\
&= \mathbf{C}(\mathbf{1}_q - \mathbf{XA})^{-1}(\delta\mathbf{X})(\mathbf{1}_p - \mathbf{AX} - \mathbf{A}\delta\mathbf{X})^{-1}\mathbf{B} \\
&\approx \mathbf{C}(\mathbf{1}_q - \mathbf{XA})^{-1}(\delta\mathbf{X})(\mathbf{1}_p - \mathbf{AX})^{-1}\mathbf{B}
\end{aligned} \tag{7.281}$$

Assuming that $\mathbf{W(X)}$, $\mathbf{X}$, $\delta\mathbf{X}$, and $\mathbf{C}$ are nonsingular with $n = m = p = q$, Eq. (7.275) becomes

$$\begin{aligned}
\mathbf{S(X)} &= \mathbf{C}(\mathbf{1}_n - \mathbf{XA})^{-1}(\delta\mathbf{X})(\mathbf{1}_n - \mathbf{AX})^{-1}\mathbf{BW}^{-1}\mathbf{(X)X}(\delta\mathbf{X})^{-1} \\
&= \mathbf{C}(\mathbf{1}_n - \mathbf{XA})^{-1}(\delta\mathbf{X})\mathbf{X}^{-1}\mathbf{C}^{-1}[\mathbf{W(X)} - \mathbf{D}]\mathbf{W}^{-1}\mathbf{(X)X}(\delta\mathbf{X})^{-1} \\
&= \mathbf{C}(\mathbf{1}_n - \mathbf{XA})^{-1}(\delta\mathbf{X})\mathbf{X}^{-1}\mathbf{C}^{-1}[\mathbf{1}_n - \mathbf{W(0)W}^{-1}\mathbf{(X)}]\mathbf{X}(\delta\mathbf{X})^{-1} \\
&= \mathbf{C}\tilde{\mathbf{F}}^{-1}\mathbf{(X)}(\delta\mathbf{X})\mathbf{X}^{-1}\mathbf{C}^{-1}[\mathbf{1}_n - \mathbf{W(0)W}^{-1}\mathbf{(X)}]\mathbf{X}(\delta\mathbf{X})^{-1}
\end{aligned} \tag{7.282}$$

where $\mathbf{W(0)} = \mathbf{D}$ and $\tilde{\mathbf{F}}\mathbf{(X)} = \mathbf{1}_n - \mathbf{XA}$. Since from (7.83), $\det \tilde{\mathbf{F}}\mathbf{(X)} = \det(\mathbf{1}_n - \mathbf{XA}) = \det(\mathbf{1}_n - \mathbf{AX}) = \det \mathbf{F(X)}$, the determinant of $\mathbf{S(X)}$ is found to be

$$\det \mathbf{S(X)} = \frac{\det[\mathbf{1}_n - \mathbf{W(0)W}^{-1}\mathbf{(X)}]}{\det \tilde{\mathbf{F}}\mathbf{(X)}} = \frac{\det[\mathbf{1}_n - \mathbf{W(0)W}^{-1}\mathbf{(X)}]}{\det \mathbf{F(X)}} \tag{7.283}$$

This expression is a direct generalization of Eq. (5.6) for the scalar sensitivity function, and it relates the sensitivity matrix, the return difference matrix, and the transfer-function matrix of a multiple-loop feedback amplifier. Like (5.6), (7.283) indicates that if $\mathbf{W(0)} = \mathbf{0}$, then $\det \mathbf{S(X)} = 1/\det \mathbf{F(X)}$, meaning that the determinant of the sensitivity matrix is equal to the reciprocal of the determinant of the return difference matrix $\mathbf{F(X)}$. As to the physical significance of $\tilde{\mathbf{F}}\mathbf{(X)}$, we recall that in Sec. 7.2.1 the return difference matrix was defined by $\mathbf{F}(\mathbf{X}) = \mathbf{1}_n - \mathbf{AX}$, where $\mathbf{AX}$ is the loop transmission matrix and is obtained by opening the loop at the input of $\mathbf{X}$, as shown in Fig. 7.37*a*. The matrix $\mathbf{XA}$ has the physical significance of loop transmission if the loop is opened at the input of $\mathbf{A}$, as indicated in Fig. 7.37*b*. Thus, $\tilde{\mathbf{F}}(\mathbf{X})$ is the return difference matrix with respect to the branch $\mathbf{A}$.

7.5.2 The Multiparameter Sensitivity

In this section, we derive formulas for the effect of change of $\mathbf{X}$ on a scalar transfer function $w(\mathbf{X})$.

Let x_k, $k = 1, 2, \ldots, pq$, be the elements of $\mathbf{X}$. The multivariable Taylor series expansion for $w(\mathbf{X})$ with respect to x_k is given by

$$\delta w = \sum_{k=1}^{pq} \frac{\partial w}{\partial x_k} \delta x_k + \sum_{j=1}^{pq} \sum_{k=1}^{pq} \frac{\partial^2 w}{\partial x_j \, \partial x_k} \frac{\delta x_j \, \delta x_k}{2!} + \cdots \tag{7.284}$$

The first-order perturbation can then be written as

$$\delta w \approx \sum_{k=1}^{pq} \frac{\partial w}{\partial x_k} \delta x_k \tag{7.285}$$

The single-parameter sensitivity function, as defined by Eq. (3.20), is given by

$$S(x_k) = \frac{x_k}{w} \frac{\partial w}{\partial x_k} \tag{7.286}$$

Combining (7.285) and (7.286) yields

$$\frac{\delta w}{w} \approx \sum_{k=1}^{pq} S(x_k) \frac{\delta x_k}{x_k} \tag{7.287}$$

This expression gives the fractional change of the transfer function w in terms of the scalar sensitivity functions $S(x_k)$.

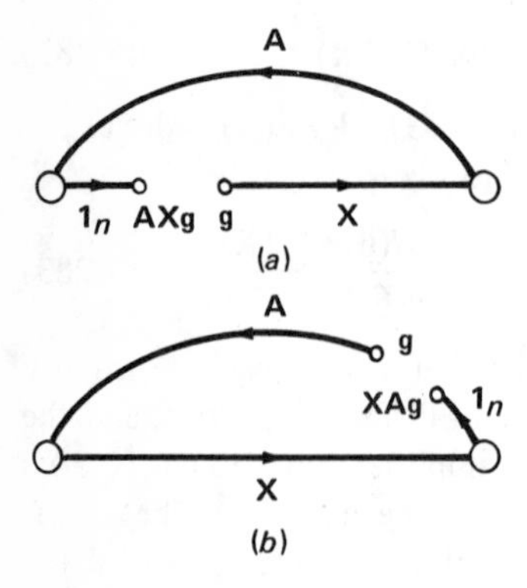

Figure 7.37 (*a*) The physical interpretation of the loop transmission **AX** and (*b*) the physical interpretation of the loop transmission **XA**.

Refer again to the fundamental matrix feedback-flow graph of Fig. 7.13. Assume that the feedback amplifier has a single input and a single output. Then from (7.281) we have

$$\delta w(\mathbf{X}) \approx \mathbf{C}(\mathbf{1}_q - \mathbf{XA})^{-1}(\delta\mathbf{X})(\mathbf{1}_p - \mathbf{AX})^{-1}\mathbf{B} \tag{7.288}$$

where $\mathbf{C}$ is a row q-vector and $\mathbf{B}$ is a column p-vector. Write

$$\mathbf{C} = [c_1, c_2, \ldots, c_q] \tag{7.289a}$$

$$\mathbf{B}' = [b_1, b_2, \ldots, b_p] \tag{7.289b}$$

$$\widetilde{\mathbf{W}} = \mathbf{X}(\mathbf{1}_p - \mathbf{AX})^{-1} = (\mathbf{1}_q - \mathbf{XA})^{-1}\mathbf{X} = [\widetilde{w}_{ij}] \tag{7.289c}$$

The increment $\delta w(\mathbf{X})$ can be expressed in terms of the elements of (7.289) and those of $\mathbf{X}$. In the case where $\mathbf{X}$ is diagonal with

$$\mathbf{X} = \text{diag}\,(x_1, x_2, \ldots, x_p) \tag{7.290}$$

where $p = q$, the expression for $\delta w(\mathbf{X})$ can be greatly simplified and is found to be

$$\delta w(\mathbf{X}) = \sum_{i=1}^{p}\sum_{k=1}^{p}\sum_{j=1}^{p} c_i\left(\frac{\widetilde{w}_{ik}}{x_k}\right)(\delta x_k)\left(\frac{\widetilde{w}_{kj}}{x_k}\right)b_j$$

$$= \sum_{i=1}^{p}\sum_{j=1}^{p}\sum_{k=1}^{p} \frac{c_i\widetilde{w}_{ik}\widetilde{w}_{kj}b_j}{x_k}\,\frac{\delta x_k}{x_k} \tag{7.291}$$

Comparing this with (7.287), we obtain an explicit form for the single-parameter sensitivity function as

$$S(x_k) = \sum_{i=1}^{p}\sum_{j=1}^{p} \frac{c_i\widetilde{w}_{ik}\widetilde{w}_{kj}b_j}{x_k w(\mathbf{X})} \tag{7.292}$$

Thus, knowing (7.289) and (7.290), we can calculate the multiparameter sensitivity function for the scalar transfer function $w(\mathbf{X})$ immediately.

Example 7.14 Consider the same voltage-series feedback amplifier of Fig. 7.14, as discussed in Example 7.2. Assume that V_s is the input and V_{25} the output. The transfer function of interest is the amplifier voltage gain V_{25}/V_s. The elements of main concern are the two controlling parameters of the controlled sources. Thus, from Example 7.2 we have

$$\mathbf{X} = \begin{bmatrix} \tilde{\alpha}_1 & 0 \\ 0 & \tilde{\alpha}_2 \end{bmatrix} = \begin{bmatrix} 0.0455 & 0 \\ 0 & 0.0455 \end{bmatrix} \tag{7.293a}$$

$$\mathbf{A} = \begin{bmatrix} -90.782 & 45.391 \\ -942.507 & 0 \end{bmatrix} \tag{7.293b}$$

$$\mathbf{B}' = [0.91748 \quad 0] \tag{7.293c}$$

$$\mathbf{C} = [45.391 \quad -2372.32] \tag{7.293d}$$

yielding
$$\tilde{\mathbf{W}} = \mathbf{X}(\mathbf{1}_2 - \mathbf{AX})^{-1} = 10^{-4}\begin{bmatrix} 4.85600 & 10.02904 \\ -208.245 & 24.91407 \end{bmatrix} \tag{7.294}$$

Also, from (7.100) we have

$$w(\mathbf{X}) = \frac{V_{25}}{V_s} = 45.39 \tag{7.295}$$

To compute the sensitivity functions with respect to $\tilde{\alpha}_1$ and $\tilde{\alpha}_2$, we apply (7.292), as follows:

$$S(\tilde{\alpha}_1) = \sum_{i=1}^{2}\sum_{j=1}^{2} \frac{c_i \tilde{w}_{i1} \tilde{w}_{1j} b_j}{\tilde{\alpha}_1 w(\mathbf{X})}$$

$$= \frac{c_1 \tilde{w}_{11}\tilde{w}_{11} b_1 + c_1 \tilde{w}_{11}\tilde{w}_{12} b_2 + c_2 \tilde{w}_{21}\tilde{w}_{11} b_1 + c_2 \tilde{w}_{21}\tilde{w}_{12} b_2}{\tilde{\alpha}_1 w}$$

$$= 0.01066 \tag{7.296a}$$

$$S(\tilde{\alpha}_2) = \frac{c_1 \tilde{w}_{12}\tilde{w}_{21} b_1 + c_1 \tilde{w}_{12}\tilde{w}_{22} b_2 + c_2 \tilde{w}_{22}\tilde{w}_{21} b_1 + c_2 \tilde{w}_{22}\tilde{w}_{22} b_2}{\tilde{\alpha}_2 w}$$

$$= 0.05426 \tag{7.296b}$$

The return differences and the null return differences with respect to $\tilde{\alpha}_1$ and $\tilde{\alpha}_2$ were computed earlier in Chap. 4 and are given by Eqs. (4.141), (4.171), and (4.172). They are repeated below:

$$F(\tilde{\alpha}_1) = 93.70 \tag{7.297a}$$

$$F(\tilde{\alpha}_2) = 18.26 \tag{7.297b}$$

$$\hat{F}(\tilde{\alpha}_1) = 103.07 \cdot 10^3 \tag{7.297c}$$

$$\hat{F}(\tilde{\alpha}_2) = 2018.70 \tag{7.297d}$$

By appealing to Eq. (5.5), the sensitivity functions are found to be

$$S(\tilde{\alpha}_1) = \frac{1}{F(\tilde{\alpha}_1)} - \frac{1}{\hat{F}(\tilde{\alpha}_1)} = 0.01066 \tag{7.298a}$$

$$S(\tilde{\alpha}_2) = \frac{1}{F(\tilde{\alpha}_2)} - \frac{1}{\hat{F}(\tilde{\alpha}_2)} = 0.05427 \tag{7.298b}$$

confirming (7.296).

Assume that $\tilde{\alpha}_1$ is changed by 4% and $\tilde{\alpha}_2$ by 6%. The fractional change of the voltage gain $w(\mathbf{X})$ is found from (7.287) as

$$\frac{\delta w}{w} = S(\tilde{\alpha}_1)\,\frac{\delta\tilde{\alpha}_1}{\tilde{\alpha}_1} + S(\tilde{\alpha}_2)\,\frac{\delta\tilde{\alpha}_2}{\tilde{\alpha}_2} = 0.01066 \cdot 0.04 + 0.05427 \cdot 0.06 = 0.003683 \quad (7.299)$$

or 0.37%.

7.6 SUMMARY

We began this chapter by reviewing the rules of the matrix signal-flow graph and presenting some of its properties. One way to obtain the graph transmission is to apply the graph reduction rules repeatedly until the final graph is composed of only one edge. The transmittance of the edge is the desired graph transmission. To avoid the necessity of repeated graph reduction, we described a topological procedure for computing the graph transmission directly from the original matrix signal-flow graph.

In the study of a single-loop feedback amplifier, we usually single out an element for particular attention. The scalar return difference and null return difference are then defined in terms of this particular element. For a multiple-loop amplifier, we pay particular attention to a group of elements and study its effects on the whole system. This leads to the concepts of return difference matrix and null return difference matrix. Like the return difference, the return difference matrix is defined as the square matrix relating the difference between the applied vector signal and the returned vector signal to the applied vector signal. Likewise, the null return difference matrix is the return difference matrix under the situation when the input excitation vector of the feedback amplifier is adjusted so that the total output resulting from this input excitation and the applied signal vector at the controlling branches of the elements of interest is identically zero. The overall transfer characteristics of a multiple-loop feedback amplifier are described by its transfer-function matrix. We showed that the determinant of the transfer-function matrix can be expressed in terms of the determinants of the return difference and null return difference matrices and its value when the elements of interest assume the zero value, thereby generalizing Blackman's impedance formula for a single input to a multiplicity of inputs.

In deriving the null return difference matrix, we assume that the effects at the output resulting from the applied signals at the controlling branches of the elements of interest can be neutralized by a unique input excitation vector of the amplifier. In the situation where the above condition is not satisfied, the null return difference matrix is not defined. To avoid this difficulty, we introduced the complementary return difference matrix. This is the return difference matrix under the situation when the input excitation vector of the amplifier is adjusted so that the sum of this input excitation and the output is identically zero. This matrix together with the return difference and null return difference matrices generally depend on the choice of the elements of interest. However, we demonstrated that their determinants are invariant with respect to the ordering of these elements.

Like the return difference and null return difference for a general reference value, the return difference and null return difference matrices were generalized for a general reference matrix. We showed that the general return difference matrix with respect to **X** for a general reference matrix **K** is equal to the product of the inverse of the return difference matrix with respect to **K** and the return difference matrix with respect to **X**, both being defined for the zero reference. A similar statement can be made for the null return difference matrix for a general reference matrix. The determinant of the transfer-function matrix can therefore be expressed in terms of its value when the elements of interest assume their reference values and the determinants of the general return difference and null return difference matrices. We also developed relations between the scalar return difference and null return difference and the return difference and null return difference matrices. Specifically, we indicated that the scalar return difference of a multiple-loop feedback amplifier with respect to an element x of **X** is equal to the ratio of the determinants of the return difference matrix with respect to **X** and that with respect to **K**, where **K** is obtained from **X** by setting x to zero. A similar relation was obtained between the scalar null return difference and null return difference matrix.

In Chaps. 4 and 5, the concepts of return difference and null return difference were introduced by means of the first- and second-order cofactors of the indefinite-admittance matrix. In this chapter, we demonstrated that the determinant of the return difference matrix can be expressed elegantly and compactly in terms of the ratio of the two functional values assumed by the determinant of the hybrid matrix under the condition that the elements of interest assume their nominal and reference values.

Finally, we introduced the sensitivity matrix and the multiparameter sensitivity function. We indicated that the sensitivity matrix is a direct extension of the sensitivity function defined for a single-input, single-output system and for a single parameter. In the particular situation, the sensitivity matrix reduces to the form similar to that defined for the scalar sensitivity function, and we obtained an expression relating the determinant of the sensitivity matrix and the determinants of the return difference matrix and a matrix involving only the transfer-function matrix. Formulas for the effects of changes of a group of elements on a scalar transfer function were also derived.

PROBLEMS

7.1 Show that Eqs. (7.8) and (7.9) are equivalent.

7.2 From (7.16), show that if $p = n$ and $\mathbf{B} = \mathbf{1}_n$, then

$$\mathbf{W}(\mathbf{X}) = \mathbf{W}(\mathbf{0})\hat{\mathbf{F}}(\mathbf{X})\mathbf{F}^{-1}(\mathbf{X}) \tag{7.300}$$

7.3 From (7.16), show that if $q = m$ and $\mathbf{C} = \mathbf{1}_n$, then

$$\mathbf{W}(\mathbf{X}) = \mathbf{W}(\mathbf{0})\mathbf{F}_1^{-1}(\mathbf{X})\hat{\mathbf{F}}_1(\mathbf{X}) \tag{7.301}$$

where

$$\mathbf{F}_1(\mathbf{X}) = \mathbf{1}_q - \mathbf{XA} \tag{7.302a}$$

$$\hat{\mathbf{F}}_1(\mathbf{X}) = \mathbf{1}_q - \mathbf{X}\hat{\mathbf{A}} \tag{7.302b}$$

7.4 Consider the matrix signal-flow graph of Fig. 7.38. By using the topological procedure, confirm that the graph transmission from node $\mathbf{X}_7$ to node $\mathbf{X}_6$ is given by

$$\mathbf{W} = \mathbf{D}[\mathbf{1} - \mathbf{C}(\mathbf{1} - \mathbf{BJ})^{-1}\mathbf{K} - \mathbf{G}(\mathbf{1} - \mathbf{FH})^{-1}\mathbf{FE}(\mathbf{1} - \mathbf{JB})^{-1}\mathbf{JK}]^{-1}\mathbf{G}(\mathbf{1} - \mathbf{FH})^{-1}\mathbf{FE}(\mathbf{1} - \mathbf{JB})^{-1}\mathbf{A}$$
$$+ \mathbf{D}[\mathbf{1} - \mathbf{C}(\mathbf{1} - \mathbf{BJ})^{-1}\mathbf{K} - \mathbf{G}(\mathbf{1} - \mathbf{FH})^{-1}\mathbf{FE}(\mathbf{1} - \mathbf{JB})^{-1}\mathbf{JK}]^{-1}\mathbf{C}(\mathbf{1} - \mathbf{BJ})^{-1}\mathbf{BA} \tag{7.303}$$

7.5 From Fig. 7.27, show that the currents I_a and I_b and voltages V_{13} and V_{45} are related by Eq. (7.131) when the input excitation V_s is adjusted so that the output voltage $V_{25} = 0$.

7.6 In Fig. 7.27, suppose that the input V_s is adjusted so that $V_s + \check{V}_{25} = 0$, where $\check{V}_{25}$ denotes the contributions to V_{25} by the currents I_a and I_b. Show that the currents I_a and I_b and voltages V_{13} and V_{45} are related by Eq. (7.134).

7.7 Demonstrate that (7.303) can be expressed equivalently as

$$\mathbf{W} = \mathbf{DC}(\mathbf{1} - \mathbf{BJ} - \mathbf{KC})^{-1}\mathbf{BA} + [\mathbf{DC}(\mathbf{1} - \mathbf{BJ} - \mathbf{KC})^{-1}\mathbf{KG} + \mathbf{DG}]$$
$$\cdot [\mathbf{1} - \mathbf{FH} - \mathbf{FEJ}(\mathbf{1} - \mathbf{BJ} - \mathbf{KC})^{-1}\mathbf{KG}]^{-1}[\mathbf{FEA} + \mathbf{FEJ}(\mathbf{1} - \mathbf{BJ} - \mathbf{KC})^{-1}\mathbf{BA}] \tag{7.304}$$

7.8 Prove the identity (7.200).

7.9 Prove the identity (7.203).

7.10 In the feedback network of Fig. 7.11, let the elements of interest be represented by the matrix equation

$$\boldsymbol{\theta} = \begin{bmatrix} V_1 \\ V_a \\ V_b \end{bmatrix} = \begin{bmatrix} Z_1 & 0 & 0 \\ 0 & \beta_1 & \beta_2 \\ 0 & \alpha_1 & \alpha_2 \end{bmatrix} \begin{bmatrix} I_1 \\ I_2 \\ I_3 \end{bmatrix} = \mathbf{X}\boldsymbol{\phi} \tag{7.305}$$

Show that the return difference matrix with respect to $\mathbf{X}$ is given by

$$\mathbf{F}(\mathbf{X}) = \frac{1}{\xi}\begin{bmatrix} (Z_1 + Z_3)(Z_2 + Z_4) + Z_1 Z_3 & -\alpha_1 Z_3 - \beta_1(Z_2 + Z_4) & -\alpha_2 Z_3 - \beta_2(Z_2 + Z_4) \\ Z_1 Z_3 & Z_3(Z_2 + Z_4 - \alpha_1) & -\alpha_2 Z_3 \\ Z_1(Z_2 + Z_4) & -\beta_1(Z_2 + Z_4) & (Z_3 - \beta_2)(Z_2 + Z_4) \end{bmatrix} \tag{7.306}$$

where $\xi = Z_3(Z_2 + Z_4)$.

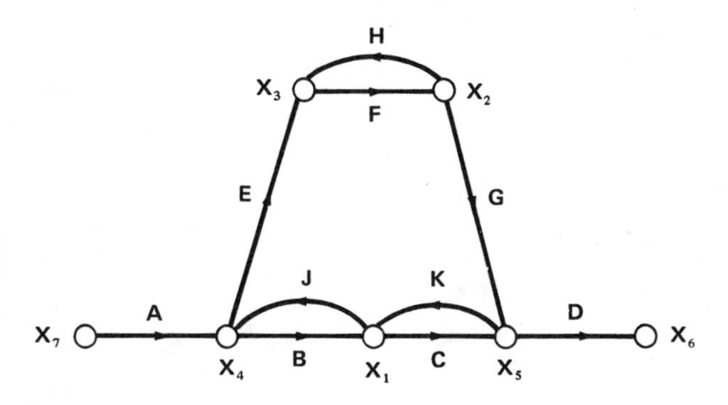

Figure 7.38 A matrix signal-flow graph.

7.11 By applying (7.225) and (7.226) in conjunction with (7.249), verify (7.254) and (7.255). Also confirm the identities (7.224).

7.12 Show that (7.281) can also be expressed in the form

$$\delta \mathbf{W}(\mathbf{X}) = \mathbf{C}(\mathbf{1}_q - \mathbf{X}^+\mathbf{A})^{-1}(\delta \mathbf{X})(\mathbf{1}_q - \mathbf{A}\mathbf{X})^{-1}\mathbf{B} \tag{7.307}$$

7.13 Repeat Example 7.14 if current I_{51} is chosen as the output variable and the input admittance I_{51}/V_s is the network function of interest.

7.14 Consider the parallel series or current-shunt feedback amplifier of Fig. 5.14, whose equivalent network is shown in Fig. 5.15*b*. Let the controlling parameters of the two controlled current sources be the elements of interest. Assume that the voltages V_{15} and V_{25} are the output variables. Compute the matrix equations (7.13), and determine the transfer-function matrix of the amplifier.

7.15 In Prob. 7.14, let the voltage V_{15} be the output variable. Compute the following with respect to the two controlling parameters:

(*a*) The return difference matrix $\mathbf{F}(\mathbf{X})$
(*b*) The null return difference matrix $\hat{\mathbf{F}}(\mathbf{X})$
(*c*) The complementary return difference matrix $\check{\mathbf{F}}(\mathbf{X})$
(*d*) The voltage gain of the amplifier
(*e*) The amplifier input impedance

7.16 In Prob. 7.14, let the voltage V_{25} be the output variable. By using physical interpretations outlined in Sec. 7.2.5, compute the matrices $\mathbf{A}$, $\hat{\mathbf{A}}$, and $\check{\mathbf{A}}$ directly from the equivalent network of Fig. 5.15*b*. Invoke (7.118) and calculate the voltage gain V_{25}/V_s.

7.17 In Prob. 7.14, assume that the reference values for the two controlling parameters are 0.01 and 0.02 mho. Compute the following:

(*a*) The matrices $\mathbf{A}_1$, $\mathbf{B}_1$, $\mathbf{C}_1$, and $\mathbf{D}_1$
(*b*) The return difference matrix $\mathbf{F}_\mathbf{K}(\mathbf{X})$
(*c*) The null return difference matrix $\hat{\mathbf{F}}_\mathbf{K}(\mathbf{X})$
(*d*) The amplifier input impedance

7.18 Refer to Prob. 7.14. By using (7.199), (7.200), and (7.209), calculate the return differences, the null return differences, and the sensitivity functions with respect to the two controlling parameters.

7.19 In Prob. 7.14, let the voltage V_{25} be the output variable. The transfer function of interest is the amplifier voltage gain V_{25}/V_s. By using (7.292), compute the sensitivity functions with respect to the controlling parameters of the two controlled sources. Suppose that the controlling parameters are each changed by 5%. Determine the fractional change of the voltage gain.

7.20 In the feedback network of Fig. 7.11, the elements of interest are represented by the matrix equation

$$\boldsymbol{\theta} = \begin{bmatrix} I_1 \\ V_a \\ V_b \end{bmatrix} = \begin{bmatrix} Y_1 & 0 & 0 \\ 0 & \beta_1 & \beta_2 \\ 0 & \alpha_1 & \alpha_2 \end{bmatrix} \begin{bmatrix} V_1 \\ I_2 \\ I_3 \end{bmatrix} = \mathbf{X}\boldsymbol{\phi} \tag{7.308}$$

Demonstrate that the return difference matrix with respect to $\mathbf{X}$ is given by

$$\mathbf{F}(\mathbf{X}) = \frac{1}{U}\begin{bmatrix} (Z_2+Z_4)(1+Y_1Z_3)+Z_3 & -\beta_1(Z_2+Z_4)-\alpha_1 Z_3 & -\beta_2(Z_2+Z_4)-\alpha_2 Z_3 \\ -Y_1 Z_3 & Z_2+Z_3+Z_4-\alpha_1+\beta_1 & \beta_2-\alpha_2 \\ -Y_1(Z_2+Z_4) & \alpha_1-\beta_1 & Z_2+Z_3+Z_4+\alpha_2-\beta_2 \end{bmatrix} \tag{7.309}$$

where $U = Z_2 + Z_3 + Z_4$. Compute the matrix $\mathbf{A}$ by its physical interpretation.

7.21 In the feedback network of Fig. 7.11, let the voltage V_2 be the output variable. Assume that $\mathbf{X}$ of (7.305) is the matrix of interest. Show that the null return difference matrix with respect to $\mathbf{X}$ is given by the expression

$$\hat{\mathbf{F}}(\mathbf{X}) = \frac{1}{Z_3}\begin{bmatrix} Z_3 & \alpha_1 - \beta_1 & \alpha_2 - \beta_2 \\ 0 & Z_3 & 0 \\ 0 & \alpha_1 - \beta_1 & Z_3 + \alpha_2 - \beta_2 \end{bmatrix} \tag{7.310}$$

Compute the matrix $\hat{\mathbf{A}}$ by its physical interpretation.

7.22 In the feedback network of Fig. 7.11, let the voltage V_4 be the output variable. Assume that the matrix $\mathbf{X}$ of (7.305) is the matrix of interest. Show that the null return difference matrix with respect to $\mathbf{X}$ is given by (7.310).

7.23 A three-stage common-emitter feedback amplifier is shown in Fig. 6.9. Assume that the three transistors are identical, with $h_{ie} = 1.1$ kΩ, $h_{fe} = 50$, and $h_{re} = h_{oe} = 0$. The values of other network elements are given below:

$$R_1 = 100\ \Omega \quad R_2 = 75\ \Omega \quad R_3 = 1\ \text{k}\Omega$$
$$R_4 = 1\ \text{k}\Omega \quad R_f = 100\ \text{k}\Omega \tag{7.311}$$

Let the controlling parameters h_{fe} of the three transistors be the elements of interest. The transfer function of interest is the amplifier voltage gain. Compute the following:

(*a*) The return difference matrix $\mathbf{F}(\mathbf{X})$
(*b*) The null return difference matrix $\hat{\mathbf{F}}(\mathbf{X})$
(*c*) The complementary return difference matrix $\check{\mathbf{F}}(\mathbf{X})$
(*d*) The voltage gain of the amplifier
(*e*) The fractional change of the voltage gain for a 10% change of each of the three controlling parameters h_{fe}

7.24 For the multiple-loop feedback amplifier of Fig. 7.39, assume that transistors Q_1 and Q_3 are identical, with $h_{ie} = 1$ kΩ, $h_{fe} = 50$, and $h_{re} = h_{oe} = 0$, and that transistors Q_2 and Q_4 are identical, with $h_{ie} = 1.2$ kΩ, $h_{fe} = 75$, and $h_{re} = h_{oe} = 0$. Compute the return difference matrix, the null return difference matrix, and the complementary return difference matrix with respect to the four controlling parameters h_{fe} of the transistors by ignoring the biasing circuitry (Ramey and White, 1971).

7.25 In the fundamental matrix feedback-flow graph of Fig. 7.13, assume that the system has a single input u and a single output y. Suppose that the output node y is connected to the input node u by a branch with transmittance k as shown in Fig. 7.40. Show that the return difference matrix, written as $\mathbf{F}(\mathbf{X}, k)$, with respect to $\mathbf{X}$ in Fig. 7.40 can be written as

$$\mathbf{F}(\mathbf{X}, k) = \mathbf{1}_p - \left(\mathbf{A} + \frac{k}{1 - kd}\mathbf{BC}\right)\mathbf{X} \tag{7.312}$$

By using this result, show that

$$\mathbf{F}(\mathbf{X}) = \mathbf{F}(\mathbf{X}, 0) \tag{7.313a}$$

$$\hat{\mathbf{F}}(\mathbf{X}) = \mathbf{F}(\mathbf{X}, \infty) \tag{7.313b}$$

$$\check{\mathbf{F}}(\mathbf{X}) = \mathbf{F}\left(\mathbf{X}, \frac{1}{d-1}\right) \tag{7.313c}$$

This indicates that the return difference matrix $\mathbf{F}(\mathbf{X})$, the null return difference matrix $\hat{\mathbf{F}}(\mathbf{X})$, and the complementary return difference matrix $\check{\mathbf{F}}(\mathbf{X})$ are special values of the return difference matrix $\mathbf{F}(\mathbf{X}, k)$.

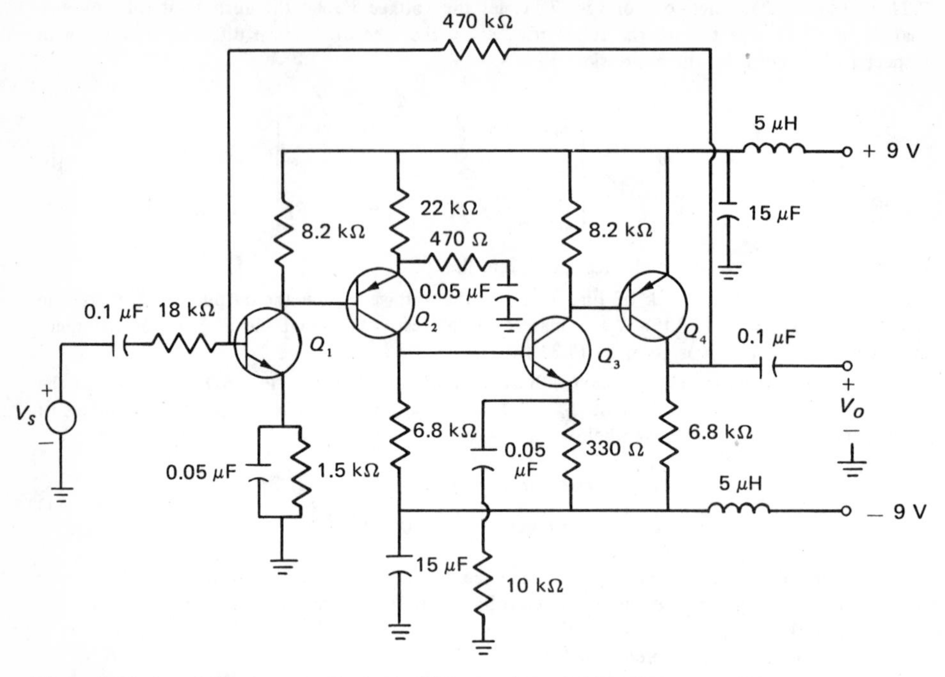

Figure 7.39 A multiple-loop feedback amplifier together with its biasing circuitry.

7.26 Instead of invoking (7.83), show that we can apply the following formulas for the determinants of the partitioned matrices to derive the identity (7.88) directly from (7.16*a*):

$$\det \begin{bmatrix} \mathbf{W} & \mathbf{X} \\ \mathbf{Y} & \mathbf{Z} \end{bmatrix} = \det \mathbf{W} \det (\mathbf{Z} - \mathbf{Y}\mathbf{W}^{-1}\mathbf{X}) \quad \text{if } \det \mathbf{W} \neq 0 \tag{7.314a}$$

$$\det \begin{bmatrix} \mathbf{W} & \mathbf{X} \\ \mathbf{Y} & \mathbf{Z} \end{bmatrix} = \det \mathbf{Z} \det (\mathbf{W} - \mathbf{X}\mathbf{Z}^{-1}\mathbf{Y}) \quad \text{if } \det \mathbf{Z} \neq 0 \tag{7.314b}$$

7.27 Refer to the block diagram of Fig. 7.36. As indicated in the paragraph following Eq. (7.270), the matrix $\mathbf{1}_m - \mathbf{X}\mathbf{H}_2\mathbf{H}_3$ can be defined as the return difference matrix, since, by

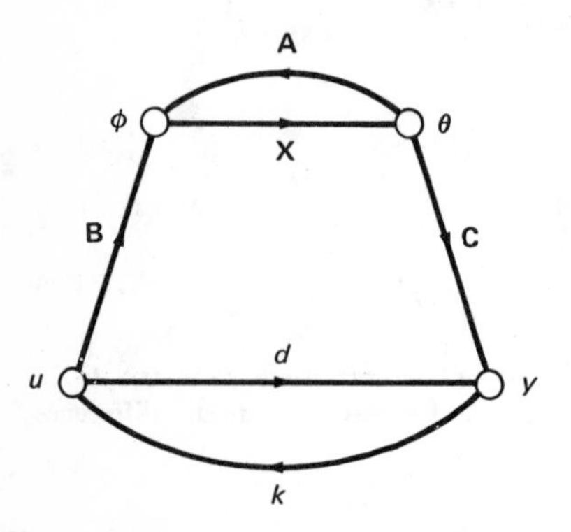

Figure 7.40 A matrix signal-flow graph associated with a new return difference matrix.

opening the output of $\mathbf{X}$, it is the matrix relating the difference between the applied vector signal and the returned vector signal to the applied vector signal. Likewise, by opening the output of $\mathbf{H}_2$, the matrix $\mathbf{1}_p - \mathbf{H}_2\mathbf{H}_3\mathbf{X}$ may be regarded as the return difference matrix, and by opening the input of $\mathbf{H}_2$, the matrix $\mathbf{1}_n - \mathbf{H}_3\mathbf{X}\mathbf{H}_2$ is the return difference matrix appropriate to this break point. Show that all three return difference matrices have the same determinant:

$$\det(\mathbf{1}_m - \mathbf{X}\mathbf{H}_2\mathbf{H}_3) = \det(\mathbf{1}_p - \mathbf{H}_2\mathbf{H}_3\mathbf{X}) = \det(\mathbf{1}_n - \mathbf{H}_3\mathbf{X}\mathbf{H}_2) \tag{7.315}$$

As a result, the determinant of the sensitivity matrix (7.270) is independent of which return difference matrix is used, and we may work any convenient return difference matrix.

7.28 In (7.260), assume that $\mathbf{XH}_2$ and $\mathbf{W}(\mathbf{X})$ are square and nonsingular. Suppose that $\mathbf{XH}_2$ is perturbed to $\mathbf{XH}_2 + \delta(\mathbf{XH}_2)$ and that $\mathbf{W}(\mathbf{X})$ is correspondingly perturbed to $\mathbf{W}(\mathbf{X}) + \delta\mathbf{W}(\mathbf{X})$. Demonstrate that

$$\mathbf{W}^{-1}(\mathbf{X})\delta\mathbf{W}(\mathbf{X}) \approx \mathbf{S}(\mathbf{X})(\mathbf{XH}_2)^{-1}\delta(\mathbf{XH}_2) \tag{7.316}$$

Compare this result with (7.275) and (7.276).

7.29 Refer to the matrix feedback-flow graph of Fig. 7.24. As in Sec. 7.2.4, we apply a signal p-vector $\mathbf{g}$ to the right of the breaking mark. Suppose that we adjust the input excitation n-vector $\mathbf{u}$ so that $\mathbf{u} + \mathbf{y} = \mathbf{0}$, assuming that $\mathbf{u}$ and $\mathbf{y}$ are of the same dimension. Show that

$$\mathbf{g} - \mathbf{h} = \mathring{\mathbf{F}}(\mathbf{X})\mathbf{g} \tag{7.317}$$

where

$$\mathring{\mathbf{F}}(\mathbf{X}) = \mathbf{1}_p - \mathring{\mathbf{A}}\mathbf{X} \tag{7.318a}$$

$$\mathring{\mathbf{A}} = \mathbf{A} - \mathbf{B}(\mathbf{1}_n + \mathbf{D})^{-1}\mathbf{C} \tag{7.318b}$$

The square matrix $\mathring{\mathbf{F}}(\mathbf{X})$ relating the difference between the applied and the returned signals to the applied signal under the condition that $\mathbf{u} + \mathbf{y} = \mathbf{0}$ may also be defined as the *complementary return difference matrix*. For a single input and single output, show that $\mathring{\mathbf{F}}(\mathbf{X})$ can be expressed as

$$\mathring{\mathbf{F}}(\mathbf{X}) = \mathbf{F}(\mathbf{X}, -1) \tag{7.319}$$

where $\mathbf{F}(\mathbf{X}, k)$ is defined in (7.312).

7.30 Refer to Prob. 7.29 and Sec. 7.2.5. Give a physical interpretation of the matrix $\mathring{\mathbf{A}}$ of Eq. (7.318*b*). By applying this interpretation to the problem considered in Example 7.7, compute $\mathring{\mathbf{A}}$ and $\mathring{\mathbf{F}}(\mathbf{X})$.

7.31 Refer to Prob. 7.29 and consider the situation where $m = n = 1$ with $w(\mathbf{0}) = D = 0$. Express the transfer function $w(\mathbf{X})$ in terms of the determinants of the matrices $\mathbf{F}(\mathbf{X})$ and $\mathring{\mathbf{F}}(\mathbf{X})$, and compare this with Eq. (7.118).

BIBLIOGRAPHY

Acar, C.: New Return Difference Matrix, *Int. J. Circuit Theory and Applications*, vol. 3, no. 1, pp. 87–94, 1975.

Arbel, A. F.: Identification of the Return Difference Matrix of a Multivariable Linear Control System, in Terms of Its Inverse Closed-Loop Transfer Matrix, *Int. J. Circuit Theory and Applications*, vol. 1, no. 2, pp. 187–190, 1973a.

Arbel, A. F.: Return Ratio and Sensitivity Computation for Multivariable Control Systems and Feedback-Stabilized Multiple Amplifier Active Circuits, *Int. J. Circuit Theory and Applications*, vol. 1, no. 2, pp. 191–199, 1973b.

Biswas, R. N. and E. S. Kuh: Optimum Synthesis of a Class of Multiple-Loop Feedback Systems, *IEEE Trans. Circuit Theory*, vol. CT-18, no. 6, pp. 582–587, 1971a.

Biswas, R. N. and E. S. Kuh: A Multiparameter Sensitivity Measure for Linear Systems, *IEEE Trans. Circuit Theory*, vol. CT-18, no. 6, pp. 718–719, 1971b.

Blecher, F. H.: Transistor Multiple Loop Feedback Amplifiers, *Proc. Natl. Elect. Conf.*, vol. 13, pp. 19–34, 1957.

Chen, W. K.: "Applied Graph Theory: Graphs and Electrical Networks," 2d rev. ed., New York: American Elsevier, and Amsterdam: North-Holland, 1976a.

Chen, W. K.: Topological Analysis of Multiple-Loop Feedback Networks, *Proc. Tenth Asilomar Conf. on Circuits, Systems, and Computers*, Pacific Grove, Calif., pp. 192–197, 1976b.

Chen, W. K.: The Hybrid Matrix in Linear Multiple-Loop Feedback Networks, *IEEE Trans. Circuits and Systems*, vol. CAS-24, no. 9, pp. 469–474, 1977a.

Chen, W. K.: Topological Evaluation of the Null Return Difference in Feedback Amplifier Theory, *Proc. Eleventh Asilomar Conf. on Circuits, Systems, and Computers*, Pacific Grove, Calif., pp. 388–395, 1977b.

Chen, W. K.: Topological Analysis of Feedback Matrices, *Proc. Twelfth International Symposium on Circuits and Systems*, Tokyo, Japan, pp. 837–840, July 17–19, 1979a. IEEE catalog no. 79CH1421-7 CAS.

Chen W. K.: Topological Evaluation of Feedback Matrices in Multiple-Loop Feedback Amplifiers, *J. Franklin Inst.*, vol. 308, no. 2, pp. 125–139, 1979b.

Chen, W. K. and H. M. Elsherif: Determinant of the Null Return-Difference Matrix, *Electron. Lett.*, vol. 13, no. 10, pp. 306–307, 1977.

Cruz, J. B., Jr. and W. R. Perkins: A New Approach to the Sensitivity Problem in Multivariable Feedback System Design, *IEEE Trans. Automatic Control*, vol. AC-9, no. 3, pp. 216–223, 1964.

Elsherif, H. M.: The Return Difference Matrix in the Multiple-Loop Feedback Systems, M.S. thesis, Ohio University, Athens, Ohio, 1974.

Elsherif, H. M. and W. K. Chen: The Return Difference Matrix in Multiple-Loop Feedback Systems, *Proc. 17th Midwest Symp. Circuits and Systems*, University of Kansas, Lawrence, Kansas, pp. 95–103, 1974.

Goldstein, A. J. and F. F. Kuo: Multiparameter Sensitivity, *IRE Trans. Circuit Theory*, vol. CT-8, no. 2, pp. 177–178, 1961.

Hakim, S. S.: Multiple-Loop Feedback Circuits, *Proc. IEE* (*London*), vol. 110, no. 11, pp. 1955–1959, 1963.

Hakim, S. S.: Evaluation of Sensitivity in Transistor Multiple-Loop Feedback Amplifiers, *Proc. IEE* (*London*), vol. 113, no. 2, pp. 219–224, 1966.

Kuh, E. S.: Some Results in Linear Multiple Loop Feedback Systems, *Proc. Allerton Conf. on Circuit and System Theory*, University of Illinois, Urbana, Illinois, vol. 1, pp. 471–487, 1963.

Kuh, E. S. and R. A. Rohrer: "Theory of Linear Active Networks," San Francisco, Calif.: Holden-Day, 1967.

MacFarlane, A. G. J.: Return-Difference and Return-Ratio Matrices and Their Use in Analysis and Design of Multivariable Feedback Control Systems, *Proc. IEE (London)*, vol. 117, no. 10, pp. 2037–2049, 1970.

MacFarlane, A. G. J.: Return-Difference and Return-Ratio Matrices and Their Use in Analysis and Design of Multivariable Feedback Control Systems, *Proc. IEE* (*London*), vol. 118, no. 7, pp. 946–947, 1971.

Ramey, R. L. and E. J. White: "Matrices and Computers in Electronic Circuit Analysis," New York: McGraw-Hill, 1971.

Riegle, D. E. and P. M. Lin: Matrix Signal Flow Graphs and an Optimum Topological Method for Evaluating Their Gains, *IEEE Trans. Circuit Theory*, vol. CT-19, no. 5, pp. 427–435, 1972.

Sandberg, I. W.: On the Theory of Linear Multi-Loop Feedback Systems, *Bell Sys. Tech. J.*, vol. 42, no. 2, pp. 355–382, 1963.

Tassny-Tschiassny, L.: The Return Difference Matrix in Linear Networks, *Proc. IEE* (*London*), vol. 100, part IV, pp. 39–46, 1953.

Truxal, J. G.: "Automatic Feedback Control System Synthesis," New York: McGraw-Hill, 1955.

APPENDIX

I

HERMITIAN FORMS

Hermitian forms and their properties are reviewed briefly in this appendix.

Let

$$\mathbf{A} = [a_{ij}] \tag{I.1}$$

be a hermitian matrix of order n and let $\mathbf{X} = [x_j]$ be a complex n-vector. Then the scalar expression

$$\mathbf{X}^*\mathbf{A}\mathbf{X} = \sum_{i=1}^{n} \sum_{j=1}^{n} a_{ij}\bar{x}_i x_j \tag{I.2}$$

where

$$\mathbf{X}^* = \bar{\mathbf{X}}' \tag{I.3}$$

$$\bar{\mathbf{X}} = [\bar{x}_j] \tag{I.4}$$

and $\bar{x}_j$ denotes the complex conjugate of x_j, is called a *hermitian form*. The matrix **A** is referred to as the *matrix of the hermitian form* (I.2). Even though **A** and **X** are complex, the hermitian form $\mathbf{X}^*\mathbf{A}\mathbf{X}$ is real. To see this, we take the conjugate of $\mathbf{X}^*\mathbf{A}\mathbf{X}$. Since $\mathbf{X}^*\mathbf{A}\mathbf{X}$ is a scalar, $(\mathbf{X}^*\mathbf{A}\mathbf{X})' = \mathbf{X}^*\mathbf{A}\mathbf{X}$. Hence, since **A** is hermitian, we have

$$\overline{\mathbf{X}^*\mathbf{A}\mathbf{X}} = (\mathbf{X}^*\mathbf{A}\mathbf{X})^* = \mathbf{X}^*\mathbf{A}^*\mathbf{X} = \mathbf{X}^*\mathbf{A}\mathbf{X} \tag{I.5}$$

showing that $\mathbf{X}^*\mathbf{A}\mathbf{X}$, being equal to its own conjugate, is real for every choice of **X**. However, for a given **A**, the sign associated with such a form normally depends on the values of **X**. It may happen that, for some **A**, its hermitian form remains of one sign, independent of the values of **X**. Such forms are called *definite*. Since definiteness of a hermitian form must be an inherent property of its matrix **A**, it is natural then to refer to the matrix **A** as *definite.* We now consider two subclasses of the class of definite hermitian matrices.

Definition I.1: Positive definite matrix An $n \times n$ hermitian matrix **A** is called a *positive definite matrix* if

$$\mathbf{X}^*\mathbf{A}\mathbf{X} > 0 \tag{I.6}$$

for all complex n-vectors $\mathbf{X} \neq \mathbf{0}$.

Definition I.2: Nonnegative definite matrix An $n \times n$ hermitian matrix **A** is called a *nonnegative definite matrix* if

$$\mathbf{X}^*\mathbf{A}\mathbf{X} \geqslant 0 \tag{I.7}$$

for all complex n-vectors **X**.

Evidently, a positive definite matrix is also nonnegative definite. Very often, a nonnegative definite matrix is also called a *positive semidefinite matrix*, but we must bear in mind that some people define a positive semidefinite matrix **A** as one that satisfies (I.7) for all **X**, provided that there is at least one $\mathbf{X} \neq \mathbf{0}$ for which the equality holds. In the latter case, it is evident that positive definiteness and positive semidefiniteness are mutually exclusive. Together, they form the class of nonnegative definite matrices. Thus, care must be taken to ensure the proper interpretation of the term "positive semidefiniteness."

In a matrix **A** of order n, define a *principal minor* of order r to be the determinant of a submatrix consisting of the rows $i_1, i_2, \ldots, i_r$, and the columns $i_1, i_2, \ldots,$ i_r. The *leading principal minor* of order r is the determinant of the submatrix consisting of the first r rows and the first r columns. We now present Sylvester's criteria for positive definiteness and nonnegative definiteness of a hermitian matrix.

Theorem I.1 A hermitian matrix is positive definite if and only if all of its leading principal minors are positive. A hermitian matrix is nonnegative definite if and only if all of its principal minors are nonnegative.

We remark that in testing for positive definiteness, the positiveness of all of its leading principal minors also implies the positiveness of all of its principal minors. However, in testing for nonnegative definiteness, the nonnegativeness of all of its leading principal minors does not necessarily imply the nonnegativeness of all of its principal minors.

An $n \times n$ matrix **U** is said to be *unitary* if and only if it has the property that $\mathbf{U}^*\mathbf{U} = \mathbf{1}_n$. We write

$$\mathbf{D}(\lambda_1, \lambda_2, \ldots, \lambda_n) = \begin{bmatrix} \lambda_1 & 0 & \cdots & 0 \\ 0 & \lambda_2 & \cdots & 0 \\ \cdots & \cdots & \cdots & \cdots \\ 0 & 0 & \cdots & \lambda_n \end{bmatrix} \tag{I.8}$$

We can prove the following theorem by induction over the order of **A**.

Theorem I.2 Let **A** be a hermitian matrix of order n. Then there exists a unitary matrix **U** such that **U*AU** is a diagonal matrix whose diagonal elements are the eigenvalues of **A**; that is,

$$\mathbf{U^*AU} = \mathbf{D}(\lambda_1, \lambda_2, \ldots, \lambda_n) \tag{I.9}$$

For a hermitian form (I.2), Theorem I.2 implies

Theorem I.3 By a suitable unitary transformation $\mathbf{X} = \mathbf{UY}$, the hermitian form $\mathbf{X^*AX}$ can be reduced to the form

$$\mathbf{X^*AX} = \mathbf{Y^*D}(\lambda_1, \lambda_2, \ldots, \lambda_n)\mathbf{Y} = \sum_{j=1}^{n} \lambda_j \bar{y}_j y_j \tag{I.10}$$

where the λ's are the eigenvalues of the hermitian matrix **A** and y_j is the jth row element of the n-vector **Y**.

It is possible to use other nonsingular transformations to reduce a given hermitian form to the diagonal form. No matter by what nonsingular transformation, a hermitian form can be reduced to a standard form given in the following theorem.

Theorem I.4 Irrespective of the nonsingular transformation used, a given hermitian form can be reduced to a form

$$d_1\bar{y}_1y_1 + d_2\bar{y}_2y_2 + \cdots + d_p\bar{y}_py_p - d_{p+1}\bar{y}_{p+1}y_{p+1} - \cdots - d_r\bar{y}_ry_r \tag{I.11}$$

where the d's are all positive and the intergers p and r remain invariant.

The number r of terms in (I.11) is called the *rank* of the form, and the number p of positive terms is referred to as the *index* of the form. The *signature* of the form is then defined to be $2p - r$. Evidently, the rank of the matrix of the hermitian form is equal to the rank of the form. Thus, we can state that a hermitian matrix is positive definite if and only if its order, rank, and the index of its hermitian form are equal, and that it is nonnegative definite if and only if the rank and index of its hermitian form are equal.

Theorem I.5 A hermitian matrix **A** is positive definite if and only if any one of the following conditions is satisfied:

1. $\mathbf{B}^*\mathbf{AB}$ is positive definite for arbitrary and nonsingular **B**.
2. $\mathbf{A}^q$ is positive definite for every interger q.
3. There exists a nonsingular matrix **B** such that $\mathbf{A} = \mathbf{B}^*\mathbf{B}$.

APPENDIX

II

CONVERSION CHART FOR TWO–PORT PARAMETERS

$\Delta_x = x_{11}x_{22} - x_{12}x_{21}$

To \ From	**Z**	**Y**	**T**	**T^{-1}**	**H**	**H^{-1}**
Z	$\begin{matrix} z_{11} & z_{12} \\ z_{21} & z_{22} \end{matrix}$	$\begin{matrix} \frac{y_{22}}{\Delta_y} & -\frac{y_{12}}{\Delta_y} \\ -\frac{y_{21}}{\Delta_y} & \frac{y_{11}}{\Delta_y} \end{matrix}$	$\begin{matrix} \frac{A}{C} & \frac{\Delta_T}{C} \\ \frac{1}{C} & \frac{D}{C} \end{matrix}$	$\begin{matrix} \frac{D'}{C'} & \frac{1}{C'} \\ \frac{\Delta_{T'}}{C'} & \frac{A'}{C'} \end{matrix}$	$\begin{matrix} \frac{\Delta_h}{h_{22}} & \frac{h_{12}}{h_{22}} \\ -\frac{h_{21}}{h_{22}} & \frac{1}{h_{22}} \end{matrix}$	$\begin{matrix} \frac{1}{g_{11}} & -\frac{g_{12}}{g_{11}} \\ \frac{g_{21}}{g_{11}} & \frac{\Delta_g}{g_{11}} \end{matrix}$
Y	$\begin{matrix} \frac{z_{22}}{\Delta_z} & -\frac{z_{12}}{\Delta_z} \\ -\frac{z_{21}}{\Delta_z} & \frac{z_{11}}{\Delta_z} \end{matrix}$	$\begin{matrix} y_{11} & y_{12} \\ y_{21} & y_{22} \end{matrix}$	$\begin{matrix} \frac{D}{B} & -\frac{\Delta_T}{B} \\ -\frac{1}{B} & \frac{A}{B} \end{matrix}$	$\begin{matrix} \frac{A'}{B'} & -\frac{1}{B'} \\ -\frac{\Delta_{T'}}{B'} & \frac{D'}{B'} \end{matrix}$	$\begin{matrix} \frac{1}{h_{11}} & -\frac{h_{12}}{h_{11}} \\ \frac{h_{21}}{h_{11}} & \frac{\Delta_h}{h_{11}} \end{matrix}$	$\begin{matrix} \frac{\Delta_g}{g_{22}} & \frac{g_{12}}{g_{22}} \\ -\frac{g_{21}}{g_{22}} & \frac{1}{g_{22}} \end{matrix}$
T	$\begin{matrix} \frac{z_{11}}{z_{21}} & \frac{\Delta_z}{z_{21}} \\ \frac{1}{z_{21}} & \frac{z_{22}}{z_{21}} \end{matrix}$	$\begin{matrix} -\frac{y_{22}}{y_{21}} & -\frac{1}{y_{21}} \\ -\frac{\Delta_y}{y_{21}} & -\frac{y_{11}}{y_{21}} \end{matrix}$	$\begin{matrix} A & B \\ C & D \end{matrix}$	$\begin{matrix} \frac{D'}{\Delta_{T'}} & \frac{B'}{\Delta_{T'}} \\ \frac{C'}{\Delta_{T'}} & \frac{A'}{\Delta_{T'}} \end{matrix}$	$\begin{matrix} -\frac{\Delta_h}{h_{21}} & -\frac{h_{11}}{h_{21}} \\ -\frac{h_{22}}{h_{21}} & -\frac{1}{h_{21}} \end{matrix}$	$\begin{matrix} \frac{1}{g_{21}} & \frac{g_{22}}{g_{21}} \\ \frac{g_{11}}{g_{21}} & \frac{\Delta_g}{g_{21}} \end{matrix}$
T^{-1}	$\begin{matrix} \frac{z_{22}}{z_{12}} & \frac{\Delta_z}{z_{12}} \\ \frac{1}{z_{12}} & \frac{z_{11}}{z_{12}} \end{matrix}$	$\begin{matrix} -\frac{y_{11}}{y_{12}} & -\frac{1}{y_{12}} \\ -\frac{\Delta_y}{y_{12}} & -\frac{y_{22}}{y_{12}} \end{matrix}$	$\begin{matrix} \frac{D}{\Delta_T} & \frac{B}{\Delta_T} \\ \frac{C}{\Delta_T} & \frac{A}{\Delta_T} \end{matrix}$	$\begin{matrix} A' & B' \\ C' & D' \end{matrix}$	$\begin{matrix} \frac{1}{h_{12}} & \frac{h_{11}}{h_{12}} \\ \frac{h_{22}}{h_{12}} & \frac{\Delta_h}{h_{12}} \end{matrix}$	$\begin{matrix} -\frac{\Delta_g}{g_{12}} & -\frac{g_{22}}{g_{12}} \\ -\frac{g_{11}}{g_{12}} & -\frac{1}{g_{12}} \end{matrix}$

$\Delta_x = x_{11}x_{22} - x_{12}x_{21}$ *(Continued)*

To \ From	Z	Y	T	T^{-1}	H	H^{-1}
H	$\begin{matrix} \frac{\Delta_z}{z_{22}} & \frac{z_{12}}{z_{22}} \\ -\frac{z_{21}}{z_{22}} & \frac{1}{z_{22}} \end{matrix}$	$\begin{matrix} \frac{1}{y_{11}} & -\frac{y_{12}}{y_{11}} \\ \frac{y_{21}}{y_{11}} & \frac{\Delta_y}{y_{11}} \end{matrix}$	$\begin{matrix} \frac{B}{D} & \frac{\Delta_T}{D} \\ -\frac{1}{D} & \frac{C}{D} \end{matrix}$	$\begin{matrix} \frac{B'}{A'} & \frac{1}{A'} \\ -\frac{\Delta_{T'}}{A'} & \frac{C'}{A'} \end{matrix}$	$\begin{matrix} h_{11} & h_{12} \\ h_{21} & h_{22} \end{matrix}$	$\begin{matrix} \frac{g_{22}}{\Delta_g} & -\frac{g_{12}}{\Delta_g} \\ -\frac{g_{21}}{\Delta_g} & \frac{g_{11}}{\Delta_g} \end{matrix}$
H^{-1}	$\begin{matrix} \frac{1}{z_{11}} & -\frac{z_{12}}{z_{11}} \\ \frac{z_{21}}{z_{11}} & \frac{\Delta_z}{z_{11}} \end{matrix}$	$\begin{matrix} \frac{\Delta_y}{y_{22}} & \frac{y_{12}}{y_{22}} \\ -\frac{y_{21}}{y_{22}} & \frac{1}{y_{22}} \end{matrix}$	$\begin{matrix} \frac{C}{A} & -\frac{\Delta_T}{A} \\ \frac{1}{A} & \frac{B}{A} \end{matrix}$	$\begin{matrix} \frac{C'}{D'} & -\frac{1}{D'} \\ \frac{\Delta_{T'}}{D'} & \frac{B'}{D'} \end{matrix}$	$\begin{matrix} \frac{h_{22}}{\Delta_h} & -\frac{h_{12}}{\Delta_h} \\ -\frac{h_{21}}{\Delta_h} & \frac{h_{11}}{\Delta_h} \end{matrix}$	$\begin{matrix} g_{11} & g_{12} \\ g_{21} & g_{22} \end{matrix}$

APPENDIX

III

OUTLINE OF A DERIVATION OF EQ. (7.224)

Let **M** denote the coefficient matrix of (7.221). We first rewrite the matrix **M** as the product of two matrices and then evaluate its determinant:

$$\det \mathbf{M} = \det \begin{bmatrix} \mathbf{1}_m & \mathbf{0} & \mathbf{0} & \mathbf{0} \\ \mathbf{0} & \mathbf{1}_r & \mathbf{0} & \mathbf{0} \\ \mathbf{0} & \mathbf{0} & \mathbf{1}_r & \mathbf{0} \\ \mathbf{0} & \mathbf{0} & \mathbf{0} & -\mathbf{H}_{\bar{t}\bar{t}} \end{bmatrix} \begin{bmatrix} -\mathbf{Q}'_{\bar{t}} & \mathbf{1}_m & \mathbf{0} & \mathbf{0} \\ \mathbf{0} & \mathbf{0} & \mathbf{1}_r & \mathbf{Q}_{\bar{t}} \\ \mathbf{H}_{tt} & \mathbf{0} & -\mathbf{1}_r & \mathbf{H}_{t\bar{t}} \\ -\mathbf{H}_{\bar{t}\bar{t}}^{-1}\mathbf{H}_{\bar{t}t} & \mathbf{H}_{\bar{t}\bar{t}}^{-1} & \mathbf{0} & -\mathbf{1}_m \end{bmatrix}$$

$$= (-1)^m \det \mathbf{H}_{\bar{t}\bar{t}} \det \begin{bmatrix} -\mathbf{Q}'_{\bar{t}} & \mathbf{1}_m & \mathbf{0} & \mathbf{0} \\ \mathbf{0} & \mathbf{0} & \mathbf{1}_r & \mathbf{Q}_{\bar{t}} \\ \mathbf{H}_{tt} - \mathbf{H}_{t\bar{t}}\mathbf{H}_{\bar{t}\bar{t}}^{-1}\mathbf{H}_{\bar{t}t} & \mathbf{H}_{t\bar{t}}\mathbf{H}_{\bar{t}\bar{t}}^{-1} & -\mathbf{1}_r & \mathbf{0} \\ -\mathbf{H}_{\bar{t}\bar{t}}^{-1}\mathbf{H}_{\bar{t}t} & \mathbf{H}_{\bar{t}\bar{t}}^{-1} & \mathbf{0} & -\mathbf{1}_m \end{bmatrix}$$

$$= (-1)^m \det \mathbf{H}_{\bar{t}\bar{t}} \det \begin{bmatrix} \mathbf{B}_f & \mathbf{0} \\ \mathbf{0} & \mathbf{Q}_f \\ \mathbf{Y}_b & -\mathbf{1}_b \end{bmatrix} \tag{III.1}$$

where $b = r + m$ and $\mathbf{Y}_b$ is defined by (7.226). It should be noted that in deriving (III.1) we have implicitly assumed that the network possesses the admittance representation, which would require that $\mathbf{H}_{\bar{t}\bar{t}}$ be nonsingular. To continue our derivation, we apply the elementary row operations to the matrices in (III.1), yielding

$$\det \begin{bmatrix} \mathbf{B}_f & \mathbf{0} \\ \mathbf{0} & \mathbf{Q}_f \\ \mathbf{Y}_b & -\mathbf{1}_b \end{bmatrix}$$

$$= \det \begin{bmatrix} -\mathbf{Q}'_{\bar t} & \mathbf{1}_m & \mathbf{0} & \mathbf{0} \\ \mathbf{H}_{tt} - \mathbf{H}_{t\bar t}\mathbf{H}_{\bar t\bar t}^{-1}\mathbf{H}_{\bar t t} - \mathbf{Q}_{\bar t}\mathbf{H}_{\bar t\bar t}^{-1}\mathbf{H}_{\bar t t} & \mathbf{H}_{t\bar t}\mathbf{H}_{\bar t\bar t}^{-1} + \mathbf{Q}_{\bar t}\mathbf{H}_{\bar t\bar t}^{-1} & \mathbf{0} & \mathbf{0} \\ \mathbf{H}_{tt} - \mathbf{H}_{t\bar t}\mathbf{H}_{\bar t\bar t}^{-1}\mathbf{H}_{\bar t t} & \mathbf{H}_{t\bar t}\mathbf{H}_{\bar t\bar t}^{-1} & -\mathbf{1}_r & \mathbf{0} \\ -\mathbf{H}_{\bar t\bar t}^{-1}\mathbf{H}_{\bar t t} & \mathbf{H}_{\bar t\bar t}^{-1} & \mathbf{0} & -\mathbf{1}_m \end{bmatrix}$$

$$= (-1)^b \det \begin{bmatrix} \mathbf{0} & \mathbf{1}_m \\ \mathbf{H}_{tt} - \mathbf{H}_{t\bar t}\mathbf{H}_{\bar t\bar t}^{-1}\mathbf{H}_{\bar t t} - \mathbf{Q}_{\bar t}\mathbf{H}_{\bar t\bar t}^{-1}\mathbf{H}_{\bar t t} + \mathbf{H}_{t\bar t}\mathbf{H}_{\bar t\bar t}^{-1}\mathbf{Q}'_{\bar t} + \mathbf{Q}_{\bar t}\mathbf{H}_{\bar t\bar t}^{-1}\mathbf{Q}'_{\bar t} & \mathbf{H}_{t\bar t}\mathbf{H}_{\bar t\bar t}^{-1} + \mathbf{Q}_{\bar t}\mathbf{H}_{\bar t\bar t}^{-1} \end{bmatrix}$$

$$= (-1)^{mr+b} \det (\mathbf{H}_{tt} - \mathbf{H}_{t\bar t}\mathbf{H}_{\bar t\bar t}^{-1}\mathbf{H}_{\bar t t} - \mathbf{Q}_{\bar t}\mathbf{H}_{\bar t\bar t}^{-1}\mathbf{H}_{\bar t t} + \mathbf{Q}_{\bar t}\mathbf{H}_{\bar t\bar t}^{-1}\mathbf{Q}'_{\bar t} + \mathbf{H}_{t\bar t}\mathbf{H}_{\bar t\bar t}^{-1}\mathbf{Q}'_{\bar t})$$

$$= (-1)^{mr+b} \det \mathbf{Q}_f\mathbf{Y}_b\mathbf{Q}'_f \tag{III.2}$$

where $\mathbf{Q}_f$ is defined in (7.219*a*). Combining (III.1) and (III.2) gives

$$\det \mathbf{M} = (-1)^{m+mr+b} \det \mathbf{H}_{\bar t\bar t} \det \mathbf{Q}_f\mathbf{Y}_b\mathbf{Q}'_f = (-1)^{mr+r} \det \mathbf{H}_{\bar t\bar t} \det \mathbf{Q}_f\mathbf{Y}_b\mathbf{Q}'_f \tag{III.3}$$

Since (7.222) is obtained from (7.221) by using only the elementary row operations to eliminate the variables $\mathbf{V}_{\bar t}$ and $\mathbf{I}_t$, it is not difficult to show that the determinants of the coefficient matrices of (7.221) and (7.222) are related by the equation

$$\det \mathbf{M} = (-1)^{rb} \det \mathbf{H} \tag{III.4}$$

Substituting (III.4) in (III.3), we obtain

$$\det \mathbf{H} = \det \mathbf{H}_{\bar t\bar t} \det \mathbf{Q}_f\mathbf{Y}_b\mathbf{Q}'_f \tag{III.5}$$

In a similar manner, if the feedback network possesses the impedance representation, $\mathbf{H}_{tt}$ is nonsingular and the determinant of $\mathbf{H}$ is related to $\det \mathbf{B}_f\mathbf{Z}_b\mathbf{B}'_f$ by

$$\det \mathbf{H} = \det \mathbf{H}_{tt} \det \mathbf{B}_f\mathbf{Z}_b\mathbf{B}'_f \tag{III.6}$$

where the fundamental circuit matrix $\mathbf{B}_f$ is defined in (7.219*b*), and the branch-impedance matrix $\mathbf{Z}_b$ is given by (7.225).

By appealing to (5.201) and (5.202), (III.5) and (III.6) can be expressed as

$$\det \mathbf{H} = \lambda_{hc}(\det \mathbf{H}_{\bar t\bar t})\, \Delta_c \tag{III.7a}$$

$$\det \mathbf{H} = \lambda_{hm}(\det \mathbf{H}_{tt})\, \Delta_m \tag{III.7b}$$

where λ_{hc} and λ_{hm} are real constants, depending only on the choices of the basis cutsets and loops.

SYMBOL INDEX

The symbols that occur most often are listed here, separated into four categories: Roman letters, Greek letters, matrices and vectors, and others.

Roman letters

Matrices and vectors

Greek letters

Other symbols

SUBJECT INDEX

Admittance model of transistor, 80
Admittance parameters:
  of n-port network, 21
  of two-port network, 127